# ELLIS HORWOOD SERIES IN
# APPLIED SCIENCE AND INDUSTRIAL TECHNOLOGY

*Series Editor:* Dr D. H. SHARP, OBE, former General Secretary, Society of Chemical Industry; formerly General Secretary, Institution of Chemical Engineers; and former Technical Director, Confederation of British Industry.

This collection of books is designed to meet the needs of technologists already working in the fields to be covered, and for those new to the industries concerned. The series comprises valuable works of reference for scientists and engineers in many fields, with special usefulness to technologists and entrepreneurs in developing countries.

Students of chemical engineering, industrial and applied chemistry, and related fields, will also find these books of great use, with their emphasis on the practical technology as well as theory. The authors are highly qualified chemical engineers and industrial chemists with extensive experience, who write with the authority gained from their years in industry.

*Published and in active publication*

**PRACTICAL USES OF DIAMONDS**
A. BAKON, Research Centre of Geological Technique, Warsaw, and A. SZYMANSKI, Institute of Electronic Materials Technology, Warsaw
**NATURAL GLASSES**
V. BOUSKA *et al.*, Czechoslovak Society for Mineralogy & Geology, Czechoslovakia
**POTTERY SCIENCE: Materials, Processes and Products**
A. DINSDALE, lately Director of Research, British Ceramic Research Association
**MATCHMAKING: Science, Technology and Manufacture**
C. A. FINCH, Managing Director, Pentafin Associates, Chemical, Technical and Media Consultants, Stoke Mandeville, and S. RAMACHANDRAN, Senior Consultant, United Nations Industrial Development Organisation for the Match Industry
**THE HOSPITAL LABORATORY: Strategy, Equipment, Management and Economics**
T. B. HALES, Arrowe Park Hospital, Merseyside
**OFFSHORE PETROLEUM TECHNOLOGY AND DRILLING EQUIPMENT**
R. HOSIE, formerly of Robert Gordon's Institute of Technology, Aberdeen
**MEASURING COLOUR: Second Edition**
R. W. G. HUNT, Visiting Professor, The City University, London
**MODERN APPLIED ENERGY CONSERVATION**
Editor: K. JACQUES, University of Stirling, Scotland
**CHARACTERIZATION OF FOSSIL FUEL LIQUIDS**
D. W. JONES, University of Bristol
**PAINT AND SURFACE COATINGS: Theory and Practice**
Editor: R. LAMBOURNE, Technical Manager, INDCOLLAG (Industrial Colloid Advisory Group), Department of Physical Chemistry, University of Bristol
**CROP PROTECTION CHEMICALS**
B. G. LEVER, International Research and Development Planning Manager, ICI Agrochemicals
**HANDBOOK OF MATERIALS HANDLING**
Translated by R. G. T. LINDKVIST, MTG, Translation Editor: R. ROBINSON, Editor, *Materials Handling News*. Technical Editor: G. LUNDESJO, Rolatruc Limited
**FERTILIZER TECHNOLOGY**
G. C. LOWRISON, Consultant, Bradford
**NON-WOVEN BONDED FABRICS**
Editor: J. LUNENSCHLOSS, Institute of Textile Technology of the Rhenish-Westphalian Technical University, Aachen, and W. ALBRECHT, Wuppertal
**REPROCESSING OF TYRES AND RUBBER WASTES: Recycling from the Rubber Products Industry**
V. M. MAKAROV, Head of General Chemical Engineering, Labour Protection, and Nature Conservation Department, Yaroslavl Polytechnic Institute, USSR, and V. F. DROZDOVSKI, Head of the Rubber Reclaiming Laboratory, Research Institute of the Tyre Industry, Moscow, USSR
**PROFIT BY QUALITY: The Essentials of Industrial Survival**
P. W. MOIR, Consultant, West Sussex
**EFFICIENT BEYOND IMAGINING: CIM and its Applications for Today's Industry**
P. W. MOIR, Consultant, West Sussex
**TRANSIENT SIMULATION METHODS FOR GAS NETWORKS**
A. J. OSIADACZ, UMIST, Manchester

*Series continued at back of book*

# MODERN BATTERY TECHNOLOGY

# MODERN BATTERY TECHNOLOGY

Editor
CLIVE D. S. TUCK B.Sc., Ph.D.
Alcan International Limited
Banbury, Oxfordshire

**ELLIS HORWOOD**
NEW YORK  LONDON  TORONTO  SYDNEY  TOKYO  SINGAPORE

First published in 1991 by
**ELLIS HORWOOD LIMITED**
Market Cross House, Cooper Street,
Chichester, West Sussex, PO19 1EB, England

A division of
Simon & Schuster International Group
A Paramount Communications Company

© Ellis Horwood Limited, 1991

All rights reserved. No part of this publication may be reproduced, stored in a retrieval system, or transmitted, in any form, or by any means, electronic, mechanical, photocopying, recording or otherwise, without the prior permission, in writing, of the publisher

Typeset in Times by Ellis Horwood Limited
Printed and bound in Great Britain
by Redwood Press Limited, Melksham, Wiltshire

---

British Library Cataloguing in Publication Data

---

Modern battery technology. —
(Ellis Horwood series in applied science and industrial technology)
I. Tuck, Clive D. S. II. Series
621.31
ISBN 0–13–590266–5

---

Library of Congress Cataloging-in-Publication Data

---

Modern battery technology / editor, Clive D. S. Tuck.
p. cm. — (Ellis Horwood series in applied science and industrial technology)
Includes bibliographical references and index.
ISBN 0–13–590266–5
1. Electric batteries — Design and construction. I. Tuck, Clive D. S., 1950– . II. Series.
TK2901.M57   1991
621.31′242–dc20                                                                                    91-11102
                                                                                                          CIP

# Table of contents

**LIST OF CONTRIBUTORS** . . . . . . . . . . . . . . . . . . . . . . . . . . . .14

**PREFACE** . . . . . . . . . . . . . . . . . . . . . . . . . . . . . . . . . . . . . . .15

**INTRODUCTION** . . . . . . . . . . . . . . . . . . . . . . . . . . . . . . . . .17
    I.1   Types of battery . . . . . . . . . . . . . . . . . . . . . . . . . . . . . . .19
    I.2   Battery design . . . . . . . . . . . . . . . . . . . . . . . . . . . . . . . .20

1. **HISTORICAL REVIEW OF BATTERY DEVELOPMENT AND COMMERCIALIZATION** (A. J. Salkind) . . . . . . . . . . . . . . . . .23

2. **THE CHEMISTRY AND ELECTROCHEMISTRY OF BATTERY SYSTEMS** . . . . . . . . . . . . . . . . . . . . . . . . . . . . . . . . . . . . .31
    (C. D. S. Tuck, 2.1 to 2.4 and A. Gilmour, 2.5)
    2.1   The thermodynamics of battery systems . . . . . . . . . . . . . . . . .31
          2.1.1   The equilibrium cell potential . . . . . . . . . . . . . . . . .31
          2.1.2   The effect of electrolyte concentration on cell voltage . . . . . .35
    2.2   The electrochemistry of battery systems . . . . . . . . . . . . . . . . .36
          2.2.1   Voltage losses of operating cells . . . . . . . . . . . . . . . .37
          2.2.2   Faraday's laws of electrolysis . . . . . . . . . . . . . . . . .38
          2.2.3   The kinetics of electrode charge transfer . . . . . . . . . . . . .39
          2.2.4   The kinetics of cell mass transport . . . . . . . . . . . . . . .42
          2.2.5   Inefficiencies of battery operation . . . . . . . . . . . . . . .43
                2.2.5.1   Self-discharge processes . . . . . . . . . . . . . . .43
                2.2.5.2   Passivation processes . . . . . . . . . . . . . . . .45
                2.2.5.3   Rechargeable battery inefficiencies . . . . . . . . . . . .46
    2.3   Electrochemical techniques used in the development of batteries . . . .48
          2.3.1   Single electrode studies . . . . . . . . . . . . . . . . . . . .49
                2.3.1.1   D.C. techniques . . . . . . . . . . . . . . . . . .49
                2.3.1.2   A.C. techniques . . . . . . . . . . . . . . . . . .53
          2.3.2   Whole unit (two-electrode) studies . . . . . . . . . . . . . . . .56
                2.3.2.1   D.C. techniques . . . . . . . . . . . . . . . . . .56

           2.3.2.2   A.C. techniques . . . . . . . . . . . . . . . . . . . . . . . . . 58
    2.4  Physical techniques used in the development of batteries . . . . . . . . 60
           2.4.1  Electron microscopy . . . . . . . . . . . . . . . . . . . . . . . . . . . . . . 61
                   2.4.1.1   Scanning electron microscopy . . . . . . . . . . . . . . 61
                   2.4.1.2   Transmission electron microscopy . . . . . . . . . . . 65
           2.4.2  X-ray diffraction and neutron diffraction . . . . . . . . . . . . . 65
           2.4.3  Extended X-ray absorption fine structure (EXAFS) . . . . . . . 67
           2.4.4  Electron spectroscopy . . . . . . . . . . . . . . . . . . . . . . . . . . . 69
           2.4.5  Secondary ion mass spectroscopy (SIMS) . . . . . . . . . . . . . . 70
           2.4.6  Infrared and Raman spectroscopy . . . . . . . . . . . . . . . . . . 70
           2.4.7  Thermogravimetric analysis (TGA) and differential thermal
                   analysis (DTA) . . . . . . . . . . . . . . . . . . . . . . . . . . . . . . . . 73
    2.5  Physical techniques used in the analysis of particulate battery
           materials (A. Gilmour) . . . . . . . . . . . . . . . . . . . . . . . . . . . . . . . . 75
           2.5.1  Measurement of surface area . . . . . . . . . . . . . . . . . . . . . . 75
           2.5.2  Porosity and density measurement . . . . . . . . . . . . . . . . . . 77
           2.5.3  Particle sizing and counting . . . . . . . . . . . . . . . . . . . . . . 80

3. **COMMERCIAL NON-RECHARGEABLE BATTERY SYSTEMS** . . . . . . 87
    3.1  Zinc–carbon batteries (B. Schumm) . . . . . . . . . . . . . . . . . . . . . . . 87
           3.1.1  Introduction . . . . . . . . . . . . . . . . . . . . . . . . . . . . . . . . . 87
           3.1.2  Manufacturers . . . . . . . . . . . . . . . . . . . . . . . . . . . . . . . . 88
           3.1.3  Chemistry of zinc–carbon systems . . . . . . . . . . . . . . . . . 89
           3.1.4  Types of zinc–carbon cell . . . . . . . . . . . . . . . . . . . . . . . 97
           3.1.5  Construction details . . . . . . . . . . . . . . . . . . . . . . . . . . . 97
           3.1.6  Cell or battery performance . . . . . . . . . . . . . . . . . . . . . 102
           3.1.7  Future developments . . . . . . . . . . . . . . . . . . . . . . . . . . . 110
    3.2  Alkaline–manganese batteries (J. C. Hunter) . . . . . . . . . . . . . . . . 111
           3.2.1  Introduction . . . . . . . . . . . . . . . . . . . . . . . . . . . . . . . . . 111
           3.2.2  History . . . . . . . . . . . . . . . . . . . . . . . . . . . . . . . . . . . . . 112
           3.2.3  Battery sizes, applications . . . . . . . . . . . . . . . . . . . . . . 112
           3.2.4  Battery construction features . . . . . . . . . . . . . . . . . . . . 113
           3.2.5  Chemistry of the alkaline–manganese battery . . . . . . . . . 115
           3.2.6  Materials used in the alkaline–manganese battery . . . . . . . 117
           3.2.7  Performance characteristics . . . . . . . . . . . . . . . . . . . . . 118
                   3.2.7.1   General . . . . . . . . . . . . . . . . . . . . . . . . . . . . . 118
                   3.2.7.2   Input capacity . . . . . . . . . . . . . . . . . . . . . . . . 118
                   3.2.7.3   Output capacity . . . . . . . . . . . . . . . . . . . . . . . 119
                   3.2.7.4   Effect of temperature on discharge performance . . . 121
                   3.2.7.5   Storage of batteries . . . . . . . . . . . . . . . . . . . . . 122
                   3.2.7.6   Resistance to leakage . . . . . . . . . . . . . . . . . . . 123
                   3.2.7.7   Battery cost . . . . . . . . . . . . . . . . . . . . . . . . . . 124
                   3.2.7.8   Summary . . . . . . . . . . . . . . . . . . . . . . . . . . . 124
           3.2.8  Likely future developments . . . . . . . . . . . . . . . . . . . . . . 124
    3.3  Small-size alkaline batteries (E. A. Megahed) . . . . . . . . . . . . . . . 125
           3.3.1  Definitions . . . . . . . . . . . . . . . . . . . . . . . . . . . . . . . . . . 125

|          |       |           |                                                                 |
| -------- | ----- | --------- | --------------------------------------------------------------- |
|          | 3.3.2 |           | History: 'Necessity leads to invention' .................125    |
|          | 3.3.3 |           | Chemistry.................................................126  |
|          | 3.3.4 |           | Cell construction ........................................133  |
|          | 3.3.5 |           | System characteristics and application performance ......137   |
|          |       | 3.3.5.1   | Hearing aid batteries...............................137        |
|          |       | 3.3.5.2   | Watch batteries .....................................147       |
|          | 3.3.6 |           | Available sizes and types ................................155  |
|          |       | 3.3.6.1   | Hearing aid batteries................................155       |
|          |       | 3.3.6.2   | Watch batteries .....................................155       |
|          | 3.3.7 |           | Further product development .............................155   |
| 3.4      | Large zinc–air batteries (V. H. Vu) ............................160 |
|          | 3.4.1 |           | History ..................................................160  |
|          | 3.4.2 |           | General characteristics..................................161   |
|          | 3.4.3 |           | Cell chemistry ..........................................161   |
|          | 3.4.4 |           | Cell components..........................................162   |
|          |       | 3.4.4.1   | Carbon cathode ......................................162       |
|          |       | 3.4.4.2   | Electrolyte ..........................................170      |
|          |       | 3.4.4.3   | Zinc anode ..........................................175       |
|          |       | 3.4.4.4   | Miscellaneous materials .............................177       |
|          | 3.4.5 |           | Cell design and construction details ......................177 |
|          |       | 3.4.5.1   | Wet air-depolarized batteries .......................177       |
|          |       | 3.4.5.2   | Air-depolarized batteries with gelled electrolyte ...179       |
|          | 3.4.6 |           | Applications ............................................185   |
|          | 3.4.7 |           | Likely future developments ..............................185   |

**4. RECHARGEABLE BATTERY SYSTEMS** ................................194
  4.1 Lead–acid batteries (K. Peters)...................................194

|          |       |           |                                                              |
| -------- | ----- | --------- | ------------------------------------------------------------ |
|          | 4.1.1 |           | Historical notes and general characteristics...............194 |
|          | 4.1.2 |           | Materials and structural features .......................196 |
|          |       | 4.1.2.1   | Negative active material............................196      |
|          |       | 4.1.2.2   | Positive active material............................197      |
|          |       | 4.1.2.3   | Grid alloys ........................................199       |
|          |       | 4.1.2.4   | Electrolyte ........................................200      |
|          |       | 4.1.2.5   | Separators .........................................202      |
|          | 4.1.3 |           | Basic characteristics ...................................203 |
|          |       | 4.1.3.1   | Discharge behaviour ................................204      |
|          |       | 4.1.3.2   | Charging behaviour.................................207       |
|          |       | 4.1.3.3   | Self-discharge behaviour ...........................210      |
|          |       | 4.1.3.4   | Gas recombination behaviour .......................211       |
|          | 4.1.4 |           | Construction and performance ...........................213  |
|          |       | 4.1.4.1   | S.L.I. (automotive) batteries.......................213      |
|          |       | 4.1.4.2   | Motive power batteries .............................219      |
|          |       | 4.1.4.3   | Standby batteries ..................................227      |
|          |       | 4.1.4.4   | Portable equipment and electrical appliances .....235        |
|          |       | 4.1.4.5   | Other applications .................................241      |
|          |       | 4.1.4.6   | Storage of renewable energy........................243       |

## Table of contents

- 4.2 Nickel–cadmium batteries .......... 244
  - Introduction (C. P. Albon and J. Parker) .......... 244
  - 4.2.1 Pocket plate cells (C. P. Albon and J. Parker) .......... 246
    - 4.2.1.1 Construction .......... 246
    - 4.2.1.2 Reaction mechanisms of nickel–cadmium cells .... 254
    - 4.2.1.3 Battery technology .......... 258
    - 4.2.1.4 Recent developments .......... 262
  - 4.2.2 Sealed cylindrical nickel–cadmium cells (R. N. Thomas) .... 263
    - 4.2.2.1 Introduction .......... 263
    - 4.2.2.2 Construction .......... 264
    - 4.2.2.3 Electrical characteristics .......... 266
    - 4.2.2.4 Applications and battery design .......... 277
    - 4.2.2.5 Future prospects .......... 283

## 5. LITHIUM BATTERY SYSTEMS .......... 287

- 5.1 Lithium–thionyl chloride batteries (H. F. Gibbard and T. B. Reddy). 287
  - 5.1.1 General characteristics .......... 287
  - 5.1.2 Manufacturers .......... 288
  - 5.1.3 Chemistry .......... 288
    - 5.1.3.1 Cell reaction .......... 288
    - 5.1.3.2 Safety .......... 289
  - 5.1.4 Construction details .......... 289
    - 5.1.4.1 Materials .......... 289
    - 5.1.4.2 Bobbin cells .......... 291
    - 5.1.4.3 Wound cells .......... 294
    - 5.1.4.4 Prismatic cells .......... 302
    - 5.1.4.5 Reserve cells .......... 307
  - 5.1.5 Likely future development .......... 310
    - 5.1.5.1 General improvements .......... 310
    - 5.1.5.2 Improvements in low-rate cells .......... 311
    - 5.1.5.3 Improvements in moderate- and high-rate cells .... 312
- 5.2 Lithium–sulphur dioxide batteries (T. B. Reddy) .......... 312
  - 5.2.1 Historical background .......... 312
  - 5.2.2 Technical advantages .......... 313
  - 5.2.3 Chemistry and electrochemistry .......... 314
    - 5.2.3.1 Electrode and cell reactions .......... 314
    - 5.2.3.2 Electrolyte composition and properties .......... 314
    - 5.2.3.3 Balanced cell concept .......... 316
  - 5.2.4 Cell construction .......... 316
    - 5.2.4.1 Spiral wound construction in hermetically sealed case 317
    - 5.2.4.2 Cell case material .......... 318
    - 5.2.4.3 Cathode construction .......... 318
    - 5.2.4.4 Anode construction .......... 318
    - 5.2.4.5 Separators and insulators .......... 318
    - 5.2.4.6 Vent designs .......... 319
    - 5.2.4.7 Glass-to-metal seals .......... 319

|  |  |  |
|---|---|---|
| | 5.2.5 | Commercially available cells and their performance . . . . . . . 320 |
| | | 5.2.5.1 Commercially available cells . . . . . . . . . . . . . . . . 321 |
| | | 5.2.5.2 Cell performance . . . . . . . . . . . . . . . . . . . . . . 323 |
| | | 5.2.5.3 Safety considerations in the use of lithium–sulphur dioxide cells . . . . . . . . . . . . . . . . . . . . . . . . . 327 |
| | 5.2.6 | Battery design and performance. . . . . . . . . . . . . . . . . . . . . 327 |
| | | 5.2.6.1 Battery design . . . . . . . . . . . . . . . . . . . . . . . . . 327 |
| | | 5.2.6.2 Fuses, diodes, and discharge circuits . . . . . . . . . . 327 |
| | | 5.2.6.3 Other design considerations . . . . . . . . . . . . . . . . 328 |
| | | 5.2.6.4 US military battery types . . . . . . . . . . . . . . . . . . 328 |
| | | 5.2.6.5 Performance of BA-5598 batteries . . . . . . . . . . . 330 |
| | 5.2.7 | Applications of lithium–sulphur dioxide cells and batteries . . 331 |
| | | 5.2.7.1 Industrial and electronic applications . . . . . . . . . 333 |
| | | 5.2.7.2 Military applications . . . . . . . . . . . . . . . . . . . . . 334 |
| | 5.2.8 | Conclusions . . . . . . . . . . . . . . . . . . . . . . . . . . . . . . . . . . . 334 |
| 5.3 | Lithium–carbon monofluoride batteries (D. Eyre and C. D. S. Tuck) 336 |
| | 5.3.1 | Introduction . . . . . . . . . . . . . . . . . . . . . . . . . . . . . . . . . . . 336 |
| | 5.3.2 | Manufacturers . . . . . . . . . . . . . . . . . . . . . . . . . . . . . . . . . 336 |
| | 5.3.3 | Chemistry and cell components . . . . . . . . . . . . . . . . . . . . . 337 |
| | 5.3.4 | Cell types and performance. . . . . . . . . . . . . . . . . . . . . . . . 339 |
| | | 5.3.4.1 Cylindrical cells . . . . . . . . . . . . . . . . . . . . . . . . 339 |
| | | 5.3.4.2 Coin cells . . . . . . . . . . . . . . . . . . . . . . . . . . . . . 340 |
| | | 5.3.4.3 Pin-type cells . . . . . . . . . . . . . . . . . . . . . . . . . . 340 |
| | 5.3.5 | Conclusion . . . . . . . . . . . . . . . . . . . . . . . . . . . . . . . . . . . 345 |
| 5.4 | Lithium–manganese dioxide batteries (S. Narukawa and N. Furukawa) . . . . . . . . . . . . . . . . . . . . . . . . . 348 |
| | 5.4.1 | General characteristics. . . . . . . . . . . . . . . . . . . . . . . . . . . 348 |
| | 5.4.2 | Manufacturers/statistics . . . . . . . . . . . . . . . . . . . . . . . . . . 349 |
| | 5.4.3 | Chemistry . . . . . . . . . . . . . . . . . . . . . . . . . . . . . . . . . . . . 349 |
| | 5.4.4 | Construction details . . . . . . . . . . . . . . . . . . . . . . . . . . . . 352 |
| | | 5.4.4.1 Flat-type batteries. . . . . . . . . . . . . . . . . . . . . . . 352 |
| | | 5.4.4.2 Cylindrical type batteries . . . . . . . . . . . . . . . . . 352 |
| | 5.4.5 | Cell types available and applications . . . . . . . . . . . . . . . . . 352 |
| | 5.4.6 | Performance characteristics . . . . . . . . . . . . . . . . . . . . . . . 353 |
| | | 5.4.6.1 Flat-type batteries. . . . . . . . . . . . . . . . . . . . . . . 353 |
| | | 5.4.6.2 Cylindrical batteries (inside-out structure) . . . . . . 356 |
| | | 5.4.6.3 Cylindrical batteries (spiral structure) . . . . . . . . . 358 |
| | 5.4.7 | Likely future developments . . . . . . . . . . . . . . . . . . . . . . . 365 |
| 5.5 | Lithium–iodine batteries (J. Jolson, S. Wicelinski, and D. Schrodt) . 365 |
| | 5.5.1 | History . . . . . . . . . . . . . . . . . . . . . . . . . . . . . . . . . . . . . . 366 |
| | 5.5.2 | Chemistry . . . . . . . . . . . . . . . . . . . . . . . . . . . . . . . . . . . . 367 |
| | 5.5.3 | Medical use . . . . . . . . . . . . . . . . . . . . . . . . . . . . . . . . . . . 368 |
| | | 5.5.3.1 Available types . . . . . . . . . . . . . . . . . . . . . . . . 369 |
| | | 5.5.3.2 Construction . . . . . . . . . . . . . . . . . . . . . . . . . . 370 |
| | | 5.5.3.3 Performance characteristics . . . . . . . . . . . . . . . 374 |
| | 5.5.4 | Industrial use. . . . . . . . . . . . . . . . . . . . . . . . . . . . . . . . . . 377 |

|  |  | 5.5.4.1 Available types ........................ 378 |
|---|---|---|
|  |  | 5.5.4.2 Construction .......................... 379 |
|  |  | 5.5.4.3 Performance characteristics ............ 380 |
|  | 5.5.5 | Likely future developments ..................... 381 |
| 5.6 | Lithium–copper oxide batteries (A. E. Brown) ............. 383 |
|  | 5.6.1 | Introduction .................................. 383 |
|  | 5.6.2 | Manufacturers ................................. 384 |
|  | 5.6.3 | Construction .................................. 384 |
|  |  | 5.6.3.1 Anode assembly ....................... 385 |
|  |  | 5.6.3.2 Cathode assembly ..................... 385 |
|  |  | 5.6.3.3 Electrolyte .......................... 385 |
|  |  | 5.6.3.4 Separator ............................ 385 |
|  |  | 5.6.3.5 Sealing .............................. 385 |
|  | 5.6.4 | Performance ................................... 385 |
|  | 5.6.5 | Storage ....................................... 387 |
|  | 5.6.6 | Chemistry ..................................... 388 |
|  | 5.6.7 | Safety ........................................ 392 |
| 5.7 | Lithium–vanadium pentoxide batteries (H. V. Venkatasetty) ..... 393 |
|  | 5.7.1 | Cell design and performance ..................... 393 |
|  | 5.7.2 | Stability and safety ........................... 399 |
|  | 5.7.3 | Lithium–vanadium pentoxide medical batteries .......... 400 |
|  | 5.7.4 | Lithium–silver vanadium pentoxide batteries .......... 401 |
| 5.8 | Lithium–iron disulphide batteries (A. Gilmour) ............ 402 |
|  | 5.8.1 | Introduction .................................. 402 |
|  | 5.8.2 | Chemistry ..................................... 403 |
|  | 5.8.3 | Cell designs .................................. 403 |
|  | 5.8.4 | Performance of the Li–FeS$_2$ system .............. 404 |
|  | 5.8.5 | Discharge mechanisms of the Li–FeS$_2$ system .......... 406 |
|  | 5.8.6 | The market's reaction to the Li–FeS$_2$ system .......... 407 |
| 5.9 | Lithium–anode thermal batteries (A. Attewell) ............. 409 |
|  | 5.9.1 | Introduction .................................. 409 |
|  | 5.9.2 | Battery construction ........................... 410 |
|  | 5.9.3 | Cell components and electrochemistry ............. 413 |
|  |  | 5.9.3.1 The lithium anode ..................... 414 |
|  |  | 5.9.3.2 Electrolytes .......................... 414 |
|  |  | 5.9.3.3 Cathodes ............................. 415 |
|  | 5.9.4 | The pyrotechnic ................................ 416 |
|  | 5.9.5 | Computer modelling ............................. 417 |
|  | 5.9.6 | The performance of thermal batteries ............. 417 |
|  | 5.9.7 | The use of thermal batteries .................... 419 |
|  | 5.9.8 | Alternatives to thermal batteries ................ 422 |
|  | 5.9.9 | Major suppliers of thermal batteries ............. 422 |
|  | 5.9.10 | The future ................................... 422 |
| 5.10 | Lithium–molybdenum disulphide rechargeable batteries (F. C. Laman) .......................................... 423 |
|  | 5.10.1 | Introduction .................................. 423 |

|  |  |  |  |  |
|---|---|---|---|---|
| | 5.10.2 | Cell chemistry and operation | 423 |
| | 5.10.3 | Construction | 425 |
| | 5.10.4 | Electrical characteristics | 426 |
| | | 5.10.4.1 | Cell voltage | 426 |
| | | 5.10.4.2 | Internal resistance | 428 |
| | | 5.10.4.3 | Energy efficiency | 429 |
| | | 5.10.4.4 | Charge requirement | 429 |
| | 5.10.5 | Performance characteristics | 430 |
| | | 5.10.5.1 | Deliverable capacity | 430 |
| | | 5.10.5.2 | Cycle life | 434 |
| | | 5.10.5.3 | Charge retention | 437 |
| | 5.10.6 | Safety | 437 |
| | | 5.10.6.1 | Electrical abuse test results | 437 |
| | 5.10.7 | Shipping and handling | 438 |
| | 5.10.8 | Reproducibility and reliability | 439 |
| | | 5.10.8.1 | Single cells | 439 |
| | | 5.10.8.2 | Multi-cell batteries | 441 |
| | 5.10.9 | Future developments | 443 |
| | 5.10.10 | Summary | 443 |

**6. SYSTEMS UNDER DEVELOPMENT** .................. 452
  6.1  Nickel–zinc batteries (A. Duffield) ........... 452
      6.1.1  The nickel–zinc system .............. 452
      6.1.2  Cell and battery construction and performance ...... 456
      6.1.3  The future for nickel–zinc .............. 458
  6.2  Sodium–sulphur batteries (M. McNamee) ........... 458
      6.2.1  Introduction ................... 458
      6.2.2  Performance characteristics ............. 461
      6.2.3  Cell construction .................. 463
      6.2.4  The battery system ................. 464
      6.2.5  Development status ................. 469
  6.3  Nickel–hydrogen batteries (J. J. Smithrick) ........ 472
      6.3.1  General characteristics ............... 472
      6.3.2  Manufacturers and experience ............ 473
      6.3.3  Chemistry ..................... 474
          6.3.3.1  Normal operation ............ 474
          6.3.3.2  Overcharge ............... 474
          6.3.3.3  Reversal ................. 475
      6.3.4  Cell construction .................. 475
          6.3.4.1  Air Force/Hughes cell .......... 475
          6.3.4.2  Comsat/Intelsat cell ........... 476
          6.3.4.3  NASA Advanced cell ........... 476
      6.3.5  Battery information ................. 477
      6.3.6  Cell performance .................. 478
          6.3.6.1  Discharge ................ 478
          6.3.6.2  Charge .................. 480

|  |  | 6.3.6.3 | State-of-charge .......................480 |
|---|---|---|---|

- 6.3.6.3 State-of-charge .......................480
- 6.3.6.4 Capacity retention ....................482
- 6.3.6.5 Life test ............................483
- 6.3.6.6 Storage .............................485
- 6.3.7 Likely future developments ..................486

6.4 Aluminium–air batteries (C. D. S. Tuck) ................487
- 6.4.1 Historical introduction. ....................487
- 6.4.2 Chemistry of the system. ...................489
- 6.4.3 Anode development .......................490
- 6.4.4 Air cathode development....................494
- 6.4.5 Electrolyte management ....................496
- 6.4.6 Commercial battery development. .............498
- 6.4.7 Future developments. .....................500

6.5 Zinc–bromine batteries (A. Leo). ...................502
- 6.5.1 Introduction ..........................502
- 6.5.2 Battery configuration .....................505
- 6.5.3 Bipolar stack shunt current management ..........506
- 6.5.4 Stack fabrication and sealing methods...........508
- 6.5.5 Materials of construction ..................509
- 6.5.6 System design ........................511
- 6.5.7 Battery performance characteristics ............512
- 6.5.8 Technology status. ......................514

6.6 Rechargeable lithium–iron sulphide batteries (H. F. Gibbard) .....516
- 6.6.1 General characteristics. ...................516
- 6.6.2 Manufacturers .........................519
- 6.6.3 Chemistry. ...........................519
  - 6.6.3.1 Negative electrode ................519
  - 6.6.3.2 Positive electrode. ................520
  - 6.6.3.3 Negative-to-positive capacity ratio ........521
  - 6.6.3.4 Separator ......................521
- 6.6.4 Construction ..........................522
- 6.6.5 Performance ..........................525
- 6.6.6 Future developments......................527

6.7 Lithium solid state batteries (R. J. Neat). .................528
- 6.7.1 Introduction .........................528
- 6.7.2 The cell components .....................529
  - 6.7.2.1 Lithium anodes ..................529
  - 6.7.2.2 Polymer electrolytes ................529
  - 6.7.2.3 Intercalation composite cathodes ..........530
  - 6.7.2.4 Component fabrication ..............531
  - 6.7.2.5 Cell fabrication ..................532
- 6.7.3 Cell performance .......................532
- 6.7.4 'Room temperature' performance .............535
- 6.7.5 Summary ...........................537

**APPENDICES** . . . . . . . . . . . . . . . . . . . . . . . . . . . . . . . . . . . . . . 544
Appendix A.  Performance characteristics of battery systems included in
this volume (in alphabetical order of common name) . . . . . . . 544
Appendix B.  Standard potentials of electrodes: reactions at 25°C
(in alphabetical order of reacting component) . . . . . . . . . . . 559
Appendix C.  Electrochemical equivalents per mass and volume of possible
battery electrode elements and compounds . . . . . . . . . . . . . 561
Appendix D.  Conversion tables and physical constants. . . . . . . . . . . . . . . 563

**INDEX** . . . . . . . . . . . . . . . . . . . . . . . . . . . . . . . . . . . . . . . . . . 567

# List of contributors

C. P. Albon, Alcad Ltd., Redditch, Worcestershire, UK.
A. Attewell, Royal Aerospace Establishment, Farnborough, Hampshire, UK.
A. E. Brown, SAFT (UK) Ltd., Hampton, Middlesex, UK
A. Duffield, Transfer Technology Ltd., Coleshill, N. Warwickshire, UK.
D. Eyre, Crompton Eternacell Ltd., South Shields, Tyne & Wear, UK.
N. Furukawa, Sanyo Electric Co. Ltd., Sumoto-City, Hyogo, Japan.
H. F. Gibbard, Duracell Research Center, Needham, Massachusetts, USA.
A. Gilmour, Battery Consultant, Henley-on-Thames, Berkshire, UK.
J. C. Hunter, Eveready Battery Co. Inc., Westlake, Ohio, USA.
J. Jolson, Catalyst Research, Owings Mills, Maryland, USA.
F. C. Laman, Advanced Energy Technologies Inc., Burnaby, British Columbia, Canada.
A. Leo, Energy Research Corporation, Danbury, Connecticut, USA.
M. McNamee, Chloride Silent Power Ltd., Runcorn, Cheshire, UK.
E. A. Megahed, Rayovac Technology Center, Madison, Wisconsin, USA.
S. Narukawa, Sanyo Electric Co. Ltd., Sumoto-City, Hyogo, Japan.
R. J. Neat, Materials Development Division, Harwell Laboratory, Oxfordshire, UK.
J. Parker, Alcad Ltd., Redditch, Worcestershire, UK.
K. Peters, Battery Consultant, Worsley, Manchester, UK.
T. B. Reddy, Power Conversion Inc., Saddlebrook, New Jersey, USA.
A. J. Salkind, Rutgers University, Piscataway, New Jersey, USA.
B. Schumm, Eagle-Cliffs Inc., Cleveland, Ohio, USA.
D. Scrodt, Catalyst Research, Owings Mills, Maryland, USA.
J. J. Smithrick, NASA Lewis Research Center, Cleveland, Ohio, USA.
R. N. Thomas, SAFT (UK) Ltd., Hampton, Middlesex, UK.
C. D. S. Tuck, Alcan International Ltd., Banbury, Oxfordshire, UK.
H. V. Venkatasetty, Honeywell Inc., Bloomington, Minnesota, USA.
V. H. Vu, SAFT, SEAP, Bordeaux, France.
S. Wincelinski, Catalyst Research, Owings Mills, Maryland, USA.

# Preface

As civilization has become more sophisticated and reliant on readily available sources of energy, so battery research and development has been stimulated. During the last twenty-five years in particular, many new systems have become commercially available, and their full potential has not yet been realized. This book offers a collection of contributions concerning both currently commercialized battery systems and those nearing commercialization, by authors who have first-hand experience of the systems in question. Thus it is hoped that the book will be of use both to those requiring batteries as a source of power, in which case ready comparisons between systems can be made, and those developing new systems.

When I was first approached by David Sharp, Editor-in-Chief of this series, to undertake the task of compiling this volume, the potential problems seemed immense. My feeling was that any likely authors would be so busy with the task of maintaining or establishing their battery technologies in the marketplace that they would have little time for writing. However, I must particularly thank my wife, Jean, as well as family and friends for encouraging me to take up the challenge and for assisting me along the way. In fact, I found that the response of the authors was extremely positive, and, in spite of their busy schedules, the promised contributions were soon forthcoming. I must express my sincere thanks to them for all the time and effort they have given to the project.

I am writing this preface during a stay in a secluded valley in the Yorkshire dales, and the surroundings very much remind me that concern for the environment seems to be having an ever-increasing part to play in the research and development of battery systems. Indeed, it was this concern, particularly with regard to mercury pollution, which fuelled the Japanese development of lithium batteries in the early 1970s, initially for lighted fishing floats. At recent battery conferences it has been noticeable that the environment has been at the forefront of discussion, and this has largely been in three main areas. Firstly, an electric alternative to the internal combustion engine is sought, and this is being assisted by legislation to install large numbers of electric vehicles in cities such as Los Angeles over the next few years. The application of this policy will undoubtedly be accelerated by the recent rises and instability in the price of oil. Secondly, a reduction in the use of toxic substances in current batteries is successfully taking place, most notably in the case of mercury. And thirdly, there is an awareness that the recycling of battery materials should be encouraged. Thus, there is an increase of general interest in battery systems, and it is intended that this book will be of use both to those developing them as well as those wishing to find the most appropriate power source for their application.

The book begins with a history of battery development and then briefly covers the science of batteries, including sections on the practical methods (electrochemical and physical) used in battery research and development. Following that, there are chapters on the present, most widely used systems, both of non-rechargeable and rechargeable type. The book then deals with batteries which have appeared relatively recently and whose time is just beginning or just about to begin, most notably the lithium batteries and those with more intractable electrode materials. At present the latter systems have their niches or expected places in the market, and their successful commercialization is in the hands of their developers together with the vagaries of the business world and public opinion. It is anticipated that this book will assist in publicizing their possibilities, as well as demonstrating the capabilities of the more traditional systems. Inevitably, some battery types could not be included, but it is hoped that the large numbers of references present will assist in overcoming that limitation.

I must express my thanks to Alcan International for allowing me to undertake this venture and for the encouragement and assistance I have received. Thanks must go to my secretary, Charlotte King, for the inevitably large quantity of typing, to Jeanette Reynolds for drawing a considerable number of the figures, and to my wife for her very efficient preparation of the index. I must also thank Ray Brown of Newcastle University for advising me on the chapter on battery chemistry and electrochemistry. Lastly, I would like to thank the staff of the Ellis Horwood publishing company, particularly Rachael Jones, for their patience, encouragement, and professionalism.

*Snaizeholme, Yorks*
*April 1991* Clive Tuck

# Introduction

A battery is defined as a power generating device which is able to convert stored chemical energy into work of an electrical nature. The stored energy is either inherently present in the chemical substances used, as in the case of a non-rechargeable battery, or is induced in those substances by application of an external source of electrical work. In the latter case, a rechargeable battery is the result.

The word 'battery' was originally applied by Benjamin Franklin as a collective term to describe the apparatus obtained when several Leyden jar capacitors were connected together, its previous use being restricted to assemblies of cannon in artillery units. It was Humphrey Davy, in 1801, who first applied the term to an array of individual galvanic couples, these being harnessed together to yield greater power. The term 'cell', used to describe each separate couple, came later, and was originally applied to the individual compartments of a battery designed by Cruikshank in 1808. When batteries were first invented they were seen as merely laboratory curiosities, and it needed the introduction of the electric telegraph in the 19th century and the development of the electric incandescent lamp at the beginning of the 20th century to give them practical value. Today, batteries are indispensable and can be found in varieties of applications, ranging from miniature forms such as button cells to vast banks of rechargeable cells occupying several hundred square metres, designed to equalize the daytime and night-time electricity production from the power stations of some countries, a process known as load-levelling.

The construction of a typical battery cell is schematically illustrated in Fig. I.1. This shows the simplest type of battery, which can be constructed from suitable electrodes connected electrically together and placed in an ionically-conducting electrolyte. If the electrodes were to be made of zinc and copper immersed respectively in dilute sulphuric acid containing zinc ions and copper sulphate solution, then the cell would be termed a Daniell cell, named after its inventor. Faraday gave the electrodes the distinguishing names of anode and cathode. These were derived respectively from the Greek words for 'way up' (ανοδος) and 'going down' (καθοδος), as Faraday supposed that the anode received the current and the cathode generated the current. It was discovered later that the electron flow is from

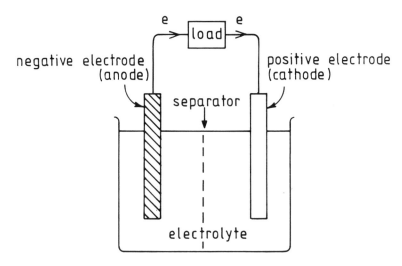

Fig. I.1 — Generalized battery cell showing its main components and the direction of electron flow when the cell is discharged through a load.

the anode to the cathode, and, although the terms are not truly descriptive, they are useful in distinguishing the two electrodes.

The invention of the internal combustion engine and its widespread use for transportation acted partly to promote and partly to hinder battery development. The lead-acid battery was adopted for powering the starting motor, lights, and ignition of vehicles, but batteries for traction ceased to be economic or appropriate. It is interesting to note that the land speed record of 1902 was held by an electric car which travelled at 104 mph. During the last forty years the introduction of solid state electronic equipment operating at low voltage has produced a revival in battery research and development. Fig. I.2 shows that the number of battery patents has increased during recent years, with a particularly dramatic rise of Japanese patents. Several new battery technologies have been successfully developed over this period, such as the lithium battery in Japan, and there is now a plethora of manufactured battery systems.

This situation allows there to be an appropriate battery for most applications, but certain considerations have to be made when selecting a suitable battery:

(i) Should the battery be rechargeable or non-rechargeable?
(ii) What are the required voltage and current levels, and what are their desired profiles for the application in mind?
(iii) What is the maximum permissible cost of the battery?
(iv) What are the required physical dimensions of the battery?
(v) If rechargeable, what types of charge/discharge duty cycle will the battery be required to undertake?
(vi) What is the temperature range over which adequate performance of the battery will be required?

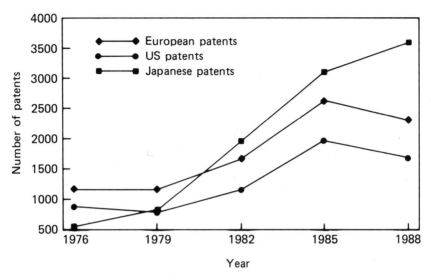

Fig. I.2 — Total numbers of battery patents published each year during the period 1976 to 1988 for the USA, Europe, and Japan.

(vii) What is the battery shelf-life? In other words, will the battery performance deteriorate significantly during storage or in the switched-off state?
(viii) How user-friendly and environmentally friendly is the battery?

## I.1 TYPES OF BATTERY

A nomenclature to describe the different general types of battery has been developed, each type having its advantages and disadvantages.

### (i) Primary batteries

These are not designed to be rechargeable, and they are convenience products which generally have high energy per unit volume (energy density) and good shelf-life (i.e. they will not significantly discharge when not in use). They are the most widely used batteries, with main applications in electric torches and electronic equipment. Common examples of this type of battery are zinc–carbon (Leclanché) and alkaline–manganese (zinc–manganese dioxide) cells.

### (ii) Secondary batteries

These are batteries designed to be electrically recharged, and they offer a cost saving once the initial outlay for the system has been made. Their most common application is in conjunction with the internal combustion engine as a rechargeable power source for starting, lighting and ignition (SLI). Common examples of this type of system are lead–acid and nickel–cadmium batteries.

### (iii) Mechanically rechargeable batteries
These batteries are rechargeable by renewing one of the electrodes of the system once it has been expended during discharge, and, in practice, this is invariably the anode. It therefore offers the advantage of increased speed of recharge over secondary batteries. Examples of this type of battery are aluminium–air and zinc–air.

### (iv) Hybrid battery
This generally describes a battery which has some features of a fuel cell in that one electrode (usually the cathode) consumes gas while the other (the anode) is itself consumed. Examples of this type of battery are zinc–bromine, aluminium—air, and zinc–air.


## I.2 BATTERY DESIGN
Battery design brings together a number of scientific disciplines in the areas of physics, chemistry, mechanical engineering, and electrical engineering, as there are a large number of factors to be considered when designing a successful battery.

### (i) Thermodynamic aspects
Reactions must be considered which are thermodynamically feasible by an extensive margin so that significant voltages can be developed. However, the components and products must remain in stable or controllable form during storage and operation. Large electrical charges should be available from low masses of reactants in order to achieve high energy per unit weight or volume.

### (ii) Electrode kinetics aspects
The chemical reactions need to take place at a high rate for the battery to yield appreciable power. Thus the rate constants of the electrode processes must be high.

### (iii) Mass transport effects
Batteries are of a finite size, which is often minimized for ease of use, and considerations have to be made about the accessibility of the electrode materials for reaction. Another important aspect is the ease of removal of the reaction products which might otherwise block continuing electrode reactions.

### (iv) Battery construction — electrical aspects
Every effort must be made to minimize internal voltage losses by maximizing both the ionic conductivity of the electrolyte and the electronic conductivity of the electrode materials. Electrode gaps must be minimized if high currents are to be drawn, and connections to external tabs must be well constructed and appropriate for the currents experienced during use.

### (v) Battery construction — mechanical aspects
The battery must be packed in an appropriately sized container of the best material for optimum mechanical, electrical, and chemical behaviour. This may be of

cylindrical, prismatic, coin cell, or button cell type if universal application is required. Cell vents, seals, and lead-through connections would also be designed in the packaging medium.

At the present time, there is a large range of battery types available, with very diverse applications, although the three major types in commercial use are the zinc–carbon (Leclanché), alkaline–manganese, and lead–acid batteries. The major characteristics of all the batteries covered in this book are summarized in Appendix A to facilitate comparisons.

# 1

# Historical review of battery development and commercialization

(A. J. Salkind)

It would appear from the archaeological work of König about 50 years ago [1] that galvanic cells were known to the Babylonians as early as 500 BC. During his excavations at a site near Baghdad, earthenware jars approximately 15 cm high were found, each containing copper tubes having an inserted central iron rod cemented at the top with asphalt. This construction was seen to be strongly reminiscent of simple battery cells, and, on filling with naturally occurring acids, they would undoubtedly have produced reasonable currents. Although their true purpose has been lost in antiquity, modern suggestions as to their use have included electroplating, medicinal therapy, and use in religious rites. However, in spite of this evidence that early electrochemical cells existed in the ancient world, it can be stated that the development of modern battery systems and technology undoubtedly began with Volta's experiments about 200 years ago [2]. He combined metals such as copper, brass, or silver (as positive electrodes) with tin or zinc as the opposite (negative) electrodes in simple salt electrolytes. Volta's discoveries were remarkable. He was able to obtain currents in the order of amperes, compared to the electrostatic generators in use at that time which could only deliver milliamperes. Volta was also able to achieve high voltages by combining cells into a series (pile). The leading scientists and engineers of the next century were intrigued by the use of batteries for laboratory work, particularly for obtaining thermodynamic and chemical data. These include: Ampère, Arrhenius, Bacon, Becquerel, Cruikshank, Daniell, Davy, Edison, Faraday, Jungner, Leclanché, Nernst, Oersted, Planté, Rutherford, and Volta. Before the invention of the dynamo by Siemens in 1866, there was little commercial interest in secondary batteries, since recharging had to be done from costly primary batteries which occupied much space.

Modifications and improvements of Volta's cell were carried out by many innovators, especially Cruikshank [3], Daniell [4], Grove [5], and Bunsen [6]. All the cells were open and flooded with electrolyte, which made them dangerous to move. In 1860, Leclanché [7] constructed the first cell design with restricted electrolyte, using manganese dioxide as the positive electrode and zinc as the negative electrode, which led to the present portable and sealed zinc–manganese dioxide battery systems. Fig. 1.1 shows antique drawings of this cell and designs by Daniell, Grove and Bunsen.

At the same time, Gaston Planté [8] built the first true secondary battery consisting of two sheets of lead in a sulphuric acid electrolyte. The sheets, held apart by a separator, were coiled to form a cylinder. The formation of the Planté cell was carried out by means of primary cells of different types and was a laborious and time consuming process. However, once dynamos became commercially available, much experimentation and rapid progress was made with the lead–acid battery system. Glasstone & Tribe [9] developed the double sulphate theory of reaction in 1881, the same year that Sellon introduced the lead–antimony alloy [10].

The origins of alkaline batteries can be found in the 1881 patent of Felix de Lalande and Georges Chaperon [11]. In the original form, their cell utilized a horizontal positive plate with copper oxide as the active material. Suspended above this was a horizontally oriented zinc negative electrode. The electrolyte consisted of potassium or sodium hydroxide. However, the e.m.f. of this couple was less than 0.9 V. Although Lalande and Chaperon attempted to develop a secondary cell, their system/construction functioned only as a primary cell. In the next decade, many variations were devised which attempted to provide a more practical secondary design. These included efforts by Desmazures [12], De Virloy, Commelin, & Bailhache [13], Boettcher [14], Schoop [15], Waddell & Entz [16], and Entz & Phillips [17]. One design was used by the French government in 1887–1888 to power a submarine boat, which excited much interest at the time. A battery of zinc–copper oxide cells, known as the Waddell–Entz battery, was used in New York in 1893–1894 to power a trolley car line [18]. However, practical difficulties with the Lalande type cells in storage battery applications were caused by the high solubilities of copper oxide and zinc and the low cell voltage.

Edison patented [19] an improved version of the Lalande–Chaperon primary cell in 1899. Modern versions of the cell, often called Edison–Lalande cells, are still used in some railway signalling applications. Another modification of the Lalande–Chaperon type cell which substituted mercury oxide for copper oxide, was initially proposed by Aron in 1886 [20], but found no immediate use. Over fifty years later, it was modified by Samuel Ruben [21] and became an extremely successful primary battery system. It found extensive use in high-reliability electronic and medical equipment with high energy density and very flat voltage performance curves.

In an attempt to make a more reversible cell by elimination of the soluble zinc material in his Lalande type cell, Edison proposed using cadmium as the negative in a 1900 patent [22]. Soon this was followed by his work with nickel electrode positive systems, especially nickel–iron.

Desmazures had previously mentioned nickel oxide as a possible alternative to copper oxide in one of his earlier alkaline cell patents. In 1887, nickel oxide was again mentioned as a possible active material by Dun & Haslacher [23]. However, no one seems to have made any serious experiments with the oxides or hydroxides of nickel until Waldemar Jungner noted the suitability of these compounds as active materials based on his experiments with the formation of nickel sheet in 1897–1898 [24].

Ch. 1]    **Historical review of battery development and commercialization**    25

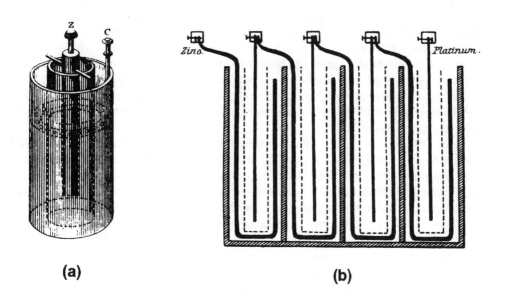
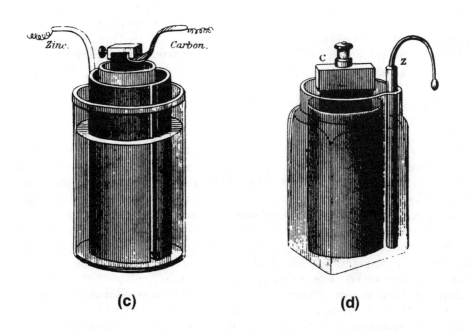

Fig. 1.1 — Nineteenth century drawings of battery cells. (a) Daniell cell; (b) Grove cells; (c) Bunsen cell; (d) Leclanché cell.

Independently, T. de Michalowski described methods for the preparation of nickel active material in an 1899 patent [25]. It is evident that work was being carried out in many different locations on alkaline accumulators with relatively insoluble electrodes during the last decade of the 19th century. Included were the activities of the French Lieutenant, V. Darrieus, who described the electrode combinations bismuth–copper oxide and cadmium–copper oxide in an 1893 patent [26]. These systems, however, had very low cell potentials. Another interesting report was the 1898 patent of Pollack [27] who disclosed a cell in which both electrodes consisted of iron or iron oxides.

However, the fundamental inventions and commercialization of the nickel electrode secondary battery are mainly due to the independent activities of Waldemar Jungner in Sweden and Thomas Alva Edison in the USA. They were both motivated by their interests in electric vehicles, and they researched systems in which the electrolyte did not enter into the chemical reaction. This ruled out the lead–acid battery system. Jungner carried out research on many positive and negative active materials including the compounds of silver, manganese, copper, nickel, iron, and cadmium. In parallel independent studies, Edison studied the systems nickel oxide (hydroxide)–cadmium, iron–cobalt oxide, zinc–mercuric oxide, and on other combinations. His group carried out extensive work on the effects of additives, such as mercury to iron electrodes, as well as manufacturing methods. Eventually Jungner and his successors developed the nickel–cadmium battery and Edison the nickel–iron battery. After many early technical problems, both systems became successful articles of commerce; the redesigned Edison cell in 1908 [28] and the improved Jungner type cell in 1910 under the management of Robert Ameln and the engineering of A. Estelle [29].

Although the three systems above described — zinc–manganese dioxide, lead–acid, and nickel–cadmium — dominated the commercial marketing of battery systems, there were a great number of other systems used for speciality applications. Although the early systems had metallic electrodes and liquid aqueous electrolytes, many new systems often had non-aqueous liquid or solid electrolytes. Electrode materials could be solids, liquids, or gases.

Captain Charles Renard used a high specific energy zinc–chlorine battery as a power source on his dirigible (airship) 'La France' in 1884 [30]. The chlorine was made *in-situ* by reacting $CrO_3$ with HCl. H. H. Dow, the founder of the Dow Chemical Company, proposed a zinc–bromine storage battery in a 1902 article in the Transactions of The American Electrochemical Society [31]. Some eighty-five years later this system became a commercial reality [32]. In the interim, conducting plastics and bipolar type cell designs became common, which made the new design practical. This underlines the point that battery systems depend upon materials science and materials availability in order to be commercially viable. Over thirty chemical elements are used in battery compounds, and in some cases, for instance lead, fabrication of batteries is the major use for that element.

There are more than two dozen primary battery systems in commercial use [33] and a somewhat smaller number of storage battery systems. Specific systems and designs find application because they are the most economical product that is readily available for a particular performance requirement.

The lithium anode battery systems evolved in the 1970s because of the need for long shelf life in heart pacers and other electronic devices that did not require high currents. Nickel–hydrogen battery systems are being developed to replace nickel–cadmium in those applications where cycle life requirements exceed 10 000 cycles and for special environmental situations. Zinc–silver oxide designs have been used to provide very high power density systems and, at times, to fabricate a non-magnetic battery. Thermal batteries provide a type of power source with almost infinite storage life. Thus, it can be seen that battery technology is constantly evolving as new devices and applications appear. Some systems are also being modified to provide methods for convenient materials recovery from spent batteries. Fabrication techniques are also changing, in order to make battery assembly environmentally safer.

Table 1.1 lists the significant events which have occurred during the history of battery development.

**Table 1.1** — Chronology of battery developments and related events

| Year | Event |
|---|---|
| 1791 | Galvani frog leg experiment |
| 1792 | Volta pile battery |
| 1834 | Faraday laws of electrochemical reactions |
| 1836 | Grove cell |
| 1859 | Planté lead–acid battery |
| 1860 | Leclanché, Zinc–$MnO_2$ system (dry cell) |
| 1866 | Siemens dynamo |
| 1875 | Practical $MnO_2$ dry cell |
| 1881 | Glasstone & Tribe, double sulphate theory of reaction for the lead–acid battery |
| 1881 | Sellow, lead–antimony alloy |
| 1885 | Bradley, zinc–bromine battery patent |
| 1899 | Jungner, alkaline accumulators including nickel–cadmium |
| 1905 | Edison, commercial production of nickel–iron batteries |
| 1927 | André, zinc–silver oxide cells |
| 1928 | Pflerder, Spoon, Gimelin, & Ackerman, sintered electrode |
| 1935 | Haring & Thomas, lead–calcium alloy |
| 1950 | Ruben, sealed zinc–mercuric oxide cell |
| 1956 | Bacon, alkaline fuel cell |
| 1966 | Kummer & Weber, sodium–sulphur battery |
| 1970 | Tobias aprotic solvent research |
| 1980s | Sealed lead–acid batteries become common |
| | Practical zinc–bromine secondary battery becomes available |
| | Hydride type hydrogen electrodes go into commercial production |
| | Lithium–$MnO_2$ primary batteries capture a large market |
| | Sealed small nickel–cadmium cells exceed 10% of the value of all storage battery system marketing. |

The commercial applications for traditional and new battery systems are continuously expanding on a worldwide basis. Batteries are produced locally in almost every inhabited region of the World, and are traditionally not shipped long distances because of high shipping costs relative to the battery value for most systems.

The manufacturers' value for battery production in 1989 was approximately $19 thousand million (over £10 thousand million). The retail value of products would probably be three times as large. The manufactured values for the different battery system/design segments are listed in Table 1.2.

**Table 1.2** — Estimated production values for different battery system/design segments (1989)†

|  | US dollars (millions) |
|---|---|
| *Primary batteries (consumer types)* | |
| Leclanché (zinc–$MnO_2$) | 3 000 |
| Alkaline–Manganese | 2 000 |
| Lithium | 900 |
| Metal–air, zinc–silver oxide, zinc–mercury oxide | 900 |
| Other | 800 |
|  | 7 600 |
| *Secondary batteries* | |
| Lead–acid (SLI) | 6 000 |
| Lead–acid (industrial) | 1 400 |
| Lead–acid (small sealed) | 400 |
| Nickel–cadmium (small sealed) | 1 100 |
| Nickel–cadmium (industrial) | 700 |
| Other secondary | 300 |
|  | 9 900 |
| *Special* | |
| Reserve, and water activated | 100 |
| Military, aerospace (including thermal) | 500 |
| Implanted biomedical | 100 |
| Other | 300 |
|  | 1 000 |
| Total: | 18 500 |

†Worldwide not including the USSR, China, and other countries for which data are not available.

## NEW TRENDS AND FUTURE GROWTH

In the last decade, the fastest growing segments of battery technology and markets have been in small sealed cells. The market for small sealed-cell nickel–cadmium cells has grown by a factor of five, and that for small sealed maintenance-free (MF) lead–acid by a similar amount. New research and developments are being reported in the journals to extend these technologies to much larger size batteries, especially in the case of lead–acid.

In the last year, there has been a much more concerted effort towards energy storage systems. This has included utility load-levelling and customer side of the meter efforts to reduce peak current charges. New concepts such as horizontal electrode sealed lead–acid are being developed for these applications which also entails more sophisticated charge control circuits.

Lastly, the effect of new chemical, analytical and instrumental techniques on the basic understanding of electrode structures and mechanisms has been apparent. Further improvement of battery technology is expected, owing to these influences, and this will be assisted by the increasing availability of new materials for cases, separators, electrode structures, active materials, and electrolytes. However, in many countries there appears to be a diminishing amount of industrial research and development and fewer academic institutions where battery research is carried out. These problems will undoubtedly retard the rate of introduction for some new battery technologies.


## REFERENCES

[1] König, W. *Forsch. Forschr.*, **14**(1), 8, (1938).
[2] Volta, A. *Phil. Trans. Roy. Soc.*, **90,** 403 (1800).
[3] Cruikshank, W. *Tilloch's Phil. Mag.*, **7,** 337 (1800).
[4] Daniell, J. F. *Phil. Mag.* III **8,** 421 (1836).
[5] Grove, W. *Phil. Mag.* III, **14,** 388 (1839), and **15,** 287 (1839).
[6] Bunsen, R. W. *Pogg. Ann. Physik.*, **54,** 417 (1841).
[7] Leclanché, G. French Patent 69980 (1866).
[8] Planté, G. *Compt. Rend. Acad. Sci. Paris,* **50,** 640 (1860); *Recherches sur l'Electricite,* Gauthier-Villars, Paris (1883).
[9] Glasstone, J. H., & Tribe, A., *Nature* **25**, 221, 461 (1881) and **26**, 251, 342, 602 (1882).
[10] Sellow, British Patent 3987 (1881).
[11] de Lalande, F. & Chaperon, G. US Patent 274,110 (1883); French Patent 143,644 (1881).
[12] Desmazures, C. French Patent (1887), US Patent 402,006 (1889).
[13] DeVirloy, A., Commelin, E. & Bailhache, G. French Patent 164,681 (1884); U.S. Patent 345,124 (1886).
[14] Boettcher, E. German Patent 57,188 (1890).
[15] Schoop, P. British Patent 7711 (1893).
[16] Waddell, J. & Entz, J. B. US Patent 461,858 (1891).
[17] Entz, J. B. & Phillips, W. A. US Patent 421,916, 440,023, 440,024 (1890).
[18] Discussion In: Wade, E. J. *Secondary batteries,* the Electrician Printing and Publishing Co., London (1902); and in Falk, S. U. & Salkind, A. J. *Alkaline storage batteries,* John Wiley, New York (1969).

[19] Discussion in Vinal, G. W. *Primary batteries,* John Wiley, New York (1950).
[20] Aron, H. German Patent 38,220 (1886).
[21] Ruben, S. US Patent 2,422,045 (1947).
[22] Edison, T. A. British Patent 20,960 (1900).
[23] Dun, A. & Haslacher, F. British Patent 1,862 (1887).
[24] Jungner, W. Swedish Patent 10,177 (1899).
[25] de Michalowski, T. British Patent 15,370 (1899).
[26] Darrieus, V. French Patent 233,083 (1893).
[27] Pollack, C. German Patent 107,727 (1898).
[28] The Edison Storage Battery Co., Orange, N.J., Commercial literature (1934–1960).
[29] Estelle, A. British Patent 9,964 (1910).
[30] Renard, C. *The light batteries of the dirigible La France,* Aeronautics Review Library, 1890; and in *History of battery technology* Salkin, A. ed., *Proc. Electrochemical Soc.,* **87**-14 (1987) p. 268.
[31] Dow, H. H. *Trans. Am. Electrochem. Soc.* **1**, 120 (1902).
[32] Grimes, P. Exxon Corp., Private Communication.
[33] Salkind, A. J. & Brodd, R. Primary and secondary batteries, in electrochemistry and solid state science education, *Electrochem. Soc. Proceedings* **87**-3 (1987).

# 2

# The chemistry and electrochemistry of battery systems

(C. D. S. Tuck, 2.1 to 2.4 & A. Gilmour, 2.5)

In 1791 it was discovered by Luigi Galvani of the University of Bologna that two dissimilar metals, when contacted by a moist substance, would allow the passage of a direct current between them. This principle was developed further by Alessandro Volta of Pavia University, who arranged stacked pairs of zinc and silver disks which were separated by cardboard soaked in sodium chloride solution. By connecting wires to the top and bottom of the stack, he produced the first battery power source, the 'voltaic pile'. In this case there were chemical reactions taking place at both the zinc anode and the silver cathode such that electrons were generated at the anode and discharged at the cathode, these two reactions being known as half-cell reactions. The two reactions of the voltaic pile are:

(i) Anodic reaction at the zinc anode

$$Zn \rightarrow Zn^{2+} + 2e^- \qquad (2.1a)$$

(ii) Cathodic reaction at the silver cathode

$$2 H_2O + 2 e^- \rightarrow H_2 + 2OH^- \qquad (2.1b)$$

The voltaic pile is termed a corrosion cell because the positive electrode (silver) does not actually take part in the cathodic reaction as this is simply the reduction of water. Zinc itself would undergo both these reactions on its surface if immersed in an aqueous electrolyte, owing to its inherent self-corrosion, although the water reduction reaction equation (2.1b), occurs more readily on silver than on zinc.

## 2.1 THE THERMODYNAMICS OF BATTERY SYSTEMS

### 2.1.1 The equilibrium cell potential

A battery in which anode and cathode participate in the overall cell reaction is the Daniell cell of 1836 which is named after its inventor, J. F. Daniell. Fig. 2.1 shows

the case of the Daniell cell this becomes the equation

$$E_{th} = E_{Cu^{2+}/Cu} - E_{Zn/Zn^{2+}} \qquad (2.8)$$

When the cell components possess activities of unity

$$E_{th} = 0.36 - (-0.76)$$
$$= 1.12 \text{ V} .$$

If the direction of equation (2.2c) is reversed, denoting a cell charging reaction, the cell voltage, by this convention, would be $-1.12$ V.

Values of voltages for various half-cell reactions are given in Appendix B, and these enable overall theoretical cell voltages to be derived if the anodic and cathode equations are combined. It is apparent that the highest voltages would be achieved by coupling the most reactive metals with the most powerful oxidizing agents, but this may not be practically possible. From the theoretical cell potentials and equation (2.5), theoretical energies of the battery components can be derived, and these are useful when comparing different systems, particularly when expressed in terms of the weight or volume of the components.

They are defined theoretically as:

(i) Gravimetric energy density,
$$(E.D.)_G^{Th} = nFE_{th}/(\text{total molar mass of active components}) \qquad (2.9)$$

(ii) Volumetric energy density,
$$(E.D.)_V^{Th} = nFE_{th}/(\text{total molar volume of active components}) \qquad (2.10)$$

In practical batteries, the energy content of the actual systems are measured in terms of watt-hour (Wh) of output and compared with the battery weight and volume. Practical energy densities are defined as

(i) Gravimetric energy density,
$$(E.D.)_G^{Pr} = \text{Wh of discharge/battery mass (kg)} \qquad (2.11)$$

(ii) Volumetric energy density,
$$(E.D.)_V^{Pr} = \text{Wh of discharge/battery volume (litres or dm}^3) \qquad (2.12)$$

The volumetric energy density would obviously be influenced by the overall shape and construction of the battery. Needless to say, the practically measured energy densities are somewhat less than the theoretical ones, particularly for the higher theoretical energy density cases, and Fig. 2.2 shows that there is an approximate relationship

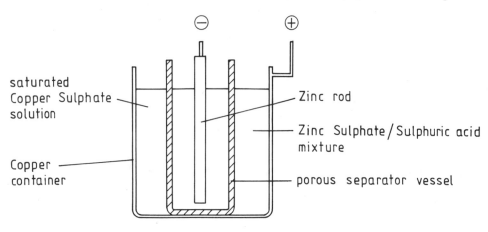

Fig. 2.1 — The components of a Daniell cell.

schematically how the cell is constructed and demonstrates its two compartments which house the zinc anode and copper cathode respectively and allow them to undergo their separate half-cell reactions. The reactions are:

(i) Anodic reaction at the zinc anode

$$Zn \rightleftarrows Zn^{2+} + 2e^- \qquad (2.2a)$$

(ii) Cathodic reaction at the copper cathode

$$Cu^{2+} + 2e^- \rightleftarrows Cu \qquad (2.2b)$$

The basic design of the cell comprises a zinc electrode immersed in an electrolyte containing zinc ions and a copper electrode in the form of the cell container, filled with a solution of copper ions. In the situation where no current was flowing through the cell, the reactions denoted by equations (2.2a) and (2.2b) would both be at equilibrium. The overall cell at equilibrium would thus be given by the reaction equation

$$Zn + Cu^{2+} \rightleftarrows Zn^{2+} + Cu \qquad (2.2c)$$

For this reaction, the standard enthalpy change at 25°C is $\Delta H° = -227$ kJ. mol$^{-1}$. The quantity of heat which would be released if the reaction occurred by simple mixing of reactants would be equal to $-\Delta H$. However, when a spontaneous reaction occurs electrochemically in a cell the heat released in the cell ($-q$) is less than $-\Delta H$ because energy is also dissipated in the load. This can be expressed as the equation

$$-\Delta H = -q + \int V dQ$$

Here, $V$ is the cell voltage and $Q$ is the electrical charge p situation occurs when the load becomes vanishingly small. In th becomes zero such that equilibrium is established (i.e. the rea reversibly). At equilibrium, $V$ assumes its maximum value which is equilibrium e.m.f., and $-q$ reaches its minimum value of $T\Delta S$ absolute temperature and $\Delta S$ the entropy change in the reaction. Thi be expressed by the equation

$$-\Delta H = -T\Delta S + E_{th} nF$$

where the charge, $Q$, is given as the product of $n$, the number of mol of el passed, and the Faraday, $F$ (the charge per mol of electrons). The latter has the of 96 487 C. mol$^{-1}$, and $E_{th}$ for the Daniell cell is 1.12 V if all the cell component at unit activity. The maximum amount of electrical work available from the react is given by

$$-\Delta G = nE_{th} F \qquad (2.5)$$

where $\Delta G$ is the Gibbs free energy change of the reaction, in accordance with the familiar relationship

$$\Delta G = \Delta H - T\Delta S \ . \qquad (2.6)$$

The potential of 1.12 V generated at equilibrium by the Daniell cell can be expressed as the difference between the individual potentials associated with the electrode reactions given by equations (2.2a) and (2.2b). As it is impossible to measure individual electrode potentials in an absolute sense, they are each measured with reference to another electrode which is used as a standard. The electrode normally chosen for this purpose is the standard hydrogen electrode (SHE) which in practice is usually composed of a platinum black foil immersed in a solution of hydrogen ions of unit activity and saturated with hydrogen gas at a partial pressure of 1 atm. With this electrode as reference and assuming the electrolytes are at unit activity, the zinc electrode reaction, equation (2.2a), produces an electrode potential $E_{Zn/Zn^{2+}}$ of $-0.76$ V, and the copper electrode reaction, equation (2.2b), gives a potential $E_{Cu^{2+}/Cu}$ of $+0.36$ V. It is possible to calculate the overall theoretical cell potential, $E_{th}$, by subtracting the anode electrode potential, $E_{anode}$, from the cathode electrode potential, $E_{cathode}$ giving the expression

$$E_{th} = E_{cathode} - E_{anode} \ . \qquad (2.7)$$

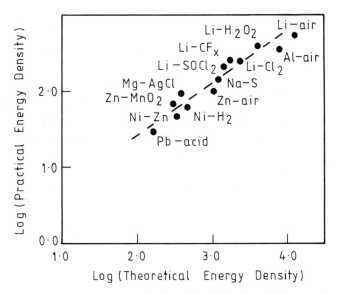

Fig. 2.2 — The relationship between the logarithm of the theoretical energy density and the logarithm of the energy density achieved in practice for a number of different battery systems.

$$(E.D.)_G^{Pr} \propto [(E.D.)_G^{Th}]^{0.7} \qquad (2.13)$$

The existence of this seemingly ubiquitous equation suggests that there may be quantifiable theoretical constraints on the dissipation of the chemical energy stored in batteries, caused by the need to ensure that it occurs in a safe manner.

### 2.1.2 The effect of electrolyte concentration on cell voltage

The Daniell cell reaction, equation (2.2c), represents a chemical reaction at equilibrium. Thus it is possible to employ the Van't Hoff isotherm to the reaction to produce an equation using the activities of the different reaction components, $a_c$, together with the equilibrium constant of the reaction, $K$, to obtain the reaction free energy, $\Delta G$. This equation is

$$\Delta G = -RT \ln K + RT \ln \left\{ \frac{[a_{Zn^{2+}}][a_{Cu}]}{[a_{Zn}][a_{Cu^{2+}}]} \right\} . \qquad (2.14)$$

If the activities are all unity, application of equation (2.5) gives

$$\Delta G° = -RT \ln K = -nFE° \qquad (2.15)$$

were $E°$ is defined as the standard e.m.f. of the cell. A generalized equation for the observed cell voltage, $E_{th}$, involving activity terms can be derived from equations (2.14) and (2.15), and the resulting expression is

$$E_{th} = E° + \frac{RT}{nF} \ln \left[ \frac{\text{product of activities of the reactants}}{\text{product of the activities of the reaction products}} \right]. \quad (2.16)$$

Thus it can be seen that, generally, as a battery discharge proceeds so that the activities of the products increase and the activities of the reaction products decrease, the open circuit cell voltage will fall. This is primarily the case when the reactants or products are soluble in the electrolyte or form a solid with the electrodes. An example of the latter occurs when lithium ions, formed by discharge of a lithium battery, form intercalation compounds with certain transition metal oxide cathode materials. Fig. 2.3 shows the sloping open circuit voltage observed during discharge

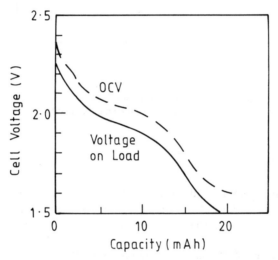

Fig. 2.3 — The constant load discharge behaviour and corresponding open circuit voltage (OCV) during the discharge of a cell which employed a Li–Al alloy anode and a transition metal oxide (intercalation type) cathode.

of such a cell, together with its discharge voltage when under load, which follows a similar pattern. If the particular battery has reactants or products which do not form solutions during discharge or which have reactants of constant activity supplied to the system, then the open circuit voltage will remain constant during discharge. For instance, in the case of the Al–air battery operating with a sodium chloride electrolyte, a flat open circuit behaviour during discharge is observed, owing to the rapid precipitation of aluminium hydroxide, thus avoiding the accumulation of $Al^{3+}$ ions in solution which would cause the cell potential to fall. Its behaviour on discharge, together with the correspondingly measured open circuit voltage, is shown in Fig. 2.4.

## 2.2 THE ELECTROCHEMISTRY OF BATTERY SYSTEMS

Electrochemistry defines the practical usefulness of couples which seem thermodynamically suited as battery materials, because the rates of electron transfer reactions

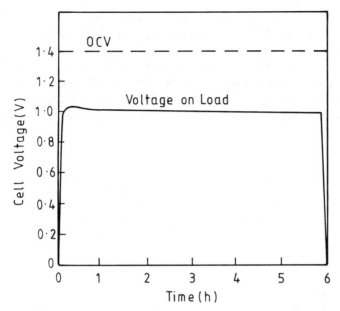

Fig. 2.4 — The discharge behaviour under constant load and corresponding open circuit voltage (OCV) of an aluminium–air battery having an electrolyte of 12% NaCl.

at the electrode–electrolyte interfaces determine the possible power available. Studying the electrochemical performance of individual battery electrodes separately is a valuable technique towards understanding the overall behaviour observed when the electrodes are coupled together in a battery.

### 2.2.1 Voltage losses of operating cells

As discussed in section 2.1, if two dissimilar metals are connected together and immersed in an appropriate battery electrolyte under equlibrium conditions, a voltage, $E_{th}$, will be developed between them. However, if this situation is disturbed by the application of a load which causes current drain, the voltage between the electrodes will begin to fall from its equilibrium value, usually fairly rapidly. This process is termed polarization. It will take place more readily if the electrolyte is diluted or of lower conductivity, showing that the effect is somewhat dependent on ohmic factors within the cell. These could be associated with either the electrodes or the electrolyte. Another factor controlling polarization is found to be the rates of charge transfer reactions at the electrodes.

The overall practical cell voltage, $E_{cell}$, is found to be given by the expression

$$E_{cell} = E_{th} - E_{oe} - E_{os} - E_p , \qquad (2.17)$$

where $E_{th}$ has been defined previously as the equilibrium potential (see section 2.1), $E_{oe}$ is the ohmic drop in the electrodes, $E_{os}$ is the ohmic drop in the solution (and any separators) and $E_p$ is the voltage drop due to reactions at the electrode/solution interface. The magnitude of $E_p$ can relate to a number of effects, including charge transfer, nucleation and growth of phases or availability of reacting species. The terms $E_{oe}$ and $E_{os}$ can be minimized by altering the physical design and construction of the battery; that is, by lessening the anode/cathode gap, by ensuring adequate electrode conductivity (particularly if constructed from powders), and by optimizing the connections to the electrodes. Fig. 2.5 shows the effect of changing such physical

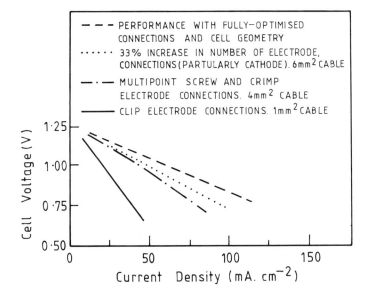

Fig. 2.5 — The improvements which can be made in the relationship between cell voltage and operating current density for a saline aluminium-air battery by optimization of the cell engineering.

factors on the voltage–current behaviour of an aluminium-air battery operating in a saline electrolyte, showing that substantial improvements can be made by judicious engineering.

### 2.2.2 Faraday's laws of electrolysis

Faraday's two laws of electrolysis, first stated in 1833, are of fundamental importance in the electrochemistry of batteries. The laws are as follows:

(i) The quantity of primary product formed by electrolysis is directly proportional to the quantity of electricity flowing.
(ii) The passage of a given quantity of electricity causes the masses of the primary products formed by electrolysis to be in the ratios of the atomic or molecular

weights of those products, each being first divided by the number of electrons involved in their respective formation.

Thus, if a current $i$ flows through a battery for time $t$ and the anode is composed of an active metal of atomic weight, $A_w$, having a valency, $n$, the application of Faraday's laws results in the equation:

$$m = itA_w/nF \tag{2.18}$$

where $m$ is the mass of anode which directly takes part in the battery discharge and $F$ is Faraday's constant (96487 C.mol$^{-1}$). For battery considerations, the charge is more usually expressed as amp-hour (Ah), and Appendix C gives values of Ah.g$^{-1}$ and Ah.cm$^{-3}$ for a number of possible battery materials. When such substances are coupled as battery electrodes, the overall battery capacity relates to the combined mass of both anode and cathode materials, and a theoretical Ah capacity of the system can be determined. For example, in the case of the alkaline–manganese battery (see section 3.2.5), the overall reaction is

$$Zn + 2MnO_2 + H_2O \rightarrow ZnO + 2MnOOH \ . \tag{2.19}$$

A battery of 1 Ah capacity would consume 1.22 g Zn and 3.24 g MnO$_2$. Thus, as the observed cell voltage is 1.5 V, the theoretical gravimetric energy density, $(E.D.)_G^{Th}$, would be calculated as:

$$(E.D.)_G^{Th} = 1.5 \times 1000/(1.22 + 3.24)$$

$$= 336 \ Wh.kg^{-1} \ .$$

In practice, the energy density is much lower. This is because there are supplementary masses due to the electrolyte, separator materials, and case, as well as their being a small excess quantity of zinc included to compensate for off-load corrosion. Also, the MnO$_2$ is made electronically more conductive by the addition of carbon, thereby adding to its bulk. The practical energy density is thus found to be approximately 95 Wh.kg$^{-1}$: about 30% of the theoretical figure.

### 2.2.3 The kinetics of electrode charge transfer

For the metal immersed in a solution of its ions, an equilibrium will be established at the metal/solution interface which is represented by the equation:

$$M \underset{c}{\overset{a}{\rightleftharpoons}} M^{n+} + ne^- \ , \tag{2.20}$$

where $a$ denotes the anodic (dissolution) reaction and $c$ denotes the cathodic (deposition) reaction. To preserve electroneutrality, the resulting metal surface charge will be balanced by an equal and opposite charge distribution in the electrolyte region adjacent to the electrode. This process is known as the generation of an electrical double layer. If the equilibrium is disturbed by the passage of current through the electrode, then electrons will be lost from the metal and an excess of positive ions will be generated at the surface. The electrode will display an overvoltage, η, which is a measure of the degree of voltage disturbance from the equilibrium situation. The electrode is said to be anodically polarized, and it is found that

$$\eta = E_a - E_r ,\qquad(2.21)$$

where $E_a$ is the electrode potential under such polarization conditions and $E_r$ is the electrode equilibrium potential. It is possible to derive an equation for the rate of the anodic reaction by using Faraday's first law, making the rate of the forward reaction of equation (2.20) proportional to the current passed, $i_a$, by the relationship

$$i_a = k_a n\, FA\, [M]\qquad(2.22)$$

Where

$$k_a = k_o \exp(\alpha n \eta F/RT) .\qquad(2.23)$$

In these expressions, $k_a$ is a potential-dependent rate constant, $k_o$ is a standard rate constant, and $A$ is the area of the specimen surface over which the reaction occurs.

In the case of the reverse (cathodic) reaction of equation (2.20), a similar equation is derived:

$$i_c = k_c \text{nFA}\, [M^{n+}]\qquad(2.24)$$

where

$$k_c = k_o \exp(-(1-\alpha)\eta n F/RT) ,\qquad(2.25)$$

where $i_c$ is the cathodic current and $k_c$ is a potential-dependent rate constant.

In general, the observed external current, $i$, is equal to the difference between anodic and cathodic currents, as given by the expression

$$i = i_a - i_c .\qquad(2.26)$$

At equilibrium, $i = 0$, and $i_a$ becomes equal to $i_c$, this value being described as $i_o$, the exchange current. From equations (2.23), (2.25), and (2.26), the so-called Butler–Volmer equation can be derived. This has the form

$$i = i_o [\exp(\alpha n \eta F/RT) - \exp(-(1-\alpha)n\eta F/RT)] \tag{2.27}$$

where

$$i_o = k_o nFA \tag{2.28}$$

The exchange current is a measure of the rate of exchange of charge at the equilibrium potential. It is advantageous for each battery electrode to possess a high $i_o$ value so that $i_a$ and $i_c$ are high for any value of overvoltage.

Fig. 2.6 shows plots for aluminium, zinc, and iron in potassium hydroxide

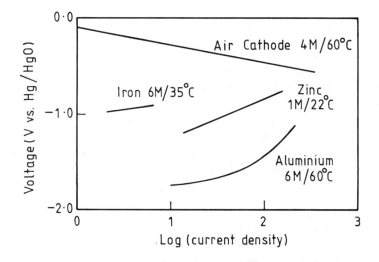

Fig. 2.6 — Tafel plots of oxygen reduction on an air cathode (manufactured by Alupower, Inc.) and the oxidation of iron, zinc, and aluminium anodes in potassium hydroxide at the indicated molarities and temperatures.

solutions of the relationship between their electrode potentials measured against a reference electrode and the logarithm of imposed current density.

Plots of these types are generally termed polarization curves or Tafel plots. Also shown in Fig. 2.6 is the polarization behaviour of an air electrode. Thus, possible metal–air cell voltages at specified current densities would be given by the differences between potential values of the air cathode at those current densities and the corresponding potential values of the particular metal selected by use of equation (2.7). However, equation (2.17) has indicated that, in practice, this voltage is lowered because of the cell internal resistances which add to the electrode polarization losses apparent in Fig. 2.6. These factors need to be considered when calculating the practical cell voltage during discharge, but it is clear that the best electrodes for batteries would be those displaying a low value of d$E$/d(log $i$), that is, low polarization behaviour.

### 2.2.4 The kinetics of cell mass transport

The successful operation of a cell depends not only on the kinetics of the charge transfer reactions at the electrodes but also on the ease with which ionic species and battery reaction products can be transported in the electrolyte either toward or away from electrodes. If the electrolyte is flowing during cell operation by pumping or convection, this will facilitate mass transport. However, if the cell experiences low electrolyte flow or stagnant conditions, then diffusion phenomena become more apparent, these being either normal diffusion due to concentration gradients developed in the cell or electromigration of charged species due to the applied electric field. These two cases will now be considered.

### (i) Normal diffusion mechanism

In this case, the normal equations derived by Fick can be used to calculate the relationship of the concentration of species in solution both with its distance from the electrode surface and with time. To simplify the procedures, it is assumed that, in stirred solutions, there is a thin electrolyte layer of thickness $\delta$ (known as the Nernst diffusion layer) close to the electrode which remains stagnant, and that beyond this layer, the solution is of completely uniform concentration. Thus, if [S] is the bulk concentration of active species, S, and $[S]_s$ is the corresponding concentration immediately adjacent to the electrode surface, then the steady-state diffusion flux of ions towards the electrode, $j_{s,d}$, will be given by:

$$j_{s,d} = D\{([S] - [S]_s)/\delta\} \qquad (2.29)$$

where $D$ is the diffusion coefficient of S.

The reaction rate, measured as the modulus of current $|i|$ as S may be oxidized or reduced, will be given by a similar equation to equation (2.22), namely:

$$|i| = nFAj_{s,d} \, . \qquad (2.30)$$

In such a case, $|i|$ will have a diffusion-limited value, $i_{l,d}$. From equations (2.29) and (2.30), this limiting current is given by

$$i_{l,d} = nFAD[S]/\delta \, . \qquad (2.31)$$

For example, an oxygen cathode operating submerged in stagnant seawater has a limiting current density of approx. 40 $\mu$A.cm$^{-2}$ ($D \sim 10^{-5}$ cm$^2$.s$^{-1}$, [S] $\sim 2 \times 10^{-7}$ mol.cm$^{-3}$, $\delta \sim 10^{-2}$ cm). Also, from equation (2.29) and (2.31), the ratio of $[S]_s$ to [S] can be derived as:

$$[S]_s/[S] = (i_{l,d} - |i|)/i_{l,d} \, . \qquad (2.32)$$

This can be substituted in a modified form of equation (2.16), to give a value for the concentration overvoltage, $|\eta_{conc}|$, according to:

$$|\eta_{conc}| = -\frac{RT}{nF} \ln\left[\frac{i_{1,d} - |i|}{i_{1,d}}\right]. \qquad (2.33)$$

*(ii) Diffusion with electromigration*
In this case the migrating species will carry a fraction of the ionic current termed the transport number, $t_s$, and the flux of the species under electromigration, $j_{s,e}$, will relate to $i$ by a similar equation to the diffusion-only case, equation (2.30); allowing $j_{s,e}$ to be given by

$$j_{s,e} = t_s i/nF. \qquad (2.34)$$

The total flux of species will be the sum of the diffusion flux and electromigration flux, and rearrangement of equation (2.29) added to equation (2.34) produces a modified form of equation (2.31), giving the limiting current density, $i_{1,e}$, under these conditions, as:

$$i_{1,e} = \frac{nFD\,[S]}{\delta(1-t_s)} \qquad (2.35)$$

It should be noted that the concept of single ion diffusion coefficients is meaningful only in a vast excess of inert electrolyte. However, there is no problem with diffusion of an electrolyte which overall is neutral. For combined diffusion and migration of an ion it is possible to consider two cases:

(a) There is a considerable excess of inert electrolyte, in which case the transport number of the ion is zero.
(b) The ion is derived from a binary salt, in which case the transport number can be measured and used, but the ion cannot diffuse independently of its counter-ion.

It is probable that any other cases are too complicated to deal with satisfactorily.

### 2.2.5 Inefficiencies of battery operation

#### 2.2.5.1 *Self-discharge processes*
For a battery to operate efficiently, the electrode materials should be only used during the discharge period and should take part only in reactions which generate external current. In reality, side reactions occur, and, in aqueous electrolytes, the production of hydrogen from the anode is the normal cause of battery inefficiency. This hydrogen is often generated by local corrosion caused by the presence in the

material of noble element impurities which have low overvoltages for the hydrogen evolution reaction. For instance, an aluminium electrode may contain iron-bearing second phase particles, as shown schematically in Fig. 2.7. In this case, the operating

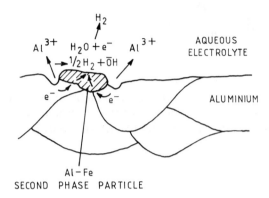

Fig. 2.7 — A local cathodic site produced at an Al–Fe second phase particle on an aluminium battery anode surface during discharge.

polarization curves for anodic and cathodic processes are shown in Fig. 2.8, where

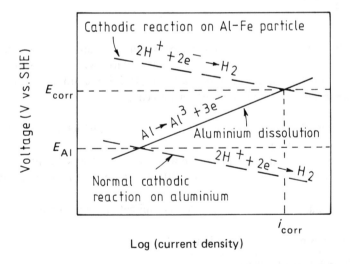

Fig. 2.8 — Plots of the Tafel behaviour of the normal polarization of aluminium in aqueous electrolytes and the hydrogen reduction (cathodic) reaction on Al–Fe second phase particles present due to contamination of the aluminium with iron. $E_{Al}$ is the normal open circuit electrode potential observed on pure aluminium and $E_{corr}$ is the potential observed on aluminium containing Al–Fe particles, $i_{corr}$ being the corrosion current density in the latter situation.

current density is plotted against voltage. It is found that the open circuit voltage displayed by the pure aluminium electrode is shifted in a positive direction to a corrosion potential $E_{corr}$. This results in a lower battery voltage caused by the more positive operating potential of the aluminium and a residual anodic and cathodic current, $i_{corr}$, on switch-off. Also, the iron phases will occupy some of the electrode surface area and may accumulate during discharge. Thus the operating anode area will be lower than expected, resulting in a higher current density for this area and a subsequent further lowering of the battery operating voltage.

Some metals in certain aqueous electrolytes display an increased generation of hydrogen during their anodic discharge. Fig. 2.9 shows this in the case of aluminium

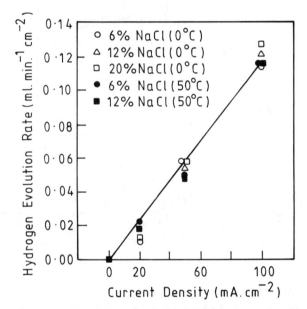

Fig. 2.9 — The hydrogen evolution rate during aluminium dissolution under galvanostatic conditions in NaCl of various concentrations at temperatures of 0°C and 50°C.

in saline electrolytes which displays a linear increase in hydrogen evolution rate as the anodic current density increases (true Tafel behaviour predicting a decreasing hydrogen rate).

Magnesium, iron, chromium, nickel, cobalt, and zinc can also display this phenomenon, which has been termed the 'negative difference effect' [1]. Its explanation is as yet unknown, but it has been suggested that it is caused by the reduction of water by singly charged positive metal ions [2] which exist for a short time before being oxidized to the normally charged state of the metal. If this process occurs, then even the use of ultra-pure metal electrodes will result in a coulombic efficiency of less than 100%.

### 2.2.5.2 Passivation processes

Reactions of the electrode surfaces with the battery electrolyte may cause a blocking or passivation of the electrode surface which will result in a reduced battery voltage and difficulties in achieving adequate voltage levels immediately upon switch-on. In magnesium batteries, the magnesium reacts with the aqueous battery electrolyte and is coated with a film of magnesium hydroxide. This layer is an advantage for long storage periods off load, but causes a 'voltage delay' effect when the battery is activated. This is shown in Fig. 2.10 [3], and the effect is enhanced as the temperature of operation falls.

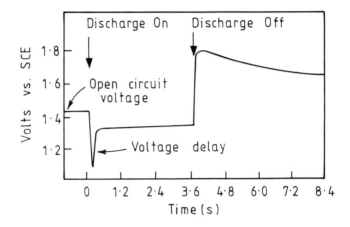

Fig. 2.10 — The voltage transient behaviour of an AZ21 magnesium alloy (Mg 2%, Al 1%, Zn 0.2%, Mn 0.15%) discharged at 20 mA. cm$^{-2}$ in an aqueous electrolyte containing 4M Mg(ClO$_4$)$_2$,Li$_2$ CrO$_4$ [3]. (Courtesy of Prof. R. Das, Editor: *Annual Battery Conference, CSULB*).

In lithium batteries, the unexpected stability of lithium in the non-aqueous electrolytes used has been found to be due to a rapidly formed porous film, this generally being 1.5–2.5 nm thick [4]. Table 2.1 [5] gives a summary of the present understanding of the composition of some of these films. It is thought that the thickness of such films is limited by electron-tunnelling, and it has been found that the film displays properties generally possessed by solid electrolytes, their having high resistivity. As the films act as separating media between the metal and electrolyte they have been given the term 'solid electrolyte interphase' or SEI. As with the previously mentioned magnesium battery, a voltage delay problem can occur because of the passive layer, and these delays are particularly severe for Li/SOCl$_2$ batteries after prolonged storage at high temperature.

### 2.2.5.3 Rechargeable battery inefficiencies
*Dendritic deposition on recharge*
Efficient recharging of a metal electrode requires the redeposition of atoms on that electrode which will reproduce the topography of the original metal surface as closely as possible. In practice this is rarely achieved, and dendritic deposits are often formed [14]. These result from:

**Table 2.1** — The surface films developed on lithium in several different battery electrolytes [5]

| Electrolytes | Components of surface films | Ref. |
|---|---|---|
| $LiAlCl_4$–$SOCl_2$ | $LiCl$ | [6]–[7] |
| $LiBr$–$SO_2$ | $Li_2S_2O_4$ | [6] |
| | $Li_2SO_3$, $Li_2S_xO_3$, $Li_2S_2O_3$ | [8] |
| $LiClO_4$–PC | $Li_2CO_3$ | [6] |
| | Polymer | [9] |
| | $CH_2CHCH_2OCO_2Li$, $CH_3CHCHOCO_2Li$ | [10] |
| | $CH_2CH_2CH_2OCO_2Li$, $(CH_2CHCH_2OCO_2Li)_2$ | |
| $LiClO_4$–DOL | $Li_2O$ | [11] |
| $LiClO_4$–THF | Polymer | [6] |
| $LiAsF_6$–2MeTHF | $(-O-As-O-)R$ | [12] |
| | $As(OR)_xF_{n-x}$ (n = 3,6) | [13] |

| | | |
|---|---|---|
| PC | : | Propylene carbonate |
| DOL | : | Dioxolane |
| THF | : | Tetrahydrofuran |
| 2MeTHF | : | 2-methyl tetrahydrofuran |

(i) partial blocking of a proportion of the surface by inhomogeneities such as defects present after fabrication or partial coverage by oxide or foreign materials.

(ii) a combination of ohmic drop and diffusion effects causing any unevenness of the deposit to be emphasized.

The presence of dendrites has an adverse effect on battery discharge performance as they tend to react more readily with the electrolyte and can cause internal battery shorts if they become long enough. In some cases they can even perforate the battery separator materials. For instance, owing to dendritic lithium deposition in rechargeable lithium batteries [5], only 40% of the lithium plated during recharge is available for subsequent discharge. However, when aluminium is used as a substrate for the lithium deposition in such batteries, this lithium charge-discharge cycling efficiency rises to near 100%, as, in this case, dendrites do not occur as readily [15].

Zinc electrodes are subject to the related phenomenon of electrode shape change during charge-discharge cycling (see Section 6.1.1).

*Charge retention*
To be effective, rechargeable batteries need to be able to retain charge as long as possible, and this requires a good deal of maintenance either in recharging the batteries frequently or keeping them in a permanently charged state by means of

constant connection to the charger. NiCd batteries can be stored in a discharged state without harm, whereas lead–acid batteries degrade irretrievably if stored in that way. The ability to retain charge decreases logarithmically with temperature, with zinc–silver oxide batteries showing the best overall behaviour in this respect, still possessing 85% of their capacity after three months' storage at room temperature (loss of over 10% of charge in 100 h). The poor charge retention of nickel–iron batteries has resulted in their being rejected as possible electric vehicle batteries [16].

*Overcharge reactions*
During charging of secondary batteries, side reactions often occur which are associated with the electrolysis of the battery electrolyte. In the case of aqueous electrolytes these are the oxygen evolution reaction and the hydrogen evolution reaction at the positive and negative electrodes respectively. These reactions will reduce the efficiency of the battery and will also cause the production of potentially hazardous gases and depletion of the battery electrolyte volume.

In sealed nickel–cadmium batteries a solution to the problem has been found [17]. A chemical shuttle has been devised such that the oxygen evolved from the nickel electrode is consumed at the cadmium electrode after diffusing across the electrode gap (see section 4.2.1.4). In a similar way, ferrocene additions can operate as a chemical shuttle species for overcharge protection in certain lithium batteries [18].

## 2.3 ELECTROCHEMICAL TECHNIQUES USED IN THE DEVELOPMENT OF BATTERIES

Electrochemistry, broadly defined as the study of the chemical effects produced by electricity, has strong links with the science associated with battery development. In electrochemical studies, a potential or current is externally impressed on a particular substance to be investigated while it is immersed in a suitable ion-conducting medium. In a battery the opposite situation occurs in that current is produced by the system itself. Although classical electrochemical methods are truly applicable only to planar solid electrodes or spherical amalgam electrodes of known area, techniques which have been developed for electrochemical investigations can be used very effectively to determine the properties of likely battery electrode materials. Once these properties have been defined for single electrodes, then possible battery couples can be assembled and tested as power sources, although the resulting cells may produce secondary electrochemical effects that are due to their size, three-dimensional geometry, electrode morphology, or non-uniform internal potential distribution.

It is possible effectively to investigate these secondary phenomena by the application of electrochemical techniques to partly or fully assembled batteries. For instance, Fig. 2.11 shows the overall discharge behaviour of an experimental calcium/thionyl chloride cell [19] together with individual measured potentials (against an Ag/AgCl reference electrode) for the calcium anode and carbon cathode recorded *in situ* during the discharge. It is apparent that the performance of the cathode dominates the overall cell characteristics, and subsequent analysis reveals

### Sec. 2.3] Electrochemical techniques used in the development of batteries

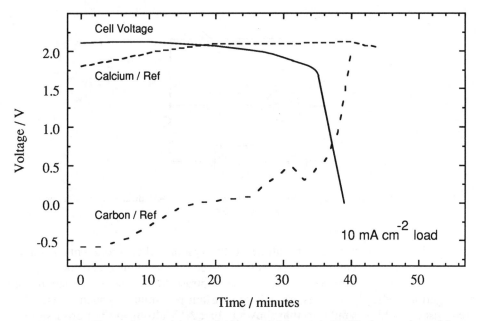

Fig. 2.11 — The anode potential and cathode potential (vs. an Ag/AgCl reference electrode), together with the cell voltage, of a calcium–thionyl chloride cell during a 10 mA. cm$^{-2}$ discharge at 20°C. The electrolyte used was 1 M Na (AlCl$_4$)$_2$ in thionyl chloride, and the cathode was composed of Shawinigan acetylene black [19].

that the failure of the cell is due to blocking of the cathode by the deposition of calcium chloride reaction product within its structure.

Individual investigations of the polarization behaviour of each electrode in a thionyl chloride electrolyte before coupling them together in a cell would allow the calculation of the expected cell voltage by the use of equation (2.7). This would enable a comparison to be made between the expected value and the experimentally observed voltage. Any differences found between the two values would form the subject of an investigation which would aid further cell development. Thus, research is initially carried out on the electrochemistry of single electrodes, and this is related to whole-unit (2-electrode) studies.

#### 2.3.1 Single electrode studies

##### 2.3.1.1 D.C. techniques

For single electrode studies, the simplest type of circuit employs a cell which consists of the electrode to be investigated (working electrode), a counter electrode (usually of platinum, although pure carbon may be preferred), and a reference electrode. The latter is a non-polarizing electrode such as calomel, silver–silver chloride, or mercury–mercury oxide for aqueous electrolytes, or Li/Li$^+$ for non-aqueous electrolytes. These three electrodes are immersed in the particular electrolyte appropriate to the battery in question, and the electrical circuit shown in Fig. 2.12 is

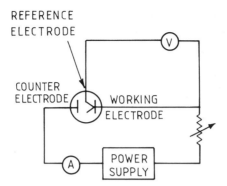

Fig. 2.12 — A three-electrode galvanostatic electrochemical cell. 'A' indicates an ammeter and 'V' a voltmeter.

connected to it. If the system is operated in a battery mode, then a constant current is passed through the circuit and the potential between the working electrode and the reference electrode is recorded. This is in contrast to normal electrochemical investigations, when specimens are held at constant potentials. In such a case, the circuit used would be similar to that shown in Fig. 2.12, although the power supply and variable resistor components of the circuit would be replaced by a single potentiostat, allowing constant potentials to be set and maintained.

The reference electrode is not normally placed directly on the specimen surface, but is connected to a narrow solution-filled tip, known as a Luggin capillary, which is more able to be placed close to the electrode surface. As this capillary would tend to shield part of the electrode surface, an alternative procedure would be to place it through the specimen from the back face. These two geometries are shown in Fig. 2.13. If the Luggin capillary is taken as a parallel planar configuration external to the electrode then the voltage drop $V_{iR}$ experienced between it and the specimen surface can be calculated [20] as:

$$V_{iR} = i\vartheta/\kappa , \qquad (2.36)$$

where $i$ is the operating current density, $\kappa$ is the specific conductivity of the solution, and $\vartheta$ is the effective distance at which the potential is measured, it being slightly less than the specimen-to-tip distance.

Compensation for this voltage drop can be made by the application of a number of well known techniques [21] which will result in a calculated potential measurement as close as possible to the true specimen surface potential.

Fig. 2.14 shows galvanostatic measurements made on some aluminium alloys in 2 M NaCl, exhibiting the advantages of Al–Ga and Al–In binary alloys as possible battery negative electrodes over pure aluminium, owing to their more electronegative behaviour [22]. Correction for the solution resistance in the case of the aluminium produces the negatively sloped dashed line of Fig. 2.14 which has been explained by suggesting [23] that the oxide film is progressively thinned as current

## Sec. 2.3] Electrochemical techniques used in the development of batteries

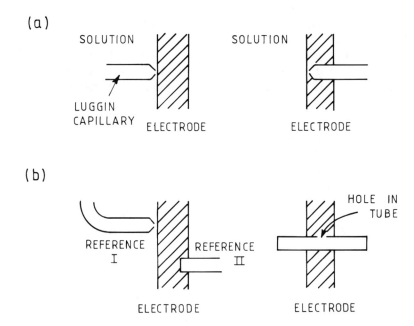

Fig. 2.13 — Possible locations for the Luggin capillaries of reference electrodes. (a) For the examination of solid electrodes (e.g. metals). (b) For the examination of porous electrodes (e.g. metal oxides).

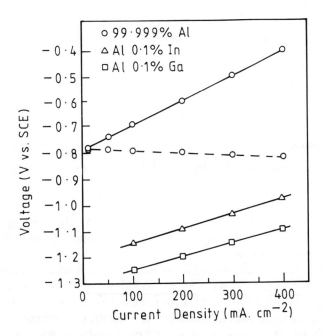

Fig. 2.14 — The potential vs current density of pure aluminium, Al 0.1% Ga, and Al 0.1% In in 2 M NaCl at ambient temperature. The dashed line gives the aluminium behaviour corrected for the voltage drop caused by solution resistance [22,23]. (Courtesy of Pergamon Press).

density increases, causing the aluminium to undergo depassivation and enabling it to display the increasingly negative electrode potentials observed.

Plots of electrode potentials against $t^{1/2}$ are often made, particularly if it is thought that diffusion of an active electrode species takes place during the discharge of an electrode. Fig. 2.15 shows such a plot at various impressed currents [24]. These

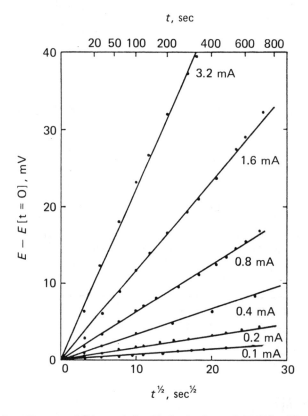

Fig. 2.15 — The potential/time relationship in the form of $(E-E\,[t=0])$ vs $t^{1/2}$ at various galvanostatic currents during the discharge of lithium from an Al–Li alloy electrode [24]. (Reproduced by permission of the publisher, The Electrochemical Society, Inc.)

pertain to the diffusion of lithium out of a discharging Al–Li alloy in a propylene carbonate electrolyte, and, from the slopes of the lines, a lithium diffusion coefficient of $7 \times 10^{-9}$ cm$^2$.s$^{-1}$ can be derived.

In rechargeable systems, cyclic voltammetry is a useful technique in determining oxidation and reduction reactions. As an example of this, Fig. 2.16 shows the cyclic voltammogram of pure lead in 5 M $H_2SO_4$ compared to that taken in 0.1% $H_3PO_4$/5 M $H_2SO_4$, both after cycling one hundred times [25]. In the positive current region the anodic peak (labelled A) is due to the reaction $PbSO_4 \rightarrow PbO_2$ and is the charging reaction. The cathodic peak in the negative current region (labelled B) is the discharge reaction $PbO_2 \rightarrow PbSO_4$. This work shows that the addition of phosphoric

### Sec. 2.3] Electrochemical techniques used in the development of batteries

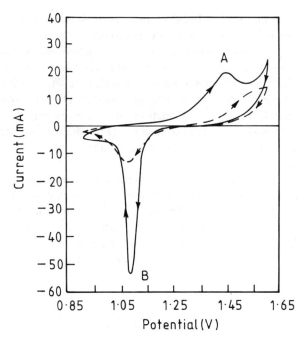

Fig. 2.16 — The cyclic voltammograms of pure lead cycled in 5 M $H_2SO_4$ and 0.1% $H_3PO_4$/5 M $H_2SO_4$ [25]. (Courtesy of International Power Sources Symposium Committee).

acid to the sulphuric acid causes an increase in the potential required to effect the formation of lead dioxide from lead sulphate, and consequently this reaction becomes more difficult. This is contrary to the observations that the addition of phosphoric acid to sulphuric acid has generally been found to be advantageous in terms of cycle life, owing to minimization of the formation of refractory lead sulphate [26].

#### 2.3.1.2 A.C. techniques

The main A.C. technique used is commonly known as 'A.C. impedance' but it is more correctly named electrochemical impedance spectroscopy (E.I.S). It relies on the supposition that electrochemical systems behave linearly under sufficiently small potential perturbations. They can thus be tested, and their A.C. behaviour can be modelled in a similar way to conventional electrical circuits consisting of passive components. At any one frequency, the behaviour of an electrode towards small amplitude electrical perturbations can be described alternatively in terms of

(i) a modulus of impedance, $Z$, and a phase angle, $\phi$
or (ii) a series resistance, $R_s$, and a capacitance, $C_s$
or (iii) a parallel combination of resistance, $R_p$, and capacitance, $C_p$.

In general, the resistive and capacitative components can possess positive or negative values. The dependence of the values of these pairs of parameters upon frequency further characterizes the behaviour of an electrode.

In simple cases the relationship between the electrode and its A.C. equivalent circuit is straightforward. Thus a perfectly polarizable electrode (that is, one involving no Faradaic process), free from effects of adsorption, will behave as a pure, frequency-independent capacitance, corresponding to the value of the double layer capacitance. If the electrode supports a one-step Faradaic reaction, with no adsorption or blockage due to insoluble species, then, provided that the electrolyte solution is sufficiently well stirred, the electrode reaction can be represented by a resistance $R_p$ in parallel with the double layer capacitance $C_p$. Here, $R_p$ will be independent of frequency but will be dependent upon the mean value of the potential or current density being applied.

However, many electrode reactions have several mechanistic steps, and are rate limited by mass transfer or include insoluble species. In such cases the equivalent circuit cannot easily be predicted or interpreted, and will in general require more than two passive components. The simplest circuit proposed for a planar electrode supporting a simple one-step redox reaction of the type $Ox + ne^- = Rd$ is the so-called Randles circuit, shown in Fig. 2.17(a), the $Fe^{2+}/Fe^{3+}$ exchange on

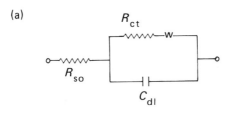

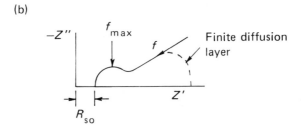

Fig. 2.17 — (a) A simple Randles circuit simulating an active electrode. (b) A schematic representation of the complex plane impedance spectrum resulting from the Randles circuit shown in (a), typical of the $Fe^{2+}/Fe^{3+}$ exchange reaction on platinum. $R_{so}$, $R_{ct}$ and $C_{dl}$ are the solution resistance, charge transfer resistance and double layer capacitance respectively and $2\pi R_{ct} \cdot C_{dl} f_{max} = 1$. W is the Warburg impedance component.

platinum in perchloric acid being such a system. Fig. 2.17(b) shows a typical impedance spectrum of this system, the individual points defining the spectrum being derived at frequencies, $f$, which decrease from left to right. The axes show the real and imaginary components of the impedance ($x$-axis and $y$-axis respectively) and the plot is variously termed a Nyquist plot, Sluyters plot, or Cole-Cole plot. The battery electrode which comes closest to producing Randles circuit behaviour is the zinc

### Sec. 2.3] **Electrochemical techniques used in the development of batteries**

electrode in a solution of 5 M KOH containing $10^{-2}$ M $Zn(OH)_4^{2-}$ ions. Here, the equation

$$R_{ct} = RT/nFi_o \qquad (2.37)$$

is appropriate, where $R_{ct}$ is the charge transfer resistance and $i_o$ is the exchange current density for the zinc oxidation reaction. At low frequencies, a Warburg component, $Z_w$, which has real and imaginary impedance contributions and which is indicative of diffusion control, is evident, and the diffusion coefficient of the zincate ions can be derived from the equation

$$Z_w = \sigma \, \omega^{-1/2} \qquad (2.38)$$

where $\omega$ is the angular frequency and $\sigma$ the Warburg coefficient. The value of $\sigma$ for the above case is given by

$$\sigma = RT/[n^2 F^2 \sqrt{2} \, (C_o \sqrt{D_o})] \qquad (2.39)$$

where $C_o$ is the concentration of zincate ions at the electrode/solution interface, $D_o$ their diffusion coefficient, and $n$ is the number of electrons transferred in the electrochemical reaction.

If a zinc electrode in KOH solution is polarized at a potential slightly negative of that required for zinc dissolution as zincate ions, then it behaves as a perfectly polarizable electrode. The impedance behaviour of a flat zinc electrode under such conditions is shown in Fig. 2.18(a), and it is compared in Fig. 2.18(b) with the behaviour of a sintered (porous) zinc electrode polarized similarly. For the latter, the high frequency limit corresponds to the normal flat electrode behaviour, as only the top surface of the electrode is operating, whereas the low frequency limit depends on the area of the pores. The behaviour at intermediate frequencies depends on the pore sizes and geometry. The penetration distance of the A.C. diffusion wave into the solution $l_s$, is given by

$$l_s = (D^{1/2}/2\omega)^{1/2} \qquad (2.40)$$

where $D$ is the diffusion coefficient of the diffusing species and $\omega$ is the angular frequency. The penetration distance, $l_p$, of the signal down a pore which is purely capacitative is given by

$$l_p = (\kappa r/4\pi f C_{dl})^{1/2} , \qquad (2.41)$$

where $\kappa$ is the solution conductivity, $C_{dl}$ is the double layer capacitance (typically 20 $\mu F \, cm^{-2}$), $f$ is the frequency, and $r$ is the pore radius.

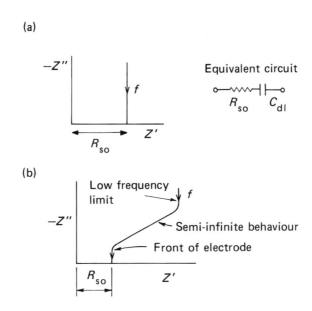

Fig. 2.18 — Complex plane plots of impedance spectra, $Z''$ being the imaginary impedance component and $Z'$ being the real impedance component. (a) The spectrum of a flat zinc electrode in KOH solution, polarized just negative of the potential required for its discharge as zincate. The corresponding equivalent circuit is shown. (b) The spectrum of a sintered zinc electrode polarized under the same conditions. $R_{so}$ and $C_{dl}$ are the solution resistance and double layer capacitance respectively.

Non-Randles behaviour is very common, and Fig. 2.19 shows such behaviour in the Li/V$_2$O$_5$ cell. Here the charge transfer resistance seen is due to Li$^+$ ion transfer at the V$_2$O$_5$ interface.

### 2.3.2 Whole unit (two-electrode) studies

#### 2.3.2.1 D.C. techniques

In whole unit studies, full cells containing both anode and cathode are subjected to electrical testing of various kinds in an attempt to obtain parameters which can be used to judge cell performance and energy availability. In the most elementary kind of test the cell is simply discharged, and this is carried out either through a constant load or with an applied constant current.

Fig. 2.20 shows the effect of these two modes of discharge on a typical cell. Both curves exhibit three stages, denoted by A, B, and C. In region A, during the initial discharge period, there is a rapid voltage fall due to electrochemical polarization of the electrodes. As the electrode materials dissolve, this causes the conductivity of the battery electrolyte to fall steadily, consequently the overall internal resistance rises. This produces the gradual decline of voltage in region B. When the active material in the battery becomes depleted and the internal resistance very high, the voltage falls off at a rapid rate, as seen in region C. It is evident that the constant load drain produces a lower initial voltage but a longer-lived battery, owing to the high initial

Sec. 2.3] **Electrochemical techniques used in the development of batteries** 57

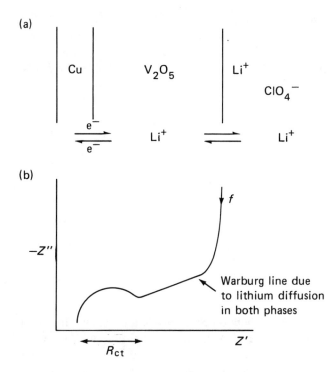

Fig. 2.19 — (a) The configuration of a Li/V$_2$O$_5$ cell cathode with copper current collector and lithium perchlorate electrolyte. (b) The impedance spectrum obtained from a Li/V$_2$O$_5$ cell.

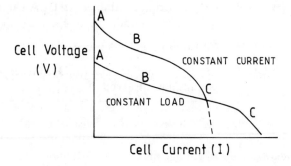

Fig. 2.20 — The typical voltage versus current behaviour observed for cells discharged under constant load and constant current conditions.

current and low final currents obtained. For some discharges, for example for a Zn–AgO battery, a two step discharge behaviour is observed, as shown in Fig. 2.21. This is due to the two-electron transfer of the AgO reduction to silver taking place in two distinctly separate stages, $Ag_2O$ being formed after the first stage (see section 3.3.3).

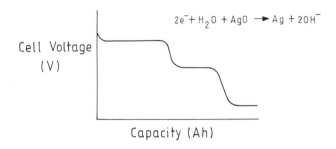

Fig. 2.21 — The voltage versus cell capacity behaviour (discharge curve) for a Zn–AgO cell.

Other systems may show different discharge behaviour from that described in Fig. 2.20 above, because of such factors as rising electrolyte conductivity during discharge (for example, the initial discharge period of Mg/AgCl cells) or a constant electrolyte conductivity during discharge (for example, Ni–Cd cells). Table 2.2 shows some battery couples and their discharge behaviour.

Table 2.2 — The discharge behaviour observed for different battery electrode reactions

| Battery electrode reaction | Discharge behaviour |
| --- | --- |
| $Li \rightarrow Li^+ + e^-$ | If $Li^+$ is removed as it is formed, then a constant potential is observed |
| $Ni(OH)_2 \rightarrow NiOOH + H^+ + e^-$ | $Ni^{3+}$ is formed within $Ni(OH)_2$. A variable potential is seen, as both phases change composition |
| $Cd + 2OH^- \rightarrow Cd(OH)_2 + 2e^-$ | Constant potential observed |
| $H^+ + e^- + MnO_2 \rightarrow MnOOH$ | $Mn^{3+}$ formed in the $MnO_2$ phase. A variable potential is observed |

### 2.3.2.2 A.C. techniques

The application of electrochemical impedance spectroscopy to full cells is much more meaningful if the behaviour of one electrode dominates. Often, this situation is caused by the larger size of the dominant electrode, but it is present in some cells even if the electrodes are of a similar size. Impedance data obtained from full cell

### Sec. 2.3] Electrochemical techniques used in the development of batteries

tests have yielded some useful results. For two electrodes in series, which is the case in a typical cell, the equivalent circuit is shown in Fig. 2.22(a).

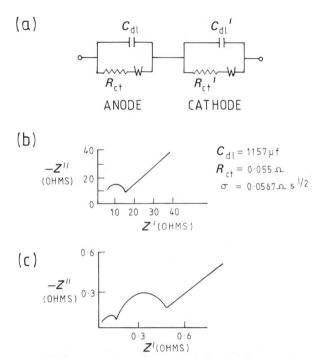

Fig. 2.22 — (a) The simplified equivalent circuit of a complete battery cell. (b) The impedance spectrum resulting from the circuit shown, if the component values are identical in both electrodes. (c) The impedance spectrum for the circuit shown, if there is gross imbalance between the electrode areas and component values.

If the time constants (the products $C_{dl}.R_{ct}$ and $C'_{dl}.R'_{ct}$ in Fig. 2.22(a)) are identical, then the complex plane response is that shown in Fig. 2.22(b), which is broadly the shape of the single electrode impedance spectrum except that the component magnitudes are increased. More commonly, however, the time constants differ, as in the case of Fig. 2.22(c). A decrease in the area of one electrode causes an increase in its current density and so a decrease in its charge transfer resistance. This lowers its time constant, therefore its semicircle appears at high frequencies.

Other features of impedance spectra arise from adsorption of reactive intermediates, roughness or porosity, and factors induced by cell geometry. However, in spite of these complicating phenomena, a semi-empirical approach can allow the A.C. technique to be used to monitor battery state-of-charge in a number of different systems.

#### (i) Nickel–cadmium cells

The impedance spectra of 23 Ah Ni–Cd cells with sintered electrodes is shown at various states of charge in Fig. 2.23. There is a distinct absence of a well-defined high

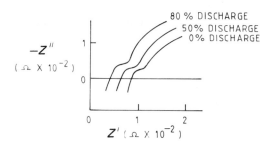

Fig. 2.23 — Impedance spectra obtained from nickel–cadmium cells after various degrees of discharge, discharging at a C/5 rate.

frequency charge transfer process for all the states of charge, and the spectra cannot easily be modelled in terms of an electrical equivalent circuit. However, it has been found possible to empirically relate the impedance behaviour to battery state of charge by plotting the real and imaginary impedance components determined at each frequency against state of charge and searching for correlations. By this method it has been found that the imaginary component of impedance observed at 0.39 Hz correlated well and consistently with battery state of charge. This observation has enabled the development of a Ni–Cd battery state of charge indicator.

(ii) *Zinc–mercury oxide alkaline cells*
The behaviour of this cell is dominated by the zinc electrode, and, as previously mentioned, it displays good Randles circuit behaviour. It has been found that the growth in diameter of the high frequency semi-circle (see Fig. 2.17(b)) can be correlated with the degree of discharge of the cell, making this parameter an effective state of charge indicator.

(iii) *Lead–acid batteries*
The previous two examples show the use of impedance to monitor state-of-charge in a mode which requires the battery to be switched off during measurement. For lead–acid batteries the measurements can be made with the battery switched on, as the parameter to be recorded is the impedance at high frequency. This measurement allows the derivation of the electrolyte resistance ($R_{so}$ of Fig. 2.17(b)) which can be shown to be indicative of the specific gravity of the acid. This, in turn, can be related to the battery state of charge. Thus the variation of ohmic resistance with depth of discharge can be tracked by continuous tuning of the frequency to allow a single point measurement at high frequency to be sufficient to measure state of charge.

## 2.4 PHYSICAL TECHNIQUES USED IN THE DEVELOPMENT OF BATTERIES

Battery systems are composed of electrodes which primarily undergo chemical changes during charge and discharge. However, as the electrodes are almost always solids, they undergo physical changes associated with their use which to a large extent take a controlling role in the battery performance, principally because the chemical reactions take place at or near the electrode surfaces. Thus physical techniques, particularly those designed to study surface morphological and compositional changes, have had an increasing part to play in the development of batteries, especially as these techniques have become more refined and more widely available. In this section, some of these techniques will be described in relation to particular battery systems. However, they are usually generally applicable, and a number of battery systems have yet to be investigated fully by the use of these methods.

### 2.4.1 Electron microscopy

Electron microscopy as a whole is probably the most useful of surface analysis techniques as well as being the most generally available. The technique relies on the fact that the degree of image resolution is inversely proportional to the wavelength of the incident beam. A beam of electrons, which has a wavelength similar to that of crystal lattice dimensions, is easily manipulated electrically and electronically to produce images analogous to the behaviour of light in an optical microscope.

The two main methods used in electron microscopy, which require different instrumentation, are transmission electron microscopy (TEM) and scanning electron microscopy (SEM), although the hybrid, scanning transmission electron microscopy (STEM) is now fairly common. For battery development, SEM has a more widespread application than the other forms.

#### 2.4.1.1 Scanning electron microscopy

In the scanning electron microscope, a finely focused electron beam is electronically scanned across the surface of a specimen held in the vacuum chamber of the instrument. Fig. 2.24 is a schematic diagram of the effects produced by this primary

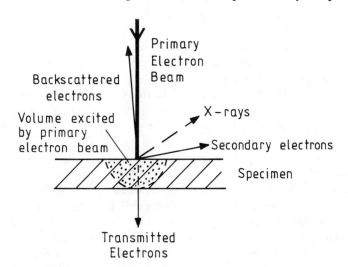

Fig. 2.24 — The resultant species produced by electron-beam bombardment of an electron microscope specimen.

electron beam and shows the resulting species which are produced. Two main types of electron are generated when the beam interacts with the specimen, and there is both a penetration and internal scattering of the beam [27]. The beam penetration depth is related to the electron beam energy by the approximate expression [28].

$$x_r = \frac{1.44 \times 10^{-11} A_w E_o^2}{2Z\rho \ln(0.174 E_o/Z)}, \qquad (2.42)$$

where $x_r$ is the penetration depth, $A_w$ the specimen atomic weight, $\rho$ the specimen density, $E_o$ the initial electron beam energy, and $Z$ the specimen atomic number. For instance, for electrons of energy 20 kV, their penetration in gold is around 0.4 $\mu$m and that in aluminium is approximately 2 $\mu$m.

The highest energy electrons produced after specimen bombardment are largely those reflected by recoil from the atoms of the specimen, and these have energies between that of the primary beam and 50 eV. Imaging of these electrons is termed back-scattered electron imaging (BSI), and, as their energy is dependent on collision with the specimen atoms, the backscattering coefficient, $\eta_b$, is related to the atomic number, $Z$, of those atoms by the equation [29]

$$\eta_b = \frac{(\ln Z)}{6} - 0.25, \qquad (2.43)$$

which is found to fit the observed data above atomic number 10, but is still useful for applying to elements of lower atomic number. Thus the main use of BSI is to determine segregation of materials of different atomic number on a surface, and Fig. 2.25(a) shows an image of an Al–Sn alloy which has been discharged in sodium hydroxide [30]. As tin has a higher atomic number than aluminium the tin particles present show up as bright patches on the aluminium surface. As a further refinement of the technique, use of an accurate brightness meter can enable separation and identification of several different elements or compounds if they are sufficiently segregated in the specimen.

As well as the back-scattered electrons, the incident electron beam causes electronic transitions to take place in the electron orbitals of the specimen atoms such that electrons are emitted from them. These are the secondary electrons which have a lower energy than the back-scattered electrons, this value being less than 50 eV. Their intensity is proportional to the reciprocal of cos $\sigma_e$, where $\sigma_e$ is the angle between the primary beam and the normal to the object surface, and they tend to be emitted from the upper 10 nm layer of the surface, close to their point of generation.

Thus, by secondary electron imaging (SEI), surface morphology and structure can be examined. Fig. 2.25(b) shows the same area as Fig. 2.25(a) taken under secondary electron imaging conditions. It can be seen that the bright particles of tin are widely dispersed over an extensively pitted specimen surface, the pit morphology depending on the grain orientation. The topographical information can be greatly

Sec. 2.4]     **Physical techniques used in the development of batteries**     63

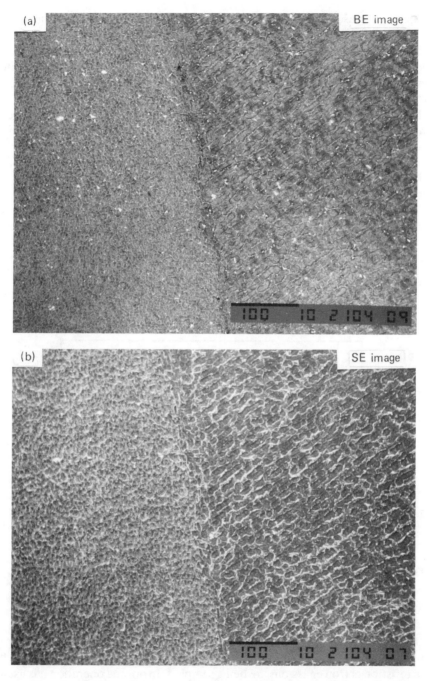

Fig. 2.25 — Electron micrographs of an Al 0.22% Sn alloy after discharge in an aluminium–air battery for several minutes, the electrolyte being 4 M NaOH at 60°C [30]. (a) Backscattered electron (BE) image showing the distribution of tin particles in the surface region. (b) Secondary electron (SE) image showing topographical information and the presence of a grain boundary running from top to bottom.

enhanced by the use of stereo imaging, where two micrographs of the same area are taken at two different tilt angles. For normal SEM studies it is found that a tilt difference of between 10° and 15° at 10 000 × magnification will give good 3-D images. This technique is fairly simple to carry out as most instruments have tilting specimen stages, although use of a stereo viewer or red/green photographic imaging to produce an anaglyph is required to analyse the results.

The addition of X-ray analysis facilities to scanning electron microscopes in recent years has considerably enhanced their capability. These techniques either allow energy-dispersive spectroscopy (EDS), using a solid-state detector, or the similar wavelength dispersive technique, using a diffractometer. Both yield compositional data, as the electron beam incident on (and within) the specimen will cause generation of X-rays characteristic of the atoms encountered. The latter technique is more accurate in terms of quantitative compositional data, but is more time consuming and less appropriate if an unknown composition is present [31]. Thus it is the EDS technique which has proved more popular, although this is not able to analyse for elements with atomic number less than 8, unlike the wavelength dispersive technique. Fig. 2.26 shows a typical EDS spectrum of a second-phase particle in an impure Al–In binary alloy. It demonstrates that the particle contains the contaminants iron and lead.

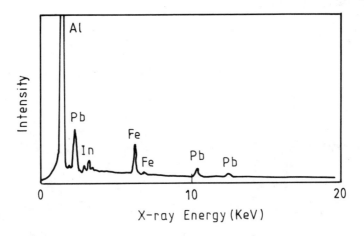

Fig. 2.26 — An X-ray energy dispersive spectrum (EDS) produced from a second phase particle of size 0.2 $\mu$m in a contaminated Al–In binary alloy.

Resolution of the scanning electron microscope image is directly related to the beam diameter and energy. It is also dependent on the electron optics and on which characteristic electron emissions are being sampled. In normal operation it is usual to achieve a resolution of about 2 nm by using a fine beam with low energy, although the best instruments can be made to achieve less than 1 nm resolution. It is necessary for the specimen to have an electronically conductive surface in order to obtain good image clarity (hence avoiding a distortion of the secondary electron emission by

Sec. 2.4]     **Physical techniques used in the development of batteries**     65

specimen 'charging'), and it is customary to achieve this by the deposition of thin coatings of gold, platinum, or a platinum/palladium mixture on the specimen surface. This is normally carried out by vacuum evaporation or sputter-coating.

### 2.4.1.2  *Transmission electron microscopy*

Transmission electron microscopy (TEM) employs a beam of high energy electrons to penetrate the specimen, these commonly having energies in excess of 100 keV, and the interaction of this beam with the atoms of the sample allows the user to determine something of its internal structure. Image contrast is achieved by reflection of the beam away from the main viewing direction, as it encounters the specimen crystal lattice and is scattered because of the specimen thickness. Discontinuities in the structure such as grain boundaries or crystal dislocations thus become clearly visible. Fig. 2.27 shows electron micrographs of various chromium oxides prepared in order to study their properties as lithium battery cathodes [32]. The specimens are supported for viewing on carbon microgrids, and the amorphous nature of $CrO_x$ is suggested by its less crystalline morphology. A disadvantage of the technique is that specimens need to be thin enough for electron transmission, that is below about 0.1 $\mu$m, and techniques to produce thin specimens such as electropolishing, ion-beam etching, or microtomy need to be carefully refined before definitive observations can be made. Care must be taken, as such preparation techniques can sometimes irreversibly alter the specimen structure and produce image artefacts.

As with scanning electron microscopy, the X-rays generated by interaction of the electron with the specimen can be used to yield analytical information, and EDS analysis coupled with TEM can be made more quantitative than that carried out in the SEM [33]. Also, present instruments allow a simple conversion of the image from that characteristic of real crystallographic space to that present in reciprocal crystallographic space, and electron diffraction patterns of the imaged areas can readily be produced. Accurate small area structural analysis can thus be achieved, as demonstrated by the electron diffraction patterns of the various chromium oxides in Fig. 2.27. The more amorphous morphology of $CrO_x$ is displayed by the evident ring-like nature of its diffraction pattern.

### 2.4.2  X-ray diffraction and neutron diffraction

X-ray diffraction is widely used for substance and phase identification, as the diffraction pattern obtained is characteristic of atomic structure both in terms of its spatial pattern and intensity distribution. Computer assistance in both controlling the diffractometer and analysis of the data has greatly simplified the application of the technique. In the normal method used, analysis of a finely divided powder of the unknown substance is normally carried out, and the random nature of the crystallites present enable quantitative information to be gained concerning the concentrations of the different phases observed. Information about the crystallite size can be obtained from the degree of diffraction line broadening, a process which generally occurs when the crystallites are less than 20 nm in diameter.

Fig. 2.28 [34] shows X-ray diffraction patterns for $Ni(OH)_2$ in the positive electrode of a Ni–Cd battery. In this case, the crystallite size of $Ni(OH)_2$ was observed to be relatively small, and its pattern was compared to that of $Ni(OH)_2$ with

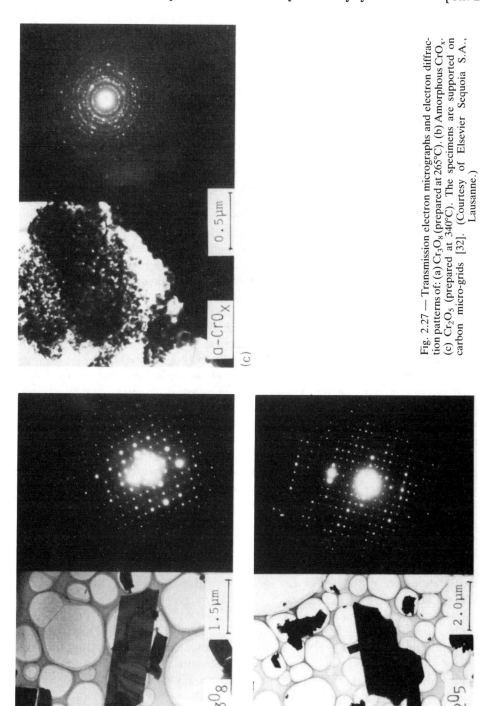

Fig. 2.27 — Transmission electron micrographs and electron diffraction patterns of: (a) $Cr_3O_8$ (prepared at 265°C). (b) Amorphous $CrO_x$. (c) $Cr_2O_5$ (prepared at 340°C). The specimens are supported on carbon micro-grids [32]. (Courtesy of Elsevier Sequoia S.A., Lausanne.)

Sec. 2.4]     **Physical techniques used in the development of batteries**     67

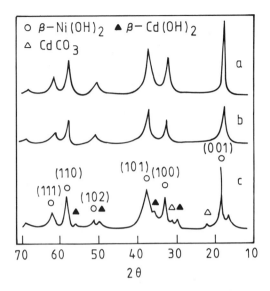

Fig. 2.28 — X-ray diffraction patterns produced on a diffractometer of Ni(OH)$_2$ with a Cd$^{2+}$ ion or Cd(OH)$_2$ addition [34]. (a) No addition. (b) Cd$^{2+}$ at 5 wt%, (c) Cd(OH)$_2$ at 5 wt%. (Courtesy of the International Power Sources Symposium Committee).

added Cd$^{2+}$ and Cd(OH)$_2$. No evidence of the presence of cadmium oxides could be observed in the Ni(OH)$_2$ with Cd$^{2+}$ additions, and, from the sharp diffraction angle characteristic of Ni(OH)$_2$ and the fact that the crystal lattice of Ni(OH)$_2$ was seen to be somewhat enlarged, it was assumed that cadmium formed a solid solution in the Ni(OH)$_2$.

In the neutron diffraction technique [35], thermal neutrons of about 4000 m.s$^{-1}$ velocity and wavelength approx. 10 nm are normally used, and it is more appropriate than X-ray diffraction for analysis of substances containing light elements (e.g. hydrogen) and magnetic materials. Also, the inelastic scattering produced allows the analysis of vibrational modes (phonons) and other dynamic effects of solids. A disadvantage of the technique is that it requires specimens at least two orders of magnitude greater in volume than those necessary for X-ray diffraction to produce the equivalent line intensity.

### 2.4.3 Extended X-ray absorption fine structure (EXAFS)
This technique enables bond distances and coordination numbers to be obtained from materials having a variety of forms, including crystalline, amorphous, liquid, or glassy states [36]. The incident X-ray beam is of high energy and is usually provided from a synchroton storage ring. Its absorption by the sample is recorded over an energy band from approximately 100 eV below to about 1000 eV above an absorption edge which is usually the K-edge of the chosen target element. The EXAFS spectrum is seen as a series of oscillations on the high-energy side of the edge, and it gives information concerning the number, distance, and geometric

arrangements of the first three nearest-neighbour shells within a spherical region of up to 0.6 nm of the target atom.

EXAFS has been very useful in studying polymer electrolytes which cannot be investigated by normal X-ray diffraction, owing to their amorphous nature. Fig. 2.29

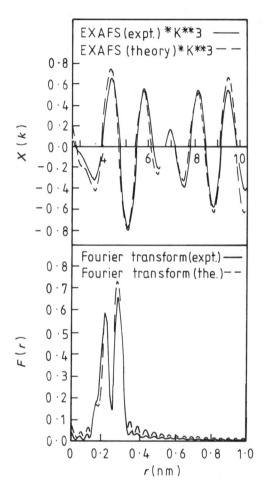

Fig. 2.29 — EXAFS spectra of $PEO_{12}:ZnI_2$ and their corresponding Fourier transforms [37]. (Courtesy of Springer-Verlag.)

shows a spectrum exhibiting the Zn K-edge of EXAFS obtained during a study of the polyethylene oxide (PEO) compound $(PEO)_{12}\cdot ZnI_2$, together with its Fourier transform [37]. Also shown is a theoretical spectrum obtained by assuming that the zinc ions in the iodide compound have nearest neighbour shells of four oxygen and two iodide ions. The fit is good, indicating that the structure $(PEO)_{12} ZnI_2$ is of this type. It is distinguishable from $(PEO)_{12} Zn Br_2$ which, in the same study, was shown to have six oxygen and two bromide ions surrounding the zinc ions.

### 2.4.4 Electron spectroscopy

Electron spectroscopy is a technique in which the surface of a solid *in vacuo* is subjected to a flux of near-monochromatic electromagnetic radiation which interacts with atoms of the sample and causes them to emit photoelectrons. As these can escape without collision only from a depth of approximately 6 nm, the technique is very surface specific. There are three main fields of electron spectroscopy; ultraviolet photoelectron spectroscopy (UPS), X-ray photoelectron spectroscopy (XPS), and Auger electron spectroscopy (AES). In the latter technique, which is closely related to photoelectron spectroscopy, a vacancy caused by photoelectron emission from a core level is filled by an electron descending from a higher level. Energy from this process is transferred to a second high-level electron which is ejected as the Auger electron. Auger electron peaks are normally found to be present during the recording of XPS spectra, but can be distinguished from photoelectron peaks by their independence from the energy of the primary X-ray source. Because of the latter property, Auger electrons can be emitted from a specimen by applying an electron beam which is not monochromatic. Thus AES can be carried out in modified SEM instruments to give highly spatially-resolved surface analysis of down to 50 nm resolution and this enables surface element mapping. The best spatial resolution which has been achieved with XPS instruments is 10 $\mu$m, and most instruments can achieve 150 $\mu$m resolution.

During the chemical bonding process, the elements bound together undergo changes in the energy states of the electrons involved in bonding, and XPS is able to gain information about these binding energy changes by a study of the 'chemical shift' or change in position of a spectral peak for different displayed oxidation states or types of bonding. For instance, the carbon 1s photoelectron peak exhibits a larger binding energy in $CF_4$ than $CCl_4$, which is in turn higher than the energy of the same peak observed in $CH_4$. Fig. 2.30 shows XPS spectra obtained for the Cl2p peak from

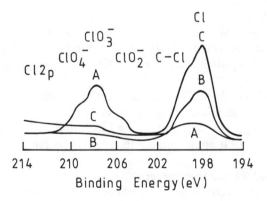

Fig. 2.30 — XPS spectra of Cl 2p in the surface layer formed on Li, electrochemically deposited from a solution of 1 M $LiClO_4$ in propylene carbonate indicating the decomposition of $ClO_4^-$ ions. A: Top surface, B: After 10 min sputtering, C: After 20 min sputtering. Sputtering rate 1 nm. min$^{-1}$ [38]. (Reproduced by permission of the publisher, The Electrochemical Society, Inc.)

the surface layer formed on lithium which had been electro-deposited on silver from a solution of $LiClO_4$ in propylene carbonate [38]. It shows the presence of five different chlorine compounds, particularly after bombardment of the surface with high energy Argon ions.

Quantitative data can be obtained from peak intensity measurement, and depth profiling is also commonly used. This can be by non-destructive methods such as varying the take-off angle of the X-rays, by the use of varying energy X-ray sources [39], or by repeated Argon–ion etching of the surface with subsequent analysis. XPS and AES are both used for this type of analysis, although the Auger peaks in an XPS spectrum can sometimes display much greater sensitivity to the chemical environment of particular atoms and are thus of greater value in chemical state identification [40]. Fig. 2.31 shows two AES depth profiles of metal hydrides produced by Ovonic containing V, Zr, Ti, and Ni [41]. Fig. 2.31(a) is the depth profile obtained from uncycled material, showing an oxide thickness of 7.5 nm, and Fig. 2.31(b) shows the depth profile after cycling the same material for 69 cycles. In the latter case the oxide thickness is 75 nm. It is apparent that, after cycling, the elements with soluble oxides are not present at the surface but begin to appear more towards the bulk of the specimen. In the same study, XPS analysis at 50 nm depth after cycling showed that Ti was present as $TiO_2$, nickel was metallic, zirconium was present as $ZrO_2$, and chromium was in the form of $Cr_2O_3$.

### 2.4.5 Secondary ion mass spectrometry (SIMS)

In this technique [42] fragments emitted from a surface during its bombardment by a beam of energetic particles are mass analysed. Ions of argon, gallium, oxygen, or caesium are most commonly used as the irradiating sources, but neutral beams have also been used. SIMS is divided into dynamic SIMS, where intense ions beams are used to etch away surfaces at rates between 1 and 30 $\mu m\,h^{-1}$, or static SIMS where the erosion rate is far less (less than 2 nm $min^{-1}$). By rastering the incident beam over the surface it is possible to produce maps of element or fragment distributions, and, with the aid of computer systems, these can be stored at different positions during surface erosion, allowing mapped depth profiles to be obtained.

Recently, improved spatial resolution has been achieved by the use of liquid metal sources which are usually of gallium [43]. Ion beams of liquid metals can be focussed to a spot of approximately 20 nm in diameter, giving a spatial resolution of around 50 nm. The spectra from SIMS are recorded as negative ion spectra and positive ion spectra, and the technique is particularly useful in the case of low atomic number elements such as lithium. Fig. 2.32 shows a SIMS spectrum from the same system as that of Fig. 2.30 [38], and strong Li (7) and $Li_2$ (14) peaks are visible, showing the inclusion of lithium metal in the passivating film.

### 2.4.6 Infrared and Raman spectroscopy

Infrared and Raman spectroscopies have been used to identify species in battery electrolytes and polymeric battery materials and to study electrode surfaces *in situ*. Raman spectroscopy [44], which arises from the scattering of an incident laser beam by the sample molecules, has been found to be more versatile for such studies than infrared spectroscopy. The main reason for this is that, generally, *in situ* reflectance

### Sec. 2.4] Physical techniques used in the development of batteries 71

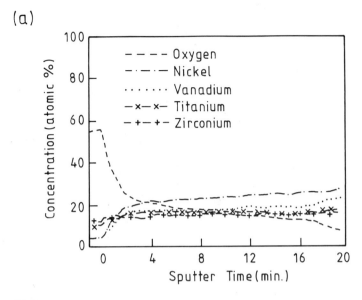

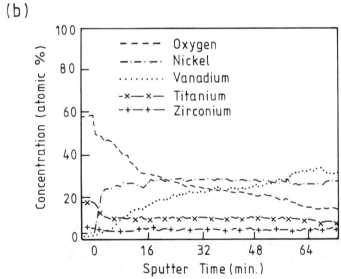

Fig. 2.31 — Auger electron spectra depth profiles of the surface of a metal hydride containing nickel, vanadium, titanium, and zirconium with sputter rate 4.16 nm. min$^{-1}$. (a) Uncycled electrode with thickness (defined as 50% oxygen concentration) of 33 nm. (b) Electrode cycled 69 times with thickness (defined as 50% oxygen concentration) of 75 nm [41] Courtesy of the International Power Sources Symposium Committee).

techniques encounter problems when used on rough surfaces such as reacting electrode materials, and Raman spectroscopy suffers less in this respect than infrared spectroscopy. Raman spectroscopy of non-aqueous battery-type electrolytes has

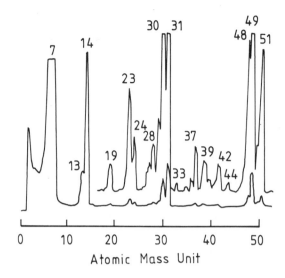

Fig. 2.32 — SIMS spectrum of the surface layer produced on lithium electrodeposited from a 1 M LiClO$_4$ solution in propylene carbonate. Two different sensitivities are shown in the low mass range [38]. (Reproduced by permission of the publisher, The Electrochemical Society, Inc.)

been studied by using silver electrodes, and investigations of LiI in both acetonitrile [45] and propylene carbonate [46] has been reported. Signals from both adsorbed solvent, and solute molecules have been detected as well as those from the decomposition products.

Fourier transform infrared spectroscopy (FTIR) has been used to identify surface films on lithium *ex situ* in the solvents 1,2 dimethoxyethane (DME) and tetrahydrofuran (THF) [47]. Fig. 2.33(a) shows spectra obtained in the DME case. The spectra show the presence of lithium methoxide. For the similar investigation in THF, it was found that lithium butoxide was formed. If water was present in the solvent, then spectra due to LiOH were observed.

In more recent studies, attempts have been made to get more information from this type of system by using Raman spectroscopy *in situ*, and investigations have been carried out on LiAsF$_6$ dissolved in THF and 2-methyl THF [48]. Fig. 2.33(b) illustrates some of the findings. In the trace marked 'C', Raman bands can be seen in the spectral region 900 to 1380 cm$^{-1}$ which arise from the THF, the bands relating to the AsF$_6^-$ ions appearing at a lower frequency. In the trace marked 'B', which was obtained from the electrode *in situ* at open circuit after lithium charge/discharge cycling, new signals at 1293 cm$^{-1}$ and 1104.5 cm$^{-1}$ appear. These signals were found to arise from polymeric THF. Raman spectroscopy of polymeric THF studied *ex situ* produces the spectrum marked 'A', and the similarity of this to the spectrum marked 'B' indicates that polymeric THF is formed in the operating cell. It has been found that the polymer develops in the electrochemical cell owing to species associated with the solute, and it is not due to electrochemical phenomena. To date, the passifying layers formed on lithium electrodes have prevented the use of the more powerful technique of surface enhanced Raman spectroscopy (SERS).

### Sec. 2.4] Physical techniques used in the development of batteries

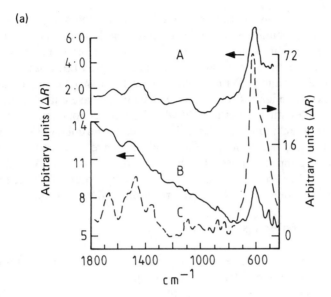

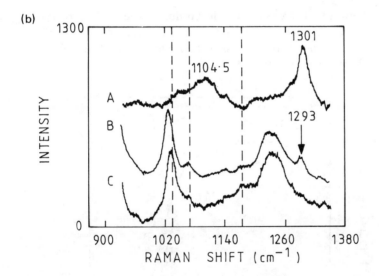

Fig. 2.33 — (a) FTIR spectra from silver plates after lithium bulk deposition from 1,2 Dimethoxyethane (DME) solutions [47]. A: DME/LiClO$_4$ (0.2 M); B: DME/LiAsF$_6$ (0.2 M); C: Spectrum of LiOCH$_3$ loaded on a silver plate from methanol solution. (b) Raman spectra in the THF ring-stretch modes region [48]. A: Dried polymerized THF; B: Lithium electrode *in situ* in a solution of 0.6 M LiAsF$_6$/THF at open circuit after lithium charge/discharge cycling; C: 0.6 M LiAsF$_6$/THF electrolyte spectrum.

### 2.4.7 Thermogravimetric analysis (TGA) and differential thermal analysis (DTA)

These two techniques analyse the thermal properties of substances with particular reference to temperature dependent reactions. In thermogravimetric analysis

(TGA) a specimen is simply heated or cooled in a controlled manner during which time the mass of the specimen is recorded. Differential thermal analysis (DTA) is a more refined technique in which a recording is made of temperature differences between the material to be analysed and a reference sample while both are subjected to identical thermal treatment. With DTA, if a thermal change occurs in the sample, that is, if it melts, solidifies, or undergoes a structural phase change, then a peak occurs at the characteristic temperature at which the effect appears. DTA is closely related to the similar technique of differential scanning calorimetry (DSC) where the temperature difference between specimen and reference is forced to zero by an energy extraction or input imposed by the instrument. Both DTA and DSC are quantitative in that the recorded peak area produced by a thermal change is related to the quantity of substance involved in the reaction by the equation

$$\Delta H = A_p M_w (60 B \varepsilon \Delta qs) \times 10^{-6}/m \ , \tag{2.44}$$

where $\Delta H$ = heat of reaction (J. mole$^{-1}$); $A_p$ = area of peak (cm$^2$); $m$ = sample mass (g); $B$ = min/cm of scan (dependent on the heating rate); $\Delta qs$ = mV/cm of recorded output voltage; $\varepsilon$ = cell calibration coefficient (mW. mV$^{-1}$); $M_w$ = molecular weight of the sample material.

Fig. 2.34 shows the TGA curve produced during the heating of β-manganese dioxide showing oxygen adsorption and release to occur at about 300°C [49], a process which was found to be reversible and which was observed to affect the electrochemical properties of the material.

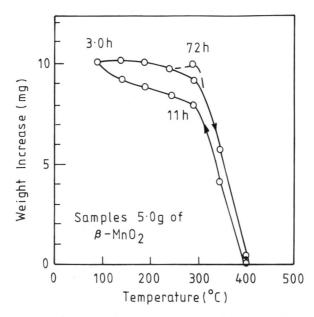

Fig. 2.34 — Thermogravimetric behaviour of β-MnO$_2$ heat-treated in air between 400°C and 100°C starting after a steady-state weight had been achieved at 400°C and dwelling for the number of hours indicated at each temperature until no further weight change was observed [49]. (Courtesy of Independent Battery Manufacturers Association).

Fig. 2.35 shows the DSC curve for a sample of nickel iron hydroxide produced as an iron-immune positive nickel hydroxide electrode [50]. It exhibits the release of water from the specimen in two distinct temperature ranges; between 150°C and 200°C and between 220°C and 500°C.

## 2.5 PHYSICAL TECHNIQUES USED IN THE ANALYSIS OF PARTICULATE BATTERY MATERIALS
(A. Gilmour)

In practical situations a finely divided solid usually consists of particles whose size spans a range of values; that is, there is a particle size distribution. A plot of numbers of particles present against particle size gives a distribution curve which is characteristic of the sample concerned. It is assumed throughout that samples for analysis are consistent in composition and that no segregation has occurred either in the process of taking samples or in the preparation for the analysis technique itself. Because of the wide range of particle sizes and shapes usually present in commercially used powders, no single physical parameter is adequate to describe its character. It is therefore convenient to approach the study of particulate materials under three headings, all of which give measured parameters with optimum values dependent on the materials function or application.

### 2.5.1 Measurement of surface area

The surface area of a finely divided powder, whether it is electrochemically active or serves as an electronic conductor, has a major influence on the performance of the composite electrode. This parameter is usually expressed in square metres per gram ($m^2.g^{-1}$).

The most useful and generally applicable measurement technique is based on adsorption of gases by the particulate material, which is referred to as the adsorbent. The gas to be adsorbed, termed the adsorbate, has an associated heat of adsorption, the value of which depends to some extent on the nature of the adsorbent. Langmuir's kinetic theory of monolayer adsorption [51] was extended to take account of multilayer adsorption by Brunauer, Emmett, and Teller [BET] in 1938 [52]. The BET equation for adsorption at a free surface is:

$$n^s = \frac{n_m^s C P}{(P_o - P)\left[1 + (C-1)\frac{P}{P_o}\right]} \quad (2.45)$$

Where $n^s$ is the quantity of gas adsorbed at an equilibrium pressure $P$, $P_o$ is the vapour pressure of the adsorbate in the condensed (liquid) state of the adsorption temperature, and $C$ is a constant related to the heat of adsorption. $n_m^s$ is the quantity of gas adsorbed which corresponds to a monolayer on the entire adsorbent surface. The surface area of the adsorbent, $A_a$, is given by:

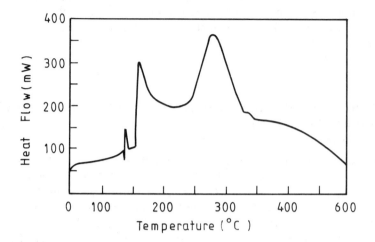

Fig. 2.35 — The DSC behaviour of iron-immune nickel hydroxide electrode material containing $Ni^{2+}$, $Fe^{3+}$ and $NO_3^-$ ions during heating from 50°C to 500°C [50]. (Courtesy of International Power Sources Symposium Committee).

$$A_a = a_m n_m^s, \qquad (2.46)$$

where $a_m$ is the effective area per molecule in the monolayer.

The BET theory has gained universal acceptance in that it enables an experimental determination of the number of molecules required to form a monolayer, despite the fact that exactly one monomolecular layer is never actually formed. This derives from the assumption that multilayer molecules in adsorbed states are in dynamic equilibrium with the vapour and are equivalent energetically to the liquid state of the adsorbate.

For practical purposes, the BET equation is generally fitted to data over a range where $P/P_o = 0.05$ to $0.35$. This gives a condition where very high energy sites are occupied but extensive multilayer adsorption has not yet commenced. A minimum of two values for $P/P_o$ is required in order to derive a value for $n_m^s$.

The gases argon and krypton are appropriate adsorbates since they approach 'ideal' criteria, but they can be used only for BET analysis below their critical temperatures. Nitrogen as an adsorbate has a higher critical temperature. It has the unusual property that, on almost all surfaces, its critical temperature is sufficiently small to prevent localized adsorption and yet adequately large to prevent the adsorbed layer from behaving as a two-dimensional gas. These unique properties of nitrogen have led to its wide acceptance for BET measurement. The effective cross-sectional area per molecule, $a_m$ has been accurately determined by numerous workers to be $16.2 \times 10^{-2}$ nm² at 77° K. However, it should be recognized that $a_m$ is often somewhat dependent on the nature of the adsorbent. The problem has been summarized by Young & Crowell [53], Anderson & Baker [54], and Gregg & Sing [55].

Experimental techniques can be broadly classified into three adsorption methods. These are:

(i) Volumetric
(ii) Continuous flow
(iii) Gravimetric.

In volumetric methods, a quantity of gas, which is initially contained in a calibrated glass tube, is expanded into a larger volume containing the sample, the initial and final gas pressures being measured. Ideal gas behaviour is assumed, and it is necessary to maintain a constant temperature throughout the experiment. The apparatus required consists of vacuum-tight glassware and taps, manometers, a vacuum pump, gas supplies, and good thermostatic control of the sample containers.

Continuous flow or dynamic methods use a technique whereby nitrogen is adsorbed from a mixture of helium and nitrogen. In the method of Loebenstein & Deitz [56] the gas mixture is passed back and forth over the sample until equilibrium is established by noting no further change in pressure. The quantity adsorbed is determined by the pressure decrease at constant volume. Successive data points are acquired by adding more nitrogen to the system. The results obtained show good agreement with vacuum volumetric measurements on a large variety of samples with a wide range of surface areas. An extension of this technique was developed by Nelson & Eggerton [57] who used a thermal conductivity detector to sense the change in effluent gas composition during adsorption by the sample.

The most widely used gravimetric method is that developed by Cahn & Schultz [58], in which a microbalance of 1 $\mu$g sensitivity has one of its arms suspended in a chamber containing the sample, and which can be evacuated. It has the advantage that the mass of the adsorbed species is directly measured.

Surface area measurements on a wide range of particulate materials have been reported by Crowl [59], using low temperature nitrogen adsorption. Examples of the results obtained include 0.34 $m^2.g^{-1}$ for zinc dust, 9.33 $m^2.g^{-1}$ for zinc oxide and results for two different types of carbon black which gave values of 123.7 $m^2.g^{-1}$ and 644 $m^2.g^{-1}$ respectively.

A comparison of experimental adsorption methods is given in Table 2.3.

### 2.5.2 Porosity and density measurement

In many of the materials used in battery manufacture, porosity and density need to be specified, these two parameters having a broadly inverse relationship for a given material. A good example is particulate carbon, where the presence of a wide range of particle sizes, shapes, and degrees of aggregation can have a dramatic effect on both physical and chemical/catalytic properties. Owing to the compressibility factor and the presence of air in more highly porous materials, porosity can generally be measured more reproducibly than density.

The measurement of pore size and pore size distribution generally makes use of the Kelvin equation which relates the equilibrium vapour pressure of a liquid on a curved surface such as that experienced in a capillary or pore, to the equilibrium pressure of the same liquid on a plane surface. The equation is given by

**Table 2.3** — Critical comparison of the experimental methods for determining adsorption of gases on solid surfaces

| Volumetric method | Continuous flow method | Gravimetric method |
|---|---|---|
| Ideal gas behaviour assumed | Independent of ideal gas requirements | |
| Void volume must be measured | Independent of void volume | |
| Constant liquid nitrogen level required in sample tube | Independent of liquid nitrogen level | |
| Accurate thermostat needed | No thermostating required | |
| Non-automatic recording | Electronic output signals obtained for permanent record | |
| No buoyancy correction needed | | Must correct for buoyancy |
| Needs vacuum | No vacuum required | Needs vacuum |
| Fragile glass apparatus | All metal except sample cell | Fragile apparatus |
| Less sensitive resolution (°0.01 cm$^3$) | Sensitivity better than 0.001 cm$^3$ via transducer | Sensitivity around 1 $\mu$g |

$$\ln \frac{P}{P_o} = -\frac{4\gamma\bar{v}}{d\,RT} \cos\theta ,  \tag{2.47}$$

where $P$ is the equilibrium vapour pressure of the liquid contained in a narrow pore of diameter $d$, and $P_o$ is the equilibrium vapour pressure of the same liquid on a large plane surface, $\gamma$ is the surface tension of the liquid, $\bar{v}$ is the molar volume of the liquid, and $\theta$ is the contact angle at which the liquid meets the pore wall.

This equation implies that, in a small pore, adsorbate molecules will condense at a lower pressure than is normally required on an open or plane surface. Thus, as the relative pressure $P/P_o$ is increased, condensation will occur first in pores of smaller diameter and will progress into the larger pores until, at a value of unity, condensation will have taken place on surfaces where the radius of curvature is essentially infinite. Using the Kelvin equation, an assessment of pore diameter and pore volumes can be made over a range of different relative pressures by fitting in known values for $\gamma$, $\bar{v}$, $\theta$, $R$ and $T$. The method is most useful for studies on materials with very small pores [60]. The term 'micropore' is commonly used to refer to pores whose main dimension is under about 50 nm. Micropore volume, derived from the Kelvin equation, must assume a spherical or other regular shape for the pores, and this is seldom found with practical materials. By measuring total sorption and knowing the weights of both adsorbate and adsorbent, the skeletal information derived gives some understanding of the pore structure. The largest pores amenable to analysis from the Kelvin equation have diameters of about 100 nm, which corresponds to relative pressures of approximately 0.99.

Mekhail, Brunauer, & Bodor [61] have devised a method for the analysis of micropores which offers several advantages. By plotting the volume of nitrogen adsorbed per g of adsorbent against $P/P_o$, they were able to obtain the micropore volume, surface, and distribution from one experimental isotherm.

When the particle size exceeds about 20 $\mu$m, the measurement of total pore volume is relatively simple [62]. In most battery materials, however, a considerable amount of micropore volume is usually present. For materials having particles of diameter greater than about 2 nm, filling of the micropore space without filling interparticle space can be essentially achieved by soaking in a suitable liquid and careful removal of the excess by a 'blotting' operation. Low speed centrifugal action has been used to remove interparticle space liquid without removing that in the pores [63]. Mercury porosimetry is one of the most widely used techniques for porosity determination. It is based on the fact that the contact angle between mercury and non-wetted solids exceeds 90°, and mercury can enter pores only by the application of relatively high positive pressures. For a mercury porosimeter the relationship between the required pressure difference $P_i$ and a pore diameter $d$ (assumed cylindrical) is a derivation of the Kelvin equation given by:

$$P_i = \frac{4\gamma \cos\theta}{d} \tag{2.48}$$

Where $\gamma$ is the surface tension of mercury and $\theta$ the contact angle. Taking values for $\gamma$ of 0.474 Nm$^{-1}$ and 130° as typical for $\theta$, the pore diameter which will just permit mercury penetration at a pressure of $10^5$ kPa (987 atmospheres) is 12.2 nm. A mercury porosimeter will not function reliably at pressures above around 3000 atmospheres, and this corresponds to a minimum pore diameter which can be evaluated of about 4 nm. In practice, the useful pore diameter range which can be covered is from 10 nm to about 150 $\mu$m. This just overlaps data available from physical adsorption already discussed. The determination of pore diameter distribution requires measurement of the amount of mercury forced into the pore space of a material as a function of the applied pressure. A number of pieces of commercial equipment are available, all of which record a trace of volume versus pressure (or volume versus pore diameter) more or less automatically [64]. Since mercury is slightly compressible, a blank calibration is required.

In view of the experimental difficulties with liquid wetting measurement and the dangerous high pressures required with mercury porosimetry, a novel method for density measurement of finely divided but porous particles has been developed at University College London [65]. In this technique, a weighed quantity of the particulate material is set in a low viscosity colourless resin in such a way that the volume and hence bulk density is known accurately. The solidified composite is sectioned, polished, and thinly vacuum deposited with gold. Electron microscopy is used to image this section, and the photomicrograph produced is examined by using an automatic scanning microdensitometer. Area and volume voidage can be determined from the distinction made between clear interstitial resin and particles of powder, the data being processed by a computerized image analyser.

### 2.5.3 Particle sizing and counting

Most of the materials used in the battery industry have irregular and often elongated shaped particles, and this makes the mathematical interpretation of experimentally measured parameters more difficult than with ideal spherical particles. For irregular particles, the assigned size depends on the method of measurement, hence the particle sizing technique should, wherever possible, duplicate the process which has to be controlled. The shape factor is best considered to be dealt with under the heading of the surface area and density parameters already discussed.

Microscopy is the only widely used particle sizing technique in which individual particles are observed and measured. A single irregular particle has an infinite number of linear dimensions, and it is only when they are averaged that a meaningful value results. When linear dimensions are measured parallel to some fixed direction (the microscope section plane), the size distribution of these measurements reflects the size distribution of the projected areas of the particles. These are called statistical diameters. Since a single particle can have an infinite number of statistical diameters, the latter are meaningful only when sufficient particles have been measured to give average statistical diameters in each size range.

Optical microscopy is used for particles down to about 0.8 $\mu$m, while for smaller particles it is necessary to use electron microscopy. The most severe limitation of optical microscopy is its small depth of field which is approximately 10 $\mu$m at a magnification of 100 × and approximately 0.5 $\mu$m at a magnification of 1000 × . The surface of particles larger than about 5 $\mu$m can be studied by reflected light, but, for smaller particles, transmission microscopy (see section 2.4.1.2) with which silhouettes are observed, must be used. Sample preparation is very important, as is the skill of the operator.

Electron microscopy has shown great advances in recent years to become a widely applied and powerful technique. It is necessary to distinguish between transmission electron microscopy (TEM) and scanning electron microscopy (SEM).

In the TEM technique (see Section 2.4.1.2), the specimen is deposited on a very thin (10 to 20 nm) plastic membrane such as collodion resting on a copper grid. The electron beam passing through the specimen produces an image on a fluorescent screen or photographic plate, the latter being required for detailed analysis. TEM offers a resolution down to about 0.3 nm, but there is little point in trying to measure particles of size less than about 1.0 nm; however, their presence can be detected and a rough estimate of their count made. The useful upper limit is around 5 $\mu$m.

SEM (see Section 2.4.1.1) uses a fine beam of electrons (5 to 50 keV energy) of about 10 nm cross-section which scans across the sample in a series of parallel tracks. In the normal mode of operation, it is secondary electrons emitted at the specimen

### Sec. 2.5] Physical techniques used in the analysis of particulate battery materials 81

which are detected by using a scintillator photomultiplier, and a photographic record can be made. SEM is considerably faster than TEM and gives more three-dimensional details owing to its greater depth of field, which can be up to about 100 nm with the highest voltage conditions. As a large number of measurements are required to obtain meaningful results from microscopy, operator time and skill are limiting factors. A plethora of semi-automatic and automatic aids have been developed to speed up counting and sizing analyses. They are usually designed to accept images directly from the microscopes, or use photomicrographs which are scanned by a TV camera. The information is stored and manipulated, using a computer, and displayed in either 'true' colour or false colour, the latter being a technique used to differentiate between grey levels [66].

The Coulter counter is a technique for determining the number and size of particles suspended in an electrolyte. The particles are passed through a small orifice on either side of which an electrode is immersed. The changes in resistance as particles pass through the orifice generate a series of voltage pulses whose amplitude is proportional to the volume of particles [67]. By means of a computer, a size distribution graph of the suspended phase is produced. The technique was originally developed for blood cell and bacteria counting, but was later refined to include particle sizing. This technique is suitable for particulate matter greater than about 0.6 $\mu$m diameter, and calibration is effected by using particles having a narrow size range and placed in suspension at very low concentration. The instrument response is essentially due to particle volume, and, although the manufacturers claim that particle shape, roughness, and material have little effect on the analysis, there is considerable evidence that the parameter measured is the spherical envelope which surrounds the particle. Porous materials are unsuitable for applying to this method since their effective densities are not known. Ratios in error by 1.3 to 1 have been quoted for non-extreme shapes, with higher error values for flaky particles [68]. The method has the great advantage of giving rapid results, and, although its range is limited for a given set of instrument settings, it can rapidly distinguish between two distributions with very similar peaks.

A more recent technique which is becoming widely used is that of laser diffraction. When a laser beam is passed through a suspension of particles in a liquid, the angle of diffraction is inversely proportional to particle size, and the intensity of the diffracted beam at a given angle is a measure of the mean projected areas of particles of a specific size [69]. Malvern Instruments have developed a particle analyser based on this principle. The instrument consists of a small He–Ne gas laser fitted with a spatial filter and collimating lens. This provides a parallel, monochromatic coherent incident beam. The suspended particles are ultrasonically agitated and subjected to the beam, the different light being brought to a focus on a special detector consisting of 30 concentric, semicircular, photosensitive rings. The signals are computer-controlled, and the acquisition of 200 data points for each detected ring takes less than 1 second. Each detector/lens system gives approximately a 100 to 1 particle size ratio. Unfortunately, the instrument is insensitive to particle sizes much below about 1 $\mu$m [70].

For the measurement of very small particles, X-ray methods have been widely used (see section 2.4.2). X-ray diffraction line broadening occurs when the crystallite size falls below about 100 nm, and the technique is particularly suitable for investigation of metal crystallites of size 3 to 50 nm [71]. X-ray scattering at angles less than 5° is another useful technique [72].

The principle of sedimentation of particles in a static column of air from which Stokes diameters are determined has been used for particle size analysis of $MnO_2$, by the British Ever Ready Co. [73]. With an instrument known as a Micromerograph (Sharples Corp. USA) a cloud of de-agglomerated particles settle under gravity in a vertical column 2.1 m tall at a rate dependent on their size. The accumulation of particles at the bottom of the cylinder is recorded by an automatic servo-controlled electronic torsion balance, and the graph of weight against time is converted to a cumulative size distribution chart. It is claimed to be suitable for particles in the range 1 to 100 $\mu$m with a $\pm 3\%$ accuracy. Its main disadvantage is that, since coarser material preferentially settles first, finer material can take a long time to come down, and calibration errors may arise.

## ACKNOWLEDGEMENTS

The editor would like to acknowledge and thank the various people who assisted in suggesting the content of this chapter and advised on the text. In particular, the section on electrochemical methods used in battery research was based on presentations made by Dr R. D. Armstrong of Newcastle University and Dr P. J. Mitchell of Loughborough University at the Workshop on Battery Research held in Oxford at the beginning of 1989 under the auspices of the Royal Society of Chemistry. The main consultant for the chapter was Dr O. R. Brown of Newcastle University, and those advising on the section concerning physical techniques were M. P. Amor (electron microscopy), Dr J. A. Treverton (ESCA/Auger/SIMS), Dr R. Latham (EXAFS), Dr G. Chapman (X-ray diffraction), Prof. D. E. Irish and Dr M. Odziemkowski (IR/Raman), and G. Flynn (DTA/DSC).

## REFERENCES

[1] Kolotyrkin, Y. M. & Florianovich, G. M., *Bull. Chem. Soc. Beograd*, **48** [suppl.] (1983), 413.

[2] Despic, A. R., Drazic, D. M., Purenovic, M. M. & Cikovic, N., *J. Appl. Electrochem.*, **6** (1966), 527.

[3] Oxley, J. E., Spellman, P. J., Larren, D. M. & Tarvis, L., *Proc. 5th Annual Battery Conference on Applications and Advances, Los Angeles, 1990*, University of Los Angeles, Long Beach, CA.

[4] Peled E., *J. Electrochem. Soc.*, **126** (1979), 2047.

[5] Yamaki, J., *Proc. IBA Meeting, Los Angeles, May 1989*, p. 31.

[6] Dey, A. N., *Thin Solid Films*, **43** (1977), 131.

[7] Peled, E. In: *Lithium batteries*, Gabano, J.-P., Ed., Academic Press, London, (1983), 43.

[8] Abraham, E. M. & Chaudri, S. M., *J. Electrochem. Soc.*, **133** (1986), 1307.

## References

[9] Nazri, G. & Muller, R. M., *J. Electrochem. Soc.,* **132** (1985), 1385.
[10] Aurbach, D., Daroux, M. L., Faguy, P. W. & Yeager, E. *J. Electrochem. Soc.,* **134** (1987), 1661.
[11] Cuasta, A. J. & Bumb, D. D., *Proc. Symp. on Power Sources for Bio-implantable Applications and Ambient Temperature Lithium Batteries,* Owens B. & Margali, N., Eds, Electrochemical Society, London (1980), 95.
[12] Koch, V. R., *J. Electrochem. Soc.,* **126** (1979), 181.
[13] Yen, S. P. S., Shen, D. M., Carter, B. J., & Somoano, R. B., *Proc. 31st Power Sources Symposium* (1984), 114.
[14] Duffield, A., *Chemistry and Industry,* No. 3 (1988), 88.
[15] Tuck, C. D. S., Armstrong, R. D., Brown, O. R., & Ram, R. P., *Proc. 5th Annual Battery Conference on Applications and Advances, Los Angeles, 1990,* University of Los Angeles, Long Beach, CA.
[16] Cairns, E. J. & Heitbrink, E. H., *Comprehensive Treatise of Electrochemistry,* Bockris, J. O'M., Conway, B. E., Yeager, E. & White, R. E., Eds, **Vol. 3,** (1981), 421.
[17] Dassler, A., German patent 899,216 (1951).
[18] Wilkinson, D. P. & Dudley, J., *Ext. Abs. 48, Electrochemical Society Fall Meeting, Hollywood, Florida, 1989.*
[19] Bradley, J., PhD thesis, University of Loughborough, 1991.
[20] Gileadi, E., Kirowa-Eisner, E. & Penciner, J., *Interfacial electrochemistry: an experimental approach,* Addison Wesley, Reading, MA, (1975), 207.
[21] Britz, D., *J. Electroanal. Chem.,* **88** (1978), 309.
[22] Despic, A. R., *Indian Journal of Technology,* **24** (1986), 465.
[23] Despic, A. R., Drazic, D. M., Zecevic, S. K. & Atanasoski, R. T., *Electrochimica Acta,* **26** (1981), 173.
[24] Jow, T. R. & Liang, C. C. *J. Electrochem. Soc.,* **129** (1982), 1429.
[25] Morris, G. A., Mitchell, P. J., Hampson, N. A. & Dyson, J. I., *Power Sources,* Vol. 12, Keily, T. & Baxter, B. W., Eds, International Power Sources Symposium Committee, Leatherhead, (1989), p. 61.
[26] Tudor, S., Weisstuch, A. & Davang, S. H., *Electrochem. Technol.,* **3** (1966), 90.
[27] Treverton, J. A. & Amor, M. P., *J. of Microscopy,* **140** (1985), 383.
[28] Archard, G. D. & Mulvey, T., *X-ray optics and X-ray microanalysis,* Pattee, H. H., Cosslett, V. E. & Engstrom, A., Eds, Academic press, London (1963), p. 393.
[29] Tomlin, S. G., *Proc. Phys. Soc.,* **82** (1963), 465.
[30] Hunter, J. A., D.Phil. thesis, University of Oxford (1990).
[31] Reed, S. J. B., *Electron microprobe analysis,* Cambridge University Press, London, (1975).
[32] Takeda, Y., Kanno, R., Oyabe, Y., Yamamoto, O., Nobugaya, K. & Kanamaru, F., *J. Power Sources,* **14** (1985), 215.
[33] Cliff, G. & Lorimer, G. W., *J. Microscopy,* **103** (1975), 203.
[34] Matsumoto, I., Ogawa, H., Iwaki, T. & Ikeyama, M., *Power Sources,* Vol. 12, Keily, T. & Baxter, B. W., Eds, International Power Symposium Committee, Leatherhead (1989), p. 214.

[35] Rao, C. N. R., *Materials for solid state batteries*, Chowdari, B. V. R. & Radshakrishna, S. Eds, World Scientific (1986), 3.
[36] Latham, R. J., Linford, R. G. & Schlindwein, W. S., *Farad. Dis. Chem. Soc.*, **88** (1989), 103.
[37] Cole, M., Scheldon, M. H., Glasse, M. D., Latham, R. J. & Linford, R. G., *Appl. Phys.* **A49** (1989), 249.
[38] Nazri, G. & Muller, R. H., *J. Electrochem. Soc.*, **132** (1985), 2050.
[39] Robinson, K. S. & Sherwood, P. M. A., *Surface and Interface Analysis,* **6** (1984), 261.
[40] West, R. H. & Castle, J. E., *Surface and Interface Analysis,* **4** (1982), 68.
[41] Fetcenko, M. A., Venkatesan, S., Hong, K. C. & Reichman, B., *Power Sources,* Vol. 12, Keily, T. & Baxter, B. W., Eds, International Power Sources Symposium Committee, Leatherhead (1989), p. 411.
[42] Vickerman, J. C., *Chemistry in Britain* **23** (1987), 969.
[43] Swift, A. J., *Metals and Materials* (1988), 688.
[44] Long, D. A., *Chemistry in Britain,* **25** (1989), 589.
[45] Irish, D. E., Hill, I. R., Archambault, P. & Atkinson, G. F., *J. Solution Chem.*, **14** (1985), 221.
[46] Hill, I. R., Irish, D. E. & Atkinson, G. F., *Langmuir,* **2** (1986), 752.
[47] Aurbach, D., Daroux, M. L., Faguy, P. W. & Yeager, E. *J. Electrochem. Soc.*, **135** (1988), 1863.
[48] Odziemkowski, M. & Irish, D. E., *J. Electrochem. Soc.* (to be published).
[49] Kamo, G., Horita, K., Kawakami, K. & Nakai, K., *Proc. IBA Meeting, Los Angeles, 1989,* p. 8; and *Denki Kagaku,* **57** (1989).
[50] Glemser, O., Bauer, J., Buss, D., Harms, H. J. & Low, H., *Power Sources,* Vol. 12, Keily, T. & Baxter, B. W., Eds, International Power Sources Symposium Committee, Leatherhead (1989), p. 165.
[51] Langmuir, I., *J. Am. Chem. Soc.,* **40** (1918), 1361.
[52] Barnauer, S., Emmett, P. H. & Teller, E., *J. Am. Chem. Soc.,* **60** (1938), 309.
[53] Young, D. M. & Crowell, A. D., *Physical adsorption of gases*, Butterworth, London (1962).
[54] Anderson, J. R. & Baker, B., *J. Phys. Chem.,* **66** (1962), 482.
[55] Gregg, S. J. & Sing, K. S. W., *Adsorption, surface area and porosity*, Academic Press, London (1967).
[56] Loebenstein, W. V. & Deitz, V. R., *J. Res. Nat. Bur. Stan.,* **46** (1951), 51.
[57] Nelson, F. M. & Eggertsen, F. T., *Anal. Chem.,* **30** (1958), 1387.
[58] Cahn, E. & Schultz, H. R., *Vac. Microbalance Tech.,* **3** (1962), 29.
[59] Crowl, V. T., *Proc. Conf. organised by the Society of Analytical Chemistry at Loughborough Univ., (Sept. 1966),* p. 288.
[60] Innes, W. B., *Anal. Chem.,* **23**, 759 (1951).
[61] Mekhail, R. H., Brunauer, S. & Bodor, E. E., *J. Colloid Interface Sci.,* **26** (1968), 45.
[62] Innes, W. B., *Anal. Chem.,* **28** (1956), 332.
[63] Innes, W. B., *Anal. Chem.,* **29** (1957), 1069.
[64] Instrument handbooks, Micrometerics Instrument Corp., Georgia 30071, U.S.A. & Quantachrome Corp., New York 11548, U.S.A.
[65] Knight, M. J., Rowe, P. N., MacGillivray, H. J. & Cheeseman, D. J., *Trans. Inst. Chem. Engrs.,* **58** (1980), 203.

[66] Jesse, A., *Microscope*, **24** (1976), 1.
[67] Coulter, W. H., *Proc. Nat. Electronic Conf.*, **12** (1956), 1034.
[68] Karuhn, R. *et al.*, *Powder Technol.*, **11** (1974), 157.
[69] Talbot, J. H., *Proc. Conf. on Particle Size Analysis, London Soc. Analyt. Chem.* (1970), pp. 95–100.
[70] Felton, P. G., *Int. Symp. In-stream Measurement of Particle Solid Prop, Bergen, Norway, August 22–23* (1978).
[71] Pope, D., Smith, W. L., Eastlake, M. J. & Moss, R. L., *J. Catal.*, **22** (1971), 72.
[72] Guinier, A. & Fournet, G., *Small angle scattering of X-rays*, John Wiley, New York (1955).
[73] Bryant, A. C., Freeman, D. S. & Tye, F. L., *Proc. Conf. organised by the Society of Analytical Chemistry at Loughborough Univ.* (September 1966), p. 154.


## FURTHER READING

### Thermodynamics
A. J. Bard, R. Parsons & J. Jordan (1985), *Standard potentials in aqueous solution*, M. Dekker, U.S.A.
M. H. Cardew (1990), *Thermodynamics for chemists and chemical engineers*, Prentice-Hall, U.K.
E. B. Smith (1990), *Basic chemical thermodynamics*, Oxford University Press, U.K.

### Electrochemistry
A. J. Bard & L. R. Faulkner (1980), *Electrochemical methods: fundamentals and applications*, John Wiley, U.S.A.
J. O'M. Bockris & A. K. N. Reddy (1970), *Modern electrochemistry* (2 vols), Macdonald, London.
D. R. Crow (1988), *Principles and applications of electrochemistry*, Chapman and Hall, London.
R. Greef, L. M. Peter, J. Robinson, R. Peat & D. Pletcher (1985), *Instrumental methods in electrochemistry*, Ellis Horwood, U.K.
P. H. Rieger (1987), *Electrochemistry*, Prentice-Hall, U.S.A.

### Electron microscopy
D. Chescoe & P. J. Goodhew (1990), *The operation of transmission and scanning electron microscopes*, Oxford University Press, U.K.
I. M. Watt (1989), *Principles and practice of electron microscopy*, Cambridge University Press, U.K.

### X-ray diffraction
B. D. Cullity (1978), *Elements of X-ray diffraction*, Addison-Wesley, U.K.
D. M. Moore & R. C. Reynolds (1989), *X-ray diffraction and the identification and analysis of clay minerals*, Oxford University Press, U.K.

### Electron Spectroscopy/SIMS
D. Briggs, A. Brown & J. C. Vickerman (1989), *Handbook of secondary ion mass spectrometry* (SIMS), John Wiley, U.K.

M. Thompson, M. D. Baker, J. F. Tyson & A. Christie (1985), *Auger electron spectroscopy*, John Wiley, U.S.A.

J. F. Watts (1990), *Introduction to surface analysis by electron spectroscopy*, Oxford University Press, U.K.

### Infra-Red/Raman spectroscopy
N. B. Colthup, L. H. Daly, H. Lawrence & S. E. Wiberly (1990), *Introduction to infrared and Raman spectroscopy*, Academic Press, U.S.A.

W. O. George & P. McIntyre (1987), *Infrared spectroscopy*, John Wiley, U.S.A.

P. R. Griffiths & J. A. DeHaseth (1986), *Fourier transform infrared spectrometry*, John Wiley, U.S.A.

### DTA/DSC
M. E. Brown (1988), *Introduction to thermal analysis: techniques and applications*, Chapman and Hall, U.K.

M. I. Pope & M. D. Judd (1977), *Differential thermal analysis and its applications*, Heyden, U.K.

# 3

# Commercial non-rechargeable battery systems

## 3.1 ZINC–CARBON BATTERIES
### (B. Schumm)

### 3.1.1 Introduction

Zinc–carbon batteries or 'dry' cells are galvanic cells which have thin solid zinc for an anode and a cathode which consists of a moist cake of manganese dioxide powder, special carbon black, electrolyte, and solution blended together. They typically provide 1.4 to 1.72 V of D.C. electric power which declines gradually to 0.9 V or less during use. From the earliest inception over 100 years ago [1] the Leclanché cell, Fig. 3.1, was successful commercially because the zinc of the anode, naturally occurring manganese dioxide ore for the cathode, and ammonium chloride electrolyte salt were plentiful and cheap. More recently a modification of the system referred to as a zinc chloride cell has appeared. In this design variant, the use of chemically stable thin separators between the anode and the cathode, and improved, yet simple, seals, has permitted the ammonium chloride to be greatly reduced in favour of commercial zinc chloride salt.

The low cost and general utility of the Leclanché cell permit it to be a popular consumer system even today. This is true in spite of numerous advanced design competitive systems. The cells are very tolerant of many impurities in the ingredients, and they provide cost effective performance on both heavy and light electrical loads. With very inexpensive constructions the cells have good shelf life and fair resistance to abuse during use. The advent of the zinc chloride 'heavy duty' and 'super heavy duty' cells has extended the range of performance on continuous loads and reduced the likelihood of cell electrolyte leakage when abused, with only modest increases in the cost of construction.

Both the Leclanché and zinc chloride cells can be made with cathode formulations and separator modifications to greatly change the amount of delivered energy on a given electrical load. One can also redesign to control the type of load on which the cell provides the best relative performance. Thus the continuous load service to

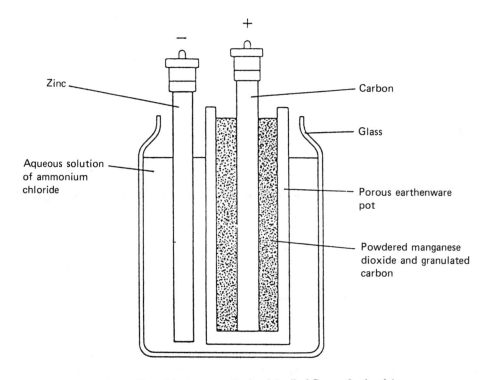

Fig. 3.1 — The original type wet Leclanché cell of George Leclanché.

0.9 V can be doubled by changing the manganese dioxide in a zinc chloride cell from a high quality, but low cost, natural manganese dioxide one to a more expensive electrolytically deposited special manganese dioxide. The intermittent service in an electric torch could similarly be increased by around 100%. If the carbon and electrolyte are increased at the expense of the manganese dioxide (and the theoretical cell capacity) then much larger currents (amperes) can be drawn from the cells for heavy drain applications. As a result of this versatility, a large and possibly confusing variety of these dry cells are available for consumer and industrial use with various capabilities and prices.

The most common cylindrical cell sizes are from about 1 cm to about 3 cm in diameter and 3 cm to 10 cm in height. However, the cells have been made and sold in sizes as small as 3 mm high and 1 cm side length, or as large as barrels (in sets of larger cells) for ocean buoys where the battery weighs several hundred kg and serves as ballast for the buoy as well as a power supply.

### 3.1.2 Manufacturers

One can find in the World today over two hundred manufacturers (both large and small) of zinc–carbon dry cells and batteries. Each has capitalized on the long tradition of the Leclanché cell. They have many methods of assembling the cells. Depending on the market, the manufacturer, labour costs, materials availability, and capital a factory may be found making cells largely by hand or employing extremely sophisticated, high speed equipment. Table 3.1 gives a partial list of manufacturers, some countries of operation, and typical brand names.

Despite competition from alkaline–manganese cells, the worldwide sales of zinc–carbon cells and batteries continue to rise. This is less true in Europe, Japan, and the United States where more expensive portable appliances and *per capita* income encourage the use of more expensive cells. But, in most countries of the World, the lower cost and competitive performance of the various grades of zinc–carbon cells are dominant factors in sales. It is worthwhile to remember that batteries for torches, radios, and automobiles are the first form of electrical power to enter developing regions, so these have a strong starting base of usage. In many countries, dry cell powered illumination is cheaper than candles or other manufactured flame-type light sources. Thus zinc–carbon cells sales are likely to continue to rise for some time to come on a worldwide basis.

Statistics of zinc–carbon cell sales growth can be presented in a number of ways. General areas of possible interest are (1) monetary value and trends of sales in various major markets, (2) number of actual units sold by size, (3) where they are sold, and (4) when they are sold.

Tables 3.2 to 3.4 [2,3,4] and Figs 3.2 to 3.6 [3,4,5,6,7] show the absolute levels or the level and the shift of sales dollar percentage in various markets between zinc–carbon cells and alkaline–manganese cells, the most popular competitor. In less developed economies this shift has not turned so obviously in favour of alkaline cells.

Because the cost and retail price of zinc–carbon cells is lower, the actual unit sales are more positive, as shown in Figs 3.2 and 3.3. From the viewpoint of suppliers to this segment of the industry, the tonnage of material required, while not huge, is maintained or is continuing to grow, particularly because large zinc–carbon cells compete more effectively against alkaline cells.

In the least developed areas of the World dry cells are sold by small general outlets or hardware shops. As the country becomes more like the most developed countries the trend seems to be quite uniform. The sales volume pattern shifts to the distribution of outlets shown in Figs 3.6 and 3.7 [4].

Another factor of interest in at least the United States is the annual cycle of sales. Holidays very definitely affect sales dollar volume, as shown in Fig. 3.8 [8].

### 3.1.3 Chemistry of zinc–carbon systems

For every unit of electrical energy produced by a galvanic cell an equivalent amount of electrode materials and the salts must move or be altered to provide the energy. As noted earlier, the electrode chemicals of the Leclanché cell are zinc as a sheet metal anode, and manganese dioxide of particular crystal types as a powder mixed with carbon powder. The electrolyte which moistens this mixture is ammonium chloride and zinc chloride in an aqueous solution. The cathode may also contain solid ammonium chloride to act as a fuel reserve for the cell during intermittent operation. Other additives such as gum karaya and ion exchange resins [9] may be added to the cathode to increase the discharge efficiency under at least some conditions. For the zinc chloride cell most, if not all, of the ammonium chloride is left out of the cathode

**Table 3.1** — The principal dry cell manufacturers of the World

| Manufacturer | Major countries of operation | Typical brand names |
| --- | --- | --- |
| Eveready Battery Co. (Ralston-Purina Co.) | USA, France, Singapore, Brazil, Mexico, Argentina, Egypt, Indonesia Nigeria | Eveready, UCAR, Wonder, Mazda, Marin, Cipel |
| Duracell, Int'l | USA, England, Belgium, | Duracell, Mallory, Daimon, Volta |
| Matsushita Battery Co., Ltd | Japan, Indonesia, Phillippines, Brazil Belgium (w/Philips) | National, Panasonic |
| Rayovac Corp. | USA, Mexico | Rayovac |
| Varta Battery Co. | W. Germany, Mexico | Varta, Aquila Negra |
| Hanson Trust Ltd. | England | Ever Ready, Berec |
| Toshiba Battery Co. | Japan | Toshiba |
| Oy Airam | Finland | Oy Airam, Panda |
| Chung Pak, Ltd. | Hong Kong | Vinnic |
| Kaha Battery Co. | Turkey, Egypt | Kaha, Butterfly |
| White Elephant Co. | China | White Elephant, Winco |
| HiWatt (w/555) | Hong Kong, China | HiWatt |
| Five Rams | Thailand, China | Five Rams |
| Poon Trading Co., Ltd. | Taiwan | Various |
| Nippo Battery Co. | India | Nippo |
| Leclanché, Ltd | Switzerland | Leclanché |
| Cegasa, Ltd | Spain, USA | Cegasa |
| Croatia | Yugoslavia | Croatia |
| Fuji Electrochemical Co. | Japan | Novel, Lamina |
| Hitachi-Maxell, Ltd. | Japan | Maxell |
| SONY Energytec | Japan | SONY |
| Union Carbide Corp. | India | Eveready |
| Williams Electric Co. | Taiwan | — |
| Gold Peak Industries | Taiwan, Hongkong, Peoples Republic of China | Gold Peak, Sylva, GPI |
| ABC Battery Co. | Indonesia | ABC |
| Sanyo Battery Co. | Japan | Sanyo |
| 555 Brands | China, Hong Kong | 555 |
| STC Corp | Korea | STC |
| Rocket Electric Co. | Korea | Rocket |
| Philips | Belgium | Philips, National |
| Toyo Takosago Co. | Japan | Other brands |
| Tudor Battery Co.'s | Spain, Sweden | Tudor |

**Table 3.2** — World distribution of battery production in 1981, with 1987 estimated total [2,3]. Numbers given are in millions of US $

|  | Carbon–zinc | Alkaline Zn–MnO$_2$ | Minature | Total |
|---|---|---|---|---|
| United States | 273 | 390 | 184 | 847 |
| Japan | 298 | 78 | 88 | 464 |
| W. Europe | 705 | 101 | 94 | 900 |
| Africa | 325 | — | — | 325 |
| Asia (ex. Japan) | 537 | — | 18 | 555 |
| Other | 614 | 45 | 6 | 665 |
| Total worldwide (1981)* | 2752 | 614 | 390 | 3756 |
| % — 1981 zinc–carbon | 73 | | | |
| Estimate 1984 % | 67 | | | |
| Estimate 1987† | | | | 8000 |

\* George Int'l Consultants Aug. 1981.
† D. Bradshaw.

**Table 3.3** — Retail battery sales by size and grade in the US

| Size | General purpose | Heavy duty | Alkaline | Total |
|---|---|---|---|---|
| D (R20) | 37% | 26% | 18% | 21% |
| C (R14) | 24 | 22 | 17 | 19 |
| AA (R6) | 26 | 38 | 48 | 44 |
| AAA (R03) | | | 18 | 5 |
| 9 V | 10 | 11 | 9 | 10 |
| 6 V | 3 | 3 | | 1 |

Source: Rayovac Corp.; A.C. Nielsen, 1988.

formulation. Leclanché cells may have a separator up to 3.5 mm thick, made of cereal paste and electrolyte solution. At the high thicknesses the separator serves as an electrolyte reservoir as well as a membrane between the electrodes. In zinc chloride cells and many Leclanché cell designs the separator is a piece of coated paper serving primarily as an ion transfer membrane with less of a solution reservoir function.

**Table 3.4** — Distribution of devices using dry cells [6]

| Type of device | Million devices | Percentage of total |
|---|---|---|
| Torches/lanterns | 180 | 16.4 |
| Radios | 90 | 8.0 |
| Camera equipment | 124 | 11.3 |
| Toys/games | 69 | 6.3 |
| Watches | 144 | 13.1 |
| Tape recorders and recorder-radios | 110 | 10.0 |
| Calculators | 88 | 8.0 |
| Smoke alarms | 76 | 6.9 |
| Clocks | 122 | 11.1 |
| Other | 97 | 8.9 |
| Total | 1100 | 100.0 |

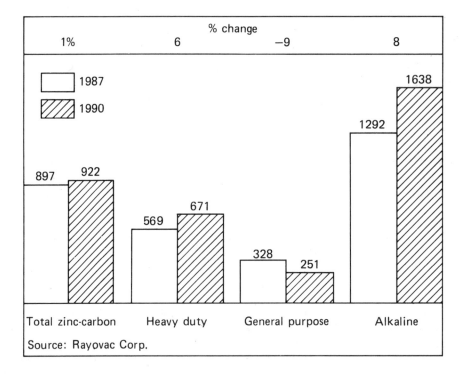

Fig. 3.2 — US 1987 and estimated 1990 primary battery sales by grade or type [4].

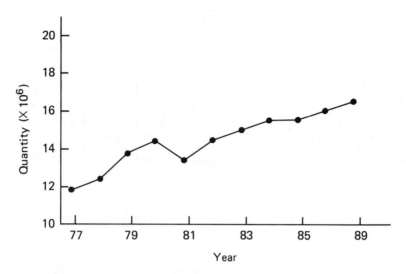

Fig. 3.3 — Quantity of zinc–carbon dry cells produced in Japan for domestic use, 1977–1989 [5].

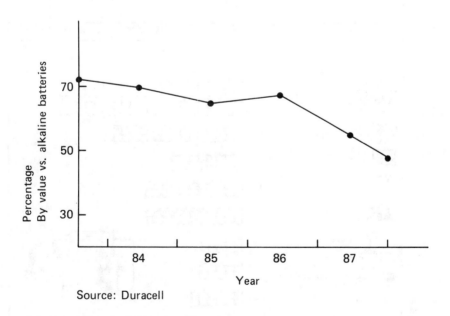

Source: Duracell

Fig. 3.4 — Sales trend in dollar value percent of UK market for zinc–carbon batteries 1984–1987 (Balance Alkaline Batteries).

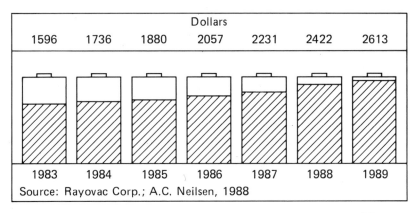

Fig. 3.5 — Estimated retail dry cell market growth in the US 1983–1989 [4].

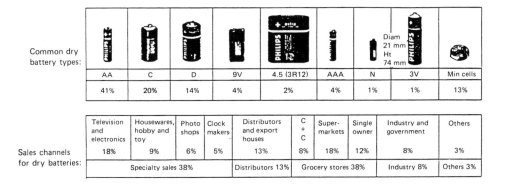

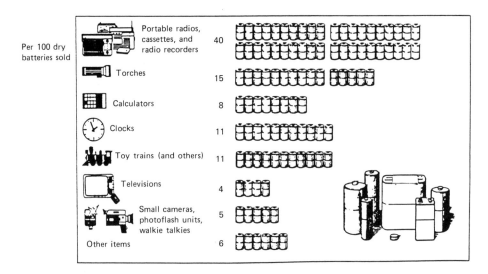

Fig. 3.6 — Retail zinc–carbon sales distribution and intended use pattern — Europe [7].

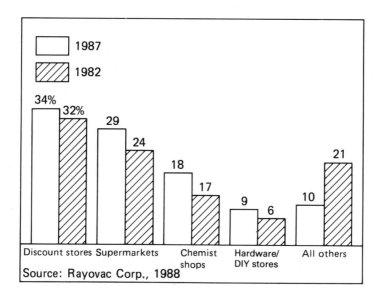

Fig. 3.7 — Where batteries are purchased, 1982 versus 1987 (US) [4].

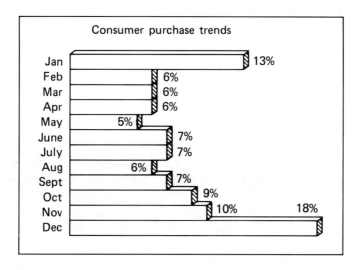

Fig. 3.8 — Purchases month by month — US [8].

Despite the long history of the Leclanché cell, the chemistry of the system is actually more complex than that of many newer systems. At the anode the first reaction is highly efficient:

$$Zn^0 \rightarrow Zn^{2+} + 2e^- \ . \tag{3.1}$$

This reaction is immediately followed by combination with the chloride ion of the electrolyte to yield a mixture of species, principally the anion, zinc tetrachloride:

$$Zn^{2+} + 4Cl^- \rightarrow ZnCl_4^{2-} \ . \tag{3.2}$$

At the manganese dioxide cathode, at least two reactions are observed. The initial reaction is:

$$MnO_2 + NH_4^+ \rightarrow MnOOH + NH_3 - e^- \ . \tag{3.3}$$

The ammonia available from this reaction usually reacts with the zinc chloride present to form the partly dissolved salt, zinc diammino chloride:

$$ZnCl_4^{2-} + 2NH_3 \rightarrow Zn(NH_3)_2Cl_2 + 2Cl^- \ . \tag{3.4}$$

As the cell discharge continues, another reaction which may occur is the more direct combination of the dissolved zinc in some form with the manganese dioxide or with manganese ions:

$$Zn^{2+} + 2MnO_2 \rightarrow ZnO:Mn_2O_3 - 2e^- \ . \tag{3.5}$$

The overall principal reaction, however, proceeds through the path:

$$Zn + 2MnO_2 + 2NH_4Cl \rightarrow Zn(NH_3)_2Cl_2 + 2MnOOH \ . \tag{3.6}$$

Alternatively, the reaction forming zinc manganese oxides (heterolite) more directly could be considered with substitution of zinc metal for the ion in the equation above. If the ammonium chloride is depleted or absent, another reaction commonly observed is the formation of zinc oxychloride, precipitating principally in the cathode body of the cell:

$$5ZnCl_2 + 9H_2O \rightarrow ZnCl_2:4Zn(OH)_2:H_2O + 8HCl \ . \tag{3.7}$$

This reaction probably occurs in some stepwise fashion with the final identified product occurring as the last step in precipitating a complex ion.

The reactions of the zinc chloride cell are those without ammonia present. When ammonia is present in small amounts the cell will behave a little more like a

Leclanché cell but with perhaps a little higher voltage on intermittent tests and in the early part of continuous tests.

While these reactions are usually presented as occurring in the aqueous phase in the cell it should be observed that the ammonia formed in the reactions together with water vapour are capable of transferring through the porous body of the cathode mix in the gas phase and redissolving to react at a point away from the origin. This means that the rate of reaction and diffusion in the cells is not strictly limited by liquid phase diffusion, which is relatively slow compared to the typical drain rates demanded of dry cells.

### 3.1.4 Types of zinc–carbon cell

Zinc-carbon dry cells are sold in two main classes (1) cylindrical cells either singly or with two more in a battery, and (2) flat cells which are nearly always sold as a battery of from four to three hundred or more cells in a stack or set of stacks.

The typical family of several grades of cylindrical cells or batteries of cells is illustrated in Fig. 3.9 [10]. A group of higher voltage or larger batteries of a prominent worldwide manufacturer is shown in Fig. 3.10 [11].

### 3.1.5 Construction details

While numerous constructions exist, they differ very little in principle. Every cell should meet three criteria for acceptable performance: (1) Excellent appearance, (2) an expected level of performance, and (3) an expected level of reliability. In competitive markets, high quality cell appearance is a mark of high quality manufacture. Since the purchase of batteries is often an impulse response at the vendor's shop, a brightly decorated jacket aids in creating the desire to buy. This seems to be true in both affluent and primitive societies. In many areas of the World a metal jacket has become a symbol of relatively high quality ingredients, whereas paper or paper plastic composites are indicative of lower quality. Beyond the visual appearance most buyers expect the dry cell to have a rock-hard character when held in the hand, and not to be soft or mushy. Again, it is supposed that the hardness represents a mark of careful, more expensive, construction.

A dry cell contains a variety of electrode and packaging materials. Each material must have a certain level of quality, or the performance of the cell or its appearance will be degraded to some extent.

The anode is zinc metal in the form of a zinc can or cup in the common cylindrical cells. These cans are produced by one of three methods. The first, extrusion, is the most common and the most recently developed. To make an extruded can a circular disk is cast or stamped out of a sheet of rolled zinc. Such a disk, sometimes called a calot, is slightly smaller than the diameter of the final cell can. It is made thick enough, (usually about 2 to 4 mm) to provide the zinc weight necessary. This calot is coated with with a lubricant such as graphite and is placed in a die in a heavy press running at 60 or more strokes per minute. About ninety tonne of weight is applied per stroke. When the chamfered punch is rammed into the die containing the calot, the zinc nearly liquefies from the impact energy dissipated and appears to jump up the sides of the punch well beyond the sides of the die. The punch must be cooled, often by water; the can is trimmed on the punch to give a smooth opening and

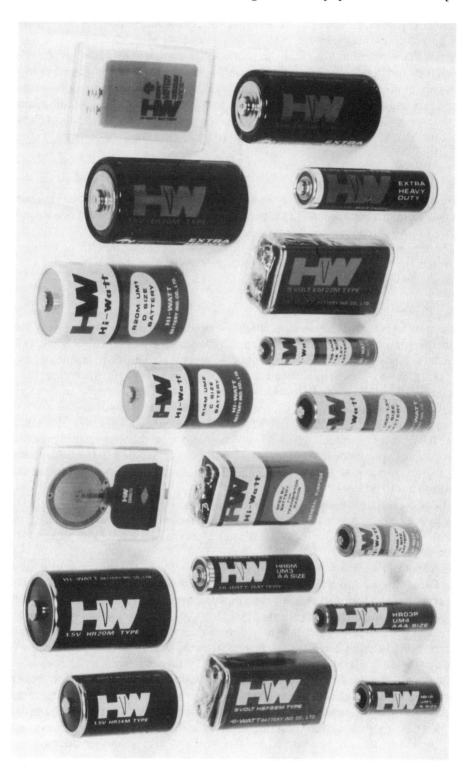

Fig. 3.9 — Dry cell product of an East Asian manufacturer, HiWatt Co. [10].

Sec. 3.1]  Zinc–carbon batteries  99

Fig. 3.10 — Larger or higher voltage battery products of a major manufacturer, Eveready Battery Co [11].

stripped from the punch before the next stroke. The thickness of the can is determined by the clearance between the punch and the die. It requires considerable skill and machine adjustment to obtain a can with the sidewall thickness the same as the bottom, the latter being usually about ten to twenty per cent thicker.

The second method of can manufacture is practised chiefly in China and the United States. This operation is called drawing. For this method a disk of rolled zinc of the desired final can thickness is made into a shallow dish shape. Then this is progressively pushed through a series of drawing orifices which force the zinc over an increasingly smaller mandrel. This action deepens the can while reducing the diameter of the part. Five to seven stages, side by side in a press, with transfer from one to the other, complete the process. Finally, the cans are trimmed to achieve a clean cut-off at the opening, of the proper height. The thickness of the entire can, side and bottom, is essentially the same as the original sheet thickness.

The third method, the oldest, is at present practised only in China or for large cells. Here rolled zinc is cut into a rectangular piece, scrolled, and the edges soldered together to form a hollow cylinder. A zinc disk is then soldered in to form the can bottom and complete the assembly. Again the rolled zinc thickness determines the thickness of the can wall. In some cells a steel disk is used for the can bottom rather than a zinc disk.

In most dry cells the zinc is amalgamated with about 0.1% mercury to significantly improve its resistance to corrosion over time. The zinc may contain about 0.05% cadmium. The cadmium serves as a grain refiner, and it makes the alloy harder and also more corrosion resistant. It also may contain about 0.25% lead to improve its drawing qualities or extrusion quality so as to allow the typical can to be formed more readily without defects. Despite these rather low levels it can be expected that these elements will be reduced in percentage to meet increased environmental concerns for both cell disposal and occupational safety in the experimental and manufacturing workplaces. The zinc must also not contain much iron (<0.007%), or splits may appear in the can during forming or spots could be present with potential high corrosion rates. Other impurities of special concern are cobalt and antimony in trace amounts as these also accelerate local spot corrosion of the can. This type of corrosion may lead to rapid formation of perforations and loss of seal, inviting electrolyte loss from the cell and short storage life.

The next chemical component of prime importance is the manganese dioxide cathode material. In many Leclanché and zinc chloride cells the manganese dioxide originates from mines with special high purity ore and with particular crystal forms which occur naturally in Mexico, Gabon, China, and Brazil. The content of the 'gamma' crystal form and other electrochemically active forms of manganese dioxide in these ores approaches 70% to 80% of the ore, and impurities such as nickel, copper, arsenic, and cobalt are either very low or insoluble. Iron, which occurs in these ores, is generally not harmful in the fully oxidized state in which it is usually found. If free iron were introduced by milling, then this iron would reduce some of the manganese dioxide and be harmful to battery performance. The manganese dioxide is always mixed with either graphite or acetylene (carbon) black to provide better conductivity and absorption of electrolyte. Usually, only a minor portion of graphite is used, with the major portion of carbon as acetylene black. This highly conductive, stable form of finely divided carbon absorbs a greater volume of electrolyte when comparing its properties to graphite, and probably contains a surface of particular value on which to precipitate secondary reaction products of the cell reactions. When acetylene black began to be used in dry cells in the 1930s the achievable service doubled. It is important that the carbon black or graphite contains very little impurities such as nickel. Such impurities, just as in the case of manganese dioxide ore impurities, can dissolve and migrate to the surface of the zinc where they may be reduced and cause local corrosion centres. This will both consume zinc and generate hydrogen gas which could disrupt cell seals and dry out the cell through chemical consumption of water and moisture evaporation. Similarly, the ammonium chloride and the zinc chloride salts used in the electrolyte must be free of soluble iron and the same undesirable impurities as listed for manganese dioxide. The electrodes are separated by a flour-starch or polymer-coated paper membrane which permits ions to pass but prevents electrical short-circuits between the electrodes. The flour starch or polymers used in the separator must be specially chosen and of tested quality to assure that the material will not degrade to an unacceptable degree during high temperature storage or the various cycles of cell usage and temperature variation. Similarly, the paper base for such separators must be tested carefully before full use in commercial dry cells.

The seals and outer construction materials of dry cells must be similarly carefully chosen and tested. Even the asphalt used for a seal in less expensive cells must have a proper balance of melting point, adhesive quality, and surface tension to permit ease of manufacture and long term stability in warehouses, shops, and customers' devices. Plastics chosen for construction parts must resist ammonia vapours and strong acid solutions which may be driven around the cells when the cell is abused or heavily discharged. Only baked carbon or graphite (as found in the collector rods of almost all round cells) has been found to be an effective conductive material which can withstand the extremely corrosive and chemically active electrolyte. Titanium could also serve in this role. As will be noted in the construction sketches, Figs 3.11 to 3.13, each cell is generally closed with some type of plastic or asphaltic cover, often combined with or covered by another jacketing system to provide both double containment and a better base for high quality decorative printing or labelling.

The construction of a cell dictates the capacity that can be built into the cell. The cell outer dimensions and total volume are fixed by international standards. Some of these standards are outlined in the IEC recommendations, as shown in Fig. 3.14 and Tables 3.5 and 3.6 [12]. The thickness of the jacket, containment parts, and cover assemblies occupies a certain amount of the standardized volume. The volume remaining is then available for active ingredients and the conductors essential to complete the electrical circuit to the manganese dioxide cathode material. Similarly, the thickness of the separator, which is an ion-conducting membrane between the zinc metal anode cup and the cathode mixture, restricts the volume available for the cathode material to a large degree. The available space for the cathode mixture differs for different cell designs in commercial practice by as much as 81% for the same size cell, with a commensurate effect on performance if the same cathode formula is used.

The zinc–carbon cell chemicals are fundamentally very stable [13]. However, the cell must have a sealing system which allows most of the gaseous corrosion products to escape. The seals must also largely exclude air or oxygen from the cell. If both of these somewhat contradictory requirements are not adequately balanced the cell will be either less safe if abused or will allow oxygen to eventually combine with and render unusable all of the zinc. At the same time, much of the water in the cell would evaporate, shutting off the cathode activity as well. This venting to sealing balance is easier to achieve in larger cells than in smaller ones. In most commercial cells the ingenuity of materials usage and placement provides an acceptable balance of closure versus openness, so that cell capacity is preserved for years in a safe manner.

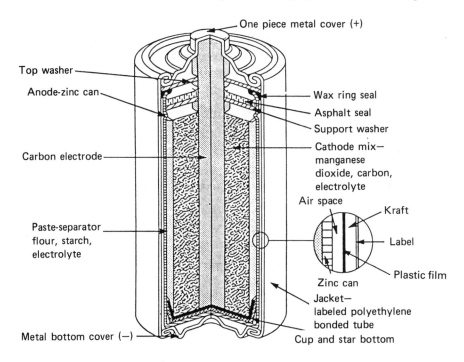

Fig. 3.11 — Construction details of a general purpose Leclanché cell [11].

### 3.1.6 Cell or battery performance

As materials and methods of construction have improved over the years, the service likely from the highest performance commercial dry cells has markedly increased, as shown by Fig. 3.15.

Several examples are given in Figs 3.16 to 3.19 of the range of service which can be purchased for cells classified as general purpose or 'ordinary' and super heavy duty, or premium, cells. Two general types of test are illustrated. The first, a continuous test, is one in which the cell is discharged without interruption until a chosen lower voltage is reached. The other type is any one of a number of possible intermittent tests. An intermittent test is one in which the discharge usually lasts for a period and then is stopped for a period before repeating the cycle, perhaps once a day. Other intermittent tests attempt to simulate periods of use of a torch during the day with a sixteen hour rest before repeating a series of pulses of, for instance, four minutes every fifteen minutes. These tests are run until the voltage drops below a chosen end voltage level. Each discharge curve of voltage versus time exhibits the typical shape for zinc–manganese dioxide cells. At the start of cell operation the electrolyte of the cell is in contact with fresh manganese dioxide or zinc surface with the highest chemical activity. As this material reacts, the concentration of reaction products on, or in, the surface of the electrodes increases, causing the cell to be 'polarized' and to exhibit lower voltage. For the zinc, this concentration effect is at the surface of the zinc sheet, and it tends to be small. For the metal oxide cathode,

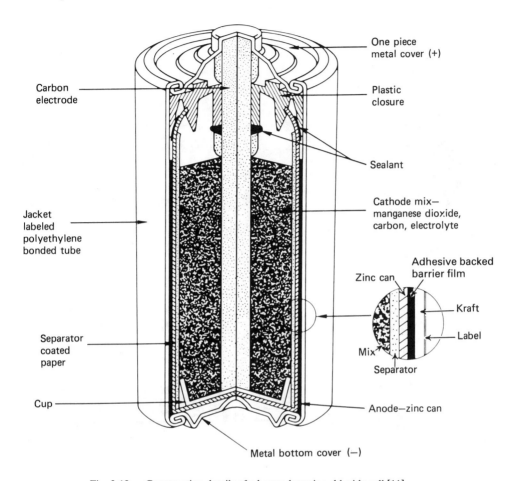

Fig. 3.12 — Construction details of a heavy duty zinc chloride cell [11].

however, there is a concentration gradient of protons induced in the pore solution of the cathode cake and in the solid state structure of each particle, which also causes the electrode to exhibit the typical lower voltage. Thus the cell activity is sensitive to the rate at which energy is drawn from the cell. At high rates the cell voltage will fall more rapidly as the electron demand exceeds the rate at which essential chemicals can diffuse through the cell to maintain chemical equilibrium. When the cell is off-load the chemicals tend to come to equilibrium throughout the cell, and the voltage and rate capability recover markedly. As a result, the total energy delivered on intermittent loads can be two or three times that delivered on common continuous tests. For the continuous data, the premium cells are zinc chloride cells. For the other tests the electrolyte system makes less difference.

Another test commonly used to screen dry cells for quality is called a 'flash current' or amperage test. Here the cell is short-circuited and the current measured which is flowing at, for example, 200 ms later. Such a test provides a check on the

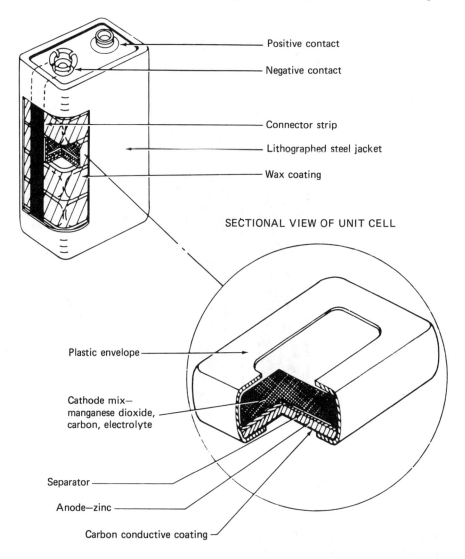

Fig. 3.13 — Construction details of a Leclanché or zinc chloride 9 V battery with flat bipolar cells [11].

internal resistance of the cell and to some extent its performance on tests simulating torch or motor driven toy applications. It does not, however, predict intermittent or light load performance well. It is more useful as a quality surveillance procedure to observe if the cell assembly is reasonably uniform from an internal electrical control viewpoint.

Apart from the essential performance aspect of service, a zinc–carbon dry cell is expected to have a relatively low price compared to other common commercial consumer cells, and it is expected to be safe and reliable. The aspect of cost is not

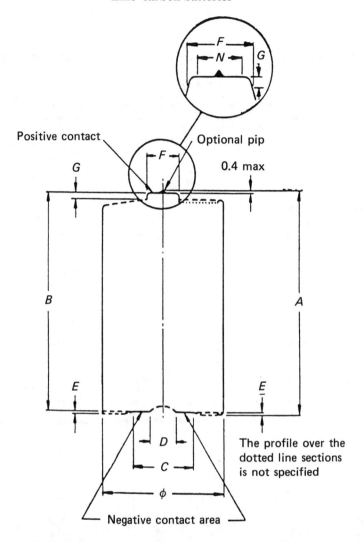

Fig. 3.14 — Dimensions which have been assigned minima or maxima per IEC Standards [12].

addressed here other than to note that the merchandizing practices of manufacturers may render the sale price of a more expensive system quite low on certain occasions, so that price comparisons between systems become difficult.

Safety and reliability are interacting performance aspects, as noted in the construction discussion. To discuss this performance area it can be divided somewhat differently. Reliability factors can be considered to be (1) capacity retention during storage, (2) absolute containment of corrosive electrolyte, and (3) no risk of the cell building up internal gas pressure and bursting, even if intentionally abused. This could be called a fail-safe criterion.

The variety of constructions available and formulae for the cells create a great deal of confusion in the area of estimating the reliability of a given set of dry cells.

**Table 3.5** — Assigned cylindrical cell sizes (IEC/ANSI Standards) [12]

| Designation | | Nominal cell dimension (mm) | | Maximum battery dimensions (mm) | |
|---|---|---|---|---|---|
| IEC | ANSI | Diameter | Height | Diameter | Height |
| R06 |  | 10 | 22 | — | — |
| R03 | AAA | — | — | 10.5 | 44.5 |
| R01 |  | — | — | 12.0 | 14.7 |
| R0 |  | 11 | 19 | — | — |
| R1 | N | — | — | 12.0 | 30.2 |
| R3 |  | 13.5 | 25 | — | — |
| R4 | R | 13.5 | 38 | — | — |
| R6 | AA | — | — | 14.5 | 50.5 |
| R8 | A | 16.0 | 47.8 | — | — |
| R9 |  | — | — | 16.0 | 6.2 |
| R10 |  | — | — | 21.8 | 37.3 |
| R12 | B | — | — | 21.5 | 60.0 |
| R14 | C | — | — | 26.2 | 50.0 |
| R15 |  | 24 | 70 | — | — |
| R17 |  | 25.5 | 17 | — | — |
| R18 |  | 25.5 | 83 | — | — |
| R19 |  | 32 | 17 | — | — |
| R20 | D | — | — | 34.2 | 61.5 |
| R22 | E | 32 | 75 | — | — |
| R25 | F | 32 | 91 | — | — |
| R26 | G | 32 | 105 | — | — |
| R27 | J | 32 | 150 | — | — |
| R40 | 6 | — | — | 67.0 | 172.0 |
| R41 |  | — | — | 7.9 | 3.6 |
| R42 |  | — | — | 11.6 | 3.6 |
| R43 |  | — | — | 11.6 | 4.2 |
| R44 |  | — | — | 11.6 | 5.4 |
| R45 |  | 9.5 | 3.6 | — | — |
| R48 |  | — | — | 7.9 | 5.4 |
| R50 |  | — | — | 16.4 | 16.8 |
| R51 |  | 16.5 | 50.0 | — | — |
| R52 |  | — | — | 16.4 | 11.4 |
| R53 |  | — | — | 23.2 | 6.1 |
| R54 |  | — | — | 11.6 | 3.05 |
| R55 |  | — | — | 11.6 | 2.1 |
| R56 |  | — | — | 11.6 | 2.6 |
| R57 |  | — | — | 9.5 | 2.7 |
| R58 |  | — | — | 7.9 | 2.1 |
| R59 |  | — | — | 7.9 | 2.6 |
| R60 |  | — | — | 6.8 | 2.15 |
| R61 |  | 7.8 | 39 | — | — |
| R62 |  | — | — | 5.8 | 1.65 |
| R63 |  | — | — | 5.8 | 2.15 |
| R64 |  | — | — | 5.8 | 2.70 |
| R65 |  | — | — | 6.8 | 1.65 |
| R66 |  | — | — | 6.8 | 2.60 |
| R67 |  | — | — | 7.9 | 1.65 |
| R68 |  | — | — | 9.5 | 1.65 |
| R69 |  | — | — | 9.5 | 2.10 |

IEC: International Electrotechnical Commission.
ANSI: American National Standards Institute.

## Table 3.6 — Assigned flat cell sizes [12]†

| Designation IEC | Dimensions in millimetres | | | |
| --- | --- | --- | --- | --- |
| | Diameter | Length | Width | Thickness |
| F15 | | 14.5 | 14.5 | 3.0 |
| F16 | | 14.5 | 14.5 | 4.5 |
| F20 | | 24 | 13.5 | 2.8 |
| F22 | | 24 | 13.5 | 6.0 |
| F24 | 23 | — | — | 6.0 |
| F25 | | 23 | 23 | 6.0 |
| F30 | | 32 | 21 | 3.3 |
| F40 | | 32 | 21 | 5.3 |
| F50 | | 32 | 32 | 3.6 |
| F70 | | 43 | 43 | 5.6 |
| F80 | | 43 | 43 | 6.4 |
| F90 | | 43 | 43 | 7.9 |
| F92 | | 54 | 37 | 5.5 |
| F95 | | 54 | 38 | 7.9 |
| F100 | | 60 | 45 | 10.4 |

† The complete dimensions of batteries are given in the relevant specification sheets.

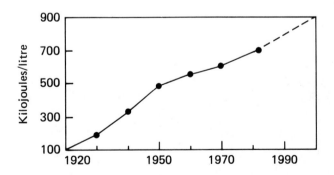

Fig. 3.15 — Increase of R20 dry cell energy density over the years, with future projection.

Beyond this, the storage qualities of the cells are dependent on the size of the cell, with smaller cells being more difficult to design with long shelf life. A classic study [14] estimated the shelf life of cells at different temperatures to be about four years at

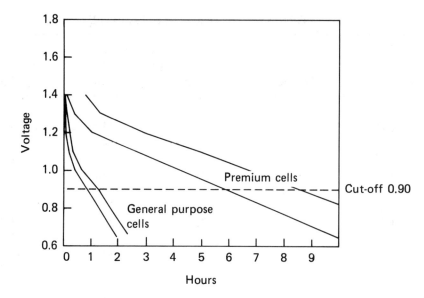

Fig. 3.16 — Performance ranges available in commercial R20 zinc–carbon on 2.0 ohm per cell continuous drain (toy or torch application).

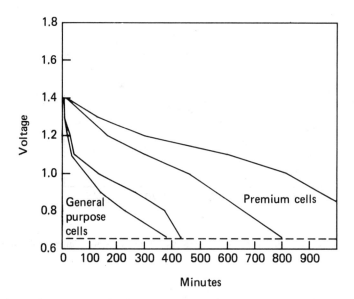

Fig. 3.17 — Performance ranges available in commercial R20 zinc–carbon cells on 2.2 ohm per cell intermittent tests of approximately thirty minutes per day.

# Sec. 3.1] Zinc–carbon batteries

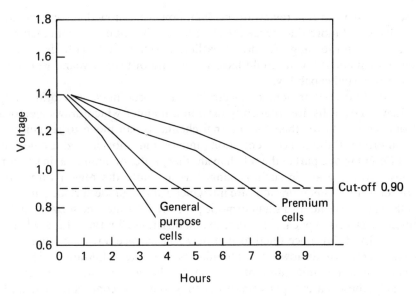

Fig. 3.18 — Performance ranges available in commercial R6 zinc–carbon cells on 10 ohm per cell test, one hour on load per day, cassette application.

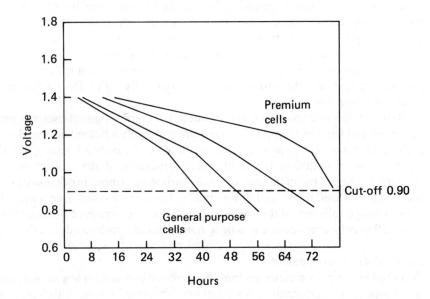

Fig. 3.19 — Performance ranges available in commercial R6 zinc–carbon cells on 75 ohm per cell test, four hours on load per day, radio application.

room temperature. A review of the data, however, show that some manufacturers' products had more than twice the shelf life of that of other manufacturers', so the average is misleading. As noted earlier, another study [13] demonstrated that specially made dry cells could have a shelf life of twenty years, which exceeds all other reported capability.

Absolute containment of electrolyte, leakproofness, is a difficult quality to achieve. Dry cells are frequently used in a manner which draws electrical energy from the cell faster than the internal chemicals can move to remain in chemical equilibrium. If the salt concentration (ammonium chloride or zinc chloride or both) is different in one part of the cell than another, water or ammonia, in vapour or liquid phase, will migrate in such a manner as to dilute the region with a higher salt concentration. Since zinc ions are introduced at the anode surface during discharge, cells typically build up a slight osmotic pressure near the zinc surface, and as a result liquid electrolyte appears. To some extent this liquid fills the volume where solid zinc once existed, but once filled to that new zinc surface any continuing pressure will cause electrolyte to slowly move to other parts of the cell. Many designs contain one or more features which attempt to confine this liquid movement inside the cell jacket. Usually, these features perform successfully, but if the conditions of abuse happen to be just right, the cell may 'leak'. If this happens, the very corrosive, low pH electrolyte may attack metal or electronic parts nearby. If the electrolyte gets on human skin, especially that of young children, a surface burn could result. The greater danger, however, occurs if the child rubs his eyes with hands smeared with electrolyte and tiny salt crystals. Whereas the electrolyte itself may not be so harmful, the salt crystals if rubbed on the surface of the eyeball will irritate it severely. As a result of this concern for safety at least one manufacturer [15] includes a warning on the cell package noting that small children should not be allowed to play with dry cells without adult supervision. Similarly, adults should not treat dry cells without respect for both their electrical energy content and their chemical content. This, of course, applies to all cell systems, not just zinc–carbon cells. A typical standard warning notes that the cells should not be disposed of in fire, recharged, or put in an appliance in the wrong orientation especially with cells of different age and remaining capacity.

To finish this discussion, it must be stated that in some limited circumstances a dry cell may burst from inside pressure build-up, especially if heated or mixed with other cells of different age or capacity. This is almost exclusively a bursting like that of a balloon, but nonetheless the rigid jacket structures of dry cells may contain a pressure of several bar in the case of a zinc–carbon cell (more in other systems) which will produce a loud report and throw small parts across a room. Every manufacturer tries to design cells so that this will not happen, but the variety of treatments to which a dry cell may be exposed in customers' hands makes absolute safety a difficult goal.

### 3.1.7 Future developments
As noted in the performance section, the zinc–carbon system has steadily improved in all aspects of performance for nearly one hundred years. Although one might think that all avenues to improvement must have been exhausted this is not the case.

Already, in the recent past, sample cells with more efficient jacketing have appeared for the Japanese market, which may provide up to 25% more service on many load cycles. New carbon blacks and more active manganese dioxide forms seem likely. Thinner containment structures and improved separators (thinner and more stable) may also be possible. Therefore it is likely that the projected growth in performance of Fig. 3.15 is possible or even conservative in the near term and hence highly probable for the foreseeable future.

Another performance possibility where there has been very little commercialization is in the area of rechargeable zinc–carbon cells. Work such as that of Kozawa & Powers [16] twenty years ago indicated that the phase changes in the manganese dioxide of the acid electrolyte dry cell may not occur for some considerable depth of discharge. If the manganese dioxide electrode is relatively stable during cycling, it is a key component in these systems when paired with a zinc electrode which cycles very well in the acid electrolyte. Other work by Wroblowa [17] and Kordesch et al. [18] at BTI, Inc. and at the University of Graz has also shown manganese dioxide to be rechargeable in alkaline electrolyte or other media. Electronic control systems of the future with sophisticated 'smart' power conversion may further support concepts using what is today considered a poorly rechargeable system. In general, a system with poorer performance can be improved a great deal more by sophisticated control than the more robust performers such as lead–acid batteries and nickel–cadmium systems. The low cost of the zinc–carbon systems and their relatively high energy density for a rechargeable system make the possibility well worth pursuing.

## 3.2 ALKALINE–MANGANESE BATTERIES
(J. C. Hunter)

### 3.2.1 Introduction

The batteries discussed in this section are commonly called 'alkaline–manganese batteries', 'alkaline zinc–manganese dioxide batteries', 'cylindrical alkaline batteries', or sometimes just 'alkaline batteries'. They are high quality power cells particularly well suited for use in devices such as toys, audio devices, and cameras. They have excellent shelf life, very low leakage rates, and are capable of supplying high currents. They provide very good service in demanding applications, at a relatively low cost. Alkaline–manganese batteries differ from zinc–carbon batteries in cell chemistry and construction features. They differ from miniature alkaline batteries in construction characteristics (sizes, shapes), and in the fact that miniature alkaline batteries most often use other cathode materials in place of manganese dioxide.

Many of the ingredients in the alkaline–manganese battery are similar to those in a zinc–carbon battery, including zinc as anode material and manganese dioxide as cathode material. But the electrolyte, instead of being a mildly acidic solution of zinc chloride and ammonium chloride, consists of an alkaline solution of potassium hydroxide. This solution has very high conductivity, resulting in increased ability to carry electric current. However, making an alkaline–manganese battery entails more than just substituting one electrolyte solution for another. Along with the change in electrolyte, there are important changes in the chemical purity of the ingredients, as well as in their physical form. There are also important differences in the construction of the battery. Thus, while the zinc–carbon battery was commercially developed by the late 1800s, it was not until the late 1950s that a commercial, general use alkaline–manganese battery was developed.

### 3.2.2 History

The carbon–zinc battery was developed in the 1800s, based on the pioneering work of Georges Leclanché [19] in the 1860s, with further development by Gassner [20] in the 1880s. In the early days some work was also done on batteries with alkaline rather than acidic electrolytes, including the work done by de Lalande & Chaperon [21] on a zinc–copper oxide battery with a potassium hydroxide electrolyte. Such batteries were used for years in railway switching applications. In the 1940s Ruben [22] did significant work on zinc–mercury oxide alkaline batteries. He solved many of the problems of the use of zinc in small alkaline batteries, showing the need for high surface area zinc powder, defining the proper range of electrolyte compositions, and demonstrating means of controlling zinc corrosion. In the late 1940s and early 1950s, Herbert [23] developed this technology further, to produce small zinc–manganese dioxide alkaline batteries which were used in portable radios. However, these batteries did not provide good performance at high discharge rates, and thus were not successful competitors to the general purpose zinc–carbon batteries. In the late 1950s further development work was done, and a practical general purpose alkaline zinc–manganese dioxide battery was developed and commercialized by the Battery Products division of Union Carbide Corporation under the 'Eveready-ENERGIZER' brand name. In the following years, similar batteries were developed by major manufacturers around the World. Since then there has been intense competition to improve the capacity and discharge performance of these batteries, with resulting benefits to the consumer. This is illustrated in Fig. 3.20 which shows the improvement in performance of 'Eveready-ENERGIZER' alkaline–manganese batteries over the past nine years.

Alkaline–manganese batteries are currently made by a large number of companies throughout the World. Some of these manufacturers include Eveready Battery Company, Duracell, Rayovac, Eastman Kodak, Ever Ready Ltd, CIPEL, Philips, VARTA, Imatra, Matsushita, Fuji, Hitachi-Maxell, Sony, Sanyo, Toshiba, and Rocket.

Alkaline–manganese batteries are taking an ever increasing share of the total primary battery market, as electronic devices are developed which make more severe demands on the battery. On a worldwide basis, alkaline–manganese batteries make up about 36% of the primary consumer battery market [24]. Moreover, these percentages are increasing each year.

### 3.2.3 Battery sizes, applications

Cylindrical alkaline–manganese batteries are made in the same sizes as the zinc–carbon general purpose batteries (see Tables 3.5 and 3.6). Sizes range from the large 'F' size through the 'D', 'C', and 'AA' sizes, to the small size 'AAA', 'AAAA', 'N', and 'E1' batteries. Smaller size batteries are also made, using the somewhat different

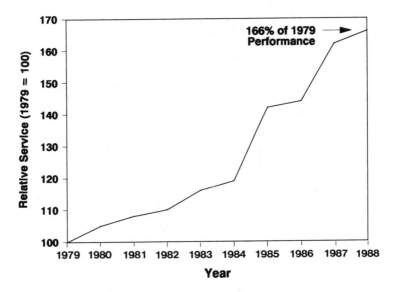

Fig. 3.20 — Improvement in service of 'Eveready-ENERGIZER' alkaline–manganese batteries from 1979 to 1988; averaged performance of 'D', 'C', 'AA', and 'AAA' batteries. Courtesy of Eveready Battery Company.

construction of miniature alkaline batteries (discussed in the next section of this book).

'F' cells are used typically in the large four cell, six cell, or eight cell lantern batteries, and are not sold individually. 'D' size batteries are widely used in torches as well as in audio devices (e.g. radios, cassette players). Their advantage in torches is their long shelf life and resistance to leakage. In audio devices, their ability to deliver high sustained currents is important. 'C'-size batteries are used in torches and audio devices, as well as in toys. 'AA'-size batteries are widely used in audio devices (personal stereo sets), cameras, and toys. In these applications they are called on to provide very high currents with high reliability and leakage resistance. 'AAA'-size batteries are used in audio and photo applications. 'AAA' cells are mainly used in 9 V batteries for a variety of photo and audio applications, as well as in smoke detectors. 'N' and 'E1' batteries are used in devices such as calculators and pagers, which require small sized batteries with the higher performance of the alkaline–manganese system.

### 3.2.4 Battery construction features

The construction features of the alkaline–manganese battery are shown in Fig. 3.21 (a typical 'D'-size battery). The battery is contained in a steel can which also serves as current collector for the cathode. The inner surface of the can is often covered with a layer of nickel plate to reduce corrosion and provide good contact. Some manufacturers use a layer of conductive carbon paint as well. Inside the can is a cathode containing manganese dioxide (as active material), carbon (as a conductor), and

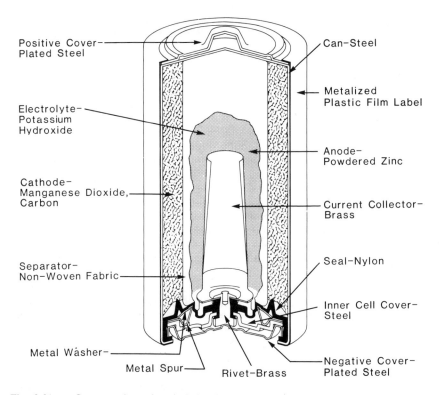

Fig. 3.21 — Cutaway view of typical 'D'-size alkaline–manganese battery. Courtesy of Eveready Battery Company.

possibly some binder material. The cathode may be formed directly in the can, or it may be made externally and then inserted into the can. The cavity formed by the hollow cathode is lined with separator, typically a type of paper material, porous enough to allow liquid to penetrate through it, while keeping the anode and cathode from physically touching each other and thus shorting out the battery. Within the separator-lined cathode cavity is the zinc anode. This consists mainly of zinc powder (solid sheet zinc cannot sustain the high currents needed in an alkaline battery), together with electrolyte solution, and in many cases a gelling material which makes the electrolyte thicker and which helps to support the zinc powder. An anode collector is inserted into the anode, to provide a current path for the electricity. The collector passes through a seal, or is connected to another piece of metal which passes through the seal. The seal assembly fits tightly against the cell container and ensures that electrolyte does not leak out. The entire cell is covered with a finish, including a label, and positive and negative covers. The covers are necessary because the alkaline battery is built 'inside-out' compared to a zinc–carbon battery. In the zinc–carbon battery, the positive (+) cover is placed over the carbon rod, and therefore has a bump, while the negative end is flat. But in an alkaline–manganese battery, the positive end is the bottom of the steel container and is flat, while there is a bump at the anode end where the collector goes through the seal. Therefore, a flat cover is placed at the negative end, and a cover with a bump is placed at the otherwise flat positive end. In this way, the alkaline battery looks on the outside like a conventional zinc–carbon battery, and can be used in the same devices without confusion to the consumer as to which end is positive or negative. Some manufacturers mould the bump directly into the battery container, eliminating the need for a positive cover; these batteries still need the flat negative cover, however.

A fail-safe mechanism is a vital part of the alkaline–manganese battery. The battery is hermetically sealed, which contributes to its exceptionally good shelf life and leakage resistance, but if a battery is charged, there can be internal generation of gas, which could cause excessive pressure to build up inside. The alkaline–manganese battery is not intended to be recharged, but a battery user might nonetheless attempt to charge it. Or, if one cell is placed backwards in a multicell battery compartment, the remaining cells could charge the one cell while they are discharging. The fail-safe mechanism provides for a safe method of pressure-release in such cases.

### 3.2.5 Chemistry of the alkaline–manganese battery

The power for the alkaline–manganese battery comes from the electrochemical reactions of the zinc and the manganese dioxide. During discharge, the zinc is oxidized, forming zinc oxide. At the same time, the manganese dioxide in the cathode is reduced to manganese oxyhydroxide (MnOOH).

$$Zn + 2OH^- \rightarrow ZnO + H_2O + 2e^- \quad \text{anode reaction} \quad (3.8)$$

$$2e^- + 2MnO_2 + H_2O \rightarrow 2MnOOH + 2OH^- \quad \text{cathode reaction} \quad (3.9)$$

-----------------------

$$Zn + 2MnO_2 + H_2O \rightarrow ZnO + 2MnOOH \quad \text{overall reaction} \quad (3.10)$$

The reaction of manganese dioxide is unusual, in that the product (MnOOH) has the crystal structure of the original manganese dioxide. As the discharge proceeds, an intimate crystalline mixture (solid solution) of the two materials is formed [25,26]. As a result, the voltage level of the cathode does not stay constant during discharge, but, rather, decreases continually. Thus an alkaline battery (or a zinc–carbon battery for that matter) has a sloping discharge voltage (Fig. 3.22), in contrast to batteries such as lead–acid or nickel–cadmium, which have nearly constant voltage during discharge. The initial voltage of an alkaline zinc–manganese dioxide battery is typically 1.5–1.6 V, dropping to 1 V or less at the end of discharge, depending on the type of use.

Zinc is more easily passivated in alkaline electrolyte than in acidic electrolyte. That is, the discharge reaction proceeds more slowly in alkaline than in acidic electrolyte, and when one attempts to discharge the zinc at a sufficiently high rate, the discharge reaction cannot keep up. Therefore, the alkaline–manganese battery needs to use zinc powder as the anode material [22] rather than the sheet used in

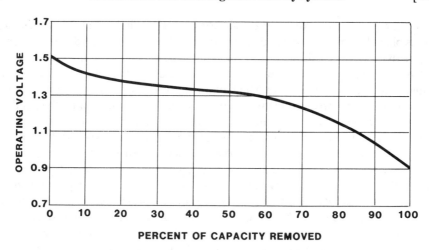

Fig. 3.22 — Typical alkaline–manganese battery operating voltage as a function of remaining capacity. Courtesy of Eveready Battery Company.

zinc–carbon batteries. The high surface area of the zinc powder makes up for the more sluggish discharge reaction and allows the zinc anode to function well even at very high discharge currents. The zinc discharge reaction [27] is considerably more complicated than the reaction given above, which is just a summary of the overall zinc reaction. Discharge of the zinc entails the formation of soluble zinc in the form of zincate ions:

$$Zn + 4OH^- \rightarrow Zn(OH)_4^{2-} + 2e^- \:. \tag{3.11}$$

Zincate is quite soluble in the electrolyte. However, the point is eventually reached when the zincate solubility is exceeded, and solid zinc oxide is then precipitated in the battery:

$$Zn(OH)_4^{2-} \rightarrow ZnO + H_2O + 2OH^- \:, \tag{3.12}$$

and the overall reaction is then that given in equation (3.8) above. If the concentration of the caustic electrolyte is low, then the zinc will tend to passivate at the point where zincate saturation is reached. This phenomenon was common in some alkaline zinc–air batteries produced in the past, which used electrolytes of relatively low concentration. In such cases, battery capacity could be extended by using larger amounts of electrolyte, allowing more solution to contain the zincate ions, and thus delaying passivation. However, this resulted in low energy density because of the excess electrolyte being used. The alkaline–manganese battery uses caustic electrolyte of higher concentration (typically in the range of 30% to 45% potassium hydroxide). In this concentrated electrolyte, the zinc discharge reaction can proceed efficiently even after electrolyte saturation is achieved and zinc oxide starts to precipitate. Therefore, alkaline–manganese batteries have proportionately much less electrolyte, and greatly improved energy density compared to that which would be possible otherwise.

Zinc is thermodynamically unstable in contact with water. That is, it has a tendency to corrode in water, reacting with water to form oxidized zinc and hydrogen gas. If this were to happen to any great extent, the zinc would be consumed without producing any energy, large amounts of hydrogen gas would be produced in the batteries, and zinc would be a very inefficient battery material. However, it is possible that the rate of zinc corrosion can be made very low under certain conditions, thus allowing zinc to be more useful as a battery material.

### 3.2.6 Materials used in the alkaline–manganese battery

The correct specification of materials that go into the alkaline–battery is of great importance. The availability of sufficiently pure substances of the correct physical and chemical form is a necessary requirement for the production of commercially successful alkaline–manganese batteries.

Zinc is used in the form of fine powder to have high current carrying capability. The powder is formed by atomization of molten zinc metal, and the composition of the zinc is critical. Impurities such as arsenic or antimony, which would promote the corrosion of zinc, must be kept at very low levels. Other materials, which reduce zinc corrosion, may be added to the zinc. Numerous additives are described in battery patents. Lead is one of the more common, and is helpful even when present in small amounts. One of the most effective materials for keeping zinc corrosion at low levels is mercury. Mercury can lower the corrosion rate of zinc itself, and it can help prevent the detrimental effects of other impurities. In the past, mercury was commonly added to the zinc in alkaline–manganese batteries in amounts of 6–8% versus zinc. In recent years there has been a trend toward lowering mercury levels, with some manufacturers producing batteries with less than 1% mercury in the zinc. Even at these lower levels mercury can help reduce zinc corrosion.

Other types of corrosion inhibitors can be used to further reduce zinc corrosion. There are different materials described in the battery patent literature for use in this regard. They are typically organic materials which are surface-active towards zinc and affect the corrosion reactions at the metal/solution interface.

In many cases the zinc powder anode contains gelling agents based on natural materials (starches, cellulose) or synthetic materials. These gelling agents help to support the zinc powder and partly immobilize the electrolyte. Important attributes of these materials include their ability to gel the strongly alkaline electrolyte, and to resist degradation in the electrolyte. Significant improvements in alkaline–manganese battery performance in recent years have been in part due to use of newer, more effective, synthetic gelling agents.

Manganese dioxide is a key ingredient for alkaline–manganese batteries. Whereas zinc–carbon batteries may use natural manganese dioxide, alkaline–manganese batteries are made with only high purity synthetic materials. The manganese dioxide must be of the correct crystal structure (known as $\gamma$-$MnO_2$) in order to operate in alkaline electrolyte. Moreover, it must be of high purity and must not contain harmful impurities that could affect the zinc and cause increased corrosion. Some naturally occurring manganese dioxide minerals are of the correct crystal form but are too low in purity, and they contain impurities that would promote zinc gassing. Manganese dioxide for use in alkaline–manganese batteries is, therefore, obtained synthetically, using either a strictly chemical process to produce CMD (chemical manganese dioxide) or an electrochemical process to produce EMD (electrolytic manganese dioxide).

The can used as container for the alkaline–manganese battery is formed from steel. It does not take part in the battery discharge reactions but is an inert container (in contrast to the zinc–carbon battery where the container is also the anode of the battery). Although it does not take part in the battery discharge reactions, some chemical reactions can occur at the can surface, such as corrosion. Such corrosion could introduce iron into the electrolyte, and can cause formation of resistive deposits on the can surface which would be detrimental to battery performance. Therefore, the can is commonly coated with nickel, using any one of a variety of methods, to improve contact. Some manufacturers use other conductive coatings as well.

### 3.2.7 Performance characteristics

#### 3.2.7.1 General

The performance characteristics of the alkaline–manganese battery can best be described in terms of comparison to the performance characteristics of other types of battery that can be used in the same applications. The main competitor for the alkaline battery is the zinc–carbon battery, either in the form of the standard Leclanché battery, or the premium Leclanché or heavy-duty battery. In certain applications alkaline–manganese batteries compete with nickel–cadmium batteries as well. Currently, there is increasing availability of lithium batteries. Many of these are of special size, shape, and voltage, and thus do not compete on a cell-for-cell basis with alkaline–manganese batteries, but rather are used in new applications where a completely new size of battery can be used. However, other lithium batteries are being introduced which will directly compete with standard size alkaline–manganese batteries, and performance comparisons with such batteries are of interest.

There are many aspects of battery performance to be considered: input capacity, output capacity under a variety of conditions (rate, intermittency, temperature); effect of long-term storage on output capacity; leakage resistance; and cost. Input capacity is a measure of how much energy can be put into a battery of a given size or weight. This capacity depends largely on the chemical reactants in the battery; some reactants can give higher energy densities than others. The input capacity also depends on the physical construction of the battery, since a battery contains more than just the chemical reactants. Space in the battery is taken up by separators, seal assemblies, conductive materials, and other items that are necessary but do not contribute to the capacity. Some types of construction are desirable in terms of improving output capacity on certain tests, but result in lower total input capacity in the battery.

#### 3.2.7.2 Input capacity

An alkaline–manganese battery in "AA" size has typically an input capacity of 2.5 to 3 Ah. Zinc–carbon batteries have lower capacities, ranging from below 1 Ah for some standard 'AA' zinc–carbon batteries, up to about 1.4 Ah for premium or heavy-duty 'AA' zinc–carbon batteries. Nickel–cadmium batteries generally have even lower capacities, typically 0.5 to 0.75 Ah for an 'AA'-size battery. Lithium batteries may have significantly higher energy density than alkaline–manganese batteries, depending on the specific lithium system, and the construction features of the batteries. Direct comparisons are somewhat difficult since most current lithium batteries are of special size and/or of different voltage than standard alkaline–manganese batteries. Thus, an alkaline–manganese battery has two to three times the input capacity of a zinc–carbon battery, about five times the capacity of a nickel–cadmium battery, but somewhat less capacity than a lithium battery.

*3.2.7.3 Output capacity*

More important to the battery user than the input capacity is the actual output of the battery in normal use. Under the most favourable discharge conditions one can get most of the input capacity out of the battery. But under more demanding conditions the output can be substantially less than the input capacity. Generally, one obtains higher efficiency (output capacity nearer to input capacity) when the battery is discharged slowly, that is, at low currents, or when it is discharged intermittently with periods of rest between periods of discharge. Lower efficiency results when the battery is discharged continuously and/or at high rates. One reason for this is the internal resistance of the battery. On low current discharges, the internal battery resistance is small relative to the resistance of the device being powered, and so that power loss in the battery is small. But at higher currents, the internal resistance can be significantly large relative to the device resistance, and the power loss in the battery can be large. This effect will be greater for a battery with high internal resistance, and lower for a battery with low internal resistance. Polarization is a second reason why discharge performance may be worse on high rate continuous discharge. This term refers to effects occurring in the battery during discharge which hinder the discharge reaction. Often, polarization leads to build-up of reaction products in the battery (concentration polarization), resulting in lower battery voltage. If this occurs, allowing the battery to rest will lead to reestablishment of equilibrium conditions in the battery, and the battery may be able to provide additional discharge capacity.

Fig. 3.23 shows a comparison of output capacity on a light drain test for an alkaline–manganese battery and a zinc–carbon battery. Both discharge curves show the sloping discharge shape typical of batteries that use manganese dioxide as cathode. However, the alkaline–manganese battery provides about three times the capacity of the zinc–carbon battery. The performance advantage of the alkaline–manganese battery shown in this case is directly attributable to its higher input capacity. On this light drain test, both types of battery discharge with high efficiency.

Fig. 3.24 shows a similar comparison on a heavy drain continuous test. In this case, the alkaline–manganese battery has a very large performance advantage over the zinc–carbon battery. Part of the reason for this advantage is the lower internal resistance of the alkaline–manganese battery: the electrolyte solution in the alkaline–manganese battery has higher conductivity than the electrolyte in zinc–carbon batteries or the electrolyte in lithium batteries. Also, the alkaline–manganese

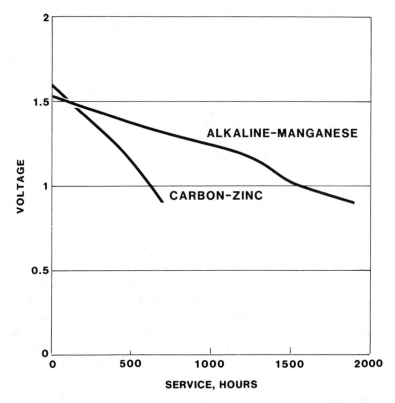

Fig. 3.23 — Performance comparison of 'D'-size alkaline–manganese versus zinc–carbon batteries on typical light drain test (150 ohm continuous test at 21°C). Courtesy of Eveready Battery Company.

battery exhibits much lower polarization than the zinc–carbon battery. These two effects combine to produce the large performance advantage for alkaline–manganese batteries on heavy drain continuous tests.

The alkaline–manganese battery is influenced very little by concentration polarization; it performs about the same on either continuous or intermittent tests. The zinc–carbon battery, on the other hand, is susceptible to rather large concentration polarization effects. For this reason, the zinc–carbon battery performs poorly on high rate continuous tests, but performs better on intermittent tests. Therefore, alkaline–manganese batteries have the the biggest advantage over zinc–carbon batteries on high rate continuous discharges. The advantage is not as great on high rate intermittent discharges, and the advantage is smaller yet on light drain discharges. However, even on the lightest drain tests, the alkaline–manganese battery has about two to three times the performance of the zinc–carbon battery.

The nickel–cadmium battery has the same type of alkaline electrolyte as the alkaline–manganese battery, and, moreover, has a special type of construction ('jelly-roll') that provides for even lower internal resistance and polarization than

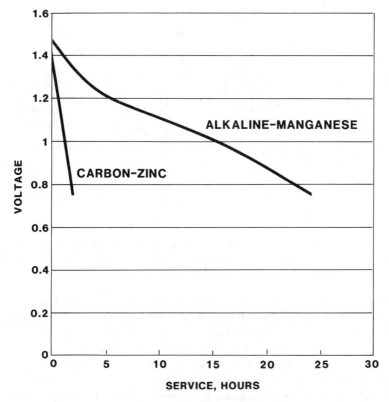

Fig. 3.24 — Performance comparison of 'D'-size alkaline–manganese versus zinc–carbon batteries on typical heavy drain test (2.2 ohm continuous test at 21°C). Courtesy of Eveready Battery Company.

the alkaline–manganese battery. Therefore, the alkaline–manganese battery does not have an efficiency advantage over the nickel–cadmium battery on high rate tests, and, in fact, has a disadvantage. But the alkaline–manganese battery does have higher input capacity (by about 5 times), so that on essentially all types of tests the alkaline–manganese battery delivers more output than the nickel–cadmium battery. The strong point of the nickel–cadmium battery is that it can be recharged.

### 3.2.7.4 Effect of temperature on discharge performance

Chemical reactions, in general, tend to occur more rapidly as the temperature is increased, and more slowly as it is decreased. The conductivity of an electrolyte varies similarly with temperature. Therefore, batteries tend to perform better at higher temperatures, and worse at lower temperatures. Of course, there are limits to the temperature range in which batteries can be used, owing to such factors as boiling or freezing of electrolyte, stability of the battery components, etc. The alkaline–manganese battery operates well over a relatively wide range of temperatures. Fig. 3.25 shows the effect of temperature on the discharge efficiency of alkaline–

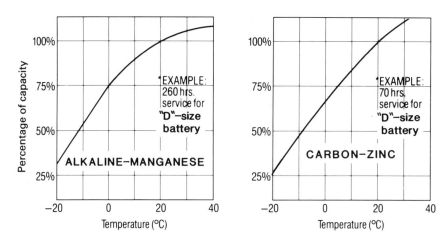

Fig. 3.25 — Effect of temperature on relative discharge performance of alkaline–manganese and zinc–carbon batteries (typical fresh 'D'-size battery service on simulated radio use, 25 ohm 4 hour/day test). Courtesy of Eveready Battery Company.

manganese and zinc–carbon batteries. Both types of battery have decreased relative performances as the temperature is decreased. However, the alkaline–manganese battery has less performance loss than the zinc–carbon battery, owing in large part to higher electrolyte conductivity. Nickel–cadmium batteries have even less performance loss than alkaline–manganese batteries, because they have the same high conductivity electrolyte as alkaline–manganese batteries together with a construction ('jelly-roll') that further reduces battery internal resistance. Lithium batteries vary in their performance, depending on the chemistry of the specific system. Electrolytes used in lithium batteries have lower conductivity at room temperature than the electrolyte in the alkaline–manganese battery, but generally, the conductivity does not decrease with lowered temperature as rapidly for the alkaline–manganese battery. Thus lithium batteries tend to have higher internal resistance than alkaline–manganese batteries at normal temperatures, but the lithium batteries may function better than the alkaline–manganese batteries at very low temperatures.

Both the alkaline–manganese and zinc–carbon batteries perform more efficiently as temperature is increased above room temperature. The alkaline–manganese battery shows less of this effect, mainly because its internal resistance is already very low and its discharge efficiency high. Based on battery performance data, an operating temperature range of −30°C to 55°C is recommended for the alkaline–manganese battery.

### 3.2.7.5 Storage of batteries

As a battery is stored for long periods, it may lose some of its useful discharge capacity. This effect may be accentuated at high temperatures. The service loss may be due to chemical and/or physical changes in the battery. There may be corrosion, as well as reaction of one battery component with another. There may be decomposition of some of the battery reactants. Incomplete sealing may allow escape of electrolyte from the battery, or allow entrance of oxygen from the atmosphere into the battery. All of these effects must be minimized if one is to have a battery that maintains good performance after long periods of storage.

The alkaline–manganese battery is very good in this regard. The battery reactants are of very high purity, resulting in very little wasteful corrosion. There is very little self-discharge and little reaction of battery components with each other. Moreover, the battery has a very good seal, which minimizes degradation due to electrolyte loss or oxygen ingress. Fig. 3.26 shows a comparison of the service maintenance of alkaline–manganese and zinc–carbon batteries, as a function of time and when stored at various temperatures.

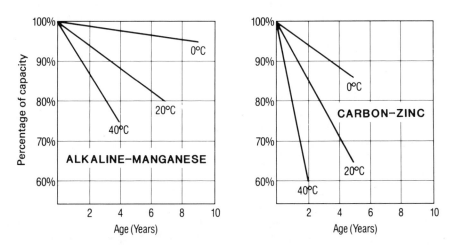

Fig. 3.26 — Effect of time and temperature on relative discharge performance of alkaline–manganese and zinc–carbon batteries (fresh and aged 'D'-size cells on simulated radio use, 25 ohm 4 hour/day test). Courtesy of Eveready Battery Company.

Alkaline–manganese batteries stored for five years at 20°C retain 85% of their original performance. Storage at 40°C leads to more rapid degradation of performance, while storage at 0°C results in much longer shelf life. Zinc–carbon batteries show poorer service maintenance than alkaline–manganese batteries, regardless of temperature. The advantage of alkaline–manganese batteries is due, in part, to the high purity of materials used in the battery, and, in part, to the 'hermetic' seal. Nickel–cadmium batteries have quite poor service maintenance, owing mainly to self-discharge reactions occurring in the battery. 'AA'-size nickel–cadmium batteries can lose 5–7% of their capacity per month of storage at 20°C. Lithium batteries generally have better service maintenance than alkaline–manganese batteries, owing to different chemistry, and a very good seal mechanism. A ten-year service life is often claimed for lithium batteries.

### 3.2.7.6 Resistance to leakage
Batteries of all types have made much progress in leakage resistance. The alkaline–manganese battery, because of its type of seal and the purity of its ingredients, has

very low leakage. It is superior to zinc–carbon batteries, even though zinc–carbon batteries have been greatly improved in recent years. Lithium batteries have leakage resistance that is comparable to or better than alkaline–manganese batteries.

### 3.2.7.7 Battery cost

The battery user should take into account the performance characteristics, as well as the cost, of the various battery types for the type of device used. The alkaline–manganese battery provides very good performance at a reasonable cost. Zinc–carbon batteries give less performance, but also cost less.

Nickel–cadmium batteries are more expensive, have less capacity on a single discharge basis, but can be recharged many times, so that the total cost of energy can be much lower than for alkaline–manganese batteries. Lithium batteries are more expensive than alkaline–manganese batteries, offer better performance in some applications, and have longer shelf life.

### 3.2.7.8 Summary

The choice of which battery to use will depend on the relative importance of the various performance factors, and will depend very much on the type of device being used. There is no single best battery for all devices.

Alkaline–manganese batteries are most appropriate for applications with high currents and long periods of discharge. Devices that use motors (such as toys, cassette players, etc.) are typical examples. Zinc–carbon batteries typically do not have the ability to perform so well in such devices. Cameras often contain motors for film-winding and focusing, often have electronic flash units which have very high current requirements, and must be free from battery leakage. For these reasons, alkaline–manganese batteries are strongly recommended for use in cameras. Audio devices often require high currents, both for operation of cassette players or compact disc players, as well as for providing high levels of sound with low distortion. Again, alkaline–manganese batteries are highly recommended for these devices.

In some high current applications, nickel–cadmium batteries may have advantages over alkaline–manganese batteries. This will tend to be the case when the device is used often and tends to consume large numbers of batteries; where the device is used according to a predictable schedule; and where it is feasible to keep the batteries charged. Such applications might include radio-controlled cars and planes, camcorders, etc. However, if one requires the device to sit for an unknown period up to several years and then be ready to function, the alkaline–manganese battery is the better choice.

## 3.2.8 Likely future developments

The cylindrical alkaline–manganese battery has been a high performance, premium battery from its inception. Since then, there has been intense competition among manufacturers, with the result that the alkaline batteries of today provide greatly increased capacity and improved performance, compared to batteries made just a few years ago. This competition is expected to continue unabated. In many cases, the specific areas of improvement being explored by battery manufacturers are highly proprietary. However, some general directions for future development can be suggested. It appears likely that there will be continued effort toward performance improvements, so that batteries will run a little longer than before, or handle a little more current than before. Efforts will probably be expanded to improve the already excellent storage life of batteries. Moreover, owing to worldwide interest in environmental concerns, there is interest in reducing the mercury content of batteries, which various battery manufacturers have taken steps towards. Thus the alkaline–manganese battery should continue to offer the battery user a continually improving product, that provides excellent performance at a reasonable cost.

## 3.3 SMALL SIZE ALKALINE BATTERIES
(E. A. Megahed)

### 3.3.1 Definitions

The definition of small size alkaline batteries will be restricted to cell sizes of the button type. Button cells may range in size from a small button (5.8 mm diameter × 1.2 mm height) to a large ice-hockey puck (37.0 × 10.5 mm). The term 'cell' and 'battery' may be used interchangeably, even though batteries are commonly referenced to multiples of cells connected in series or parallel. Alkaline battery systems will be restricted to those having commercial success and availability. Mercuric oxide–zinc (HgO–Zn) will be discussed first, followed by the recently developed air–zinc ($O_2$–Zn) for hearing aid applications. Monovalent silver oxide–zinc ($Ag_2O$–Zn) will be characterized, followed by divalent silver oxide-zinc (AgO–Zn) and the more recently developed 'PLUMBATE'–Zn and silver nickel oxide–Zn ($AgNiO_2$–Zn) systems for watch and calculator applications. Although these battery systems are widely used for hearing aid and watch applications, many improvements have been made in their chemistry in the last 10 years to warrant proliferation of the applications to new frontiers — pager, medical electronics, and a few implants.

### 3.3.2 History: 'Necessity leads to invention'

During World War II, the US Signal Corps required a high capacity per unit volume battery for radio transceivers 'handy talkie', 'walkie talkie', and other military purposes. The zinc–carbon battery available at that time suffered from low energy density and poor high temperature performance. The mercuric oxide (HgO) cathode was developed by Samual Ruben and introduced in various sizes with zinc (Zn) anode and alkaline electrolyte (KOH or NaOH) to provide the high capacity, high drain, constant voltage level, and good storage characteristics required for these devices [28,29]. With the introduction of the transistor for hearing aids in 1952 and beyond, miniaturized HgO–Zn cells were developed. Within a span of three years (1952–1954), 97% of the valve (vacuum tube) hearing aids were converted to transistor aids using miniaturized button cells. During the next 20 years (1950–1970), miniaturized HgO–Zn and $Ag_2O$–Zn batteries powered transistorized hearing aids with integrated circuit (IC) circuitry made with either silicon (1.3–1.5 V) or germanium (0.9–1.3 V) substrates.

Optimization of battery systems in the past 10 years has resulted in various brands of 'air cells' that met the electrical requirements of many types of hearing aid, e.g.

Class A, Class B and Class D type amplifiers. In the past five years, work on 'air' batteries was intensified to obtain twice the service life of mercury batteries under various environmental discharge conditions.

Similarly, the need for longer operating time before battery replacement for electronic watches has necessitated the invention of long life silver batteries. The introduction of light-emitting diode (LED) watches in 1972 by the Hamilton watch company Pulsar™ required a long life $Ag_2O$–Zn cell. Improvements in watch circuitry and the switch to liquid crystal display (LCD) watches with added functions resulted in improved cathode and barrier formulations of silver batteries. The introduction of AgO–Zn batteries under the 'DITRONIC' trade name and more recently 'The new formula plumbate' resulted in long life batteries that can operate an analogue watch for 5 years before replacement.

### 3.3.3 Chemistry

Zinc is selected for the negative electrode in aqueous alkaline batteries because of its high half-cell potential ($-1.25$ V vs SHE), low polarization and high limiting current density (up to $40\,\text{mA cm}^{-2}$) in a cast electrode. In addition, its equivalent weight is fairly low, thus resulting in high theoretical capacity ($820\,\text{mAh gm}^{-1}$). Because of low polarization, its discharge efficiency (useful capacity/theoretical capacity) is fairly high (85–95%). Although zinc is thermodyanamically unstable in aqueous alkali, a small amount of mercury (amalgamation) brings the corrosion rate within a tolerable limit. A study [30,31] of the effect of various ions on the corrosion rate showed that ions containing Cu, Fe, Sb, As, and Sn increased the zinc corrosion rate, while ions with Cd, Al, and Pb decreased the rate. As a result of these properties, zinc has proved useful in combination with other cathode materials in the battery industry. Kinetically, the zinc reaction is a complicated phenomenon. It is generally accepted that the overall negative reaction is [32,37]

$$Zn + 4OH^- = Zn(OH)_4^{2-} + 2e^- \qquad (3.13)$$

$$Zn(OH)_4^{-2} = Zn(OH)_2 + 2OH^- \qquad (3.14)$$

or

$$ZnO + H_2O + 2OH^-$$

---

$$Zn + 2OH^- = ZnO + H_2O + 2e^- \qquad (3.15)$$

Theoretical capacity $= 49.2\,\text{amp·min gm}^{-1}$.

It has been shown [31] that $Zn(OH)_2$ forms zincate ion rapidly on discharge when the mean activity coefficient of hydroxyl ion in the KOH solution exceeds 1.0. This is the case for high KOH concentrations used with HgO, $Ag_2O$, and/or AgO cathodes. When the activity coefficient is less than about 0.9, the rate at which this hydroxide goes into solution is low, and the hydroxide is present as a solid film. Similarly, at low temperature or at a high rate of discharge, the solubility of $Zn(OH)_2$ is reduced, resulting in a zinc surface covered with the hydroxide film. The presence of a solid phase (ZnO) or a hydroxide film will interfere with the discharge efficiency of the zinc electrode. Such film must be removed by dissolution; for example, high KOH concentration, high temperature, low rate of discharge, or by additives such as gelling agents, fillers, or organic siliconate, for the reaction to continue. In small alkaline cells, amalgamated powdered zinc with a small amount of additives — organic siliconate, Carbopol™, carboxymethyl cellulose (CMC), gums, Waterlock™, etc. — is needed for large surface area and electrolyte accessibility. Such construction will permit high current density and maximum capacity per unit volume of the anode space [33].

The mercuric oxide cathode operates at a substantially constant voltage, since there is only one reduction step. Its solubility in alkaline electrolyte is quite low. The predominant species present in alkaline solution would appear to be non-ionic $Hg(OH)_2$. The solubility of HgO in alkaline electrolytes is smaller than that of other cathode materials such as $Mn(OH)_2$ or $Ag_2O$ [34,35]. This is one of the reasons for the good shelf life of HgO–Zn cells.

The HgO reaction in 9–10 M solution is

$$HgO + H_2O + 2e^- = Hg + 2OH^- \tag{3.16}$$

$$E° = -0.098 \text{ V}$$

Theoretical capacity $= 14.8$ amp·min. gm$^{-1}$

The overall electrochemical reaction of an HgO–Zn cell is

$$HgO + Zn = ZnO + Hg \tag{3.17}$$

Voltage: 1.343 V

The air cathode is the most efficient electrode system in terms of high energy density and oxygen reduction. The very thin air cathode permits the use of twice as much zinc in the anode compartment of the cell as can be used in the mercuric oxide equivalent. Since the air cathode has theoretically infinite life, the electrical capacity of the cell is determined only by the zinc capacity.

With active carbon, the oxygen reduction is

$$O_2 + 2H_2O + 4e^- = 4OH^- \tag{3.18}$$

$$E° = +0.400 \text{ V}$$

Theoretical capacity $= 201.4$ amp·min. gm$^{-1}$

The overall electrochemical reaction of air–Zn cell is

$$2Zn + O_2 + 2H_2O = 2Zn(OH)_2 \tag{3.19}$$

Voltage $= 1.645$ V .

Despite the high theoretical voltage of an air-Zn cell, most practical cells have voltages between 1.42 and 1.47 V. The low practical voltage value is attributed to inefficiency of the oxygen electrode and is related to peroxide build-up at the surface of the electrode.

Fig. 3.27 shows the cathode of a Rayovac air–Zn cell and the oxygen reduction process occurring at the catalytic layer interface. The cathode structure includes the barrier/separator, catalyst layer, metallic mesh, hydrophobic membrane, and an air diffusion membrane. The catalytic layer contains activated carbon, and the metallic

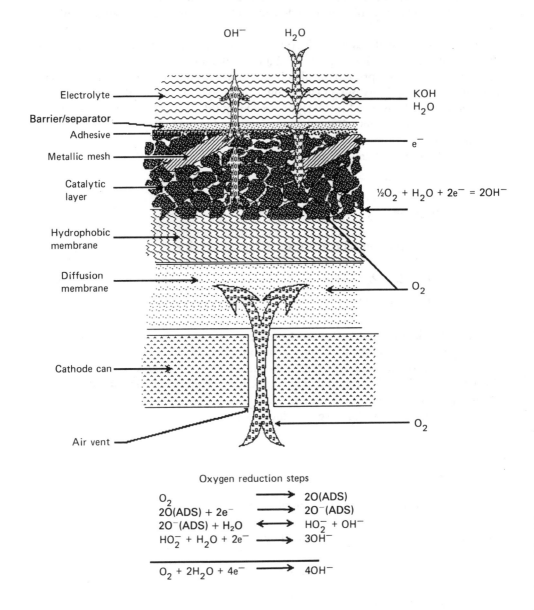

Fig. 3.27 — Oxygen reduction in the air cathode of a Rayovac air–Zn cell (courtesy of Rayovac Corp.).

mesh provides structural support and current collection. The hydrophobic membrane maintains the gas–permeable waterproof boundary between the air and the cell electrolyte. The diffusion membrane regulates gas diffusion rates and distributes oxygen evenly over the cathode surface.

An air-access hole(s) in the positive can provides a path for atmospheric oxygen to enter the cell, and this diffuses through the membrane to the cathode reaction sites. Oxygen reduction takes place in the presence of electrons provided by the metallic mesh with the production of hydroxyl ions which react with the zinc anode. Current capability increases with increasing hole size and/or the membrane porosity until the current density (reaction rate) at the catalytic layer interface becomes limiting. The improvements made in recent years in the current capability of the air cathode was facilitated by better understanding of the oxygen reduction reactions at the catalytic layer [36,37,38,39]. Additional improvements in the selection of waterproof Teflon$^{TM}$-type membranes have facilitated the introduction of practical cells without concern for leakage or parasitic cell reactions [38,39,52].

The majority of silver-zinc batteries are made with $Ag_2O$ depolarizer with a small amount of carbon (3–4% of cathode weight) for conductivity and a binder or a lubricant (0.5–1.0%) for uniform cathode pelleting. In some cases $MnO_2$ (10–15% by weight of the cathode) is added to the cathode for cost reduction. Because of the high solubility of the $Ag_2O$ cathode in alkaline electrolyte, little work was done until 1941 when André [40] suggested the use of a Cellophane barrier to reduce silver migration from the cathode to the anode. His work brought about a revival of interest in silver batteries as a primary and secondary power source.

The monovalent silver oxide reduction is

$$Ag_2O + H_2O + 2e^- = 2Ag + 2OH^- \tag{3.20}$$

$E° = +0.344 \text{ V}$ .

Theoretical capacity = 13.9 amp·min gm$^{-1}$

The overall electrochemical reaction of $Ag_2O$–Zn cell is

$$Ag_2O + Zn = 2Ag + ZnO \tag{3.21}$$

Voltage = 1.589 V .

Divalent silver oxide depolarizer is used by a few battery manufacturers as a depolarizer for long life batteries [31,37,41]. The gravimetric (433 mAh gm$^{-1}$) and volumetric (3.22 Ah cm$^{-3}$) energy density of AgO are superior to other practical depolarizers; for example $Ag_2O$, HgO and $MnO_2$ = 232, 248, 308 mAh gm$^{-1}$ and 1.67, 2.76, 1.55 Ah cm$^{-3}$; respectively.

The divalent silver oxide reduction is

$$2AgO + H_2O + 2e^- = Ag_2O + 2OH^- \tag{3.22}$$

$$Ag_2O + H_2O + 2e^- = 2Ag + 2OH^- \tag{3.23}$$

$$2AgO + 2H_2O + 4e^- = 2Ag + 4OH^- \qquad (3.24a)$$

or

$$AgO + H_2O + 2e^- = Ag + 2OH^- \qquad (3.24b)$$

$E° = 0.570 \, V$

Theoretical capacity $= 26.0 \, amp \cdot min \, gm^{-1}$.

The overall electrochemical reaction of an AgO–Zn cell is

$$AgO + Zn = Ag + ZnO \qquad (3.25)$$

Voltage: $1.815 \, V$.

Typically, AgO–Zn batteries have a two-step discharge curve corresponding to the reduction of AgO to $Ag_2O$ (1.8 V to 1.60 V) and the reduction of $Ag_2O$ to Ag (1.60 V to 0.90 V). The two-step working voltage at very low drain is not preferred for applications where one-step (voltage regulation) is required, for example, watch applications. To discharge AgO at the $Ag_2O$ voltage level, two successful approaches have resulted in commercializing AgO–Zn batteries. Megahed [42,43, 44,45] disclosed a two-step reduction process of AgO cathodes commonly defined as the 'double treatment approach'. In the first step, an AgO pellet is treated with a mild reducing agent, such as methanol, to form $Ag_2O$ around the pellet. Then, after consolidation, a strong reducing agent, such as hydrazine, is applied to form a silver layer on the pellet surface. Cells containing such cathodes display voltage characteristics similar to $Ag_2O$–Zn cells with the capacity advantage of AgO–Zn cells. The thickness of $Ag_2O$ layer around the AgO pellet and Ag layer on the top of the consolidated pellet is important to obtain voltage values typical of AgO–Zn cells initially and after delayed storage.

Fig. 3.28 shows the 'double treatment approach' in the 392 size cell with a 32.5% capacity advantage over a conventional $Ag_2O$ cathode at the same operating voltage. By contrast, AgO cells without a cathode treatment produced a two-step operating voltage.

Another approach to producing AgO–Zn cells with a single voltage level is via silver 'plumbate' [46]. Silver 'plumbate' is defined as a mixture of AgO, $Ag_2O$, $Ag_5PbO_6$, and $Ag_2PbO_2$. The process of producing silver 'plumbate' comprises AgO reaction with 5–17% by weight of finely divided lead sulphide (PbS) in a hot caustic, for example 30% NaOH solution, for several hours. The reaction product after washing and drying is a mixture of AgO, $Ag_2O$, and silver 'plumbate' with residual PbS according to the reaction:

$$3PbS + 27AgO + 6NaOH = Ag_2PbO_2 + Ag_5PbO_6 + 10Ag_2O + 3Na_2SO_4 + 3H_2O \, . \qquad (3.26)$$

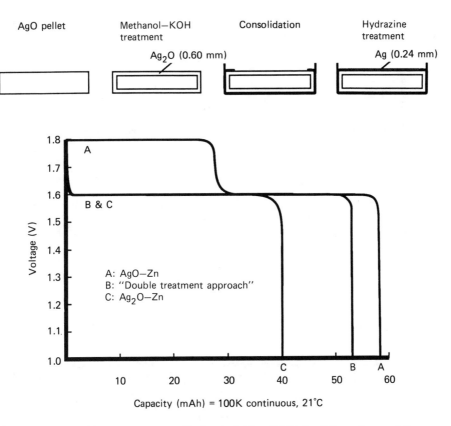

Fig. 3.28 — 'Double treatment approach' of an AgO–Zn cell (392 size; 7.9 mm diam. × 3.6 mm height).

| $Ag_2O$ thickness (mm) around each pellet surface | Ag thickness (mm) on the consolidation | Cathode capacity mAh gm$^{-1}$ | Voltage level at months | | |
|---|---|---|---|---|---|
| | | | 1 | 3 | 6 |
| 0.20 | 0.12 | 372 | 1.73 | 1.80 | 1.77 |
| 0.60 | 0.12 | 360 | 1.61 | 1.63 | 1.77 |
| 1.00 | 0.12 | 326 | 1.60 | 1.59 | 1.59 |
| 0.20 | 0.24 | 360 | 1.60 | 1.59 | 1.59 |
| 0.60 | 0.24 | 348 | 1.60 | 1.59 | 1.59 |
| 1.00 | 0.24 | 315 | 1.60 | 1.59 | 1.59 |

Fig. 3.29 shows the discharge behaviour of the silver 'plumbate' containing material and the corresponding chemical reactions versus a Zn electrode. The major portion of the curve is similar to $Ag_2O$–Zn discharge, but at a higher coulombic capacity. Further reaction of the silver 'plumbate' mixture with silver powder eliminates the high voltage step associated with AgO reduction. A model has been

developed to summarize the reactions (Fig. 3.30). Divalent silver is completely covered with $Ag_2O$ and conductive Ag and $Ag_5Pb_2O_6$.

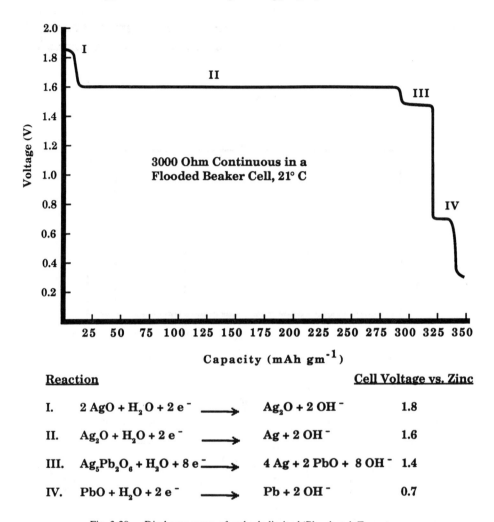

| Reaction | | | Cell Voltage vs. Zinc |
|---|---|---|---|
| I. | $2\,AgO + H_2O + 2\,e^-$ → | $Ag_2O + 2\,OH^-$ | 1.8 |
| II. | $Ag_2O + H_2O + 2\,e^-$ → | $Ag + 2\,OH^-$ | 1.6 |
| III. | $Ag_5Pb_2O_6 + H_2O + 8\,e^-$ → | $4\,Ag + 2\,PbO + 8\,OH^-$ | 1.4 |
| IV. | $PbO + H_2O + 2\,e^-$ → | $Pb + 2\,OH^-$ | 0.7 |

Fig. 3.29 — Discharge curve of cathode limited 'Plumbate'–Zn system.

Fig. 3.31 shows the discharge behaviour of 392 size cells (7.9×3.6 mm) with the 'Plumbate' cathode as compared to $Ag_2O$ and AgO cathodes. The 'Plumbate' cathode results in similar capacity to the 'double treatment approach', but with simplified processes of cathode reduction.

Still another approach to *reducing* cost reduction of the silver cathode is by reacting nickel oxyhydroxide (NiOOH) with $Ag_2O$ in the presence of aqueous alkaline solution at high temperature [47,48]. The reaction product ($Ag_2O + 2NiOOH = 2AgNiO_2$) has low specific resistance and more coulombic capacity (263 mAh gm$^{-1}$) than conventional $Ag_2O$ (232 mAh gm$^{-1}$).

# Sec. 3.3] Small size alkaline batteries

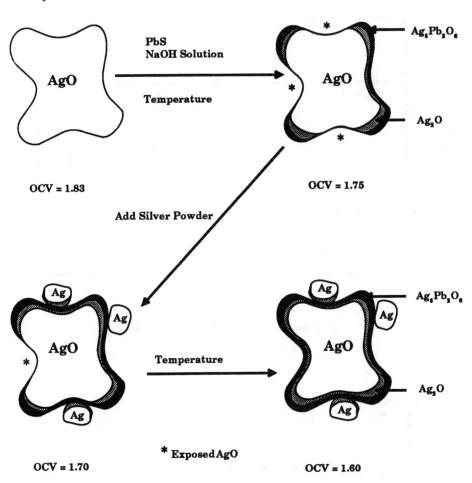

Fig. 3.30 — Model for observed open circuit voltages of Ag$_2$O covered with AgO and Ag/Ag$_5$Pb$_2$O$_6$.

Fig. 3.32 shows a 30% capacity increase in the 392 size cell over monovalent silver cells at 0.1 V lower operating voltage level. This material may be used as the cathode depolarizer in silver cells or as a conductive diluent (15–20% by weight) to Ag$_2$O cathodes.

### 3.3.4 Cell construction

Button cells with solid cathodes are designed as anode-limited (5–10% more cathode capacity than anode capacity). The anode is amalgamated powdered or gelled zinc. The amount of gelling agent is different with each manufacturer, and is considered proprietary by most battery makers. The anode is housed in a triclad metal top (nickel/stainless steel/copper) with the copper in direct contact with the zinc. The anode is separated from the cathode by a separator or absorbent (cotton-like

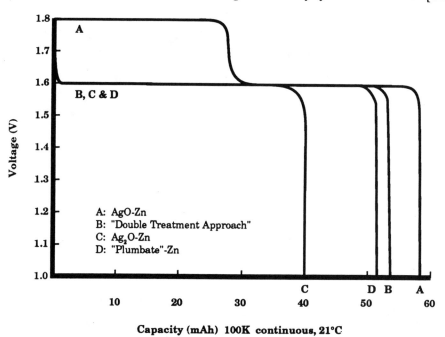

Fig. 3.31 — Silver 'Plumbate' in the Silver Cell (392 size; 7.9 mm diam. × 3.6 mm height).

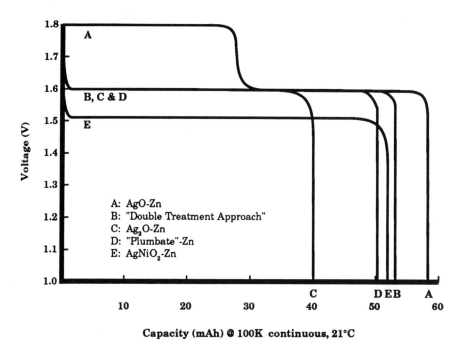

Fig. 3.32 — AgNiO$_2$ in the Silver Cell (392 size; 7.9 mm diam. × 3.67 mm height).

material) and a barrier (semi-permeable membrane). The complexity of the absorbent–separator system depends on the chemistry of the cathode and its intended use. For example, an $Ag_2O$–Zn cell for hearing aid application may use a Cellophane-type membrane to slow silver migration for a period of one or two years. Such a membrane is not sufficient to hold silver migration for 3–5 years for watch applications. Another type of membrane is grafted and irradiated polyethylene (Permion™) used in silver batteries for watch applications [49,50]. This type of membrane extends application use to last 5 years or more.

Fig. 3.33 shows the constructional details of HgO–Zn and air–Zn cells for hearing aid applications. The HgO cathode usually consists of a mixture of HgO and graphite for conductivity. A small amount of $MnO_2$ is added to disperse and imbibe mercury beads upon cell discharge. The $MnO_2$ also raises the cell voltage to 1.45 V instead of

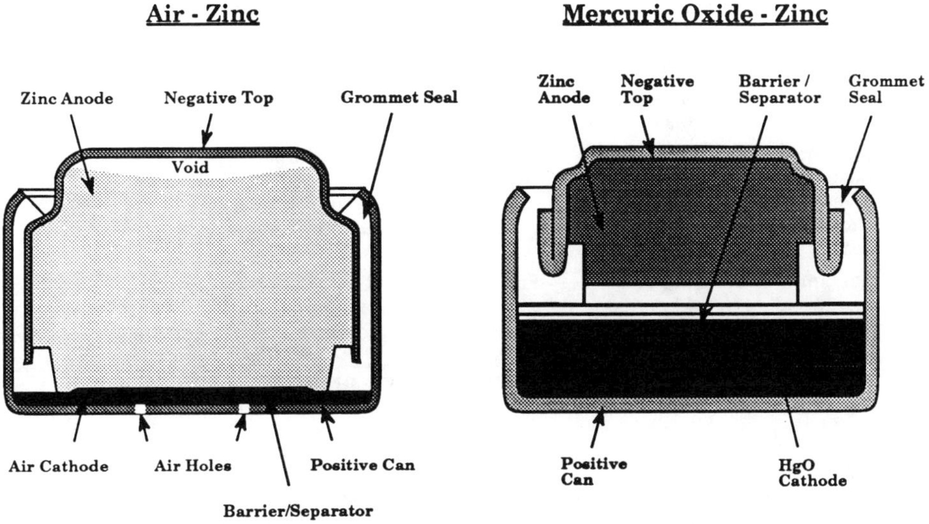

Fig. 3.33 — Air–zinc vs mercuric oxide–zinc cell construction. (Courtesy of Rayovac Corporation.)

1.35 V. Depending on the manufacturer, a small amount of Teflon™ (0.5–1.0% of cathode weight) is added to ease cathode pelleting. The cathode is pelleted and consolidated into a nickel plated steel can (preplated or postplated). The cathode can (positive) and the anode top (negative) are separated by a grommet seal material, usually nylon 6/6 with a polyamide sealant for improved seal and to prevent electrolyte leakage at the seal surfaces. Materials inside the cell are wetted with KOH electrolyte (9–10 N) and contain a small amount of zinc oxide (ZnO) to reduce zinc gassing.

The construction of an air–Zn cell is similar to the HgO–Zn cell except for the air cathode, with the corresponding air holes in the positive can in place of the $H_gO$ cathode.

A standard air–Zn cell will use a stepped can with a diffusion pad between the air cathode and the air hole (Fig. 3.34; designs b, c, d, e). A 'Proline' air–Zn cell (Fig. 3.34; design a) has an air distribution system via can gridding to allow for maximum anode volume inside the cell and no stepped can [51]. In addition, thinner hardware

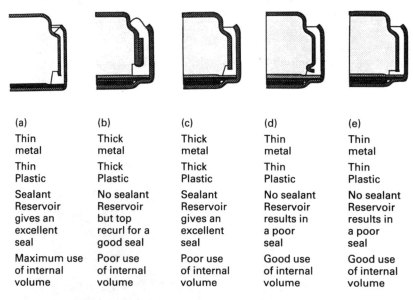

| (a) | (b) | (c) | (d) | (e) |
|---|---|---|---|---|
| Thin metal | Thick metal | Thick metal | Thin metal | Thin metal |
| Thin Plastic | Thick Plastic | Thick Plastic | Thin Plastic | Thin Plastic |
| Sealant Reservoir gives an excellent seal | No sealant Reservoir but top recurl for a good seal | Sealant Reservoir gives an excellent seal | No sealant Reservoir results in a poor seal | No sealant Reservoir results in a poor seal |
| Maximum use of internal volume | Poor use of internal volume | Poor use of internal volume | Good use of internal volume | Good use of internal volume |

Fig. 3.34 — Internal volume and seal quality of various commercial designs.

(can, top, grommet) permits adding more zinc in the anode volume, thus resulting in increased capacity over a standard air–Zn cell. Capacity and seal quality will depend on the hardware design selected by the manufacturer. In many instances the selection of a given design will depend on the manufacturer's machine and processing capability. Fig. 3.34 shows examples of the 675 design from various manufacturers. The use of thin metal and sealant reservoir will result in a maximum use of internal cell volume (capacity) and excellent seal quality (minimum or no leakage during storage or use).

The constructional details for 'Plumbate'–Zn and $Ag_2O$–Zn for watch applications are shown in Fig. 3.35. Like hearing aid cells, the hardware (can, top, grommet, etc.) houses the active components (cathode, anode, electrolyte, etc.) except for the variations in cathode thickness and barrier type. With the $Ag_2O$–Zn cell, the cathode occupies more volume thus less capacity (less zinc) as compared to the 'Plumbate' cathode. The 'Plumbate' cell has a composite barrier system made of laminated layers of Cellophane and grafted methacrylic acid crosslinked polyethylene membrane [37] to extend its life up to 5 years of storage at 21°C without silver shorting.

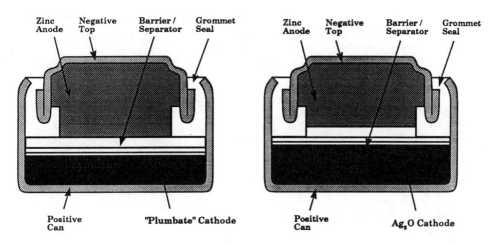

Fig. 3.35 — Construction of 'Plumbate'–Zn vs $Ag_2O$–Zn.

### 3.3.5 System characteristics and application performance

#### *3.3.5.1 Hearing aid batteries*

Table 3.7 shows the characteristics of HgO–Zn and air–Zn systems (standard and premium types) for hearing aid applications. Both systems have high capacity per unit volume, constant voltage output, high pulse capability, good service life, and reasonable cost. The HgO–Zn has been used as a reliable power source for many types of hearing aids, such as European Class B and US Class A, until recently, when the air–Zn system gained acceptance. The air–Zn system will be widely used because of its increased capacity and environmental acceptability for Class A, B, and most recently, Class D amplifiers.

**Energy density**

Mercury and air batteries have high energy density as compared to the traditional alkaline–manganese batteries. Premium air–Zn batteries will achieve up to 1200 $WhL^{-1}$ at the 1 mA discharge rate in the 675 size cell. This value is the highest among all commercial battery systems. To achieve such high values, careful selection of the cell hardware, for example, can, top, and grommet thickness, the barrier-separator thickness, and the profile of the can and top must be considered [38,51]. In addition, air–Zn batteries are designed with an anode free volume (typically 15–20% of the total anode compartment) to allow for the expansion which occurs when zinc is converted to zinc oxide during cell discharge. A disregard for these design elements may lead to an oversized discharged cell that is difficult to remove from a hearing aid compartment. Such a cell may show vent hole leakage and/or short service life.

**Cell voltage and discharge characteristics**

The open circuit voltage (OCV) of HgO–Zn and air-Zn systems will vary with cathode additives. Manganese dioxide is customarily added to HgO and air cathodes

**Table 3.7** — HgO–Zn and air–Zn characteristics for hearing aid applications; for example, e.g., 675 size; 11.6 mm diameter × 5.4 mm height

| Characteristic | Premium HgO–Zn | Standard air–Zn | Premium air–Zn |
|---|---|---|---|
| Energy density | | | |
| Wh/L | 600 | 900 | 1200 |
| Wh/Kg | 110 | 300 | 400 |
| Voltage (V) | | | |
| Nominal | 1.35 | 1.40 | 1.40 |
| Working | 1.25 | 1.23 | 1.23 |
| Nominal rate capability | | | |
| mA/cm$^2$ (initial) | 80 | 16 | 28 |
| Pulse current (mA/cell) | 50 | 20 | 35 |
| LC during discharge (mA/cell) | 30 | 15 | 20 |
| Life | | | |
| Capacity (mAh/cell) | 270 | 400 | 540 |
| Service (625Ω–16h/day) | 8 days | 12 days | 16 days |
| Shelf (years) | 3 | 5 (sealed) | 5 (sealed) |
| Temperature | | | |
| Storage — °C | −40/+60 | −40/+50 | −40/+50 |
| Operation — °C | −10/+55 | −10/+55 | −10/+55 |
| Capacity loss per year at 21°C | 4% | 3% | 3% |
| Safety | | | |
| Abuse tests | Bulging | No bulging | No bulging |
| Disposal | (reclaim) | landfill | landfill |
| Relative cost | | | |
| Material | 1.0 | 1.25 | 1.30 |
| Total | 1.0 | 1.40 | 1.50 |

to increase cell OCV above 1.40 V (up to 1.50 V). Gamma-type $MnO_2$ is added to HgO cathode (5–15% of the cathode weight), and Beta-type $MnO_2$ is added (5–10%) to the air cathode mix. The disappearance of the high OCV value may signify partial cell shorting.

Although cell OCV will not affect the operation of a hearing aid, the closed circuit voltage (CCV) level will. A drop of cell CCV below 1.0 V will stop some hearing aids from working. The discharge profiles of a 675 size cell is presented in Fig. 3.36. Curves for $MnO_2$–Zn and $Ag_2O$–Zn cells are provided for comparison. Premium air–Zn cells will provide twice the capacity of HgO–Zn cells at about the same working voltage.

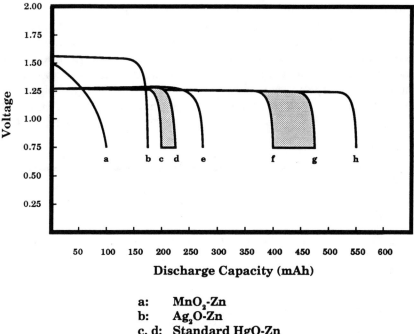

a: MnO$_2$-Zn
b: Ag$_2$O-Zn
c, d: Standard HgO-Zn
e: Premium HgO-Zn
f, g: Standard Air-Zn
h: Premium Air-Zn

Fig. 3.36 — Discharge profile of primary button cells for hearing aid applications. Cell size; 11.6 mm diam. × 5.4 mm height. Cells discharged at 625 Ω, 16 H/D, 21°C, to 0.90 V.

**Rate capability**

The ability of a hearing aid battery to deliver a maximum peak pulse drain without polarization will determine its rate capability. Hearing aids with class B amplifiers (push/pull type) will draw very high currents (up to 50 mA/cell for few milliseconds) at high volume settings. Poorly designed batteries will have a rate capability problem causing the hearing aid to enter a 'feedback' mode which results in unusual sounds, for example muffling, buzzing, distortion, or scratchiness. Typical current drains for various hearing aids are presented in Table 3.8.

Table 3.8 — Typical drains for various hearing aids (mA)

| Aid type | Idle drain | Peak continuous | Peak pulse |
|---|---|---|---|
| In the canal (ITC) | 0.5–1.0 | 2.0– 3.0 | 0.7– 5.0 |
| In the ear (ITE) | 0.5–1.0 | 2.0– 4.0 | 8.0– 10.0 |
| Behind the ear (BTE) | 1.0–2.0 | 5.0–20.0 | 20.0– 50.0 |
| Eyeglass aid | 1.0–2.5 | 5.0– 8.0 | 20.0– 50.0 |
| Body aid | 3.0–5.0 | 15.0–20.0 | 50.0–100.0 |

The maximum continuous operating current of a hearing aid cell is defined as the limiting current ($I_L$). It is determined by measuring the steady-state current from the cell to 0.90 V for one minute. In a mercury cell the $I_L$ is limited by the ionic diffusion through the barrier, while in an air–Zn cell it is regulated by oxygen diffusion through the air hole and the air cathode. Most air–Zn cells are designed with an $I_L$ at least twice the maximum expected continuous operating current. Such a design feature will safeguard against hearing aid power failure, especially in push/pull types, that is, Class B amplifiers. Fig 3.37 shows the $I_L$ for a 675 size mercury and air cell. The $I_L$

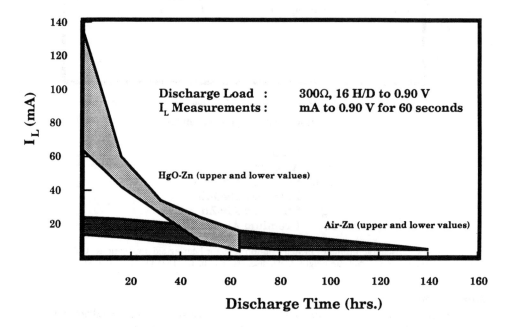

Fig. 3.37 — Limiting current during discharge of HgO–Zn and air–Zn cell (675 size; 11.6 mm diam. X 5.6 mm height).

from a mercury cell is initally high, but it degrades rapidly with cell discharge, unlike that from an air cell. The voltage of an A675H size cell falls rapidly when continuous currents above the limiting current are applied (Fig. 3.38), because the cell has become oxygen-starved. It is consuming oxygen at a faster rate than the rate at which oxygen is entering the cell. The voltage will fall until the load current is reduced to an equilibrium condition. Air–zinc cells can handle pulse currents much higher than the limiting current. This capability results from a reservoir of oxygen that builds up within the cell when the load is below the limiting current. Figure 3.39 shows the voltage profiles for various pulse waveforms. In Fig. 3.39(a), a pulse current double the magnitude of $I_L$ is applied. However, the pulses are spaced so that the average current is less than $I_L$. Therefore, the average rate of oxygen ingress is sufficient to sustain the average load current. Furthermore, the pulses are of short duration, so

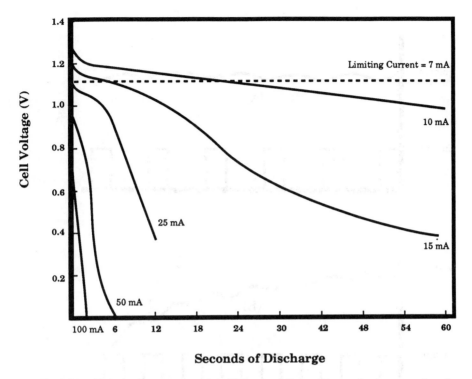

Fig. 3.38 — Voltage–time response of air–Zn button cell to continuous loads greater than the limiting current, A675H size cell. (Courtesy of Linden, D., *Handbook of Batteries and Fuel Cells*).

that the cell does not become oxygen-starved before the end of a single pulse. The cell's voltage profile shows the ripple effect of the pulse current, but the cell maintains a continuous useful average operating voltage. If the peak pulse current is increased so that the average load current is greater than $I_L$, the cell will eventually become oxygen-starved, and voltage will decline as shown in Fig. 3.39(b).

When choosing a cell for a particular pulse application, the designer must ensure that the average current for the duration of the pulse is less than the $I_L$. The designer must also check for the possibility of the cell becoming oxygen-starved during the application of a single pulse. This condition is illustrated in Fig. 3.39(c), where the average current is less than the $I_L$, but the cell becomes oxygen-starved and the voltage declines sharply during the time that a pulse is applied (because of the long pulse duration). The voltage recovers between pulses, maintaining an average close to the voltage of Fig. 3.39(a).

## Cell impedance

Cell impedance is usually measured at a frequency of 1 kHz, with or without an external load on the cell. In most types of hearing aid circuits it is desirable to have cells with low impedance values. The low impedance results in better sound quality.

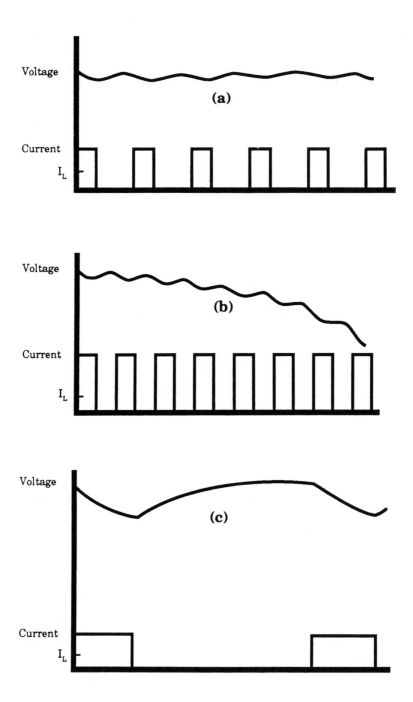

Fig. 3.39 — Pulse load performance of air–Zn button cells. (a) $I_{ave} < I_L$, short pulse duration; (b) $I_{ave} > I_L$, short pulse duration; and (c) $I_{ave} < I_L$, long pulse duration. (Courtesy of Linden, D., *Handbook of batteries and fuel cells*).

Fig. 3.40 compares the impedance values of a 675 size cell in three chemical systems at 100 Hz and 1000 Hz. Air–zinc impedance is the lowest because of low internal resistance. The combination of a conductive electrolyte (30–35%– KOH), low barrier resistivity, and conductive cathode usually results in a low impedance cell.

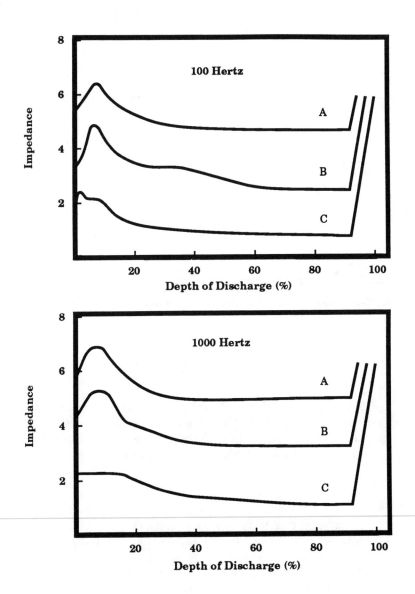

Fig. 3.40 — Cell impedance ($\Omega$) at 100 Hz and 1000 Hz (675 size cell; A: $Ag_2O$–Zn, B: HgO–Zn, C: air–Zn).

At normal operating conditions, the impedance curve usually mirrors the operating voltage curve of the cell (Fig. 3.41(a)). Sometimes, however, cell impedance will rise sharply at 10–20% depth of discharge, causing a phenomenon called 'impedance hump'. The causes of the phenomenon are not well understood. It has been postulated that a tenacious ZnO film may deposit around the zinc particles near the barrier, causing a temporary increase in cell resistivity. The addition of materials such as soluble silicates, for example, Cab-O-Sil™, will accelerate the phenomenon (Fig. 3.41(b)). At other times, cell impedance will rise at the end of cell discharge, for

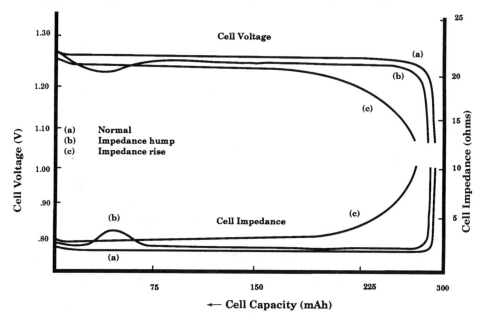

Fig. 3.41 — Voltage and impedance of HgO–Zn cells, 625V, 16 H/D to 0.90 V, 21°C (675 size; 11.6mm diam. × 5.4mm height).

example, above 80% depth of discharge. The impedance rise has been attributed to severe ZnO precipitation (Fig. 3.41(c)). Changes in the barrier system and the addition of materials that interfere with the precipitation of ZnO around the zinc particles have resulted in steady impedance curves with minimum or no humps [33,52].

### Environmental effects

Temperature and humidity effects on cell capacity are widely reported [37,52,53,54,55] for the HgO–Zn and air–Zn systems, with dry-out being the most influential factor on reducing the capacity or raising the impedance of air–Zn cells. Occasionally, with poorly designed cells, shock and vibration treatments that simulate shipping and handling of air–Zn batteries will cause loosening of the zinc from contacting the separator, thus increasing cell impedance. Fig. 3.42 shows temperature effects [59] on volumetric energy density for various primary battery systems. The alkaline MnO$_2$—Zn system is reported for comparison. The capacity degradation at low temperature is caused by a reduction in ionic diffusion through the electrolyte, and that at high temperature is caused by zinc corrosion.

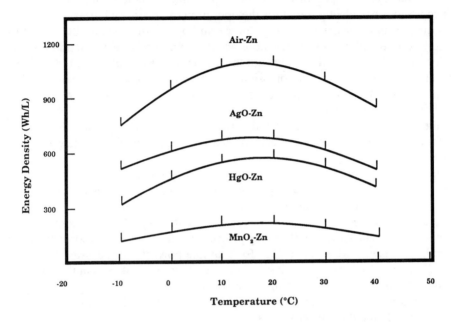

Fig. 3.42 — Volumetric energy density comparison of primary battery systems.

To relate the temperature and humidity effects on actual hearing aid use, a study was designed by Bruner [56] to measure the temperature and relative humidity within the human external auditory canal under controlled ambient conditions. The data in Table 3.9 show that the human canal temperature and humidity vary between

**Table 3.9** — Temperature and humidity within the human external auditory canal (from Bruner, R. C. & House, L. R. [56])

| Canal temperature (°C) | | | | | Canal relative humidity | | | | |
|---|---|---|---|---|---|---|---|---|---|
| Ambient temp. | Ambient humidity | | | | Ambient temp. | Ambient humidity | | | |
| | 30% | 50% | 70% | 90% | | 30% | 50% | 70% | 90% |
| 37.8°C | 35.8 | 36.3 | 37.1 | 38.2 | 37.8°C | 87.0 | 87.7 | 99.0 | 98.0 |
| 32.2°C | 35.6 | 35.7 | 35.9 | 36.0 | 32.2°C | 64.7 | 78.7 | 86.7 | 85.7 |
| 26.7°C | 35.1 | 35.2 | 35.3 | 34.4 | 26.7°C | 75.3 | 74.0 | 77.0 | 89.7 |
| 21.1°C | 35.1 | 34.0 | 34.0 | 34.4 | 21.1°C | 63.7 | 71.3 | 69.0 | 76.7 |

33–36°C and 60–100% relative humidity under ambient exposures of 20–38°C at 30–90% relative humidity. In a way, the human canal will condition the temperature and humidity around the battery to a more benign and acceptable environment. In Bruner's experiment, canal relative humidities below 60% were not found. A drop of relative humidity below 60% caused human skin to dry out and to develop cracks and fissures. Additionally, an increase of the relative humidity above 85% is a rare occurrence within the human auditory canal. In normal living conditions, the human auditory canal has a temperature range of 34–37°C and a humidity range of 60–80%. Such conditions are ideally suited for the operation of an air–Zn battery.

Reduction in cell capacity during storage and operation is generally caused by zinc corrosion reactions. These reactions are the combined effect of oxidation and self-discharge. Zinc oxidation will result in a capacity loss of 1.5% of rated capacity per year at 25°C [37] while zinc corrosion will result in 2.5% loss per year [38]. The combined loss of 4% per year may be reduced via the selection of high purity zinc and reduction of contaminants.

Other important factors connected with air–Zn batteries are the effect of gas transfer on oxidation of the zinc, electrolyte carbonation, and electrolyte water gain or loss.

During storage the air access hole(s) of an air–Zn cell are sealed to prevent oxygen transfer decay as well as moisture ingress or egress. A typical material for sealing a cell is a polyester tape coated with rubber based adhesive(s). A better sealing composite laminate has been disclosed by Oltman [57]. In this composite, a face stock of biaxially-oriented three-ply polypropylene paper is interposed between an acrylic adhesive and a plastic film, for example polyester or acetate. The bond formed between the acrylic adhesive and the metal face is weaker than the adhesive to polypropylene paper bond and the cohesive strength of the polypropylene paper. Cell activation will take place immediately upon tape removal with neither voltage delay nor residual adhesives problems.

Two phenomena are associated with the quality of the tape used in air–Zn cells. Both are detected by voltage measurement while the cell is sealed with the tape. The first is defined as *underactivation* when the cell voltage is measured below 0.9 V per cell. This voltage will reach a workable level of 1.10 V or higher in 10 seconds or less after tape removal. Such a phenomenon is usually caused by oxygen starvation inside the cell due to tight adhesion of the tape or by partial zinc shorts. It was found that voltage levels of 0.5 V to 0.90 V/cell are recoverable, while voltages below 0.5 V/cell are not recoverable. The phenomenon of underactivation may be a cause of rejection by hearing aid consumers even though cell electrical performance is not impaired.

The second phenomenon is defined as *overactivation* when the cell voltage is measured above 1.40 V. This voltage is usually caused by moisture and $CO_2$ reactions with the zinc via a loose tape or by oxygen enrichment of the air cathode. Such phenomena are harmful, as evidenced by a capacity loss or a high impedance caused by cell drying. Carbon dioxide, which is present in the atmosphere at a concentration of approximately 0.04%, reacts with cell electrolyte to from an alkali metal carbonate and bicarbonate. Extreme carbonation will increase the vapour pressure of the electrolyte, thus aggravating water vapour loss in low humidity conditions. It also increases crystal growth of carbonate in the cathode, thus inducing cathode blockage and leakage. The carbonate of the electrolyte, however, must be extreme to be detrimental to cell performance in most applications [38,39].

Water vapour transfer (humidity sensitivity) is the most important decay mechanism affecting useful service life from an air–Zn cell. This phenomenon plays an important role in low rate applications requiring more than 30 days of continuous use or high rate intermittent use for less than 30 days. It occurs when a partial pressure difference exists between the vapour pressure of the electrolyte and the surrounding environment. The equilibrium relative humidity of an air–Zn cell with 30% concentration of KOH is around 60%. The cell will lose water when the humidity at room temperature is below 60%, and will gain water at humidities above 60%. Humidity sensitivity is affected by hole size, cathode porosity, and KOH concentration. Larger hole size and porous cathodes are more sensitive to humidity variations than smaller size and low porosity cathodes. Optimization of these parameters and the selection of an appropriate sealing tape, proper KOH concentration, and low corrosion rate zinc will result in a shelf life of more than 5 years at 21°C with an average loss of 4% per year or smaller (Fig. 3.43).

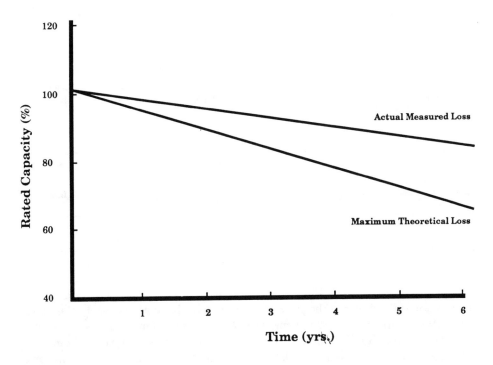

Fig. 3.43 — Actual vs theoretical capacity retention from air–Zn cells (675A size cell; 11.6 mm diam. × 5.4 mm height).

### 3.3.5.2 *Watch batteries*

Table 3.10 shows the characteristics of silver batteries, e.g. monovalent, divalent, 'Plumbate', and silver nickel oxide, for watch applications. These batteries have high energy density per unit volume, good operating voltage, high pulse capability, constant voltage output, and good service life. Their major disadvantage is the high

**Table 3.10** — $Ag_2O$–Zn, AgO–Zn, and silver 'Plumbate'–Zn characteristics for watch applications; e.g., 392 size, 7.9 mm diameter × 3.6 mm height

| Characteristics | $Ag_2O$–Zn | AgO–Zn | 'Plumbate'–Zn | $AgNiO_2$–Zn |
|---|---|---|---|---|
| Energy density | | | | |
| Wh/L | 500 | 680 | 660 | 600 |
| Wh/Kg | 120 | 160 | 155 | 145 |
| Voltage (V) | | | | |
| Nominal | 1.60 | 1.70 | 1.70 | 1.60 |
| Working | 1.50 | 1.50 | 1.50 | 1.45 |
| Nominal rate capability | | | | |
| mA/cm² (initial) | 25 | 25 | 30 | 25 |
| Backlight pulse at 80% DOD | 10 | 15 | 15 | 10 |
| Life | | | | |
| Capacity (mAh/cell) | 38 | 52 | 50 | 40 |
| Service (yrs) | | | | |
| LCD | 2.17 | 2.97 | 2.86 | 2.28 |
| LCD/backlight pulse | | | | |
| Analogue | 1.52 | 2.08 | 2.00 | 1.50 |
| Shelf (yrs.) | 5 | 5 | 5 | 5 |
| Temperature | | | | |
| Storage — °C | −40/+60 | −40/+60 | −40/+60 | −40/+60 |
| Operation — °C | −10/+55 | −10/+55 | −10/+55 | −10/+55 |
| Capacity loss per year | | | | |
| at 21°C | 3% | 4% | 3% | 5% |
| Safety | | | | |
| Abuse test | Bulging | Bulging | Bulging | Bulging |
| Disposal | Landfill | Landfill | Landfill | Landfill |
| Relative cost | | | | |
| Material | 1.0 | 0.85 | 0.80 | 0.60 |
| Total | 1.0 | 0.85 | 0.75 | 0.70 |

- LCD drain: 2 µA continuous (17.5 mAh/yr)
- LCD/backlight pulse: 1 µA background, 10 mA backlight, 8 pulses/day (25 mAh/yr)
- Analogue: 1 µA background, 0.4 mA stepping motor, 7.8 m sec./sec. (36 mAh/yr)

cost and unpredictability of the price of silver metal. The work on AgO, 'Plumbate' and $AgNiO_2$ cathodes has been intensified in the past ten years as a low cost alternative to the $Ag_2O$ cathode. The increase in coulombic capacity from these cathodes can be used either as added cell capacity for long service life or as lower cathode weight for low cell cost.

**Energy density**

Table 3.11 shows the energy density of four battery systems. The theoretical capacity from $AgNiO_2$, 'Plumbate', and AgO materials are 16%, 55%, and 98% over $Ag_2O$. These materials, unfortunately, have to go through a series of processing steps to make them useful as cathode depolarizers. The net result of the processes is a reduction in cathode capacity. Useful capacity from $AgNiO_2$, AgO, and 'Plumbate', cathodes are still 30%, 42%, and 55% higher than the $Ag_2O$ cathode. Balancing cathode capacity with zinc capacity in a given volume (357 size cell at 0.56 cm³) will

**Table 3.11** — Energy density of silver batteries; e.g., 357 size cell; 11.6 mm diameter × 5.4 mm height

| Characteristics | $Ag_2O$–Zn | AgO–Zn[†] | 'Plumbate'–Zn[‡] | $AgNiO_2$–Zn[§] |
|---|---|---|---|---|
| ● Cathode | | | | |
| Material type | $Ag_2O$ | AgO | 'Plumbate' | $AgNiO_2$ |
| Theoretical capacity (mAh/gm) | 232 | 432 | 360 | 270 |
| Actual capacity (mAh/gm) | 227 | 416 | 346 | 263 |
| Cathode conductor | Carbon | None | None | None |
| Cathode binder | Teflon™ | Teflon™ | Teflon™ | Teflon™ |
| Cathode capacity before reduction (mAh/gm) | 210 | 320 | 310 | 260 |
| Cathode capacity after reduction (mAh/gm) | 200 | 285 | 310 | 260 |
| Relative capacity | 1.00 | 1.42 | 1.55 | 1.30 |
| ● Anode | | | | |
| Material type | Zinc | Zinc | Zinc | Zinc |
| Theoretical capacity (mAh/gm) | 820 | 820 | 820 | 820 |
| Actual capacity (mAh/gm) | 760 | 760 | 760 | 760 |
| ● Energy density | | | | |
| Cell capacity (mAh) | 190 | 250 | 260 | 230 |
| Wh/L | 500 | 680 | 710 | 600 |
| Relative energy density | 1.00 | 1.36 | 1.42 | 1.20 |

† Sold commercially under the 'Ditronic' trade name from Rayovac.
‡ Sold commercially under the 'New formula' silver from Rayovac.
§ Sold commercially under the 'Energizer' trade name from Eveready and improved silver from Sony.

result in practical energy densities of 500, 600, 680, and 710 WhL$^{-1}$ for the $Ag_2O$–Zn, $AgNiO_2$–Zn, AgO–Zn and 'Plumbate'–Zn systems, respectively. The increase in energy density of $AgNiO_2$–Zn (20%), AgO–Zn (36%), and 'Plumbate'–Zn (42%) over $Ag_2O$–Zn will be affected by cell size. Smaller cell sizes will have less of a capacity advantage than larger cell sizes because of the relative increase of inactive cell components in small batteries.

### Cell voltage and discharge characteristics

The OCV of an $Ag_2O$–Zn cell is about 1.60 V. It will change slightly (1.595–1.605 V) with electrolyte concentration, zincate content in the electrolyte, and high temperature exposure [58]. Carbon dioxide reaction with silver oxide during battery manufacturing will raise the cell OCV to 1.65 V owing to the formation of silver carbonate. The increase in cell OCV, however, is temporary and will drop to the operating voltage level of 1.58 V in a watch within seconds.

Unlike $Ag_2$–Zn, the modified silver systems ($AgNiO_2$–Zn, AgO–Zn, and 'Plumbate'–Zn) will have slightly different OCV values from $Ag_2O$–Zn (Fig. 3.44). Except for AgNiO$_2$–Zn, the CCV values will stabilize at the Ag$_2$O–Zn level at or below 2% of the removed total capacity from the cell.

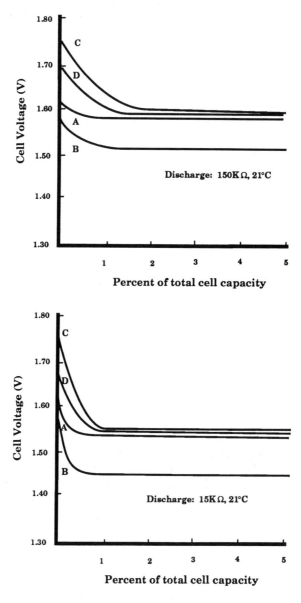

Fig. 3.44 — OCV and CCV of various silver cell chemistries A: Ag$_2$O–Zn, B: AgNiO$_2$–Zn, C: AgO–Zn (double treatment), D: 'Plumbate'–Zn (392 size cell; 7.9 mm diam. × 3.6 mm height).

Two phenomena are associated with AgO–Zn batteries (44) namely 'voltage–up' and 'impedance–up'. 'Voltage–up' represents a situation of voltage instability where the potential of a treated cathode in an AgO–Zn cell (Fig. 3.45) will show its true (not

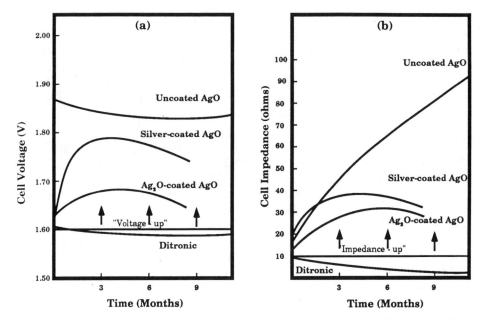

Fig. 3.45 — 'Voltage-up' and 'Impedance-up' of AgO–Zn cells (392 size) during one year storage at 21°C.

suppressed) value. In some instances, similar to those in Fig. 3.45(a), an OCV of 1.60–1.64 V appears initially, but upon ageing reverts upward to values between 1.65–1.89 V. Similarly, an 'impedance-up' represents a situation of impedance instability after a few months of storage at 21°C (Fig. 3.45(b)). 'Impedance-up' is noted to be always associated with 'voltage-up' in AgO–Zn cells but 'voltage-up' is not always associated with 'impedance-up'. Both phenomena are connected with the thickness of $Ag_2O$ and Ag layers around AgO pellets. An ideal thickness of 0.60 mm of $Ag_2O$ around the pellet and 0.24 mm of Ag on the top of the consolidation result in cells with OCV value of 1.59 V after 6 months of storage at 21°C (Fig. 3.28). A thinner Ag layer or disappearance of such layer via oxidation in the cell will cause a voltage rise above 1.65 V. Both of these phenomena are not acceptable to watch manufacturers since high operating voltages will cause ghosting of some LCD displays or time keeping inaccuracy in some analogue watches.

The 'Plumbate' formulation has resolved the problems of 'voltage-up' and 'impedance-up' via particle treatments in bulk rather than pellet treatments [46]. An OCV of 1.60 V is obtained upon complete coverage of an AgO particle with a layer of $Ag_2O$ with Ag and/or $Ag_5Pb_2O_6$. The $Ag_2O$ layer gives the cathode the necessary voltage stability typical of monovalent cells, while the $Ag_5Pb_2O_6$ and/or Ag results in low and steady cell impedance on shelf stand. These types of cell are commercially available [59].

### Rate capability

When designing batteries for watch applications, some manufacturers prefer to use NaOH electrolyte rather than KOH electrolyte for low rate applications such as

LCD watches, calculators, and some analogue watches. The NaOH cells are apt to leak less than KOH cells. This characteristic, however, has lost importance in recent years because of improvements made in seal technology by most manufacturers. Fig. 3.46 shows the pulse performance of a treated divalent cell with NaOH and KOH electrolytes. The low rate cell (IEC R41) is better suited for analogue watches (2000 ohms, 7.8 m sec/sec) even at $-10°C$, while the high rate cell (IEC SR41) is better suited for LCD watches with backlight (100 ohms, 2 sec/hr-8 h/day).

In some applications, a heavy current pulse of short duration is required in addition to the low background current, e.g. electronic shutter mechanism of a still camera, LED watch, LCD watch with a backlight, and some analogue watches. In these cases, the minimum voltage at the end of the pulse must be met to ensure proper device function. Failure to obtain the minimum voltage level has been attributed to a ZnO blockage in the cathode upper surface. A study was conducted at Rayovac with 386 silver cells (11.6 mm diameter × 4.2 mm height) used in a traditional T.I.–LED watch (steady background drain of $5 \mu A$ and an LED pulse of 40 mA for 1.25 s) for one year. With fresh cells no problem appeared upon activating the LED. After three months, however, dimming of the LED display was shown with repeated pulsing. Failure analysis of batteries removed from dim watches showed a cell voltage under pulse of 1.00 V or less. The low voltage was caused by ZnO blockage in the cathode upper surface facing the zinc. Improving the electronic conductivity of the cathode via conductive diluents, e.g. silver metal, PbO, carbon, or the ionic conductivity via reduction of ZnO dissolution and precipitation, e.g. 35% KOH, high resistance barrier, eliminated the problem. These improvements may be used in applications where high rate pulse capability is needed.

**Cell impedance**
Silver cell impedance shows a steady value or slight decline during cell discharge, since the discharge product is metallic silver. The impedance level is most influenced by conductive diluents in the cathode, barrier resistivity electrolyte type, and electrolyte concentration. A balance of these factors is exercised by battery manufacturers to obtain the desired values to meet the application.

**Environmental effects**
To extend the shelf life of watch batteries to five years, major improvements were needed in seal technology and in cell stability. The effect of temperature and humidity on leakage of button cells was reported by Hull [60]. Leakage was caused by mechanical means, e.g. improper seal, fibres in the seal, scratches, or electrochemical means, e.g. high oxygen content, high humidity. Properly designed cells are now available to operate watches for 5 years without leakage.

Stability of silver cells after high temperature storage or prolonged storage at room temperature is influenced by cathode stability and barrier selection. With an $Ag_2O$ cathode, gassing in aqueous KOH or NaOH at 74°C is not a problem. With modified cathodes, e.g. AgO, 'Plumbate', $AgNiO_2$, however, gas suppression is necessary. Megahead & Buelow [43] found that CdS, HgS, $SnS_2$, or $WS_2$ reduced $O_2$ evolution, while BaS, NiS, MnS, and CuS increased $O_2$ evolution from AgO in

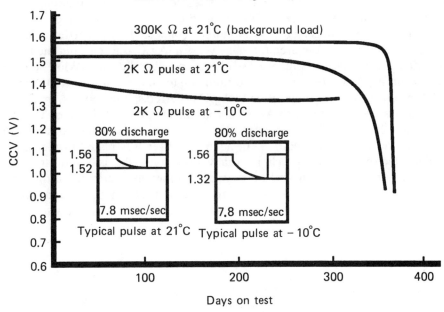
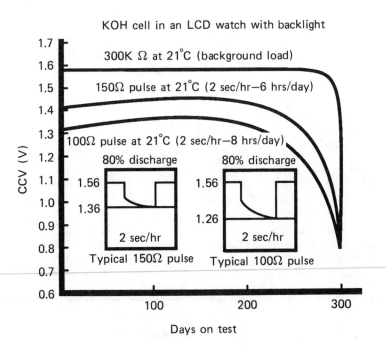

Fig. 3.46 — CCV curves for AgO–Zn cell with NaOH and KOH electrolytes. (Cell size; 7.9 mm diam. × 3.6 mm height).

386 size cells (11.6×4.2 mm). When using an AgO material with a gassing rate of 100 μl gm$^{-1}$ h$^{-1}$ in 18% NaOH+1.25% ZnO electrolyte, non-treated cells showed 0.32 mm expansion after 7 days at 74°C, while the control cells ruptured.

Failure on shelf of silver cells is closely connected with barrier selection. Cellulosic membranes were used for many years in Ag$_2$O–Zn cells, but their use in AgO–Zn cells was unsuccessful because of massive silver diffusion. While solubility of AgO and Ag$_2$O was reported by Thornton [50] to be the same (4.4×10$^{-4}$ mol l$^{-1}$ in 10N NaOH), AgO decomposition to Ag$_2$O occurred spontaneously, thus resulting in more silver diffusion with AgO–Zn cells than with Ag$_2$O–Zn cells. The small amount of soluble silver, probably Ag(OH)$^{-2}$, reaching the zinc caused accelerated corrosion and hydrogen evolution. In addition, silver was plated in the barrier, forming electronic shorts which internally discharged the cell. To stop silver migration to the zinc, laminated Permion$^{TM}$ membranes (water based or solvent based) have been used. Fig. 3.47 shows Arrhenius plots of various low and high rate silver systems. The data show that 10 years of storage at 21°C is now possible.

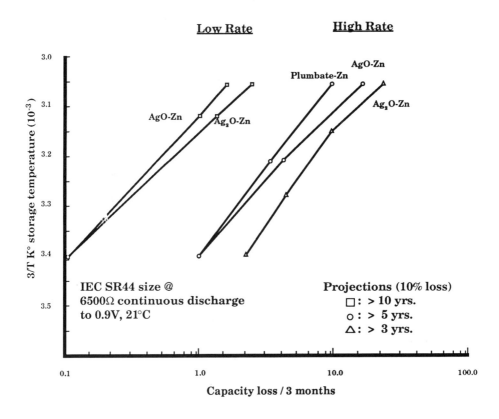

Fig. 3.47 — Arrhenius plot of various silver battery chemistries.

### 3.3.6 Available sizes and types

#### 3.3.6.1 *Hearing aid batteries*
Table 3.12 shows the commercially available mercury and air cells for hearing aid applications. Within the last five years (1985 to 1990), the ITE (in-the-ear) aids have grown to almost 80% of the market. These aids are usually powered by smaller size batteries, e.g. 13 and 312. These two sizes make up more than 75% of the total market. Recently, air–Zn batteries exceeded HgO–Zn batteries for this application. More than 60% of hearing aid batteries are now of the air–Zn type. Even smaller sizes such as the 10A are increasing rapidly for high efficiency canal aids. Silver batteries represent less than 1% of the total hearing aid batteries.

#### 3.3.6.2 *Watch batteries*
Table 3.13 shows the commercially available silver batteries for watch and calculator applications. Japanese manufacturers of watch batteries (Table 3.13(a)) have designated the SW for low rate watches, e.g. analogue LCD, and W for high rate watches, e.g. LCD with backlight and analogue watches with added functions. The numbering system designates the diameter and height of the cell. US and European manufacturers, on the other hand, have selected a numbering system without merit (Table 3.13(b)). The same is true of the major watch makers (Table 3.13(c)).

Watch batteries have been improving recently in reliability, size, and long life. A tiny watch battery ($4.8 \times 1.2$ mm) may operate an LCD watch for five years without failure.

### 3.3.7 Further product development
Future battery development will be influenced by the application requirement, as has been demonstrated in the past 50 years. Innovations in the electrical requirement of hearing aids, watches, calculators and medical devices will direct the development effort toward either longer life or higher rate capabilities. In some cases, higher voltage level and lower impedance value may bring some application advantages. The following trends are apparent.

**A switch from HgO–Zn to air–Zn for hearing aid applications**
This switch has been justified by the environmental concerns toward mercury as well as the service life advantages of air–Zn batteries. With a recent maximum allowable limit of 25 mg of mercury per cell in Europe, it is impossible to comply with environmental regulations without establishing a battery collection programme. Such programmes are costly and not without faults. The air–Zn system is offering major performance advantages over HgO–Zn without any environmental concern. Additionally, more improvements in capacity, rate, impedance, and pulse capability will be likely in future years for the air–Zn system than for the HgO–Zn system.

**Air–zinc optimization**
Performance improvements in the air–Zn system will be needed in moisture sensitivity and longer shelf life in the activated state. Both of these phenomena will

**Table 3.12** — Hearing aid batteries

| Type | Dimension (mm) Dia.×Ht. | IEC | Rayovac Type | Activair mAh | Activair Type | Duracell mAh | Duracell Type | Eveready mAh | Eveready Type | Panasonic mAh | Panasonic Type | Varta mAh | Varta Type |
|---|---|---|---|---|---|---|---|---|---|---|---|---|---|
| Mercuric oxide–zinc | 11.6×5.4 | NR44 | HR675 | 265 | | | RM675 | 235 | EP675E | 270 | NR44 | 220 | V675PX | 210 |
| | 11.6×5.4 | NR44 | RP675 | 220 | | | RM675H | 235 | E675E | 200 | NP675 | 260 | V675HP | 250 |
| | 11.6×4.2 | NR43 | | | | | RM41 | 130 | E41E | 175 | | | | |
| | 7.9×5.4 | NR48 | R13 | 95 | | | RM13 | 95 | E13E | 100 | NR48 | 90 | V13HM | 85 |
| | 7.9×3.6 | NR41 | R312 | 55 | | | RM312 | 45 | E312E | 63 | NR41 | 60 | V312HM | 54 |
| | 11.6×29.2 | NR1 | RP401 | 1200 | | | MP401H | 1150 | EP401E | 160 | NR1 | 1100 | V1PX | 1100 |
| | 15.7×6.1 | NR9 | | | | | RM625R | 300 | E625E | 260 | NR9 | 400 | V625HM | 450 |
| Zinc–air | 11.6×5.4 | AR44 | 675A | 540 | 675HP | 400 | | | | 520 | | | V675HPA | 250 |
| | 11.6×5.4 | AR44 | 41A | 390 | 675HPX | 520 | DA675X | 520 | AC675E | 390 | 675 | 540 | V675A | 400 |
| | 11.6×4.2 | | | | | | | | AC416 | | | | | |
| | 7.9×5.4 | AR48 | 13A | 230 | 13HPX | 200 | DA13X | 200 | AC13E | 220 | 13 | 230 | V13A | 170 |
| | 7.9×3.6 | AR41 | 312A | 110 | 312HPX | 90 | DA312X | 90 | AC312E | 120 | 312 | 110 | V312A | 70 |
| | 5.8×3.6 | | 10A | 60 | 230HPX | 50 | DA230 | 50 | AC230E | 50 | 536 | 50 | | |
| | 15.5×6.2 | | | | | | DA630 | 950 | | | | | | |

**Table 3.13** — Silver watch/calculator batteries

(a) Japan

| Drain type | Dimension (mm) Dia.×Ht. | Seiko Type | mAh | Maxell Type | mAh | Toshiba Type | mAh | Panasonic Type | mAh | Sony Type | mAh |
|---|---|---|---|---|---|---|---|---|---|---|---|
| Low rate | 11.6×5.4 | SR44SW | 165 | SR44SW | 165 | SR44SW | 160 | SR44SW | 170 | SR44SW | 165 |
| | 11.6×4.2 | SR43SW | 110 | SR43SW | 125 | SR43SW | 140 | SR43SW | 110 | SR43SW | 110 |
| | 11.6×3.1 | SR1130SW | 80 | SR1130SW | 82 | SR1130SW | 72 | SR1130SW | 80 | SR1130SW | 80 |
| | 11.6×2.1 | SR1120SW | 53 | SR1120SW | 55 | SR1120SW | 45 | SR1120SW | 45 | SR1120SW | 55 |
| | 11.6×1.6 | SR1116SW | 30 | SR1116SW | 29 | SR1116SW | 22 | | | SR1116SW | 33 |
| | 9.5×3.6 | SR936SW | 70 | SR936SW | 70 | SR936SW | 65 | | | SR936SW | 70 |
| | 9.5×2.7 | SR927SW | 55 | SR927SW | 55 | SR927SW | 55 | SR927SW | 55 | 927SW | 55 |
| | 9.5×2.1 | SR920SW | 39 | SR920SW | 39 | SR920SW | 38 | SR920SW | 36 | 920SW | 40 |
| | 9.5×1.6 | SR916SW | 26.5 | SR916SW | 26.5 | SR916SW | 26 | SR916SW | 26 | 916SW | 26 |
| | 7.9×3.6 | SR41SW | 45 | SR41SW | 45 | SR41SW | 40 | SR41SW | 40 | SR41SW | 42 |
| | 7.9×3.1 | SR731SW | 36 | | | | | | | | |
| | 7.9×2.6 | SR726SW | 34 | SR726SW | 33 | SR726SW | 28 | SR726SW | 30 | SR726SW | 33 |
| | 7.9×2.1 | SR721SW | 24 | SR721SW | 25 | SR721SW | 24 | SR721SW | 21 | SR721SW | 30 |
| | 7.9×1.6 | SR716SW | 21 | SR716SW | 19 | SR716SW | 19 | SR716SW | 19 | SR716SW | 19 |
| | 7.9×1.2 | SR712SW | 11 | SR712SW | 10 | | | SR712SW | 10 | | |
| | 6.8×2.6 | SR626SW | 26 | SR626SW | 26 | SR626SW | 26 | SR626SW | 26 | SR626SW | 29 |
| | 6.8×2.1 | SR621SW | 21 | SR621SW | 18 | SR621SW | 18 | SR621SW | 16 | SR621SW | 20 |
| | 6.8×1.6 | SR616SW | 15 | SR616SW | 15 | SR616SW | 11 | SR616SW | 14 | SR616SW | 15 |
| | 5.8×2.7 | SR527SW | 20 | SR527SW | 16 | SR527SW | 16 | SR527SW | 16 | SR527SW | 20 |
| | 5.8×2.1 | SR521SW | 14 | SR521SW | 14 | SR521SW | 11 | SR521SW | 13 | SR521SW | 14 |
| | 5.9×1.6 | SR516SW | 9.5 | SR516SW | 11.5 | SR516SW | 7.5 | | | SR516SW | 11.5 |
| | 5.8×1.2 | SR512SW | 5.5 | SR512SW | 5.5 | | | | | | |
| High rate | 11.6×5.4 | SR44W | 180 | SR44W | 165 | SR44W | 160 | SR44W | 180 | SR44W | 85 |
| | 11.6×4.2 | SR43W | 120 | SR43W | 125 | SR43W | 140 | SR43W | 120 | SR43W | |
| | 11.6×3.1 | SR1130W | 80 | SR1130W | 80 | SR1130W | 72 | SR1130W | 80 | SR1130W | 55 |
| | 11.6×2.1 | SR1120W | 53 | SR1120W | 55 | SR1120W | 45 | SR1120W | 45 | SR1120W | 55 |
| | 9.5×2.7 | SR927W | 55 | SR927W | 55 | SR927W | 55 | SR927W | 55 | SR927W | 40 |
| | 9.6×2.1 | SR920W | 39 | SR920W | 39 | SR920W | 38 | SR920W | 38 | SR920W | 26 |
| | | | | | | | | SR916W | 26 | SR916W | 75 |
| | 7.9×5.4 | SR48W | 75 | SR754W | 75 | SR48W | 75 | SR48W | 75 | SR48W | 45 |
| | 7.9×3.6 | SR41W | 45 | SR41W | 39 | SR41W | 45 | SR41W | 40 | SR41W | 30 |
| | 7.9×2.6 | SR726W | 34 | SR726W | 28 | SR726W | 28 | SR726W | 30 | SR726W | 24 |
| | 7.9×2.1 | SR721W | 24 | SR721W | 25 | SR721W | 24 | SR721W | 25 | SR721W | 26 |
| | 6.8×2.6 | SR626W | 26 | SR626W | 26 | SR626W | 26 | SR626W | 26 | SR626W | |

(b) USA/Europe

| Drain type | Dimension (mm) Dia.×Ht. | Rayovac Type | Eveready Type | mAh | Duracell Type | mAh | Renata Type | mAh | Varta Type | mAh |
|---|---|---|---|---|---|---|---|---|---|---|
| Low rate | 11.6×5.4 | 303 | 303 | 190 | D303 | 165 | 301 | 110 | V303 | 160 |
|  | 11.6×4.2 | 301 | 301 | 120 | D301 | 120 |  | 100 | V301 | 115 |
|  | 11.6×3.6 | 344 | 344 | 100 |  |  | 344 |  | V344 | 100 |
|  | 11.6×3.1 | 390 | 390 | 85 | D390 | 70 | 390 | 80 | V390 | 85 |
|  | 11.6×2.1 | 381 | 381 | 48 | D381 | 40 | 381 | 48 | V381 | 45 |
|  | 11.6×1.6 | 366 | 366 | 30 | D366 | 27 |  |  |  |  |
|  | 9.5×3.6 | 394 | 394 | 75 | D394 | 60 | 394 | 75 | V394 | 67 |
|  | 9.5×2.7 | 395 | 395 | 55 | D395 | 40 | 395 | 52 | V395 | 42 |
|  | 9.5×2.1 | 371 | 371 | 39 | D371 | 30 | 371 | 31 | V371 | 32 |
|  | 9.5×1.6 | 373 | 373 | 28 |  |  | 373 | 28 | V373 | 22 |
|  | 7.9×3.6 | 384 | 384 | 41 | D384 | 38 | 384 | 45 | V384 | 45 |
|  | 7.9×3.1 | 329 | 329 | 40 |  |  | 329 | 37 | V329 | 30 |
|  | 7.9×2.6 | 397 | 397 | 36 | D397 | 24 | 397 | 30 | V397 | 30 |
|  | 7.9×2.1 | 362 | 362 | 22 | D362 | 18 | 362 | 20 | V362 | 22 |
|  | 7.9×1.6 | 315 | 315 | 19 |  |  | 315 | 19 | V315 | 18 |
|  | 7.9×1.2 |  | 346 |  |  |  |  |  |  |  |
|  | 6.8×2.6 | 377 | 377 | 27 | D377 | 20 | 377 | 27 | V377 | 24 |
|  | 6.8×2.1 | 364 | 364 | 19 | D364 | 15 | 364 | 17 | V364 | 20 |
|  | 6.8×1.6 | 321 | 321 | 16 | D321 | 12 | 321 | 13 | V321 | 13 |
|  | 5.8×2.7 | 319 | 319 | 20 |  |  |  |  | V319 | 15 |
|  | 5.8×2.1 | 379 | 379 | 16 |  |  | 379 | 13 | V379 | 12 |
|  | 5.8×1.6 | 317 | 317 | 12 |  |  | 317 | 9 | V317 | 8 |
|  | 5.8×1.2 |  | 335 |  |  |  | 751 | 5 |  |  |
| High rate | 11.6×5.4 | 357 | 357 | 190 | D357 | 165 |  |  | V357 | 155 |
|  | 11.6×4.2 | 386 | 386 | 120 | D386 | 120 | 386 | 110 | V386 | 105 |
|  | 11.6×3.1 | 389 | 389 | 85 | D389 | 70 |  |  | V389 | 85 |
|  | 11.6×2.1 | 391 | 391 | 48 | D391 | 40 |  |  | V391 | 43 |
|  | 9.5×2.7 | 399 | 399 | 53 | D399 | 40 | 399 | 52 | V399 | 42 |
|  | 9.5×2.1 | 370 | 370 | 35 |  |  | 370 | 30 | V370 | 30 |
|  | 7.9×5.6 | 393 | 393 | 70 | D393 | 70 |  |  | V393 | 70 |
|  | 7.9×3.6 | 392 | 392 | 41 | D392 | 38 | 392 | 45 | V392 | 38 |
|  | 7.9×2.6 | 396 | 396 | 33 | D396 | 24 |  |  | V396 | 25 |
|  | 7.9×2.1 | 361 | 361 | 22 | D361 | 18 |  |  | V361 | 18 |

(c) Watch Makers

| Drain type | Dimension (mm) Dia.×Ht. | Timex (type) | Bulova (type) | Citizen (type) | Orient (type) | I.E.C. (designation) |
|---|---|---|---|---|---|---|
| Low rate | 11.6×5.4 | A | 226 | 280-08 | | SR44 |
| | 11.6×4.2 | D | 603 | 280-01 | 080-002 | SR43 |
| | 11.6×3.1 | | 317 | 280-24 | 080-015 | SR54 |
| | 11.6×2.1 | | | 280-27 | 080-014 | SR55 |
| | 11.6×1.6 | | | 280-46 | | |
| | 9.5×3.6 | | | 280-17 | | |
| | 9.5×2.7 | | 610 | 280-48 | 080-025 | SR57 |
| | 9.5×2.1 | | 605 | 280-31 | 080-033 | |
| | 9.5×1.6 | | 617 | 280-45 | 080-030 | |
| | 7.9×3.6 | | 247 | 280-18 | 080-001 | SR41 |
| | 7.9×3.1 | | | | | |
| | 7.9×2.6 | N | 607 | 280-28 | 080-020 | SR59 |
| | 7.9×2.1 | S | 601 | 280-29 | 080-022 | SR58 |
| | 7.9×1.6 | | | 280-56 | | |
| | 7.9×1.2 | | | | | |
| | 6.8×2.6 | BA | 606 | 280-39 | | |
| | 6.8×2.1 | T | 602 | 280-34 | 080-029 | SR60 |
| | 6.8×1.6 | DA | 611 | | | |
| | 5.8×2.7 | | 615 | 280-60 | | |
| | 5.8×2.1 | | | 280-59 | 080-039 | |
| | 5.8×1.6 | | | 280-58 | | |
| High rate | 11.6×5.4 | J | 228 | 280-41 | 080-003 | SR44 |
| | 11.6×4.2 | H | 260 | 280-15 | 080-004 | SR43 |
| | 11.6×3.1 | M | | 280-30 | 080-011 | SR54 |
| | 11.6×2.1 | L | 609 | 280-44 | 080-028 | SR55 |
| | 9.5×2.7 | W | 613 | 280-51 | 080-019 | SR57 |
| | 9.5×2.1 | Z | | | | |
| | 7.9×5.4 | F | 255 | | | |
| | 7.9×3.6 | K | 247 | 280-13 | 080-005 | SR48 |
| | 7.9×2.6 | V | 612 | 280-52 | | SR59 |
| | 7.9×2.1 | X | | 280-53 | 080-032 | SR58 |

be addressed via cathode modifications and gas diffusion selectivity. Adding catalysts to the cathode and inventing optimum semi-permeable membranes for selective gas diffusion will be the key to product success.

**Silver battery optimization**
Work on future silver battery development will concentrate on alternative mix formulations to combat the potential rise in silver prices. Also, lower current requirements will result in smaller cell sizes capable of operating watches for many years before replacement.

## 3.4 LARGE ZINC–AIR BATTERIES
(V. H. Vu)

### 3.4.1 History

Early recognition of atmospheric oxygen as an effective depolarizer in cells of various kind was discovered shortly after Volta's discovery of his famous pile (see Chapters 1 and 2). Van Marum in 1801 noticed the increase in energy of his pile when it was placed in an atmospheric oxygen environment, and Leclanché obtained better results with his battery when the upper portion of the maganese dioxide and carbon mixture was moist, but without being entirely wet. This was obviously the effect of air depolarization [61].

The search for an effective oxygen electrode had been carried out since the early nineteenth century, but it was not until 1920 that independent investigations in Europe and in the United States led the way to commercial success [64], Heise producing a workable zinc–air battery in 1932. The air-depolarized battery was developed originally for railway use, but early applications also included electric fences and the operation of radio sets in places where A.C. power lines were not yet available.

The air-depolarized primary battery of today is available in a wide range of sizes, and is adaptable to many new applications such as the miner's lamp, hazard warning lights, off-shore and on-shore markers, lighted buoys, as aids for navigation, and railway signal installations.

The air-depolarized primary battery is ideal for applications requiring low to moderate current output at a stable voltage. A long operating life and low cost are also of importance.

The market need for large air-depolarized primary batteries is not as strong today as in the past, and it has been declining for at least a decade owing to the availability of A.C. power. There are four known major manufacturers of large air-polarized primary batteries around the World today, sharing a market estimated to be near 20 million US dollars. It is expected that the need for large air-depolarized primary batteries will continue to decline, their continued operation being limited to remote locations and applications which are routinely affected by adverse weather or transmission line troubles.

### 3.4.2 General characteristics

The zinc–air depolarized primary cell consists of a caustic electrolyte, an anode of zinc metal, and a block of carbon capable of using oxygen from the air. The theoretical voltage of this cell is 1.64 V under standard conditions, and the observed open circuit voltage is in the range 1.45 V to 1.52 V. The cells are usually intended for low current drain. In commercial applications, discharge rates in the range $0.0001\,C$ to $0.001\,C$ are normally used, where $C$ is the rate capacity of the cell. They have the advantage of being relatively cheap, have a flat voltage-time curve, and a high energy density.

### 3.4.3 Cell chemistry

The general process occurring in the cell is described in Fig. 3.48. Oxygen is activated

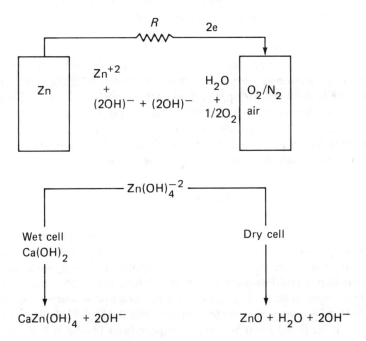

Fig. 3.48 — Zinc–air cell chemistry.

at the carbon cathode with electrons from the external circuit and water from the electrolyte to form hydroxyl ions as a net reaction, given by:

$$O_2 + 2H_2O + 4e^- = 4OH^- \ . \tag{3.27}$$

The hydroxyl ions move through the alkaine electrolyte to the anode, where zinc is oxidized with the release of electrons to the external circuit, according to

$$Zn+4OH^-=Zn(OH)_4^{-2}+2e^- . \qquad (3.28)$$

In a wet alkaline cell, the zincate ion diffuses away from the anode and migrates to a lime bed. It reacts there with the calcium hydroxide to form zincate of low solubility, and it releases hydroxyl ions back to the bulk electrolyte by the reaction

$$Ca(OH)_2+Zn(OH)_4^{-2}=CaZn(OH)_4+2OH^- . \qquad (3.29)$$

The described processes continue so long as the oxygen is supplied to the cathode and there is sufficient available lime to remove the zincate reaction products. When the lime is fully exhausted, the zinc consumes the caustic, the electrolyte alkalinity decreases and the zinc becomes passive or goes into a low performance mode. Thus, the design-limiting feature of commercial wet cells is the availability of the electrolyte.

The dry version of the zinc-air depolarized primary cell is made with gelled electrolyte. It was developed to meet the energy needs already provided for in various applications by dry saline cells. It has the advantage of being spill-proof, therefore is favoured for applications such as offshore lighted aids for navigation. Similarly to the wet cell, zinc metal is oxidized and goes into the electrolyte as zincate ions, thus reducing the effective concentration of alkali in the electrolyte. When the saturation limit is reached, zincate precipitates as zinc oxide, at the same time adding hydroxyl ions back to the electrolyte by the process

$$Zn(OH)_4^{2-}=ZnO+2OH^-+H_2O \qquad (3.30)$$

The porous and amorphous zinc oxide will continue to separate from the zinc surface until the gelled electrolyte void between zinc particles becomes filled. Thus, in this case, there is no change in electrolyte composition during and after discharge. The zinc air-depolarized primary cell made with gelled electrolyte gives much greater energy output in term of electrolyte usage per ampere-hour of capacity. The values are 8–10 ml per ampere-hour for cells without the lime, 2–3 ml per ampere-hour for cells with the lime, and 0.5–1 ml per ampere-hour for the gelled type cells [67].

### 3.4.4 Cell components

#### 3.4.4.1 *Carbon cathode*

During discharge of the cell, oxygen from atmospheric air is reduced at the air breathing carbon cathode [65], according to the equation

$$O_2+2H_2O+4e^-=4OH^- , \qquad (3.31)$$

this representing the net overall reaction. In reality, peroxide ions are formed during an intermediate step. At the exposed surface of the porous carbon electrode the oxygen partial pressure is 0.2 atm. Progressing into the carbon block, the oxygen partial pressure decreases and reaches a minimum at the gas/electrolyte interface. It is this decrease in oxygen partial pressure which provides the driving force for oxygen diffusion. At any point within the carbon block, the sum of the oxygen and nitrogen partial pressures is constant and equal to the outside atmospheric pressure. A three phase contact zone between gas, electrolyte solution, and solid carbon exists in the outer 0.5 to 1.5 mm of the carbon block submerged in the electrolyte. From the chemical reaction equation (3.31) it is clear that there has to be such a region for the cell to operate: oxygen being the gas, water and hydroxyl ions being present in the electrolyte, and the electrons travelling through the conductive medium of the carbon.

Except for the wet outer skin region, the rest of the cathode is water repellent, the dry interior providing tunnels through which the oxygen can diffuse.

Carbon cathodes of a variety of configurations and compositions have been developed, ranging from simple moulded blocks to the 'high tech' versions of today. The latter are cathodes which are made with porous polymer, the electrode thickness varying from a few hundred $\mu$m to several centimeters, depending on the type of cell.

Many materials and processes have been used in developing a workable carbon cathode capable of using oxygen from the air, and two commercial processes are currently used to produce carbon cathodes for large air depolarized cells. One process uses a high temperature to carbonize the binder into a porous structure, and, in the other, a low temperature is preferred which is used to fuse thermosetting resin together with absorbent type carbon particles. These two processes produce distinctive non-equivalent structures. In the fired electrode, the pore structure is developed through thermal decomposition at a temperature greater than 900°C, and it consists of a dispersed soft carbon structure linking the petroleum coke and carbon black grains together by continuous carbon bridges, thus guaranteeing a material with low electrical resistance. The extent of sintering of the soft carbon structure will largely define the ultimate strength of the block, and the chemistry of the decomposition and erosion process during firing is complex and rather poorly understood.

The pore structure in the plastic-bonded electrode is formed by a combination of the interstices within a packed bed with the natural porosity of the carbon particles. The particle bed is bonded together by a thermosetting plastic which is distributed as small individual particles. The electrical conductivity of this structure will depend upon a combination of the carbon particle conductivity and the number of the carbon to carbon particle contacts which can be maintained. However, the mechanical strength of the structure depends upon the continuity of the plastic bonding contacts existing throughout, the pore structure consisting of interconnecting interstices within and between particles.

The void volume of the plastic-bonded electrode can be varied by mixing carbons which are either porous in themselves or consist of aggregates capable of functioning as single particles during the fabrication step. The formation of these electrodes must be accomplished without disintegrating the porous particles, and this is part of the art of making successful air cathodes.

The mechanical strength of the structure depends largely on the number and the strength of interparticulate bonds, and, for the plastic-bonded electrode, would depend upon several factors: the adhesion of the polymer to the carbon, the degree of interlocking, as well as the degrees of polymerization and plastization of the polymer and the extent of polymer flow during fusion. Thus, the mechanical strength can be altered by varying the carbon mix composition, the plastic content, the extent of plasticization, and the fusion temperature.

With regard to the electrochemical performance of each type of block, the fired electrode with its sequence of series-parallel uniform bore capillaries would appear to have some advantage due to its low resistance. However, the series–parallel non-uniform bore capillaries of the plastic bonded electrode provide a wide range of surface tension forces, thus limiting the absorption of electrolyte into the block more effectively than the fixed structure. The distribution of voids inherent to the structure of the plastic-bonded electrode also makes it more difficult to form dead pores as often seen in the fired type.

Both types of electrodes use absorbent granular carbon such as charcoal or petroleum coke agglomerated either with a carbonizable pitch binder for the fired type, or with a thermosetting resin for the plastic-bonded type. Electrical conductivity and porosity of the cathode are improved by addition of graphite or carbon black. The processing steps are similar for both types, and include:

(i) blending,
(ii) size reduction and screening,
(iii) extruding and sintering.

The basic advantages of the plastic-bonded electrode compared to the fired type are:

(i) Elimination of the high temperature firing step, thus eliminating the decomposition product disposal problem.
(ii) Provision of a method for continuous production with a minimum process holdup time between steps.
(iii) Provision of methods for changing block sizes without major interruption of production schedule.

The pore size distribution and characteristic of typical carbon cathodes used in large commercial zinc–air depolarized cells are shown in Fig. 3.49 and Table 3.14.

The fired type of electrode is characterized by good conductivity, though it may have permeability limitations due to the presence of dead pores. At first glance, the resistivity may show the plastic-bonded electrode to be at a disadvantage compared to the fired structure, but in reality this parameter is not critical. The block resistivity is controlled by the number and area of carbon to carbon contacts, whereas porosity and pore size would tend to minimize the particulate contact. Similarly, the mechanical strength depends on the number and distribution of bonded regions, whereas increased porosity would tend to minimize that.

The performance of air-depolarized carbon cathodes is generally limited by various parameters operating singly or in combination. These are [63]:

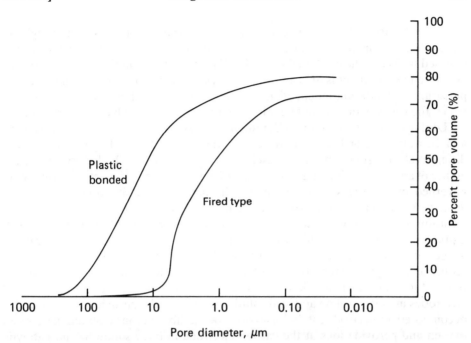

Fig. 3.49 — Pore size distribution in commercial air cathodes (Edison Batteries Laboratory data).

Table 3.14 — Physical characteristics of large air cathodes [67]

|  | Fired | Plastic bonded |
|---|---|---|
| Apparent density (g.cm$^{-3}$) | 0.82–0.86 | 0.84–0.87 |
| Porosity (%) | 55–60 | 45–55 |
| Permeability (cm$^3$ of air.cm$^{-3}$ min$^{-1}$) | 25–10 | 30–100 |
| Resistivity (ohm-cm) | 0.03–0.05 | 0.10–0.30 |

(i) Catalytic activity for oxygen reduction.
(ii) Catalytic activity for peroxide decomposition.
(iii) Electrical resistivity.
(iv) Depth of the wetted region.
(v) Gas flow characteristics of the block.

For high current performance, all parameters should be optimized, although the electrode performance is limited by the rate at which oxygen can diffuse from the exposed surface of the block to the interfacial area where it is utilized. Therefore, gas flow rate measurement is of importance and can be used to define the high current performance limitation independent of other parameters. For a low current electrode, the gas flow rate should be minimized to reduce electrolyte evaporation.

During operation, pressure differentials are developed between the atmosphere and the wetted region because of the utilization of oxygen in the latter. The pressure differentials are stable at steady discharge currents, but the high current performance of an electrode is limited by the rate of flow of air into the electrode, this being governed by the pressure differential. Alternatively, the situation may also be described as a limitation imposed by the necessity to maintain a counter nitrogen diffusion current out of the electrode. In pure oxygen, the flow of oxygen under such a pressure differential would be directly proportional to the current. The rate of oxygen utilization in the wetted zone is related to the catalytic activity for the reduction of oxygen and peroxide. It has been known for some time that oxygen is reduced to hydrogen peroxide with almost 100% current efficiency at noble metal electrodes in both alkaline and acid solutions, and that the peroxide so produced is decomposed by electrode catalytic activity. An equilibrium can be obtained between oxygen and peroxide ions at the carbon electrode in basic solution, and catalytic decomposition of peroxide is likely to be rapid. This equilibrium is given by:

$$\tfrac{1}{2}O_2 + H_2O + 2e^- = OH^- + HO_2^- \ . \tag{3.32}$$

The rate constant for the peroxide ion decomposition is shown in Table 3.15. It appears to be first order over a wide range of peroxide concentrations. When a cell is

**Table 3.15** — Rate of decomposition of a peroxide ion in a discharging air-depolarized cell [62]

| Current density $(mA.cm^{-2})$ | $HO_2^-$ $(mole.l^{-1})$ | Rate constant $(s^{-1})$ |
|---|---|---|
| 0.33 | $1.77 \times 10^{-4}$ | $29 \times 10^{-4}$ |
| 3.9 | $2.14 \times 10^{-3}$ | $28 \times 10^{-4}$ |
| 6.2 | $3.98 \times 10^{-3}$ | $24 \times 10^{-4}$ |
| 13.8 | $1.15 \times 10^{-2}$ | $19 \times 10^{-4}$ |

run at various rates of discharge, the concentration of peroxide increases, but the rate constant for the production and the decomposition remains the same. When pulverized carbon or traces of copper are added to the solution, the rate of decomposition increases several-fold. A typical set of data collected for a model air-depolarized cell with NaOH electrolyte is shown in Fig. 3.50.

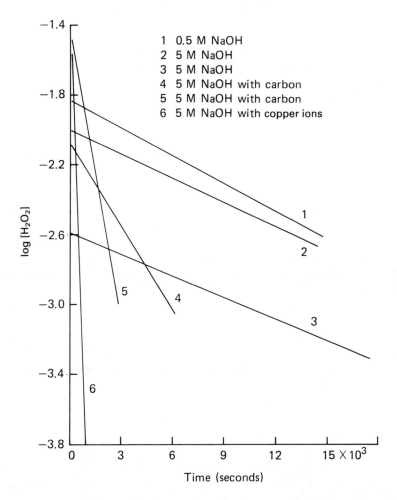

Fig. 3.50 — Rate of decomposition of hydrogen peroxide.

Present day air cathodes incorporate a peroxide decomposition catalyst, various catalytic metals being used. The catalysis of hydrogen peroxide decomposition by manganese has often been observed and reported. The mechanism proposed to account for this catalytic activity involves free radical intermediates, and such a mechanism has also been suggested for reactions involving permanganate. However, the formation of manganese dioxide in the reaction of permanganate with hydrogen peroxide is well known, and it is possible that it is this manganese dioxide which is the true catalyst, the mechanism involving an oxidation-reduction cycle such as [62]:

$$MnO_2 + H_2O_2 + 2H^+ = Mn^{2+} + 2H_2O + O_2 \qquad (3.33)$$

$$Mn^{2+} + 2H_2O = Mn(OH)_2 + 2H^+ \tag{3.34}$$

$$Mn(OH)_2 + H_2O_2 = MnO_2 + 2H_2O . \tag{3.35}$$

At high current density and in the absence of a specific decomposition catalysts, the concentration of peroxide in the bulk electrolyte will tend to increase, and non-electrochemical oxidation of the zinc will occur under these conditions, leading to lower current efficiency for the cell by the reaction equation:

$$Zn + H_2O_2 = Zn(OH)_2 \tag{3.36}$$

The electrochemical activity of air cathodes is considered to be concentrated in the wetted regions of the electrode submerged in the electrolyte. The electrode polarization is proportional to the current, and the electrode current is proportional to the concentration of the reactants and products in the wet zone. The practical limits of electrode operation arise from variations in peroxide decomposition efficiency and the rate of oxygen migration, firstly through the porous structure in the gas phase, then across the gas/liquid barrier, and finally through the liquid to the reaction sites.

Gas flow to the reaction sites depends upon the presence of pressure differentials between the wetted region and the atmosphere. This can be increased or decreased by changing the pore size within the electrode. It has been suggested that pore size in the range of 0.1–1 $\mu$m operate predominantly by a Poiseuille's viscous flow model, while those below 0.1 $\mu$m operate by a Knudsen molecular flow model [63]. Therefore, if the pore size is mostly in the smaller size range, the establishment of a pressure differential would not increase the gas flow. If short diffusion paths are used to increase the net gas flow rate, the formation of a large pressure differential can lead to increased absorption by the electrolyte.

A model of the wetted region is presented in Fig. 3.51. The model shown consists of a series of carbon grains forming a normal type of section. At the interface in the wetted zone, there are portions of the grains located in the gas phase and other portions in the electrolyte.

Generally, the electrolyte reaction is described as the sorption of oxygen on carbon followed by its reaction with water to form the hydroxyl ion and hydrogen peroxide ion. The hydrogen peroxide is subsequently removed from the surface region by diffusion, by electrolytic transport, or by decomposition to oxygen and a hydroxyl ion. A more detailed description of the reaction sequences starting from atmospheric oxygen would be as follows [63]:

$$O_2(g)(air) = O_2(g)(carbon/gas/electrolyte\ interface) \tag{3.37}$$

$$O_2(interface) = O_2(dissolved) \tag{3.38}$$

$$O_2(dissolved) = O_2(adsorbed) \tag{3.39}$$

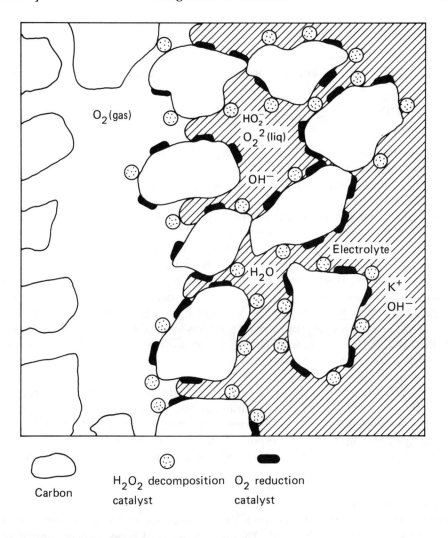

$[H^+] + [K^+] = [HO_2^-]$
$[O_2] = \text{constant} \times [HO_2^-]$

Fig. 3.51 — Model of wetted region of the carbon electrode [63].

$$O_2(\text{adsorbed}) + 2e^- = O_2^{2-}(\text{adsorbed}) \quad (3.40)$$

$$O_2^{2-}(\text{adsorbed}) + H_2O = OH^- + HO_2^-(\text{wetted region}) \quad (3.41)$$

$$HO_2^-(\text{wetted region}) + \text{catalyst} = OH^- + \tfrac{1}{2}O_2(\text{dissolved}) \quad (3.42)$$

$$O_2(\text{dissolved}) = O_2(\text{gas}) \tag{3.43}$$

$$HO_2^- + OH^- (\text{wetted region}) = HO_2^- + OH^- (\text{bulk}) \tag{3.44}$$

$$HO_2^- (\text{bulk}) + \text{catalyst} = \tfrac{1}{2}O_2 + OH^- \tag{3.45}$$

The peroxide decomposition reaction produces oxygen which raises the overall electron yield from 2 to 4 electrons per oxygen molecule. The rate of peroxide decomposition reaction controls the ratio of $O_2$ to $HO_2^-$ in the wetted region, thereby determining the electrode potential. The term 'electrode activation', frequently referred to in the literature, is associated with increasing this reaction rate, thereby improving the overall electrode performance.

Oxygen gas is formed from the peroxide decomposition reaction, thus, if the rate of reaction is sufficiently large, gas bubbles would force electrolyte into the interior part of the block, leading to a serious soaking problem. This condition can be a significant limitation during operation at high current densities, owing to the generation of high peroxide concentrations near the gas/electrode interface.

As the cell discharges, the reaction products must be transported away from the wetted region. The increased peroxide and hydroxide concentration helps to establish a concentration gradient to enable transportation, although this causes a slight drop in electrode potential.

It is interesting to note the the generation of hydroxyl ions in the wetted region of the cathode should raise the surface tension of the electrolyte, thereby decreasing the rate of wetting. Peroxide also decomposes on the carbon surface in the bulk electrolyte, and this process is responsible for gas bubble formation on the exterior surface of the carbon electrode, this providing a visual indication of the efficiency of the catalyst.

### 3.4.4.2 Electrolyte

The physical properties of alkaline hydroxide solution with and without dissolved zinc are shown in Fig. 3.52 to 3.55. The electrolytes used in zinc–air cells are alkaline hydroxides, and, in the early days, sodium hydroxide was used owing to its lower cost and the greater zinc solubility in that electrolyte. Today, potassium hydroxide is used as it gives better cell performance. Initially, sodium hydroxide was used in the range of 5 M to 6.5 M as, at these concentrations, satisfactory low temperature service was achieved and zinc solubility was high enough to ensure adequate capacity for the normal range of operation. All wet air depolarized primary cells are designed with capacity limited by the exhaustion of the electrolyte and, in older cells, the volume of electrolyte was sufficiently large to ensure that zinc oxide would not precipitate on the electrodes and thus cause operating problems.

The solubility of zinc oxide in potassium and sodium hydroxide solutions is shown in Figs 3.56 and 3.57.

Potassium hydroxide solutions saturated with zinc oxide can accept more zinc through an electrochemical oxidation process, and a zinc concentration of up to four

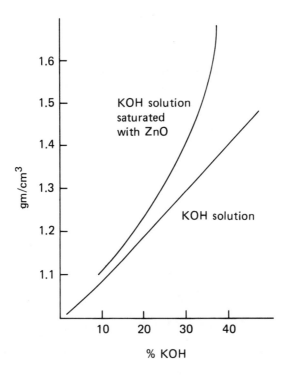

Fig. 3.52 — Density of KOH solution at 25°C [62].

times the chemical saturation is possible in discharged cells [62]. The solutions will precipitate zinc oxide and hydroxide with time, and the final concentration will evidently reach the chemical saturation point, the rate of precipitation of oxide depending on the zinc ion and hydroxide ion concentrations.

If a zinc–air cell is operated with a liquid electrolyte of 5 M KOH or 5 M NaOH, 7–8 ml of electrolyte must be provided for each Ah of rated capacity [66]. This is a large volume for a cell size of interest, therefore it is important to reduce the electrolyte volumetric requirement. Lime (calcium hydroxide) is essentially insoluble in hydroxide solution and can react slowly with the zincate ion to regenerate the hydroxyl ion [62] according to equation (3.29), and, in theory, 1.00 g of lime is capable of removing 0.88 g of zinc from the solution. This is equivalent to 0.72 Ah of battery operation. In large air-depolarized cell, lime is used in a quantity of $0.5 \, \text{g} \, \text{l}^{-1}$ of electrolyte to reduce the electrolyte requirement from 8 ml to 2 ml per Ah [66]. Thus, higher cell energy densities can be realized.

Granulated lime with high surface area is generally preferred in large air-depolarized primary cells. Table 3.16 shows some important characteristics of the lime granules used in large commercial zinc–air cells.

Zincate absorption by particles of lime could be increased by using a smaller particle size, but the formation of encrusted layers of crystalline calcium zincate would interfere with the diffusion of the used electrolyte into the unconverted lime.

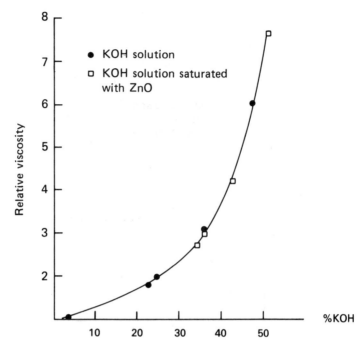

Fig. 3.53 — Relative viscosity of KOH solutions at 25°C [62]. Ratio of absolute solution viscosity to that of water at 25°C (0.8904 cP).

Therefore, an optimum particle size distribution must be selected for the chosen lime bed depth. When the lime has been used up, the solution becomes saturated with discharge products, and zinc oxide begins to precipitate according to equation (3.30).

Operation in the zinc oxide mode is less certain than when unreacted lime is available to clean the solution. Depending on discharge rate and temperature, the zinc oxide reaction may not continue until the total quantity of zinc metal is oxidized, as a layer of zinc oxide can form on the zinc anode and cause a sudden failure.

The electrolyte regeneration, in an air-depolarized cell with a gelled electrolyte is very different from that of the wet cell. Zinc oxide is precipitated from zincate directly, giving the overall cell reaction:

$$2Zn + O_2 = 2ZnO .\tag{3.46}$$

Much greater output in terms of electrolyte usage per Ah has been observed for dry air-depolarized cells with gelled electrolyte. Values in the range of 0.3–0.5 ml per Ah are common [67]. The electrolyte shows no composition change during, and even after, the discharge. Porous and amorphous blue to white coloured zinc oxide continuously separates from the zinc particle surface until the electrolyte cavity between zinc particles is filled. High gravimetric energy densities up to 330 Wh.kg$^{-1}$ are possible with gelled electrolyte air-depolarized cell. The comparison of performance of various commercial air-depolarized cells is shown in Table 3.17.

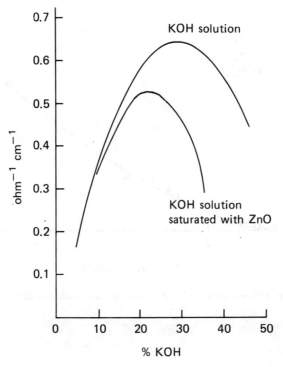

Fig. 3.54 — Specific conductance of KOH solution at 25°C [62].

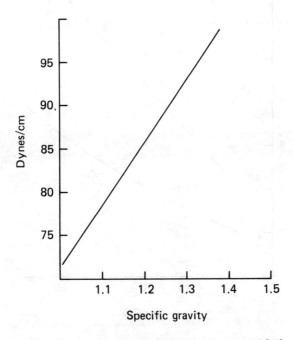

Fig. 3.55 — Surface tension of KOH solution at 25°C [62].

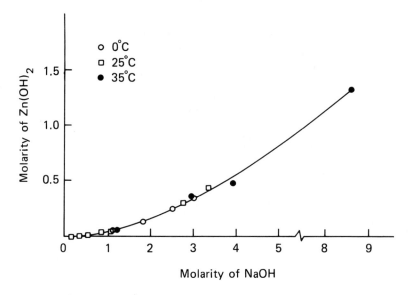

Fig. 3.56 — Zinc solubility in sodium hydroxide solution [62].

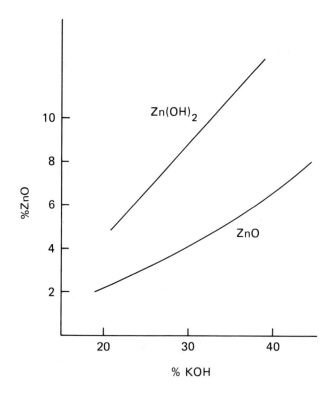

Fig. 3.57 — Solubility of ZnO and Zn(OH)$_2$ in KOH solution [62].

**Table 3.16** — Physical characteristics of granulated lime for large wet zinc–air depolarized cell (from Edison Batteries laboratory data)

| | |
|---|---|
| Absorption ($cm^3.g^{-1}$) | 0.65–0.85 |
| Bulk density ($g.cm^{-3}$) | 0.50–0.40 |
| Particle size distribution 6–25 mesh | +95% |

**Table 3.17** — Performance characteristics of commercial air-depolarized cells [70], [71], [72]

| | Nominal C.C.V. (V) | Maximum continuous current (A) | Rated capacity (Ah) | Energy density | |
|---|---|---|---|---|---|
| | | | | $Wh.cm^{-3}$ | $Wh.kg^{-1}$ |
| Dry saline | 1.1 | 1.0 | 1200 | 28 | 115 |
| Wet alkaline | 1.2 | 1.0 | 1000 | 55 | 170 |
| Dry alkaline | 1.2 | 1.0 | 1200 | 86 | 275 |

### 3.4.4.3 Zinc anode

The standard potential and pH diagram of zinc, hydrogen, and oxygen electrodes are shown in Fig. 3.58. The potential of the zinc electrode depends on many factors.

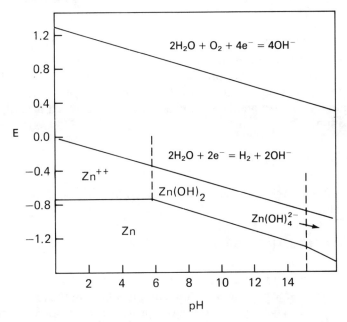

Fig. 3.58 — Potential-pH diagram for zinc electrode at 25°C [62].

These are the composition and metallurgical history of the zinc itself, the type and concentration of the electrolyte, and the type, quantity, and age of dissolved species of zinc in the electrolyte. Potential differences as great as 50 mV have been observed between single crystal and polycrystal zinc or cold worked and annealed zinc [62].

Cast zinc anodes can support very high current density before passivation occurs at 30 to 50 mA.cm$^{-2}$ [68]. During most of the discharge, zinc loses weight according to the theoretical 2e$^-$ oxidation of equation (3.28). As the caustic electrolyte becomes heavily loaded with dissolved zincate ions toward the end of the discharge, the zinc electrode goes into a passive mode indicated by a sharp drop in voltage. It is found that such passivated zinc anodes can be reactivated by placing them in a fresh electrolyte solution.

A typical chemical analysis of zinc used in commercial air-depolarized cells is given in Table 3.18. Apart from the chemical composition, the physical characteristics which are important to the battery include the grain size and the surface condition.

**Table 3.18** — Chemical composition of zinc anode for wet air-depolarized alkaline cells (from Edison Batteries laboratory data)

| | |
|---|---|
| Pb | <0.005% |
| Al | <0.001% |
| Fe | <0.002% |
| Cu | <0.005% |
| Sn | <0.001% |
| Zn | >99.95% |

When zinc metal is in contact with caustic electrolyte as in an air-depolarized cell, the dissolution of zinc and the evolution of hydrogen gas occur simultaneously. This is understood as the self-discharge process (see section 2.2.5.1). Zinc metal corrodes on open circuit stand by the reaction given by equation (3.28).

At the cathodic or less anodic sites hydrogen gas is evolved according to the reaction equation.

$$2H_2O + 2e^- = H_2 + 2OH^- . \tag{3.47}$$

The overall self-discharge reaction can be obtained by combining equations (3.28) and (3.47), giving

$$Zn + 2OH^- + 2H_2O = Zn(OH)_4^{-2} + H_2 . \tag{3.48}$$

Corrosion is detrimental to the battery operation not only because of the resulting loss of discharge capacity with time, but, in addition, the need to incorporate a hydrogen vent makes the battery more prone to deterioration through the evaporation of the electrolyte solvent.

Various alloying elements can be added to the zinc to reduce its self-corrosion, but they tend to affect the electrode potential. Several experimental zinc compositions with indium, gallium, lead, or cadmium have been tried, but amalgamation of the zinc with mercury is still the most common practice. Formerly, electrodes have contained as much as 4% mercury, but economic and environmental concerns have forced battery manufacturers to lower the mercury content. Cold working the zinc to produce grain sizes less than 10 $\mu$m has also been found to reduce the self-discharge. This has been used in commercial zinc–air depolarized cells [69].

#### 3.4.4.4 Miscellaneous materials
Brass and nickel plated copper have been found to be suitable materials for the current collectors, and zinc or nickel-plated steel is generally preferred for the external hardware.

### 3.4.5 Cell design and construction details
Large air-polarized primary batteries were originally developed for railway use and have been found to be ideal for applications requiring frequent or continuous use at low drain rates. Both wet and dry cells are available commercially.

#### 3.4.5.1 Wet air-depolarized batteries
Primary batteries of this type are manufactured in models ranging from one thousand to several thousand Ah [70]. All intercell connections are external, and the cells can be connected in series or parallel. Fig. 3.59 shows construction details of a preactivated two cell unit with nominal rating of 1000 Ah at 2.50 V. The lime granules are held inside a perforated plastic cup which keeps them from disintegrating during transportation. This type of battery is preactivated by the manufacturer and shipped wet to the users.

The wet cell type is manufactured with a non-spillable vent cap or with a reservoir top which can hold the electrolyte temporarily when the battery is in a tilted position. This is necessary to ensure that there is no spillage of the electrolyte in applications requiring the battery to withstand vigorous physical motion. Air-depolarized batteries with this built-in feature are often selected for off-shore lighted buoys used as aids for navigation. This type of battery can also be manufactured dry to ensure complete freedom from deterioration during storage. Details of construction of a two cell unit with nominal rating of 1100 Ah at 2.50 V are shown in Fig. 3.60. The batteries are shipped dry from the factory with removable adhesive tape covering the vent holes. The dry caustic is readily cast or pressed into a block, using granular flakes, and is placed into the cells during their manufacture. The user then needs to add water to the fill-line to activate the battery, the dry caustic dissolving rapidly without stirring. During this process, the heat of solution may cause the electrolyte temperature to rise as high as 90°C, and there is a noticeable shrinkage in electrolyte volume as dissolution proceeds, an electrolyte top-up normally being required before the battery is put into service.

This dry construction offers many years of shelf life without any performance deterioration. Discharge characteristics of wet air–depolarized primary cells are shown in Figs 3.61 to 3.64.

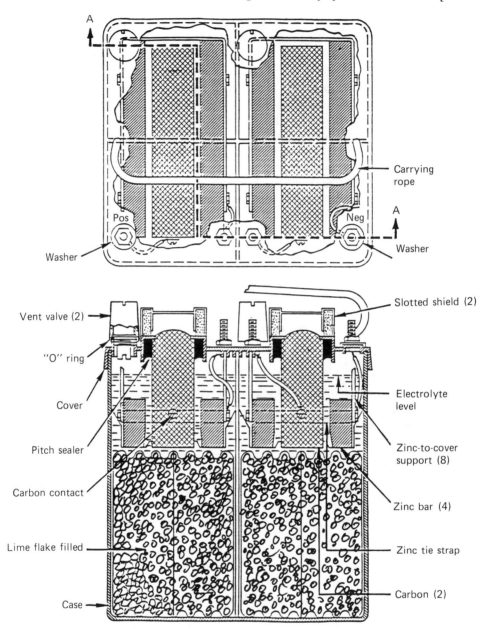

Fig. 3.59 — Construction of a deactivated wet zinc–air depolarized battery. Two cells can be connected in series (1000 Ah at 2.5 V) or parallel (2000 Ah at 1.25 V).

Wet air-polarized primary cells require very little maintenance or inspection, but when the service life is well over a year, an annual inspection of the electrolyte level is normally recommended. In hot and dry climates, more frequent inspection may be necessary to avoid the solution level dropping below that of the zinc.

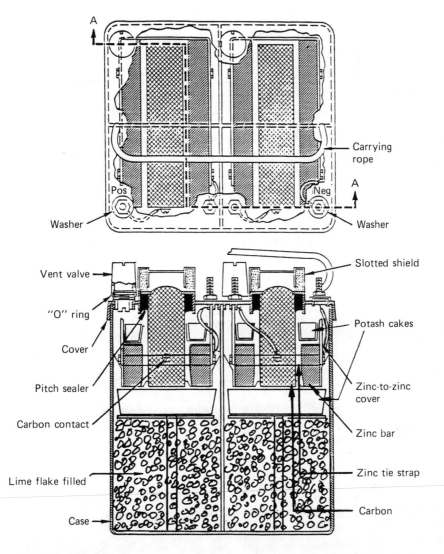

Fig. 3.60 — Construction of an *unactivated* (user activated) wet zinc–air depolarized primary battery. Two cells can be activated with tap water and connected in series (1100 Ah at at 2.5 V) or parallel (2200 Ah at 1–25 V).

### 3.4.5.2 *Air-depolarized batteries with gelled electrolyte*

The so-called dry air-depolarized batteries are manufactured with gelled electrolyte [71,72]. They were developed to meet the energy needs provided in various applications by wet air-depolarized cells and large dry saline cells. The dry saline cell is preactivated with immobilized electrolytes, therefore it is safe and convenient for the user, but it has capacity limitations and undesirable discharge characteristics. The wet air–depolarized cell offers high capacity per unit volume and desirable

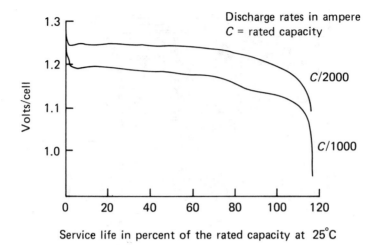

Fig. 3.61 — Typical life of wet air-depolarized cell (Edison Batteries Laboratory data).

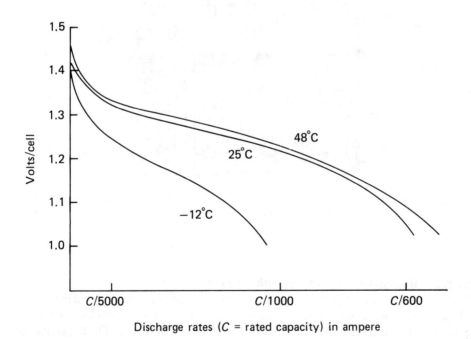

Fig. 3.62 — Current–voltage characteristics of wet air-depolarized cell at various temperatures (Edison Batteries Laboratory data).

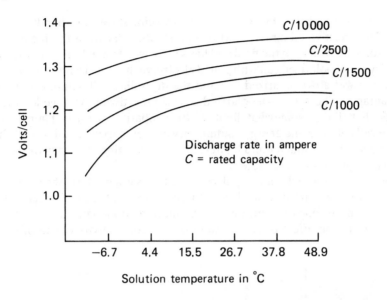

Fig. 3.63 — Voltage characteristics of wet air-depolarized cell at various temperatures (Edison Batteries Laboratory data).

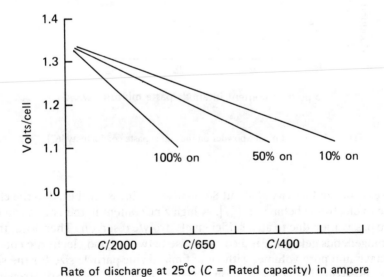

Fig. 3.64 — Intermittent discharge characteristics of wet air-depolarized cell (Edison Batteries Laboratory data).

discharge characteristics, but the user has to ship, store, and handle a corrosive liquid. The unactivated air-depolarized cell shipped dry from the factory requires troublesome activation in the field and has severe weight and bulk penalties. The so-called dry air-depolarized cell with gelled electrolyte has the desirable characteristics of both the wet air-depolarized cell and the dry saline cell without their inherent disadvantages. The dry air-depolarized cells is preactivated with gelled electrolyte thus its shelf life is somewhat limited. In comparison with wet air-depolarized alkaline cells of the same energy content, dry air-depolarized alkaline cells have the advantages of being smaller and lighter, although the zinc efficiency of the anode is limited at high rates of discharge.

The anode in these dry air-depolarized batteries is a mixture of zinc powder and gelled electrolyte. The mixture has both electronic and ionic characteristics, and depending on the zinc content, one can be more predominant than the other. Fig. 3.65 illustrates the effect of zinc content on the conductivity of the mixture. The

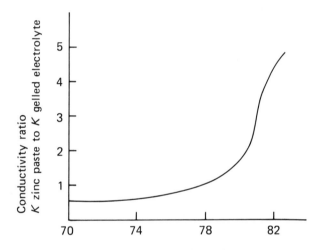

Fig. 3.65 — Effect of zinc powder on the anode paste conductivity [67].

sharp increase in conductivity at about 80% (w/w) of zinc is attributed to the change from ionic to electronic behaviour [67]. A high zinc content in the anode can cause premature passivation due to an insufficient electrolyte reservoir. Therefore, the so-called homogeneous gel anode is a compromise between good electronic conductivity and maximum pore volume, with an optimized zinc particle size for the size of electrolyte reservoir. The concentration of gelling agent and the caustic strength are also important parameters for obtaining satisfactory efficiency [67].

Chemical and physical characteristics of a commercial zinc powder used in air-depolarized cell with gelled electrolyte are shown in Table 3.19.

**Table 3.19** — Chemical and physical characteristics of zinc powder used in commercial dry air depolarized cells with gelled electrolyte [67]

| | |
|---|---|
| Pb | <0.06% |
| Fe | <0.002% |
| Cu | <0.001% |
| Cd | <0.002% |
| ZnO | <0.5% max |
| Total Zn | >99.9 |
| Particle size distribution | |
| +30 | 0.5 max. |
| −30+60 | 35–60 |
| −60+100 | 30–45 |
| −100+150 | 10–25 |
| −150 | 2.0 max. |

Cellulose derivatives and starch have been found to be effective thickening agents for caustic electrolytes. Reticulated starch is commonly used in commercial air-depolarized cells. This type of starch is prepared by treating natural ungelatinized starch granules with a cross-linking agent to block the hydroxyl groups, thus increasing the chemical stability of the starch. The effectiveness of this treatment is dependent on the number of hydroxyl group blocked and the nature of the cross-linking bond.

The effect of gelled electrolyte composition on gel stability and viscosity as well as its conductivity and vapour pressure, are illustrated in Figs 3.66 to 3.69. High caustic

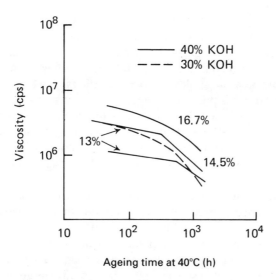

Fig. 3.66 — Stability of gelled electrolyte with time [66]. Percentage levels of starch content are given.

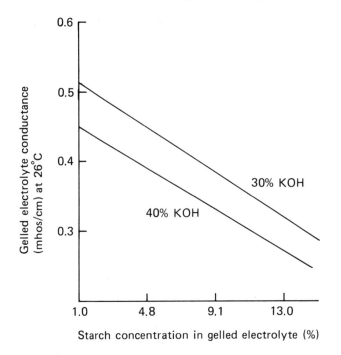

Fig. 3.67 — Effect of gelled electrolyte composition on conductivity [66].

concentrations are generally preferred for the low rate cell, whereas low caustic concentration are good for the high rate cell, where long term stability is not a critical parameter. In smaller air-depolarized cells designed with capacities less than 100 Ah, the air-breathing carbon cathode is pressed between two layers of gelled electrolyte which also serves as the seal for the next two layers of gelled zinc anode which have embedded current collectors. This is illustrated in Fig. 3.70. The cells can be stacked and connected in series or parallel to give the required voltage and Ah capacity. They are banded together and placed inside a rigid housing which has the top perforated for the entry of air. The breathing holes are covered with oxygen barrier tape which is removable when the battery is put into service.

The larger version of the gelled electrolyte zinc–air depolarized primary battery is constructed with the carbon cathode at the centre of the cell. The surrounding spaces are occupied by the gelled electrolyte and the gelled zinc anode paste. The zinc anode is placed next to the inner wall of the cell housing. This design is illustrated in Fig. 3.71.

Depending on the application, several cells can be series or parallel connected through external hardware. These cells are intended for low current drain with capacity ranging from a few hundred Ah to several thousand Ah.

The discharge characteristics of air-depolarized cells with gelled electrolyte are shown in Figs 3.72 to 3.75. Small air-depolarized cells designed with thin zinc anodes and large carbon surface areas can sustain higher rates of discharge than larger cells, with a minimum loss of zinc efficiency.

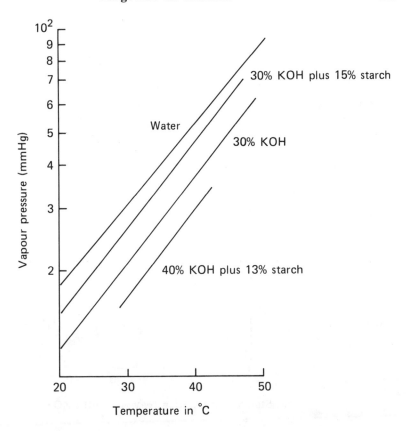

Fig. 3.68 — Effect of gelled electrode compositions on vapour pressure of the mixture [66].

### 3.4.6 Applications

With their flat discharge-voltage characteristics, high energy content, low cost, and high reliability, zinc–air depolarized primary batteries have been widely used in many signal applications. They provide separate and entirely independent power sources which are not impaired by adverse weather, A.C. power transmission line troubles, or other factors which would affect the reliability of most signal systems. Applications of large air-depolarized primary batteries include:

(i) Aids to navigation: shore aids which include all types of shore lights, bell strikers, fog horns, bridge lights, and off-shore lighted buoys.
(ii) Railway signals: track circuits, switch lamp lights, and marker lights.

### 3.4.7 Likely future developments

Today, the requirement for large zinc air-depolarized primary batteries is not great. The market demand has been declining for the past fifteen years owing to the successful development of electronic coding systems and photovoltaics for signalling

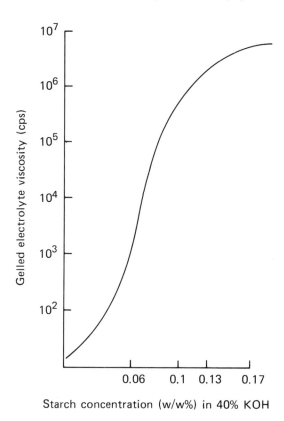

Fig. 3.69 — Effect of starch on gelled electrolyte viscosity [66].

applications. As a result, no major technological advancement is expected for large zinc air-depolarized primary batteries. However, there is growing concern regarding the safe disposal of spent batteries, thus key developments in the immediate future are expected to focus on issues related to the toxic nature of battery discharge products and possible ways to recycle battery materials.

Formerly, zinc anodes in large air-depolarized batteries contained as much as 4% mercury, but over the years, manufacturers have been reducing the level of mercury in the zinc anodes, largely for economic reasons.

Wet zinc air-depolarized batteries made today, which have cast zinc anodes, generally use less than 0.5% of mercury, but dry air-depolarized batteries made with gelled electrolyte and zinc powder still require more then 1% of mercury to have adequate shelf life. This is due primarily to the large surface area of the zinc powder. Zinc alloy compositions containing indium, gallium, lead, and cadmium have been experimented with, but amalgamation with mercury is still the most common practice. Cold worked zinc has been used as a mercury-free anode in wet air-depolarized batteries owing to its lower self-discharge rate. Mercury is very toxic,

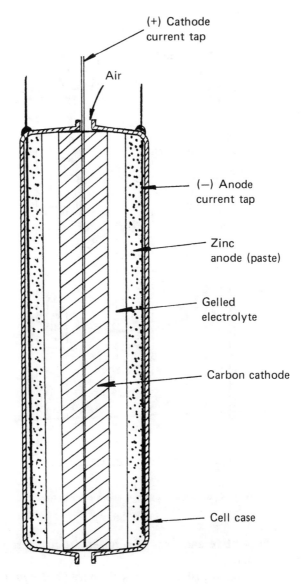

Fig. 3.70 — Cross-section of small zinc–air depolarized cell with gelled electrolyte.

and with the growing knowledge of the environmental hazards of this element, its removal from air-depolarized batteries is considered to be a desirable objective.

## ACKNOWLEDGEMENTS

J. C. Hunter would like to thank Eveready Battery Company for allowing him to contribute to this book. He is grateful to C. M. Langkau for helpful suggestions on

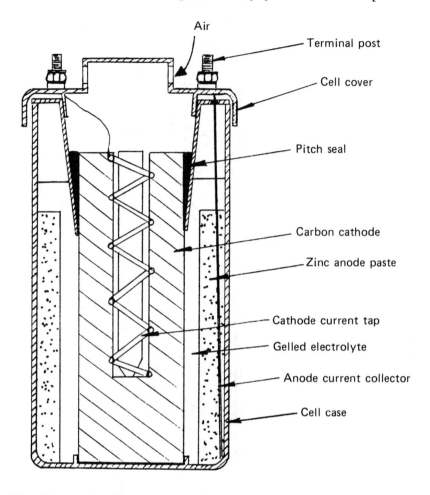

Fig. 3.71 — Cross-section of large zinc–air depolarized cell with gelled electrolyte.

the manuscript. 'Eveready' and 'Energizer' are registered trademarks of Eveready Battery Company, Inc.

E. A. Megahead wishes to thank Bernard C. Bergum and Robert B. Dopp for reviewing his manuscript. Also, many thanks to the Rayovac Corporation for supporting his work and permitting publication of the manuscript.

# REFERENCES

[1] Leclanché, G., *Les Mondes*, **16**, 532 (1868).
[2] George Int'l Consultants, *The World Battery Industry*, (1981).
[3] D. Bradshaw, Duracell Holdings Inc. publication, (1988).
[4] *Hardware Age*, August (1988), 322.

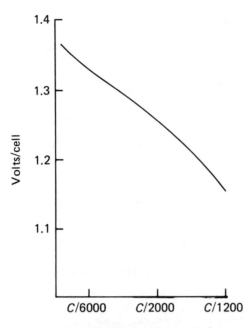

Fig. 3.72 — Current–voltage characteristics of air-depolarized cell with gelled electrolyte.

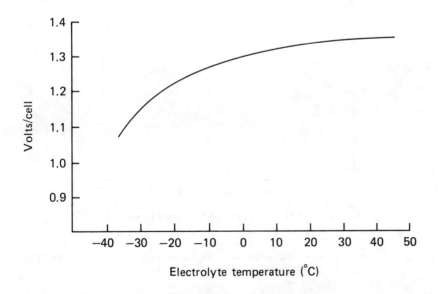

Fig. 3.73 — Variation in cell voltage of an air-depolarized cell with gelled electrolyte at various temperatures [67]. Cell discharged at C/10000.

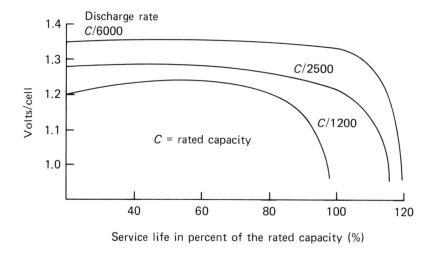

Fig. 3.74 — Service life characteristics of large air-depolarized cell with gelled electrolyte [67].

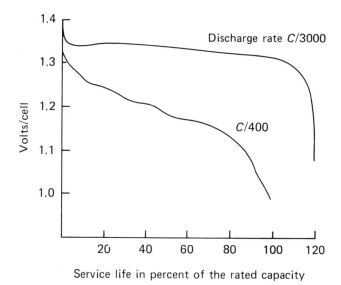

Fig. 3.75 — Service life characteristics of small zinc–air depolarized cell with gelled electrolyte (Edison Batteries Laboratory data).

[5] *JEC Battery Newsletter*, No. 3. (May–June) (1988), p. 8.
[6] Courtesy Duracell Int'l, Inc.
[7] Sales Information, Philips Ltd, Belgium.
[8] Eveready Battery Co, USA.
[9] B. Schumm, Jr., US Patent 4,608,279, August 1986.
[10] Courtesy HiWatt Battery Co., Hong Kong.
[11] Courtesy Eveready Battery Co., USA.
[12] International Standard, Int'l Electrotechnical Commission, IEC 86-1, 6th edn (1987), pp. 18–21.
[13] Schumm, B., Jr., *Proc. 33rd Int'l Power Sources Symp., 13–16 June 1988*, ECS, NJ, p. 728.
[14] Warburton, D. L., *Proc. 17th Annual Power Sources Conf., May 21–23, 1968*, p. 138.
[15] British Ever Ready Co., UK. Retail Battery Packs (1988).
[16] Kozawa, A. & Powers, R. A., *Electrochem. Technol.*, **5** (1967), 535.
[17] Wroblowa, H. S., *Ext. Abst., Fall Meeting*, The Electrochemical Society, Pennington, NJ, (1987), p. 3.
[18] Kordesch, K., Harer, W., Sharma, Y., Ujj, R., Tomantschger, K. & Freeman, D., *Proc. 33rd Int'l Power Sources Symp., June 1988*, The Electrochemical Society, Pennington, NJ, p. 440.
[19] Leclanché, G., French Patent 71,865 (1866).
[20] Gassner, C., German Patents 37,748 (1886); 42,251 (1887); US Patent 373,064 (1887).
[21] de Lalande, F. & Chaperon, G., French Patent 143,644 (1881); US Patent 274,110 (1883).
[22] Ruben, S., US Patents 2,422,045 (1947); 2,576,266 (1951).
[23] Herbert, W. S., The alkaline manganese dioxide dry cell, *J. Electrochem. Soc.*, **99** (1952), 190C–191C.
[24] *Japan Chemical Week*, 3 August (1989) 2.
[25] Bell, G. S. & Huber R., On the cathodic reduction of manganese dioxide in alkaline electrolyte, *J. Electrochem. Soc.*, **111** (1964), 1–6.
[26] Kozawa, A. & Yeager, J. F., The cathodic reduction mechanism of electrolyte, *J. Electrochem. Soc.*, **112** (1965), 959–963.
[27] Dirkse, T. P., Electrolytic oxidation of zinc in alkaline solution, *J. Electrochem. Soc.*, **102** (1955), 497–501.
[28] Ruben, S., US Patent 2,422,045 (1947).
[29] Friedman, M. & McCauley, C. E., The Ruben cell — a new alkaline primary dry cell battery, *Trans. Electrochem. Soc.*, **92** (1947), 195–215.
[30] Kober, F. & West, H., The Anodic Oxidation of Zinc in Alkaline Solutions, *Extended Abstracts*, The Electrochemical Society, Battery Division, Vol. 12 (1967), pp. 66–69.
[31] Fleischer, A. & Lander, J., *Zinc–Silver oxide batteries*, John Wiley, New York (1971).
[32] Latimer, W. M., *Oxidation potentials*, Prentice–Hall, Englewood Cliffs, NJ. (1952).
[33] Dopp, R. B., US Patent 4,617,242 (1986).

[34] Ruetschi, P., The electrochemical reactions in the mercuric oxide–zinc cell, In: Collins, D. H., Ed., *Power Sources*, Vol. 4, Oriel Press, Newcastle (1972), pp. 381–400.
[35] Ruetschi, P., The longest life alkaline primary cell, In: Thompson, J., Ed., *Power Sources*, Vol. 7, Academic Press, London (1978), pp. 533–570.
[36] Putt, R. A. & Attia, A. I., Zinc–air button cell technology, *Proc. 31st Power Sources Symp., 11–14 June (1984)*, pp. 339–349.
[37] Linden, D., *Handbook of batteries and fuel cells*, McGraw-Hill, New York (1984).
[38] Bendor, S. F., Cretzmeyer, J. W. & Hall, J. C., Longlife zinc–air cells as a power source for consumer electronics, In: *Progress in Batteries and Solar Cells*, Vol. 2, JEC Press, Ohio (1979), pp. 63–66.
[39] Cretzmeyer, J. W., Espig, H. R. & Melrose, R. S., Commercial zinc–air batteries, *Power Sources*, Collins, Ed., Vol. 6, Academic Press, London (1977), pp. 269–290.
[40] André, H., *Bull Soc. Franc. Elec.*, **6** (1941), 132.
[41] Megahed, E. A., Small batteries for conventional and specialized applications, *The Power Electronics Show and Conference, San Jose, California, 7–9 October (1986)*, pp. 261–272.
[42] Megahed, E. A., Buelow, C. R. & Spellman, P. J., US Patent 4,009,056 (1977).
[43] Megahed, E. A. & Buelow, C. R., US Patent 4,078,127 (1978).
[44] Megahed, E. A. & Davig, D. C. Long life divalent silver oxide–zinc primary cells for electronic applications, In: Thompson, J., Ed., *Power Sources*, Vol. 8, Academic Press, London (1981), pp. 141–161.
[45] Megahed, E. A. & Davig, D. C., Rayovac's divalent silver oxide–zinc batteries, *Progress in Batteries and Solar Cells*, **4** (1982), 83–86.
[46] Megahed, E. A. & Fung, A. K., US Patent 4,835,077 (1989).
[47] Nagaura, T. & Aita, T., US Patent 4,370, 395 (1981).
[48] Nagaura, T., New material $AgNiO_2$ for minature alkaline batteries, *Progress in Batteries and Solar Cells*, **4** (1982), 105–107.
[49] D'Agostino, V., Lee, J. & Orban, G., Grafted membranes, Chapter 19, *Zinc–silver Oxide Batteries*, Fleischer, A. & Lander, J., Eds, John Wiley, New York (1971), pp. 271–281.
[50] Thornton, R., *Diffusion of soluble silver–oxide species in membrane separators*. General Electric Final Report, – Schenectady, NY (1973).
[51] Oltman, J. E., Carpenter, D. D. & Dopp, R. B., US Patent 4,591,539 (1986).
[52] Weise, G. W. & Cahoon, A. C., *The primary battery*, John Wiley, New York (1971).
[53] Bagshaw, N. E., *Batteries on ships*, Research Studies Press, John Wiley, Chichester (1982).
[54] Crompton, T. R. (1983), *Small Batteries*, John Wiley, Chichester, (1983).
[55] Brodd, R. J., *Batteries for cordless appliances,* Research Studies Press, John Wiley, Chichester (1987).
[56] Bruner, R. C. & House, L. R., Thermodynamics of the external auditory canal, *Ann Otol Rhinol. Laryngol.*, **76** (1967), 409–413.
[57] Oltman, J. E., Dopp, R. B. & Carpenter D. D., US Patent 4,649,090 (1987).

[58] Hills, S., Thermal coefficients of EMF of the silver (I) and silver (II) oxide–zinc in 45% potassium hydroxide systems, *J. Electrochem. Soc.*, **108** (1961), 810–815.
[59] Duracell, Eveready, and Rayovac Product data books (1988–1990).
[60] Hull, M. N. & James, H. I., Why alkaline cells leak, *J. Electrochem. Soc.*, **124** (1977), 332–339.
[61] Vinal, G. W., *The primary batteries*, John Wiley, New York (1950), pp. 214–229.
[62] McGraw-Edison, Co., *Zinc–air cell technology report* (1970), Appendix D, P. G. Grimes.
[63] Thomas Edison Laboratory Inc., *Plastic bonded electrodes* (1957), D. Tuomi.
[64] Heise, G. W. & Cahoon, N. C. *The primary battery*, John Wiley, New York (1971), Vol. 1, Chapter 8, E. A. Schumacher.
[65] Yeager, E., *The oxygen electrode in aqueous fuel cell*, Office of Naval Research technical report 12 (1960).
[66] McGraw-Edison, Co., *Gelled electrolyte for zinc–air battery* (1982), V. H. Vu and P. H. Hettwer.
[67] McGraw-Edison Co., *Regenerative zinc–air battery* (1983), V. H. Vu.
[68] McGraw-Edison Co., *Electrochemistry and corrosion study on zinc–air cell* (1985), V. H. Vu.
[69] McGraw-Edison Co., *Mercury-free zinc anode for zinc–air battery* (1985), V. H. Vu.
[70] Edison Batteries Inc., Product literature.
[71] SAFT America Inc., Product literature.
[72] Panasonic Batteries, Product literature.

**Further reading on alkaline–manganese cells:**

Cahoon, N. C. & Holland, H. W., The alkaline manganese dioxide:zinc system, In: Heise, G. W. & Cahoon, N. C., Eds, *The primary battery*. John Wiley, New York (1971), pp. 239–263.

Kordesch, K. V., Alkaline manganese dioxide zinc batteries, In Kordesch, K. V., ed, *Batteries*. Marcel Dekker, New York (1971) pp. 241–384.

Saxe, C. & Brodd, R., History of alkaline zinc manganese dioxide cells, *Progress in Batteries and Solar Cells*, **7** (1988), 136–147.

# 4
# Rechargeable battery systems

## 4.1 LEAD–ACID BATTERIES
(K. Peters)

### 4.1.1 Historical notes and general characteristics

The commercial importance of the lead–acid battery and its effect on the lead producing industries is illustrated by the fact that the battery industry has been the major consumer of lead for a number of years. Whilst there has been a steady decline in other traditional uses of the metal, the demand for batteries has shown a generally upward trend reflecting the continuing and worldwide growth of the automobile industry and the steadily increasing number of battery powered applications.

In 1987, the battery industry in the USA used 77% (940 000 tonnes) of all lead produced. The proportion was somewhat lower in Europe, being about 50% of the total consumption, but the ratio is increasing.

This demand for lead batteries continues despite the appearance over the past twenty years of several promising new types of rechargeable battery and considerable investment in their development. Batteries such as zinc/bromine, sodium/sulphur, sodium/metal chloride, and an increasing number of rechargeable lithium systems, have been widely publicized. These all have much greater energy and power per unit weight than the lead battery. So far, none of these new battery types has been developed to the commercial stage, and when available they are unlikely to replace the lead battery in its traditional markets.

Partly, this is due to the continuing development and evolution of the lead battery which has resulted in major improvements in performance. Gains of up to 50% in energy and power density have been achieved in portable units. In some standby applications the gain in output per unit volume has exceeded 200%. In addition the service life and durability have been increased in almost every field of application.

The first practical lead–acid battery was developed in 1859 by Gaston Planté. He formed the lead dioxide active material by corroding the surface of a lead sheet. His objective was to find a replacement for primary batteries used in telegraphy . In 1880 Fauré modified the Planté process by coating the lead foil with a layer of paste made from lead oxide and sulphuric acid, and in so doing, increased the surface area of the active material and its electrical capacity.

At this time and for some years after, telecommunication and emergency power requirements produced the only market for these batteries. Toward the end of the 19th century and early in the 1900s, more compact and portable batteries became available and were used for railway train lighting, standby and regulating service, telephone exchanges, and isolated lighting plants. They were also used in submarines and for vehicle propulsion to a limited extent.

The most important event in the history of the lead battery was the invention of the electric self-starter by Kettering in 1912. As a result, the battery market has grown with the growth of the automobile market throughout this century. Engine starting batteries account for three quarters of total battery sales. In North America alone, 81 million automobile batteries were produced in 1987, and forecasts suggest that this market will continue to grow in spite of the technical improvements which have doubled the average service life in the last 20 years.

Other applications have opened up a variety of new and valuable markets for lead acid batteries throughout this century. These include batteries for motive power, for both on and off road electric vehicles, and for stationary or standby applications such as telephone and emergency power supplies. Each of these markets require batteries with specific operating features, and, in this respect, designs have been developed for particular applications. The major markets are summarized in Table 4.1, together with the specific operating requirements and battery designs.

The major developments throughout the history of the lead battery which have resulted in the present-day product include:

(i) Improved plate additives, which provide and maintain an open and porous structure within the negative active material.
(ii) Improved positive grid alloys, with optimized antimony–arsenic–selenium or calcium–tin compositions which minimize grid growth and prevent intergranular corrosion. These alloys also provide a stable overvoltage, thereby minimizing overcharge and loss of water during service.
(iii) New and continuously evolving separator designs and materials, resulting in a virtual complete absence of separator shorts or damage as a mode of failure together with much reduced internal resistance and improved power capability.
(iv) Strong and durable plastic containers which can be heat sealed thus greatly reducing the incidence of leakage of electrolyte.
(v) Designs which provide the capability for oxygen to be produced on overcharge in order to react with the negative electrode and be converted back to water. This enables designs to be produced which are spill-proof, more compact, and safer.

In addition to these material and design changes, automatic manufacturing methods and quality control procedures have been introduced progressively at every stage of production with the result that the present-day lead–acid battery has much improved energy and power capability, is smaller and lighter, and longer lasting, and more durable than its predecessors.

Table 4.1 — Major markets for lead–acid batteries

| | 1988 sales† £M | Operating duty | Design features |
|---|---|---|---|
| Starting, lighting, ignition (cars, motor cycles, etc.). | 4000 | High power discharge performance over wide temperature range. Shallow cycling duty with service life of more than 3 years. | Thin plate, polypropylene 6 V/12 V monobloc containers. Maintenance free. |
| Motive power (electric trucks, road vehicles, golf carts, etc.) | 750 | Deep cycle discharges at 1 to 5 h rate. Service life of 500 to 2000 cycles. | Tubular or flat plate designs. 2 V cells in 24/96 V assemblies. |
| Standby (telecom, UPS, emergency lighting). | 550 | Reliability over prolonged periods. Discharge duty at 1 to 6 h rate. Service life of 5 to 25 years under continuous charge conditions. Limited cycling capability. | Variety of designs, including Planté, tubular, flat plate, Bell round cell, Sealed valve regulated cells. |
| Portable (cordless tools, equipment, etc.). | 200 | Unspillability, 200 to 500 cycles, 1 to 10 h discharge rate. | Sealed, absorbed or gelled acid 6 V/12 V units upwards. |

†Estimated sales excluding eastern Europe and China.

### 4.1.2 Materials and structural features
#### 4.1.2.1 Negative active material

In the fully formed condition the negative electrode has an open porous structure of sponge lead. It is made from a mixture of lead and lead oxide, produced in a ball mill or by the oxidation of molten lead. The powder produced in these processes may contain up to 40% lead and is mixed with sulphuric acid, water, and various additives. Without these additives the charge–discharge process quickly results in densification of the active material with shrinkage, reduced porosity, and loss of electrical performance.

The additives are usually a mixture of organic lignins or lignosulphonates, barium sulphate, and carbon black. Frequently they are supplied as a proprietary mix from specialist suppliers. The organic materials act as a surfactant, lowering the surface tension of the lead surface and preventing coagulation. Barium sulphate is isomorphous with lead sulphate, and when widely distributed in the active material provides a nucleating base for lead sulphate, preventing the build-up of large crystals which would be difficult to reduce during charge. In the presence of the other additives it is doubtful if carbon black has any significant electrochemical benefit. However, its addition provides a useful colouring distinction from the positive plate during the manufacturing stages.

Fig. 4.1 shows the changes in the porosity and surface area of the negative active material during the processing stages and up to 300 charge–discharge cycles. During the paste mixing stage there is a large increase in surface area, but as the active lead is

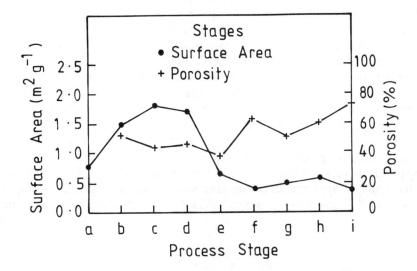

Fig. 4.1 — Negative plates: surface area and porosity changes. (a) Dry oxide; (b) Wet paste; (c) After 16 h set; (d) After 4 h drying; (e) After 4 h of formation; (f) At the end of formation; (g) After discharge; (h) After 50 cycles; (i) After 300 cycles.

produced during formation the surface area decreases to about 0.5 $m^2g^{-1}$, whilst the porosity increases to about 60%. During repeated cycling the porosity increases to about 75%, and this is accompanied by a decrease in surface area.

### 4.1.2.2 Positive active material

Lead dioxide, the positive active mass in the lead–acid battery, exists in orthorhombic and tetragonal crystalline modification ($\alpha$-$PbO_2$ and $\beta$-$PbO_2$). Both modifications are present to some degree in positive plates. The equalibrium potential of alpha lead dioxide is more positive than that of the beta form by 0.01 V. Alpha lead dioxide has a larger crystal size and is less chemically active with lower capacity per unit weight than the beta form. Positive plates which contain a high proportion of alpha generally have a longer cycle life. Neither of the two forms is fully stoichiometric, and the composition is frequently represented as $PbO_x$, where $x$ can be between 1.85 and 2.05. Various authors have reviewed these structures and their properties, including Bode & Voss [1], Burbank [2, 3], Ruetschi & Angstadt [4], and Zaslavski et al. [5].

The positive active material is prepared in a similar manner to the negative, by mixing lead oxide powder with sulphuric acid and water. Frequently, plastic fibres are added to the positive paste to improve its mechanical strength in the dried condition. However, the mixing conditions of positive pastes and the subsequent post-treatment require more careful control. In particular it is important that the temperature does not exceed 75°C at any stage, and it is desirably maintained at much lower values during the subsequent setting and curing stages. High temperatures cause paste shrinkage and cracking and produce large crystals of $4PbO.PbSO_4$ which are extremely difficult to oxidize to lead dioxide in the formation process. Positive plates made without adequate temperature control will have a much reduced electrical capacity. In addition the ability of the positive plate to provide a large number of charge–discharge cycles without losing performance is influenced by the crystal structure produced during the curing process. It is important to reduce the lead content to a very low value, usually less than 2%. Pastes with a high residual lead content will gas extensively early in the formation stage, damaging the structure and causing premature loss of active material.

The change in surface area and porosity of the positive active mass during processing and cycling is different from the negative plate. The surface area increases during paste mixing and formation, whilst the porosity decreases. During cycling the porosity remains approximately constant whilst the surface area continues to increase, and after about 200 deep cycles, is $7 m^2$ per g. This effect is illustrated in Fig. 4.2.

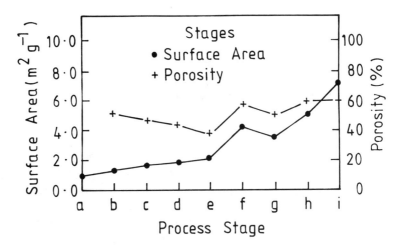

Fig. 4.2 — Positive plates: surface area and porosity changes. (a) Dry oxide; (b) Wet paste; (c) After 16 h set; (d) After 4 h drying; (e) After 4 h of formation; (f) At the end of formation; (g) After discharge; (h) After 50 cycles; (i) After 300 cycles.

### 4.1.2.3 Grid alloys

The battery grid has the dual function of active material support and current collector. The design has to satisfy both functions. The grid metal may be pure lead in some designs, but in the great majority of cases lead is alloyed with selective metals to provide both mechanical strength and corrosion resistance. Modern alloys are made with a ternary composition of either lead, antimony, and arsenic, or lead, calcium, and tin.

Lead is alloyed with antimony in various amounts. In industrial batteries which may be subject to deep cycling operation, up to 8% of antimony is used, but in the majority of other designs the antimony content has been progressively reduced to between 1% and 3%. Antimony is helpful in the manufacturing process by improving the fluidity of the metal and reducing the volume change on setting. However, under anodic conditions, the corrosion product contains soluble antimony ions which can deposit on the negative electrode as Sb atoms. This has two significant effects. Firstly, because of the different electrode potentials, the lead active material self-discharges. According to Dawson et al. [6] the reaction is of the following form:

$$2SbO^+ + 3Pb + 3H_2SO_4 \rightarrow 2Sb + 3PbSO_4 + 2H_3O^+ \tag{4.1}$$

Secondly, because antimony has a lower overpotential than lead, the charging efficiency is reduced, so that hydrogen is evolved in greater amounts. Although some antimony may be evolved as stibine the bulk of it stays in the negative active material. With a progressive build-up the overall effect is to reduce the shelf life and to increase the water losses in service. With severe poisoning of the negative plate, recharge may be difficult. Antimony also reduces the conductivity of lead. For example, the resistance of 8% antimonial alloys is about 20% higher than that of pure lead.

Arsenic is added to antimonial alloys in amounts up to 0.5%. Its presence refines the microstructure, which aids castability and provides greatly improved resistance to anodic attack and growth. With alloys containing low amounts of antimony, small amounts of other grain refining metals such as selenium are used to improve the castability and to prevent cracking.

Over the last 20 years calcium has been increasingly used as a hardening agent in place of antimony. It has a number of advantages over antimony. Firstly, it is used in much smaller amounts. Usually 0.1% is the maximum amount used, and these alloys have a lower resistance than typical antimonial alloys. Most important, however, is the lack of any poisoning effect on the negative active material. As a result the self-discharge is greatly reduced compared with cells containing antimonial grids, and the gassing overpotentials are constant throughout the life of the battery. The effect is that batteries made with calcium grids have much improved shelf life and lose less water in service.

Calcium is used in the range 0.03 to 0.1% and is frequently alloyed with between 0.5 and 1.0% tin which controls the grain size and structure and prevents intergranular corrosion. These alloys, originally developed for submarine and then telephone applications (to reduce the emission of hydrogen and stibine in confined environments) have been used increasingly in the last 20 years to produce low maintenance and maintenance-free batteries. The enhanced storage life and the stable overpotential has been particularly beneficial in SLI applications (automobile, engine starting) and batteries for telephone systems which require prolonged service without maintenance. The effect of the different grid alloys on shelf life and water loss is shown in section 4.1.3.3.

A summary of the various alloys in use in the modern day lead acid battery and their applications is given in Table 4.2. Most grids are produced by gravity casting

**Table 4.2** — Properties of lead alloys

|  | Ultimate tensile strength $Pa \times 10^{-6}$ | Elongation % | Resistivity mohm cm. (20°C) |
| --- | --- | --- | --- |
| Lead, antimony (8%) arsenic (0.35%) | 38–46 | 20–25 | 26.8 |
| Lead, antimony (2%) arsenic (0.4%) selenium (0.01%) | 33–40 | 10–15 | 22.9 |
| Lead, calcium (0.08%) tin (0.35%) | 36–40 | 25–35 | 21.4 |
| Lead, calcium (0.09%) | 34–36 | 35–40 | 22.2 |

into a book mould, but wrought production methods such as slitting sheet lead and expanding it, are now widely used. Aluminium may also be added to the calcium alloys to inhibit the oxidation and drossing of calcium in the molten state.

Grid design is particularly important in the positive plate because of the lower conductivity of lead dioxide compared with lead. Faber [7] demonstrated the effect of grid dimensions on the efficiency of the active material. Fig. 4.3 demonstrates the improvement in mass utilization which is achieved as the mesh size is reduced.

### 4.1.2.4 Electrolyte
In many types of battery the electrolyte is merely an ionic conductor. However, in the lead–acid battery, sulphuric acid is also an active material, and its concentration is reduced during discharge and increased during charge. Properties which are indicators of concentration such as specific gravity, viscosity, and refractivity can be used as a guide to the state of charge.

Most battery designs use electrolyte with a specific gravity (s.g) between 1.270 and 1.300. Planté cells are designed to operate with lower strength acid (s.g. 1.210–1.240), and batteries destined for use in hot climates contain weaker electrolyte in order to reduce the self-discharge rate. Higher strengths of electrolyte (s.g. 1.300–1.310) have been used in recent years in electric vehicle batteries to improve the

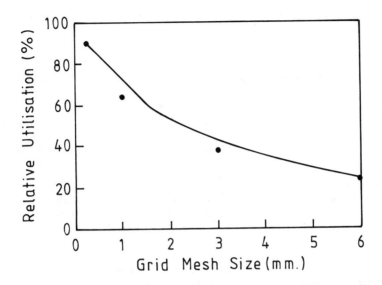

Fig. 4.3 — Positive mass utilizations (Courtesy of International Power Sources Symposium Committee).

utilization of the active material and to increase the energy density. The resistivity of sulphuric acid goes through a minimum in concentrations having a specific gravity of 1.200 to 1.300, as can be seen in Fig. 4.4.

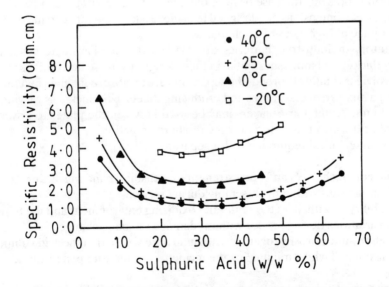

Fig. 4.4 — Resistivity of sulphuric acid solutions.

The strength of electrolyte in various designs of battery used in tropical and temperate climates is given in Table 4.3.

**Table 4.3** — Electrolyte strength (s.g.) of various types of lead–acid battery

| Application | Specific gravity | |
| --- | --- | --- |
| | Temperate climate | Tropical climate |
| SLI | 1.270–1.290 | 1.220–1.240 |
| Motive power | 1.280–1.320 | 1.240–1.280 |
| Standby (flooded) | 1.220–1.240 | 1.200–1.220 |
| Standby (absorbed) | 1.270–1.300 | 1.260–1.280 |
| Aircraft | 1.260–1.280 | 1.260–1.280 |

The purity of both the acid and the replenishing water has to be maintained to a high standard. The impurity levels are defined in various national standards [8,9].

### 4.1.2.5 Separators

Separators insulate the electrodes from those of opposite polarity but must be porous to the electrolyte and allow ionic transport. They are made from a variety of plastic and inorganic materials. Batteries for automotive and engine starting duties (SLI) contain separators made from porous polyethylene sheet, or nonwoven fibrous structures made from polypropylene, glass, or impregnated cellulosic fibres. The overall thickness of the separator in this application is between 0.8 and 1.2 mm. Where a rib is present, the base or flesh thickness of the separator can be as low as 0.25 mm. These dimensions are designed to achieve a low resistance to maximize the battery voltage at high rates of discharge.

Separators in industrial batteries are made from similiar materials but are generally thicker and more substantial in their design. In the sealed designs in which the electrolyte is totally absorbed, the separators are an absorbent glass mat made by wet laying a mixture of glass fibres and producing sheets with thickness varying from 1.0 to 2.0 mm. A number of layers may be used between the plates to absorb the volume of acid which is necessary for the discharge reaction.

The main technical requirements for separators are:

(i) Sufficient mechanical and burst strength to withstand handling during assembly and the stresses within the assembly during service.
(ii) The ability to withstand the acidic and oxidizing environment and the temperature variations during service without damage or shape change.
(iii) A high volume porosity together with small pore size so that the acid volume can be maximized whilst minimizing the incidence of dendrite penetration.

A summary of the range of separators that are now available and their specific properties is given in Table 4.4.

**Table 4.4** — Properties of separators

| | Porosity % | Pore size μm | Resistance ohm/cm2 |
|---|---|---|---|
| Cellulose | 60 | 25 | 0.30 |
| Microporous PVC | 80 | <3 | 0.18 |
| Rubber | 62 | <2 | 0.30 |
| Microporous Polyethylene | 63 | <1 | 0.15 |
| Non woven polypropylene | 60 | 12 | 0.21 |
| Non woven glass mat | 65 | 20 | 0.18 |
| Phenolic base | 68 | 0.5 | 0.15 |
| Sintered PVC | 45 | 25 | 0.25 |
| Absorbent glass mat | 90 | 24 | 0.10 |

### 4.1.3 Basic characteristics

The basic electrochemical reactions and the reversible potentials at the positive and negative electrode in the lead–acid battery are:

Negative electrode $\quad Pb \rightarrow Pb^{2+} + 2e \quad$ (4.2)
$\quad Pb^{2+} + SO_4^{2-} \rightarrow PbSO_4 \quad E_a^0 = -0.356 \text{ V} \quad$ (4.3)
Positive electrode $\quad PbO_2 + 4H^+ + 2e^- \rightarrow Pb^{2+} + 2H_2O \quad$ (4.4)
$\quad Pb^{2+} + SO_4^{2-} \rightarrow PbSO_4 \quad E_c^0 = 1.685 \text{ V} \quad$ (4.5)
Overall reaction $\quad Pb + PbO_2 + 2H_2SO_4 \rightarrow 2PbSO_4 + H_2 \quad$ (4.6)
$\quad\quad\quad\quad\quad\quad\quad\quad\quad\quad\quad\quad\quad\quad E^0 = 2.041 \text{ V}$

where $E_a^0$ and $E_c^0$ are the anode and cathode equilibrium potentials respectively.

The potentials of the electrodes and the cell given in these equations vary from the steady state values due to the overvoltage effects on each electrode and the resistive losses within the cell as follows.

$$\text{Cell voltage} \quad E_c = (E_c^0 - \eta c) - (E_a^0 - \eta a) - IR \quad (4.7)$$

where ηc and ηa are the overpotential effects on the reaction at each electrode and IR is the resistive influence.

The reaction at each electrode can be monitored with a reference electrode. A mercury/mercurous sulphate electrode is suitable for the lead–acid battery. This has a potential of +0.97 V against a standard hydrogen electrode, and it provides a suitable means for following the potential of each electrode during the charge–discharge process.

An alternative reference electrode which is more convenient for general battery monitoring, is the cadmium electrode. This consists of a piece of cadmium wire usually held in an open-ended rubber sheath. The standard potential of cadmium in sulphuric acid is approximately −0.40 V, and when immersed in the electrolyte and connected to an electrode via a high impedance meter the potential of each electrode can be followed throughout the charge and discharge process. Because of the danger of dissolution of the cadmium it should not be immersed in the acid permanently.

Electrode and cell potentials may differ from the reversible values during service. During discharge, resistive and concentration effects reduce these potentials, and during the charge stage similar effects, together with gas evolution reactions, create substantial deviations from the thermodynamic values. The ability of the lead–acid battery to provide a large number of charge–discharge cycles without appreciable change in performance is due to a number of characteristic features of the system. The high values of the hydrogen and oxygen overpotentials on the negative and positive electrodes [10,11,12,13] allow the charging process to proceed efficiently, and although the formation of lead sulphate during discharge goes through a dissolution and precipitation process, there is no significant migration of lead sulphate during cycling because of its low solubility in strong acid [14,15]. As a result, the electrodes maintain a similar structure throughout service.

The capacity of the lead–acid battery varies according to the current density and temperature at which the battery is discharged. The positive and negative electrodes behave differently in this respect, and this was demonstrated by Baikie et al. [16] and is shown in Fig. 4.5(a) and (b).

### 4.1.3.1 Discharge behaviour

The effects of current density on the discharge time of the positive and negative plates are shown in Fig. 4.6(a) and (b). Discharge currents are usually defined by the duration of the discharge, and these tests were carried out at between the 3 min and 100 min rate of discharge. At these rates, polarization and the termination of the discharge is due to diffusion limitations, particularly within the pores of the electrode. The formation of lead sulphate with a density of 6.3 g.cm$^{-3}$ compared with lead dioxide (of density 9.3 g.cm$^{-3}$) and lead (of density 11.3 g.cm$^{-3}$) at the electrode surface causes pore blocking which limits the transport of electrolyte. A further factor which limits the utilization of the active materials is the reduction in electronic conductivity caused by the insulating layers of lead sulphate reducing contact between the active sites.

Typical compositions in terms of weight of the components for SLI and motive power (tubular design) are shown in Fig. 4.7.

The relationship between battery capacity and the rate of discharge can be expressed empirically by the Peukert equation [17].

$$I^n . t = K, \qquad (4.8)$$

where $t$ is the discharge duration, $I$ is the current and $K$ and $n$ are constants. Reported values of the index $n$ vary considerably. For typical operational designs and at discharge current densities between $2.07 \times 10^2$ and $1.65 \times 10^3$ A.m$^{-2}$, values of $n=1.39$ (positive electrodes) and $n=1.41$ (negative electrodes) have been reported [18,19].

The value of $K$ in Peukert's equation is known to vary with changes in electrolyte concentration and temperature [20,21]. Temperature corrections can be included, thus:

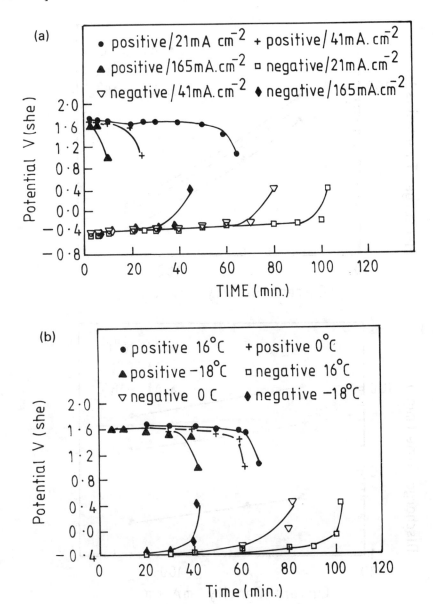

Fig. 4.5 — Discharge characteristics; (a) constant temperature, (b) constant current.

$$I^n \cdot t = K_0(1 + \alpha T) \tag{4.9}$$

where $t$ is the discharge duration (min), $T$ is the temperature (°C), and $\alpha, K_0$, and $n$ are constants with the following values in 6 M sulphuric acid and over a temperature range of $-18°C$ to $+20°C$:

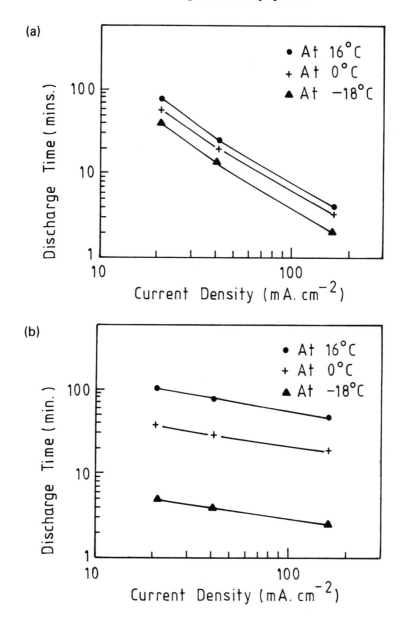

Fig. 4.6 — Discharge time, effect of current density; (a) positive plates, (b) negative plates.

$$\text{Positive} \quad 8.75 \times 10^4 \, (1 + 0.015T) = I^{1.39} \cdot t \quad (4.10)$$

$$\text{Negatives} \quad 1.40 \times 10^5 (1 + 0.024 \, T) = I^{1.41} \cdot t \quad (4.11)$$

The influence of the design of the electrode can also be accommodated by including a thickness factor as follows [22]:

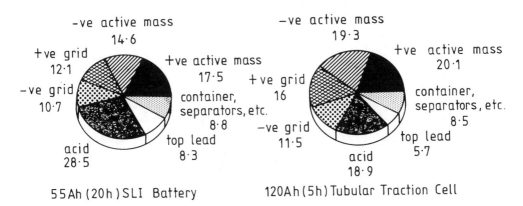

Fig. 4.7 — Composition of SLI and tubular traction cells.

$$I^n\left(\frac{t}{d^{(2-n)}}\right) = K. \tag{4.12a}$$

where $d$ is the thickness of the electrode.

For positive electrodes varying in thickness between 1.9 and 13.0 mm the following relationship holds at 21°C:

$$I^{1.4}\left(\frac{t}{d^{0.6}}\right) = 1.05 \times 10^5. \tag{4.12b}$$

Combining the two relationships gives the following equation:

$$I^n\left(\frac{t}{d^{(2-n)}}\right) = K_0(1 + \alpha T). \tag{4.13}$$

Many workers have indicated that deviations from Peukert's equation occur at extreme currents. At high current density, the value of $n$ increases, tending toward a value of 2. At lower currents the utilization increases, and the value of $n$ approaches 1. The absolute values of the coefficient and constants vary with design changes. However, the Peukert equation provides a sufficiently accurate indication of capacities at different rates of discharge for any specific design. Fig. 4.8 is a logarithmic plot of current density against discharge time for a 12 V SLI battery and this enables the capacity of the battery to be estimated at any rate of discharge. Further relationships between capacity and discharge rate and temperature are given under the specific product types.

### 4.1.3.2 *Charging behaviour*
With the appropriate design of cell and the correct charging procedure, lead–acid batteries are capable of giving many hundreds of charge/discharge cycles without

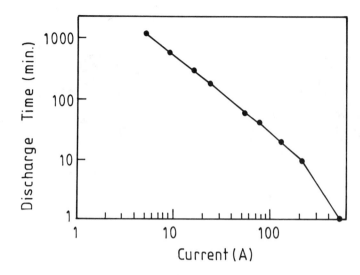

Fig. 4.8 — Current/discharge relationships for a 100 Ah (20 h) SLI battery at 25°C.

deterioration. Charging can be carried out only from a D.C. source. Most rectifiers produce D.C. with some alternating current component which causes heating of the cell. For a good cycling performance, the charging current should be limited, especially near the end of the charge. A variety of charging procedures can be used. In every case, sufficient energy must be provided to fully convert the discharged active masses to the charge state. This value must include an allowance for the inefficiency of the charge reaction which is a consequence of the simultaneous evolution of oxygen and hydrogen from the positive and negative plates respectively.

As a result of this gassing, the ampere hour efficiency (the ratio of output to input) is about 0.85 to 0.90, and the reciprocal of this, known as the charge factor, is about 1.1. These values vary according to the depth of discharge and the speed of recharge. Also, the charge factor decreases if the battery has been held in the partly or fully discharged condition for some time. The watt-hour efficiency can be defined similarly.

Partial charging is frequently used to reduce gassing and to improve the charge efficiency. In this process the charge is terminated before gassing commences, and the charge factor may be only 1.0 or marginally above. If this method is used for more than a few consecutive cycles, the capacity will decrease, and it becomes difficult to fully charge the cell and recover the capacity. The state of charge can be conveniently monitored by the cell voltage or the specific gravity of the electrolyte, which reach constant values when the cell or battery is fully charged.

During charge, although the anodic and cathodic currents must be equal, the amount of current consumed in the evolution of gas is different for each electrode. As a result, the charge acceptance, which is a measure of the useful energy accepted during charge, is different for the positive and negative electrodes. These values can be monitored by measuring the quantity of gas evolved from each electrode under different temperatures and current densities. The charge acceptance, $C_a$, for each electrode is given by

$$C_a = \frac{\text{useful charge accepted}}{\text{total charge input}} = \frac{I_c}{I_c + I_g} \tag{4.14}$$

where $I_c$ is the partial current used to charge the electrodes and $I_g$ is the partial current used to produce gas. The value of $C_a$ decreases as the charging temperature is decreased and the current density increases and as the electrodes become more charged. The gassing rates, and therefore the charge acceptance at different states of charge, are markedly different at the positive and negative plates. Fig. 4.9(a) and (b)

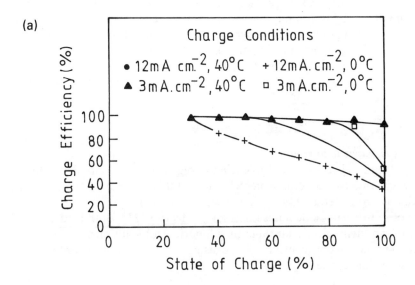

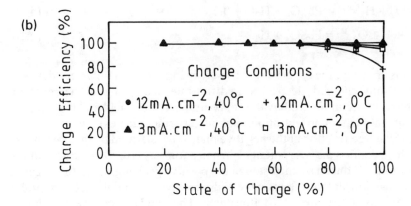

Fig. 4.9—Changes in charging efficiency; (a) positive, (b) negatives. Values relate to 10 Ah cell fully discharged with charge efficiency based on input to nominated charge level.

show the change in the charge acceptance of the positive and negative plates at different current densities and temperature as both plates become charged.

At the negative electrode $C_a$ has a value of 1.0 over a wide range of temperatures and current densities until it is nearly fully charged, whereas at the positive, $C_a$ has a value of less than 1.0 from the start of charge. As a result, oxygen is evolved from the positive in increasing amounts as the charge proceeds, and very little hydrogen is evolved until the cell is nearly fully charged.

The charge reactions proceed via dissolution and electrolytic dissociation of lead sulphate, followed by the deposition of lead on the negative plate and oxidation to lead dioxide at the positive. Whilst in the discharge state, lead sulphate crystals grow by recrystallization. After discharge the crystal size is between 0.1 and 1.0 µm, and after 8 days' storage they can be as large as 5 µm. The speed of dissolution is influenced, therefore, by the stand time before charge. The charge acceptance can also be influenced by organic additives such as expanders or anti-oxidant protectives that are occasionally used in producing dry-charged plates. The charge acceptance can be reduced substantially if the hydrogen overpotential is lowered by the contamination of the lead surface with metals possessing a low hydrogen overvoltage.

### 4.1.3.3 Self-discharge behaviour

The capacity of lead–acid cells decreases during storage in the open circuit condition. The capacity is normally recoverable on recharge, but this may not be possible if the storage period is prolonged and the retained capacity is very low.

The rate of self-discharge depends upon several factors [21]. Both lead and lead dioxide are thermodynamically unstable in sulphuric acid solutions, and on open circuit stand they react with the electrolyte, causing the slow evolution of hydrogen and oxygen according to the reactions:

$$PbO_2 + H_2SO_4 = PbSO_4 + H_2O + 0.5\ O_2 \tag{4.15}$$

$$Pb + H_2SO_4\ \ \ = PbSO_4 + H_2. \tag{4.16}$$

Both reactions are slow, the capacity loss usually being less than 0.2% per day with new cells and fully charged electrodes, and the rate decreasing as the reactions proceed. The rates are influenced by the concentration of the acid and by temperature (increasing at higher concentrations and higher temperatures).

The self-discharge reaction at the negative plate is accelerated by the presence of impurities which depress the overpotential. When antimony is used in the grid alloys, the antimony ions produced by corrosion of the positive grid migrate to the negative where the local action and the effect on the hydrogen overpotential can considerably increase the self-discharge rate as the battery ages. The trend in recent years to low or zero antimony alloys in which calcium is used as the hardening agent has reduced the self-discharge and increased the shelf life of lead–acid batteries. The capacity losses of batteries made with different grid alloys are typically shown in Fig. 4.10. This graph also shows the effect of storage temperature on the self-discharge rate.

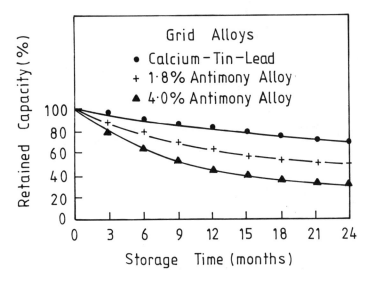

Fig. 4.10 — Capacity losses during storage for standard SLI batteries stored at 25°C and discharged at the reserve capacity rate.

Normally, this capacity loss on storage is recovered within a few charge–discharge cycles. However, after prolonged storage without any freshening charges, the lead sulphate crystals which are produced become irreversible, and the plates cannot be recharged.

### 4.1.3.4 Gas recombination behaviour

In recent years, sealed lead–acid batteries which operate on the oxygen recombination principle have been developed and are now used in a variety of applications [23–26]. These batteries contain a one-way valve which opens to release gases if the internal pressure gets too high but prevents atmospheric gases from entering. They are more properly described as valve-regulated batteries. The oxygen cycle in aqueous secondary batteries is the formation of oxygen at the positive electrode and its reduction at the negative. With proper design and controlled use, the evolution of hydrogen from the negative plate can be substantially reduced, and batteries can be operated throughout a service life of many years with little loss of water and only small amounts of gas evolved through the vent. In consequence they are particularly suitable for safe use in closed environments. In addition, the design requirements necessary to produce gas recombination allow smaller batteries with lower internal resistance.

The oxygen cycle is similar to that occurring in the sealed nickel–cadmium battery with oxygen produced at the positive electrode during charge reacting with the negative electrode and subsequently producing water via a two stage reaction as follows:

$$Pb + 0.5\ O_2 \rightarrow PbO \tag{4.17}$$

$$PbO + H_2SO_4 \rightarrow PbSO_4 + H_2O \tag{4.18}$$

with an overall reaction

$$Pb + 0.5\ O_2 + H_2SO_4 \rightarrow PbSO_4 + H_2O. \tag{4.19}$$

As these reactions proceed, the potential of the negative electrode is reduced and the evolution of hydrogen is greatly diminished. The typical voltage behaviour of a gas recombination cell charged at constant current is shown in Fig. 4.11.

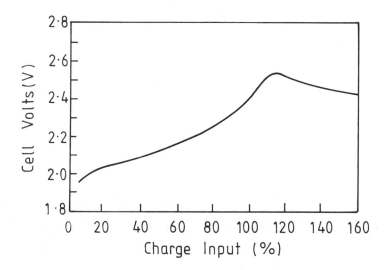

Fig. 4.11 — Voltage changes of sealed recombination cells on constant current charge at a 0.1 C rate after full discharge.

The solubility and diffusion coefficients of oxygen in 5 M sulphuric acid are 0.65 mol. dm$^{-3}$ and $0.8 \times 10^{-5}$ cm$^2$ s$^{-1}$ respectively [27]. These are higher than the similar coefficients in 7 M KOH, which suggests that transport of oxygen is more rapid in the lead–acid system than in nickel–cadmium cells. However, these rates are insufficient to support the recombination efficiencies that both systems exhibit, and it is clear from these values that oxygen transport from the positive plate to the negative is via the gaseous phase.

Lead dioxide has a low charge efficiency, and oxygen is evolved at an early stage in the charge process even at low charging rates. At similar charging rates the reduction of lead sulphate to lead at the negative electrode proceeds very efficiently with hydrogen evolution occurring only in the later stages of the charge, as demonstrated in Fig. 4.9(a) and (b).

The charge acceptance of the lead electrode remains at 100% throughout a considerable proportion of the charge. The duration of this period increases with decreasing charge rate, but is not markedly affected by temperature. As oxygen from the positive diffuses to the negative, reaction (4.17) proceeds rapidly, and the negative potential is reduced as in Fig. 4.11. As a result, the evolution of hydrogen is depressed. This condition exists as long as the charging current is less than the current equivalent of the oxygen recombination reaction.

The critical requirement in the design and manufacture of these cells is for rapid transport of oxygen to the negative. Since the electrolyte is also an active material and not merely a conductor as in nickel–cadmium cells, the design must provide a facility for gas transport without a substantial reduction in the volume of electrolyte. This can be achieved by using a highly absorbent and resilient separator which holds sufficient electrolyte for the discharge reaction and provides a tight closely packed assembly. Gas transport is achieved by operating in a partly saturated condition or by movement of the electrolyte within the separator caused by the pressure of the evolving gas. Alternative designs use silica powder to immobilize the electrolyte. In service, small cracks develop in the gel which permit the diffusion of oxygen.

Gas recombination efficiencies can be determined by monitoring the weight loss over a prolonged period of charge. With fully charged cells, electrolysis of water is the only significant reaction. The secondary reactions such as corrosion, oxidation of degradable materials, and the production of heat consume a very small proportion of the charging energy. An alternative procedure is to measure the amount of hydrogen evolved and equate this to the current equivalent, thus:

$$\text{Recombination efficiency} = \frac{I - I_H}{I} \times 100\%. \qquad (4.20)$$

where $I$ = average charge current during the test period and $I_H$ = average current equivalent to the hydrogen evolved during the same period.

With suitable designs and the correct operating procedures, gas recombination efficiencies approaching 100% can be achieved.

### 4.1.4 Construction and performance

The design and construction of lead–acid batteries varies according to the particular duty that they have to fulfil. The main categories are SLI or automotive batteries for engine starting, traction or motive power batteries for vehicle propulsion, stationary or standby batteries for emergency power supplies, and portable batteries for equipment, alarms, toys, etc. In addition, specific applications such as aircraft, submarine, and other defence systems require special designs. The primary operating features of these main categories are given in Table 4.1.

#### 4.1.4.1 SLI (automotive) batteries

The primary purpose of these batteries is to start the engine, reliably and repeatedly, but there is also a need for a reserve capacity to power services on the vehicle when the engine is stopped. The design approach in recent years has been to maximize the power by reducing the internal resistance. The resistance of lead–acid cells varies according to size, and also changes throughout a discharge, owing to the increasing polarization resistance. When a discharge commences, the voltage drops because of the ohmic resistance and polarization effects. The voltage drop across the ohmic resistance is proportional to the current density, and it appears with no delay. The internal resistance can be calculated from the difference between the rest potential and the voltage of the cell immediately after connecting the load. Its value is determined to a large extent by the resistance of the acid and the separators. As a result the improvements in the power capability for engine starting batteries have in recent years been achieved by the use of improved separators with thinner back web allowing the spacing between the plates to be reduced.

Recent improvements in the performance of SLI batteries when discharged at high current densities, together with the changes in the internal resistance, are shown in Fig. 4.12.

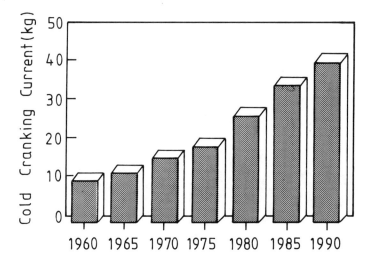

Fig. 4.12 — Improvements in SLI batteries since 1960, measured by the improvement in cold cranking current. This is the current available from a fully-charge battery at $-18°C$ for 30 s to 1.2 V per cell.

Many types of SLI battery are now made with grids produced by slitting and expanding sheet lead rather than by the traditional method of gravity casting. Fig. 4.13 shows such a design. In addition, increasingly plates of one polarity are positioned in an envelope separator to eliminate the danger of edge shorts and to allow the plate group to sit on the bottom of the container, thereby increasing the head space or allowing the overall battery height to be reduced. The practice of using an odd number of plates with negatives as the outer plates has changed in some applications. To balance the cranking and reserve performance more effectively, even plate and reverse assembly (i.e. with positive plates on the outside of the group) are becoming more common.

Most SLI batteries are now designed so that they do not require the addition of water throughout the normal service life. They are made with low concentration

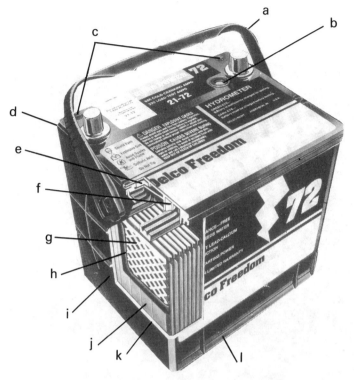

Fig. 4.13 — The construction of an SLI battery. (a) Flexible fold-down handle; (b) Built-in hydrometer; (c) Moulded symbols; (d) Heat-sealed covers; (e) Built-in flame arrestor vents; (f) Liquid-gas separator; (g) 'Small window' wrought lead-calcium grid; (h) High-density plate paste; (i) Polypropylene case; (j) Separator envelopes; (k) Reinforced case end wall; (l) Hold-down ramp. (Courtesy of AC Delco.)

antimony alloy grids or antimony-free grids. With these grid compositions and with a suitable setting on the voltage regulator, the water losses are low and the large head space accommodates sufficient electrolyte for the life time requirement.

Alternative designs have been suggested in recent years to improve specific features. The TORQUESTARTER™ battery produced by Chloride operates with gas recombination so that the electrolyte is totally absorbed within the plates and the separator. The hazard of acid leakage is eliminated, and the evolution of explosive gas mixtures during service are greatly reduced. In addition, the close-packed assembly reduces the internal resistance, providing improved performance at high rates of discharge and at low temperatures. Owing to the limited acid volume in these designs the capacity at lower discharge rates is usually lower than the equivalent size of battery containing free acid. A more recent innovation is the inclusion of a back-up power supply within the same container. Fig. 4.14 shows such a design. In addition to the main unit which provides the engine start and reserve capacity function a further part of the battery which will start the engine is kept in reserve in case the main unit fails. In the example shown, the main battery provides 525 cold cranking amperes, and in the event of an emergency or failure to start, the back-up supply will provide 275 cold cracking amperes.

Fig. 4.14 — SLI batteries with back-up power system (Courtesy Johnson Controls Inc.).

A typical range of discharge curves over periods ranging from 1 min to 20 h are given in Fig. 4.15(a) and (b). These graphs also give the mean voltages at each discharge rate and the final voltages to which the capacity is measured according to the specifications.

At the mean voltage values, the power available at different levels of discharge can be calculated. The power output at low temperatures and high currents is shown in Table 4.5. These values are most important for engine starting applications, and the reduction in cell resistance in recent years as a result of improved designs and new materials has resulted in significant improvements in engine starting performance. The charge can be estimated from the expression

$$C_2 = C_{25}[I - \alpha(25 - \theta_2)] \tag{4.21}$$

where $C_{25}$ is the capacity (Ah) at 25°C, $C_2$ is the capacity at $\theta_2$, and $\alpha$ is the temperature coefficient.

Smith [28] quotes a value of the temperature coefficient at the 5 min rate of discharge of $1.26 \times 10^{-2}$ Ah.deg$^{-1}$.

As an indication of the effect of these conditions on the engine starting capability, curves are given in Fig. 4.16, showing the engine power under various conditions.

Standard ratings and tests for SLI batteries cover the electrical performance at high current and low temperature discharge rates, the reserve capacity performance,

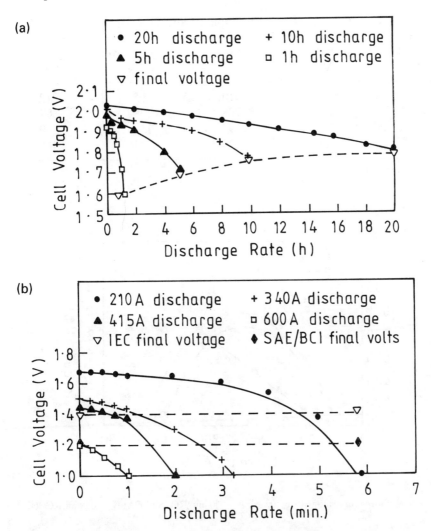

Fig. 4.15 — SLI battery performance during discharge on 70 Ah (20 h) battery. (a) low rate (25°C); (b) high rate ($-18°C$).

vibration resistance, charge acceptance, and life tests. In the United States the specifications defined by the Society of Automotive Engineers (SAE) and published by the Battery Council International (BCI) differ in some ways from the national standards defined in Europe and Japan. In addition, vehicle manufacturers have their own specifications which suppliers have to meet. The essential requirements of all these standards are summarized as follows:

(i) *Engine start capability*
This is defined as the current which the battery will deliver under cold conditions (usually at an initial temperature of $-18°C$) whilst maintaining a specified minimum

**Table 4.5** — Power output of SLI batteries

| D.C. current A | Mean volts V | Power W | Power density W/kg | W/dm$^3$ |
|---|---|---|---|---|
| 210 | 9.1 | 1910 | 95 | 212 |
| 340 | 8.2 | 2788 | 139 | 307 |
| 415 | 7.7 | 3195 | 159 | 349 |
| 600 | 6.7 | 4020 | 201 | 445 |

*Note*: All discharges at − 18°C.

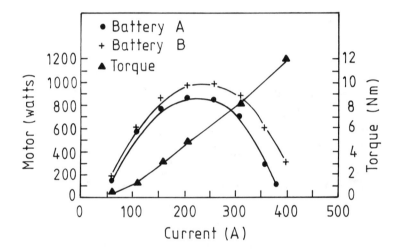

Fig. 4.16 — Power curves for SLI batteries. A — 50 min RC/280 CCA; B — 65 min RC/360 CCA.

voltage for a defined time. In Europe, the battery must have a voltage 1.4 V after discharging at the cranking current for 60 s. The cold cranking current in amperes (CCA), according to the SAE test, is the current which a fully charged battery can deliver at − 17.8°C for 30 s to a voltage of 1.2 V per cell. Automobile manufacturers' requirements and other national specifications may differ, but they are based on the same principle, and differ only in the duration of the test. In colder climates the temperature of the test discharge may be reduced.

*(ii) Reserve capacity*
The battery is discharged at a constant current of 25 A, and at 25°C. The reserve capacity is defined as the discharge duration in minutes to a voltage of 1.75 V per cell, and is a test of the battery's ability to provide power for lights, ignition, and the

auxiliaries. The 20 h capacity is also used in Europe and by certain manufacturers for the same purpose.

(*iii*) *Endurance tests*
These are designed to simulate automotive service conditions by assessing their behaviour under overcharge or cycling conditions. Some procedures consist of a combination of both. The B.S. overchange endurance test consists of charging for 100 h at 0.015 × the cranking current at 40°C followed by an open circuit period of 68 h and a discharge at the specified cranking current. This schedule is repeated until a total of 4 units of overcharge has been successfully achieved. To assess the cycling capability, batteries are placed on automatic charge/discharge units on a 1 h discharge/5 h charge at 40°C. The capacity is measured every 36 cycles at the nominal cranking current, but at the same temperature. The voltage at the end of the check discharge should exceed the specified value. The standard SLI life test specified by the SAE consists of a shallow discharge of 2 min at 25 A followed by a 10 min charge at current and voltage limits. This is repeated continuously for 96 h followed by an open circuit period of 68 h and a check discharge at a high current. This unit of test is repeated if the voltage remains above a specified minimum value. Instead of a 2 min discharge, the test is often carried out with a 4 min discharge to simulate more realistic life and failure modes.

In addition to these requirements, SLI batteries are subjected to various other tests. These include measurement of the charge acceptance rate, vibration resistance, and, in the case of batteries which are shipped in the dry condition, there is a performance requirement after storage and activation. The tests and procedures are defined in various national standards [29–31].

### 4.1.4.2 Motive power batteries
Lead–acid batteries are used as the power source in industrial trucks, electrically powered road vehicles, aircraft service vehicles, golf carts, various types of runabouts used at airports or hospital complexes, etc., and more recently in robotics and guided vehicles.

The primary requirement for this duty is to have good cycling capability. Most types of traction battery are guaranteed for 1200 cycles or four years' service. In reality they are expected to provide 1500 to 2000 deep cycles with little reduction in capacity. In applications such as electric road vehicles, where range and acceleration are important, it is desirable to have the highest possible energy density without a reduction in service life. In other duties such as industrial trucks, energy density is unimportant, and the commercial need is for maximum durability associated with the lowest cost.

Two types of design are widely used for this duty, namely the tubular and flat plate construction. Fig. 4.17(a) and (b) show both designs.

The positive plates in the tubular cell consist of a series of parallel porous tubes each having a centralized lead conductor surrounded by active material. The tubes are made from woven, braided, or felted fibres, which must be resistant to the acidic and highly oxidizing atmosphere, and they may be in the form of a one piece section converted into tubes by stitching (gauntlet type) or may be individual tubes threaded

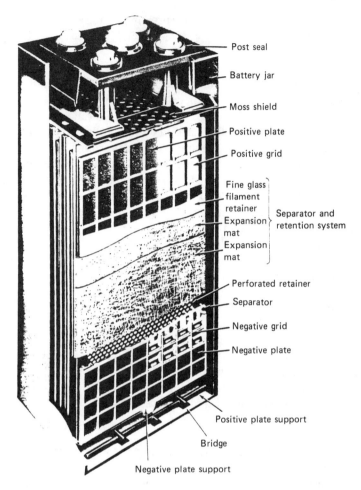

Fig. 4.17 — Lead-acid traction cells. (a) Flat plate design (Courtesy of C&D Batteries).

on the lead conductors. The tubes are sealed at the base with a plastic bottom bar. The negative plates are of standard construction. Edge shorts which may develop after extended service are prevented by an envelope separator which wraps around either the positive or negative plate.

Flat plate motive power cells are similar to SLI cells in overall construction, except that the plates are considerably thicker and more robust and the cells are built with three layers of separator. This usually consists of a perforated spacer, a glass mat, and a microporous plastic separator.

The operating characteristics of both designs differ. Because of the higher surface area and the larger quantity of acid, the utilization of the active material in terms of Ah.g$^{-1}$ is higher in tubular cells, and they generally have a higher energy density than flat plate cells. Conversely, flat plate designs usually have a higher cycle life and are more durable than tubular cells. A further beneficial feature of tubular cells is the

# Sec. 4.1] Lead–acid batteries

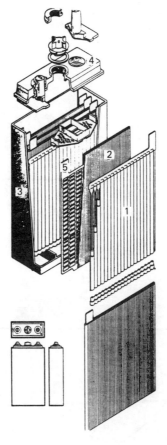

1. Gauntlet positive plates
2. Envelope separators
3. Hi-impact containers
4. Leakproof covers
5. Long life negatives

Fig. 4.17 — (b) Tubular design, (1) Gauntlet positive plates, (2) Envelope separators, (3) High impact container, (4) Leakproof covers, (5) Long life negatives. (Courtesy of CMP Batteries).

lower self-discharge rate. Because most of the conducting grid is buried within the active material, impurities, such as antimony, released as a result of corrosion during service, are adsorbed by the lead dioxide and contamination of the negative plate is reduced. A further consequence of this behaviour is that tubular cells have more stable voltage characteristics and a higher voltage at the top of charge. For this reason it is inadvisable to change from tubular to flat plate design when using the same charger without modifying the voltage settings, and the two types of cell should never be used within the same battery. In Europe the capacity of these batteries is usually quoted at the 5 h rate of discharge, whilst the 6 h capacity is used in North America. Typical performance characteristics of traction batteries at different rates of discharge are given in Fig. 4.18.

The nominal cycling capability of between 1000 and 2000 cycles is based upon discharges to 80% of the 5 h capacity and when charged according to the manufacturers recommendations. Operation at lower depths of discharge extends the cycle life substantially, as can be seen in Fig. 4.19.

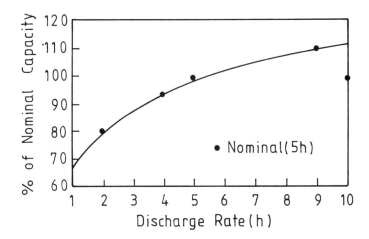

Fig. 4.18 — Traction batteries: effect of discharge rate on capacity.

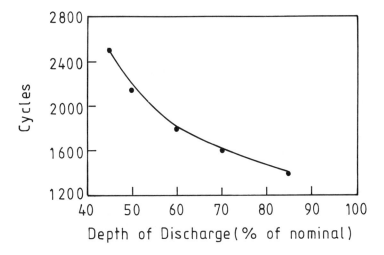

Fig. 4.19 — Traction batteries: cycle life for various discharges.

Development of these batteries in recent years has concentrated on improving the energy density and the power capability for use in advanced electric road vehicles. The general approach has been to increase the plate surface area by using thinner plates. These changes have also had the effect of reducing the plate pitch, thereby reducing the impedance of the cell and increasing the power capability. Improvements of up to 40% in the energy density have been achieved. In so doing the active material utilization is improved. At the same time, however, the cycle life is adversely affected. Although both high performance flat plate and tubular designs for electric vehicle duty have been developed, most progress has been made with the tubular design. This is because the retaining porous tube around the positive active material reduces the loss of active material which normally occurs as the utilization is increased. An indication of the improvement in both the gravimetric and volumetric energy density achieved in recent years is shown in Fig. 4.20.

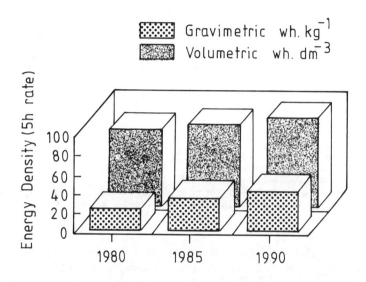

Fig. 4.20 — Electric vehicle batteries: improvements in gravimetric and volumetric energy density since 1980. 1980 — Standard motive power cell, 1985 — Electric vehicle design, 1990 — Advanced electric vehicle design.

Electric vehicle batteries are usually assembled in 6 V monobloc units as distinct from the more usual 2 V cells used in motive power battery arrays, giving a further improvement in energy density. Unlike the standard traction cell they are usually rated at the three hour rate of discharge. A comparison of the operating characteristics of two designs of electric vehicle battery is given in Table 4.6. These are both 6 V monoblocs assembled within the same cube volume. By modifying the plate configurations an improvement of about 12% is obtained.

Performance data in this table were obtained on the General Motors G-van, developed under the EPRI Transportation Programme and due to be marketed shortly (see Fig. 4.21 and Table 4.7). This electrically-powered vehicle has a range of 90 km in simulated city driving, a top speed of 52 mph, and acceleration from 0 to 30 mph in 13 seconds. Recent impetus for the EV programmes has come as a result of environmental concerns, particularly in urban areas. It is planned to introduce up to 10 000 advanced battery road vehicles into the Los Angeles area over the next five years.

**Table 4.6** — Improvements in EV batteries

|  | 6 V Electric Vehicle Monobloc Battery | |
|---|---|---|
|  | ca. 1984 | ca. 1988 |
| 3 h capacity | 165 Ah | 186 Ah |
| 5 h capacity | 184 Ah | 205 Ah |
| Energy density (5 h) | 34 Wh/kg. | 38 Wh/kg. |
| Energy output under 40 kW load per monobloc† | 0.94 kWh | 1.02 kWh |
| Peak power 1.0 V/cell fully charged | 108 W/kg | 112 W/kg |
| 80% discharged | 62 W/kg | 67 W/kg |
| Dimensions | 316 × 183 × 214 (h) mm. | |
| Weight | 31.5 kg | 32.0 kg |
| Driving range loaded van city driving‡ | 83 km | 91 km |

Note: †Results obtained with load on 216 V battery. ‡Data with General Motors G van (gross vehicle wt. 3910 kg, payload 818 kg). Courtesy CMP Batteries and Chloride EV Systems.

The power capability of the battery falls as it is progressively discharged. It decreases rapidly when discharged deeply and near the end of life. For batteries similar to those in the table the internal resistance increase by about 10% at the 50% level of discharge and by 40% when the battery is discharged to 80% depth. Under driving profiles the range and the battery performance is often limited by power and not by energy. The conventional bench tests which monitor Ah and Wh capability under either constant current or constant load conditions do not give a true indication of service behaviour. Tests have been designed to simulate service by discharging to minimum power capability. This consists of monitoring the power output at different states of discharge by imposing a high load for a short period. As an example, two EV batteries were discharged at 55 A for 2.1 h (nominally 70% depth of discharge) then by a 10 s discharge at 250 A. The discharge was continued at 55 A for 18 min followed by 250 A for 10 s and so on, until the voltage at 250 A fell below 4.44 V per monobloc. This is consistent with a vehicle battery pack having a minimum power of 40 kW. Similar tests were carried out under cold conditions (0°C) with the exception that the first 250 A pulse was given after a discharge at 55 A for 1.2 h. These tests give an indication of the useful energy to minimum power levels. Results on two types of EV battery are shown in Table 4.8.

In addition to the EPRI Programme, General Motors are continuing to develop the Impact$^{TM}$ lightweight electric car which is currently capable of accelerating from 0 to 60 in 8 s and has a top speed of 100 mph (see Fig. 4.22). It is fitted with thirty-two 10 V lead acid batteries weighing 396 kg. Each battery has a capacity of 42 Ah giving an overall output of 13.6 kWh.

Fig. 4.21 — The Electric G-Van developed as part of the EPRI Transportation Programme by Vehma International, Chloride EV Systems, and General Motors Corporation, with support from Southern California Edison. (Courtesy of Vehma International.)

**Table 4.7** — Battery and performance data for electric G-Van. (Courtesy Chloride EV Systems Ltd)

| Battery data | | Vehicle data | |
|---|---|---|---|
| Type 6 V | 3ET 205† | Gross vehicle wt (kg) | 3910 |
| Ah capacity | | Payload (kg) | 710 |
| 3 hr | 186 | Curb wt (kg) | 3200 |
| 5 h | 205 | Range, 35 mph (miles) | 90 |
| Weight (kg) | 32 | simulated city driving | 60 |
| Energy density | | Top speed (mph) | 52 |
| 3 h (Wh/kg) | 34 | Acceleration (0–30 mph) in sec. | 13 |
| 5 h (Wh/kg) | 38 | Energy consumption (kWh/mile) | 0.94 |

†Manufactured by CMP Ltd.

**Table 4.8** — Useful energy and estimated vehicle range with two types of EV battery to 40 kW minimum power

|  | At 30°C | | At 0°C | | Vehicle range | |
|---|---|---|---|---|---|---|
|  | kWh | % | kWh | % | 30°C | 0°C |
| Battery A | 34 | 100 | 24 | 100 | $x$ | $0.71x$ |
| Battery B | 42 | 123 | 31 | 127 | $1.23x$ | $0.90x$ |

Fig. 4.22 — The Impact™ Electric Car developed by General Motors Corporation. (Courtesy of General Motors Corporation.)

For optimum service life the charging system should be matched to the battery. Most chargers provide a tapering current; that is, the charge current decreases as the battery becomes progressively charged. These are usually of the single step type, but where it is necessary to charge quickly, say less than 12 h, a two stage charger can be used with a high current initially which is stepped down when the cell or battery voltage reaches a preset value. Most modern chargers contain automatic control devices which monitor the on-charge characteristics of the battery. When the voltage starts to level off as full charge is being reached, the charge is stopped and the battery receives short pulses of charge, the frequency of which is determined by the rate of decay of the voltage after each pulse. This method ensures that the battery is brought to full charge and maintained in that condition without excessive overcharge.

Charging at low temperatures is inefficient and difficult without changing the preset controls. To overcome this difficulty some equipment has the capability to pass spikes of A.C. ripple current through the cold battery to warm it up. Once the battery reaches about 20°C the normal charge proceeds. Frequent charging with substantial A.C. ripples of this kind are not normally recommended, but occasional use for short periods of this type of charging are claimed not to cause damage and to maintain the battery in a healthy condition by ensuring it receives adequate charge. Some electric vehicle systems have more sophisticated thermal management, by incorporating a combined heating and chilling system to keep the battery within given temperature limits.

During service electrolyte may become stratified, with the stronger, higher-gravity acid collecting near the base of the cell. This is normally avoided by intensive gassing at the end of the recharging period. Without thorough mixing, inhomogeneous electrolyte can result in sulphation, premature corrosion, and varying charge efficiencies to parts of the plate. To prevent this effect and to extend the service life, certain types of electric vehicle batteries are fitted with electrolyte agitation pumps [32].

### 4.1.4.3 Standby batteries

The standby market covers a variety of applications where the battery is used to power essential equipment or to provide alarms or emergency lighting if there is a breakdown in the main power supply. The uses and applications have changed significantly in recent years with an increasing demand for uninterruptable power systems (UPS) and a substantial growth in new telecommunication networks.

The paramount requirements for this application are reliability and long service life, and various types of lead–acid batteries are used. Traditionally, standby batteries have thick plates reflecting the need for durability rather than good energy or power density. For many years cells of the Planté design with a service life in excess of twenty years were extensively used in the emergency power sources of telecommunication networks, but they are being increasingly replaced by cells of the sealed design. Planté cells are still used to some extent in electricity generating stations. Both tubular and flat plate designs similar to traction cells are also widely used. These batteries have life expectations of over ten years. In the United States, the round cell developed by Bell Laboratories [33] to improve reliability and durability was introduced with a life expectation of more than thirty years. Exploded views of both the Planté cell and of the Bell round cell are shown in Fig. 4.23.

The positive plate in the Planté cell consists of a massive pure lead casting between 6 to 10 mm thick with a large surface area which is corroded by treatment with oxidizing agents to form a dense covering of lead dioxide. This layer of active material is replaced by further corrosion layers during service. The round cell developed by the Bell Telephone System and Western Electric Company contains saucer shaped pure lead grids, stacked horizontally instead of in the conventional vertical construction. The use of pure lead in place of lead–calcium is to reduce grid growth, and the circular, slightly concave shape is to counter the effect of growth and

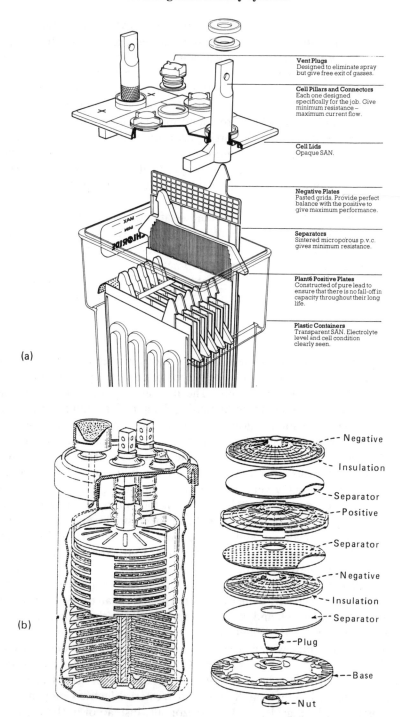

Fig. 4.23 — Two types of standby cell. (a) Planté cell design (Courtesy of Chloride Industrial Batteries Ltd); (b) The 'Bell System' cell (Courtesy of AT&T).

to ensure good contact of the active material and grid during the life of the battery. The grids are filled with a paste made from tetra basic lead sulphate. This material is difficult to oxidize to lead dioxide, and the formation process is prolonged. However, the resulting structure is less prone to shedding.

Recent years have seen the progressive introduction of sealed (valve-regulated) designs. This is particularly the case in telecom and UPS applications [34,35]. These batteries have service lives of more than ten years, they do not require watering throughout service, they give off only small amounts of gas, and they can be located in offices without special venting requirements. In addition, they are smaller and can be conveniently racked without the risk of acid spillage. The reduction in size in comparison with the more traditional flat plate and Planté cells is illustrated in Fig. 4.24, which shows the difference in volumetric energy densities at different rates of discharge. It can be seen that these new designs occupy 70% less space than Planté cells of the same capacity.

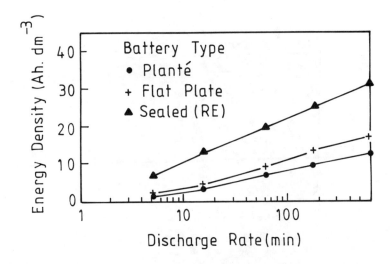

Fig. 4.24 — Energy densities of stationary batteries.

The construction of this kind of standby battery is shown in Fig. 4.25. The grids are cast in alloys free of antimony and other metals which reduce the top of charge voltage. They are usually lead–calcium or lead–calcium–tin alloys. In many designs the plates are wrapped in an absorbent separator made from glass microfibre, and the electrolyte volume is sufficient to saturate the plates and the separator but with no excess. The plates and separator are assembled into the container with slight compression, so that all the components are in intimate contact. The container and lid are welded together and the cells are sealed with a pressure relief valve which allows gas to escape if the pressure gets too high, but the valve does not let air enter. These valves are designed to exhaust at between 1 to 5 psi, and the container must be sufficiently rigid and shock resistant to withstand this pressure without deformation.

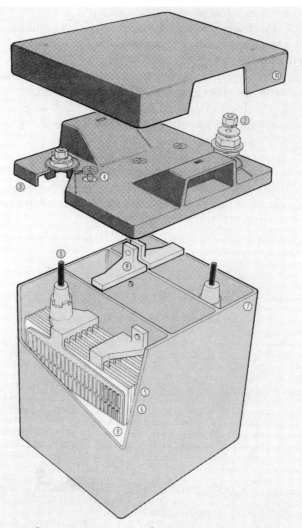

① HIGH CONDUCTIVITY PILLARS
8mm threaded brass insert for maximum conductivity and ease of installation.

② HIGH INTEGRITY PILLAR SEAL
Compression grommet designed for long life.

③ HEAT-SEALED LID
Of flame retardant ABS plastic, welded for life to the container.

④ SELF-SEALING RELIEF VALVE
Low pressure valve (operates below 1.0 p.s.i.). Prevents ingress of atmospheric oxygen.

⑤ RUGGED SUPER-THICK POSITIVE PLATES
Grids designed to resist corrosion and prolong active life.

⑥ BALANCED NEGATIVE PLATES
Ensure optimum recombination efficiency.

⑦ TOUGH FLAME RETARDANT CELL BOX
Of thick-wall V.O. rated ABS plastic, highly resistant to shock and vibration.

⑧ SEPARATORS
Of low resistance microporous glass fibre. The electrolyte is absorbed within this material.

⑨ INTER-CELL CONNECTORS
Designed to withstand ultra high currents, even short circuit.

⑩ ELECTRICAL INSULATION
Cover design provides integral insulation for each unit.

Fig. 4.25 — Sealed (RE) standby battery (Courtesy of Chloride Industrial Batteries Ltd).

In addition, many standby applications require all the components to be made from flame retardent materials.

An alternative design uses similar components, but the electrolyte is immobilized by mixing sulphuric acid with a suitable gelling agent producing a thixotropic mixture. Microcracks which form within the gel allow oxygen to diffuse to the negative plate during overcharge and recombine as in the absorbed electrolyte design.

The capacities of these batteries are usually quoted at the three hour rate of discharge. The available standby time at either constant current or constant power discharge depends upon the end of discharge voltage specified. For the same standby time at a higher finishing voltage, a larger battery is required. Tables of durations to different finishing voltages are given in manufacturers' catalogues. The voltages relate to the battery terminal values, and the voltage drop due to the cable resistance between the terminals and the load needs to be allowed for when assessing the requirements of the system. The relation between duration and output capability with respect to the nominal 3 h capacity for two finishing voltages under both constant current and constant power discharges are shown in Figs 4.26 and 4.27.

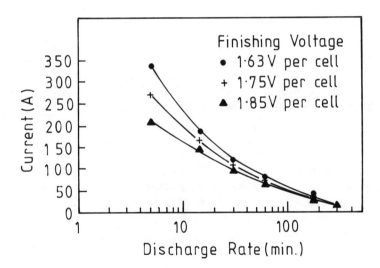

Fig. 4.26 — Sealed stationary battery constant current discharge output. Nominal capacity: 80 Ah (3 h at 20°C).

In certain applications, such as telecommunications, the load is driven directly from the D.C. output of a rectifier and stabilizing circuit. Also across this output is a bank of batteries to drive the load in the event of supply failure. The rectifier must thus fulfil both the functions of driving the load and of charging the batteries.

Normally, the lead–acid batteries are held on float at 2.15 V to 2.40 V per cell, but there is often a requirement to boost charge the battery after deep discharge.

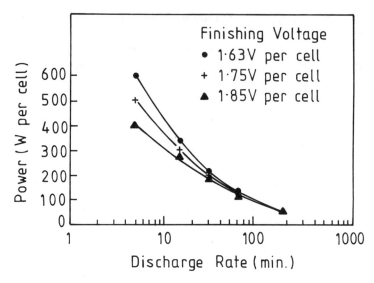

Fig. 4.27 — Sealed stationary battery constant power discharge output. Nominal capacity: 80 Ah (3 h at 20°C).

Telecommunication equipment usually operates on a nominal 24 V or 50 V D.C. supply with typical tolerances in the region of +5% to −10%. Under normal conditions the controller maintains the supply within this tolerance. For example, a $23 \times 2$ V cell battery might have a float voltage of $23 \times 2.25$ V = 51.75 V which is within the tolerance. A typical telephone exchange is likely to be in the range 45 V to 52 V.

The sealed, valve-regulated type of battery is increasingly used in telecommunication networks. For successful operation over long durations, it is important to ensure that the float and boost charge voltages are accurately controlled. These values are also dependent on the ambient temperature, and the recommended float voltage at different temperatures are shown in Table 4.9.

Table 4.9 — Recommended float voltages

| Mean operating temperature (°C) | Applied voltage per cell |
|---|---|
| 0–9 | 2.33–2.35 |
| 10–14 | 2.30–2.33 |
| 15–24 | 2.27–2.30 |
| 25–29 | 2.25–2.27 |
| 30–34 | 2.23–2.25 |
| 35–40 | 2.21–2.23 |

Where faster recharges are required, these batteries can be recharged at higher voltages, usually with a current limit of approximately 10% (in amps) of the nominal capacity. Regular operation at higher voltages will result in loss of water from the battery, and service life will be reduced. In telecommunication service, these batteries are operated in series strings or in series–parallel arrangements. The recommended charging voltage (volts per cell) is independent of the number of cells in a series string, but if more than four strings are operated in parallel, precautions must be taken to ensure charge equalization. Fig. 4.28 shows typical banks of sealed lead–acid (SLA) batteries in a telecommunications system.

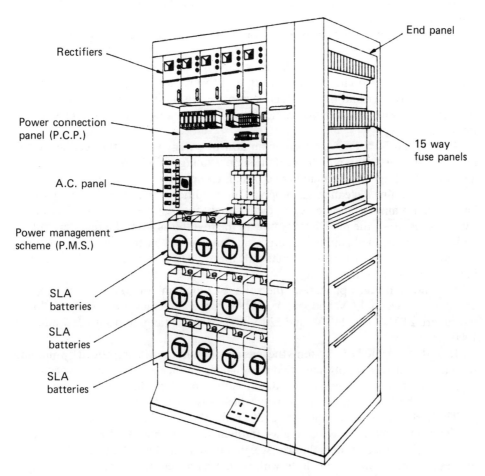

Fig. 4.28 — Power equipment rack (Courtesy of British Telecom Ltd).

Even under normal operating conditions, sealed lead–acid batteries evolve a small quantity of hydrogen gas. As a consequence, the product standard for sealed lead–acid batteries in stationary or standby systems (PB6290 Pt 4) [34] contains a test method which measures the amount of gas evolved during normal operation and expresses this as a level of the recombination efficiency. Under this procedure the ventilation requirement for the equipment housing the battery can be estimated in terms of the required number of air changes.

In recent years there has been a growing need for UPS systems to protect critical equipment from electrical supply failure. The protection often takes the form of supply cover while the system is closed down in an orderly fashion, and this may take typically 10 to 20 min. Alternatively, the battery driven system may continue to operate until a diesel generator or some alternative source of power is available. As a result, the short term discharge capability, such as the 15 min performance, is the important criterion for UPS batteries, although on occasions there may be the requirement for longer duration discharges.

Modern computer and electronic equipment requires a 'clean' and reliable supply of A.C. free of sags or surges in the line voltage, frequency variations, spikes, and transients. UPS systems achieve this by rectifying the standard mains supply, using the direct current to charge the battery and to produce 'clean' A.C. by passing through an inverter and filter system. The battery is an integral part of the supply protection system, and it is axiomatic that the battery output must be reliable and uniform.

In a typical UPS operation the main flow of energy is from the rectifier to the inverter with the batteries on float. If the supply voltage falls below the level at which a regulated voltage can be maintained or fails completely, the battery output to the inverter maintains a clean A.C. supply. This condition can last only as long as the battery can maintain an adequate power supply to the inverter. When the prime supply is restored, the main energy flow to the inverter is from the rectifier, but, in addition, the rectified supply recharges the battery. When the batteries reach full charge the charging current is automatically throttled back as a result of the steep rise in the back e.m.f. of the battery.

Depending on the application, the voltage of the bank of UPS batteries may be anywhere between 12 V and 400 V, although the extremes are not usual. More typical are 24 V, 48 V, 110 V, and 220 V with currents ranging from a few amps to 1000 A.

In both A.C. and D.C. systems the overall reliability can be improved by building redundancy into the supply lines, that is, by having duplicated UPS systems each with separate batteries and possibly a diesel generator. The dual system allows the freedom to boost charge a battery bank which is isolated from the load.

The early UPS systems operated at 50 Hz but there has been a trend, particularly with smaller units, to move to higher frequencies, for example 500 Hz and up to 25 kHz. This change reduces the size, weight, and cost of the UPS system. However, in these systems, and particularly with sealed batteries, which have lower impedance, the long term effects of high frequency ripple could have a deleterious influence on battery performance. This would be manifested by a failure to fully recharge the battery which would result in a gradual loss in capacity. Some indication of this condition can often be deduced when consistently high float-charge currents are observed, but the most reliable means of detecting this 'high frequency shallow cycling' is by monitoring the direction and amplitude of the float current.

In common with all other types of lead–acid battery, the capacity of sealed lead batteries decreases when stored on open circuit for extended periods. However, because they are made from high purity alloys and oxides, the capacity loss is lower than that of many other types of battery. Most manufacturers recommend that batteries should be given a freshening charge if stored for more than six months. Measurement of the open circuit voltage is a convenient guide to the capacity retained by the battery during storage, as can be seen in Fig. 4.29.

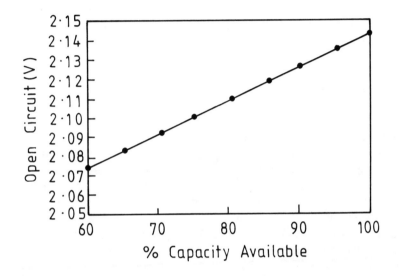

Fig. 4.29 — Open circuit voltage as an indication of state of charge. Based on 3 h capacity measurements (batteries stored at 20°C).

### 4.1.4.4 *Portable equipment and electrical applicances*
The use of small sealed lead–acid batteries in electrical applicances has increased substantially in recent years. Battery sizes in this product type range from about 1.0 Ah to 30 Ah in various voltage groupings from single 2 V cells to 12 V monoblocs. Typical applications include.

- Alarms, security systems, and emergency lighting
- Electronic cash registers, and point-of-sales back-up power supplies
- Power tools and garden equipment
- Photographic equipment
- Portable VTR/VCR/TV, record players, and recorders
- Test and measuring equipment
- Computer memory systems
- Navigation beacons, transceivers, and communications equipment

The primary requirement of these batteries is that they must be transportable and usable in any position without leakage. They are sealed with a pressure relief valve. Because of the variable nature of the duty, the batteries have designs which are capable of reliable service whether used under float or cycling conditions.

The electrolyte in these batteries is either absorbed within the separator and the plates or immobilized with suitable gelling agents. In addition, the cells are made in both cylindrical and prismatic forms. Fig. 4.30(a) and (b) illustrate both types, showing the important features.

The cylindrical cell is made by spirally winding the positive and negative electrodes which are separated by a highly absorbent glass microfibre separator. The electrodes are made from punched or expanded lead strip to which the appropriate lead oxide paste is added. The wound element is placed in a thin walled plastic container before insertion in a metal can. The grid metal is high purity lead to which a small amount of calcium is added to reduce growth and creepage. High purity levels are specified for the oxide, separator and electrolyte. The 2 V cells can be assembled into battery configurations by using shrink wrap or preformed plastic containers.

The prismatic design is made in a more conventional manner by assembling positive and negative plates with interleaved absorbent separators. The separators are often wrapped around the base of the plates as a protection from bottom edge shorts, but, owing to the resilience of the separator and the absence of free acid, there is no need for side edge sealing as in many designs with free acid. High purity materials are again used, and the grids are made by casting or expanding lead–calcium binary or lead–calcium–tin ternary alloy.

Controlled amounts of acid are added to each cell. Certain manufacturers use gelling agents such as high surface area silica powders to immobilize the electrolyte. In addition, the electrolyte in these cells frequently contains a small amount of phosphoric acid which, during discharge, produces lead phosphate within the active material and prevents the plates from being discharged too deeply. As a consequence, the cycling performance can be improved. Certain manufacturers also add sodium sulphate to the electrolyte to improve recovery after deep discharge.

An alternative design is the 'flat pack' arrangement (see Fig. 4.31). This is designed to fit into VCRs and similar equipment, and it is also a method of simplifying assembly and reducing manufacturing costs.

All these designs are sealed with a pressure relief valve located below the external cover.

The operating durations of these batteries when discharged continuously at various currents are shown in Fig. 4.32. The minimum battery size required for the prospective duty can be determined from the intersection of the current and the discharge time. The nominal capacity is usually quoted at the 20 h rate to a final voltage of 1.75 V per cell.

The voltage pattern during discharge at various rates is shown in Fig. 4.33. The C rating of this figure is the current in amperes equivalent to the 20 h rating. Thus in the case of the 30 Ah battery the 2 C discharge current is 60 A and 0.05 C is 1.5 A. This graph shows the voltage profile of 2 V, 6 V, and 12 V systems. The capacity values in these graphs are quoted at a discharge temperature of 20°C. These values vary with temperature, as shown in Fig. 4.34.

Sec. 4.1]  **Lead–acid batteries**  237

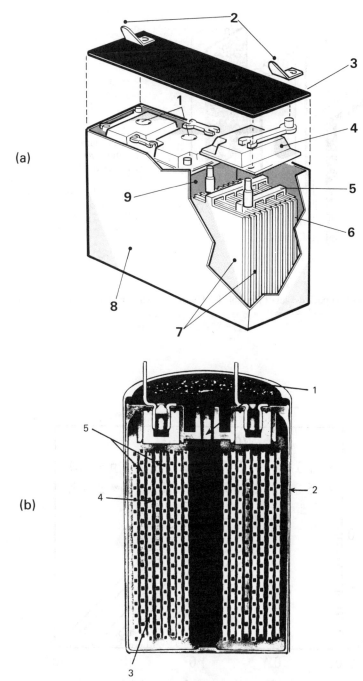

Fig. 4.30 — Portable sealed lead battery. (a) Prismatic form; (1) Self sealing vent, (2) Welded terminal, (3) Cosmetic cover, (4) Inner cover, (5) 2 V element, (6) Separator, (7) Lead/calcium grids, (8) High-impact polystyrene case, (9) Plastic cell divider. (b) Wound design; (1) Self sealing vent, (2) Plastic lined metal can, (3) Pure lead grids, (4) Absorbent separator, (5) Spirally-wound positive and negative plates. (Wound design courtesy of Gates Energy Products).

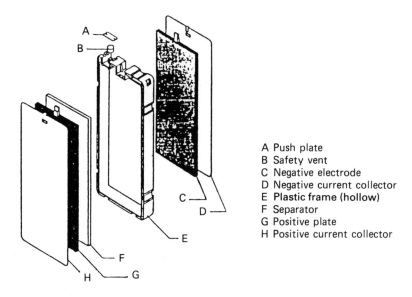

Fig. 4.31 — Flat pack design of portable lead battery (Courtesy Matsushita Electric Ltd).

A Push plate
B Safety vent
C Negative electrode
D Negative current collector
E Plastic frame (hollow)
F Separator
G Positive plate
H Positive current collector

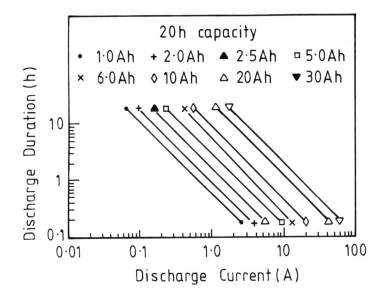

Fig. 4.32 — Battery selection chart.

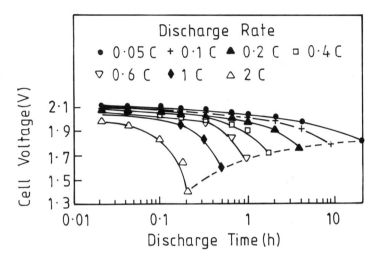

Fig. 4.33 — Discharge characteristics of the Flat Pack Battery. All discharges are at 20°C. The C rating is the current equivalent to a 20 h rating.

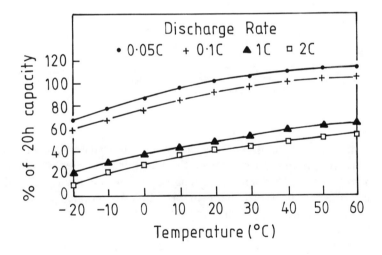

Fig. 4.34 — Capacity variation with temperature of the flat pack battery. The C rating is the current equivalent to a 20 h rating.

The service life of these batteries, in common with other lead–acid battery designs, is affected by many factors, including operational temperature, discharge rate, depth of discharge, and charging method. Recharge conditions should be controlled to give adequate input after the discharge. Recommended inputs are between 115% and 125% of the discharged capacity. Inputs in excess of this will not reduce the service life significantly so long as the finishing current is within the capability of the recombination reaction. Higher currents would result in excessive loss of water and corrosion of the positive grid. Recommended recharge conditions for optimum cycle life are provided in most manfacturers' catalogues.

Sealed batteries of this type will provide over 200 full discharge cycles when operated under optimum conditions. At lower depths of discharge, in excess of 1000 cycles are achievable, Fig. 4.35 shows the influence of depth of discharge and charging voltage on cycle life.

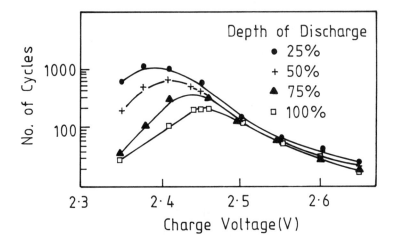

Fig. 4.35 — The charge voltage influence on cycle life of the flat pack battery. All discharges are at the 20 h rate (20°C), limited to 16 h.

This figure illustrates a situation in which the battery is used on a daily basis, that is, one cycle per day. Under these conditions the available recharge time is probably between 12 h and 16 h, and the optimum recharge voltage is in the range 2.35 V to 2.45 V per cell. If use is more infrequent, that is, one or two cycles per week (thus allowing longer charging periods), a charging voltage in the range of 2.28 V to 2.35 V per cell would increase the cycling performance.

As in most battery operations, to maintain the battery in a healthy condition it should be recharged after every discharge. Service life is much improved by recharging a partly discharged battery rather than using the battery on a number of occasions before recharge. Fig. 4.35 also demonstrates that better cycle life can be achieved by slightly oversizing the battery for the application.

Where the battery is used for emergency back-up power and is likely to be discharged infrequently, the recommended float voltage is between 2.27 V and 2.30 V per cell at ambient temperatures of 15°C to 25°C. At different temperatures the charge voltage should be modified in line with the recommended values shown in Fig. 4.36, which shows a voltage range for each temperature. If the application is

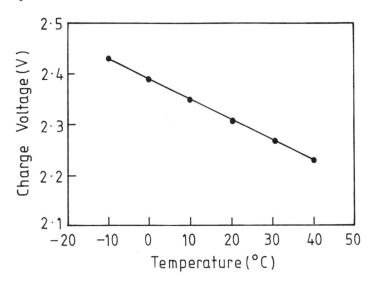

Fig. 4.36 — Charge voltages for float operation of the flat pack battery.

such that discharges are infrequent, then the lower range should be adopted. This voltage provides a full recharge from the fully discharged condition in 72 h. However, if the duty calls for a quicker recharge, then the upper voltage should be chosen. Recharge time can be increased by using step-type chargers.

To achieve satisfactory service it is important that the charge current reduces to a low value when the battery is fully charged. For example, at a charge voltage of 2.27 V per cell (25°C) the current flowing into the battery after 72 h should be between 1 and 3 mA per Ah of the nominal capacity, and should remain constant thereafter. In common with other designs these batteries should not be stored for long periods without a freshening charge. Recovery to the fully charged condition is difficult if the retained capacity is allowed to fall to less than 50% of the nominal value. This is particularly the case when the battery is installed on a low voltage charger. It is good practice to charge the battery on a constant current charger before installation.

### 4.1.4.5 Other applications

During the last ten years considerable interest has been shown by utility companies and large users of electrical power in battery energy storage schemes for load management [35,36]. A number of large systems have been installed and are undergoing proving trials. They include the 40 MWh battery in operation by Southern California, Edison which has 8256 cells, each having a capacity of 3250 Ah. Installed in 1988, and undergoing experimental trials, it is due to go into service in the near future. Further details are given in Table 4.10. Battery-based energy storage schemes of this type were widely used earlier this century, but their popularity decreased, and gas engines and turbines are now extensively used, both for back-up power supplies and to support the peak demand requirements. A recent study [37], carried out on behalf of the Electric Power Research Institute, compared the cost of

**Table 4.10** — Characteristics of Chino load levelling battery

| | |
|---|---|
| Cell capacity | 3250 Ah (5) |
| Energy | 5 kWh |
| Positive plates | 17 (lead/antimony) |
| Negative plates | 18 (lead/calcium) |
| Separator | Glass/PVC/rubber |
| Acid | 1.28–1.29 |
| Cell energy density | 24 Wh/kg |
| Number of cells | 8256 |
| Maximum storage capability | 40 MWh |
| Warranted service life | 2000 cycles (80%)/8 yr |
| Operating voltage | 2000 |
| Energy efficiency (A.C. to D.C.) | 75% |

various energy storage schemes. These include compressed air, pumped hydro, combustion turbines, and lead–acid batteries. It concluded that lead–acid batteries are the lowest-cost option for duties of about 1 h, with combustion turbines being most cost-effective for 2 h to 5 h, and compressed air the preferred option for longer periods.

Load management schemes help to improve the efficient use of energy by increasing the load factor at the generator, and they provide a strategic edge in power purchase negotiations since utilities can either store their own surplus or purchase power from other utilities and independent producers for later use. Such schemes are already practised on a large scale in many countries. Most are pumped hydro systems operating at high voltages. Battery storage schemes have the benefit that they can be virtually any size, can be located anywhere within the distribution system, and can operate at any of the system voltages. Operating within the distribution system means that load management can be localized, enabling the peak factor within a specific region to be reduced. Since equipment ratings are based on peak demand, any reduction allows these to be more conservative. Maximum technical benefit is obtained by managing the load close to the customer and at as low a voltage as possible, and there are obvious benefits with regard to the security of supply and stability.

In addition, battery energy storage schemes have many features which are commercially and environmentally favourable, including:

- Load following ability: immediate power is available to meet changing demands as generators are ramped up to operating speeds.
- Spinning reserve credit: fuel can be saved by using stored power as instantaneous reserve.
- Voltage/frequency regulation: eliminates voltage swing and avoids using large generators to control frequency
- Modular construction: battery modules can be added as needed when energy requirements rise in a local area. A battery storage scheme can be designed and installed in 1 to 2 years.

- Environmentally acceptable: units can be installed in urban areas. They are clean and quiet and have virtually no emissions.

The load management application requires long life batteries with good cycling capability. The projected service life of the large batteries now in service is 10 years, or more than 2000 cycles, and traction/motive power designs with minor modifications are being used. Whilst the energy costs increase as the service autonomy reduces, the power costs decrease, as can be seen in Fig. 4.37.

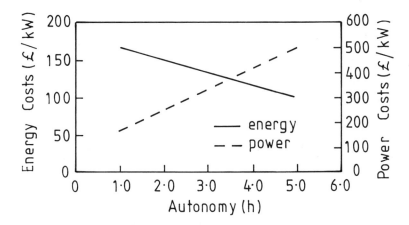

Fig. 4.37 — Costs of a load levelling battery. Initial cost £100 per kWh at a 5 h rate. 81% and 67% of nominal energy are available at the 2 h and 1 h rates respectively.

### 4.1.4.6 Storage of renewable energy

Environmental concerns have resulted in considerable interest in the use of renewable energy sources such as photovoltaic, wind, hydro, and tidal systems to generate power. Most schemes are still at the development stage, but an increasing number of applications and sites are coming into service. This is particularly true for remote areas or third world countries, where there may be no alternative energy supply.

Most renewable energy sources provide fluctuating or intermittent output, and batteries are used to store surplus energy. The use of batteries in this way normalizes the output and at the same time can provide a substantial reserve in the event of breakdown or prolonged inclement weather.

Lead–acid batteries, which have a good deep cycling capability, are generally used, and sizing together with correct choice of battery is important for extended operation. Frequently, batteries are undersized or of the wrong type, resulting in early failure. A well designed system with matched components should give up to ten years of reliable service. Installation in remote areas is not uncommon, and the battery is usually shipped in the dry charged condition and is capable of storage for two years before filling and charging. Maintenance intervals are usually a minimum of six months, and the cell design must accommodate the necessary amount of electrolyte or be fitted with an automatic watering device. A variety of battery types are used for this application, and the nature of the duty requires the battery to be discharged over long periods. Typical performance levels for discharge durations up to 500 h are given in Table 4.11. At these discharge rates, the limitations in active material and electrolyte result in only small increases in capacity as the discharge is prolonged.

**Table 4.11** — Long discharge capacities of three types of cell

| Discharge time | Planté cell Ah | Tubular cell Ah | Flat plate cell Ah |
|---|---|---|---|
| 10 h | 375 | 364 | 366 |
| 250 h | 464 | 502 | 462 |
| 500 h | 490 | 510 | 474 |

Note: All discharges to a final voltage of 1.85 V.

## 4.2 NICKEL–CADMIUM BATTERIES
### INTRODUCTION
**(C. P. Albon & J. Parker)**

Waldemar Jungner, in the year 1890, was the inventor of a form of a nickel–cadmium battery which today is still produced in large quantities. Jungner had sought a rechargeable battery system which had none of the disadvantages of the lead–acid batteries of that period. He foresaw that alkaline electrolytes, which do not change in composition during charge and discharge, could be an advantage, and that many more materials could be available as materials of construction.

Steel could be such a material. In 1897 Jungner [38] patented a design in which perforated metal sheets or wire meshes were sewn together as a container and current carrier for the many metal oxides and metals he was investigating. The forerunner of the pocket electrode for present day alkaline cells had been born. After experimenting with many oxides and metals, Jungner (in 1901) took out a patent [39] for nickel oxide and iron or cadmium couples. The nickel oxide (hydroxide) had been produced by oxidizing nickel in a dilute alkaline solution containing chloride. His early cells delivered only 10–12 Wh.kg$^{-1}$, but by using the perforated metal pocket arrangement he improved this to 22–25 Wh.kg$^{-1}$. He had found it advantageous to mix graphite with the metal oxides with which he was experimenting. Jungner had also discovered that to improve the efficiency of the cadmium electrode, the addition of iron [40] was necessary. Thus, by 1901, the present day pocket plate nickel–cadmium battery had been invented.

During the same decade, Edison had also been developing a nickel oxide electrode. The outcome of his developments was a similar nickel oxide pocket electrode [41] which he coupled with an iron electrode, and he put this cell on the market. These were withdrawn in 1904 because tests had shown that the cells lost capacity rapidly. He reintroduced the cell commercially in 1908, using a laminated tubular construction for the positive electrode. Each 10 cm tube contained 300 alternate layers of nickel hydroxide and nickel flakes. The electrolyte of the reintroduced cells had the addition of lithium hydroxide, the amount being proportioned to the amount of nickel hydroxide present. Edison had recognized the importance of lithium hydroxide for the maintenance of capacity of the nickel hydroxide electrode in the presence of iron compounds.

After a decade of experimentation and development, pocket plate nickel–cadmium cells and tubular nickel–iron cells were manufactured on a commercial scale from 1909 and 1908 respectively. The pocket plate nickel–cadmium batteries were used for many areas of application including automobile engine starting and standby requirements. The tubular nickel–iron cell was, however, mostly restricted to the traction market, as the instability of the iron electrode in potassium hydroxide made it unsuitable for standby use.

The pocket plate nickel–cadmium cells were produced in general purpose and engine starting forms. However, for some applications the high rate performance was inadequate. During the 1920s, experiments in Germany led to a patent [42] in 1928 for a nickel–iron version of a sintered plate cell. Further developments led to E. Langguth of Accumulatoren Fabrik AG (A.F.A) applying for a patent [43] on a sintered plaque of copper or nickel in combination with cadmium. During World War II, A.F.A. (now Varta) developed the first serviceable sintered nickel–cadmium batteries. The very thin flat electrodes which could be assembled close together produced cells capable of the high discharge rates required by aircraft applications. The sintered plates were produced from nickel powder made from nickel carbonyl. After the war, manufacture of sintered nickel–cadmium started in several countries, mainly for applications in the military field and for aircraft.

A. E. Lange, E. Langguth, E. Breuning, and A. Dassler of A.F.A. (now Varta) had recognized that a moist cadmium electrode could react with oxygen, and in 1933 [44] they applied for a patent for a cell that could be sealed between charges. The fully sealed cell had to wait until after the war, when the French Company, Bureau Technique de Gautrat, patented [45] (in 1952) a cell incorporating uncharged surplus negative material to suppress the evolution of hydrogen and at the same time allowing the reaction of oxygen on the charged cadmium present. Another patent [46] by the same company pointed out the advantages of a limited amount of electrolyte and of an absorbent separator which could retain electrolyte and, at the same time, allow the passage of oxygen from an overcharging positive electrode to the negative electrode.

It was about 1960 when the possibility of using the flexible property of sintered electrodes to make a wound cell was first recognized. One of the earliest on the market was produced by Societe des Accumulateurs Fixes et de Traction (S.A.F.T.) in France, later to be followed by several European, American, and Japanese manufacturers.

### 4.2.1 Pocket plate cells
(C. P. Albon & J. Parker)

#### 4.2.1.1 Construction
**Manufacture of active materials**

The positive active material consists of a mixture of nickel hydroxide and graphite. Nickel sulphate is the raw material, although this may be made by the manufacturer from nickel powder supplemented by scrap plates and sulphuric acid. The nickel sulphate is dissolved and nickel hydroxide precipitated with sodium hydroxide solution. The precipitated nickel hydroxide may be washed to remove sodium sulphate and is then removed from solution by filter pressing or centrifuging and dried at about 100°C. Manufacturers have their own techniques for precipitation and further processing, and careful control is required to produce a reproducible material. After drying, the material has to be washed again to remove further sodium sulphate and carbonates. By this stage the volume is much reduced because of shrinkage at the first drying stage, and it is easy to dry the nickel hydroxide a second time. The hydroxide has then to be ground and mixed with graphite. The techniques for this stage are unique to each manufacturer. The graphite quantity is usually 20% of the finished dry product. The graphite used is usually naturally occurring flake material, mined and purified from various countries.

The original method of manufacturing the cadmium material was an electrolytic process using a cadmium sulphate/iron sulphate electrolyte with a row of cadmium anodes and a row of iron anodes, the anodic current of each being controlled separately. The deposit was collected on a rod which had to be scraped free of deposit frequently to keep the cathode current density high. The sponge cadmium/iron deposit was recoved by dredging. It had to be washed, pressed in canvas bags, dried, and milled. This method was labour intensive and produced a material low in cadmium utilization. It has been superseded by chemical methods.

The chemical methods entail the production of cadmium oxide by burning cadmium vapour. The cadmium oxide is recovered from the 'smoke' by cyclones and is then converted to cadmium hydroxide by hydration. Iron oxide is introduced at some stage of the process, and small amounts of graphite and organic additives are added. The cadmium hydroxide based material is much higher in cadmium efficiency than the electrolytic material or materials based on cadmium oxide.

The final grinding stages of both the nickel hydroxide and cadmium oxide processes depend upon the requirements of the subsequent plate-making machinery rather than any electrochemical efficiencies. The materials must have the correct particle sizes for the flow properties, cohesiveness, and densities required by this machinery. To obtain the required densities, a compacting stage may be included in the finishing processes.

**Manufacture of perforated steel strip**

There are two main routes for perforating the strip. In either method only the central position of the strip is perforated, leaving a plain margin along each edge. The fastest and cheapest of these is roller perforation in which the strip is passed between male and female rollers. These produce indents which are made into holes by passing under a grinding wheel. The drawback with this method is that an open area of only about 12% can be obtained, since the rollers cannot be made to produce holes very close together. The maximum hole size usable by the manufacturer is in the region of 0.25 mm, otherwise too much active material is lost from the electrodes during formation and use.

The performance of a plate is dependent upon the open area of the ribbon, and to exceed 12%, needle perforation has to be used. A die of needles attached to a cam punches the steel strip. The process is much slower than the roller process, but, because the needles can be arranged in the die close together, much closer spacing of the holes can be obtained, with open areas up to 20%. Fig. 4.38 shows an enlarged

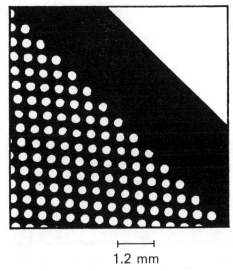

1.2 mm

Fig. 4.38 — Section of perforated strip.

view of a section of this perforated strip. Even higher open areas can be obtained by special patented means. In a patent by Ahlgren [47] the strip is punched from both sides so that the lower needles penetrate the ribbon in the space between the needles of the upper punch. In the Moffat [48] patent, two punches of needles separated by a half diameter of the needles are used sequentially. By these methods a close 'packing' of the holes is obtained, and open areas above 25% can be achieved. These ribbons are used in the ultra high performance ranges of cells.

After perforation of the strip it is next nickel plated for the positive plates to resist corrosion at the charge/discharge potentials of this electrode. To reduce environmental hazards during plate making and cell assembly, an organic or inorganic film may be applied to temporarily seal over the perforations until the formation process, which has to dissolve this film.

**Strip filling**
Two coils of perforated strip are loaded on to each filling machine to form the two faces of a rectangular pocket containing the active material. Fig. 4.39 illustrates the

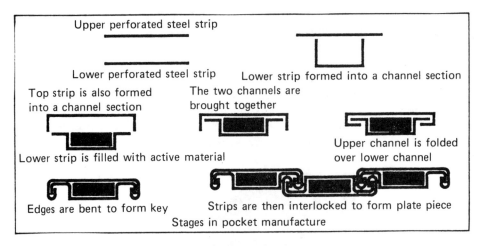

Fig. 4.39 — Stages in pocket manufacture.

metal bending processes performed by roll forming machines to produce a length of banded strip for cutting into plates. The perforated strip has smooth and rough faces due to the roller perforating or needle punching process. The rough face always faces inside the active material and helps to prevent movement of the material within the pocket. It also helps to conduct the electricity to or from the active material. Two methods of inserting the active material into the strips are employed. The cheaper and faster is known as loose powder filling in which either a controlled amount is fed by a screw into the fast moving channelled strip, or the channel is overfilled and passes under a scraper to achieve the desired weight. Although faster, this method has a limitation on the weight of active material which can be put into the strip. For larger weights the material has to be tabletted into 10–20 cm long strips which are then inserted into the strip. Although much slower at this stage, thicker plates can be made by this method, leading to fewer plates for a fixed capacity and cheaper assembly.

The interlocking of the strips is known as lacing, and a number of strips laced together constitute the height of the active part of the plate. This is limited only by the size of the subsequent rollers and the assembly table. The strip is laced in lengths of 5–10 m and then passed under rollers to secure the interlocking and size the thickness of the plates. Next a guillotine cuts the laced strip, now known as a band, into lengths which constitute the width of the plate. Thus a plate piece is obtained. This method of construction is very versatile, and plates of different heights and active mass content can be made without any expensive tooling.

**Plate construction and plate pressing**
The next stage is to construct a frame around the plate piece consisting of a lug and two side pieces of channelled steel known as beading. This is shown in Fig. 4.40. The framed plate is then pressed. During this process the beading is secured tightly in place, and the grooves for the separator ribs are pressed into the surface. Other

Sec. 4.2]  Nickel–cadmium batteries  249

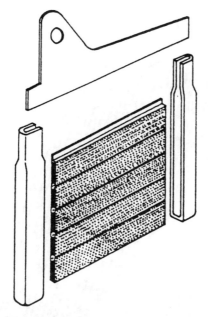

Fig. 4.40 — Expanded view of plate production.

corrugations for strength may also be made, especially in positive plates, which have to be strong enough to restrain the swelling process which takes place during use. A pressed plate is shown in Fig. 4.41.

**Group assembly and separators**
The plates in a group can be either welded or bolted to a bridge and pole. Some manufacturers use welded groups throughout their range, and others use welded assemblies only for the smaller cells where one of the lengths of beading is used as the current collector or lug. Figs 4.42 and 4.43 show a welded and bolted assembly, respectively.

The older separator known as pin and edge, shown in Fig. 4.43, has largely been superseded by frame separators shown in the photograph of a modern assembly in Fig. 4.44, or by perforated plastic sheet separators shown in Fig. 4.42.

**Containers**
Various polymers have been, and are, in use as containers. The original plastic containers were polystyrene, usually with acrylonitrile butadiene styrene (ABS) lids, but these have largely been superseded by polypropylene with its much higher impact resistance. Furthermore, it is translucent, which allows the electrolyte levels to be easily checked. Transparent containers in polyethyl sulphone (PES) or styreneacrylonitrile (SAN) can be provided, but these are expensive.

Where the application requires that the battery withstands severe vibration, for instance in mobile railway applications, the cells can be constructed in steel containers. These are usually finished in a corrosion-proof paint or may be of a stainless grade steel.

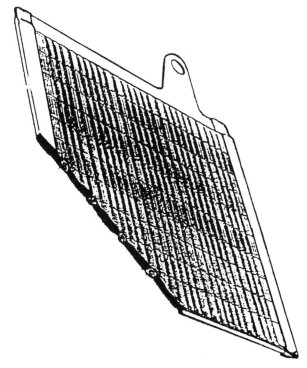

Fig. 4.41 — Pressed plate.

**Formation**
This is carried out to prepare the cell for the customer. In addition to ensuring that the cell performs to the customer's requirements, before leaving the factory, the process known as formation is required in order to do various things. It removes the carbonate and sulphate anions from the active materials, removes any dust suppressing the protective film applied to the plate, and, in general, removes traces of the active masses from the outer surface of the plates so that the debris cannot result in the development of short circuits during the life of the cell. The process may include a soaking in electrolyte before applying a charging/discharging regime, and can include changes of electrolyte. Some manufacturers prefer to form the groups of plates before inserting and sealing into the containers; others jar-form in the container.

**Electrolyte**
The ideal electrolyte should have high conductivity, low freezing point, have no reactivity with the electrodes or other constructional materials, and be inert to the atmosphere. These requirements have led to potassium hydroxide being the choice of alkali for the nickel–cadmium system. The specific conductance of potassium hydroxide is almost twice that of sodium hydroxide at the point of maximum conductance.

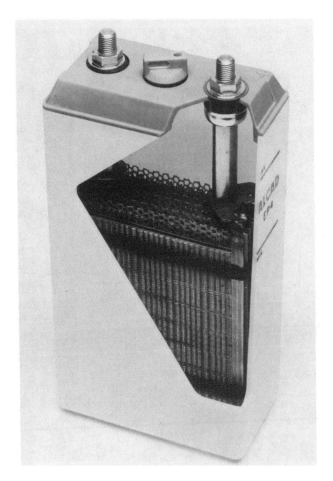

Fig. 4.42 — Weld assembly with perforated sheet separators.

At a temperature of 25°C the maximum conductance occurs with a 25 wt% solution. This solution has a specific gravity of 1.26. However, since manufacturers of pocket plate cells design cells with an appreciable difference between maximum marks to allow for the electrolytic loss of water between maintenance attention, the cells are usually filled at the onset to the maximum mark with electrolyte of density 1.20 g.ml$^{-1}$ such that the density is 1.26 g.ml$^{-1}$ by the time the electrolyte volume is at the minimum level. Some special cells designated HW (high and wide) designed for extremely long intervals between maintenance, are filled with a density of 1.16 g.ml$^{-1}$. The specific gravity range from 1.20 to 1.26 has been found to be the best compromise for performance and life of pocket plate nickel cadmium cells. Although maximum conductance occurs at density 1.26 g.ml$^{-1}$, it has been found that the positive electrode loses capacity on cycling more readily at concentrations

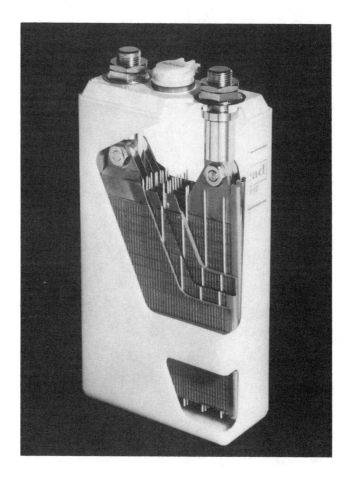

Fig. 4.43 — Bolted assembly with pin and edge separators.

above 1.26 g.ml$^{-1}$, and also the open circuit e.m.f. of the system begins to decrease significantly [49].

The exception to this gravity range is when it is known that the cells are to be used in Arctic regions, when they are filled with KOH of density of 1.25 g.ml$^{-1}$. Whilst the density range 1.20 to 1.26 g.ml$^{-1}$ has freezing points between $-25°C$ and $-45°C$, a potassium hydroxide solution of density 1.30 g.ml$^{-1}$ has the eutectic point of $-66°C$ [50].

Pocket plate nickel–cadmium cells invariably contain some lithium hydroxide in the electrolyte. The concentration is usually in the range 8 to 20 g.l$^{-1}$ LiOH. Lithium hydroxide was found by Edison [41] to have a profound effect on nickel–iron cells by increasing the life of the positive electrodes. Pocket plate nickel hydroxide electrodes are susceptible to contamination by iron species which adversely affect their capacity. This has been found to be due to the lowering of the oxygen evolution potential which reduces the charge acceptance by allowing the oxygen evolution

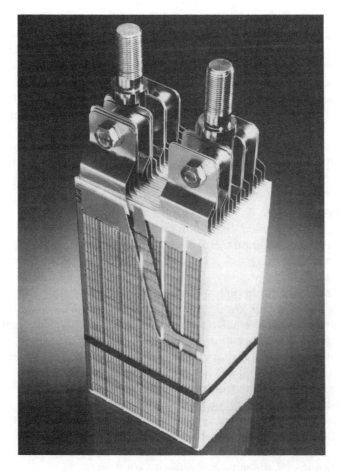

Fig. 4.44 — Bolted assembly with frame separators.

reaction to take place rather than the charging process. Lithium hydroxide has been found to raise the oxygen overpotential and thereby to nullify the iron poisoning effect. Hence lithium hydroxide is added to enable the nickel hydroxide positive electrode to maintain its capacity over 2000 cycles.

The electrolyte of pocket plate cells inevitably absorbs carbon dioxide from the atmosphere and results in the carbonate concentration slowly rising. A much lower amount of carbonate accumulation is due to anodic oxidation of the graphite present in the nickel hydroxide. The specific conductance of a 25 wt% KOH solution containing 5% of $K_2CO_3$ is 10% lower than that of the pure KOH solution. A further 10% reduction occurs if the $K_2CO_3$ increases to 10% of the KOH concentration. Thus from a performance point of view it is important to keep the carbonate concentration as low as possible. Most manufacturers recommend a change of electrolyte in the cells when the concentration reaches certain limits, generally in the range between 40 and 80 g l$^{-1}$ $K_2CO_3$.

## Battery building

Such is the versatility of the pocket plate battery manufacturer that battery building can be done in many ways to suit the needs of the customer. Small plastic cells may be taped together to form small 6 V units, and larger plastic cells can be supplied with stands and assembled at the customer's site. Protecting strips are usually clipped over the exposed terminals to guard against accidental short circuits as required by the Low Voltage Directive which requires terminals/connectors to be covered in batteries above 90 V.

Steel container cells are often built into small units in fireproofed wooden crates made from special timber or plywood. These can be supplied with handles if required.

## Product ranges

Most manufacturers market three ranges of cells — low, medium, and high performance types. In addition there may be special ranges for cycling duties or solar cell applications. The ranges differ in the number of plates for a given capacity brought about by using plates of different thicknesses. The high performance types have a large number of thin plates to provide a high plate surface area. Thus the current density ($A.cm^{-2}$) is low on the application of a high discharge current, and this, coupled with low internal resistance hardware, results in a small voltage reduction below the open circuit or rest value. Low performance types have a few thick plates and significantly higher internal resistance hardware. However, since the cost of manufacture is related to the number of plates within a cell, these types are generally of lower cost. The medium range is a compromise between the high and low performance ranges. The cost effectiveness depends upon the duty for which the cell is purchased. For applications requiring high currents for durations up to about 15 min, the high performance range will be economic; for requirements between 15 min and 3 h the medium range is recommended, and for longer than 3 h, the low performance product is the most cost effective solution.

### 4.2.1.2 Reaction mechanisms of nickel–cadmium cells

The overall reaction taking place in a nickel–cadmium cell is usually given by the reaction:

$$2NiOOH + Cd + 2H_2O \underset{\text{charge}}{\overset{\text{discharge}}{\rightleftharpoons}} 2Ni(OH)_2 + Cd(OH)_2. \tag{4.22}$$

This is arrived at by the addition of the reaction taking place at each electrode. At the positive electrode the reaction is complex and has still not been completely resolved. The simplified reaction proposed by Glemser & Einerhand [51] is

$$NiOOH + H_2O + e^- \underset{\text{charge}}{\overset{\text{discharge}}{\rightleftharpoons}} Ni(OH)_2 + OH^-. \tag{4.23}$$

The standard electrode potential for this reaction is usually given as $E° = 0.49$ V.

$Ni(OH)_2$ exists in two X-ray diffraction identifiable forms known as α and β. Both forms are layer type structures with layers of $Ni^{2+}$ ions between layers of $OH^-$

groups. The difference between the α and β forms is that, in the α form, water molecules are inserted between each layer of nickel and hydroxide groups. Bode and co-workers [60] proposed a relationship between the two forms as follows:

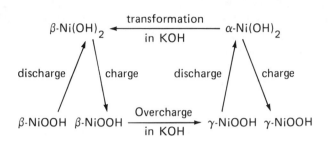

where γ-NiOOH and β-NiOOH are the charged forms of α-Ni(OH) and β-Ni(OH)$_2$ respectively.

The potential of the β–β system is 30–40 mV more positive than the α–γ system. This may explain a reduction in voltage of cells charged for long periods. In practice, the voltage of the electrode depends upon the state of charge and exhibits hysteresis between charging and discharging values, indicating a high degree of heterogeneity.

The overall reaction of the negative or cadmium electrode is generally written as:

$$Cd + 2OH^- \underset{\text{charge}}{\overset{\text{discharge}}{\rightleftharpoons}} Cd(OH)_2 + 2\ e^-. \tag{4.24}$$

This simplifies the reaction, as opinions differ as to details of the mechanism.

Some workers postulate the existence of soluble cadmium intermediates $Cd(OH)_3^-$ and $Cd(OH)_4^{2-}$, whilst other workers consider that there is a solid state reaction involving a film of CdO.

A review of the literature is given by Barnard [52]. The standard electrode potential of reaction (4.24) is $-0.809$ V. Thus, the theoretical open circuit voltage of the nickel–cadmium system is 1.299 V, although, in practice, it is dependent upon the state of charge of the nickel hydroxide electrode bcause of the heterogenous nature of the reaction. The voltage of a cell on open circuit will take a considerable time to decay or rise to this value from charged and discharged conditions respectively.

**Overcharging reactions**

In common with all types of aqueous electrolyte cells, overcharging leads to electrolysis of water into oxygen and hydrogen. In cells with alkaline electrolyte the reactions are, for the positive electrode,

$$4\ OH^- \rightarrow O_2 + 2\ H_2O + 4\ e^-, \tag{4.25}$$

and for the negative electrode,

$$2H_2O + 2\ e^- \rightarrow H_2 + 2\ OH^-. \tag{4.26}$$

The standard electrode potentials for these reactions are 0.401 V and $-0.828$ V respectively. Fortunately for cell operation, the above reactions have appreciable overvoltages. In the case of the nickel hydroxide, the overvoltage is affected by impurities such as iron oxide and cobalt hydroxide, and the presence of lithium hydroxide in the electrolyte, but its value is sufficient to raise the oxygen evolution potential into the region of the charging process of the nickel hydroxide. For the negative electrode, although the $E°$ value of the evolution of the hydrogen is only 19 mV more negative than the charging reaction of the cadmium hydroxide, the hydrogen evolution overpotential is significant. This results in the hydrogen evolution being approximately 200 mV more negative than the cadmium hydroxide reduction potential at the practical current densities of cell operation, and it does not occur until the electrode is approaching 100% charge.

**Charging**
Figs 4.45 and 4.46 show the charge curves of nickel hydroxide and cadmium pocket electrodes respectively when charged at a constant current measured against a mercury–mercuric oxide electrode with an $E°$ value of 0.098 V. The nickel hydroxide

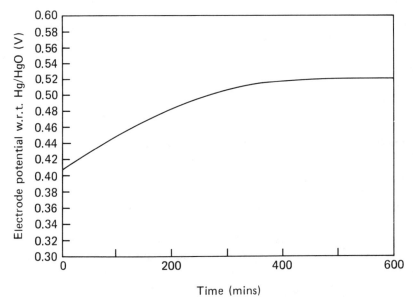

Fig. 4.45 — Charging of nickel hydroxide pocket plate.

electrode charging starts at 0.40 V. There is a gradual rise to 0.52 V with some oxygen evolution commencing as the potential approaches 0.50 V. The negative electrode potential remains constant until hydrogen evolution commences as the electrode approaches complete charge.

**Discharging**
Fig. 4.47 shows the discharge curves of the electrodes in a pocket plate cell measured against the same mercury–mercuric oxide electrode. This clearly illustrates the positive electrode limiting the capacity which is a feature of most manufacturers' cells.

## Sec. 4.2] Nickel–cadmium batteries

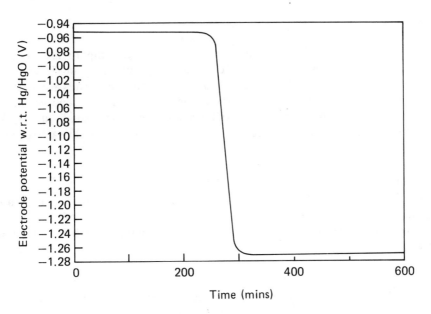

Fig. 4.46 — Charging of cadmium pocket plate.

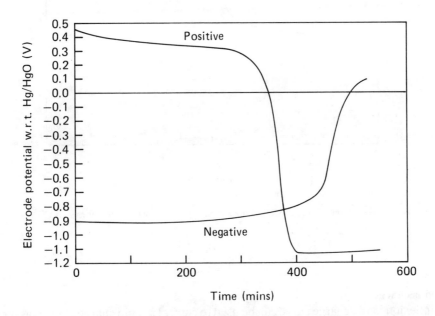

Fig. 4.47 — Discharge characteristics at C/5 of positive and negative plates in a pocket plate cell.

### 4.2.1.3 Battery technology
**Battery ratings**

Pocket plate nickel–cadmium cells are generally rated by the ampere hours capacity which can be delivered by a fully charged cell to a 1.00 V end voltage in 5 h. This value is designated as the $C_5$ capacity of the cell. Discharge and charge rates are written as multiples or fractions of this value. For instance a high rate discharge may be expressed as 10 times the numeric value of the 5 hour capacity $C_5$ usually shown as 10 $C_5$A, whilst 0.2 $C_5$ is a typical charge rate with a current of a fifth of the rated capacity of the cell.

**Battery charging — constant current**

Fig. 4.48 shows the charge characteristics at a constant current of 0.2 $C_5$. This shows the gradual rise until an input of 0.9 $C_5$ has been achieved. If the charge is cut off at this point, approximately 75% of the nominal capacity of the cell will be available on discharge.

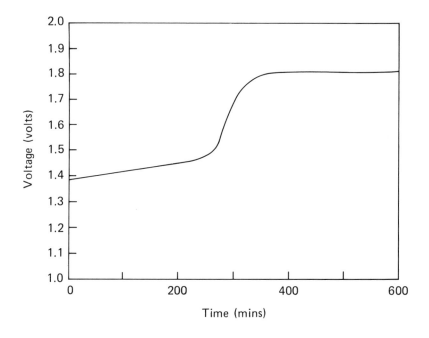

Fig. 4.48 — Charge characteristics of pocket plate cell at constant current rate of C/5.

**Rapid recharge**

The inflexion in the charge curve can be used to control a rapid charge at rates as high as C rate by sensing the point at which overcharge reactions commence, and switching to a lower current. Continuance of high charge rates will result in excessive gas evolution and hence water loss and a rise in cell temperature, since the overcharge reactions are exothermic.

The rapid charge can include a two-rate constant current charge, or a high constant voltage can be used for the initial part.

**Constant voltage charging**
It is normal practice in industry to charge batteries continuously for as long as the main power supply is available. This is achieved by using a constant potential charger which delivers a variable current. The system is self compensating, and the value of the current drawn from the charger automatically reduces as the state of charge of the battery increases. Fig. 4.49 shows the relationship between the current and time of charge of a fully discharged cell.

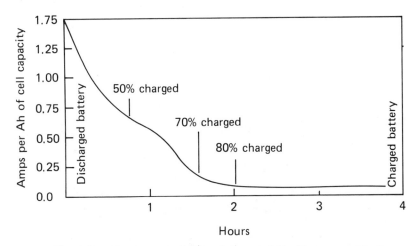

Fig. 4.49 — Constant potential charge characteristic with no current limit.

With cells of low internal resistance and voltages above 1.60 V an initial current of four times the cell capacity could be obtained which can be very high for the charger circuitry. Therefore, to limit the charger cost, it is normal practice to limit the initial current usually to C/5 or C/10 values. The current–time relationship now becomes that shown in Fig. 4.50.

Typical voltage settings in common use are given in Table 4.12. Fig. 4.51 shows the time required for a fully discharged cell to reach various states of charge at different charging voltages for the medium performance range of cells.

**Low maintenance and minimizing gassing**
If the cell is operated so that the overcharging gas evolution is avoided, the period between maintenance can be several years. This condition may be met by operating the battery at the 80% charged condition.

Sometimes environmental conditions require that in the interest of safety the presence of hydrogen and oxygen in the vicinity of the battery must be minimal. In this case the battery may be operated in the 50% charged condition by using a constant voltage charger set at 1.36 V per cell.

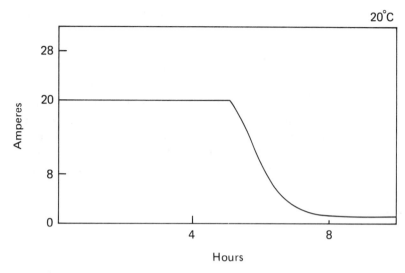

Fig. 4.50 — Constant potential charge characteristic with C/5 current limit.

Table 4.12 — Float voltages of pocket-plate nickel–cadmium batteries

| Range Requirements | High Volts/cell | Medium Volts/cell | Low Volts/cell |
| --- | --- | --- | --- |
| Minimum float voltage. Negligible water loss, but will require boost charging if full capacity is required. | 1.40 | 1.41 | 1.41 |
| Recommended float voltage for fully automatic operation, supplementary charge not required. 75% of capacity will be restored within 6 hours after emergency discharge. | 1.45 | 1.47 | 1.47 |
| Will restore full capacity within 7 days, but water consumption will be high. | 1.58 | 1.61 | 1.63 |
| Will restore full capacity within 24 hours. Water consumption very high. | 1.65 | 1.67 | 1.70 |

**Temperature effects**
The temperature range for the use of pocket plate nickel–cadmium cells is −40°C to +50°C. The only change to the cell required to cover this range is that below −25°C a stronger electrolyte of specific gravity 1.25–1.30 is used in place of the normal 1.20,

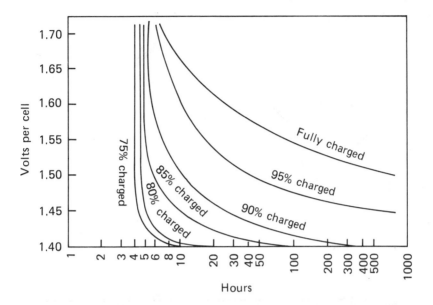

Fig. 4.51 — Time to reach state of charge at various charging voltages for medium peformance cells with current limit of C/5.

to prevent the formation of ice crystals. Complete freezing of a pocket plate cell is not harmful. On thawing it will function normally.

The capacity and voltage are decreased with temperature, but at $-40°C$ the cells will still give about 60% of their rated capacity. However, at higher rates of discharge than C/5 the effect is more pronounced. The effect of temperature is not due to the temperature coefficient of the system, which Falk reported to be almost negligible at $-0.00030$ V.°C$^{-1}$ [53]. The lowering of temperature does have a significant effect on the conductivity of potassium hydroxide, and below $-10°C$ the viscosity increases rapidly with decreasing temperature which adversely affects the diffusion controlled processes in the cell, particularly at the cadmium electrode.

At elevated temperature, the charged species of nickel hydroxide becomes unstable, resulting in a decrease in charge retention and, in cells on constant voltage charge, a higher voltage is required to maintain a given state of charge.

Fig. 4.52 shows the performance variation with temperature.

**Charge retention**
The temperature effects of the charge retention results in the capacity loss at 45°C being about three times that at 25°C. At 25°C a battery will retain 75% of its capacity for six months on open circuit. At temperatures below $-20°C$ the self-discharge is negligible.

**Life**
The pocket plate nickel–cadmium battery is renowned for its long life. This long life stems both from the reversible electrochemical reactions with very few side reactions

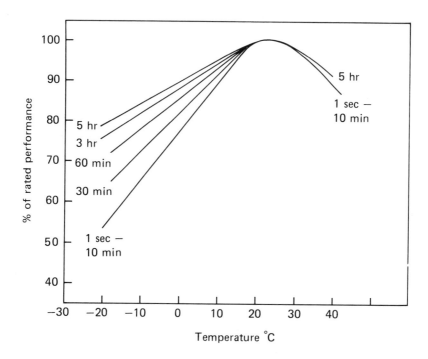

Fig. 4.52 — Variation of performance with temperature.

and from the robust construction which can withstand severe mechanical abuse, including vibration and rough handling in general. The plates are mechanically strong, and the active materials are not easily lost from them. The bolted or welded construction of the groups, the steel (or high impact plastic) containers, and the battery assemblies are all consistent with a good engineering integrity for long life products. The electrolyte does not corrode or damage any components of the cell, and the cell reactions do not lead to any significant migration of the active masses which would cause shape change of the electrodes.

The life of the battery can be thought of in terms of numbers of cycles or total period lifetime. The cycle life of a cell can be in excess of 2000 cycles, especially if cycling design plates are used in the assembly. The period lifetime varies between 8 and 25 years, depending upon the application and on the environmental conditions. The operating temperature, depth of discharge, frequency of cycle, and amount of overcharge are factors which can affect life. The presence of lithium hydroxide in the potassium hydroxide electrolyte extends the life, particularly at high operating temperatures. Cells are also able to stand in storage for long periods in any state of charge.

### 4.2.1.4 *Recent developments*
The trend for users to require batteries needing a minimum of, or no, maintenance has led to the development and introduction of ranges of cells using the oxygen

recombination cycle invented in 1952 by the Bureau Technique de Gautrat [45] to prevent water loss. This cycle uses the rapid reaction between oxygen and cadmium to prevent hydrogen evolution and loss of oxygen and therefore water from the cell.

The positive electrode evolves oxygen on overcharge by the reaction:

$$4OH^- \rightarrow O_2 + 2H_2O + 4e^-. \tag{4.27}$$

The oxygen reacts with cadmium and water to form the hydroxide:

$$2Cd + O_2 + 2H_2O \rightarrow 2Cd(OH)_2, \tag{4.28}$$

and the cadmium hydroxide is in turn reduced back to cadmium by the electrons at the cadmium electrode:

$$2Cd(OH)_2 + 4e^- \rightarrow 2Cd + 4OH^-. \tag{4.29}$$

The left and right hand sides of these three reactions balance, so that there are no net losses from the cell.

For these reactions to take place the oxygen has to diffuse from the positive electrode to the negative, and, to allow this to occur at the required rate, the electrolyte has to be contained in an absorbent fibre separator in specific quantities. The presence of the fibre separator in place of the open frame separators inevitably leads to an increase in the resistance of the cell. Thus the trade-off for a maintenance-free cell of this type is a slight loss of performance.

The pocket plate construction, being 80 years old, has led to much research and development for more modern alternatives. The post-war sintered plate was the first attempt, but the cost of the nickel carrier did not allow this to compete on cost factors. The last decade has seen the introduction of nickel fibre plates based on polypropylene felt which is nickel plated by an electroless process.

Plastic or rubber bonded plates have also been introduced, particularly for the negative electrodes, mainly in conjunction with sintered positives.

Much work has been done on Teflon™-bonded plates, but the strength required to restrain the swelling processes of the positive electrode due to the change in density of the charged and discharged forms of nickel hydroxide is difficult to obtain, and the cost of Teflon™ make these electrodes an unlikely contender to replace the pocket electrode.

### 4.2.2 Sealed cylindrical nickel–cadmium cells
(R. N. Thomas)

#### *4.2.2.1 Introduction*
Cylindrical sealed nickel–cadmium cells have been commercially available for some 30 years, and in that time have gained wide acceptance as a source of portable and standby power. Typical applications include cordless tools and household applicances, models, communications equipment, and emergency lighting.

Their major advantages are long maintenance-free life, excellent storage capability, and high power output over a wide temperature range. Although their initial cost is higher than that of conventional primary cells, their re-usability can quickly prove cost-effective.

Availability in a wide range of sizes allows the construction of many different customer designed packs to meet specific requirements, in addition to direct replacement of conventional primary cells. Whilst the following discussion is confined to cylindrical cells, many of the principles are relevant to button cells and prismatic cells.

### 4.2.2.2 Construction
**Electrodes**
Sealed cylindrical nickel–cadmium cells rely on the use of thin flexible electrodes which are wound into a spiral configuration. The most common electrode technology used is the so-called 'sinter' type. Electrode manufacture has two principal stages:

Firstly, a 'sinter' strip or 'plaque' is formed by coating a substrate which is subsequently dried and subjected to a high temperature non-oxidizing environment. The slurry consists of a mixture of low density carbonyl–nickel powder with an aqueous solution of a cellulose derivative such as carboxy methyl cellulose. The substrate is typically a perforated nickel plated mild steel strip less than 0.2 mm thick.

At the high temperatures used (800°C–1000°C) the organic materials are decomposed and the powder particles are bonded to each other and to the substrate to form a highly porous (80–90%), flexible plaque.

The sinter strip or plaque is used in the second stage of electrode manufacture, which is termed impregnation. In this process, active materials are deposited in the porous plaque, most usually by alkaline precipitation from a nitrate solution: nickel nitrate for positive electrodes or cadmium nitrate for negative electrodes. Cycles of immersion in nitrate, alkali, and washing water are repeated until the appropriate weight gain is achieved. The reactions can be summarized as:

$$\text{Negative: } Cd^{2+} + 2OH^- \rightarrow Cd(OH)_2 \qquad (4.30)$$

$$\text{Positive: } Ni^{2+} + 2OH^- \rightarrow Ni(OH)_2. \qquad (4.31)$$

After cleaning and drying, the electrodes are cut to size ready for incorporation into cells. The final electrode thickness ranges between 0.4 and 0.8 mm.

Recently, alternative manufacturing methods have been sought with the chief aims of reducing the cost of the complex 'sinter' process and of improving performance.

Electrodeposited negatives are manufactured by direct plating from an acid solution of cadmium nitrate whereby spongy metallic cadmium is deposited on to the nickel plated mild steel substrate which passes through the bath. This process leads to a precharged electrode, containing very low amounts of nickel.

Pasted negatives are manufactured by applying a mixture of cadmium oxide or hydroxide, conductive additives such as nickel or cadmium powder, and a binder directly to the nickel plated mild steel substrate.

A further method which is applicable to both positive and negative electrodes has been the subject of much attention in recent years: the sinter strip is replaced by a nickel 'foam' made by plating a plastic foam with nickel and then removing the plastic by burning. This material offers the possibility of increased porosity, and hence active material loading, while retaining the strength of sinter electrodes which is vital to positive electrode cycle life. [54].

**Electrolyte**
The electrolyte used in sealed cells is normally potassium hydroxide (e.g. 1.25–1.30), but it may have lithium hydroxide added in cells designed for high temperature, as this improves the charge acceptance of the positive electrode [55].

Electrolyte conductivity is very significant in relation to internal resistance. For a given composition, the conductivity decreases almost linearly with decreasing temperature between $+30°C$ and $-20°C$. If lithium hydroxide is added, the conductivity is further reduced, with a consequent increase in internal resistance.

The amount of electrolyte contained in a sealed cell is strictly limited. The greater part is absorbed in the separator and the electrodes, so that there is virtually no free electrolyte. Excessive quantities of electrolyte reduce the efficiency of the oxygen recombination mechanism by blocking the surface of the negative electrode.

**Mechanical**
A typical mechanical construction of a cylindrical cell is illustrated in Fig. 4.53. The positive and negative electrodes, along with a separator, are wound together to form a spiral which is inserted into a cylindrical can. The separator is normally a polyamide or polypropylene non-woven felt which prevents electronic conduction between the electrodes but allows ionic conduction via the electrolyte and oxygen transfer for recombination. The negative electrode is connected to the cover which contains a resealing safety vent. The cover is insulated from the can by a plastic gasket which also acts as a gas-tight seal. Final closure of the cell is achieved by crimping the top edge of the can over the gasket, which is then compressed between can and cover.

**Recombination**
The principle of a sealed electrochemical generator using an oxygen recombination cycle were first advanced by Lange in 1938 [56]. A sealed nickel–cadmium cell using this principle was described by Levin & Thompson in 1943 [57]. Jeannin, in 1950, defined the basic rules used in the manufacture of sealed cells [58].

Cylindrical sealed cells rely on oxygen recombination to maintain internal pressure below the relief pressure of the safety vent during charging. The cell is therefore designed so that there is an excess of negative capacity. In this way the positive electrode always becomes fully charged before the negative electrode. Oxygen is generated at the positive electrode and passes through the separator where it reacts with the negative electrode. The reaction can be summarized as:

$$2Cd + O_2 + 2H_2O \rightarrow 2Cd(OH)_2. \qquad (4.32)$$

This supresses further charging of the negative electrode, thus preventing the evolution of hydrogen which cannot be recombined to any significant degree under normal cell operating conditions.

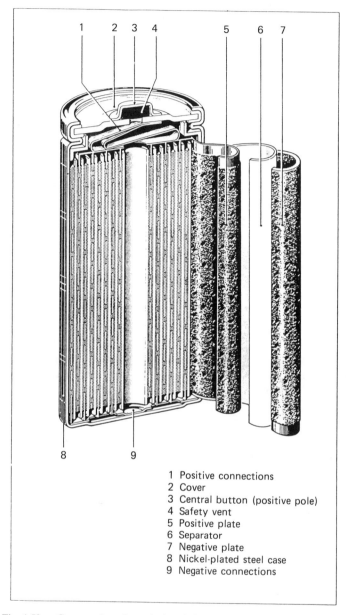

Fig. 4.53 — Construction of a typical sealed cylindrical nickel–cadmium cell.

1 Positive connections
2 Cover
3 Central button (positive pole)
4 Safety vent
5 Positive plate
6 Separator
7 Negative plate
8 Nickel-plated steel case
9 Negative connections

#### 4.2.2.3 Electrical characteristics

*Discharge capacity*

*General.* The capacity available from a sealed nickel cadmium cell is influenced by the same factors as a vented cell: discharge rate, cut-off voltage, temperature, and history. The rated capacity must therefore be measured under specific conditions defined by agreed specifications. The most commonly applied method is that defined by IEC standard 285, *Sealed nickel–cadmium cylindrical rechargeable single cells*, which leads to a measurement at the 5 h rate (0.2 $C_5A$ or $C_5$). The rated capacity is then used to define other currents. For example, a cell rated at 4 Ah must give a minimum duration of 5 h when discharged at 0.8 A (0.2 × 4). The one hour rate (1$C_5A$) is then 4 A (1 × 4), the 20 hour rate (0.05 $C_5A$) is 200 mA (0.05 × 2), and so on. This method allows comparison between different capacity cells and the use of 'generic' data in application calculations.

*Effect of discharge rate on capacity.* Typical discharge curves for a 1.2 Ah cylindrical cell at 20°C are illustrated for different discharge rates in Fig. 4.54. After an initial

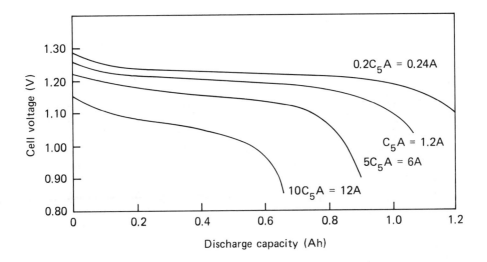

Fig. 4.54 — Discharge curves for a typical 1.2 Ah sealed nickel–cadmium cell at various discharge rates.

drop, the discharge voltage reaches a plateau region where it falls more slowly until starting to fall rapidly at the knee of the curve. It will be seen that, as the discharge rate increases the plateau voltage decreases, and the rate of voltage drop increases. Thus the capacity available at loads in excess of 0.2 $C_5A$ is less than the rated capacity.

It can also be seen that the end-point voltage is critical to the capacity available. At modest rates (less than 1 $C_5A$) most of the capacity is available above 1 V, which is the end-point used to determine rated capacity: but at higher rates a lower end-point of 0.8 to 0.9 V is required to obtain maximum capacity.

At the end of a high rate discharge, residual capacity remains available at lower discharge rates.

*Effect of temperature on capacity.* Typical discharge curves for cylindrical cells at various temperatures are illustrated in Fig. 4.55. The reduction in performance at

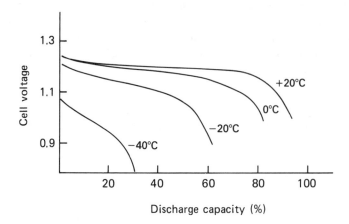

Fig. 4.55 — Discharge curves for a typical sealed nickel–cadmium cell at various temperatures.

low temperature is chiefly due to increased internal resistance arising from decreased conductivity and increased viscosity of the electrolyte: the open circuit voltage is not significantly affected. Performance can also be reduced at high temperatures, principally because of increased self-discharge which will be of particular significance at low rates of discharge.

The effects of discharge rate and temperature on capacity are summarized in Fig. 4.56.

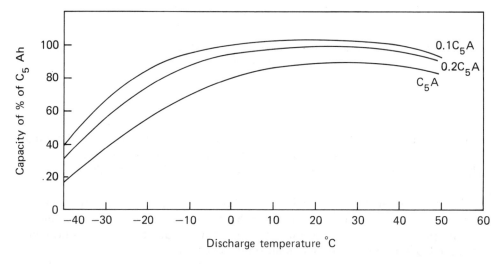

Fig. 4.56 — Effect of discharge rate and temperature on the typical sealed nickel–cadmium cell capacity, after charge at 20°C.

*Peak discharge.* Sealed nickel–cadmium cells are capable of delivering many times the C rate of discharge in short duration pulses. Fig. 4.57 shows the current available

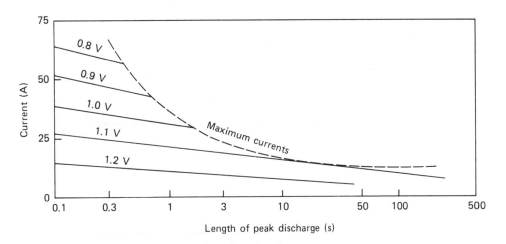

Fig. 4.57 — Current as a function of peak discharge for a typical 1.2 Ah sealed nickel–cadmium cell, for various end of discharge voltages per cell.

at +20°C from a fully charged 1.2 Ah cell as a function of pulse duration for various end of discharge voltages.

The current at any given duration must be limited to a maximum to prevent damage to the cell. In the extreme, a direct external short circuit can cause rapid overheating and cell venting, though this is normally limited by the current carrying capability of external circuitry.

*Internal resistance*
The internal resistance of cells is a complex property determined by the resistive properties of electrolyte, electrodes, separators, and conductors. Various methods are used for measuring it, but for most applications a D.C. method is appropriate.

Internal resistance is affected by the same factors as capacity, and it is also to an extent dependent on the state of charge and the rate of current flow. Fig. 4.58 illustrates the typical change in internal impedance according to depth of discharge, from which it can be seen that the effect is not pronounced until the depth of discharge exceeds 50%. This fact, coupled with changes in internal resistance with cell ageing, has rendered the use of internal resistance techniques to measure state of charge virtually useless.

*Charge characteristics*
*Constant current charging.* Typical voltage temperature and pressure curves for constant current charge are shown in Fig. 4.59.

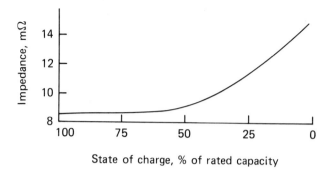

Fig. 4.58 — Typical variation of impedance with state of charge for a sealed nickel–cadmium cell.

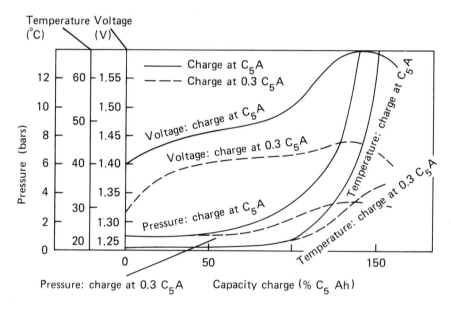

Fig. 4.59 — Voltage, temperature, and pressure curves for constant current charge of a sealed nickel–cadmium cell.

Under normal conditions, after an initial steep increase, the voltage rises more gradually until approximately 100% of the rated capacity has been charged. Thereafter, with the onset of oxygen evolution, the voltage rises more quickly. At the same time, pressure and temperature rise. The temperature rises because the oxygen recombination reaction is exothermic. Eventually the curve reaches a peak and begins to fall because of the negative temperature coefficient of the charging voltage. Ultimately a stable state is reached, depending on charge rate and thermal dissipation.

At the end of charge, almost all the charge current is being used to form oxygen which is being recombined. The recombination capability of the negative electrode is limited, but most cylindrical cells can sustain 0.1 $C_5A$ rate charging at normal temperature for long periods without excessive pressure build-up.

Since most of the energy supplied to the cell during overcharge is converted to heat, high overcharge currents can cause heat dissipation problems, particularly in larger cells.

Fig. 4.60 shows the effects of temperature and charge rate on the charging characteristic. Both the absolute voltage and the distinctiveness of the voltage peak are increased by increased current and by reduced temperature.

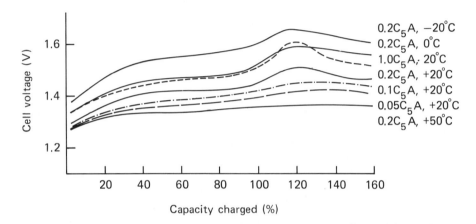

Fig. 4.60 — Charge curves for a typical sealed nickel–cadmium cell at various rates and temperatures.

At low temperature, hydrogen can be evolved at the negative electrode if the cell voltage rises too far. Since hydrogen is not recombined, this can lead to venting and loss of electrolyte. Thus charge currents must be restricted so that the cell voltage does not exceed 1.5 V to 1.55 V. With this restriction, charging can be carried out at temperatures as low as $-40°C$.

At high temperature care must be taken to ensure that the manufacturer's recommended cell case temperature is not exceeded, bearing in mind the temperature rise due to overcharge and any restriction of thermal dissipation arising from battery configuration or location.

*Constant potential charging.* This method of charging is not generally recommended, principally because it may lead to thermal runaway. At the start of charge, current decreases, mirroring the rise of voltage seen in constant current charging. However, during overcharge the increase in temperature due to oxygen recombination reduces internal resistance and the oxygen overpotential at the nickel electrode, so that the cell voltage falls. This allows more current to be driven into the cell, raising temperature further, and initiating a vicious circle.

Constant potential is used, often in memory back-up circuits, where only low rate charging is required. The current must be limited at the end of charge. This is normally achieved with a fixed resistor selected to ensure that large variations in charge current cannot occur.

*Charge efficiency.* Charge efficiency is the proportion of capacity input to a cell which is subsequently available in discharge. Fig. 4.61 illustrates typical available capacities plotted against input capacity at various rates for a conventional cell.

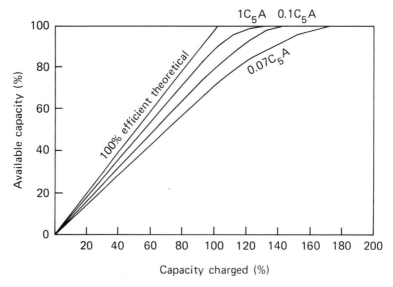

Fig. 4.61 — Charge efficiency curves for a typical sealed nickel–cadmium cell at 20°C.

It can be seen that about 140% of rated capacity input at $0.1\ C_5A$ is required to achieve full capacity output, giving an efficiency of 70%. However, at $1\ C_5A$ only 120% is required, giving an efficiency of 83%.

Apart from charge rate, elevated temperature has a significant influence on efficiency, particularly at low charge rates. Fig. 4.62 shows the typical variation of capacity available after 24 hours and 48 hours charge at $0.05\ C_5A$ with temperature, for a conventional cell. Cells with improved charge efficiency have been developed for high temperature standby applications such as emergency lighting. These cells have modified electrolyte compositions, generally incorporating lithium hydroxide, and additives to the electrode active materials (such as cobalt) which enhance high temperature chargability. Fig. 4.63 shows the variation of capacity availability for such a high temperature cell.

*Charge retention*
Like all rechargeable systems, sealed nickel–cadmium cells lose charge while standing at open circuit. The capacity loss is associated with the intrinsic instability of the charged electrodes and with the presence of impurities.

## Sec. 4.2] Nickel–cadmium batteries

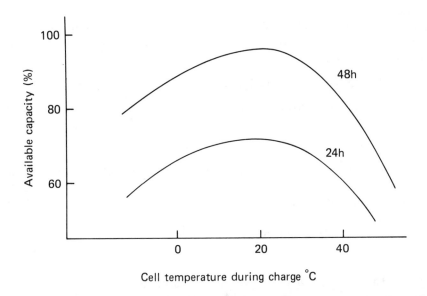

Fig. 4.62 — Available capacity after low rate charge for conventional sealed nickel–cadmium cells.

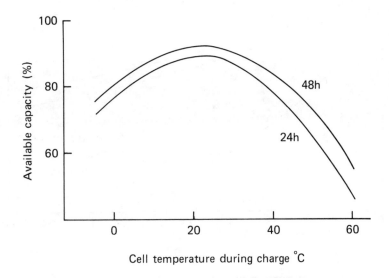

Fig. 4.63 — Available capacity after low rate charge for high temperature sealed nickel–cadmium cells.

Nitrate ions are of particular significance since they cannot be completely eliminated from sintered electrode materials produced by impregnation/precipitation from nitrate solutions. Nitrate ions are reduced at the negative electrode by a process given by the equation:

$$Cd + H_2O + NO_3^- \rightarrow Cd(OH)_2 + NO_2^-, \quad (4.33)$$

and can be transferred to the positive electrode where they are oxidized. The latter process is given by:

$$2NiOOH + H_2O + NO_2^- \rightarrow 2Ni(OH)_2 + NO_3^-. \quad (4.34)$$

The summation of these reactions is discharge while the 'shuttle' reaction continues. However, this shuttle mechanism is progressively eliminated since further reaction ultimately converts nitrate to nitrogen, which is inert. Thus the nitrate effect is greatest in early life.

Alternative electrode manufacturing processes such as pasting which do not use nitrates as starting materials can demonstrate improved charge retention characteristics.

Fig. 4.64 shows a comparison of typical self-discharge rates at various temperatures for a conventional cell and a cell with a pasted negative electrode. As might be expected, the self-discharge reactions are accelerated at elevated temperature and become insignificant below 0°C.

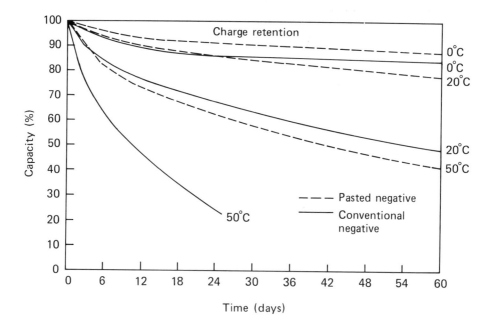

Fig. 4.64 — Charge retention curves for sealed nickel–cadmium cells.

*Operational life of cells*

Two extremes of operational duty can be distinguished: standby applications where cells are continuously charged and only infrequently discharged, and cyclic applications where cells are continuously subject to alternating discharge and charge.

In cyclic duty, life is measured in numbers of cycles. Fig. 4.65 shows the influence

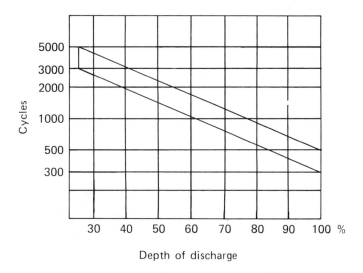

Fig. 4.65 — Cycle life as a function of depth of discharge for sealed nickel–cadmium cells.

of depth of discharge on cycle life for a conventional cell operating at +20°C. Typically at 100% depth of discharge (that is, to 1 V at moderate rates), 300–500 cycles are obtained, while at 50% depth of discharge (that is, to half the rated capacity) this increases fourfold. Effectively a halving of depth of discharge doubles the total life capacity delivered.

Operation at elevated temperature causes a reduction of cycle life: in ambient temperatures of 50°C this could be greater than 50%.

In standby operation, life is controlled by capacity overcharged and temperature. Fig. 4.66 shows life expectation for a conventional cell in terms of temperature and multiples of capacity overcharged. Particularly at elevated temperature, degradation of the normal polyamide separator is a significant factor in determining overcharge life. As a result, polypropylene separators are generally incorporated in cells designed for high temperature applications.

In reality, no application is likely to entail pure standby or pure cyclic duty, nor perfect temperature control. The prediction of actual service life is therefore difficult except in qualitative terms, since pieces of equipment, and their users, can vary greatly.

Failure modes for sealed cells follow bathtub type curves in a similar way to electronic components. High initial failure rates (infant mortality) fall to a low

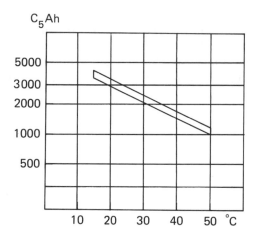

Fig. 4.66 — Overcharge life as a function of temperature for sealed nickel–cadmium cells.

constant failure rate, before climbing again at the end of service life. Typically, during initial 'burn in', lasting a few cycles or a few tens of $C_5Ah$ overcharge, the failure rate may be of the order of $10^{-5}$ per cycle or per operating hour. This would fall to between $10^{-7}$ and $10^{-8}$ in the plateau region.

The principal failure mode for sealed nickel–cadmium cells, excluding manufacturing defect and abuse, is probably internal short circuit. This arises from the growth of cadmium dendrites through the separator, and is accelerated by high temperature and overcharging. The internal short circuit may be of high resistance, but it can divert charging current from the conversion of active material, and it can increase charged stand losses. It is thus recognized in low voltage during charge, open circuit stand and discharge, and in low capacity.

Degradation of the polyamide separator under high temperature conditions along with oxidization during overcharge leads to carbonation of the electrolyte. The presence of excessive carbonate reduces the electrolyte conductivity and thus increases internal resistance. This is avoided in cells which employ a polypropylene separator. However, polypropylene separators must use special additives to ensure wettability. Breakdown of wettability can lead to maldistribution of electrolyte with a consequent effective increase in internal resistance.

Apart from irreversible deterioration of cell performance, two reversible phenomena have been identified as occurring in sealed nickel–cadmium cells. The first of these, the so-called 'memory' effect, is frequently confused with the second, known as 'alloy formation'.

'Memory' effect was originally established for 'space cells' used in geostationary orbit satellites, and can also be generated under laboratory conditions. It occurs when precisely the same amounts of charge and discharge are repeated. Part of the unused negative electrode is thought to undergo a morphological change which makes it difficult to discharge. Thus, after prolonged cycling, a step develops in the complete discharge curve at the point where discharge normally stopped. During cycling the voltage subsequent to this point declines progressively (see Fig. 4.67). Once complete discharge (that is, <1 V/cell) is carried out and the battery is recharged, the normal morphology is re-established and the voltage step disappears.

The second phenomenon, alloy formation, also results in a voltage step in the discharge voltage curve which can be confused with the 'memory' effect. It is thought to arise [59] under conditions of high temperature overcharge through the formation of an alloy ($Ni_5 Cd_{21}$) between nickel and cadmium, these being in intimate contact in the sintered electrodes. In this case the step (or about 100–120 mV) develops because the alloy discharges at a lower voltage than the cadmium normally present in the negative electrode. However, unlike the memory effect, the point in the discharge at which it occurs depends on the period of overcharge and progresses with time (see Fig. 4.68).

Like the memory effect, complete discharge causes the phenomenon to disappear, since the alloy is destroyed and pure cadmium is re-formed. Because alloy formation is most significant in high temperature continuous overcharge, cells designed for these applications make use of non-sintered negative electrodes, for example pasted electrodes. These contain low quantities of nickel, thus preventing alloy formation.

### 4.2.2.4 Applications and battery design
*Cell selection*
Table 4.13 lists the commonly available sizes of cylindrical cell and their ratings. Individual manufacturers may offer variants having enhanced performance, specific high temperature capability, or special cell sizes.

Fig. 4.69 is a typical cell selector guide: discharge duration vs discharge current are plotted on a logarithmic scale to enable an initial estimate of cell size to be made. In addition, the manufacturer, or his data, may need to be consulted to establish the de-rating necessary for the particular application requirement. This de-rating will take into account the relevant characteristics described in section 4.2.2.3 above. In particular, it is advisable to address the following:

- cycling or standby duty
- operating voltage
- capacity, peak current, duty load
- temperature range
- self-discharge
- charging

*Operating voltage*
The maximum and minimum voltages the equipment which is to be powered can tolerate will define the number of cells to be connected in series. In practice, the useful end of discharge voltage is 1 V/cell (or sometimes less at high discharge rates or low temperature). Thus, if the minimum tolerable voltage is 5 V, five cells in series are required, giving a nominal voltage of 6 V and a peak charging voltage in the range 7 V–9 V.

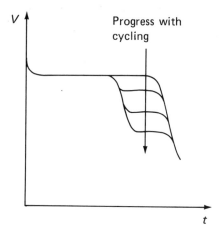

Fig. 4.67 — The progress of the 'memory' effect with cycling for a sealed nickel–cadmium cell.

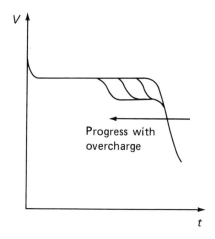

Fig. 4.68 — The progress of the 'alloy formation' effect with cycling for a sealed nickel–cadmium cell.

*Battery capacity*
In assessing the capacity required to meet the duty load, it is usually necessary to consider the peak current rather than the average load. Application loads are not normally simple constant current discharges.

If the load is resistive, the current drain changes with battery voltage. In this case it is appropriate to consider the average current, which can be approximately derived from the mid-point voltage of the discharge curve. The peak current occurs at the

**Table 4.13** — Commonly available cells

| Designation | | Typical rated capacity (Ah) | Diameter (mm) | Nominal Height (mm) | Weight (g) |
|---|---|---|---|---|---|
| ½AA | KRH 15/29 | 0.25 | 14 | 28 | 14 |
| AA | KRN 15/51 | 0.5–0.7 | 14 | 50 | 25 |
| $A_f$ | KRN 18/51 | 0.8–1.0 | 17 | 50 | 37 |
| $C_s$ | KRN 23/43 | 1.2–1.7 | 22 | 42 | 52 |
| C | KRH 27/50 | 1.8–2.2 | 25 | 50 | 75 |
| D | KRH 35/62 | 4.0–5.0 | 32 | 60 | 150 |
| F | KRH 35/92 | 6.0–7.0 | 32 | 91 | 235 |

start of discharge when the battery is best able to support it: thereafter the current is falling, and the discharge efficiency will increase.

When the application load is at constant power the current drain increases toward the end of discharge as the battery voltage falls. It is then reasonable to use the end-point voltage to calculate the worst case load. This peak current occurs at the end of discharge when the battery is least able to support it, and is therefore the limiting factor influencing discharge efficiency.

The application load may be intermittent and at different levels, for example in transceiver applications where high transmit pulses alternate with lower receive loads and standby operation. Under these conditions the available capacity is determined by the peak current load, but the average current can be applied to that capacity to give the duty duration.

*Temperature*
Because operating voltage is depressed at low temperature, particularly at higher rates of discharge, an increased proportion of capacity is available below 1 V/cell. To make use of this capacity it is necessary to work with lower end-point voltages. Thus, if the minimum tolerable voltage is fixed, then the lower end-point implies an increase in the number of series connected cells.

When considering the temperature range for discharge it is important to consider the practical situation. Although the equipment may be required to operate at −40°C ambient, the battery itself may not be exposed to that temperature for extended periods, for example if the equipment is mounted in a vehicle or carried in the pocket. The time between removal from a charger and discharger and thermal insulation between the battery and its environment may thus be significant in this context.

*Self-discharge*
Two factors relating to self-discharge are significant in battery design. Firstly, it is important to estimate how soon after discharging will the battery be applied to the

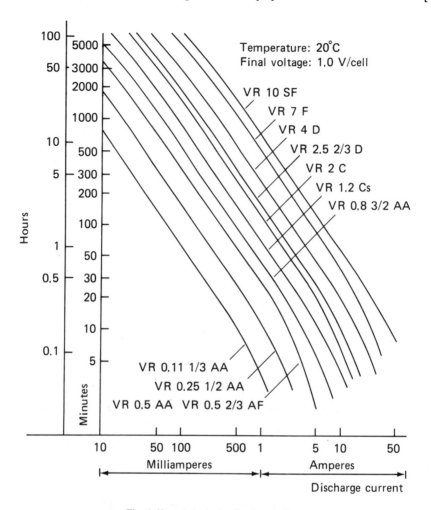

Fig. 4.69 — A typical cell selector diagram.

**Discharge duration**

This diagram facilitates the choice of the capacity of the required batteries. It is presented for type VR 023. For other types, please see the individual specifications.

### USING THE DIAGRAM
— Define the required duration and discharge (final voltage 1 V/cell).
— Read the battery type to be used, above the intersection of Duration and Current

### Example:
Discharge duration: 1 hour
Discharge current: 1 A     Choice: VR 1.2 Cs

load. Account should also be taken of intermittent use or very low rates of discharge such that the required duration is measured in days. If stand times or required durations are long, it may be necessary to select a cell type with enhanced charge retention, or to provide extra capacity to account for self-discharge losses.

**Charging methods**
The choice of battery charging method depends on battery usage and how quickly battery power needs to be restored. If the battery pack is detachable from a portable piece of equipment, then the possibility exists to carry one or more spare batteries: spare batteries can be on charge while others are in use, thus extending the permissible restoration period.

In standby applications a spare battery is not an option, since the battery is maintained in a state of readiness by continuous charging. It is therefore important to examine the permitted recovery in relation to the discharge duty load in order to set the appropriate charging current. In some cases it may be necessary to provide a two-stage charger allowing fast recharge followed by a low rate maintenance charge. More commonly, where a recovery period of 24 hours or more is permitted, a single low rate charge may be possible, the level of which will be chosen to be compatible with acceptable overcharge life and with charge efficiency.

If the charging temperature is higher than normal, charge efficiency is impaired and must therefore be taken into account in calculating the required battery capacity.

Generally, for standby applications where charge is prolonged and discharge infrequent and shallow, only very low charge rates of less than $0.1\ C_5A$ and even as low as $0.01\ C_5A$ are required, according to the permitted recovery period. These lower rates are chosen to maximize service life. However, these very low rates may require the battery to be precharged before placing it in service.

Most cylindrical nickel–cadmium cells will withstand overcharge at $0.1\ C_5A$, except at low temperature. Thus if 14–16 h is permissible for recovery, charger design for cyclic use can be very simple, since it is normally unnecessary to control either the state of charge of the battery before charging or the duration of charge. Of course, if discharge is very infrequent, a lower rate of charge should be considered to maximize life.

Many cylindrical cells can be charged at accelerated rates up to $0.3\ C_5A$, according to cell type and size, enabling full charge to be obtained in about 4 h. Usually, such cells will not withstand overcharge at these rates indefinitely but can be charged irrespective of initial state of charge with simple timer control. According to the application, the charger may either terminate charge or switch to a low rate maintenance charge when the battery is not going to be used immediately. This charging method is generally suited to cycling applications where the battery is normally used to a reasonable depth of discharge, exceeding 25%, otherwise excessive overcharge may occur during total battery life.

Some cylindrical cells have been specially designed to withstand overcharge at higher rates (between $1.0\ C_5A$ and $2.0\ C_5A$) with only simple timer control, as described above. However, normal cells, while equally capable of rapid charging, must be subject to charge control to prevent excessive overcharge and possible cell damage.

It is possible to control rapid charging by using a timed charge, but this requires that any residual charge first be discharged. Although this is a simple method, it does minimize cycle life since it has 100% depth of discharge at each cycle. Thus opportunity can be taken of a practical 'random' depth of discharge to minimize charge time and maximize cycle life by using more sophisticated charge methods.

Charge control methods normally make use of the nickel–cadmium voltage and/or temperature characteristic, and can be listed as follows:

*Absolute voltage or voltage target.* This can be set according to charge rate. It should be temperature compensated, which requires at least a third terminal for the sensor. The charger will have to lock off, or more normally switch to a top-up charge on termination which is normally set at the 100% input level. Circuitry may be required to cater for the possibility of one cell developing an internal short circuit, preventing the voltage target being reached.

*Absolute temperature or temperature target.* This can be achieved quite simply with an internal thermostat, allowing two terminal charging. It is not always favourable, because it will not operate at low ambient temperature. It may also be used in a three terminal battery as a signalling device allowing switching to a lower rate top-up or maintenance charge. A three terminal battery may also use a thermistor device to sense temperature and to allow fast charge at low temperature to be prevented.

*Differential temperature.* With this method, two sensors are used to provide information on temperature rise above ambient: one sensor is usually located at the centre of the battery pack, while the other is placed in good thermal contact with the environment. This type of control is employed in a number of military communication system batteries. It allows a higher upper temperature for charging than absolute temperature systems.

*Voltage derivatives.* The fact that the charging voltage goes through a rise, peak, and fall can be used to control fast charge. At low charge rates, especially at higher temperatures, the characteristic is not sufficiently distinct, but at rates above 1 $C_5A$ the change in battery voltage can be used. In this case voltage readings must be stored and compared to indicate $+\Delta V$, $\Delta V = 0$, or $-\Delta V$. Termination is then set on either an absolute value of $\Delta V$ or on the differentials $\Delta V/\Delta t$. It is also possible to use the second differential $\Delta^2 V/\Delta t^2 = 0$ which indicates the inflection point in the rising voltage characteristic. Each derivative places the charge termination in a different part of the charge curve which corresponds to a different state of charge. The choice will depend on the required state. In principal, the earlier the charge is terminated the less overcharge is delivered, and the longer the cycle life expectation will be.

Apart from the straightforward derivatives of the charge voltage characteristics a number of systems have been marketed which use pulse techniques, including discharge pulses, aimed at improving charge efficiency. Control or termination of charge still relies on the voltage characteristic, now modified to take into account the particular waveform, and this does not really differ from the methods just described.

Apart from fast or rapid charge capability some cell types are capable of accepting a substantial proportion of charge in very short periods from 1–10 min. Generally, this is used only in special situations where only a limited output is required. It usually necessitates pre-discharge and timing to prevent excessive internal pressure rise.

## Cell matching

It is inevitable, because of the methods used to produce nickel–cadmium sealed cells, that there will be both batch to batch and cell to cell variations in the actual capacity of cells having the same rated capacity. It is of course the manufactuer's aim to minimize this spread so as to optimize the use of active materials. Nonetheless, in batteries containing three or more cells which will be subject to deep cycling or fast charge it is generally considered desirable to ensure that cell capacities are matched as closely as possible.

In deep discharge the lowest capacity cell in a battery can be driven into reverse by the remaining cells: the larger the number of cells, the greater is the possibility of this happening. If a cell reaches less than $-1.0$ V, hydrogen can be evolved at the positive electrode. Since the hydrogen cannot be recombined it will eventually lead to venting with effective loss of water and electrolyte. Frequent and extended reversal will therefore eventually lead to premature failure of the lowest capacity cell. Thus the closer the cells in a battery are matched, the better will be the cycle life of the battery.

Cell matching is also significant in charging since the lowest capacity cell will become fully charged and enter overcharge ahead of the other cells in the battery. At low rates of charge this is not important, since the cell can withstand many hours of overcharge. However, at high charge rates the low capacity cell will reach a high temperature earlier than the others, and in extreme cases could be subject to excessive pressure build-up with consequent venting.

### 4.2.2.5 Future prospects

The progress of nickel–cadmium cylindrical cell development has been continuous. Fig. 4.70 illustrates the way in which the range of capacity available from an AA size

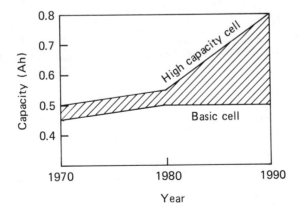

Fig. 4.70 — The growth of the Ah capacity of AA-sized sealed nickel–cadmium cells for the years 1970 to 1990.

cell has increased in the last 20 years. It seems likely that the introduction of new electrode manufacturing techniques is likely to yield further improvements during the next 5 years.

A major driving force behind these improvements has been the growth of the market in portable electronics devices, for example the mobile telephone. As these devices have developed, they have become smaller, and, although their efficiency has increased, more sophisticated facilities or greater range have been required which have not reduced the battery load requirements.

Thus, to maintain a non-disproportionate battery size it has been necessary to enhance the volumetric energy density of the battery pack. It now seems likely that further development in this area may occur in alternative chemistries, such as lithium or nickel–metal hydride [61] which offer the prospect of even greater improvements in energy density. Nonetheless, the benefits offered by the nickel–cadmium cell are likely to be with us for a good many years to come.

## REFERENCES

[1] Bode, H. & Voss, E., *Zeitschr. Electrochem.* **60**, 1053 (1956).
[2] Burbank, J., *J. Electrochem. Soc.* **106**, 369 (1959).
[3] Burbank, J., *Power sources*, Pergamon Press, Oxford, pp. 147–161 (1966).
[4] Ruetschi, P. & Angstadt, R. T., *J. Electrochem. Soc.* **104**, 406 (1957).
[5] Zaslaski, A. F., Kordnashov, Y. D. & Talkachev, U., *Doklady Akad. Nauk., S.S.S.R.* **75**, 599 (1950).
[6] Dawson, J. L., Gillibrand, M. I. & Wilkinson, J., *Power Sources*, Vol. 3, Oriel Press pp. 1–11 (1971).
[7] Faber, P., *Power Sources*, Vol. 4, Oriel Press pp. 525–540 (1973).
[8] British Standard BS3031 (1972).
[9] United States Standard OS8016.
[10] Kabanov, V., Fillipov, S., Vanyukova, L., Iofa, Z. & Prokofeva, A., *Zhurnal Fiz Khim.* **13**(3), 11 (1938).
[11] Gillibrand, M. I. & Lomax, G. R., *Electrochimica Acta*, **11**, 281–287 (1966).
[12] Ruetschi, P., Angstadt, R. T. & Cahan, B. D., *J. Electrochem. Soc.* **106**, 547 (1959).
[13] Barak, M., Gillibrand, M. I. & Peters, K., *Proc. 2nd Intnl. Symposium on Batteries*, p. 9 (1960).
[14] Craig, D. N. & Vinal, G. W., *J. Res. Natl. Bur. Standards*, **24**, 475 (1940).
[15] Bode, H., *Lead acid batteries*, John Wiley, New York p. 40 *et seq.* (1977).
[16] Baikie, P. E., Peters, K. & Gillibrand, M. I., *Electrochimica Acta*, **17**, (1972).
[17] Peukert, W., *Z. Electrochem.* **18**, 287 (1897).
[18] Gillibrand, M. I. & Lomax, G. R., *Electrochimica Acta* **8**, 693 (1963).
[19] Baikie, P. E., Gillibrand, M. I. & Peters, K., *Electrochimica Acta* **17**, 839–844 (1972).
[20] Gillibrand, M. I. & Lomax, G. R., *Electrochimica Acta* **11**, 281–287 (1966).
[21] Ruetschi, P. & Angstadt, R., *J. Electrochem Soc.* **111**, 1323–1330 (1964).

[22] McClelland, D. H. & Devitt, J., U.S. Patent 3,704,173.
[23] *The sealed lead battery handbook*, General Electric Co., Gainesville, FL (1979).
[24] Culpin, B., Peters, K. & Young, N. R. *Power Sources*, Vol. 9, Academic Press, New York p. 129 (1982).
[25] Peters, K., *J. Power Sources* **28,** 207–214 (1989).
[26] Harrison, A. I. & Wittey, B., *Power Sources*, Vol. 10, Academic Press, New York (1984).
[27] Mrha, J., Micka, J., Jindra, J. & Musilova, M., *Journal of Power Sources* **27,** 91 (1989).
[28] Smith, G., *Storage batteries*, Pitman, London, p. 211 (1964).
[29] *SAE Handbook*, Society of Automotive Engineers, Warrendale, PA.
[30] *Storage battery yearbook*, Battery Council International, Chicago.
[31] IEC Specifications, International Electrotechnical Commission.
[32] Muller, H. G., *Proc. Intnl. Conf. on Electric Vehicles*, Peter Peregrinus, London, pp. 183–192 (1977).
[33] Biagetti, R. V. & Luer, H. J., *Journal of Power Sources* 4 (1979).
[34] British Standard BS6290 Pt 4, 1987.
[35] *Proc. 1st International Conference on Batteries for Utility Energy Storage, Berlin* (1987).
[36] *Proc. 2nd International Conference on Batteries for Utility Energy Storage, Newport Beach, C.A.* (1989).
[37] Technical Bulletin GS3002, June 1989, Electric Power Research Institute, PO Box 10412, Palo Alto, CA94303, U.S.A.
[38] Jungner, W., Swedish Patent 8,558 (1897).
[39] Jungner, W., Swedish Patent 15,567 (1901).
[40] Jungner, W., Swedish Patent 11,487 (1899).
[41] Edison, T. A., US Patent 678,722 (1901).
[42] Pfleiderer, G., Spoun, F., Gmelin, P. & Ackermann, K., German Patent 491,498 (1928).
[43] Langguth, E., Swedish Patent 94,686 (1936).
[44] Lange, A. E., Langguth, E., Breuning, E. & Dassler, A., German Patent 674,829 (1933).
[45] Bureau Technique Gautrat British Patent 677, 770 (1952).
[46] Neumann, G. & Gottesmann, U., US Patent 2,571,927 (1951).
[47] Ahlgren, B. A. E., British Patent 1,232,122 (1971).
[48] Moffat, J. E., British Patent 2,088,767 (1982).
[49] Falk, S. U. & Salkind, A. J., *Alkaline storage batteries,* John Wiley, New York p. 610 (1969).
[50] Falk, S. U. & Salkind, A. J., *Alkaline storage batteries,* John Wiley, New York p. 585 (1969).
[51] Glemser, O. & Einerhand, J., *Z. Electrochem.* **54,** 302 (1950).
[52] Barnard, R., *J Appl. Electrochem.* **11,** 217–237 (1981).
[53] Falk, S. U., *J. Electrochem. Soc.*, **107,** 661 (1960).
[54] Matsumoto, I., Ogawa, H., Iwaki, T. & Ikeyama, M., *Power Sources,* Vol. 12, Keily T. & Baxter, B. W., Eds, Academic Press, New York pp. 203–220. (1989).

[55] Falk, S. U. & Salkind, A. J., *Alkaline storage batteries*, John Wiley, New York, p. 387 (1969).
[56] Lange', A. E., U.S. Patent 2,131,592 (1938).
[57] Levin, A. & Thompson, W. S., British Patent 571820 (1943).
[58] Jeannin, R., French Patent 102,709 (1950).
[59] Barnard, R., Crickmore, G. T., Lee, J. A. & Tye, F. L., *Power sources*, Vol. 6, Collins, D. H., Ed. pp. 161–179 (1977).
[60] Bode, H., Dehmelt, K. & Witte, T., *Electrochim. Acta* **11**, 1079 (1966).
[61] Fetcenko, M. A., Venkaten, S., Ovshinsky, S. R., Kajita, K., Hirota, M. & Kidon, H., *Power sources*, Vol. 13, Keily, T. & Baxter, B. W., Eds, Academic Press, London p. 149.

# 5

# Lithium battery systems

## 5.1 LITHIUM–THIONYL CHLORIDE BATTERIES
(H. F. Gibbard and T. B. Reddy)

### 5.1.1 General characteristics

The specific energy and volumetric energy density of lithium–thionyl chloride cells are among the highest of all practical battery systems. At low rates small cells have achieved more than 500 Wh.kg$^{-1}$ and 1000 Wh.dm$^{-3}$, and large cells have achieved more than half the theoretical values of 1489 Wh.kg$^{-1}$ and 2000 Wh.dm$^{-3}$. With an open-circuit voltage of 3.66 V and an almost flat voltage profile during discharge, the lithium–thionyl chloride system finds many electronic applications. The storage characteristics of the system are outstanding; low-rate cells stored for five years at room temperature have demonstrated 97 per cent of their original capacity [1]. The operating temperature range is exceptionally wide: −55°C to more than 150°C.

Thionyl chloride serves as both the positive active material and the medium for transport of ionic current through the cell. This feature strongly contributes to the high energy density of the system, but it also brings several disadvantages which have limited the usage of lithium–thionyl chloride cells in a number of fields. One problem is the phenomenon of voltage delay. Lithium is thermodynamically unstable in contact with thionyl chloride, but it reacts on immersion to form an insoluble, thin film of lithium chloride which passivates the lithium surface and prevents the highly energetic reaction between the two active materials. Excessive growth of the passivating film during storage causes a large drop in cell voltage when a load is applied; the time required for the thick film to break up and permit the voltage to recover to a minimum acceptable level is called the 'voltage delay'. Cell manufacturers have adopted various strategies to minimize the effects of voltage delay, but the user generally must test cells on the actual load profile and in the intended environment to verify that voltage delay will not be a problem in his application.

The intimate contact between lithium and thionyl chloride also can lead to safety hazards in high-rate cells. In the early development of high-rate lithium–thionyl chloride technology, cells shorted or discharged at high current commonly exploded. Similar problems were encountered with charging and voltage reversal of cells. The introduction of safety vents, internal and external electrical fuses, and diodes has done much to improve the safety of high-rate lithium–thionyl chloride cells and batteries. Nevertheless, the large-scale use of these batteries in applications such as military communications, which was predicted a few years ago [2], has not yet occurred.

Lithium-thionyl chloride cells are made in small prismatic and cylindrical cells with capacity under 1 Ah, in a variety of flat, cylindrical, and prismatic configurations in capacities of a few ampere-hours, and in low-rate prismatic cells as large as 20 000 Ah. Reserve cells with a 20-year storage life have been produced in large quantity in small, low-rate sizes and have been designed and demonstrated for large, high-rate applications [3].

### 5.1.2 Manufacturers

Major manufacturers of lithium–thionyl chloride cells and batteries include the following corporations: Battery Engineering, Crompton Eternacell, Eagle Picher Industries, Electrochem Industries, Honeywell, Hitachi–Maxell, Power Conversion, Saft, Tadiran, and Yardney. Additionally, Eveready has a strong position in the technology but negligible sales in lithium–thionyl chloride batteries.

### 5.1.3 Chemistry

#### 5.1.3.1 Cell reaction

The lithium–thionyl chloride primary cell typically consists of a lithium metal foil anode, a porous carbon cathode, a porous non-woven glass or polymeric separator between them, and an electrolyte containing thionyl chloride and a soluble salt, usually lithium tetrachloroaluminate. The carbon cathode serves as a catalytic surface for the reduction of thionyl chloride, as a current-collecting medium in conjunction with a metal support, typically nickel, stainless steel or Alloy 52, and as the repository for the insoluble products of the discharge reaction.

Although the detailed mechanism for reduction of thionyl chloride at a carbon surface is rather complicated and has been the subject of much controversy [4–6], the overall cell reaction is accurately described as follows [7]:

$$4Li + 2SOCl_2 \rightarrow 4LiCl + S + SO_2 \:. \tag{5.1}$$

Sulphur dioxide is soluble in the electrolyte; sulphur is soluble up to about 1 mol.dm$^{-3}$ but can precipitate in the cathode pores near the end of discharge. Lithium chloride is essentially insoluble and precipitates on the surfaces of the pores of the carbon cathode, forming an insulating layer which terminates operation of cathode-limited cells.

The distribution of lithium chloride throughout the cathode strongly influences the operating lifetime and the capacity of lithium–thionyl chloride cells [8]. The capacity decreases markedly at low temperature or at high rate, as, in these cases, reduction of thionyl chloride occurs primarily near the surface of the cathode. Maximum capacity of the cathode can be attained under the conditions of moderate to high temperature and low current, where the current density throughout the cathode is relatively uniform [9].

The capacity of the cathode can be substantially extended by including excess aluminium chloride in the electrolyte. In this case, the cathode reaction can be written as

$$4AlCl_3 + 2SOCl_2 + 4e^- \rightarrow 4AlCl_4^- + SO_2 + S. \qquad (5.2)$$

Electromigration of lithium ions into the cathode then produces soluble lithium tetrachloroaluminate rather than insoluble lithium chloride. The presence of excess aluminium chloride thus postpones the onset of cathode passivation and extends cell capacity. The use of this technique is limited to high-rate reserve cells, since aluminium chloride, as a Lewis acid, dissolves the passivating coating of lithium chloride from the anode and increases its corrosion rate.

Various types of catalyst have been found to improve the performance of cathodes subjected to high-rate discharge. These include metals, metallic salts and oxides, and heat-treated monomeric and polymeric transition-metal coordination complexes. The use of these catalysts is still in the development stage for active cells, because the dissolution of the catalyst and subsequent deposition of transition metals at the anode generally increases the corrosion rate and leads to excessive thickness of the protective anode film.

### 5.1.3.2 Safety

Because of their extremely high energy density and the toxicity of their contents, the usual safety rules for lithium batteries must be followed carefully in the use of lithium–thionyl chloride cells and batteries. These include the following:

- Do not recharge.
- Use diodes to protect cells from charging currents in any parallel configurations.
- Do not short circuit cells or store them loosely or in dump bins, or place them on metal surfaces where they may short.
- Observe polarity at all times when installing cells.
- Do not open, puncture, or crush cells.
- Do not ineinerate.
- Do not assemble cells into batteries without consulting the manufacturer.
- Do not use cells in unvented containers, or block or interfere with the functioning of the safety vent.

### 5.1.4 Construction details

#### 5.1.4.1 Materials
*(i) Electrolyte*
The purity of all of the components, and particularly that of the electrolyte, is critical to the proper performance of lithium–thionyl chloride cells. Various impurities can affect shelf life, voltage delay, capacity, and safety. Limits must be set for the

maximum concentration of water and of transition metals, especially iron; and these concentrations must be carefully monitored during the manufacturing process. These limits vary according to the manufacturer, but maximum concentrations of 1–5 ppm iron and 20–50 ppm 'water' in the electrolyte with which the cells are filled are typical. Water is unstable in the electrolyte, which undergoes hydrolysis reactions as shown in equations (5.3) and (5.4)

$$H_2O + SOCl_2 \rightarrow 2HCl + SO_2 \tag{5.3}$$

$$H_2O + AlCl_4^- \rightarrow HCl + AlCl_3OH^- . \tag{5.4}$$

Common analytical methods for the determination of hydrolysis products and iron are IR and UV–visible spectroscopy, respectively. Iron may be present in the thionyl chloride or lithium tetrachloroaluminate salt, or it may be introduced during the manufacturing process. Iron and other heavy metals, as well as hydrolysis products, can be removed from the electrolyte by refluxing the electrolyte in contact with elemental lithium, which scavenges the impurities.

The concentration of lithium tetrachloroaluminate in lithium–thionyl chloride cells varies with the manufacturer. The conductivity of the electrolyte at room temperature reaches a maximum at about 1.8 mol litre$^{-1}$, and this concentration is frequently employed. Since the concentration of the salt increases during discharge, lower values of 0.5–1.5 mol litre$^{-1}$ are also selected.

*(ii) Anode*

Lithium used as the anode is the commercial battery grade, which is low in iron, sodium, and nitrogen. It is important that the manufacturing process does not introduce nitrogen, because lithium nitride can undergo undesirable side reactions leading to thermal runaway [10,11].

The lithium chloride layer covering the anode is a critical feature in that it enables the kinetic stability of a thermodynamically unstable system. It is important, however, that the buildup of this film during storage does not impair cell performance during discharge. The behaviour of the interfacial lithium chloride layer depends on several factors, including the concentration of lithium tetrachloroaluminate and impurities in the electrolyte, the temperature, and the presence of various additives to the electrolyte, the separator, or the anode surface [12]. Several approaches have been taken to control the properties of the protective anode film. These include variations in the electrolyte, such as the addition of lithium oxide or sulphur dioxide, and the substitution of lithium–boron–halogen compounds like $Li_2B_{10}Cl_{10}$ for lithium tetrachloroaluminate. One of the most effective methods is the addition to the cell of polymeric materials such as polyvinyl chloride, which change the morphology of the surface film of lithium chloride so that voltage delay is ameliorated [13,14]. Coating the lithium surface with a cyanoacrylate adhesive film has also been used to reduce the effects of film growth on cell performance [15].

*(iii) Separator*

The separator for lithium–thionyl chloride batteries must be thin, highly porous, and an electronic insulator. Its function in the cell is to prevent contact of the positive and negative active materials leading to an internal short, while permitting an ionically conductive path between the electrodes. It must be inert to all the cell components with which it is in contact, both under normal conditions of storage and use and under abuse conditions. The most commonly used material is non-woven glass, which is highly porous and inexpensive but lacks mechanical strength. For high-rate batteries which are exposed to conditions of shock and vibration, porous Tefzel™ separators have been used with good results [16]. Tefzel™ separators are chemically stable in cells under normal and abuse conditions [17] and are considerably stronger than separators of non-woven glass, but the capacity at high rates at or below $-20°C$ has been found to be lower than that of cells with non-woven glass separators.

*(iv) Cathode*
The cathode is composed of carbon black and Teflon™ binder. Various compositions have been used, but cathodes containing approximately 10% by weight Teflon™ give adequate mechanical properties as well as good rate capability and capacity. Many different types of carbons have been evaluated, individually and as blends. The industry standard is Chevron Shawinigan acetylene black, which has been employed at levels of compression from 50% to 100%. Significant improvements in performance at high current density have been reported [18] by the use of carbon fibre cathodes and cathodes catalyzed with cobalt tetramethoxyphenylporphyrin (CoTMPP).

The current collector for the cathode varies with the type of cell construction. In high-rate cells the carbon–Teflon™ mixture is deposited on a nickel substrate, typically expanded metal, perforated foil, or woven screen. The carbon mixture is applied to the substrate either by deposition from an aqueous slurry or as sheets extruded from a 'dough' containing a nonaqueous liquid component, e.g. methanol or isopropanol. An alternative technology involves curing a carbon–Teflon™ mixture, grinding the resultant mixture, and pressing it into a current collector. In bobbin cells the cathode is extruded from a dough in a cylindrical form, and the current collector is a coaxial metal pin inserted in the centre of the cathode material. In large bobbin cells, an annular bobbin with a cylindrical current collector on the inside wall has been used.

*5.1.4.2 Bobbin cells*
*(i) Configuration, sizes*
Lithium–thionyl chloride bobbin cells are made in a cylindrical configuration as shown in Fig. 5.1, usually with a central, cylindrical cathode, a lithium anode swaged against the stainless steel or nickel-plated steel can, and a separator in between. They are hermetically sealed, and the axial terminal exits the cell through a glass-to-metal seal. The cells are available in ANSI sizes of $\frac{1}{2}$ AA to DD, corresponding to a capacity range of 800 mAh to 30 Ah, as well as in non-standard sizes ranging in capacity up to 150-Ah (#6 cell). Cells are available with a variety of standard terminals (round wire, pressure contacts, flat strip, connectors); custom terminals and battery assemblies can be obtained directly from the manufacturers or their distributors.

*(ii) Discharge characteristics*
The long ionic path from the anode to the active sites in the cathode limits this type of cell to low rates, typically C/100 or lower. The specific energy is high, approximately

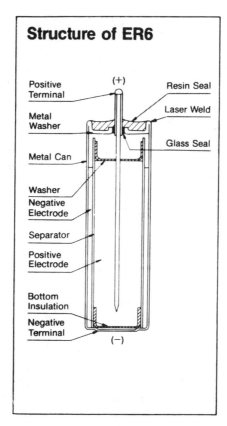

Fig. 5.1 — ER6 bobbin lithium–thionyl chloride cell in cross section (Maxell).

400 Wh kg$^{-1}$ and up to 1.0 Wh cm$^{-3}$. The open-circuit and operating voltages at low rates are nearly flat during most of the discharge, as shown for the discharge of AA cells at several temperatures and currents in Fig. 5.2. Examination of this figure shows that the capacity at a fixed current is higher at 20°C than at −40°C or 60°C. The capacity as a function of temperature thus passes through a maximum in this temperature range. This is seen more clearly in Fig. 5.3. The maximum in capacity occurs because the discharge process is limited at low temperature by mass transport in the pores of the cathode, and by the increased rate of corrosion of the anode at elevated temperatures. The rapid increase in the rate of self-discharge with increasing temperature is shown in Fig. 5.4. This increase in the rate of corrosion of the anode with increasing temperature, and also with disruption of the surface film by current flow, has been demonstrated by microcalorimetry and impedance measurements [19–21] (see section 2.3.1.2).

(*iii*) *Applications*
Because of their long life and high energy density, lithium–thionyl chloride cells have found many applications in the electronics industry. The largest single usage is probably as the power back-up for volatile random access memory. Rapidly growing

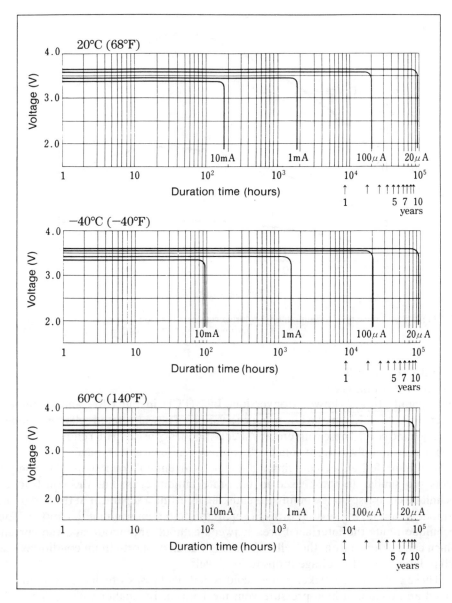

Fig. 5.2 — Discharge characteristics of lithium–thionyl chloride AA cells (Maxell).

applications include utility metering gauges and security/alarm systems. Other applications are: automobile electronic components, down-hole oil well logging, industrial process controllers, measuring instruments, pushbutton telephones, car telephones, pocket pagers, small communications equipment, security/alarm systems, and sensors.

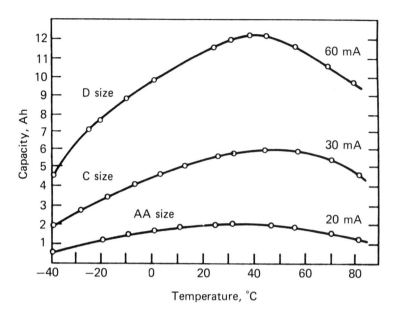

Fig. 5.3 — Capacity of lithium–thionyl chloride cells as a function of temperature at various currents.

### 5.1.4.3 Wound cells

#### (i) Low-rate wound cells

Two manufacturers, Power Conversion, Inc. (PCI) in the USA and Crompton Eternacell Ltd (CEL) in the UK produce a variety of power-limited cells containing a spiral wound core. These cells employ sheet cathodes consisting of Teflon™-bonded carbon on a metallic current collector, a non-woven separator (typically glass fibre), and a lithium foil anode containing a metallic current collector strip. The components are similar to those used in high-rate lithium cells; but the construction technique comprises first winding the cathode, inserting a separator at the end of the cathode wrap, taking several turns of separator, and then winding the anode. This technique limits the interfacial area between cathode and anode and the current which can be drawn from the cells. Under high drain or short-circuit conditions, the cell polarizes and the voltage drops to low levels.

These cells employ nickel-plated, cold-rolled steel cases which are hermetically sealed and contain a high-pressure vent mechanism. In smaller cells ($\frac{1}{2}$ AA, $\frac{2}{3}$ A, AA), these are double convolution vents located in the bottom of the cell case, while larger C and D cells use a coined side-vent mechanism. Direct external short circuits or forced discharge do not lead to venting of these cells, some of which have been approved by Underwriters Laboratories Component Recognition Program. The vent mechanism in the cell case will activate only in the event of extreme abuse conditions such as incineration of the cell.

The characteristics of the cells available from PCI and CEL are given in Table 5.1. These cells are commercially available with a variety of terminations: button tops, solder tabs, radial leads, and axial leads.

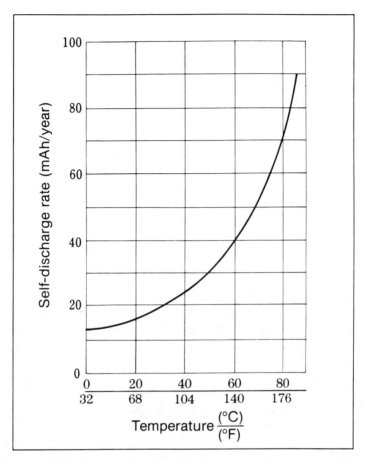

Fig. 5.4 — Self-discharge rate as a function of temperature for AA cells (Maxell).

**Table 5.1** — Characteristics of power-limitted wound cells

| Model No. | Standard size designation | Weight (g) | Nominal capacity (Ah) | Maximum discharge rate (mA) |
|---|---|---|---|---|
| T04 | ½ AA | 9 | 0.80 | 8.0 |
| T06 | AA | 16 | 1.70 | 17.0 |
| T32 | ⅔ A | 17 | 1.35 | 13.5 |
| T52 | C | 52 | 5.00 | 50.0 |
| T20 | D | 99 | 10.00 | 100.0 |

Although these cells will support higher drain rates for short periods, such as a series of high-current pulses on a duty cycle, they are generally recommended for applications where the drain rate is C/100 or less; and typical applications are C/1000 to C/100 000. Room-temperature discharge data at the C/500, C/1000, C/10 000, and C/100 000 rates are given for the AA cell of this design in Fig. 5.5. Voltage delay is seen not to be a significant factor at these rates at room temperature.

Fig. 5.6 shows the polarization curve for the T06/4 cell, giving discharge voltages as a function of current at 54°C, 21°C and −29°C. The discharge voltage is only slightly reduced by increasing the current up to 5 mA at the two higher temperatures. At −29°C, however, the operating voltage decreases significantly with increasing current. At very low drain rates, the voltage depression is only of the order of 0.1 V to 0.2 V at very low temperatures. In fact, these cells have been employed for low-drain aircraft applications at temperatures to −54°C. Shelf life under ambient conditions is excellent, as shown [1] by the loss of 3% capacity for the T06/4 cell after five years' storage at room temperature when discharged at the 1000 hour rate (1800 ohms).

In general, the applications for cells with the PCI power-limited design are similar to those for bobbin cells. The largest use is in random-access memory circuits where a single cell will serve to hold the information in memory in the event of a loss of line voltage. A schematic of a circuit showing such an application is shown in Fig. 5.7. Diode protection of the cell is required in this application to prevent high-rate charging of the cell. Diodes should be specified to have a leakage current of 30 $\mu$A or less. An added feature to limit the charging current if the protective diode fails in a shorted condition is to insert a current-limiting resistor, typically 470 $\Omega$ or greater, in series with the diode as shown in Fig. 5.8. Parallel arrays of cells should also employ a diode in each string to prevent charging.

Another significant application for this product line is in industrial process controller circuits where the T20/4 D cell has been widely employed. In two-cell, 6 V battery packs, these batteries have been used to replace the alkaline batteries supplied by the manufacturers of personal computers to power clock or memory circuits. Because of their wide operating temperature range, these cells have also found applications in avionics systems for military aircraft, such as their use in a battery pack in the fire-control computer for the US F-14 fighter aircraft.

*(ii) High rate wound cells*
The largest market for high-energy lithium batteries is military communications. For more than a decade this field has been dominated by the lithium–sulphur dioxide system. The intention of the US Army, the world's largest purchaser of lithium batteries, to replace lithium–sulphur dioxide batteries by a combination of lithium–thionyl chloride primary batteries and lithium rechargeable batteries [2], has not been realized. Although a BA-6598 lithium–thionyl chloride replacement for the BA-5598 sulphur dioxide battery was demonstrated [14], and 10 000 units were delivered to the US Army for field testing, large-scale purchases have not materialized. The higher cost of the thionyl chloride system and continuing concerns over the safety of these high-rate, high-energy batteries are two factors which probably contribute to this hiatus between development and deployment.

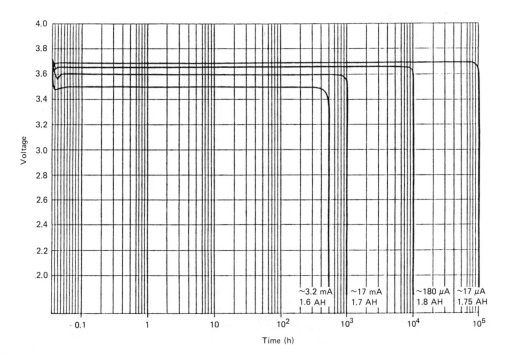

Fig. 5.5 — Discharge curves for low-rate AA cells (Power Conversion, Inc.).

More development work has been done on high-rate F cells than on any other size, because this geometry is well suited to the BA-6598 battery, for which the US Army let three production contracts, to Eveready, Altus, and Power Conversion. Tens of thousand of cells and batteries have been built and tested to the exacting specifications of the US Army. This includes electrical capacity as a function of temperature, with and without storage at 54°C for one month, and electrical, mechanical, and thermal abuse safety testing. Other high-rate cell sizes include D and C cells, but little demand has existed for smaller cells such as the AA or 1/2 AA.

Figs 5.9 to 5.11 show the discharge characteristics of Eveready BA-6598 batteries [14] composed of four-F cells, discharged at 2.0 A at three temperatures of 21°C, −29°C, and 54°C. These batteries met the requirements of the military standard MIL-B-49461, except for the capacity at 21°C (7.0 A h) and the voltage delay characteristic at −29°C (less than 60 s to 10.0 V). The voltage delay at −29°C was marginal on fresh batteries and did not meet the specification after storage for four weeks at 54°C, but the authors stated that 'adjusting the PVC additive' and 'controlling process contaminants' had solved this problem.

The capacity of the battery can be seen to depend strongly on temperature; about 40% of the capacity to the cutoff voltage of 10.0 V is lost when the temperature is decreased from 21°C to −29°C. The low-temperature loss in capacity is even more pronounced when individual cells are discharged. The resistance to heat flow of the battery package causes significant self-heating, which increases the cell temperature

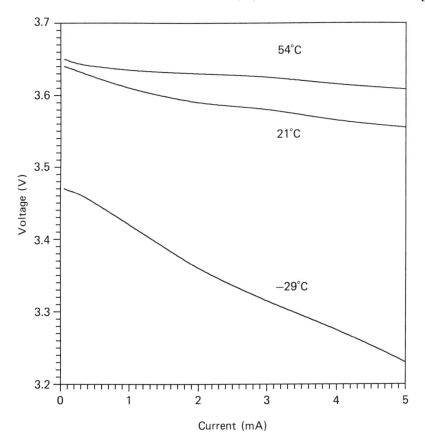

Fig. 5.6 — Polarization curves for low-rate AA cells (Power Conversion, Inc.).

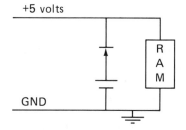

Fig. 5.7 — Random access memory circuit using a low-rate lithium–thionyl chloride cell protected from charging by a diode.

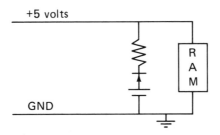

Fig. 5.8 — Random access memory circuit using a low-rate lithium–thionyl chloride cell protected by a diode and a resistor.

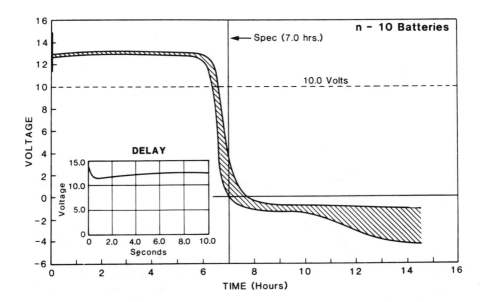

Fig. 5.9 — Discharge curve envelope at 21°C for BA-6598 batteries (Eveready first-article samples).

and thus capacity. The dissipation of heat from lithium–thionyl chloride cells is an important design consideration for high-rate applications, particularly if the cells are to be discharged to an operating voltage of less than 2.0 V. The rate of generation of thermal energy rises dramatically as the voltage drops at the end of life, and this can lead to internal pressure sufficient to open the safety vent. In this connection, it is interesting to note that the surface temperature of the cells in the Eveready batteries exceeded 90°C without opening of the safety vent. In fact, the location of the required safety thermal switch in their design was on the inner case surface and not near the warmer cell surfaces near the geometric centre of the battery.

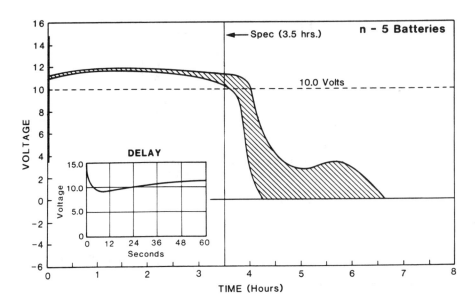

Fig. 5.10 — Discharge curve envelope at −29°C for BA-6598 batteries (Eveready first-article samples).

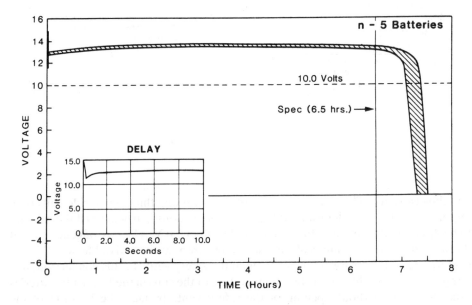

Fig. 5.11 — Discharge curve envelope at 54°C for BA-6598 batteries (Eveready first-article samples).

Lithium–thionyl chloride cells are well suited to discharge regimes requiring high current pulses and high specific energy. Recently, Gibbard & Nadkarni [16] described a battery of seven C cells, specially designed for high-power operation. This battery was tapped between the third and fourth cells, and the two sections were discharged at different loads. Fig. 5.12 shows the voltage of the three-cell section

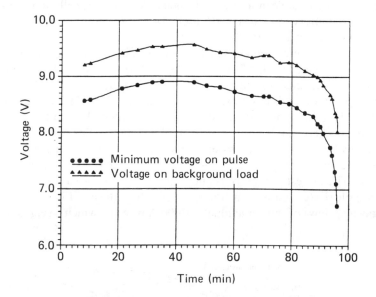

Fig. 5.12—Voltage of high-rate, three-cell battery section on pulse load during shock testing at room temperature.

during its discharge on a pulsed duty cycle of 3.40 A for 18 ms and 6.16 A for 2 ms. The capacity on this regime was in the range of 4.9 A h to 5.9 A h. Termination of the discharge of the battery when the voltage of the three-cell section fell below 7.0 V completely avoided the rapid rise in temperature which otherwise occurs at the end of discharge.

At this time there is no significant usage of high-rate lithium–thionyl chloride batteries in industrial or consumer applications. Military applications tend to be for expendable systems, where concerns about heat generation at the end of life and storage and disposal of expended batteries are minimal. Examples include missiles, fuses, and several types of underwater systems. Many of these military uses are classified, and details are not available. Moreover, the cells themselves are not widely available; potential users are advised to contact the manufacturers, including the following: Battery Engineering, Electrochem Industries, Honeywell Power Sources Center, Power Conversion, SAFT, and Yardney. A notable omission from this list is Altus Corporation, a company primarily known for its lithium–thionyl chloride technology, which terminated its manufacturing operations in the summer of 1990.

### 5.1.4.4 Prismatic cells

Although the prismatic design offers significant advantages in terms of volumetric utilization, particularly in multi-cell batteries, relatively few prismatic designs of lithium–thionyl chloride cells have been developed. Small prismatic cells have been commercialized by Eagle Picher Industries for use on printed circuit boards with dual in-line pin (DIP) connectors. These are marketed under the Keeper and Keeper II trade names and are available in sizes of 350–1600 mA h as single cells and in nominal 3.5 V and 6.8 V battery modules with 1600 mA h capacity. These designs are intrinsically low-rate and are intended for application in memory back-up and clock applications. The only intermediate-sized prismatic cell known to have been commercialized is a prismatic D-cell developed by GTE Power Systems Operations, now a part of Yardney Technical Products. It is marketed as the 'Pris D' cell.

Very large prismatic cells have also been manufactured by GTE Power Systems Operations and by Altus Corp. for standby power to activate U.S. Intercontinental Ballistic Missiles. These batteries have been employed in the silos of the Minuteman and MX ('Peacekeeper') Missile Systems as auxiliary power which would allow launch to occur even after loss of the lead–acid battery which acts as the main launch power system. The largest lithium cells and batteries ever constructed are believed to be the Missile Extended System Power (MESP) batteries. Fig. 5.13 shows a partial cutaway view of an individual 10 000 A h cell, which weighs 68 kg.

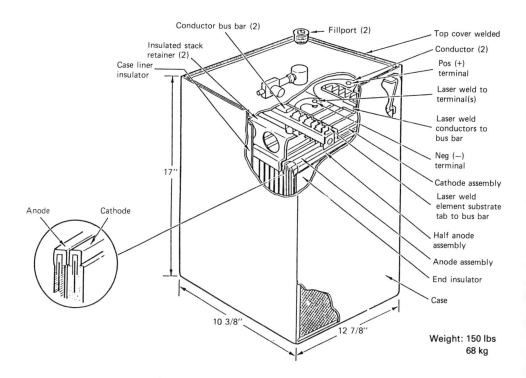

Fig. 5.13 — Missile Extended System Power (MESP) cell cutaway view.

Sec. 5.1]  **Lithium–thionyl chloride batteries**  303

A prismatic array of electrodes with non-woven glass fibre separators is seen to comprise the case-neutral design. Feedthroughs insulate both positive and negative terminals from the stainless steel case. Each cell is equipped with a pressure relief valve for the event of an over-pressure condition, which could occur at the end of discharge when the pressure of sulphur dioxide gas becomes appreciable. Fig. 5.14

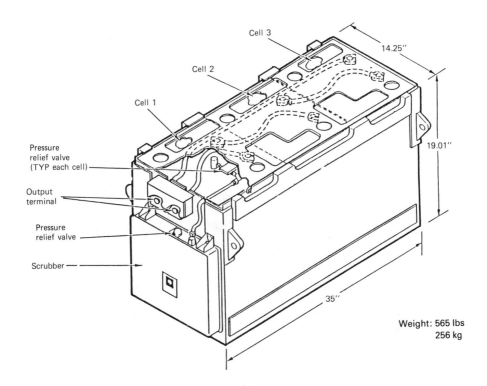

Fig. 5.14 — Missile Extended System Power (MESP) three-cell submodule.

shows the lithium power submodule in which three cells are connected in series and the cell vents are connected to a scrubber unit to contain vapours of thionyl chloride and sulphur dioxide resulting from cell over-pressure. The total weight of the subassembly is 256 kg. Fig. 5.15 shows the complete 33 V, 10 000 A h lithium power module, which has a maximum weight of 798 kg. An energy density of 905 W h dm$^{-3}$ and a specific energy of 480 W h kg$^{-1}$ have been achieved in these systems, which are designed to operate at the C/300 rate.

(*i*) *Configurations and sizes of prismatic cells and batteries*
The design of Eagle Picher's prismatic cells employed in their computer memory back-up batteries is shown in Fig. 5.16. Rectangular cathodes and anodes are pressed on opposite sides of the case with separator interposed, in this case-positive design. A spring-loaded current collector keeps this assembly in compression and connects both anodes to the pin of the glass-to-metal seal which serves as the negative terminal

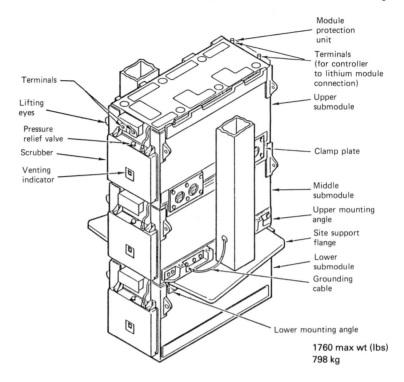

Fig. 5.15 — Missile Extended System Power (MESP) module.

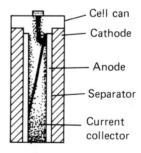

Fig. 5.16 — Prismatic cell in cross-section (Eagle Picher).

of the cell. These 1.6 A h cells are assembled into computer memory back-up batteries with working voltages of 3.5 V and 6.8 V.

Another novel design is that used for Eagle Picher's Keeper II cells, as shown in an exploded view in Fig. 5.17. A prismatic cathode fits inside a formed separator. The anode consists of a rectangular piece of lithium foil folded over three sides of the

# Sec. 5.1] Lithium–thionyl chloride batteries

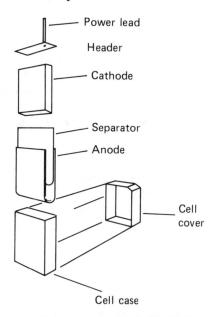

Fig. 5.17 — Prismatic cell exploded view (Eagle Picher, Keeper II).

cathode. The entire assembly is inserted into the prismatic case, which is welded to a header assembly containing a glass-to-metal seal. The central pin in the seal is inserted in the porous carbon cathode to form the positive termination of the cell.

Table 5.2 lists the specifications for the Keeper II cells and batteries. The capacity is given for the 350 h rate.

**Table 5.2** — Specifications for Keeper II cells and batteries

| Battery number | Working voltage, V | Capacity mA h | Length mm | Width mm | Thickness mm | Weight, g |
|---|---|---|---|---|---|---|
| LTC-3PN | 3.5 | 350 | 15.2 | 15.5 | 6.4 | 4.0 |
| LTC-7P | 3.5 | 750 | 30.5 | 17.8 | 8.4 | 9.0 |
| LTC-7PN | 3.5 | 750 | 24.5 | 16.5 | 6.4 | 6.8 |
| LTC-7PMP | 3.5 | 1500 | 38.1 | 30.5 | 8.4 | 19.0 |
| LTC-7PMS | 7.0 | 1750 | 38.1 | 30.5 | 8.4 | 19.0 |
| LTC-16P | 3.5 | 1600 | 36.6 | 13.5 | 13.5 | 16.0 |
| LTC-7P-MP-F-52 | 3.5 | 1500 | 38.1 | 30.5 | 11.4 | 25.0 |
| LTC-16P-CO-F-S11 | 3.5 | 1600 | 47.2 | 16.8 | 16.1 | 25.0 |

GTE developed a prismatic D cell with a parallel-plate design in a stainless steel case having the same volume as a cylindrical D cell [22]. Both platinum-catalyzed and uncatalyzed cathodes were employed during the development phase. The final design consisted of 11 plate-type cells with cathodes which were catalyzed with platinum at the level of 2.0%. The cells were designed to operate at a continuous drain of 1.1 A (30 ohm load), but rates up to 7.5 A were tested. Continuous operation at 10 A was not recommended, although short-circuiting the cell gave a maximum current of 120 A, which reportedly led only to bulging of the cells. The case terminals were designed to act as safety vents in the case of abuse of the cell.

*(ii) Discharge characteristics of prismatic cells*

Typical discharge data for the Eagle Picher LTC-16P computer memory back-up battery are shown at two rates (1 mA and 10 $\mu$A) and three temperatures ($+75°C$, $+20°C$, and $-20°C$) in Fig. 5.18. This model is rated at 1600 mA h and is seen to provide 1500 h of service at 1 mA and $1.6 \times 10^5$ h at 10 $\mu$A. Fig. 5.18 indicates that the operating temperature has a more pronounced effect on operating voltage than on service life.

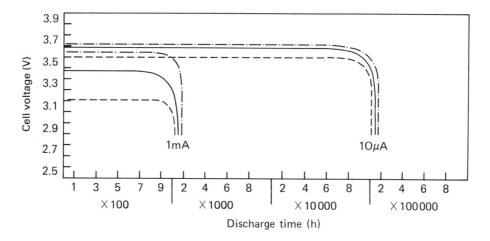

Fig. 5.18 — Typical discharge curves for prismatic cells (Eagle Picher LTC-16P) at three temperatures. Dash-dot line, 75°C; solid line, 20°C; dashed line, $-20°C$.

Fig. 5.19 shows typical discharge curves for the Eagle Picher Keeper II Model LTC-7P at the same two rates and temperatures. This model is rated at 750 mA h and 3.5 V, and is seen to provide 740 h of service at the 1 mA rate $+20°C$ and 66 000 h at 10 $\mu$A at $+20°C$. As in the case of the computer back-up batteries, temperature has a significant effect on operating voltage. The manufacturer claims 80% capacity retention for this hermetically sealed unit after 15 years' storage on the basis of microcalorimetric data.

Discharge curves for the GTE prismatic D cell on a 3.0 $\Omega$ load approximating a 1.1 A discharge are shown at three temperatures (0°C, $+22°C$, and $+50°C$) in Fig. 5.20. This cell employed four catalyzed cathodes and five anodes but was not the final design [22]. Capacities of 10.4 Ah, 11.4 Ah, and 12.4 Ah to a 2.0 V cut-off were

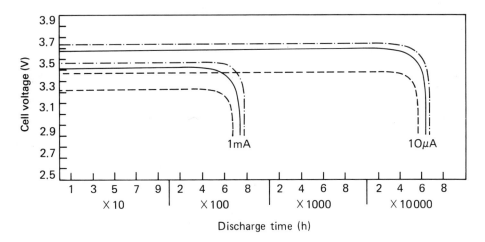

Fig. 5.19 — Typical discharge curves for prismatic cells (Eagle Picher LTC-7P) at three temperatures. Dash-dot line, 75°C; solid line, 20°C; dashed line, −20°C.

reported for the 3.0 Ω load at the three temperatures cited for Fig. 5.20. Pulse polarization data showed that this cell gave a 60 s voltage above 3.0 V at a current density of 40 mA cm$^{-2}$. Discharge on a regime of 13.3 A pulses for 10 s, four times per hour, gave a capacity of 6.67 Ah, or 60% of the design goal for this cell. Short-circuit currents of 62 A, which were observed with this cell, led only to distortion of the case and leakage through the glass-to-metal seals.

Fig. 5.21 shows the discharge voltage and temperature for a 10 000 Ah cell from the MESP battery discharged at 40 A. Approximately 280 h of discharge was obtained to a 3.0 V cut-off, which exceeded the nominal capacity. The cell temperature is seen to be essentially constant during discharge of this large, prismatic cell.

### 5.1.4.5 Reserve cells

Until recently the need for large reserve batteries has been met by the zinc–silver oxide system, and lead–acid or lithium–vanadium pentoxide batteries have filled the need for small reserve batteries. Now reserve lithium–thionyl chloride batteries are beginning to be developed for a wide range of power and capacity, and have begun to displace other systems in some applications. The most important feature of lithium–thionyl chloride reserve batteries is the separation of the electrolyte from the anode and cathode until the battery is activated. In this way the side-reactions which tend to degrade capacity during storage are avoided, and shelf life of 10–20 years can be achieved.

### (i) Configurations, sizes

Probably the smallest reserve cell manufactured in large volume is Power Conversion's RT03/40 cell, shown in Fig. 5.22. The dimensions of this cell are 12.8 mm (diameter) and 21.3 mm (height), and its weight is 5.1 g. The liquid electrolyte is held in a glass ampoule. When the ampoule is fractured by a mechanical or explosive shock, the electrolyte wets the cathode, separator, and anode; and the cell voltage

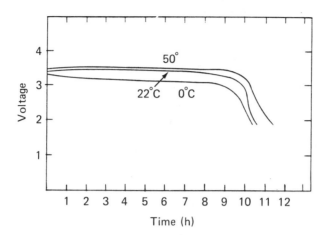

Fig. 5.20 — Discharge curves for prismatic D cell (GTE) at three temperatures (C/10 rate, 3Ω load).

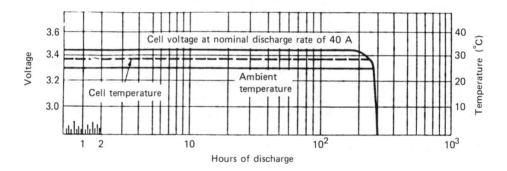

Fig. 5.21 — Voltage and temperature for 10 000-A h MESP battery discharged at 40 A.

rises to 3.66 V in less than 0.5 s. The nominal capacity of the cell is 280 mA h, and its operating temperature range is $-40°C$ to 75°C. The cell is designed to withstand specified, high levels of shock and vibration without activating. It can be used individually or in series or parallel configurations for reserve batteries.

Another type of reserve cell is activated by injecting the electrolyte, held in a reservoir, into the electrode assembly. An interesting example of this type of cell was developed by Altus for the power source of the Small Intercontinental Ballistic Missile [3]. The cell, shown in Fig. 5.23, employs a spiral wound core contained in a tube which also contains the electrolyte. The electrolyte is separated from the cell core by a burst disk. Upon activation, gas pressure from a manifold bursts the disk and forces the electrolyte into the cell core. Since each cell in a battery can contain an electrolyte of different composition, the electrical performance of the battery can be tailored for particular requirements. For example, extremely tight voltage regulation tolerances can be held, as shown by Chang *et al.* [3].

Sec. 5.1]     **Lithium–thionyl chloride batteries**     309

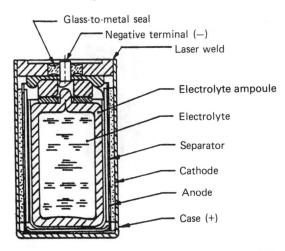

Fig. 5.22 — Small reserve lithium–thionyl chloride cell (Power Conversion Model RT03/40).

More conventional reserve batteries employ stacks of bipolar cells and a single, external electrolyte reservoir. A pyrotechnic device generates gas pressure which forces electrolyte through a manifold and into the spaces between cells. Alternatively, in small pile-type batteries the electrolyte can be held in a frangible glass ampoule until activation. Large batteries of this type have been evaluated as replacements for zinc–silver oxide missile batteries [23,24]; these batteries face formidable competition from thermal reserve batteries, except for applications which require long active or 'wet-stand' life.

(*ii*) *Discharge characteristics*
Fig. 5.24 shows the discharge curve for Power Conversion's RT03/40 reserve cells at −40°C and 75°C. Approximately one million of these cells have been constructed for use in a family of scatterable mines, and other applications for this and similar reserve cells are under investigation. The larger reserve batteries tend to be designed specifically for a single use, and the variety of types precludes the presentation of a typical example. The interested reader is referred to proceedings of the bienniel International Power Sources Symposium held at Cherry Hill, New Jersey [23–25].

(*iii*) *Applications*
Current and projected applications for reserve lithium–thionyl chloride batteries are almost exclusively military. They include fuses, missiles, mines, sonobuoys, torpedoes, and emergency power for crew rescue in a space station. Reserve batteries with flowing electrolyte have been proposed for torpedo propulsion [25]; although the system would develop extremely high specific power and energy, the technical difficulties are formidable.

Reserve lithium–thionyl chloride batteries are usually ordered on a custom basis, but several manufacturers have 'off-the-shelf' designs which could be adapted for applications other than the one for which they were developed. The potential user should contact the developers and manufacturers, which include the following: Eagle Picher, Honeywell, Philips USFA B.V., Power Conversion, SAFT, Silberkraft GmbH, Sonnenschein Lithium GmbH, and Yardney.

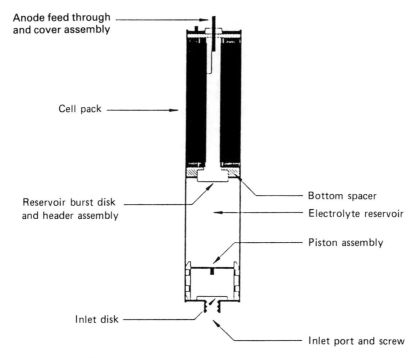

Fig. 5.23 — Spirally wound, reserve, high-rate lithium–thionyl chloride reserve cell (Altus).

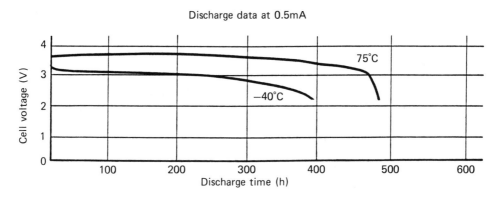

Fig. 5.24 — Discharge curves for small reserve cell (Power Conversion Model RT03/40 at −40°C and 75°C.

### 5.1.5 Likely future development
#### 5.1.5.1 General improvements
(*i*) *Electrocatalysts*

The development of catalysts for the electroreduction of thionyl chloride is an active area of research which is likely to result in general improvement in the performance of lithium–thionyl chloride cells. The addition of heat-treated, transition metal macrocyclic compounds such as CoTMPP to the cathode has been shown to increase not only the average discharge voltage but also the discharge capacity [26]. The latter effect is not, as yet, fully explained. In early experiments such catalysts were found to be partially soluble in the electrolyte, and migration of the soluble transition metal ions to the anode had a deleterious effect on the lithium chloride film covering the lithium. It is likely that treatments will be developed which will virtually eliminate the solubility of such catalysts, at which time their use in active cells may become common.

*(ii) Carbons*
Another area of research which promises general improvements in lithium–thionyl chloride technology is the use of mixtures of carbon and new forms of carbon in the cathode. The performance of many carbon blacks and mixtures thereof has been studied [12], and no consistent improvement over Shawinigan acetylene black (SAB) has been observed. Recently, Blomgren *et al.* [18] reported increases of nearly 70% in capacity of cathodes made from carbon microfibres, over controls of SAB. Advances in the very active area of the science of carbon materials can be expected to be translated into improvements in the capacity of lithium–thionyl chloride cells of all types.

*(iii) Anode film*
Work on improvement of the impedance characteristics of the lithium anode in lithium–thionyl chloride cells continues at an active pace. The most effective 'antipassivation' additives, such as polyvinyl chloride, have been found to affect the morphology of the lithium chloride crystals in the passivating film [13], and this suggests research on other compounds such as surfactants which affect the surface energy of crystalline deposits. Fundamental understanding of the surface chemistry of the interface between the anode film and the electrolyte can be expected to result in further improvements in the voltage-delay properties and shelf life of all types of lithium–thionyl chloride cell.

### 5.1.5.2 *Improvements in low-rate cells*
The most significant near-term trend in low-rate cells will probably be the extension of shelf and operating life to more than 10 years. It is likely that cell designs will be tailored to meet the requirements of individual, large-volume applications. For example, it has been noted [19] that the rate of corrosion of the anode can be minimized by operation below a critical current which is characteristic of the cell design. For many applications, such as the recording and periodic reporting of utility meter data, a low, steady current is punctuated by periodic pulses. Each pulse typically disrupts the film on the anode, and repair of the film occurs at the expense of anode capacity. If the area of the anode is increased, so that the critical current density for film disruption is not exceeded during the pulses, then the capacity loss due to film repair can be greatly reduced. On the other hand, the increase of anode area will increase the intrinsic, steady-state rate of anode corrosion, which is proportional to area. It appears that for each application in which current pulses cause film disruption, there may be an optimum choice of anode area which will achieve maximum cell capacity and active life.

Other means of lengthening cell life undoubtedly will be employed. These include additives, purification of materials, alternative electrolyte salts, and conditioning regimes for the completed cells. Storage at elevated temperature to thicken the anode film, for example, may substantially reduce the rate of anode corrosion but not impair cell performance at extremely low current drains.

### 5.1.5.3 *Improvements in moderate- and high-rate cells*
Users of high-performance batteries for military systems continually push the state-of-the-art to obtain more energy per unit mass or per unit volume; the present hiatus in the widespread use of lithium–thionyl chloride high-rate batteries is thus unlikely to persist. The introduction of these batteries will require advances in the safety and reliability of high-rate systems. One method to improve battery safety is the incorporation of modern solid-state electronic circuits into the battery design. A first step in this direction is an electronic circuit which monitors cell voltage and disconnects the battery from its load when the voltage of one of the cells falls below a predetermined value [27]. Although this circuitry could be included in the end-use item, miniaturized application-specific integrated circuits could be incorporated in the battery itself at an increase in cost of only a few per cent. An on-board microprocessor could monitor cell voltage and temperature at several locations, as well as the time derivatives of these quantities, and could employ sophisticated algorithms to control battery output or to provide information to the host equipment.

Another area which may be expected to provide substantial improvements in performance and safety of high-rate batteries is that of thermal management. The limiting factor for the rate of discharge of lithium–thionyl chloride batteries has been the increase in temperature of the cells due to internal generation of thermal energy. A few of the techniques which have been suggested or used to decrease the internal temperature of high-rate batteries include the following: circulating electrolyte, phase-change heat sinks, heat pipes, and augmentation of heat-transfer area (fins). Effective use of such techniques will yield batteries capable of sustained specific power of the order of kilowatts per kilogram.

## 5.2 LITHIUM–SULPHUR DIOXIDE BATTERIES
(T. B. Reddy)

### 5.2.1 Historical background

The lithium–sulphur dioxide primary battery system was developed as a viable product in the 1960s by a research group working in the Central Research Division of the American Cyanamid Company in Stamford, Connecticut, USA. Earlier work in Switzerland [28] had produced a rechargeable sodium battery with aluminium trichloride dissolved in a liquid sulphur dioxide electrolyte. Concurrent with the work at American Cyanamid, a group working for the Livingston Electronic Corporation, now the Honeywell Power Sources Center, showed that the addition of 'ligand gases' such as sulphur dioxide to nonaqueous electrolytes enhanced the conductivity of these solutions.

The work at American Cyanamid culminated in the issuance of several relevant patents [29,30,31], but this firm elected not to develop this product and licensed its technology to the Mallory Battery Company, now Duracell, Inc. During the 1970s, several other firms including Power Conversion, Inc. (PCI) and the Honeywell Power Sources Center also developed this technology, which has now emerged as the principal high-energy, primary battery system used to power military electronics and communications equipment in the World. For example, the BA-5590 battery, consisting of ten lithium–sulphur dioxide (Li–$SO_2$) D cells, is employed to power some forty different types of electronics equipment employed by the US Military establishment.

## 5.2.2 Technical advantages

As at present manufactured, the Li–$SO_2$ system possesses numerous technical advantages over conventional aqueous primary battery systems:

(1) *Higher energy density* Lithium–sulphur dioxide cells have an energy density of 275 Wh.$kg^{-1}$ (125 Wh.$lb^{-1}$). This is approximately twice the energy density of the Mg–$MnO_2$ systems and three times that of carbon–zinc Leclanché cells. Multi-cell Li–$SO_2$ batteries achieve energy densities of 176 Wh.$kg^{-1}$ (80 Wh.$lb^{-1}$).

(2) *High power density* Lithium–sulphur dioxide cells provide high power levels as a result of their spiral wound construction. A typical Li–$SO_2$ D cell is rated at 2 A for constant current applications, but can put out 30 A pulses for use in applications such as sonobuoys. High-rate D cell designs can produce 8 A, but will vent on continuous use at this rate.

(3) *Wide operating temperature* Because of the low viscosity of its electrolyte, Li–$SO_2$ batteries are operational over the range from −54°C to +71°C (−65°F to +160°F). US Military batteries are rated more conservatively from −29°C to +55°C (−20°F to +130°F).

(4) *High cell voltage and flat discharge voltage* The open circuit voltage (OCV) of the lithium–sulphur dioxide cell is 2.95 V, allowing batteries to be constructed from one half the number of cells needed using aqueous cell technology. Thus, two BA-5598 batteries may be carried in the AN/PRC-77 military radio whilst only one BA-4386 (magnesium) or one BA-3386 (alkaline) battery will fit in the same volume. This allows the soldier to carry one active Li–$SO_2$ battery and a spare for emergency use. The flat discharge curve (see section 5.2.5.2) allows electronics equipment to operate under constant power conditions over most of the useful battery life.

(5) *Long shelf life* Owing to the passive film on the lithium anode, long shelf life is obtained from lithium–sulphur dioxide batteries. US Military batteries are rated for five years' life, and 99% capacity retention has been obtained on discharge at +21°C (+70°F) after five years' ambient temperature storage. The actual shelf life is in excess of ten years for ambient storage and one year at +71°C (+160°F).

Experience has shown that the batteries should not be tested during storage, particularly if high drain rate performance is required, since intermittent testing contributes to the build-up of the passive film on the anode [42].

The hermetic cell design required by the pressurized electrolyte contributes substantially to the long shelf life since it prevents the ingress of reactive contaminates and the egress of active material from the cell.

### 5.2.3 Chemistry and electrochemistry

The lithium–sulphur dioxide cell was the first practical galvanic cell to be discovered with an active wet-stand in which both active materials are in direct physical contact. The anode material is lithium metal foil and the cathode reactant is liquid sulphur dioxide dissolved in the electrolyte. These materials will react directly, but the stability of the system is provided by a passive film of lithium dithionite [33] formed on the lithium surface. This film is stable under normal storage conditions and provides the remarkably long shelf life afforded by this battery type.

#### 5.2.3.1 Electrode and cell reactions

The electrode reactions have been defined as follows:

$$\text{Anode: Li} \rightarrow \text{Li}^+ + e^- \tag{5.5}$$

$$\text{Cathode: } 2\,SO_2 + 2\,Li^+ + 2e^- \rightarrow Li_2S_2O_4 \downarrow \tag{5.6}$$

$$\text{Overall cell: } 2\,Li + 2\,SO_2 \rightarrow Li_2S_2O_4 \downarrow \tag{5.7}$$

On activation, the lithium metal anode acts as an electrode of the first (simplest) kind, producing soluble lithium ions. Concurrently, the sulphur dioxide in the electrolyte is electro-reduced and reacts with the lithium ions in solution to produce an insoluble precipitate of lithium dithionite, $Li_2S_2O_4$, at the positive electrode. Owing to the need to accommodate this precipitate, highly porous electrode structures are employed in this cell.

At high temperatures, cathodic side reactions occur which reduce the capacity of the cell as shown by Bowden et al. [32]. These authors identified sulphur, lithium dithionate ($Li_2S_2O_6$), and possibly lithium pyrosulphite ($Li_2S_2O_5$) as well as the lithium dithionite, in the porous cathodes of Li–$SO_2$ cells discharged on 0.4 ohm loads at $+72°C$.

#### 5.2.3.2 Electrolyte composition and properties

The electrolyte employed in lithium–sulphur dioxide cells is composed of a mixture of liquid sulphur dioxide with acetonitrile as co-solvent and lithium bromide as the conductive salt. This electrolyte typically contains about 70 wt. % $SO_2$, and the vapour pressure over the liquid phase is primarily due to sulphur dioxide. Figure 5.25 shows the vapour pressure of sulphur dioxide as a function of temperature. It is apparent that the vapour pressure increases dramatically with temperature. At $+20°C$, the pressure is in the order of $3 \times 10^5$ Pa (43.5 p.s.i.) whilst at $+71°C$ (a storage temperature required by US Military specifications) it has risen to $14 \times 10^5$ Pa (203 p.s.i.). This increase makes obvious the need to maintain hermeticity of the cell case over a wide temperature range. The conductivity of the electrolyte is $5 \times 10^{-2}$ ohm$^{-1}$ cm$^{-1}$ at $+20°C$ and is $2.2 \times 10^{-2}$ ohm$^{-1}$ cm$^{-1}$ at $-50°C$. This

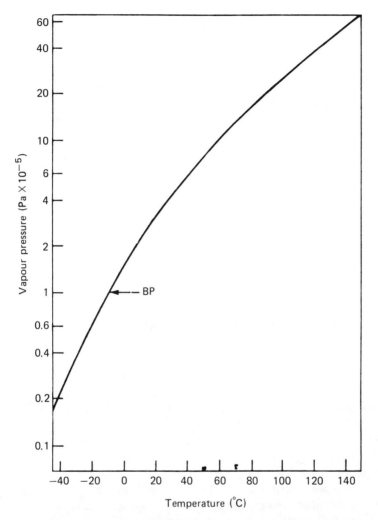

Fig. 5.25 — Vapour pressure of sulphur dioxide vs temperature.

relatively high conductivity of the electrolyte at low temperature is a major factor in the ability to use lithium–sulphur dioxide cells in low temperature applications where cells containing aqueous electrolytes would be inoperative.

The stability of the electrolyte is another factor to be considered since storage at +71°C (+170°F) is required by US military specification [34] and US Government contracts require five years' storage under ambient conditions. At this temperature, the following reaction has been found to occur [35]:

$$4\ LiBr + 4SO_2 \rightarrow 2\ Li_2SO_4 + S_2Br_2 + Br_2 \ . \tag{5.8}$$

This reaction, which is catalyzed by water, can lead to lowered capacity due to the removal of $SO_2$ and LiBr from the electrolyte. The presence of bromine in the electrolyte produces an abnormally high open circuit voltage and can lead to the corrosion of internal components which are normally stable. The use of LiAs $F_6$ as a replacement for LiBr has been proposed for reserve cells to provide greater electrolyte stability. However, lithium bromide continues to be the electrolyte salt employed for active cells.

Because of the high coefficient of thermal expansion of the $SO_2$–$CH_3CN$–LiBr electrolyte, it is necessary to underfill cells at room temperature to allow for electrolyte expansion of about 5% between +20°C and +71°C.

### 5.2.3.3 Balanced cell concept

Early designs of lithium–sulphur dioxide cells were constructed with excess lithium metal, and whilst they provided both high capacity and high rate capability, they were prone to overheating and to venting of electrolyte, sometimes with the concomitant combustion of the vented gases. This was found to be due to the reaction between acetonitrile and lithium metal when the sulphur dioxide levels were reduced to below 5% in the electrolyte, removing the passivity of the film on the anode. Lithium is known to react with acetonitrile, producing a variety of products as shown by the following reaction:

$$Li + CH_3CN \rightarrow CH_4 + LiCN + \text{other organic products} . \tag{5.9}$$

The ejection of methane from hot cells leads to the combustion reactions seen in early unbalanced cells. The lithium cyanide formed is also a concern from a disposal standpoint. These reactions were particularly common in multi-cell batteries where one or more cells with a low capacity can be driven into a reverse voltage condition at the end of discharge.

DiMasi et al. [36,37] showed that cells where the coulombic ratio of lithium to sulphur dioxide (Li–$SO_2$ ratio) was close to 1.0 were less prone to venting and ignition than unbalanced cells where this ratio could be as high as 1.5. In balanced cells (i.e. Li–$SO_2$ ratio near 1.0), the lithium metal remains passive owing to the equivalent amount of sulphur dioxide remaining in the cell so that the lithium–acetonitrile reaction (5.9) is unlikely to occur. The current US military specifications for lithium–sulphur dioxide batteries [34] calls for a Li–$SO_2$ ratio in the range of 0.90–1.05. The same specification requires that cells and batteries can be capable of forced discharge into reversal at half the discharge rate at +21°C (+70°F) in order to verify the ability of the cells and batteries to withstand this abusive condition.

At present, only cells constructed for special applications such as sonobuoys employ an unbalanced design. In these cases, the high pulse capability required can be provided only by an unbalanced design, whilst venting and disposal problems are not significant factors owing to the nature of the application.

### 5.2.4 Cell construction

Early lithium–sulphur dioxide cells were constructed with crimp seals, but storage life was limited by $SO_2$ loss through the seals. This led to the development of the hermetic design currently employed by all major manufacturers. The details of this construction are given below.

Sec. 5.2]    **Lithium–sulphur dioxide batteries**    317

### *5.2.4.1 Spiral wound construction in hermetically sealed case*
Lithium–sulphur dioxide cells of this type employ a case negative design with a spiral wound configuration in which cathode, separator, and anode are wound in a 'jelly-roll' and inserted into the case. Fig. 5.26 shows a partial cut-away view of this design. The design features illustrated are as follows:

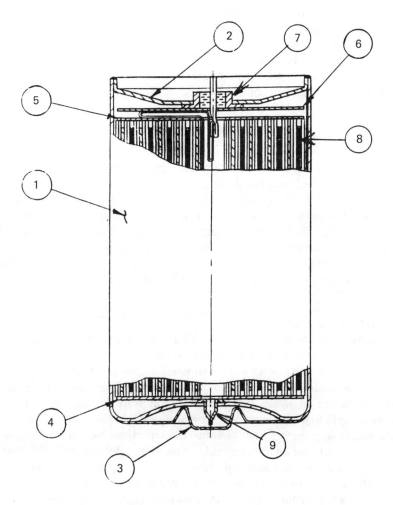

Fig. 5.26 — Partial cut-away view of lithium–sulphur dioxide cells.

(1) The cell case.
(2) The top shell assembly.
(3) Protective bottom cap over electrolyte filling tube.
(4) Bottom core insulator.
(5) Top core insulator.
(6) Top shell insulator.

(7) Glass-to-metal seal welded into top shell. Pin is positive terminal of cell.
(8) Spiral wound core.
(9) Electrolyte filling tube which is welded shut.

### 5.2.4.2 Cell case material

The cell case and top shell material is normally nickel-plated cold rolled steel. Silberkraft GmbH in Germany offers lithium–sulphur dioxide cells in stainless steel cases, and Power Conversion, Inc. has produced cells in stainless steel cases in certain sizes and for special applications. The polarity is normally case negative, and a tab connects the case to the lithium anode. The lithium metal polarizes the case to a potential at which it is not attacked by the corrosive electrolyte. If the lithium metal is entirely consumed or if lithium contact to the case is lost, the electrolyte will attack the cell case.

### 5.2.4.3 Cathode construction

The cathode is electrochemically inert and acts as a site for the electrochemical reaction (5.7) given above, and as a host for the lithium dithionite precipitate during cell discharge. It must be highly conductive, have a high surface area, and be highly porous. The need for these properties has led to the development of a Teflon™-bonded acetylene black structure which is laminated to an aluminium expanded metal grid which acts as a current collector. This cathode material is produced in sheet form which is subsequently cut to the desired size before the mix is removed from one end to allow an aluminium tab to be welded to the expanded metal grid. This tab is welded to the pin of the glass-to-metal seal which acts as the positive terminal of the cell.

### 5.2.4.4 Anode construction

The anode consists of lithium metal foil which is extruded to the desired thickness which is typically 0.15 mm–0.30 mm (0.006–0.012 in) and width which is dependent upon the cell size. A metallic strip is laminated in the centre of this by certain manufacturers, to maintain electrical continuity during discharge. Earlier designs employed a structure in which two pieces of foil were laminated to a central expanded metal grid to which a tab was previously welded [38].

An alternative approach to maintaining the structural integrity of the anode during discharge is to fuse an impermeable stripe in the porous separator material (see next section) before cell assembly [39]. This stripe prevents discharge from occurring in this area and serves to maintain continuity during discharge.

In any event, a tab is cold-welded to the anode and electrically welded to the cell case to provide the necessary electrical continuity. Tabs of copper, nickel, and nickel plated cold rolled steel have been successfully employed for this application.

In a few instances where very high surface area electrodes are required, lithium metal, a few hundredths of a mm thick, has been coated on either side of a copper metal foil.

### 5.2.4.5 Separators and insulators

Both non-woven and microporous separator materials have been employed in lithium–sulphur dioxide cells. The microporous polyolefins are very thin, typically

25 μm thick, which minimizes separator volume and allows a greater quantity of active material to be incorporated in the cell case. Polypropylene has a higher melting point than polyethylene and thus provides greater thermal stability, making it the material of choice. Microporous polypropylene separators which are commercially available are 45% porous and have very fine channels of approximately 200 nm diameter which allow electrolyte to flood the pores and provide good conductivity, but prevent carbon particles from the cathode shorting to the anode. The microporous polyethylene separator also is slightly elastic, a property desirable during corewinding.

Non-woven polypropylene separators are cheaper than the microporous material, but since they contain pores of the same diameter as the carbon particles in the cathode, they must be much thicker than the microporous variety. Non-wovens also tend to be employed in lower-rate cells because of the increased interelectrode distance associated with their use.

Internal insulators are also employed in the cells to prevent the development of short circuits within the cell case, as illustrated in Fig. 5.26. These are also fabricated from polypropylene for the same reasons as the separators employed in Li–$SO_2$ cells. Whilst fluorocarbons are more inert under most circumstances, the highly energetic reaction which can occur between molten lithium and these materials has restricted their use.

### 5.2.4.6 Vent designs

Two types of vent are commonly employed in lithium–sulphur dioxide cells. Power Conversion, Inc. (PCI) and its licensees employ a coined side vent mechanism in their large cells, i.e. D size and larger. A die is used to form an indentation in the side wall of the cell case, thinning the wall thickness to a few hundredths of a mm (Fig. 5.27). These mechanisms are designed to expel the electrolyte from the cell in the range of 2.4–3.1 MPa (350–450 p.s.i.) internal pressure.

An alternative design employs a double convolution vent mechanism coined in the bottom of the cell case. When pressure builds up within the cell, the ends of the cell bulge in a 'drum head' effect. This causes a crack to develop at the bridge between the convolutions, resulting in a loss of electrolyte from the cell.

Although sulphur dioxide is a noxious material, the loss of electrolyte associated with a venting mechanism serves to deactivate the cell and prevent further discharge. This provides a major safety feature which is absent from the oxyhalide type of high-rate lithium battery where a much lower vapour pressure electrolyte is employed. In high-rate lithium–thionyl chloride cells, the loss of electrolyte on venting is not complete, and the cell can continue to discharge. This may result in a thermal runaway situation which is not the case with lithium–sulphur dioxide cells.

### 5.2.4.7 Glass-to-metal seals

Glass-to-metal seals are employed in all hermetically sealed Li–$SO_2$ cells. They may be discrete units which are welded into the top shell assembly, or they may be an integral part of the top shell.

Corrosion of the glass was a common problem in early Li–$SO_2$ cells. This phenomenon is associated with the 2.95 V potential difference across the glass which results in replacement of the alkali metal ions in the glass with lithium ions from the

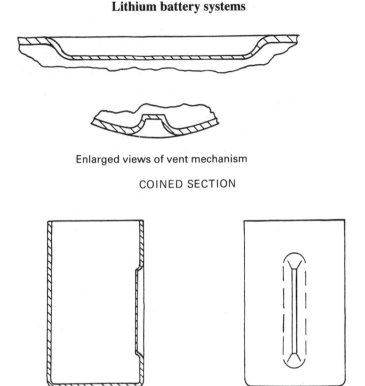

Fig. 5.27 — Safety vent mechanism.

electrolyte via an ion exchange mechanism. This is followed by reduction of the lithium ions to lithium metal at the edge of the glass where it is in contact with the case. This process results in a loss of the structural integrity of the glass, termed 'glass corrosion'. Two different approaches have eliminated this problem in the current generation of Li–SO$_2$ cells. One solution has been to coat the glass with a polymeric coating which prevents contact with the electrolyte [40]. This type of seal normally employs a tantalum pin. A second approach uses glasses with increased corrosion resistance such as type TA-23 or CABAL-12 glasses. The TA-23 glass is normally employed with a molybdenum pin. Both approaches provide excellent stability, and the 'glass corrosion' problem is no longer a significant cause of failure in Li–SO$_2$ cells.

### 5.2.5 Commercially available cells and their performance

Lithium–sulphur dioxide cells are available in a large variety of sizes for use as cells or for fabrication into battery packs. The principal manufacturers of these products are Power Conversion, Inc. (PCI) and SAFT America, Inc. in the US, Ballard Battery Systems in Canada, Crompton Eternacell Ltd in the UK, and and Silberkraft GmbH in Germany. Jian Zhong Chemicals Corp., in the People's Republic of China, is also beginning the production of Li–SO$_2$ cells. The latter three organizations are all

Sec. 5.2]    **Lithium–sulphur dioxide batteries**    321

licensees of PCI in the US, and their products are similar to those of the latter organization. The main difference in the products manufactured by Silberkraft is that their Li–$SO_2$ cells are packaged predominantly in stainless steel cases. Since the products of Crompton Eternacell, Silberkraft, and Jian Zhong Chemicals are similar to those of PCI, only the latter products will be described in detail.

### 5.2.5.1 Commercially available cells

Table 5.3 lists the cell sizes and characteristics of products available from PCI.

**Table 5.3** — Lithium–sulphur dioxide cells available from Power Conversion, Inc.

| Standard size designation | Model No. | Rated drain (mA) | Rated capacity (Ah) | Ht. (mm) | Dia. (mm) | Wt. (g) |
|---|---|---|---|---|---|---|
| ½ AA    | G04 | 17   | 0.50 | 27.9  | 14.2 | 8   |
| AA      | G06 | 46   | 1.1  | 50.3  | 14.2 | 14  |
| ½ A     | G31 | 25   | 0.59 | 26.7  | 16.3 | 9   |
| ⅔ A     | G32 | 3    | 0.96 | 34.5  | 16.3 | 12  |
| ¾ C     | G50 | 100  | 2.5  | 40.1  | 25.7 | 36  |
| C       | G52 | 125  | 3.4  | 49.5  | 25.7 | 44  |
| 1¼ C    | G54 | 180  | 4.4  | 60.5  | 25.7 | 54  |
| D       | G20 | 1750 | 8.3  | 60.2  | 33.3 | 80  |
| 'squat' D | G70 | 160 | 9.0  | 50.3  | 38.4 | 90  |
| —       | G58 | 420  | 10.0 | 49.8  | 41.6 | 105 |
| —       | G60 | 1000 | 28.0 | 116.6 | 41.6 | 230 |
| —       | G62 | 1250 | 35.0 | 139.7 | 41.6 | 280 |

As seen in Table 5.3, PCI offers Li–$SO_2$ cells in twelve sizes ranging from 0.5 to 35.0 Ah. Nonstandard sizes for specialized applications are also available. Many of the current ratings of the PCI products given in Table 5.3 are conservative. The G70 cell, which is rated at 160 mA, operates safely at 2.0 A in the BA5598 battery. These designs are all hermetically sealed in a case containing a safety vent and are of balanced construction, i.e. the lithium–sulphur dioxide ratio does not exceed 1.1. These cells also contain a current collector in the anode which prevents lithium fragmentation during discharge and allows the cells to be force discharged safely. PCI's ⅔ A cell (model G32) is approved by Underwriters Laboratories Component Recognition Program.

Table 5.4 contains information on the Li–$SO_2$ cells currently available from SAFT America, Inc.

SAFT currently offers nine different sizes with capacities ranging from 0.500 to 8.00 Ah. The C and D sizes are available in both standard-rate and high-rate versions. The latter models contain an H designation in the model number and offer

**Table 5.4** — Lithium–sulphur dioxide cells available from SAFT America, Inc.

| Standard size designation | Model No. | Rated drain (mA) | Rated capacity (Ah) | Ht. (mm) | Dia. (mm) | Wt. (g) |
|---|---|---|---|---|---|---|
| $\frac{1}{2}$ A | LX 1622 | 100 | 0.500 | 27.6 | 16.8 | 10.0 |
| — | LX 2317 | 100 | 0.675 | 19.0 | 23.0 | 12.0 |
| $\frac{2}{3}$ A | LX 1634 | 225 | 0.850 | 35.1 | 16.8 | 14.0 |
| $\frac{2}{5}$ C | LX 2618 | 225 | 0.900 | 18.0 | 26.2 | 13.0 |
| — | LX 1978 | 1000 | 3.00 | 80.0 | 19.0 | 37.0 |
| C | LX 2649 | 1000 | 3.50 | 50.0 | 26.2 | 43.0 |
| C | LXH 2649 | 2000 | 3.00 | 50.0 | 26.2 | 43.0 |
| — | LXH 2958 | 2500 | 4.40 | 58.0 | 29.0 | 60.0 |
| D | LX 3457 | 2000 | 7.50 | 61.5 | 34.2 | 85.0 |
| D | LXH 3457 | 2500 | 7.00 | 61.5 | 34.2 | 85.0 |
| 'squat' D | LX 3851 | 2000 | 8.00 | 48.0 | 38.0 | 87.0 |

higher drain rate at the expense of reduced capacity. The model numbers LX 1634 ($\frac{2}{3}$ A), LX 2649 (C), and LX 3457 (D) are all approved by Underwriters Laboratories Component Recognition Program. The LX1634 and LX2649 meet US military specification requirements and are, therefore, of balanced design. With some models, the height given in Table 5.4 are those of unfinished cells. SAFT Li–SO$_2$ cells employ TA-23 glass and molybdenum pins in the glass-to-metal seal assembly. The seals must be protected to prevent external corrosion during use.

Table 5.5 lists the cells produced by Ballard Battery Systems Corp. of Canada on which information is currently available. Ballard is also known to produce a $\frac{1}{2}$ D cell for battery use, but no information is available on the characteristics of this product.

**Table 5.5** — Lithium–sulphur dioxide cells available from Ballard Battery Systems

| Standard size available | Rated drain (mA) | Rated capacity (Ah) | Ht. (mm) | Dia. (mm) | Wt. (g) |
|---|---|---|---|---|---|
| $\frac{1}{3}$ C | 50 | 0.800 | 18.3 | 25.4 | 20 |
| D | 2000 | 7.6 | 58.0 | 33.0 | 85 |

At present, information is obtainable on only two cells produced by Ballard primarily for military use. Little additional information is available on the characteristics of these cells.

### 5.2.5.2 Cell performance

Data on PCI's Model G52 Li–SO$_2$ C size cell will be presented as typical of an intermediate sized cell. Figs 5.28, 5.29, and 5.30 show the room temperature discharge data for C cells on 150, 22.4, and 3.0 ohm loads. Figs 5.31, 5.32 and 5.33 indicate the discharge characteristics of C cells on 150, 22.4, and 3.0 ohm loads at $-29°C$ ($-20°F$). The performance characteristics on these and three other loads are given in Table 5.6.

The exceedingly flat discharge characteristic of the Li–SO$_2$ cell is evident in the room temperature discharge curves at 150 and 22.4 ohms (Figs 5.28 and 5.29). At the higher rates, more curvature is observed as seen in Fig. 5.30. The room temperature capacity is relatively unaffected by drain rate from C/200 to C/10 and remains in the range of 3.30 to 3.40 Ah. At higher rates, the capacity is reduced, but a higher capacity is obtained on a 1.50 ohm load where the average current is 1.613 A compared to a 3.00 ohm load where the average current is 0.857 A. This effect is explained by an increased cell efficiency at the higher rate due to a higher operating temperature.

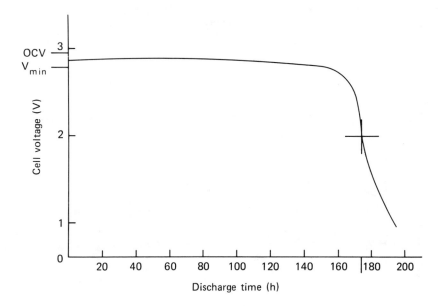

Fig. 5.28 — Discharge curve for lithium–sulphur dioxide C-cell on 150 ohm load at room temperature.

At $-29°C$ ($-20°F$), the discharge curve is still very flat on 150 ohm load as seen in Fig. 5.31, and the capacity is 93.6% of that obtained at room temperature. At higher drain rates, more curvature is observed in the discharge curve as seen in Figs 5.29 and 5.30. The capacity is also reduced to 83.3% of the room temperature value on 22.4 ohm load and 44.0% on 3.0 ohm discharge. The low temperature discharge

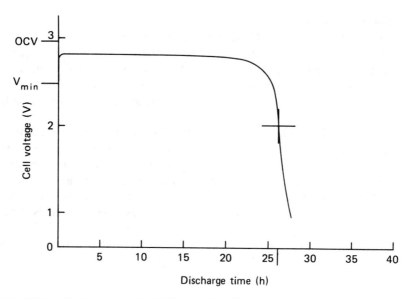

Fig. 5.29 — Discharge curve for lithium–sulphur dioxide C-cell on 22.4 ohm load at room temperature.

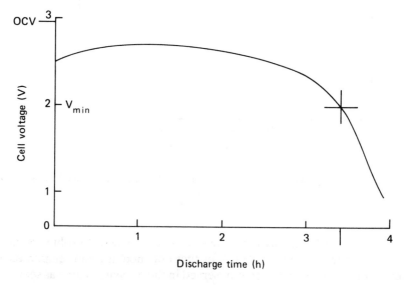

Fig. 5.30 — Discharge curve for lithium–sulphur dioxide C-cell on 3.0 ohm load at room temperature.

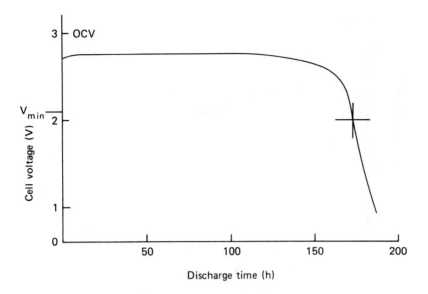

Fig. 5.31 — Discharge curve for lithium–sulphur dioxide C-cell on 150 ohm load at −29°C (−20°F).

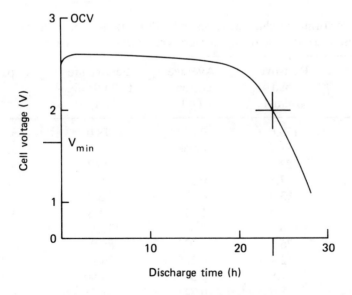

Fig. 5.32 — Discharge curve for lithium–sulphur dioxide C-cell on 22.4 ohm load at −29°C (−20°F).

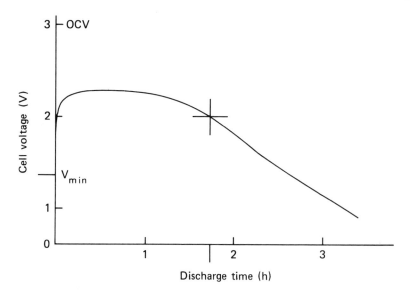

Fig. 5.33 — Discharge curve for lithium–sulphur dioxide C-cell on 3.0 ohm load at −29°C (−20°F).

Table 5.6 — Performance characteristics of PCI's lithium–sulphur dioxide C-cell at six discharge rates and two temperatures. Cut-off voltage = 2.00 V

| Temp. (°C) | Resistive load (ohms) | Average current (A) | Service life to 2.00 volts (h) | Capacity (Ah) |
|---|---|---|---|---|
| R.T. | 150.0 | 0.019 | 178.0 | 3.39 |
| R.T. | 76.7 | 0.038 | 88.5 | 3.36 |
| R.T. | 22.4 | 0.126 | 26.2 | 3.30 |
| R.T. | 7.8 | 0.355 | 9.3 | 3.30 |
| R.T. | 3.0 | 0.857 | 3.4 | 2.91 |
| R.T. | 1.5 | 1.613 | 1.87 | 3.01 |
| −29 | 150.0 | 0.018 | 172.5 | 3.17 |
| −29 | 76.7 | 0.036 | 85.0 | 3.04 |
| −29 | 22.4 | 0.115 | 23.9 | 2.75 |
| −29 | 7.8 | 0.315 | 6.60 | 2.08 |
| −29 | 3.0 | 0.741 | 1.73 | 1.28 |
| −29 | 1.5 | 1.369 | 0.617 | 0.84 |

data given in the figures and in Table 5.6 demonstrate the excellent low temperature performance of Li–SO$_2$ system in single cell applications. Multi-cell batteries perform significantly better than single cells at low temperatures because of intercell heating effects which produce higher actual operating temperatures, particularly in sealed containers, as demonstrated in the next section.

### 5.2.5.3 Safety considerations in the use of lithium–sulphur dioxide cells
Lithium–sulphur dioxide cells can be used safely for numerous applications, but should be fused if very high currents are anticipated and diode protected if any possibility of charging exists. Battery packs assembled from cells should explicity follow the manufacturer's directions, and battery design is best left to experts in the field. Cells should not be short-circuited since they are designed to vent under such conditions. Cells should also be stored and used below the manufacturer's recommended temperature maximum to avoid any leakage or venting. Incineration will also lead to venting and possible fire due to the flammability of lithium metal. Neither cells nor batteries should be incorporated into a completely sealed housing or container unless there is a pressure relief mechanism employed in the assembly in case the cell vents. Other safety considerations are detailed below in section 5.2.6.

## 5.2.6 Battery design and performance
Lithium–sulphur dioxide cells are highly energetic devices, and their principal use to date has been as components for the assembly of high-rate batteries to power a multitude of military communications and electronic devices. The design and performance characteristics of Li–SO$_2$ batteries will be considered in this section, while their applications will be described in section 5.2.7. This section will be devoted primarily to US military batteries. The basic specification for such Li–SO$_2$ batteries is MIL-B-49430(ER) [34], but each battery type has its own specification sheet as well.

### 5.2.6.1 Battery design
Since batteries are series and/or parallel arrays of individual cells, additional features must be employed in battery design to ensure that the performance and safety requirements of any particular application are met. These are described below.

### 5.2.6.2 Fuses, diodes, and discharge circuits
Battery packs employ electrical fuses normally of the slow-blow design, usually incorporated in the negative leg of the circuit. Many US military batteries are designed to operate at 2.00 A and employ a 2.25 A non-replaceable slow-blow type fuse to prevent cell venting in the event of an over-current condition. In addition, most batteries employ a thermal fuse designed to open at $88 \pm 2°C$ ($190 \pm 5°F$). Such a fuse is located at the centre of the cell stack. For batteries which can be used with two-cell strings in parallel, each leg must contain both an electrical and a thermal fuse. The lithium–sulphur dioxide cells employed for battery use are primary cells and are not capable of being recharged safely. For this reason, each battery contains a blocking diode in the positive leg of the circuit to prevent charging. The US military requirement [34] calls for a diode with a leakage current not to exceed 2.0 mA when 40 V is applied in opposition to the battery voltage. The use of parallel arrays of cells

hard-wired together has been discontinued to prevent charging if diode failure occurs.

The US Environmental Protection Agency (EPA) has ruled that lithium–sulphur dioxide batteries are non-hazardous if fully discharged to deplete the reactive components to low levels. To achieve this goal, all multi-cell US military batteries now incorporate a complete discharge circuit consisting of a switch and a resistor placed across each cell stack. After the useful life of the battery is completed, the user activates the switch which is accessed through a slot in the case to complete the discharge process. The resistors in these circuits are sized to achieve complete discharge in five days, so that batteries should be held for this period before disposal.

### 5.2.6.3 Other design considerations

US military specifications require that each cell be sleeved with a 4.76 mm ($\frac{3}{16}$ in) roll-over on the top and bottom. Once a positive lead is welded to the pin of the glass/metal seal, the cell must be potted with an insulating polymeric compound to prevent the positive tab from shorting to the cell case. The tab must be covered with an insulating sleeve at least 0.2 mm (0.008 in) thick which does not soften until 150°C (302°F). Battery cases are required to be of a material which will not support combustion, and are typically made of polymeric material. Once circuits are assembled from the individual cells, the battery must be potted with an elastomeric polymer designed to allow the unit to pass the shock, drop, and vibration requirements of MIL-B-49430 [34]. After final sealing, all units must be checked for open circuit and closed circuit compliance with the specification as well as for dimensional conformance. Further details on the requirements for these batteries may be obtained by consulting the military specification document [34].

### 5.2.6.4 US military battery types

A compilation of the battery types employed for military electronics and communication equipment throughout the World is given in Table 5.7.

The voltages shown are the nominal open circuit values except for the BA-5598, BA-5590 and BA-5557 where the operating voltages under the rated load as indicated on the battery specification sheets (known as the slash sheet) are given. In the case of the BA-5598 battery, one five-cell string is employed, but a three-volt tap between the negative leg and the first cell constitutes a second circuit. Both BA-5590 and BA-5557 have two five-cell circuits which may be employed in parallel or in series connections.

The service life requirements on the five standard military tests are shown. The H, I, and L tests are carried out on fresh batteries at the rated load, whilst the HT and LT tests are performed after storage for 28 days at +71°C (+160°F). H and HT tests are carried out at +54°C (+130°F) normally on a 50% duty cycle (5 min on/5 min off) to prevent activation of the thermal fuse. In the case of the BA-5567, H and HT tests are continuous. The I test is performed at +21°C (+70°F), whilst both L and LT tests are carried out at −29°C (−20°F).

The allowable initial voltage delay time in seconds at −29°C (−20°F) is also given in Table 5.7, since this is an important operating parameter. These times are applicable to the L and LT tests and represent the allowable time to reach operating

**Table 5.7** — US military batteries

| BA-number | Nominal voltage (V) | Current (A) | Parallel circuits | Cell size | No of cells | Battery weight (kg/lbs.) | Initial voltage delay at $-29°C/-20°F$ (s) | H-test (h) | Service life requirements I-test (h) | L-test (h) | HT-test (h) | LT-test (h) |
|---|---|---|---|---|---|---|---|---|---|---|---|---|
| 5599 | 9.0 | 2.0 | No | D | 3 | 0.45/1.0 | 60.0 | 5.6 | 3.0 | 1.6 | 5.4 | 1.4 |
| 5598 | {14.4 | 2.0 | 3V Tap | Squat D | 5 | 0.68/1.5 | 60.0 | 6.5 | 3.6 | 1.6 | 6.3 | 1.5 |
|      | 3.0  | 2.0 | 3V Tap | Squat D | 5 | 0.68/1.5 | 60.0 | 6.5 | 3.6 | 1.6 | 6.3 | 1.5 |
| 5590 | {12.0 | 4.0 | Yes | D | 10 | 1/2.33 | 60.0 | 5.6 | 3.2 | 1.8 | 5.4 | 1.6 |
|      | 24.0 | 2.0 | Yes | D | 10 | 1/2.33 | 60.0 | 5.6 | 3.2 | 1.8 | 5.4 | 1.6 |
| 5567 | 3.0 | 0.05 | No | ⅔C | 1 | 19 g/0.7 oz. | 10.0 | 14.0 | 16.0 | 9.0 | 12.25 | 8.0 |
| 5847 | 6.0 | 2.0 | No | D | 2 | 0.24/0.53 | 60.0 | 5.6 | 3.0 | 1.6 | 5.4 | 1.4 |
| 5557 | {12.0 | 0.65 | Yes | Non-Standard | 10 | 0.5/1.1 | 60.0 | 5.5 | 3.0 | 1.5 | 5.3 | 1.3 |
|      | 24.0 | 1.30 | Yes | Non-Standard | 10 | 0.5/1.1 | 60.0 | 5.5 | 3.0 | 1.5 | 5.3 | 1.3 |
| 5588 | 15.0 | 0.50 | No | Standard | 5 | 0.3/0.65 | 10.0 | 5.8 | 5.8 | 2.8 | 5.0 | 2.6 |
| 5600 | 9.0 | 2.0 | No | D | 3 | 0.36/0.8 | 60.0 | 5.6 | 3.0 | 1.8 | 5.2 | 1.6 |
| 5112 | 12.0 | 0.65 | No | Non-Standard | 4 | 0.18/0.4 | 60.0 | 5.50 | 3.0 | 1.50 | 5.30 | 1.30 |

voltage, which is typically 2 V per cell with a few exceptions (BA-5599, BA-5847, BA-5588, for tests other than L and LT, and BA-5600). The cut-off voltage for the determination of the service life is also typically 2.0 V per cell except for the BA-5588 where the cut-off voltage is 11.0 V.

In the case of the BA-5590 and BA-5557 where both series tests and parallel tests are carried out, the service life requirements are the same on both types of test.

Batteries are discharged to zero volts on the H, HT, L, and LT tests. In the case of the I-test, the battery is discharged to zero volts at the rated load and then force discharged into reversal, normally at one-half the rated load for a time equal to twice the service life requirement on the I test. These batteries are not allowed to vent, leak, rupture, or burn during or within 24 hours after any of these tests, otherwise they fail the requirement.

### 5.2.6.5 Performance of BA-5598 batteries

The BA-5598 is a typical lithium–sulphur dioxide battery which is widely used to power AN/PRC-25 and AN/PRC-77 military transceiver radios throughout the World. The circuit diagram for this battery is shown in Fig. 5.34. The battery is constructed with five 'squat' D cells (38 mm/1.5 in diameter by 51 mm/2.00 in high). The negative leg contains a 2.25 A slow-blow fuse as does the $+A_1$ leg of the 3.0 V circuit which is separately fused.

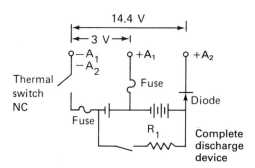

Fig. 5.34 — Circuit diagram for the BA-5598 lithium–sulphur dioxide battery.

The battery contains a thermal fuse in the negative leg and a blocking diode in the $+A_2$ (14.4 V) leg. The complete discharge device consisting of a switch and resistor (125–175 ohms) is connected across all five cells for use after the battery has been discharged. Battery dimensions are $119 \times 90.5 \times 52.4$ mm ($4\frac{11}{16} \times 3\frac{9}{16} \times 2\frac{1}{16}$ in), and the maximum weight is 0.68 kg (1.5 lb). These units are half the size and less than half the weight of the conventional BA-3386 alkaline manganese and BA-4386 magnesium batteries used to power the AN/PRC-25 and AN/PRC-77 radios, allowing the user to carry both an operating unit and a spare in the battery container. The service life of the BA-5598 is slightly greater than that of the BA-4386 at room temperature, but is four times as great at −17°C (0°F). The BA-5598 will operate these radios at temperatures down to −40°C (−40°F), whilst conventional batteries are non-functional at these temperatures. Discharge curves of PCI's BA-5598 batteries on the H, I and L tests are shown in Figs 5.35, 5.36, and 5.37. These figures show service lives of 7.73 h on the H test, 4.02 h on the I test and 2.34 h on the L test, exceeding the requirement given in Table 5.7 for this battery by 19%, 12%, and 46% respectively.

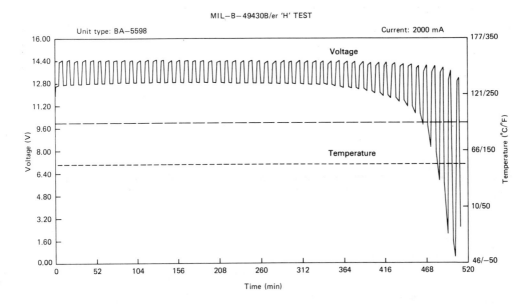

Fig. 5.35 — Discharge curve and temperature profile for the BA-5598 battery being discharged on the H-test (2.0 A, 50% duty cycle at +54°C/ +130°F).

Fig. 5.38 shows the start-up voltage characteristic on the L test for the same battery. This test is carried out after the battery has been equilibrated at −29°C (−20°F) for 16 h, and the data shows that the operating voltage of 10.0 V is reached in 28.67 s, less than half the 60 s requirement for this battery type. This data demonstrates the technical superiority of lithium–sulphur dioxide batteries over conventional technologies.

### 5.2.7 Applications of lithium–sulphur dioxide cells and batteries

As the earliest practical lithium power source to be commercialized, the Li–$SO_2$ system has found a wide variety of applications in the twenty years of its existence. This section will discuss both industrial and electronic applications as well as military and naval uses. In addition, mention will be made of reserve and rechargeable Li–$SO_2$ cells.

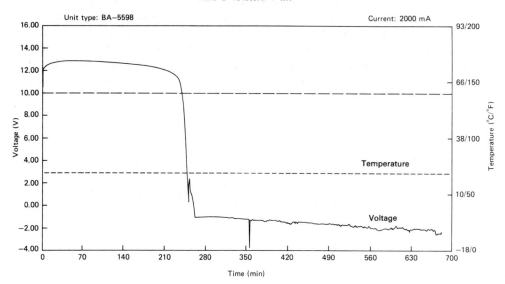

Fig. 5.36 — Discharge curve and temperature profile for the BA-5598 battery being discharged on the I-test (2.0 A, at $+21°C/+70°F$).

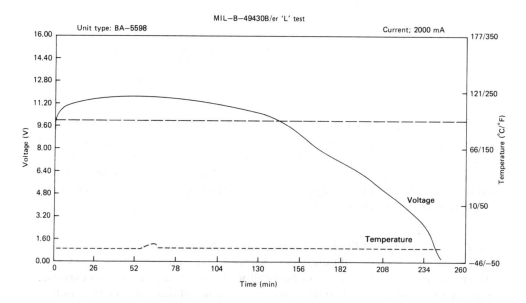

Fig. 5.37 — Discharge curve and temperature profile for the BA-5598 battery being discharged on the L-test (2.0 A, at $-29°C/-20°C$).

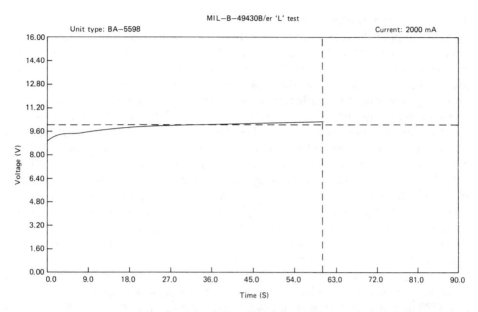

Fig. 5.38 — Initial voltage delay characteristic of the BA-5598 battery being discharged on the L-test (voltage vs time for first 90 s of discharge on 2.0 A at $-29°C/-20°C$).

### 5.2.7.1 Industrial and electronic applications

Lithium–sulphur dioxide cells and batteries have found a wide variety of use in industrial and electronic applications [40]. These include utility meters where the wide operating temperature is a major consideration. In such uses, the cell acts as a power source to hold information in memory in case of a loss of line voltage. Although somewhat overshadowed by their higher voltage competitior, lithium–thionyl chloride, Li–$SO_2$ cells find use in other random access memory (RAM) applications where wide operating temperature or pulse capability is required. The low temperature capability of Li–$SO_2$ cells has made them the preferred choice for assembling batteries used in weather balloons for telemetry. Battery packs assembled from large cells such as PCI's G60 (28.0 Ah) and G62 (35.0 Ah) cells have found wide use in this application where operation to $-53°C$ ($-65°F$) is required. At the opposite extreme, custom-made batteries have also found wide use in oceanographic experimentation. Industrial uses include applications in process controllers, switchgear, and test equipment.

Lithium–sulphur dioxide cells, which are US Federal Aviation Agency approved, are used in avionics including on-board computers for the Boeing 747 aircraft and emergency locator transmitters (ELTs). Four Li–$SO_2$ cells are employed in the inertial upper stage (IUS) rocket propulsion system which is used to place satellites in orbit from the US Space Shuttle Orbiter. In this case, the control system includes two redundant computers, each with two memory units. One of PCIs model G56 cells is used to hold the programmed information in each of the four memory units. Here again the remarkable low temperature performance of the Li–$SO_2$ cell has made it the most appropriate system for this application.

### 5.2.7.2 Military applications

Lithium–sulphur dioxide batteries are the most widely used lithium batteries for military applications in the world. The family of Li–SO$_2$ batteries used by the US military establishment and by numerous other countries as well has been described in section 5.2.6.4. Table 5.8 lists the applications for these batteries. The BA-5590 is used to power some forty different types of electronic and communications equipment, including the single channel ground and air radio system (SINCGARS), the most modern military radio in use by the US military forces. In fact, the BA-5590 is the only deployed battery which will operate SINCGARS. It also powers the AN/PRC-104, 113, and 119 radios as well as equipment such as VINSON, PLRS, BANCROFT, the global positioning system (GPS), and the battery computer system (BCS). The BA-5112 was designed to operate the AN/PRC-112 rescue radio and beacon. The BA-5567, the only single cell battery, operates several night vision devices. The BA-5599 operates the AN/PAS-7 test equipment. The BA-5800 was designed to operate the chemical agent monitoring (CAM) device, and it runs the Trimble navigation global positioning system as well. The BA-5598 replaces the conventional BA-3386 and BA-4386 batteries in operating the AN/PRC-25, 74, and 77 radios, the KY-38 scrambler, and the MOPENS mine deployment system. The BA-5588 operates several 'walkie talkie' type radios such as the AN/PRC-126 and 127 systems, whilst the BA-5557 operates the digital message device. It is clear that Li–SO$_2$ batteries have become the principal power source for the current generation of military electronics/communications equipment.

Lithium–sulphur dioxide batteries are also the chosen power source for the active sonobuoys used in anti-submarine warfare. The batteries are typically twelve-cell batteries using D or mini D cells designed for very high rate applications. Active sonobuoys require 27–30 A pulses for operation, and the high-rate capability combined with excellent low-temperature performance have made Li–SO$_2$ batteries very suitable for these applications.

Lithium–sulphur dioxide batteries have also been employed for specialized applications such as the anti-satellite (ASAT) missile where four different Li–SO$_2$ batteries powered the electronics, telemetry, and stage separation mechanism [41,42]. The electronics application for this missile required continuous operation of the battery composed of twelve very high rate D cells at an 8 A discharge rate, although venting was allowable after completion of the mission requirements. The other batteries for this missile were composed of twelve $1\frac{1}{4}$ D cells which were required to operate continuously at the 4.0 A rate.

### 5.2.8 Conclusions

The lithium–sulphur dioxide system has found wide acceptance in military electronics/communications applications and in other specialized applications where its high-rate capability, low-temperature performance, and long shelf life make it the system of choice. Its utility in other applications has been limited by safety considerations. The pressurized electrolyte enhances the tendency to vent the sulphur dioxide-rich electrolyte when a cell overheats or if a short-circuit condition develops. The development of additional applications including consumer use will depend upon increased safety. The use of a fusible separator, which deactivates the

**Table 5.8** — US military batteries and applications

| Battery No. | Application |
|---|---|
| BA5590/U | Radios (SINCGARS) |
| | Communications equipment |
| | Satellite radios |
| | Scramblers |
| | Radar |
| | Loudspeakers |
| | VHF radios |
| | Laser equipment |
| | Range finders |
| | Laser markers |
| | Counter measures |
| | Weather equipment |
| | Jammers |
| | Cooling systems |
| BA5112/U | Rescue radio/beacon |
| BA5567/U | Night vision equipment |
| BA5599/U | Test equipment |
| | Night vision equipment |
| BA5600/U | Data terminals |
| BA5800/U | Chemical agent monitors |
| | Global positioning equipment |
| BA5847/U | Test equipment |
| | Antennas |
| | Night vision equipment |
| BA5598/U | Radios (PRC-77, PRC-25), Scramblers |
| BA5588/U | Handheld radios |
| | Gas masks |
| BA5557/U | Digital message device |

cell under over-temperature or short-circuit conditions, may serve to open up additional areas of use. Similarly, the use of a cell monitoring circuit within a battery incorporating a cell voltage, temperature, or pressure monitoring device [43] holds the promise of enhancing system safety. This is particularly the case if the cell voltage is monitored, since Li–SO$_2$ cells rarely vent above 2.00 V, the normal cut-off voltage.

Another possible area for future improvement with this technology would be to develop a means of lowering the vapour pressure of the electrolyte to reduce the tendency of cells to vent during use.

In principle, Li–SO$_2$ cells can be recharged. Considerable research has been carried out in this area. The most recent attempts in this area involved the elimination of the organic co-solvent and the use of a lithium tetrachloroaluminate––sulphur dioxide electrolyte [44]. Although such cells can achieve significant cycle life, safety considerations associated with the development of highly reactive high-surface area lithium metal have limited their practical utility. This has also been true of other secondary lithium technologies, so that if this problem can be solved, the lithium–sulphur dioxide battery (which has achieved significant utility as a primary system) may achieve further use in secondary applications as well.

## 5.3 LITHIUM–CARBON MONOFLUORIDE BATTERIES
(D. Eyre and C. D. S.Tuck)

### 5.3.1 Introduction

The combination of the most energy-dense metal with the highest energy content oxidizing agent would result in a lithium–fluorine battery of theoretical energy density 6250 Wh.kg$^{-1}$. At present, the closest attainment of this system as a commercial battery has used lithium with a combination of fluorine and carbon, the latter being in the form of the compound, carbon monofluoride ($CF_x$).

The theoretical energy density of this system is approximately 2000 Wh.kg$^{-1}$, and such batteries were first developed in Japan by Matsushita (Panasonic) in the early 1970s. They began to be mass produced in 1976, Matsushita also licensing production to a number of other battery companies throughout the World.

Applications of the lithium–carbon monofluoride (Li–$CF_x$) battery are in powering electronic watches, calculators, and cameras (a $\frac{2}{3}$ A size being used in the Kodak disk camera). They are also used as back-up power for electronic devices. A major application in Japan is as a battery for lighted night-time fishing floats using a thin pin-type battery configuration. In Europe there is considerable competition with Li–$MnO_2$ cells in the electronics memory back-up market, which includes such items as 'keep alive' electronics security tagging. Li–$CF_x$ batteries have safety advantages over Li–$MnO_2$ batteries, owing to the more stable nature of $CF_x$, particularly with regard to the organic battery electrolytes used. Also, Li–$CF_x$ has a higher capacity density (2379 Ah.dm$^{-3}$) than Li–$MnO_2$ (1549 Ah.dm$^{-3}$). $CF_x$ batteries have energy densities of almost 350 Wh.dm$^3$ (200 Wh.kg$^{-1}$) and have achieved a power density of 263 W.dm$^{-3}$, with the Panasonic BR $\frac{2}{3}$ A cell. Of significant application benefit for Li–$CF_x$ in comparison within its main competitor, the Li–$MnO_2$ system, is the more typical level voltage profile as the discharge proceeds. It also has the ability to be miniaturized whilst still maintaining excellent pulse and current drain performance, this being over the temperature range $-40°C$ to $85°C$. Cells and batteries of the Li–$CF_x$ system have demonstrated high reliability and good shelf life (10 years).

### 5.3.2 Manufacturers

The original patent for Li–$CF_x$ batteries is held by the Matsushita Electronic Company of Japan, the parent company of the Panasonic Industrial Company. This company produces batteries in the form of cylindrical cells, coin cells, and pin type cells. Fig. 5.39 shows the construction of these three cell types. Among the cylindrical cells manufactured by Panasonic is the $\frac{2}{3}$ A size developed originally and selected for the Kodak disk camera, which is able to deliver currents of up to 1.6 A during the $1\frac{1}{2}$ seconds required to take a photograph. The time of operation includes

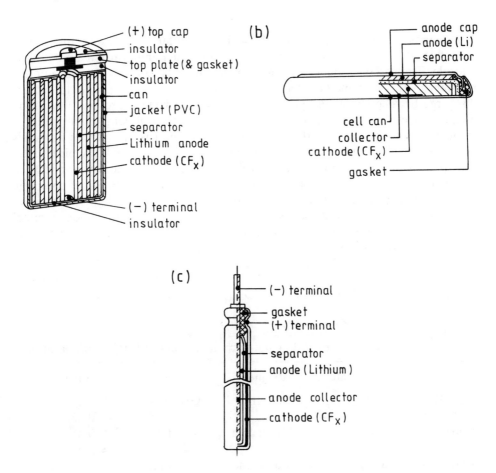

Fig. 5.39 — Cross sectional views of the different types of commercial Li–CF$_x$ battery. (a) Cylindrical type, (b) Coin type, (c) Pin type. (Courtesy of Matsushita Electric Industrial Co. Ltd.)

setting the exposure, activating the flash, taking the picture, advancing the film, and recharging the flash.

Other manufacturers produce similar types of cell, although not as complete a range as Panasonic, and they have individual agreements with Matsushita to do so. Eagle Picher, Rayovac, Eveready, Wilson Greatbatch, Duracell, and, from 1987, Crompton Eternacell Ltd (formerly Crompton Parkinson Ltd) have been granted patent access to produce these cells, with some agreements including technology exchange as well.

### 5.3.3 Chemistry and cell components

The basic Li–CF$_x$ battery reaction is given by:

$$x\mathrm{Li} + \mathrm{CF}_x \rightarrow x\mathrm{LiF} + \mathrm{C} \qquad E^\circ = 3.2\ \mathrm{V}\ . \tag{5.10}$$

This equation shows that, as the lithium in the battery reacts with the carbon monofluoride, the original cell material is converted into carbon and lithium fluoride. The carbon produced is amorphous, and acts to lower the cell internal resistance. This is a major advantage of the process when applied to battery discharge behaviour manifested as the level on-load voltage profile. Polycarbon monofluoride, $CF_x$ (where $x$ is 0.9 to 1.2) is an interstitial compound produced through direct reaction of carbon powder and fluorine gas, and it was originally developed as a lubricant. It is manufactured in the USA by Allied Chemical under the trade name of Accuflor™ specifically for the lithium battery application, although it has several other uses — a high temperature lubricant, a corrosion resistant coating material, and a colour display component. It is a light grey powder of melting point 500°C which is insoluble in water, although it undergoes thermal decomposition with water to produce hydrogen fluoride. It also reacts with alkali metals, sulphuric acid above 200°C, bases, amines, and polyacetal resins. Its density is 2.7 g.cm$^{-3}$ and its equilibrium water content is very low (0.15%). Table 5.9 gives a list of patents concerning $CF_x$ showing that material with different values of $x$ may be produced along with different carbon bases (graphitic or amorphous).

**Table 5.9** — US patents regarding polycarbon monofluoride ($CF_x$) preparation

| Patent No. | Issue date | Assignee | Carbon | $x$ range |
|---|---|---|---|---|
| 3514337 | 5/70 | US Army | Graphite | $>0.13$ $<0.28$ |
| 3536532 | 10/70 | Matsushita | 'Crystalline' | $>0.5$ $\leq 1.0$ |
| 3700502 | 10/72 | Matsushita | 'Amorphous' | $>0$ $\leq 1.0$ |
| 3892590 | 7/75 | Yardney | No restriction | $>1.0$ $\leq 2.0$ |
| 4247608 | 1/81 | Watanabe | Graphite | C(2)F |
| 4271242 | 6/81 | Matsushita | $d_{002}$ $>3.4$ $<3.5$ | No restriction |

The particular cathode formulation produced for lithium batteries is a mixture of $CF_x$ with actylene black and a binder such as PTFE in the ratios of 0.8:0.1:0.1. For its manufacture, the $CF_x$ and carbon are slurried in a solution of 50/50 isopropanol/water to which a small proportion of ammonia has been added. The latter prevents the precipitation of PTFE which is added as a suspension. The slurry is well mixed and then heated to coagulate the PTFE, a process which precipitates all the solids. These are separated and dried at 100°C before granulation and pressing into electrode pellets. It is usual to further dry the cathode pellets at 250°C to remove the surfactants used in the PTFE emulsion, as these affect the wetting characteristics of the cathode. Titanium foil is used as a current collector.

Some manufacturers, such as Crompton Eternacell Ltd, in their coin cells, use a styrene butadiene rubber binder for the cathode mix rather than PTFE. Fig. 5.40 shows X-ray diffraction patterns taken of $CF_x$ decomposition during a Li–$CF_x$ battery discharge. It can be seen that the $CF_x$ peak disappears after 25% of charge has been lost, shortening the carbon atom interlayer distance and producing amorphous carbon with crystalline lithium fluoride. The latter precipitates in the cathode during cell operation.

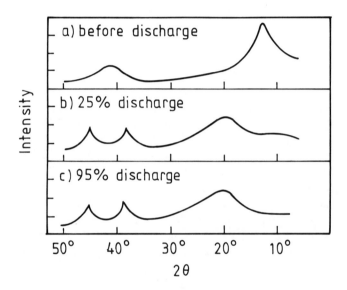

Fig. 5.40 — X-ray diffraction patterns from a diffractometer of $CF_x$ before and during decomposition during discharge. (a) Before discharge, (b) After 25% discharge, (c) After 95% discharge. (Courtesy of Matsushita Electric Industrial Co. Ltd.)

The electrolyte used in Li–$CF_x$ cells is usually a 1:1 mixture of dimethoxyethane (DME) and propylene carbonate containing 1 M $LiBF_4$, although $LiAsF_6$ in γ-butyrolactone is also used. The separator is a thin porous insulator with high ion permeability, and it is usually a polyolefin made into non-woven cloth, or microporous polypropylene.

### 5.3.4 Cell types and performance [45,46,47]
#### 5.3.4.1 *Cylindrical cells*
The cylindrical cell configuration is shown in Fig. 5.39(a), and employs the spiral winding of lithium separator and cathode, the latter bound on to an expanded titanium metal grid by pressure rolling. Sealing is achieved by use of a polypropylene disk containing the positive contact, and this disk also acts as a safety vent.

Table 5.10 gives data on the different cylindrical cells available and Fig. 5.41 shows the discharge of a R14 size BR-C cell under a constant load of 20 Ω. This is compared with a similar current discharge of a comparably sized zinc–carbon battery. A flat discharge profile for the BR-C cell is obtained, and almost 100% of the cell energy is utilized during discharge. At $-10°C$, the voltage is slightly lower and less plateau-like, but the capacity of the cell falls by only approximately 20%.

Owing to the extreme stability of the $CF_x$ the long-term storage capability of these cells is very good, with 10 years of storage being achieved without significant performance deterioration. Also, as self-discharge reactions are almost non-existent, the batteries are well-suited to long discharge at very low currents, and Fig. 5.42 shows a current of 20 μA being supplied by a BR-⅔A cell for 7 years. Pulse discharge behaviour is also very good, as shown in Fig. 5.43. Battery packs containing two cylindrical cells in series, suitably packaged together, are also available, and are able to give a cell voltage of 6 V.

### 5.3.4.2 Coin cells

A schematic cross-section of the coin cell construction is shown in Fig. 39(b). It employs a carbon-coated titanium screen as a cathode current collector. The lithium is pressed to the inner surface of the stainless steel casing. Table 5.11 lists the different types and sizes available.

Fig. 5.44(a) shows discharge curves for the BR 2330 cell under different loading conditions, again showing flat voltage discharge profiles. This behaviour is not significantly different over the temperature range $-10°C$ to $+70°C$, as Fig. 5.44(b) shows, and the batteries operate satisfactorily in more extensive temperature ranges of $-40°C$ to $+85°C$. Typical discharge loads are from ~1 nA to >10 mA impulse regimes. Although a slight voltage delay may exist, caused by an initial surface insulating layer on the $CF_x$ electrode, this is of no significance either in magnitude or duration, as can be seen in Fig. 5.45. The storage capability of the system is well demonstrated, as is shown in Fig. 5.44(d) and Fig. 5.46.

Applications of the coin cell are for powering electronic watches (digital and analogue), calculators, cameras, electronic translators, memory back-up devices, and low power cordless equipment.

### 5.3.4.3 Pin-type cells

Fig. 5.39(c) shows the configuration of this slim-line battery which is lightweight owing to its having an aluminium case. Its voltage of 3 V and small size make it ideal for coupling with light-emitting diodes, and a major application is as a power source for lighted fishing floats. It is also used in microphones and toys. Table 5.12 gives the dimensions and properties of the types available.

The constant-load discharge behaviour of the BR-425 cell is shown in Fig 5.47(a), with temperature characteristics under a 5 kΩ load discharge being shown in Fig. 5.47(b). Self-discharge is negligible, and cell utilization is high (80%–90%) even under an imposed drain of several mA. Fig. 5.47(c) and Fig. 5.47(d), which show the temperature characteristics of the operating voltage vs load resistance and capacity vs load resistance respectively, demonstrate that these cells can be used effectively over a wide temperature range ($-10°C$ to $+60°C$).

**Table 5.10** — Dimensions and properties of lithium–$CF_x$ cylindrical cells

| Model No. | | BR-C | BR-A | BR-⅔A | BR-½A | BR-⅔AA |
|---|---|---|---|---|---|---|
| Nominal Voltage (V) | | 3 | 3 | 3 | 3 | 3 |
| *Nominal Capacity (mAh) | | 5000 | 1800 | 1200 | 650 | 600 |
| Standard Drain (mA) | | 150 | 2.5 | 2.5 | 2 | 2 |
| Continuous Maximum Drain (mA) | | 300 | 250 | 250 | 120 | 80 |
| Pulse | (mA) | max. 1000 | max. 1000 | max. 1000 | max. 500 | max. 500 |
| dimension | Diameter (mm) | 26.0 | 17.0 | 17.0 | 17.0 | 14.5 |
| (Max.) | Height (mm) | 50.0 | 45.5 | 33.5 | 23.0 | 33.5 |
| Approx. Weight (g) | | 47.0 | 18.0 | 13.5 | 9.5 | 9.5 |

*Nominal Capacity shown above is based on standard drain and cut off voltage down to 1.8V at 20°C. For low (microampere) drains, capacity is significantly higher.

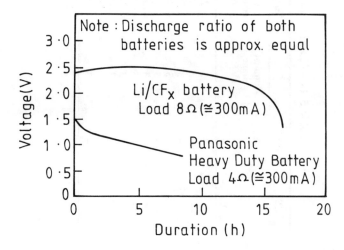

Fig. 5.41 — Comparison between the discharge behaviour under constant load conditions of an R14-size Li–$CF_x$ battery and the same size of zinc-carbon battery. (Courtesy of Matshushita Electric Industrial Co. Ltd.)

Table 5.11 — Dimensions and properties of lithium–CF$_x$ coin cells

| Model No. | JIS | IEC | Electrical characteristics (20°C) | | | | | | Dimensions | | Weight (g) |
|---|---|---|---|---|---|---|---|---|---|---|---|
| | | | Nominal voltage (V) | Nominal capacity (mAh) | Recommended Drain | | | | Diameter (mm) | Height (mm) | |
| | | | | | Pulse (mA) | Standard (mA) | | Low (μA) | | | |
| BR1216 | — | — | 3 | 25  | 5  | 0.03 | | 1  | 12.5 | 1.60 | 0.6 |
| BR1220 | — | — | 3 | 35  | 5  | 0.03 | | 1  | 12.5 | 2.00 | 0.7 |
| BR1225 | — | — | 3 | 38  | 5  | 0.03 | | 1  | 12.5 | 2.50 | 0.9 |
| BR1616 | — | — | 3 | 48  | 8  | 0.03 | | 1  | 16.0 | 1.60 | 1.0 |
| BR2016 | — | — | 3 | 75  | 10 | 0.03 | | 1  | 20.0 | 1.60 | 1.5 |
| BR2020 | — | — | 3 | 100 | 10 | 0.03 | | 2  | 20.0 | 2.00 | 2.0 |
| BR2032 | — | — | 3 | 190 | 10 | 0.03 | | 4  | 20.0 | 3.20 | 2.5 |
| BR2320 | — | — | 3 | 110 | 10 | 0.03 | | 2  | 23.0 | 2.00 | 2.5 |
| BR2325 | — | — | 3 | 165 | 10 | 0.03 | | 3  | 23.0 | 2.50 | 3.0 |
| BR2330 | — | — | 3 | 255 | 10 | 0.03 | | 5  | 23.0 | 3.00 | 3.2 |
| BR3032 | — | — | 3 | 500 | 10 | 0.03 | | 10 | 30.0 | 3.20 | 5.5 |

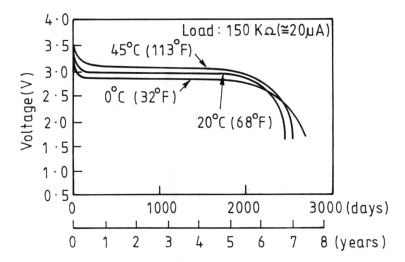

Fig. 5.42 — The discharge curves at 0°C, 20°C, and 45°C of BR-2/3 A Li–CF$_x$ batteries at approximately 20 μA drain over a period of seven years. (Courtesy of Matsushita Electric Industrial Co. Ltd.)

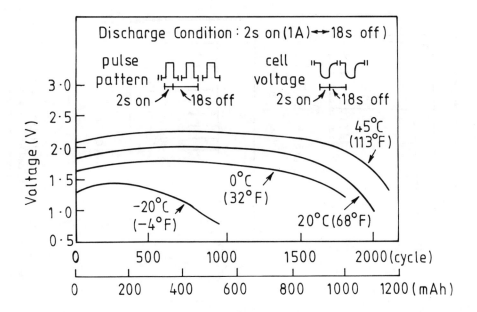

Fig. 5.43 — The 1 A pulse discharge characteristics at −20°C, 0°C, 20°C and 45°C of a BR-2/3A Li–CF$_x$ battery. (Courtesy of Matsushita Electric Industrial Co. Ltd.)

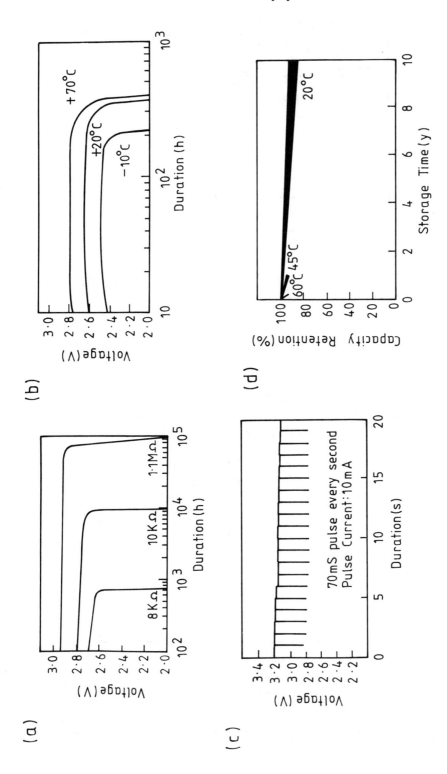

Fig. 5.44 — Various characteristics for the BR 2330 Li–CFx coin cell. (a) Constant load discharge performance for loads of 8 kΩ, 100 kΩ, and 1.1 MΩ. (b) Temperature performance −10°C, 20°C, and 70°C during discharge through a 4Ω load. (c) Pulse discharge performance at 20°C for a 70 ms pulse of current 10 mA every second. (d) Storage performance over a 10 year period for storage at 20°C, 45°C, and 60°C. (Courtesy of Crompton Eternacell Ltd.)

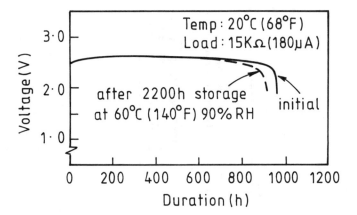

Fig. 5.45 — Storage characteristics of a BR 2325 Li–CF$_x$ coin cell under conditions of high temperature (60°C) and high humidity (90% RH). (Courtesy of Matsushita Electric Industrial Co. Ltd.)

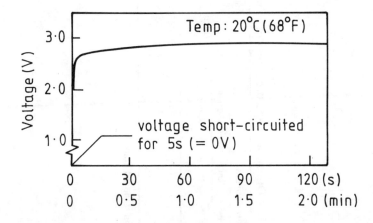

Fig. 5.46 — Voltage recovery characteristics of a BR 2325 Li–CF$_x$ coin cell after a short circuit of 5 s duration at 20°C. (Courtesy of Matshushita Electric Industrial Co. Ltd.)

### 5.3.5 Conclusion

Li–CF$_x$ cells and batteries have been extremely successful in many applications. A significant number of these have been in the consumer marketplace, with photographic applications being a principal and widespread use. The development of Li–CF$_x$ batteries for the electronics industry have consequently resulted in many of the same batteries being used in industrial applications, particularly in uses associated with memory circuitry. This has been due to their reliable and consistently safe behaviour even under arduous environmental conditions. They are continually being further developed in terms of materials, and work is progressing to produce larger

Table 5.12 — Dimensions and properties of lithium–$CF_x$ pin-type cells

| Model No. | Electrical characteristics | | | | Dimensions | | Weight (g) | Operating Temperature range |
|---|---|---|---|---|---|---|---|---|
| | Nominal voltage (V) | Nominal capacity (mAh) | Recommended drain | | Diameter mm (inch) | Height mm (inch) | | |
| | | | Pulse (mA) | Standard (mA) | | | | |
| BR211 | 3 | 5.4 | — | 0.05 | 2.2(0.087) | 11.5(0.45) | 0.09 | −10°C∼−60°C(14°F∼140°F) |
| BR425 | 3 | 25 | 4 | 0.5 | 4.2(0.165) | 25.9(1.02) | 0.6 | −20°C∼−60°C(−4°F∼140°F) |
| BR435 | 3 | 50 | 6 | 1 | 4.2(0.165) | 35.9(1.41) | 0.9 | −20°C∼−60°C(−4°F∼140°F) |

Note: Nominal capacity shown above is based on standard drain.

# Sec. 5.3] Lithium–carbon monofluoride batteries

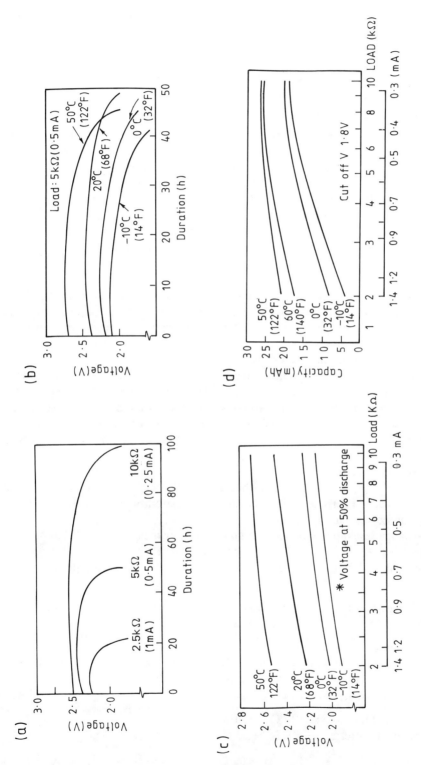

Fig. 5.47 — Various characteristics of a BR 245 Li-CFx pin-type cell. (a) Constant load discharge performance at 20°C for loads of 2.5 kΩ and 10 kΩ. (b) Temperature performance at −10°C, 0°C, 20°C, and 50°C during discharge through a 5 kΩ load. (c) The operating voltage as a function of the load resistance. (d) The cell capacity as a function of the load resistance. (Courtesy of Matsushita Electric Industrial Co. Ltd.)

batteries with superior discharge performance. Improvements in sealing are being made, and the crimped seals of present cells are being replaced by welded ones.

## 5.4 LITHIUM–MANGANESE DIOXIDE BATTERIES
(S. Narukawa and N. Furukawa)

### 5.4.1 General characteristics

Lithium–manganese dioxide (Li–$MnO_2$) batteries use $MnO_2$ as the active cathode material. It has been used for a long time as the active cathode material in zinc–carbon batteries and alkaline manganese dioxide batteries.

In 1975, Sanyo identified a novel reaction between lithium and $MnO_2$ and succeeded in exploiting this as the Li–$MnO_2$ battery [48]. Recently, the use of this battery has expanded remarkably, and now more than fifteen companies are producing it worldwide. Sanyo has also granted the manufacturing technology for Li–$MnO_2$ batteries to major battery manufacturers around the World.

The general advantages of Li–$MnO_2$ battery system are as follows:
(1) High voltage, high energy density
Lithium–manganese dioxide batteries have a high discharge voltage of 3 V, about twice that of conventional dry cell batteries. Because of this advantage, a single lithium battery can be used to replace two, and in practice even three, conventional dry cell batteries.
(2) Excellent discharge characteristics
Since lithium batteries are capable of maintaining stable voltage levels throughout long periods of discharge, a single battery can be used as the internal power source throughout the operational lifetime of a given item of equipment, eliminating the need for battery replacement. In addition, batteries using a crimp-sealed system with spiral electrodes can be used to provide high current discharge for a wide variety of applications.
(3) Superior leakage resistance
The use of an organic liquid rather than an alkali for the electrolyte results in significantly reduced corrosion and a much lower possibility of electrolyte leakage.
(4) Superior storage characteristics
Lithium–manganese dioxide batteries employing manganese dioxide, lithium, and a stable electrolye exhibit a very low tendency toward self-discharge. The degree of self-discharging exhibited by Li–$MnO_2$ batteries stored at room temperature is as follows:

Crimp-sealed batteries:1% per annum
Laser-sealed batteries: 0.5% per annum.

(5) A wide operating temperature range
Because they use an organic electrolyte with a very low freezing point, lithium batteries operate at extremely low temperatures. Moreover, they demonstrate superior characteristics over a wide temperature range from cold to hot, as follows:

Crimp-sealed batteries: $-20°C \sim +70°C$
Laser-sealed batteries: $-40°C \sim +85°C$.

(6) A high degree of stability and safety.

Since Li–MnO$_2$ batteries do not contain toxic liquids or gases, they pose no major pollution problems.

### 5.4.2 Manufacturers/statistics

Recently, the use of these batteries has expanded remarkably. More than fifteen companies in six countries around the World are producing them (see Table 5.13).

**Table 5.13** — Manufacturers of Lithium–manganese dioxide batteries

Sanyo Electric Co. Ltd (Japan)
Matsushita Micro Battery Co. Ltd (Japan)
Hitachi Maxwell Ltd (Japan)
Toshiba Battery Co. Ltd (Japan)
Seiko Electronic Components Ltd (Japan)
Sony Energytec Inc. (Japan)
Fuji Electrochemical Co. Ltd (Japan)
Toyotakasago Dry Battery Co. Ltd (Japan)
Yuasa Battery Co. Ltd (Japan)
Duracell International Inc. (United States)
RAYOVAC International Co. (United States)
Eveready Battery Co. (United States)
VARTA Batterie AG (West Germany)
SAFT (France)
BEREC (United Kingdom)
RENATA S.A. (Switzerland)

Over 90% of all Li–MnO$_2$ batteries are manufactured in Japan, and more than 80% of all lithium batteries are Li–MnO$_2$ batteries.

Fig. 5.48 shows the growth of lithium battery production in Japan. This has increased more than 20% per year over the past few years, and the lithium battery market is expected to expand further.

### 5.4.3 Chemistry

By using atomic absorption spectroscopy, ion microanalysis, and X-ray diffraction methods, the following reaction mechanism in Li–MnO$_2$ batteries has been suggested;

Anode reaction; $Li \rightarrow Li^+ + e^-$ (5.11)

Cathode reaction; $Mn^{IV}O_2 + Li^+ + e^- \rightarrow Mn^{III}O_2(Li^+)$ (5.12)

Overall battery reaction; $Mn^{IV}O_2 + Li \rightarrow Mn^{III}O_2(Li^+)$ , (5.13)

$Mn^{III}O_2(Li^+)$ signifies that lithium ion is introduced into the MnO$_2$ crystal lattice.

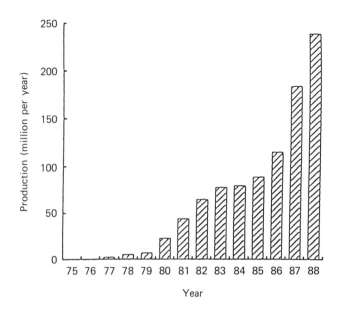

Fig. 5.48 — Growth of production of lithium batteries in Japan.

The battery voltage is approximately 3 V, and as there is no gas generation during discharge, the internal cell pressure does not increase.

Fig. 5.49 shows a schematic presentation of the solid phase during the discharge

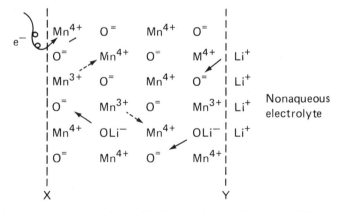

Fig. 5.49 — Schematic presentation of the solid phase during the discharge of $MnO_2$. The arrows show the directions of movement of the electrons and lithium ions: ———lithium ion movement; – – – – – – electron movement; X, $MnO_2$ — electronic conductor interface; Y, $MnO_2$ — solution interface.

of $MnO_2$. Lithium ions in a non-aqueous electrolyte are introduced into the $MnO_2$ crystal lattice, and tetravalent manganese is reduced to trivalent manganese. Various Li–$MnO_2$ batteries have been developed on the basis of this reaction mechanism.

In Li–MnO$_2$ batteries, lithium perchlorate (LiClO$_4$) is widely employed as an electrolytic salt, and propylene carbonate (PC) and 1,2-dimethoxyethane (DME) as a mixed solvent. This PC-DME-LiClO$_4$ electrolyte shows high conductivity ($>10^{-2}\Omega^{-1}$ cm$^{-1}$) and low viscosity ($<3$cP).

The requirements for the MnO$_2$ active material in Li–MnO$_2$ batteries are as follows:

(1) It must be almost anhydrous,
(2) It must have an optimized crystal structure suitable for diffusion of Li$^+$ ion into the MnO$_2$ crystal lattice.

Fig. 5.50 shows the relationship of the MnO$_2$ heat treatment temperature with the water content of MnO$_2$. Although it is important that there is no water in the

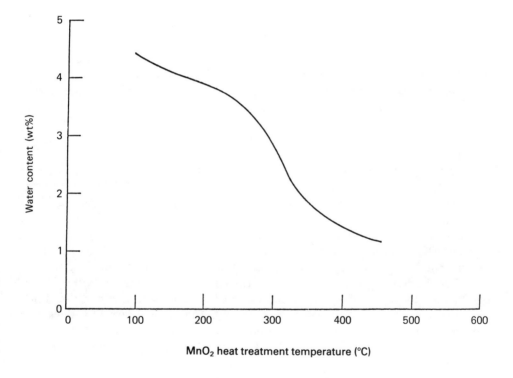

Fig. 5.50 — Relation of MnO$_2$ heat treatment temperature and water content in MnO$_2$.

cathode materials of non-aqueous batteries, the presence of a little water is unavoidable when MnO$_2$ is used as the active material. It is believed that this water is bound in the crystal structure, and that it has no effect on the storage characteristics as shown in Fig. 5.51, which shows the relation of the MnO$_2$ heat treatment temperature to the residual capacity ratio after 11 months storage at 60°C.

Fig. 5.52 shows the discharge characteristics, at a current density of $1.2 \text{ mA/cm}^2$, of electrolytic $MnO_2$ heat treated at various temperatures. From the characteristics shown, it may be concluded that the optimum heat treatment temperature range for stable discharge is between 375°C and 400°C, which agrees with the data of Fig. 5.51.

### 5.4.4 Construction details
Lithium–manganese dioxide batteries are classified according to their shape and structure [49].

#### *5.4.4.1 Flat-type batteries*
The battery construction is shown in Fig. 5.53. The cathode electrode consists of $MnO_2$ with the addition of a conductive material and binder. The anode is a disk made of lithium metal, which is pressed onto the stainless steel anode can. The separator is non-woven cloth made of polypropylene, which is placed between the cathode and the anode.

#### *5.4.4.2 Cylindrical type batteries*
The battery construction is shown in Fig. 5.54. The cylindrical batteries can be classified into two basic types: one with a spiral structure, and one with an inside-out structure [50]. The former consists of a wound, thin cathode and lithium anode with a separator in between. The latter is constructed by pressing the cathode mixture into a high-density cylindrical form. The spiral construction batteries are suitable for high-rate drain, and the inside-out construction batteries are suitable for high energy density.

The sealing system can be classified into two types: crimp sealing and laser sealing. A comparison of these sealing methods is shown in Fig. 5.55, the degree of air-tightness with laser sealing being equivalent to a ceramic based hermetic seal [50].

Figure 5.56 shows the construction of Li–$MnO_2$ battery 2CR5, which is used as the central power source for fully automatic cameras [51]. The 2CR5 is composed of two CR15400 batteries connected in series. It is encapsulated in plastic material and designed in shapes that will prevent misuse. The nominal voltage of the 2CR5 is 6 V.

When the 2CR5 is short-circuited, a thermal protector prevents the battery from overheating by substantially increasing the protector resistance. When the short circuit is removed, the 2CR5 operates normally. The thermal protector does not impede the ability of the 2CR5 to deliver high current. When the discharge current is depleted, the user can easily remove the 2CR5 from the camera and replace it with a new one.

### 5.4.5 Cell types available and applications
As mentioned above, lithium–manganese dioxide batteries are available in a variety of shapes and constructions in accordance with their particular use. Details of Sanyo Li–$MnO_2$ cells are shown in Table 5.14.

Fig. 5.57 shows the various applications of lithium batteries based on their drain currents. Flat-type batteries are generally used for low-rate drain. Spiral structured cylindrical batteries are suitable for high-rate drains such as strobe lights and cameras.

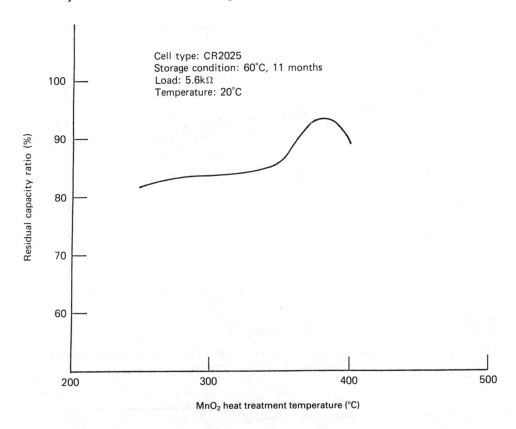

Fig. 5.51 — Relation of $MnO_2$ heat treatment temperature and residual capacity ratio after 11 months at 60°C.

Fig. 5.58 shows the applications of lithium batteries and quantities involved. In the early days, lithium batteries were mainly used for watches and calculators. They are now also used for cameras, strobe lights, memory back-up, and measuring instruments.

### 5.4.6 Performance characteristics

#### 5.4.6.1 Flat-type batteries

Fig. 5.59 shows the load characteristics of the CR2032. The cell voltage of the battery is approximately 3 V. The operating voltage varies slightly according to the discharge current.

Fig. 5.60 shows the temperature characteristics of the CR2032. The battery discharges at a stable voltage over a wide temperature range from −20°C to 70°C. Fig. 5.61 shows the pulse discharge characteristics of the CR2016. The battery can be used as a power source for alarms, LEDs, and a variety of lamps such as emergency lighting switches.

Recently, flat-shaped manganese dioxide–lithium secondary batteries have been developed. The positive active material is the lithium-containing manganese dioxide

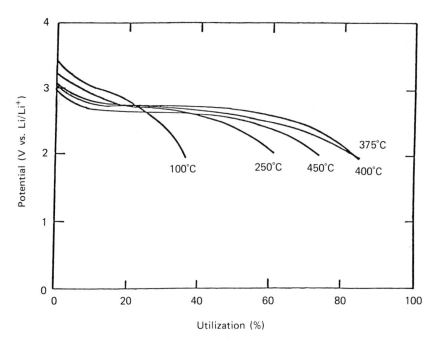

Fig. 5.52 — Discharge characteristics at a current density of 1.2 mA.cm$^{-2}$ of electrolytic MnO$_2$ heat treated at various temperatures.

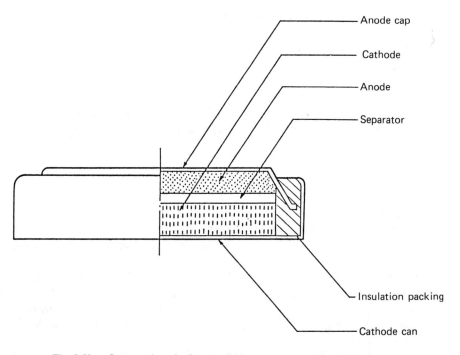

Fig. 5.53 — Construction of a flat type lithium–manganese dioxide battery.

## Sec. 5.4] Lithium–manganese dioxide batteries

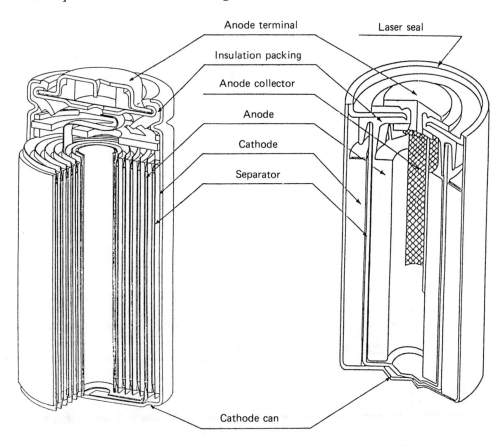

Fig. 5.54 — Different constructions of cylindrical type lithium–manganese dioxide batteries.

| Model | Crimp sealed cell | Laser sealed cell |
|---|---|---|
| Structure | 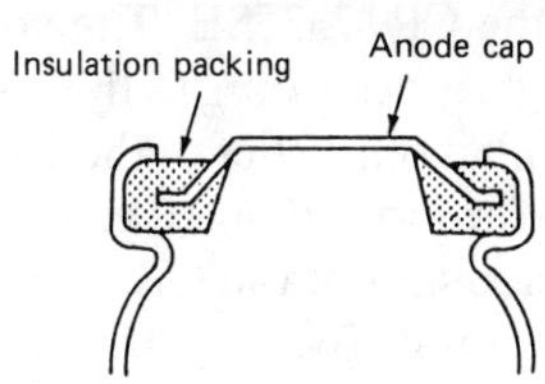 Insulation packing, Anode cap | 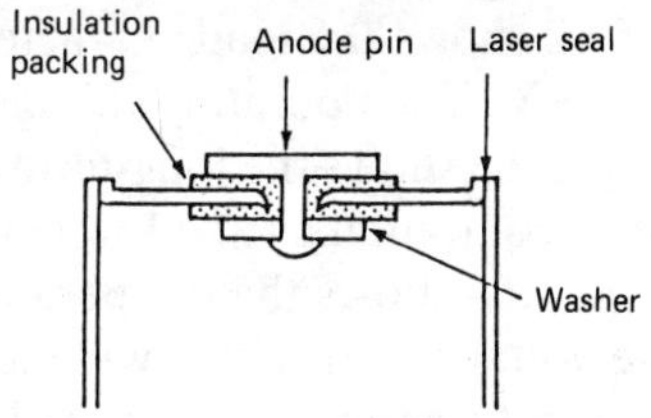 Insulation packing, Anode pin, Laser seal, Washer |
| Leak rate of helium atm cc/sec | $10^{-7}$ | $10^{-9}$ |

Fig. 5.55 — The relationship of the seal type with the leak rate of helium for cylindrical type lithium–manganese dioxide batteries.

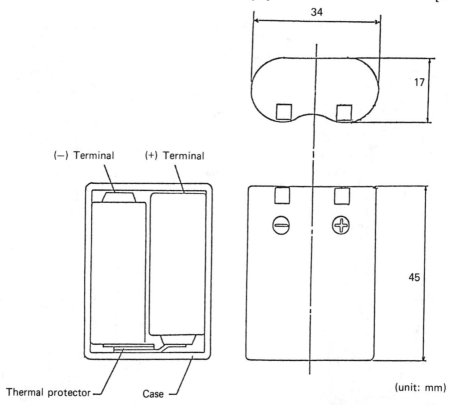

Fig. 5.56 — The construction, shape, and dimensions of the 2CR5 lithium–manganese dioxide battery for fully automatic cameras.

CDMO. CDMO stands for composite dimensional manganese dioxide, and it is prepared from LiOH and $MnO_2$ by heat treatment (see ref. [49] of Chapter 2). The batteries are mainly used for memory back-ups. Cylindrical-shaped manganese dioxide–lithium secondary batteries are expected to be developed in the near future.

### 5.4.6.2 Cylindrical batteries (inside-out structure)

Fig. 5.62 shows the load characteristics of the CR17335SE. The voltage is approximately 3 V. The operating voltage varies slightly according to the discharge current.

Fig. 5.63 shows the long-term discharge characteristics of the CR17335SE. It can be used continuously at a low current level for more than 10 years.

Fig. 5.64 shows the temperature characteristics of the CR17335SE. It delivers a stable voltage over a very wide temperature range from $-40°C$ to $85°C$.

Fig. 5.65 shows the self-discharge characteristics of the CR17335SE. The self-discharge rate of this battery is low during long-term storage, not exceeding 0.5% per year at room temperature.

Fig. 5.66 shows the storage characteristics of the CR17335SE. This battery demonstrates extremely good storage characteristics; storage for 100 days at 70°C, being equivalent to 10 years at room temperature. The operating voltage and discharge time are almost unchanged after storage.

**Table 5.14** — Specifications of Sanyo lithium–manganese dioxide batteries

| Classification | Model | Nominal voltage (V) | Nominal capacity (mAh) | Standard discharge current (mA) | Max. discharge current (mA) | | Max dimensions (mm) | | Weight (g) |
|---|---|---|---|---|---|---|---|---|---|
| | | | | | Continuous | Pulse | Diameter (D) | Height (H) | |
| *Flat-type lithium batteries* | | | | | | | | | |
| | CR1220 | 3 | 35 | 0.1 | 1 | 5 | 12.5 | 2.0 | 0.8 |
| | CR1620 | 3 | 60 | 0.2 | 2 | 6 | 16.0 | 2.0 | 1.2 |
| | CR2016 | 3 | 80 | 0.3 | 4 | 15 | 20.0 | 1.6 | 1.7 |
| | CR2025 | 3 | 140 | 0.3 | 3 | 15 | 20.0 | 2.5 | 2.5 |
| Crimp-sealed cells | CR2032 | 3 | 190 | 0.3 | 3 | 15 | 20.0 | 3.2 | 3.0 |
| | CR2430 | 3 | 270 | 0.3 | 3 | 20 | 24.5 | 3.0 | 4.0 |
| | CR2450 | 3 | 500 | 0.2 | 2 | 10 | 24.5 | 5.0 | 6.2 |
| *Cylindrical-type lithium batteries* | | | | | | | | | |
| | CR-1/3N | 3 | 160 | 2.0 | 20 | 80 | 11.6 | 10.8 | 3.3 |
| | 2CR-1/3N | 6 | 160 | 2.0 | 20 | 80 | 13.0 | 25.2 | 9.1 |
| | 2CR5 | 6 | 1300 | 10 | 250 | 3000 | 34(L)×17(W)×45(H) | | 40 |
| Crimp-sealed cells | CR14250SE | 3 | 800 | 0.5 | 3 | 7 | 14.5 | 25.0 | 9.0 |
| | CR12600SE | 3 | 1400 | 1 | 5 | 30 | 12.0 | 60.0 | 15 |
| | CR17335SE | 3 | 1700 | 1 | 4 | 15 | 17.0 | 33.5 | 17 |
| Laser-sealed cells | CR17450SE | 3 | 2500 | 1 | 5 | 20 | 17.0 | 45.0 | 22 |
| | CR23500SE | 3 | 5000 | 1 | 6 | 30 | 23.0 | 50.0 | 42 |

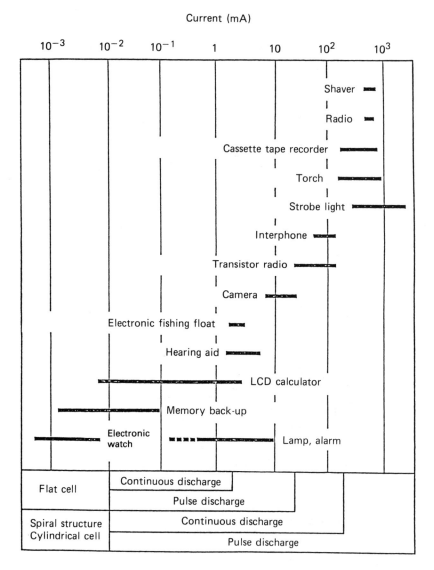

Fig. 5.57 — Various applications of lithium–manganese dioxide batteries based on their drain currents.

### 5.4.6.3 Cylindrical batteries (spiral structure)

Fig. 5.67 shows the load characteristics of the 2CR5 [52]. This battery is composed of two CR15400 cells connected in series. It discharges at a stable voltage, even at very high current levels.

Fig. 5.68 shows the temperature characteristics of the 2CR5. The operating voltage is stable over a wide temperature range from $-20°C$ to $60°C$.

Sec. 5.4]  **Lithium–manganese dioxide batteries**  359

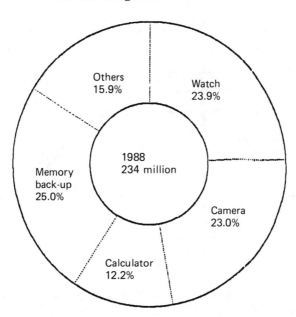

Fig. 5.58 — Major fields of application of lithium–manganese dioxide batteries.

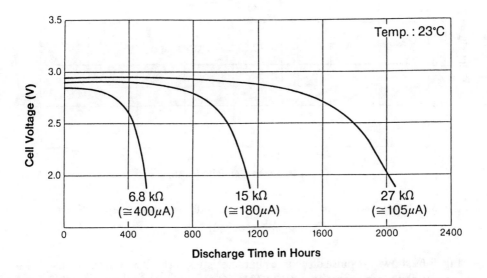

Fig. 5.59 — Load characteristics of the CR2032 lithium–manganese dioxide battery.

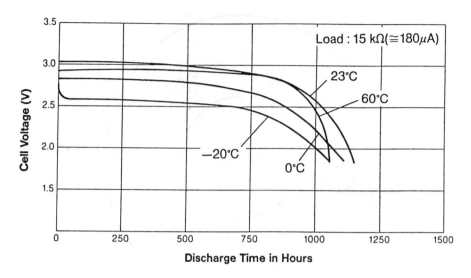

Fig. 5.60 — Temperature characteristics of the CR2032 lithium–manganese dioxide battery.

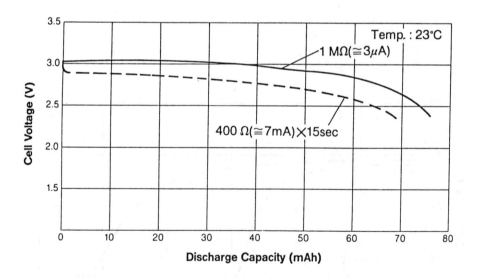

Fig. 5.61 — Pulse discharge characteristics of the CR2016 lithium–manganese dioxide battery.

Fig. 5.69 shows the pulse discharge characteristics of the 2CR5. It can be used as a power source for tape recorders, LCD TVs, camera motors for film rewinding, and camera flash systems.

## Sec. 5.4] Lithium–manganese dioxide batteries

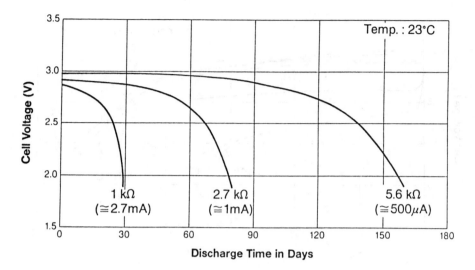

Fig. 5.62 — Load characteristics of the CR17335SE lithium–manganese dioxide battery.

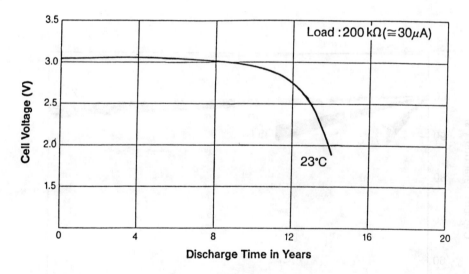

Fig. 5.63 — Long-term pulse discharge characteristics of the CR17335SE lithium–manganese dioxide battery.

Fig. 5.70 and Fig. 5.71 show practical tests of the 2CR5 in a fully automatic camera at 23°C and −40°C, respectively [53]. When the shutter is released, the current discharged powers the exposure meter and the electromagnetic shutter, and

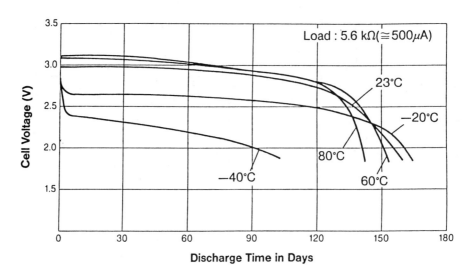

Fig. 5.64 — Temperature characteristics of the CR17335SE lithium–manganese dioxide battery.

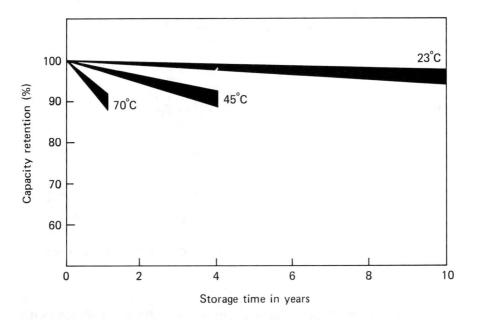

Fig. 5.65 — Graphs showing the self-discharge characteristics of the CR17335SE lithium–manganese dioxide battery.

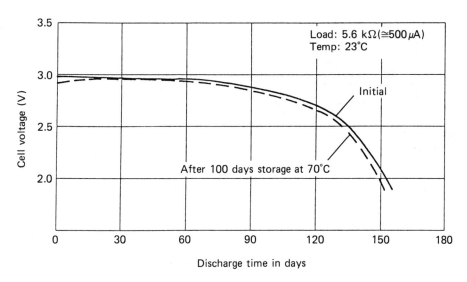

Fig. 5.66 — Graphs showing the storage characteristics of the CR17335SE lithium–manganese dioxide battery.

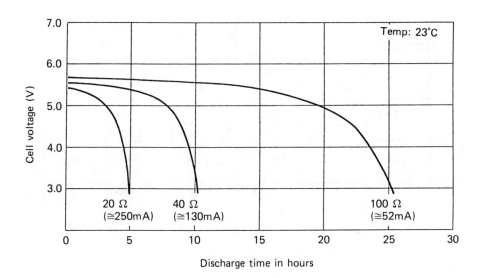

Fig. 5.67 — Load characteristics of the 2CR5 lithium–manganese dioxide battery.

it is also used for winding the film and charging the strobe light for the next photograph. Since the strobe light can be charged within 2 s, continuous photographs can be taken with the strobe light at short time intervals, as Fig. 5.70 shows. Fig. 5.71

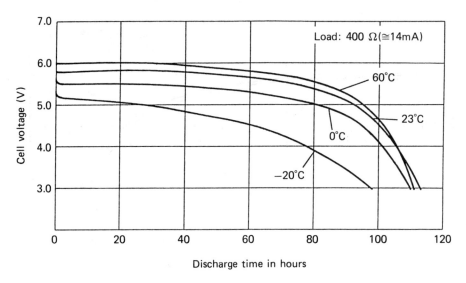

Fig. 5.68 — Temperature characteristics of the 2CR5 lithium–manganese dioxide battery.

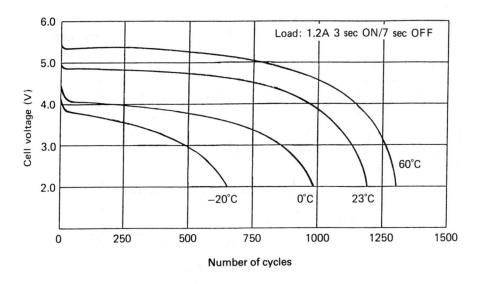

Fig. 5.69 — Pulse discharge characteristics of the 2CR5 lithium–manganese dioxide battery.

shows that continuous photographs can be taken with the strobe light even at −40°C. Moreover, there is no voltage delay during the initial discharge stage even at low temperatures and high pulse rates.

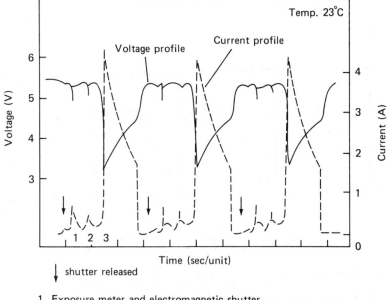

Fig. 5.70 — Practical test results of a 2CR5 lithium–manganese dioxide battery in a fully automatic camera at 23°C.

### 5.4.7 Likely future developments

Lithium batteries of various types have been developed, and Li–MnO$_2$ batteries have been commercialized by Sanyo on the basis of original technology. Major battery manufacturers around the World are now producing them.

Lithium–manganese dioxide batteries will spread into various fields as high-power sources for strobe lights and camera autowinding systems, cassette tape recorders, high performance lights, 8 m/m VTRs, LCD TVs, handy telephones, transceivers, and other highly portable electronic equipment.

## 5.5 LITHIUM–IODINE BATTERIES
### (J. Jolson, S. Wicelinski, and D. Schrodt)

**Introduction**

The lithium–iodine battery was the first commercially successful lithium battery, and it remains the only commercially successful solid state battery. Despite the recent success of other lithium battery chemistries, the lithium–iodine battery remains the preferred system for low power applications which require high energy density, high operating voltage, wide operating temperature range, and the highest possible reliability over long periods.

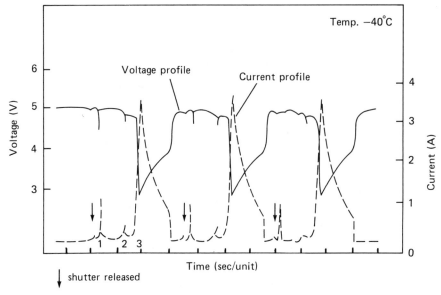

Fig. 5.71 — Practical test results of a 2CR5 lithium–manganese dioxide battery in a fully automatic camera at −40°C.

### 5.5.1 History

Portable power sources available during the mid-1950s tended to have high power capability, low open circuit voltage, low energy density, limited operating temperature range, short shelf life, and low reliability. A major contributor to the reliability problems was electrolyte leakage. Not only did this leakage lead to premature battery failure, but the release of corrosive fluids often destroyed the device housing the battery.

These characteristics were barely tolerable when the major portable devices powered by electrochemical cells were inexpensive torches and valve radios. After the transistor and integrated circuit came into widespread use, the functionality and reliability of portable electronic devices, as exemplified by today's calculators, cameras, watches, and television sets, increased their demand, and also increased the requirement for power sources with improved characteristics.

Researchers of the time understood that energy density could be maximized by choosing anode materials from the upper left-hand corner of the periodic table and cathode materials from the upper right-hand corner of the periodic table. They also understood that phase changes in the electrolyte limited the operating temperature range and that the thermodynamic stability of the electrolyte limited the choice of electrode materials. It followed naturally for them to conclude that the maximum energy density and voltage which could be obtained were determined by the electrolyte, and that attempts to work outside the electrolyte's stable range would lead to short shelf life. Solid electrolyte based systems were seen as a way of overcoming these limitations.

Early attempts to develop solid state batteries with silver anodes, silver halide electrolytes, and iodine-rich cathodes [54] proved that solid state batteries could deliver current densities of the order of 1–2 mA.cm$^{-2}$. This work also promised cells capable of a wide operating temperature range and long shelf life. It encouraged developmental activities aimed at overcoming the low energy density and voltage obtained with the silver based systems.

Some of this later work showed that a solid state magnesium battery could be constructed [55,56] which delivered high operating voltage, moderate energy density, and power capabilities of several mA.cm$^{-2}$. A novel feature of this battery was its use of an iodine-rich charge transfer complex cathode. During assembly of the battery the anode was pressed directly against the cathode, thereby forming a thin electrolyte layer via a thermochemical reaction. The self-healing nature of such electrolyte layers was to play a major role in the later success of the lithium–iodine battery. The work of Gutmann, Hermann, and Rembaum had come close to the development of a useful solid state battery. However, the shelf life of the system they developed was not acceptable because of their use of high permittivity liquids to increase current capability.

It was left to Schneider & Moser [57,58] to put the remaining pieces of the puzzle together. These workers realized that to get long shelf life from a solid state battery all traces of water and other high permittivity liquids needed to be eliminated. This was done by vacuum drying individual cell components, manufacturing cells in a dry room, and hermetically sealing them. They also realized that the only anode material which could provide high energy density and usable power under anhydrous conditions was lithium. In addition, they discovered that the use of a charge transfer complex containing poly-2-vinylpyridine (P2VP) and excess iodine resulted in unexpectedly superior power capability.

### 5.5.2 Chemistry

The basic reaction occurring in the lithium–iodine battery is:

$$2Li + I_2 \rightarrow 2LiI . \tag{5.14}$$

In a lithium–iodine/P2VP cell a more complete description of the reaction can be shown as [59]:

$$\text{Anode } 2Li \rightarrow 2Li^+ + 2e^- \tag{5.15}$$

$$\text{Cathode } 2Li^+ + 2e^- + P2VP.nI_2 \rightarrow P2VP.(n-1)I_2 + 2LiI \tag{5.16}$$

$$\text{Cell } 2Li + P2VP.nI_2 \rightarrow P2VP.(n-1)I_2 + 2LiI . \tag{5.17}$$

Before cell assembly the surface of the lithium anode is scraped clean. The iodine and P2VP are ground, mixed together, and pressed into a pellet [60]. As the anode and cathode are brought together, a very thin lithium iodide layer is formed *in situ* by the direct thermochemical reaction of lithium and iodine. On open circuit storage this layer grows thicker as iodine from the cathode diffuses through the lithium iodide layer to the lithium anode, forming more lithium iodide. This self-limiting reaction occurs very slowly, resulting in self-discharge rates of less than 5% over a ten-year period at room temperature. It is the only self-discharge reaction known to occur in lithium–iodine batteries.

The lithium iodide layer which forms during cell assembly is both electrolyte and separator, and it is self healing [61]. If a crack occurs, additional lithium iodide will form at the crack site, immediately sealing the flaw. This provides the lithium–iodine battery with intrinsic reliability and ability to withstand extreme environmental abuse, unmatched by any other commercially available active cell chemistry.

The addition of a small amount of P2VP to the cathode material, generally 5% or less by weight, makes it semiconductive. This allows current to flow in the battery and around the external circuit when the battery is loaded. As the cell is discharged, the iodine content of the cathode decreases, the lithium iodide layer becomes thicker, and the lithium anode becomes thinner. Theoretical calculations show that the cell should shrink as it is discharged. However, under typical discharge rates, little volume change occurs. Under heavier loads, expansion of the package by up to 5% has been observed. This has been related to the formation of a lithium iodide layer with less than theoretical density [62].

The theoretical open circuit potential of the lithium–iodine cell is 2.794 V at 25°C. With this system, open circuit potentials close to this value are seen until all free iodine has been consumed. Upon application of a typical load to a fresh cell a small initial decline in voltage is noted. This is due to the resistance of the cathode material. The voltage of the cell then declines almost linearly through most of its life, owing to the linear build-up of impedance in the lithium iodide layer as it gets thicker. At end-of-life a 'knee' is observed in the discharge curve. This signifies that the available iodine in the cathode has been exhausted, and is a direct result of a dramatic rise in cathode impedance.

The theoretical energy density of the lithium–iodine couple is 1.93 Wh cm$^{-3}$. In practice this is somewhat reduced by the addition of P2VP to the cathode. Commercially available batteries designed for the medical market deliver up to 1.3 Wh cm$^{-3}$. This ratio of usable-to-theoretical energy density of 0.67 is particularly high for an electrochemical power source because there is no need to add discrete separators, barriers, or electrolytes to lithium–iodine cells. The lithium–iodine battery combines this excellent energy density with exceptional reliability, safety, abuse resistance, and shelf life. Commercially available versions of the lithium–iodine battery are capable of operating over the temperature range of $-55$ to $+125$°C. The maximum rate capability of this system is several hundred $\mu$A cm$^{-2}$ [63].

### 5.5.3 Medical use
The first lithium–iodine battery was implanted in 1972 [64]. Since then the lithium–iodine battery has become the most popular system for powering implantable devices, with nearly two million cells implanted to date. Medical versions of the lithium–iodine battery have accumulated 17 years of data demonstrating excellent reliability and safety.

In addition to Catalyst Research (CR), the original developer, lithium–iodine batteries are manufactured for pacemaker applications by Medtronic, Inc. and Wilson Greatbach Limited (WGL). Today, more than 300 000 batteries are used by the pacemaker industry annually. Even though the lithium–iodine battery has become the accepted standard for implantable biomedical devices, each manufacturer continues to strive toward reliability enhancement and size reduction. These efforts along with construction and performance characteristics will be discussed in the following sections.

### 5.5.3.1 Available types

Medical versions of the lithium–iodine battery are custom designed in a variety of sizes and shapes to meet the needs of the pacemaker industry. Battery shapes range from rectangular to semicircular. Since their primary application is in cardiac pacemakers, the batteries are designed to have thicknesses of less than 10 mm and to be as compact as possible. The various manufacturers' specifications are summarized in Tables 5.15 to 5.18.

**Table 5.15** — Catalyst research 900 series specifications

| Model | Capacity (Ah) | Dimensions, mm (Thickness × height × length) | Weight (g) |
|---|---|---|---|
| 906 | 1.8 | 6.6 × 19 × 45.1 | 19 |
| 908 | 1.9 | 9.1 × 15.5 × 44.8 | 19 |
| 909 | 3.1 | 9.4 × 22.9 × 44.8 | 28 |
| 910 | 1.8 | 6.6 × 22.3 × 45.2 | 18 |
| 911 | 2.3 | 6.6 × 27.2 × 45.1 | 23 |
| 912 | 2.0 | 7.8 × 30.6 × 27.3 | 20 |
| 913 | 1.2 | 4.8 × 30.6 × 27.3 | 11.5 |
| 914 | 1.6 | 7.8 × 25 × 27.3 | 16 |
| 915 | 2.2 | 8.6 × 26 × 33 | 21 |
| 916 | 2.3 | 8.8 × 22 × 45 | 23 |
| 917 | 2.6 | 8.6 × 30.6 × 33 | 25 |
| 918 | 1.0 | 5.1 × 23.4 × 32.3 | 10.5 |
| 919 | 1.3 | 5 × 22.1 × 45 | 13 |
| 920† | 2.8 | 9.1 × 22.9 × 44.9 | 26 |
| 921† | 2.2 | 8.8 × 22 × 45 | 23 |
| 922† | 2.5 | 8.6 × 30.6 × 33 | 25 |
| 930 | 1.9 | 7.1 × 22.2 × 44.8 | 18.8 |
| 931 | 1.0 | 4.8 × 25.5 × 27.3 | 9.6 |

†Double anode.

Tables 5.15 and 5.16 contain specifications for the Catalyst Research 900 and 9000 series batteries. The 9000 series is mechanically similar to the 900 series, and both contain cathodes with a 20:1 iodine to polymer ratio. However, the 9000 series has a higher energy density because it incorporates some recent chemical and process

**Table 5.16** — Catalyst research 9000 series specifications

| Model | Capacity (Ah) | Dimensions, mm (Thickness × height × length) | Weight (g) |
|---|---|---|---|
| 9006 | 1.9 | 6.6 × 19 × 45.2 | 19 |
| 9008 | 2.0 | 9.1 × 15.5 × 44.9 | 19 |
| 9009 | 3.4 | 9.4 × 22.9 × 44.9 | 28 |
| 9010 | 1.9 | 6.6 × 22.3 × 45.2 | 18 |
| 9011 | 2.6 | 6.6 × 27.2 × 45.1 | 23 |
| 9012 | 2.3 | 7.8 × 30.6 × 27.3 | 20 |
| 9013 | 1.5 | 4.8 × 30.6 × 27.3 | 11.5 |
| 9014 | 1.8 | 7.8 × 25 × 27.3 | 16 |
| 9015 | 2.6 | 8.6 × 26 × 33 | 21 |
| 9016 | 2.6 | 8.8 × 22 × 45 | 23 |
| 9017 | 3.1 | 8.6 × 30 × 33 | 25 |
| 9018 | 1.1 | 5.1 × 23.4 × 32.3 | 10.5 |
| 9019 | 1.4 | 5 × 22.1 × 45 | 13 |
| 9020† | 3.2 | 9.1 × 22.9 × 44.9 | 26 |
| 9022† | 2.8 | 8.6 × 30.6 × 33 | 25 |
| 9023† | 2.2 | 6.6 × 30.7 × 33 | 19 |
| 9030 | 2.1 | 7.1 × 22.2 × 44.8 | 18.8 |
| 9031 | 1.2 | 4.8 × 25.5 × 27.3 | 9.6 |

†Double anode.

improvements. In addition to the 900 and 9000 series, a 9100 series is being introduced by Catalyst Research. The 9100 series contains cathode material with a 30:1 iodine to polymer ratio further increasing energy density.

Specifications for lithium–iodine batteries manufactured by Promeon, a division of Medtronic, Inc., are listed in Table 5.17. Three types of battery, the Alpha family, the Beta 263, and the Zeta family, are manufactured by Promeon. In general, Medtronic builds these batteries for in-house use.

Table 5.18 contains the lithium–iodine battery specifications for cells manufactured by Wilson Greatbatch Limited. These batteries contain a cathode which is typically 30:1 iodine to P2VP by weight. Projected capacity of the various models under a 140 k$\Omega$ resistive load to a 1.8 V cut-off at 37°C is provided.

### 5.5.3.2 Construction

Lithium–iodine batteries are assembled in controlled, dry room environments of less than 1% relative humidity and are hermetically sealed to maintain cell integrity. Cell designs currently in production use a central lithium anode which is surrounded by the cathode. Since the discharge product also functions as the electrolyte, no discrete separator is added between the anode and cathode. Case and cover blank material is typically 304L or 316L stainless steel. Either glass-to-metal or ceramic-to-metal seals

**Table 5.17** — Promeon battery specifications

| Model | Capacity (Ah) | Dimensions, mm (Thickness × height × length) | Weight (g) |
|---|---|---|---|
| Alpha 28 | 1.7 | 7.9 × 28.4 × 27.4 | 17 |
| Alpha 283 | 2.2 | 7.9 × 28.4 × 27.4 | 19 |
| Alpha 333 | 2.7 | 7.9 × 33.4 × 27.4 | 27 |
| Alpha 335 | 2.2 | 7.9 × 33.4 × 27.4 | 27 |
| Beta 263 | 2.6 | 8.3 × 26.2 × 36.6 | 23 |
| Zeta 203 | 1.1 | 5.2 × 20.0 × 36.6 | 10 |
| Zeta 265 | 1.3 | 5.2 × 26.0 × 36.6 | 14 |

**Table 5.18** — Wilson Greatbatch Ltd, battery specifications

| Model | Capacity† (Ah) | Dimensions, mm (Thickness × height × length) | Weight (g) |
|---|---|---|---|
| 7905 | 1.7 | 8.6 × 15 × 45 | 19 |
| 7906 | 3.0 | 8.6 × 23 × 45 | 29 |
| 7911 | 2.5 | 7 × 28 × 45 | 26 |
| 8031 | 1.9 | 7 × 19 × 45 | 20 |
| 8041 | 2.2 | 8.6 × 26 × 33 | 22 |
| 8074 | 2.4 | 8.6 × 23 × 45 | 24 |
| 8077 | 1.8 | 7 × 23 × 45 | 21 |
| 8082 | 2.2 | 8.6 × 18 × 45 | 24 |
| 8206 | 2.0 | 7.8 × 30.6 × 27.3 | 22 |
| 8207 | 1.5 | 7.88 × 25 × 27.3 | 18 |
| 8304 | 1.4 | 7 × 16.5 × 45 | 17 |
| 8402 | 1.2 | 5 × 23 × 45 | 13 |
| 8431 | 1.26 | 5 × 30.6 × 27.3 | 13 |

†Capacity to 1.8 V under a 140 kΩ constant resistive load

are employed to electrically isolate the anode lead from the case which forms the positive lead. Laser and/or tungsten inert gas welding techniques are used to weld the cover to the case.

Two main types of medical lithium–iodine batteries are now being produced. The type shown in Fig. 5.72 contains a ribbed lithium anode which is centrally located in a stainless steel enclosure. Within the anode is a nickel screen current collector which is brought out via a glass-to-metal feedthrough. To enhance cell performance, the anode is precoated with P2VP before final cell assembly [65]. Batteries of this type

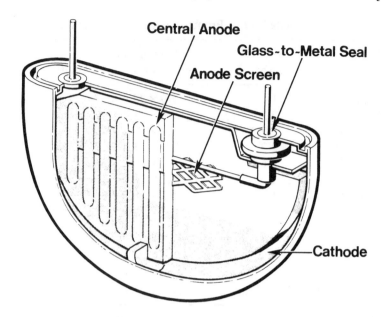

Fig. 5.72 — Cutaway view of a corrugated anode pacemaker battery. (Courtesy of Wilson Greatbatch, Ltd.)

are assembled with their header welded to the case before addition of the thermally reacted cathode. Because of this, the battery cover also contains a filling ferrule which is sealed after cathode addition by two discrete sealing mechanisms [66]. The case serves as the cathode current collector in this design.

As shown in Fig. 5.73, the second main type of medical lithium–iodine battery has a pelletized cathode which reduces cell impedance and precludes the migration of cathode material to the anode current collector after cell assembly [60]. To form these cathodes, finely ground iodine and P2VP are mixed together and then pellets are pressed. Two such cathode pellets are placed around an anode subassembly. The outer surfaces of the cathode pellets are covered with a current collector and the entire unit is slipped into a stainless steel case. The header assembly is then welded to the case to ensure hermeticity. A subsequent heat treatment at temperatures of less than 90°C allows the iodine and polymer in the cathode to react and assures the establishment of good interfacial contacts.

Despite significant differences in manufacturing details, the batteries illustrated in Figs 5.72 and 5.73 exhibit a good deal of similarity. Both are case positive and employ a central lithium anode surrounded by cathode material to reduce cell impedance by a factor of four. The current collector materials and the feedthroughs are also similar. In addition, both batteries employ techniques which cause disruption of the lithium iodide layer to produce cells with low internal resistance [67,68].

Batteries of the type illustrated in Fig. 5.72 use a ribbed anode to maximize active anode surface area and minimize cell impedance. This has been taken one step further with cells containing pelletized cathodes with the incorporation of two

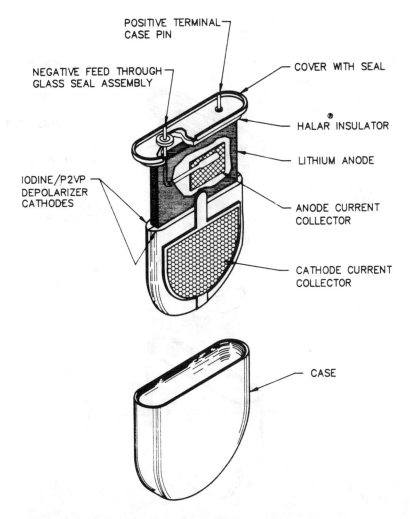

Fig. 5.73 — Exploded view of a pelletized cathode pacemaker battery.

anodes within one enclosure. As illustrated in Fig. 5.74, this design allows for the delivery of the high rate pulses required by some newer pacemaker models.

To achieve and maintain the high reliability necessary for implantable medical batteries, a stringent quality control programme which requires a large number of inspection steps has been put in place. Each manufacturer's quality programme is designed to control the materials and components used, control the assembly and manufacturing processes employed, maintain records of the battery's history and behaviour, and predict future product performance.

After final assembly, each serialized battery undergoes hermeticity and X-ray examination along with workmanship and dimensional inspection. Next, all batteries undergo electrical tests at 37°C for several weeks under loads which simulate pacemaker drains. During this time, periodic measurement of cell voltage and

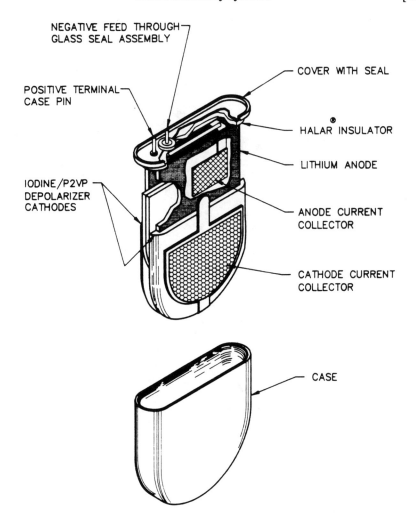

Fig. 5.74 — Exploded view of a pelletized cathode pacemaker battery with a double anode.

impedance are made so that upon test completion the electrical data can be mathematically analyzed and statistically compared with acceptance criteria. Only those cells passing all quality control requirements are prepared for shipment. Random samples are selected from each production lot and are retained for in-house testing protocols. Though the stringent quality control process used during the manufacture of implantable batteries is expensive and labour-intensive, it is necessary when one considers that a pacemaker battery is expected to operate for up to 20 years.

### 5.5.3.3 *Performance characteristics*
Battery drain rates of 20 to 50 $\mu$A are typically required for cardiac pacemaker applications. At these rates battery discharge begins near the open circuit voltage.

As discharge progresses, the lithium iodide layer gradually increases in thickness, causing cell resistance to increase and the cell voltage to decrease. At higher drain rates polarization effects become increasingly important. Fig. 5.75 illustrates the effect of increasing the drain rate on the discharge behaviour of the lithium–iodine battery. Thermodynamic and kinetic factors exert a strong influence on the form of the discharge curve [69] and must be taken into account when evaluating and predicting battery performance.

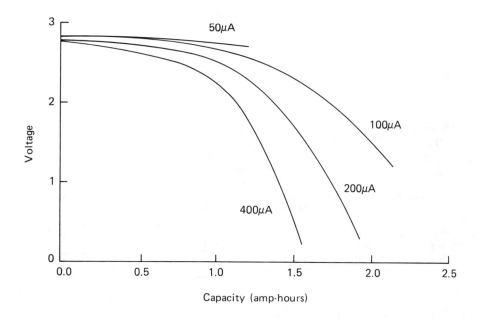

Fig. 5.75 — Discharge data for an Alpha 333 battery at a variety of drain rates. (Courtesy of Promeon, Division of Medtronic, Inc.)

The shape of the discharge curve is of importance to pacemaker designers as it must provide adequate warning of impending battery exhaustion. Desirable discharge curves exhibit a gradual knee to a chosen replacement voltage. The replacement voltage varies among pacemaker manufacturers, but typically ranges from 2.0 V to 2.3 V. Implantable lithium–iodine cells are designed to be cathode limited because cathode limited cells exhibit a less abrupt end-of-life than anode limited cells.

Fig. 5.76 shows a projected voltage vs capacity curve for a WGL 8077 battery discharged under a 140 k$\Omega$ resistive load at 37°C. This battery is typical of batteries made with precoated anodes which show lower internal resistance and higher voltage during discharge than their older uncoated counterparts. Fig. 5.77 illustrates a

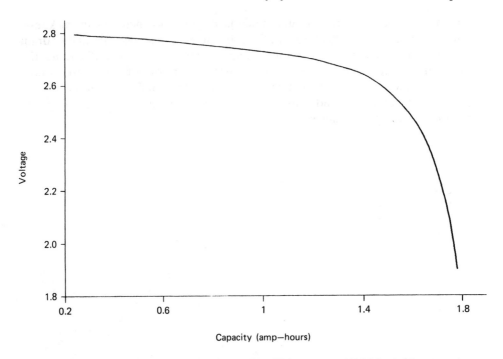

Fig. 5.76 — Projected discharge curve for a WGL 8077 battery at a 140 kΩ load. (Courtesy of Wilson Greatbatch, Ltd.)

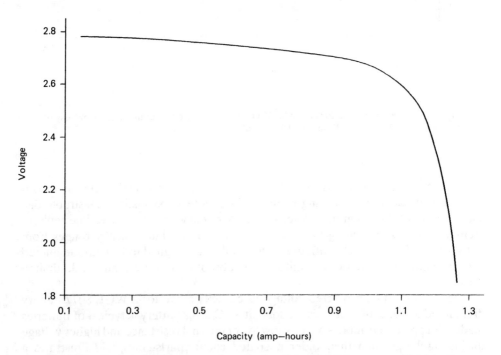

Fig. 5.77 — Projected discharge curve for a Catalyst Research 919 battery at a 150 kΩ load.

projected voltage vs capacity curve for a CR 919 battery discharged under a 150 kΩ load at 37°C. A resistance vs capacity curve for the CR 919 battery is shown in Fig. 5.78. Figs 5.77 and 5.78 are typical of the pelletized cathode batteries which exhibit lower internal resistance and higher voltages during life than uncoated anode cells made with thermally reacted cathode material.

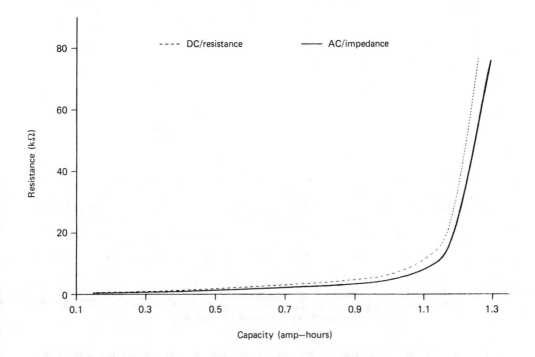

Fig. 5.78 — Projected resistance vs capacity curve for a Catalyst Research 919 battery.

### 5.5.4 Industrial use

The characteristics of high voltage, high energy density, high reliability, exceptional safety, and long life under low current drain conditions which make the lithium–iodine/P2VP system well suited for powering cardiac pacemakers also make it ideal for computer memory and clock back-up applications [59,63,70,71,72]. Today's device designer is fortunate in having this chemistry available in a variety of package configurations and sizes which have been optimized for the electronics industry.

Electronic components are often required to operate over the military specification temperature range of −55°C to +125°C or the commercial specification range of −40°C to +90°C. By using a lithium–iodine cell, the battery powering the board will have the same temperature tolerance as the rest of the package. In addition, the required current drain for CMOS-RAM data retention increases with temperature and so does the current capability of the lithium–iodine battery. This means that current capabilities selected for the RAM's 25°C requirement will be met at higher and lower temperatures as illustrated in Fig. 5.79. As current drains required by

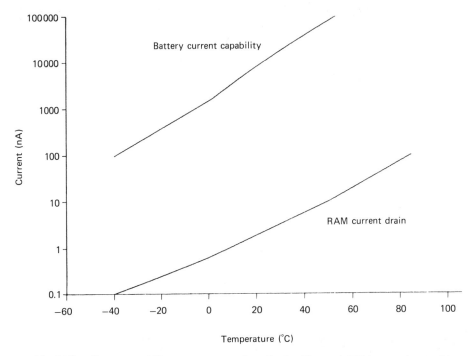

Fig. 5.79 — Current capability vs temperature for a Catalyst Research B35 battery shown with current required by a Toshiba TC5516APL RAM.

modern CMOS-RAM devices are low relative to battery current carrying capability, self-discharge generally limits battery life. Since lithium–iodine cells show less than a 5% capacity loss over a ten year period they typically have more capacity available for CMOS back-up than any other battery chemistry.

The hermetically sealed solid state nature of the lithium–iodine battery results in no gas generation, no weight change, and no appreciable volume change during discharge. Also, there is no danger of chemical explosion during short circuit, reverse charge, forced overdischarge, or puncture as is the case with some other lithium batteries. In fact, reverse current of a few microamps will prolong battery life by inhibiting the self-discharge reaction.

Finally, the battery's package should allow it to be handled just like other electronic components. Thus, the battery designer should maximize use of PC board area and allow for the use of standard insertion, wave soldering, cleaning, and burn-in protocols.

### 5.5.4.1 Available types

Early versions of industrial lithium–iodine batteries looked like coins with leads protruding from one face of the cell as shown in Fig. 5.80. These designs offered

Fig. 5.80 —Catalyst Research model 2736 battery.

many advantages over standard configurations, but the round shape did not optimize board space and the hidden leads made mounting and cleaning difficult. Therefore, a new generation of batteries, which are now available was designed. The larger capacity versions of these cells are rectilinear, while the smaller capacity versions are shaped like conventional resistors. The cells, some of which are shown in Fig. 5.81, have exposed leads which make automatic mounting and cleaning easy. Table 5.19 contains specifications for available models including some of the older coin shaped versions. Catalyst Research remains the only manufacturer of lithium–iodine batteries designed specifically for the industrial marketplace.

### 5.5.4.2 Construction

Cathode pellets for coin shaped batteries are manufactured with automatic equipment, but hand labour is used to place them in stainless steel cases and for most other operations. Cover assemblies contain glass-to-metal seals which are flush with the cover. Fabrication of the anode assembly involves placing an insulator over the inside of the cover. Through a small hole in the insulator an anode current collector screen is spot welded to the feedthrough. A lithium disk is then swaged to the current collector. The anode assembly and cathode assembly are fitted together and the perimeter is laser welded, ensuring hermeticity. Positive leads are subsequently spot welded to the case.

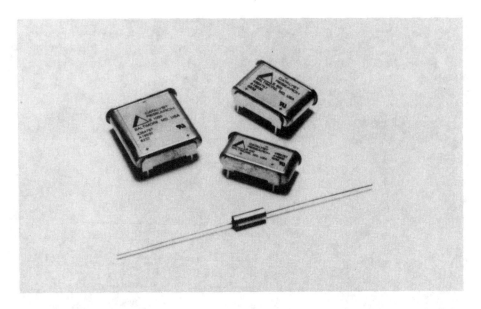

Fig. 5.81 — Current generation of Catalyst Research industrial batteries.

The rectilinear and resistor type batteries are assembled on a fully automated line. This is accomplished by pressing cathode powder directly in a case, leaving a cavity. After lithium is inserted into the cavity an insulating sheet is placed on top of the can. A cover containing a glass-to-metal seal with integral negative lead is then brought down on the can, piercing the anode. The assembly is automatically moved to the next work station where it is hermetically sealed and leads are welded to the case.

All of the batteries that are built undergo heat treatment followed by extensive leak checking, electrical testing, and visual inspection. The environmental and material controls employed are similar to those used in the manufacture of medical batteries.

### 5.5.4.3 *Performance characteristics*
Figs 5.82 and 5.83 show projected performance curves for the B-1000 battery. Within certain limits, the internal impedance of the battery is independent of drain rate and depends only upon capacity removed. Therefore, the internal impedance vs capacity curve shown in Fig. 5.82 can be used to predict discharge performance under a variety of load conditions. The curves shown in Fig. 5.83 are derived from Fig. 5.82 by Ohm's law.

Fig. 5.84 shows predicted shelf life as a function of temperature for the B-1000 battery. These curves are based on measurement of internal impedance as a function of time under open circuit storage conditions at 90°C and on microcalorimetry measurements made at 25°C. To make these projections it was assumed that self-discharge decreases linearly with the square root of time [61].

**Table 5.19** — Catalyst Research industrial battery specification

| Model | Projected capacity (mAhr) | Nominal dimensions (mm) | Nominal weight, (g) | Max current* ($\mu A$) |
|---|---|---|---|---|
| Button cell | | diam × thickness | | |
| 3440 | 650 | 53 × 4 | 12.6 | 140 |
| 2736 | 400 | 27 × 5 | 7.5 | 75 |
| 1935 | 200 | 20 × 4.5 | 4.3 | 40 |
| Rectilinear | | length × width × thickness | | |
| B-1000 | 1000 | 25 × 26 × 8 | 11.9 | 250 |
| B-600 | 600 | 18 × 26 × 8 | 8.4 | 160 |
| B-400 | 400 | 13 × 26 × 8 | 6.2 | 100 |
| Axial lead | | diam × length | | |
| B-35 | 35 | 4 × 11 | 0.7 | 25 |
| B-10 | 8 | 2.5 × 7 | 0.2 | 10 |

*25°C, ¼ capacity, 2 V cut-off.

### 5.5.5 Likely future developments

As a mature system the lithium–iodine battery is not likely to undergo dramatic improvement in performance or capability in the near future. However, it will continue to evolve in response of the needs of the marketplace. These changes will include improvements in chemistry and packaging techniques and reductions in cost.

The energy density of the active materials within a lithium–iodine battery has increased significantly over the years. Comparison of achievable values with theoretical values suggests that further improvements in energy density of the active materials will be relatively small. In time, however, the optimized chemistries which have been developed recently, such as those used in Catalyst Research's 9000 and 9100 series batteries, will replace the older chemistries. This will increase the average energy density of available lithium–iodine batteries significantly.

Great improvements in packaging density have also been made in the last two decades. Although there is still room to improve the efficiency of the best packages, the major gains have already been made. It is likely, however, that less efficient designs will be replaced by more efficient ones as time goes on, resulting in continued improvement in average packaging efficiency for some years to come. The 1990s will see increased emphasis on low costs, and there will be a growth in demand for very

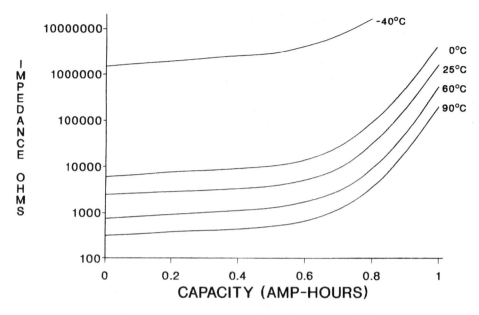

Fig. 5.82 — Internal impedance vs capacity for a Catalyst Research B-1000 battery at various temperatures.

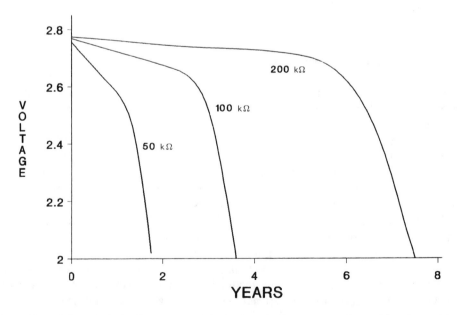

Fig. 5.83 — Projected discharge curves for a Catalyst Research B-1000 battery under various loads at 25°C.

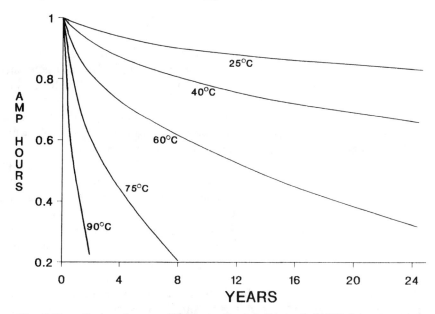

Fig. 5.84 — Projected storage life for a Catalyst Research B-1000 battery at various temperatures.

small (<50 mAh), very thin (<0.2 cm), and surface mountable packages. Driving these trends will be the need to power ID tags, smart cards, and other leading edge devices. The increased interest in watches with analogue displays will also have an impact on the market for lithium–iodine batteries.

Although continued improvements in rate capability have been made over the years by modifying the chemistry, it is not known what the limits of the system are. Certainly, rates of 1 mA/cm$^2$ may be possible. However, a more predictable improvement in power density could be envisaged by the use of stacked arrays of thin lithium–iodine cells. As described earlier, a beginning has already been made in supplying such batteries to the medical market. A 2 Ah lithium–iodine battery, capable of delivering several milliamperes continuously could be designed when the need arises.

## 5.6 LITHIUM—COPPER OXIDE BATTERIES
(A. E. Brown)

### 5.6.1 Introduction

Lithium–copper oxide cells and batteries have been available since 1969 and are now firmly established as a versatile and reliable power source suitable for a wide range of applications. Each cell comprises an anode of pure lithium, a solid cathode of cupric oxide/graphite, a separator of porous non-woven glass fibre, and an electrolyte of lithium perchlorate dissolved in an organic solvent, 1,3-dioxolane. Whilst the nominal cell operating voltage is 1.5 V, in practice the working voltage is strongly dependent upon the rate of discharge and the ambient temperature. The inherent high energy density of this system results in extremely high capacity cells, for example the AA size (14.5 mm diameter, 50 mm long) has a nominal 3.7 Ah (based on the weight of lithium).

It is ideally suited to low loads and has found widespread usage in memory support applications in the 2-cell, 3 V format. The excellent shelf life characteristics have led to use in very low rate, long life requirements often extending beyond ten years.

The system has also proved suitable for downwell logging applications together with the more recently developed lithium–copper oxyphosphate range of cells which was specifically aimed at these extremely high temperature conditions.

When suitably packaged, cells are capable of operation over an exceptionally wide temperature range: $-55°C$ to $+150°C$. The polypropylene grommet seal, however, restricts the use of bare cells to the range $-40°C$ to $+55°C$. Encapsulation or the use of glass-to-metal hermetic sealing permits operation over the wider range. Care must be exercised when selecting cells for use at such extremes of temperature, as performance, in terms of voltage and capacity, is very strongly dependent upon operating temperature, and additional capacity may have to be built in.

Storage results have demonstrated excellent capacity retention and lack of start-up problems after 14 years of storage. In addition, very low rate discharges have been in progress for 10 years and are still continuing. As a solid cathode system, lithium–copper oxide cells are relatively benign amongst available lithium electrochemistries. A programme of abuse tests has been carried out at the SAFT laboratories both in France and in the UK, and the cells have shown remarkable resilience to both electrical and mechanical forms of abuse.

### 5.6.2 Manufacturers

Lithium–copper oxide cells are currently manufactured by SAFT (France) in a bobbin, cylindrical cell format and by SANYO (Japan) in button construction. Further development into C and D size spiral wound construction cells has been reported by Jumel et al. [73], but mass production of these cells has not occurred, since the marketplace has shown a preference for large cells of other electrochemistries with better low temperature current capability. In addition, requirements for 1.5 V single cells are less common than the 3 V provided by the competing lithium–sulphur dioxide, lithium–thionyl chloride, and lithium–manganese dioxide cells.

In the main, cylindrical Li–CuO cells are built up into battery packs with the relevant number of series connected cells to provide the required voltage and/or parallel branches where higher capacities are demanded. Suitable protective devices (diodes, fuses, connectors) are then included by the battery manufacturer as appropriate for the application.

### 5.6.3 Construction

Most of the work carried out on this system relates to results obtained from cylindrical bobbin cells. This format is essentially a low-rate construction. In such a design, as illustrated in Fig. 5.85, a central tube is wrapped in a separator and surrounded by an annular cathode. This construction provides a central well to accommodate an optimum quantity of electrolyte and permits maximum cathode surface area with a minimum interelectrode distance, facilitating efficient discharge.

### 5.6.3.1 Anode assembly
A rectangle of lithium sheet is rolled onto a metallic central collector tube which gives mechanical support and acts as a 'current collector' by means of connection to the negative 'cap'. The lithium used must be of a very high purity, since the presence of impurities can be responsible for parasitic reactions which can reduce shelf life and affect performance.

### 5.6.3.2 Cathode assembly
Cold pressed annular rings of cathode mix are stacked inside a metal can. These cathode rings comprise copper oxide to which is added graphite to provide electrical conductivity. The can, therefore, has positive polarity. The copper oxide powder is produced by the controlled oxidation of high purity copper powder, finally producing cupric oxide with the required particle size and shape.

### 5.6.3.3 Electrolyte
The electrolyte comprises lithium perchlorate salt dissolved in the organic solvent, 1,3-dioxolane. All materials used in electrolyte manufacture have to be 'dried' to minimize water levels since lithium will react with any water present. Such a reaction will reduce capacity and form undesired products.

### 5.6.3.4 Separator
The specially selected separator is a porous, non-woven glass fibre material. To function correctly, a separator must provide reliable physical separation of the anode and cathode whilst permitting current to be drawn when required. To ensure physical separation the material must be mechanically rugged and stable over the long storage periods required. If the cell current capability is not to be impaired, a highly porous structure, with good wicking properties, is necessary to ensure that sufficient electrolyte is available to permit the discharge reaction to take place.

### 5.6.3.5 Sealing
The cell is sealed with a crimped polypropylene joint which also electrically isolates the cathode and anode assemblies. For use at moderate temperatures ($-40°C$ to $+55°C$) this type of seal is extremely reliable. To permit use outside these limits, two options are available:

(1) cells fitted with glass-to-metal seals,
(2) fully encapsulated cells.

The drawback of the ultra-reliable hermetic sealing with glass-to-metal technology is the high cost associated with the special glass required.

## 5.6.4 Performance
Considerable work has been reported on the performance evaluation of the AA and ½AA size cylindrical, bobbin construction cells. Takashi et al. [74] and Furukawa et al. [75] have also reported work on a Li–CuO button cell. The AA size cell produced by SAFT has a nominal capacity, based on the weight of lithium, of 3.7 Ah. Fig. 5.86 illustrates the performance achieved under near-optimum discharge

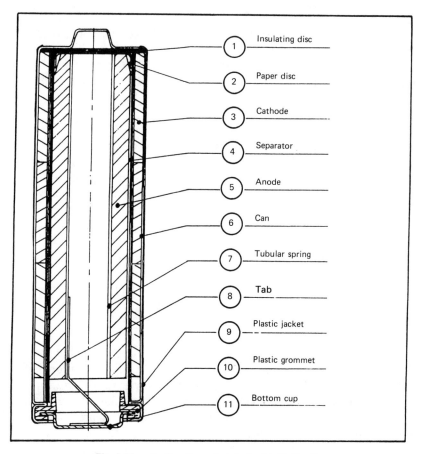

Fig. 5.85 — Section through AA size Li–CuO cell.

conditions of 1 mA cm$^{-2}$ (based on lithium surface area) at 20°C to an end-point voltage of 0.9 V.

The smaller ½AA cell has been the subject of an extensive evaluation programme by Broussely et al. [76]. Performance data obtained for this size of cell, over the current density range 1 to 10 mA cm$^{-2}$, were analysed in terms of the performance domain analysis technique of Bro et al. [77], and the following equation was used to fit performance data to a curve:

$$C = 1.24 \frac{\tanh\left[\dfrac{\bar{i}}{3.25}\right]^{1.66}}{\left[\dfrac{\bar{i}}{3.25}\right]^{1.66}} \qquad (5.18)$$

Where $C$ = capacity in Ah and $\bar{i}$ = average current density in mA cm$^{-2}$.

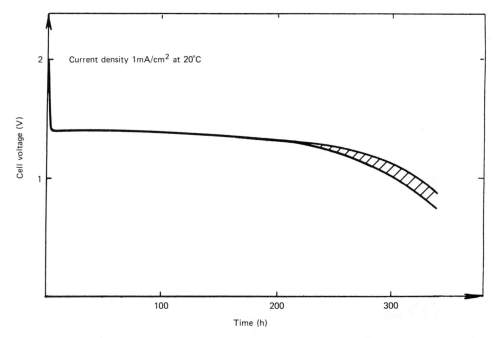

Fig. 5.86 — Performance of bobbin type cells at 1 mA cm$^{-2}$ (20°C).

The variation of volumetric energy density with current at 20°C is summarized in Fig. 5.87.

As previously discussed, cells can be discharged over a wide range of operating temperatures. Figs 5.88 and 5.89 illustrate the voltage/time characteristics for AA cells discharging at $+125°C$ and $-55°C$, respectively. The significant effect of operating temperature on cell voltage is clearly demonstrated.

An interesting feature is the appearance of two voltage plateaux under certain conditions. Fig. 5.90, which shows the variation of voltage with temperature for AA cells discharged through a 1 kΩ load, clearly illustrates that at $+70°C$ two voltage plateaux exist at approx. 1.6 V and 1.3 V, whilst at $+20°C$ and 0°C only single voltage plateaux are observed.

Interrupted discharge has been shown to have minimal effect on the overall capacity achieved [78].

Lithium–copper oxide cells do not exhibit the 'voltage delay effect' often found in other lithium systems, for example liquid cathode types (see Section 2.2.5.2). As illustrated in Fig. 5.91, on the application of a short circuit the current rises to its peak level within about 1 ms of load application. Supporting results have also been reported after 2 and $7\frac{1}{2}$ years' storage at 20°C [78] and after 6 months' storage at 60°C.

## 5.6.5 Storage

Both accelerated and real time storage tests demonstrate that the system has exceptionally good storage characteristics.

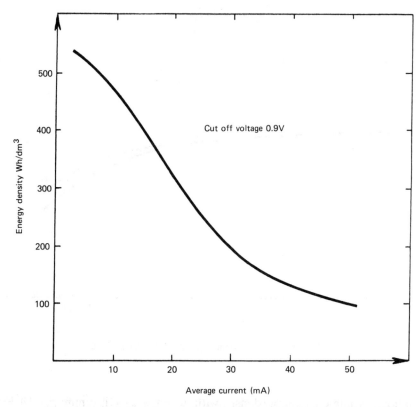

Fig. 5.87 — Variation of energy density with current for ½AA size cells at 20°C.

Results are now available after 14 years' room temperature storage as described by Broussely et al. [79]. After such prolonged storage, performance is comparable to that of fresh cells (Fig. 5.92). Cells fitted with glass-to-metal seals have been stored for 2 years at +90°C and subsequently discharged at room temperature, giving 93.5% of the capacity of a fresh cell. Similarly, the rate of self-discharge during discharge has been found to be extremely low; AA size cells have been on discharge for periods in excess of 10 years at very low rates, and continue without failure.

**5.6.6 Chemistry**
In common with many other battery systems, the electrochemistry of lithium–copper oxide cells is not fully understood.

The overall cell reaction can be written as:

$$2Li + CuO = Li_2O + CuO \qquad (5.19)$$

with a reversible potential ($E°$) of 2.24 V and a theoretical specific energy density, based on the Gibbs free energy change for the reaction, of 1285 Wh. kg$^{-1}$ and 3140 Wh dm$^{-3}$.

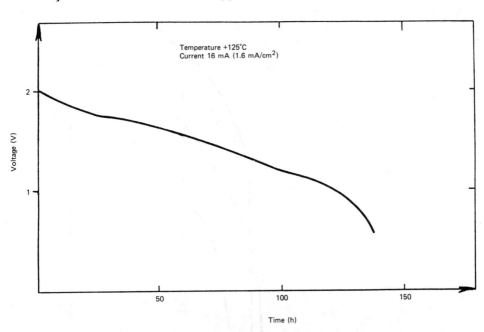

Fig. 5.88 — Performance of AA size cell at +125°C.

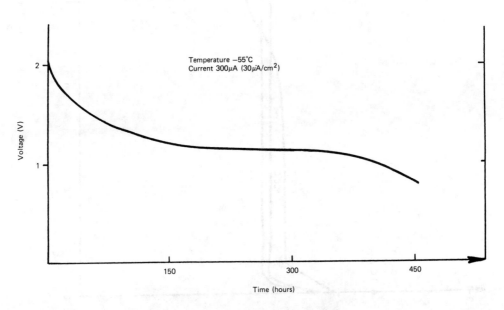

Fig. 5.89 — Performance of AA size cell at −55°C.

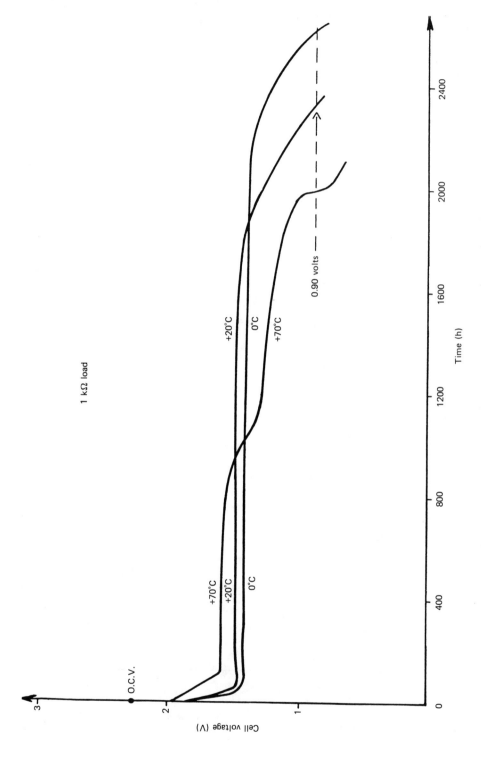

Fig. 5.90 — Performance of AA size cell under 1 kΩ load.

## Sec. 5.6] Lithium–copper oxide batteries

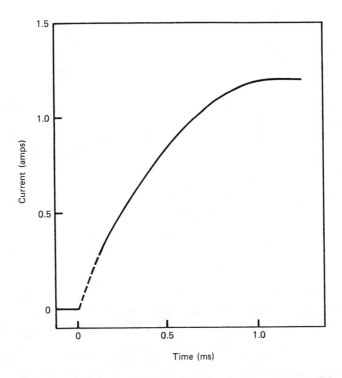

Fig. 5.91 — Current profile of AA size cell under short circuit conditions.

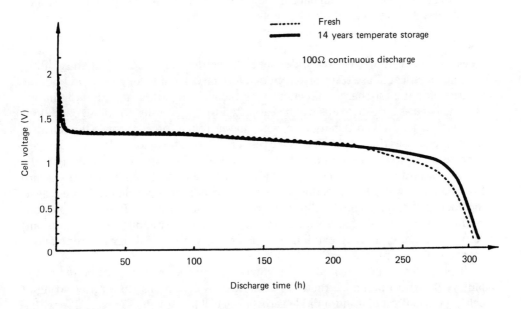

Fig. 5.92 — Performance of AA size cell after 14 years' temperate storage.

Observations of the open circuit and on-load voltages [80] support a more complex discharge mechanism than that suggested by the overall cell reaction above. A relatively recent study of AA size bobbin cells by Bates *et al.*, [81] involved modifying, in turn, the cathode and electrolyte of lithium limited cells.

Cathode studies included variation of graphite percentage, copper oxide particle size, and forming pressure. Subsequent performance evaluation showed little dependence on either cathode composition or structure.

A range of six electrolytes was prepared and evaluated in cells. Two of these maintained the standard lithium perchlorate salt and replaced the 1,3-dioxolane solvent with a binary and with ternary solvent mixtures. In other mixtures, lithium perchlorate was replaced with lithium trifluoromethane sulphonate dissolved in the same binary and ternary solvent mixtures. Performance evaluation of cells containing these various electrolytes demonstrated that the standard electrolyte gave superior performance despite its relatively low conductivity.

The results of analysis of cathodes in various states of discharge failed to provide clear evidence to support an explanation of the discharge mechanism. Cathodes from anode limited cells demonstrated the formation of cuprous species as discharge proceeded at the expense of cupric species. The excess of cathode did not permit analysis beyond 85% cathode utilization. It is probable that beyond this stage copper itself will begin to be formed at the expense of both the intermediate cuprous and the starting cupric species.

The ambiguity in results obtained from cathode limited cells tested in a button configuration test rig described in the same work failed to assist in selecting a mechanism which fits the evidence.

In conclusion, workers have put forward several theories for the discharge mechanism, but experimental data are conflicting and do not manage to fully explain the observed performance of cells.

### 5.6.7 Safety

Cells have been subjected to a wide range of electrical and mechanical abuse [82].

The electrical abuse tests have included forcing current through cells in both the 'charge' and the 'discharge' directions with the cells in various states of discharge. Results have shown that cells can be subjected to a wide range of these conditions without incident. Overdischarge, i.e. when the current is forced through the cell after discharge, causes the voltage across the cell terminals to reverse, and violent cell case rupture can follow. Attempted recharge of fresh and discharged cells only generally leads to hazardous conditions at relatively high currents such as 200 mA. No hazard has been encountered under the short circuit conditions. Mechanical abuse tests included crushing, piercing, shock, vibration, bump, etc. No fires or explosions resulted. Exposure to temperatures in excess of 180°C (melting point of lithium) resulted in molten lithium exuding via the softened crimp seal. Exposure to 500°C caused violent rupture.

The results of electrical abuse tests have led to the evolution of guidelines in battery design to minimize the risks of hazardous conditions. In a series string of cells, the inclusion of a shunt diode across each cell limits voltage reversal to within safe limits. To restrict battery charging currents, both external and internal (i.e.

## 5.7 LITHIUM–VANADIUM PENTOXIDE BATTERIES
(H. V. Venkatasetty)

In attempts to develop high energy density and low cost primary lithium batteries, metal oxide cathode materials have been the logical choice. Of the many transition metal oxides investigated, lithium–manganese dioxide (Li–MnO$_2$) and lithium–vanadium pentoxide (Li–V$_2$O$_5$) have been the most promising. The Li–V$_2$O$_5$ battery system is very attractive because the cathode, vanadium pentoxide (V$_2$O$_5$), with its high oxidation state, can provide a high open circuit voltage (OCV) in a lithium battery. In suitable aprotic electrolyte solutions, lithium can intercalate into the cubo-octahedral cavities of the perovskite-like structure of V$_2$O$_5$. Therefore, the Li–V$_2$O$_5$ cell, in addition to high OCV, has high energy and power density capability [83,84]. The disadvantage of V$_2$O$_5$ cathode material is its relatively low electronic conductivity, therefore it requires processing with about 10 w/o of carbon powder and about 5 w/o of a binder (PTFE).

### 5.7.1 Cell design and performance

V$_2$O$_5$ cathode material with its perovskite-like structure, during the discharge of the Li–V$_2$O$_5$ cell, incorporates lithium, forming Li–V$_2$O$_5$ with a predominantly uniaxial expansion of the perovskite-like cavity. Owing to the high cell voltage of approximately 3.4 V, the cell calls for aprotic solvents with high resistance to oxidation. The Li–V$_2$O$_5$ battery system, both in the active and in the reserve configuration, was one of the earliest battery systems developed by Honeywell for military and medical applications. Vanadium pentoxide is insoluble in aprotic organic electrolytes, and it is chemically and electrochemically stable. A Li–V$_2$O$_5$, cell with 1 M lithium perchlorate (LiClO$_4$) in propylene carbonate or in a 1:1 mixture (by volume) of propylene carbonate and 1,2 dimethoxyethane, has an open circuit potential of 3.4 V [85]. A laboratory cell on discharge with a 100 k$\Omega$ load gives four distinct plateaux at 3.4 V, 3.2 V, 2.4 V, and 2.0 V (Fig. 5.93a). On the other hand, if the cell is discharged with 3.3 K$\Omega$ load, it gives rise to only two plateaux at 3.0 V and 1.8 V (Fig. 5.93b). These discharge results are similar to those reported by Dey with prismatic cell configuration [86] and by Horning & co-workers using the cylindrical reserve cell configuration [87]. The discharge reactions and their mechanism are not completely understood. The postulated reduction reaction of V$_2$O$_5$ from vanadium oxidation state V to vanadium oxide VO$_2$ of oxidation state IV, should give a cell potential of 2.4 V [88]. The observed Li–V$_2$O$_5$ cell potential is 3.4 V [89,90,91]. Therefore, a topochemical reaction involving the intercalation of lithium into the cathode V$_2$O$_5$ has been proposed on the basis of X-ray diffraction studies of discharged cathodes [92,93]. The formation of an intercalation compound LiV$_2$O$_5$, a ternary phase material, may account for the first plateau appearing during the cell discharge at low rates.

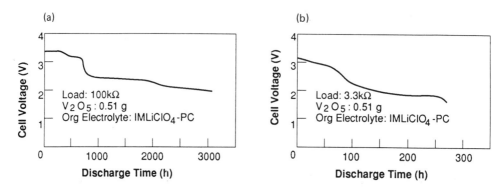

Fig. 5.93 — (a) Discharge curve of Li–$V_2O_5$ cell with a 100 kΩ load at 25°C [85]. (Reprinted with permission) (b) Discharge curve of Li–$V_2O_5$ cell with a 3.3 kΩ load at 25°C [85]. (Reprinted with permission.)

During the early 1970s, several design concepts were used to fabricate high energy density Li–$V_2O_5$ cells. A new cell design about double that of a standard 'D' cell of alkaline $MnO_2$, called the 'DD' cell, was fabricated and tested by a Honeywell group for commercial and industrial use [94]. The schematic of the cell configuration (Fig. 5.94) consists of two lithium metal anode disks pressed to completely cover the stainless steel grid current collector which is welded to a central rod attached to the pin of the glass-to-metal seal. Similarly, two cathode disks consisting of $V_2O_5$ and

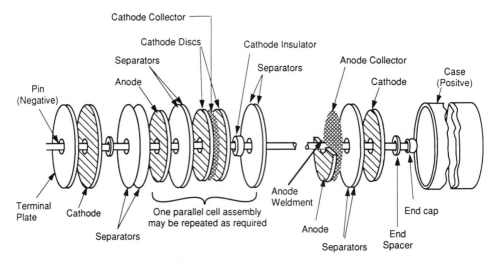

Fig. 5.94 — Schematics of an active 'DD' size Li–$V_2O_5$ cell design [94]. (Reprinted with permission.)

graphite powder, are mixed and pressed over stainless steel (316L) grids and assembled to fit into a stainless steel case. The two cathodes are isolated from the central rod by insulating rings and are connected to the case. The anodes are connected to the central rod, and Celgard™ 2400 separates the cathode from the anode. A double salt mixture of 2 M lithium hexafluoroarsenate ($LiAsF_6$) and 0.4 M lithium tetrafluoroborate ($LiBF_4$) in pure methyl formate is the electrolyte solution. A glass-to-metal seal forms the cell closure, and the 'DD' cell has dimensions of 11.89 cm by 3.4 cm with an active electrode area of 79 cm$^2$. Since the cell has a fairly high internal resistance, the short circuit shows a maximum value of 3 A with a maximum temperature of 54°C (130°F). A number of these cells were tested at different rates varying from 20 mA to 5 mA over a temperature range of $-23$°C (10°F) to $+48$°C ($+120$°F) to a 2.0 V cut-off with no significant variation. All these cells yielded capacities from 20 Ah to about 25 Ah. Cells stored for 60 days at 60°C (140°F), and tested at 24°C (75°F) at 20 mA and 10 mA yielded about 20 Ah capacity. The cell at the low rate of 10 mA with first electron discharge operates to a plateau of 2.6 V and at 20 mA to a plateau of 2.4 V, after which the voltage drops to the next plateau at about 2.2 V, corresponding to the second electron discharge (Fig. 5.95). When continuing the discharge, the curves drop sharply, indicating the loss of energy. The energy densities calculated from these curves at room temperature have values of about 244 Wh.kg$^{-1}$.

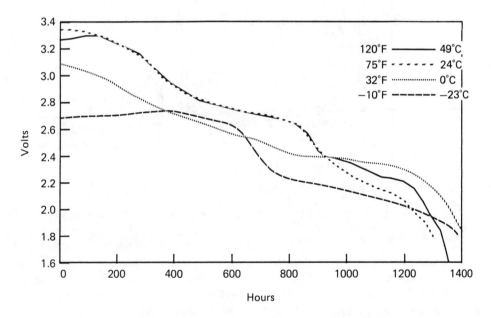

Fig. 5.95 — Typical discharge characteristics of 'DD' size Li–$V_2O_5$ cell at 20 mA discharge at different temperatures [94]. (Reprinted with permission.)

Lithium reserve cells of capacities varying from 100 mAh to 500 mAh have been designed and tested [87]. These are designed to be a single discharge power source which remains inactive for many years without suffering any capacity loss due to the electrolyte degradation or the electrode material deterioration. This reserve feature is achieved by isolating the electrolyte solution in special glass ampoules from the electrodes (Fig. 5.96). These small reserve cells are incorporated successfully into

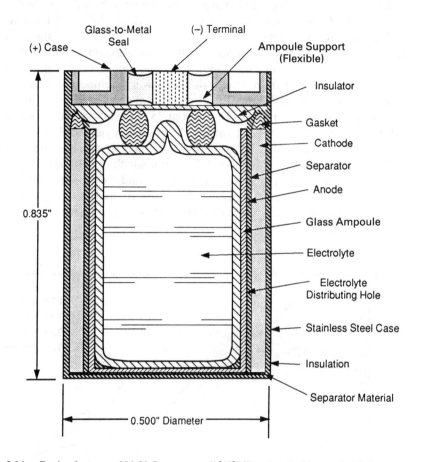

Fig. 5.96 — Design features of Li–$V_2O_5$ reserve cell [87]. (Reprinted with permission.)

several munition systems, and they probably represent a large fraction of the batteries manufactured today. The reserve cell is made of a cylindrical stainless steel (316L) case and it has an individual Li–$V_2O_5$ cell with an open circuit voltage of 3.4 V and an output voltage of 3.2 V. In addition to the case, the header is also of 316L joined to the case by a projection weld at the case flange. The header serves as the cover for the cell and incorporates a glass-to-metal seal for the centre terminal metal pin. The terminal pin has negative polarity and the case has positive polarity. The glass ampoule containing the electrolyte reservoir is centrally located with the lithium electrode pressed to the stainless steel current collector and placed with a binder around it. The cathode, made from a mixture of dry $V_2O_5$ and 10% carbon, is placed directly adjacent to the cell case with a glass fibre separator separating the electrodes. This design is cathode limited and has insulating components

Sec. 5.7]    Lithium–vanadium pentoxide batteries    397

strategically placed to prevent internal short circuiting. The electrolyte is a solution of $2\,M\,LaAsF_6$ and $0.4\,M\,LiBF_4$ in methyl formate in a glass ampoule. The electrolyte solution is of very high purity and very high stability and therefore achieves in excess of ten years of storage life and good performance upon activation. The glass ampoule containing the electrolyte solution is firmly fixed inside the cell by an ethylene propylene rubber ring support. A sharp mechanical activation force directly applied to the bottom of the can breaks the glass ampoule, releasing the electrolyte into the cell. The activation time is taken as the time required for the cell voltage to reach the cut-off voltage of 2.0 V and occurs in less than five seconds [87]. A cross-sectional view of the cell structure activation mechanism is shown in Fig. 5.97. The discharge efficiency of these reserve cells is in the range of 75 to 85%. Typical discharge characteristics of small capacity cells are shown in Fig. 5.98. The performance characteristics of these cells are suitable for many military applications.

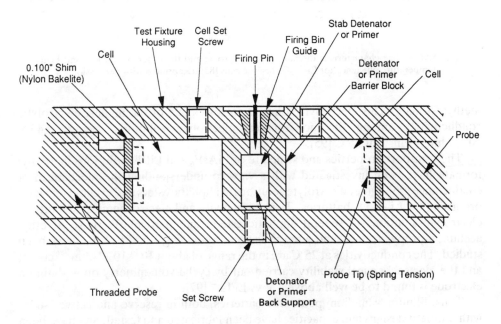

Fig. 5.97 — Cross-section view of stab-type activation mechanism for Li–$V_2O_5$ reserve cell [87]. (Reprinted with permission.)

Transport properties, electrochemical stability, and structure of the electrolyte solutions used in lithium batteries play an important role in the storage and performance characteristics of these batteries. Therefore, different electrolyte solutions are used in different lithium cathode couples to meet the requirements of the various applications. In Li–$V_2O_5$ cells, the electrolyte solution used is a double salt, a mixture of lithium hexafluoroarsenate ($LiAsF_6$) and lithium tetrafluoroborate ($LiBF_4$), usually with the composition of $2\,M\,LiAsF_6$ and $0.4\,M\,LiBF_4$ in

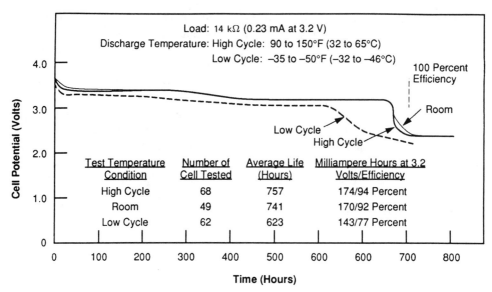

Fig. 5.98 — The characteristic discharge voltage profile and the difference in low and high temperature efficiency for Li–V$_2$O$_5$ reserve cells [87]. (Reprinted with permission.)

methyl formate. This electrolyte solution possesses one of the highest electrolytic conductivities among the non-aqueous electrolyte solutions, being $\sim 45 \times 10^{-3}$ ohm$^{-1}$ cm$^{-1}$ at 25°C [95].

The transport properties and structure of LiAsF$_6$ and LiBF$_4$ solutions in methyl formate have been investigated with a view to understanding the mechanism of conductance [96]. Other electrolyte solutions in aliphatic esters and cyclic esters look promising for Li–V$_2$O$_5$ batteries. The conductivity and electrochemical stability of electrolyte solutions of several aliphatic esters, and particularly those of ethyl acetate, n-propyl acetate, and γ-butyrolactone containing 1.5 M LiAsF$_6$, have been studied. The conductivities at 25°C are in the range of about $10 \times 10^{-3}$ ohm$^{-1}$ cm$^{-1}$, and the electrochemical stability carried out by cyclic voltammetry on a platinum electrode is found to be well above 4 V vs Li/Li$^+$ [97].

Thus, lithium–vanadium pentoxide batteries, both in reserve and active mode, with different designs and capacities have been fabricated and tested, and have been used for the last fifteen years as effective high energy density power sources. Whilst the reserve cells have a proven storage life of more than ten years, active cells under normal conditions lose about 3% of their capacity per year at 24°C [87,95,98]. The excellent stability of the electrolyte solution in active cells is demonstrated in cells stored for one and two years before discharging them at 1 kΩ load at 25°C (Fig. 5.99). The primary failure mode of Li–V$_2$O$_5$ batteries has been attributed to sporadic positive grid corrosion. The presence of impurities in the solvent and/or the electrolyte can produce corrosive products such as hydrogen fluoride [99]. Corrosion can be controlled or eliminated by a thorough purification of the solvent and the electrolyte and also by using corrosion resistant grid materials such as 316L stainless steel.

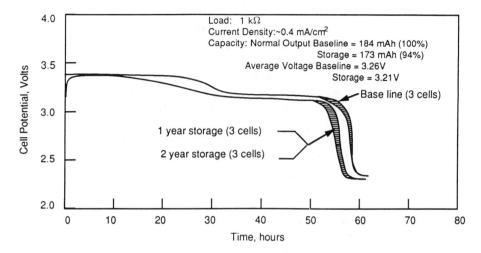

Fig. 5.99 — Discharge performance of active Li–$V_2O_5$ cells after one and two years' storage at 25°C [95]. (Reprinted with permission.)

Levy conducted electrical and environmental testing of Li–$V_2O_5$ cells (32). He investigated the Li–$V_2O_5$ batteries containing 2 M $LiAsF_6$ and 0.4 M $LiBF_4$ in methyl formate for performance over a wide temperature range (−54°C to +71°C) after long-term storage. The cells used were double that of the standard 'D' cell called 'DD', designed by Honeywell [94]. Cells were discharged at the 1 year rate, 5 year rate, 10 year and 20 year rate, at six temperatures to a cut-off voltage of 2 V. Cell performances were rated in energy density and compared to those of controlled units discharged fresh under the same conditions. The maximum life expectancy of cells stored and discharged at the 1 year rate at 71°C was found to be 40 to 70 days. Also, the cell performance was found to drop with increased storage temperature, and cells stored at 71°C vented through the fill hole seal. Li–$V_2O_5$ cells storage and discharge tests have shown that cells stored at high temperature produce a worse performance than those stored at low temperature and subsequently discharged at high temperature. However, Li–$V_2O_5$ cells of 'DD' design were found to survive up to one year storage at room temperature or below, and up to 100 days at +57°C, operating subsequently within the temperature range of −32°C to +57°C. These Li–$V_2O_5$ batteries were also found to withstand shock tests of up to 17 000 g under load at room temperature.

### 5.7.2 Stability and safety

There are only three types of primary active Li–$V_2O_5$ battery using pure vanadium pentoxide cathode materials. These are the double 'D' cells [94], button cell type [100], and the prismatic cell type used for medical applications [101]. All these cells use multiple electrodes connected in parallel to provide larger electrode areas for high-rate applications. They use lithium hexafluoroarsenate ($LiAsF_6$) and lithium tetrafluoroborate ($LiBF_4$) in methyl formate as the battery electrolyte. The major impurity in the solvent and the salt is water. Water, when present in the solvent, can result in hydrolysis of methyl formate to methanol and formic acid. The products of hydrolysis react with lithium, forming lithium formate, lithium methoxide, and hydrogen. Lithium hexafluoroarsenate in the presence of water also forms the hydrolysis product $LiAsF_5OH$ which destabilizes the electrolyte. Therefore, it is very important to thoroughly purify the solvent and to use very pure $LiAsF_6$. In reserve cells, where the electrolyte solution is to be stable up to ten years or more before the cell is activated, electrolyte solution stability is assured by thorough purification and storing it in a glass ampoule with lithium metal. The vapour pressure of methyl formate is relatively high, and proper design parameters are necessary to use these cells at high temperature. The use of vents is necessary, particularly in larger cells, and diode protection is useful in cells to protect them against being driven to reverse polarity by stronger cells or from cells driven to overdischarge. In recent years, improvements in materials quality and cell design have substantially minimized the safety hazards of lithium batteries, and many types of these improved lithium battery systems are finding an increased market for commercial applications.

### 5.7.3 Lithium–vanadium pentoxide medical batteries

Recent advances in lithium battery technology and the availability of reliable high performance microelectronic circuitry have led to the development of many implantable medical devices such as the insulin pump, versatile cardiac pacemakers, and the automatic defibrillator. The power source requirements for implantable cardiac defibrillators are rather demanding. The cell must have high energy density, high rate capability, reliability, and good active storage life, and it must meet the safety standards for implantable grade batteries. The cell must be capable of providing low rate 'background' loads in the microampere range and should also be capable of delivering high-rate pulses having an amplitude from about 1 A to about 3 A with energy from about 25 to about 50 joules for a duration of about ten seconds from a low background. The cells must be hermetically sealed and have high capacity to provide many years of operating life with a very low self-discharge rate. They must also have high voltage and high energy density so that appreciably smaller size cells can be designed for implantable applications. The function of the implantable defibrillator, for which the present $Li-V_2O_5$ and $Li-AgV_2O_5$ power sources are being developed, is to prevent sudden death from lethal arrythmia. The defibrillator device with associated electronics has the ability to monitor the heart rate of the patient, to recognize ventricular fibrillation, and deliver high current pulses to defibrillate the heart. Horning and co-workers at Honeywell Power Sources Center have designed $Li-V_2O_5$ cells with multiple cathode plates and anode material to provide large electrode areas for high rate pulse capability and have placed these in hermetically sealed packages, using glass-to-metal seals and laser welding [87,101,102]. These cells used an electrolyte of $LiAsF_6$ and $LiBF_4$ in methyl formate. The cell construction details are shown in Fig. 5.100. This $Li-V_2O_5$ medical cell has an open circuit voltage of 3.4 V and an operating voltage of 2.8 V at 45 mA/cm$^2$. The nominal cell capacity is about 800 mAh, and the power density is 80 Wh.kg with 0.24 Wh.cm$^{-3}$. During discharge of $Li-V_2O_5$ cells, the open circuit voltage of 3.4 V shifts to 3.2 V, passing through a transition, and it shows two plateaux; at 2.9 V and 2.7 V (Fig. 5.101). Several $Li-V_2O_5$ cells have been discharged under pulsing conditions through 2 A constant current pulses, and the cell utilization is 87% of its capacity.

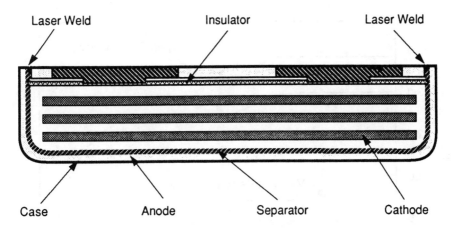

Fig. 5.100 — High capacity and high efficiency Li–$V_2O_5$ defibrillator cell construction details [101,102]. (Reprinted with permission.)

### 5.7.4 Lithium–silver vanadium pentoxide batteries

A modification of the Li–$V_2O_5$ battery is demonstrated by the development of a conductive cathode material of the composition $AgV_2O_5$. The resulting battery, Li–organic electrolyte–$AgV_2O_5$, has a superior volumetric energy density and a high rate capability. It can provide high current pulsing ability and has a sloping discharge behaviour which helps the system designer to predict the end of life of the battery. These batteries possess excellent shelf life, a very low self-discharge rate of 0.4% per year, and good safety characteristics [103]. Therefore, they are being developed for use in implantable cardiac defibrillators and drug-delivery systems [103,104]. The Li–$AgV_2O_5$ cell with $LiAsF_6$ in propylene carbonate and dimethoxyethane (50:50 by volume) on electrochemical discharge, shows five well characterized reduction steps with 3.5 equivalents of lithium from about 3.5 V to 1.5 V. The sequence of reduction is from V(V) to V(IV) followed by Ag(I) to Ag(0) and finally from V(IV) to V(III). The first three steps at 3.2 V, 3.0 V, and 2.8 V are attributed to V(V) to V(IV) reduction, the fourth step at 2.6 V is attributed to the reduction of Ag(I) to Ag(0), and the fifth step at 2.0 V is attributed to the reduction of V(IV) to V(III) [105].

Thus, Li–$V_2O_5$ and Li–$AgV_2O_5$ organic electrolyte based battery systems have been packaged into configurations compatible with implantable devices. The energy density of these systems at an average current drain of 500 $\mu$A is 0.87 Wh.cm$^{-3}$. Self-discharge appears to be very low both under shelf life conditions and during discharge. The sloping discharge curve and gradual approach to end-of-life voltage of Li–$AgV_2O_5$ make them attractive for implantable applications.

A lithium–silver vanadium pentoxide primary battery containing a different electrolyte, 1.2 molar solution of lithium trifluoromethane sulphonate ($LiCF_3SO_3$) in a mixed solvent system of propylene carbonate and dimethoxyethane (50:50 by volume), has been developed for defibrillators. The cathode plates contain 94 w/o of silver vanadium oxide ($AgV_2O_{5.5}$), 2 w/o graphite powder, 1 w/o carbon, and 3 w/o of Teflon™ powder. A plurality of these cathode plates and a large number of anode

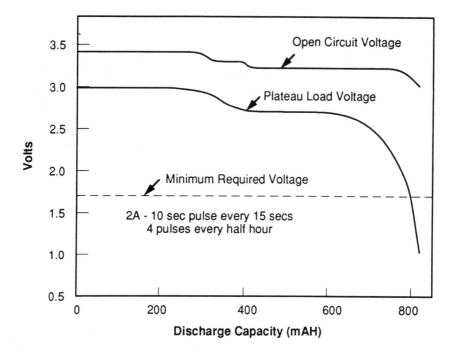

Fig. 5.101 — Discharge characteristics of high efficiency Li–$V_2O_5$ cells for implantable medical devices [101,105]. (Reprinted with permission.)

plates interposed between the cathodes (with separators) have provided a high surface area for high rate capability. They have been packaged in a stainless casing as a power source for an implantable cardiac defibrillator [106]. This cell has high volumetric capacity, high rate capability, and good current pulsing behaviour.

## 5.8 LITHIUM–IRON DISULPHIDE BATTERIES
(A. Gilmour)

### 5.8.1 Introduction

Exploratory work on the use of iron disulphide ($FeS_2$) as a positive active electrode material in conjunction with a lithium metal anode and a non-aqueous electrolyte was pioneered by Eveready (UCAR) in the early 1970s. $FeS_2$, also known as iron pyrites or 'fools gold', is one of the most abundant naturally occurring minerals and cheap bulk raw materials of commerce. It was soon established that this cell couple had an operating load voltage in the 1.5 V region, and its potential use for high capacity 'dry' cells to outperform the alkaline–manganese system became apparent. Its development was spurred on by the rising price of silver which (as the oxide) was used in conjunction with zinc and an alkaline electrolyte as the preferred watch cell electrochemistry (see sections 3.3.5.2 and 3.3.6.2). For watch cells the Li–$FeS_2$ system showed a substantial materials cost advantage over the silver oxide system.

### 5.8.2 Chemistry
The overall cell reaction is as follows:

$$4\,Li + FeS_2 \rightarrow 2\,Li_2S + Fe\,. \qquad (5.20)$$

The origin and purity of the $FeS_2$ have a significant effect on the open circuit voltage (OCV) of the cells. Traces of the oxides and sulphides of metals such as nickel, cobalt, and chromium will give an elevated OCV as high as 3.65 V. On application of a load, however, the voltage will rapidly fall to around 1.5 V, this value depending on the current density at the $FeS_2$ electrode.

Since $FeS_2$ is a fairly good electronic conductor and the finely divided iron formed on discharge is a good conductor, a constant voltage discharge curve is observed with a high percentage utilization of the active material.

As with all lithium couples, the electrolyte used in Li–FeS$_2$ cells must be completely anhydrous (<20 ppm $H_2O$). Much development has gone into investigation of a wide range of polar solvents stable to lithium, and also into the effect of using a number of different lithium salts as solutes. In the earlier studies on $FeS_2$ [107], mixtures of solvents, particularly tetrahydrofuran, 1:2 dimethoxyethane, and 3-methyl-2-oxazolidone, were used with lithium tetrafluoroborate ($LiBF_4$), lithium hexafluoroarsenate ($LiAsF_6$), and lithium trifluoromethane sulphonate ($LiCF_3SO_3$) as solutes. The solvent 1:3 dioxolane was later found to be more stable than tetrahydrofuran and to give better low temperature performance [108]. The use of $LiCF_3SO_3$ dissolved in a mixture of propylene carbonate (PC) and 1:2 dimethyoxyethane (DME) as the electrolyte for a Li–FeS$_2$ cell, has also been the subject of a patent application [109]. A solution of lithium perchlorate $LiClO_4$ in a 50/50 mixture of PC and DME was used in the Li–FeS$_2$ system by Venture Technology (UK).

### 5.8.3 Cell designs
Eveready (UCAR) first announced, in 1978, a range of coin cells using the Li–FeS$_2$ system. The six sizes listed all had existing silver oxide–zinc equivalents, three having the then popular 11.56 mm diameter. The capacities of these cells were rated some 35% higher than those of their corresponding silver monoxide ($Ag_2O$)–zinc system equivalents, the Li–FeS$_2$ system thus seeming poised for extensive market penetration and growth.

Venture Technology (UK) developed an AAA (IEC R03) size Li–FeS$_2$ cell which was put on the market in 1982. This design (Fig. 5.102) has a stainless steel can (A) with lithium foil (B) in contact with the inside surface. The positive electrode is a composite of $FeS_2$, graphite, and PTFE binder coated from a slurry onto an expanded aluminium grid. This is bound on itself to form a bobbin (C), and a porous polypropylene separator (D) surrounds the bobbin, the sub-assembly being a push fit into the can. The electrolyte is a solution of $LiClO_4$ in a PC/DME mixture as described above. The cathode is tab welded to the cap (E) and the cells are sealed by crimping around the polypropylene gasket (F).

Eveready (USA) announced their intention in mid-1988 to market an AA (IEC R6) high-rate cell which was developed especially for photographic use. This cell is

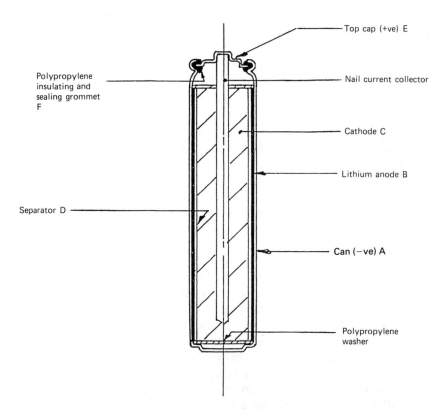

Fig. 5.102 — A cross-section through a lithium–iron disulphide LiFe 104 cell.

made up of a very thin $FeS_2$ composite cathode, a thin microporous polyethylene separator, and lithium foil spirally wound together and inserted into the nickel plated steel can. The lithium foil is in electrical contact with the can, and the cathode makes contact with the cap by means of a 'crown of thorns' pressed on the top of the spiral. This is held in place by the crimp which forms an insulating seal between the cap and the can. The electrolyte is a solution of lithium trifluoromethane sulphonate in the mixed solvents already discussed. Fig. 5.103 shows a cross-section of the Eveready AA cell.

### 5.8.4 Performance of the Li–$FeS_2$ system

Fig. 5.104 shows the discharge characteristics of the Eveready No. 803 (R44 size) button cell at three different current drains. On the 30 kΩ load corresponding to around 50 μA drain, it is seen that the running voltage starts off around 1.7 V and shows a stepped fall to around 1.6 V half way through its discharge life. On the much heavier drain of 230 μA (6.5 kΩ load) the running voltage is around 1.5 V and the step in the voltage curve is not perceptible. The capacity obtained from this cell is around 180 mAh to a 1.0 V end-point. This compares favourably with the R44 silver

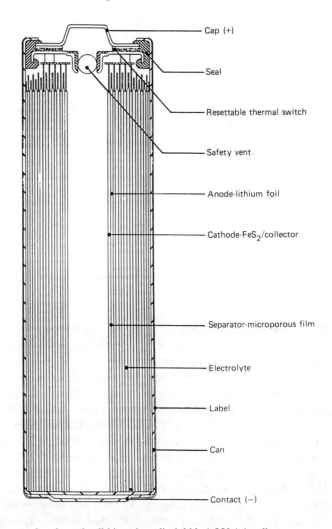

Fig. 5.103 — A cross-section through a lithium–iron disulphide 1.5 V AA-cell.

monoxide cell which delivers around 130 mAh, but the latter will deliver much higher currents and the load voltage is substantially independent of the current drain over a wide range of discharge rates.

The AAA (R03) size cylindrical cell was originally developed by Venture Technology for a paging application. With a quiescent current of 1 mA, pulses of over 80 mA were needed from the battery for up to 10 s duration. Fig. 5.105 shows the results obtained on the Li–FeS$_2$ cells known as LiFe 104, in comparison with those observed with the equivalent size of alkaline–manganese cell, the Mn2400 from Duracell. The LiFeS$_2$ cells not only give around twice the service life of the alkaline cells, but the running voltage is substantially higher and stays constant over a much longer period.

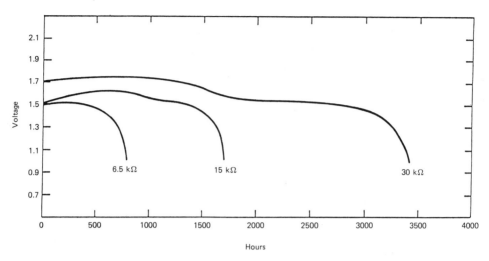

Fig. 5.104 — The effect of drain rate on cell discharge at 35°C of fresh lithium–iron disulphide button cells No. 803.

### 5.8.5 Discharge mechanisms of the Li–FeS$_2$ system

The four-electron reduction of FeS$_2$ to elemental iron and lithium sulphide involves the breaking of the polysulphide anion bonds and the valency change from the ferrous (II) to metallic (0) state. There are at least two steps in the total reduction of FeS$_2$ as evidenced by the step or 'knee' in the discharge curves of Fig. 5.104. As already indicated, this step is sensitive to rate of discharge, temperature, and source of FeS$_2$. Although several discharge mechanisms have been proposed for the Li–FeS$_2$ system, a major problem in verification of a hypothesis is the inability to positively identify the reaction products formed on discharge. X-ray diffraction, Mössbauer spectroscopy, and quantitative chemical analysis have all been used [110]. Neither of the instrumental methods was able to discriminate between metallic iron and FeS$_2$. Chemical analysis, in the presence of water, showed that the finely divided iron, formed on discharge, reacted to form hydrogen gas. The presence of metallic iron early on in the discharge was confirmed, in fact, by simply using a magnet.

The formation of ferrous sulphide FeS$_2$ as an intermediate has been dismissed, since the voltage curves on cells made from this material run at about 0.15 V lower than the second step of the Li–FeS$_2$ discharge curve.

The only viable mechanism, albeit less specific, is the formation of a lithiated iron sulphide intermediate, this species having several different formulae depending on the value of $x$ in Li$_x$FeS$_2$ and the mechanism proposes that both the FeS$_2$ and Li$_x$FeS$_2$ are reducible at the same operating potential [111].

Another useful characteristic of the LiFeS$_2$ system is the dependence of the open circuit voltage (OCV) on the degree of lithiation. The OCV is usually no higher than 1.80 V early in the discharge, and it follows the general shape of the discharge curve, confirming that more than one species is responsible for the discharge mechanism.

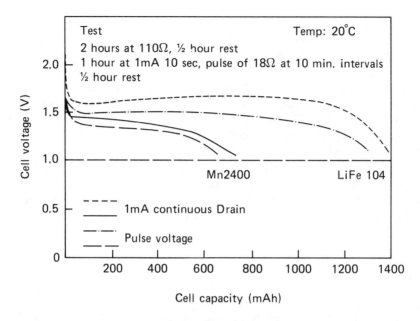

Fig. 5.105 — Continuous drain discharge characteristics and pulse discharge characteristics of Dowty AAA lithium–iron disulphide cells.

In view of the observed volumetric expansion of the $FeS_2$ cathode during discharge, it is inferred that a lattice structural change must occur when the $FeS_2$ is converted into $Li_xFeS_2$. Since $FeS_2$ has a cubic rock salt type structure with very little space to accommodate lithium ions, it is likely that the reaction is a heterogeneous process involving the rupture of the disulphide anion bonds, an extensive lattice expansion, and the insertion of a number of lithium ions corresponding to the value of $x$ in $Li_xFeS_2$. The transformation is likely to involve a hexagonal structure like the $TiS_2$ type of lattice, but this has not yet been clarified.

### 5.8.6 The market's reaction to the Li–FeS$_2$ system
(a) *Coin cells*
Since their launch in 1978 by Eveready (USA), lithium iron disulphide coin cells have not achieved the expected market success because of two basic technical deficiencies inherent in the system:

(1) On discharge, the $FeS_2$ undergoes a significant increase in volume which results in cells swelling and exceeding the height tolerance on the button cell dimensions.
(2) The system has a much lower rate capability than that of the alkaline electrolyte based systems, for example silver oxide–zinc cells as used in watches.

From a purely economic point of view, the market price of silver has not risen to the alarming heights that had been predicted, and silver oxide–zinc button cells still play a predominant role in the 1.5 V watch cell market.

(b) *Cylindrical cells*

The technical problems encountered with coin cells as listed above are minimized by using a cylindrical spirally wound cell design. The swelling of thin $FeS_2$ sections in the spiral is accommodated largely by the dissolution of the lithium foil, with the added advantage that a more consolidated interface is obtained on discharge. The high surface interfacial area between lithium and cathode overcomes the low-rate limitation, but at the cost of lower capacity and more expensive processing with more separator and current collectors needed.

The Venture Technology AAA Li–$FeS_2$ cells were technically well received in the particular application for which they were designed. The manufacturing cost in the relatively small numbers being produced (20 000 per month) was high, and the selling price proved not to be competitive with the equivalent size alkaline cell which was being made in very large numbers internationally.

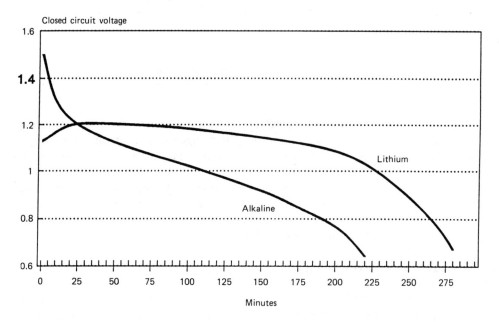

Fig. 5.106 — The continuous drain discharge characteristics (2.2 $\Omega$ load) of a lithium–iron disulphide cell compared to those of an alkaline–manganese cell, both being designed for toy or high-tech flashlight (torch) applications.

It is premature to gauge the market reaction to the Eveready Super Plus AA size Li–$FeS_2$ cell. With a two-to-one performance advantage over alkaline–manganese on constant drain (see Fig. 5.106) a faster cycling time, coupled with the long shelf life, its success will be very largely determined by whether the price charged to the consumer is profitable to the manufacturer.

Environmentally, iron disulphide in the undischarged state is acceptable since it occurs naturally as a major source for both sulphur and iron and their derivatives. Lithium sulphide, which is the discharge product, is, however, much less friendly since it reacts instantly with water to form hydrogen sulphide. The disposal of large numbers of batteries based on the lithium–iron disulphide system could therefore be a cause for concern to the environmentalist.

## 5.9 LITHIUM ANODE THERMAL BATTERIES
(A. Attewell)

### 5.9.1 Introduction

The term 'thermal battery' is used to describe a class of reserve-type molten salt electrolyte primary battery which is inert until the electrolyte is melted by the heat generated from the burning of an internal charge of pyrotechnic. Thereafter they remain active for as long as one hour, or for only a few seconds, depending upon size, thermal insulation, electrochemical system, and the rate at which power is withdrawn. Working at around 500°C and with an operating temperature 'window' of about 120°, the performance of thermal batteries is affected little by the ambient temperature.

The development of thermal batteries began in Germany toward the end of World War II. Intensive work in the USA during the 1950s [112] and slightly later in Europe, established them as a near-ideal electrical supply for guided missiles as well as suitable power sources for some emergency duties in military aircraft. This was because of a unique combination of properties which included complete inertness during storage, ability to work at very low temperatures without preheating, resistance to mechanical vibration and shock, and an ability to deliver very high powers for a short period.

The electrical output from modern designs can vary from a watt for ten milliseconds to several kW for tens of minutes depending on their size and internal layout. A range of these hermetically-sealed, steel-cased batteries is shown in Fig. 5.107.

Several reviews have traced the evolution of this type of battery [113,114,115,116,117], while Singh & Sachan [118] compiled a comprehensive literature study of cell systems. Early designs used either calcium or magnesium as the negative electrode with such materials as alkaline-earth chromates, or oxides of tungsten, iron, or vanadium as cathodes. However, these systems had severe disadvantages, mainly due to unwanted chemical side reactions, and these limited electrochemical efficiencies to about 20% of the theoretical output. Calcium types have, except for a few specialized designs, been superseded by ones using lithium (or a lithium-rich alloy) for the negative electrode with iron disulphide as the cathode. Lithium-based systems do not suffer to any major extent from efficiency-reducing and exothermic side reactions, and are now allowing batteries to remain active longer, and to be made in much larger sizes.

The most common electrolyte continues to be a eutectic, or near-eutectic mixture of lithium and potassium chlorides (60 mol% LiCl: m.pt. 352°C). Its high decomposition potential allows the use of alkali or alkaline earth electrodes, and at the

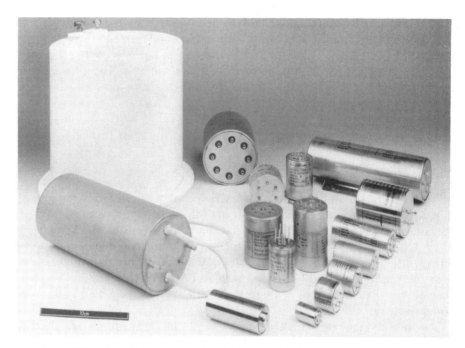

Fig. 5.107 — Typical thermal batteries. (Courtesy MSA (Britain) and SAFT (UK).)

working temperature of between 400° and 600°C, it is almost fully ionized, requiring no solute to give it a conductivity of 1.86 ohm$^{-1}$ cm$^{-1}$ (cf: 30 wt% KOH; 0.6 ohm$^{-1}$ cm$^{-1}$; 1.35 sg H$_2$SO$_4$: 1.6 ohm$^{-1}$ cm$^{-1}$ at +20°C).

Cells using the Li (or Li-alloy)–LiCl:KCl–FeS$_2$ system have an on-load voltage of about 1.6 V. They can be worked at current densities up to several amperes per cm$^2$, allowing thermal batteries to be discharged at very high currents, and giving power densities at the ten second or less rate which are better than any other complete multi-cell system, weighing a few kg (Fig. 5.108). For example, a 6 kg battery, designed to deliver its energy in a one-second discharge, will have a specific power output of about 5 kW.kg$^{-1}$.

Because of the hygroscopic nature of the chemicals used, batteries have to be made under very dry conditions. Assembled in dry rooms (Fig. 5.109) held typically at 2 to 3% RH (500 ppm water vapour), their steel containers remain hermetically sealed throughout storage and discharge.

### 5.9.2 Battery construction

The most common assembly method uses powder technology to produce thin disks of the anode, electrolyte/separator layer, the cathode, and the pyrotechnic. These are assembled into a stack of cells as shown in Fig. 5.110. The diameters of the components can vary from under 10 to over 150 mm. The thicknesses of anode and cathode layers may vary from 0.2 to several mm, depending on (a) the amount of chemicals required to give the electrical output, (b) the speed with which power is

Sec. 5.9]  **Lithium anode thermal batteries**  411

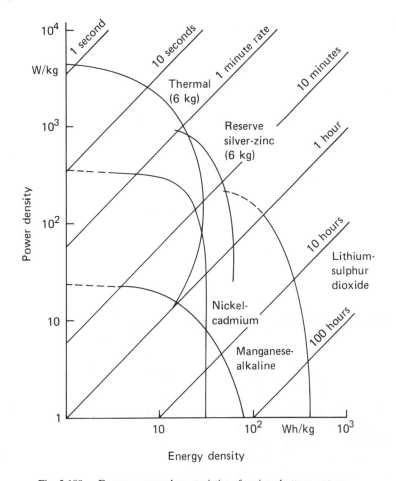

Fig. 5.108 — Energy–power characteristics of various battery systems.

Fig. 5.109 — Assembly of thermal batteries. (Courtesy MSA (Britain).)

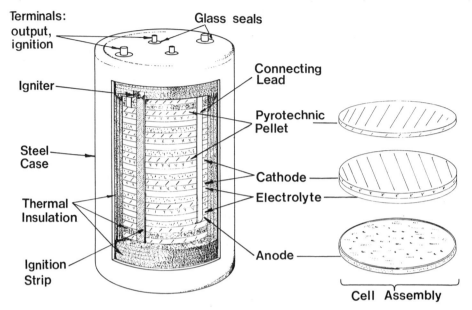

Fig. 5.110 — Internal arrangement of a thermal battery.

required after the pyrotechnic is ignited — the thinner the pellet, the more rapid will be the melting of the electrolyte layer in the wake of the flame-front of the burning pyrotechnic, and (c) handling constraints for the finished pellet.

When an output is required, the pyrotechnic disks, made from iron and potassium perchlorate powders, are ignited by a 'heat paper' fuse train made from a rapidly burning, easily ignited, mixture of powdered zirconium, barium chromate and ceramic fibres. This fuse train can either extend down the periphery of the cell stack, or, if annular components are used, be placed in the central hole.

Initiation of this sequence can either be by the electrical firing of an igniter, ('fuzehead', 'electric match') or by striking a percussion cap or stab detonator fitted into the case of the battery. Where a very high acceleration is applied to the battery, such as in gun-launched missiles, this can be used to trigger a device internal to the battery which, like the igniter or cap, will emit a flame for ignition of the fuse train.

Alternative cell shapes to the commonly used disk or annulus are emerging. Batteries using large rectangular cells have been reported [119], while 'D'-shaped cells may be used in two side-by-side stacks to halve the length of a high-voltage battery.

To emphasize the near-complete shift to lithium-based systems, only one calcium anode type is sometimes used. It employs the electrochemical system Ca–LiCl:KCl–$K_2Cr_2O_7$ and is used when very fast 'activation' or 'rise' times of tens of ms are needed — times which cannot yet be matched by comparable sized lithium types [116]. One surface of a disk of a dumbbell-shaped piece of nickel foil is coated with calcium by vacuum deposition. This is folded around a disk of heat paper. The dumbbell serves as the connector in a series chain of cells. A disk of glass cloth is impregnated with electrolyte containing dissolved dichromate. The cell components are annular, and the flame from the igniter or cap is directed down the central hole to ignite the pads of heat paper. Cells of this type are thin, the soluble cathode material lowers the melting point of the electrolyte, and these factors, combined with a pyrotechnic which burns at approximately 250 mm s$^{-1}$, give activation times which are faster than the pellet system described earlier. Active life of this system is measured in seconds, owing to the rapid chemical oxidation of the calcium by the dissolved dichromate.

Whatever the design, the cell stack is held under compression, either by an arrangement of internal clamps, or by pressure exerted from the lid of the cylindrical case.

The amount and type of thermal insulation separating the cell stack from the case will determine the rate of cooling and hence the active life of a thermal battery. In the small spaces available, insulating wraps based on glass or silica fibres are commonly employed, while high-performance materials based upon titanium dioxide ('Microtherm$^{TM}$', Micropore Ltd, UK or 'MIN K$^{TM}$', Johns Manville, USA) can be used as case liners.

A long stack of cells, with the same weights of cells and pyrotechnic disks throughout its length, will tend to cool more rapidly at the ends than in the centre. Several techniques are available to even out the temperature decay profiles. Pyrotechnic pellets with more heat content may be put between the end cells. Heat 'buffers' may be added at either end of the stack. These can be either pyrotechnic disks, separated from the main stack by thermal insulation, or heat reservoirs which use the latent heat of fusion of salt mixtures (such as $Li_2SO_4$:NaCl, m.pt 500°C) to delay heat loss through the end cells [120].

Electrical connections, for power output, electric igniter, or activation detector (see section 5.9.6) are brought through the lid of the case by glass-to-metal seals.

These design principles give batteries which are rugged, inert during storage (electrolyte solid and non-conducting with hermetically sealed assembly), and which have an ability to operate over a wide temperature range without auxiliary heating. Their internal operating temperature is between 400 and 600°C. These important, advantageous properties are achieved at the expense of short active lives, which can vary from seconds to tens of minutes, according to battery size and internal construction, and they are determined either by the time the electrolyte remains molten after the one-shot heating, or by exhaustion of the active materials.

### 5.9.3 Cell components and electrochemistry

A simple representation of the two-step discharge of the lithium–iron disulphide couple to zero volts is:

$$2\,Li + FeS_2 = Li_2S + FeS \tag{5.21}$$

$$2\,Li + FeS\ = Li_2S + Fe\ . \tag{5.22}$$

In practice, various lithiated iron sulphides are formed and reduced as discharge progresses [121], and these produce falling, rather than stable discharge plateaux as suggested by reactions (5.21) and (5.22). At a current density of 500 mA.cm$^{-2}$, the voltage of the first step (5.21) is about 1.6 V, and that of the second, 1.2 V. To obtain a reasonably regulated voltage output, the system is discharged only to the end of reaction (5.21), a reaction which gives a specific energy density of 640 Wh kg$^{-1}$ from the active chemicals. This is a high value, as compared with lead–acid at 250 Wh kg$^{-1}$ but it is lower than some other lithium anode systems (for example 1200 Wh kg$^{-1}$ for Li-sulphur dioxide). All practical battery systems have energy densities which are much less than these purely theoretical values for the active ingredients (see section 2.1.1).

For example, lead batteries are eight times as heavy as the theoretical mass needed to store and produce their energy content. The positive and negative chemicals in a sulphur dioxide primary cell account for only a quarter of its weight. Because of the need to carry the pyrotechnic, thermal insulation, and a substantial case, the specific energy of thermal batteries is, at best, about one-tenth of the theoretical.

### 5.9.3.1 The lithium anode

Lithium is used either as the metal or as a solid alloy [115]. As it is molten at the working temperature, immobilization of pure Li is required. One method is to retain it within the pores of a foam–metal disk, but this is uncommon as only a limited range of substrate thicknesses is available. The most widely used Li anode (trademark: 'LAN') is prepared by stirring iron powder into a bath of molten lithium. After cooling, this mixture, which contains about 20 wt% Li (46 A.min g$^{-1}$), can be rolled to any thickness. When the battery is operating, the iron matrix retains the molten metal [122].

Of those lithium-rich alloys which are solid at the working temperature, the most commonly used is 20 wt% Li-Al, m.pt.c 690°C. It is usable to 12 wt%, giving some 23 A min g$^{-1}$. Its open circuit potential is some 300 mV less than that of pure Li, but commercially available powders pelletize easily. A 44 wt% Li–Si alloy is harder and less easy to pelletize, and the usable discharge is in two steps, as the composition goes from an initial $Li_{13}Si_4$ either to $Li_7Si_3$, giving 35 A min g$^{-1}$ [117], or to a lower Li% alloy, giving some 88 A min g$^{-1}$ [123].

A technically interesting material with the appearance and malleability of lead, but one which has not entered production because of manufacturing difficulties, is 'lithium–boron'. This behaves as a liquid lithium electrode, being, most probably a solid 'cage' of $Li_7B_6$ permeated by 37 wt% of elemental Li (85 A min.g $^{-1}$) [124].

In practice, and given consumption of the active materials before the electrolyte freezes, anode efficiencies of up to 85% have been reported, but anode weight is only a minor consideration, being about 5% of total battery weight [125,126].

### 5.9.3.2 Electrolytes

Most batteries continue to use a eutectic, or near-eutectic, mixture of LiCl–KCl (m pt. 352°C). The UK Admiralty Marine Technology Establishment pioneered the use of a ternary mixture of lithium halides which has a melting point of 430°C [127]. This has the advantages of an invariant cation composition during discharge and better conductivity, but it needs a higher operating temperature than the binary salt.

Other alkali halide combinations are being investigated. For example, the eutectic mixture 25 mol% LiCl- 37 mol% LiBr- 38 mol% KBr mixture melts at about 310°C, so allowing operation at a lower temperature. It also has a wider liquidus/compositional range above the original m. pt, than does the binary, so delaying the precipitation of complex salts during discharge. Various workers have found that $FeS_2$ is more reversible (higher working potential) in this electrolyte [128,129].

The ternary mix of LiF:NaF:KF ('FLINAK') appeared to be a promising material [130], but little information has been published recently on its use in thermal cells. Although lithium metal is thermodynamically unstable in the melt, a layer of $Li_2O$ protects against dissolution when the metal is in practical cells [131].

All of the foregoing refers to salt mixes which are molten at the operating temperature of the battery. Immobilization of these highly mobile liquids within the pressed pellet is imperative, and this is done by adding about 35 wt% of MgO powder to the electrolyte mix.

As an alternative to a liquid, solid electrolytes have been tried. One example is the use of the solid electrolyte LiI/43 mol% $Al_2O_3$. Reported performances show promise [132].

### 5.9.3.3 Cathodes

The commonly used $FeS_2$ cathode suffers from two disadvantages: it has a low on-load voltage (about 1.6 V vs Li) and it is thermally unstable at battery temperatures. However, it is electronically conductive and is able to support current densities of several A cm$^{-2}$, and, when discharged only to the end of the $FeS_2$ stage, gives good voltage regulation.

Because $FeS_2$ is electrically conductive, a separate electrolyte layer is needed between anode and cathode, a change from earlier calcium designs using non-conducting calcium chromate, where a single pellet containing both electrolyte and cathode ingredients could be used.

An initial high-voltage peak, due to impurities, can be removed by the addition of $Li_2O$ [133] or by other, proprietory, methods. $FeS_2$ cathodes account for about 8% of battery weight.

$FeS_2$ thermally decomposes around 550°C into FeS and sulphur, but good efficiencies are obtained unless the battery is required to remain hot for long periods, when the rate of chemical reaction between Li and sulphur vapour may become significant.

A cathode with a higher working voltage and a similar or better specific capacity would cut battery size and weight in a multi-cell assembly by reducing the number of cells and associated pyrotechnic layers. One such cathode has been announced by RAE [134]: lithiated vanadium oxide ('LVO'). Its performance in comparison to $FeS_2$ is shown in Fig. 5.111. Owing to its higher voltage, it may reduce the size of some batteries by 20%. The search for other materials continues. Some, such as $V_2O_5$, have a high cathode potential but react with Br-containing electrolytes.

A more radical approach, that of using gaseous chlorine as the cathode, is being investigated [135]. If the severe engineering problems of distributing the gas to a multi-cell stack can be overcome, then a doubling of the energy density of thermal batteries may be possible.

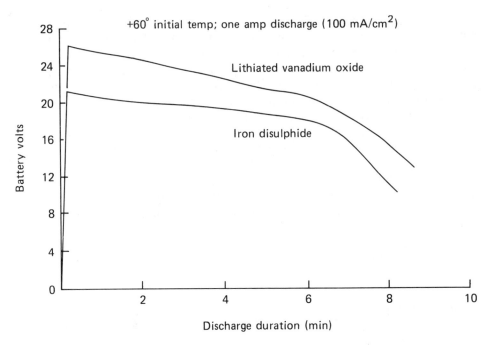

Fig. 5.111 — Performance of batteries using $FeS_2$ and lithiated vandium oxide (LVO) cathodes.

### 5.9.4 The pyrotechnic

No other area of pyrotechnic technology has such demanding criteria. The exothermic mixture used to activate a thermal battery must burn at a controlled rate, have a good thermal conductivity, not melt, not produce gas, have a calorific output which can be closely controlled, be insensitive enough to be pelleted without igniting, then be able to be lit reliably over a wide temperature range. Also, after ignition, it must be electrically conductive, as the disks of burnt pyrotechnic form the inter-cell connectors (Fig. 5.110).

The calorific output and the position of each pyrotechnic layer have to be carefully calculated (by 'heat balancing'), and the battery requirements closely maintained during production, otherwise the batteries will reach temperatures outside their operating range.

Thin pellets are used, pressed from a mixture of $KClO_4$ and high-surface-area iron powder [136]. In the presence of excess iron, $Fe_{0.947}O$ is the thermodynamically stable oxide above 600°C. Near stoichiometric ratios of the components (i.e. 63 wt% Fe; 37 wt% $KClO_4$) produce temperatures high enough to melt iron, so, to moderate the reaction, control the rate of burning and produce a conductive ash, mixtures containing between 10–16% of perchlorate are used. A typical 86:14 wt% mix gives about 1100 J g$^{-1}$ and burns evenly, but fairly slowly, at 100 mm s$^{-1}$. The pyrotechnic can account for 10 to 20% of battery weight.

The energy required to ignite a heat pellet is tens of mJ. This relative insensitivity means that, to ensure that all pellets are ignited reliably, the much more sensitive 'heat paper' is used as an intermediate between the initiating device and the pellets, as described previously in section 5.9.2.

### 5.9.5 Computer modelling

Traditionally, the evolution of a new design of battery is an iterative process, using experience and enthalpy calculations to provide a prototype, the next stage being to build, test, and adjust the design, depending on results. This procedure is expensive (particularly for large batteries) and time consuming, so the use of computer modelling of the temperature/time profiles to predict the behaviour of any design is an increasingly valuable tool. The most sophisticated model announced so far is one from RAE [137]. By its use, the burning of the pyrotechnic layers between cells can be followed, and the times at which chemicals in the cell melt can be plotted.

These data, together with the detailed temperature/time profiles at different places within the battery and on the outside case which a model can predict, can be used to reduce significantly the time spent in designing for new applications.

### 5.9.6 The performance of thermal batteries

Acknowledgement is made to MSA (Britain) Ltd, PCI Inc., SAFT (America) and SAFT (UK) Ltd, who provided much of the information presented in this section.

Early designs, with calcium as the anode material, gave a falling discharge characteristic with time with a steadily increasing internal resistance as discharge progressed. All systems based on lithium–iron disulphide exhibit a more stable on-load voltage with an internal resistance which remains substantially constant during useful life.

Fig. 5.112 shows a typical voltage/time characteristic during activation and discharge. It also defines various terms used to describe the behaviour of thermal batteries. The performance illustrated is that of a 90 mm diameter × 160 mm long battery, weighing 2.5 kg and discharging into 3.5 ohm.

Fig. 5.113 summarizes the performance of various sizes of battery, ranging from a fairly small example (30 mm dia. × 75 mm long) to one of the largest batteries available (10 $dm^3$). All data are for a voltage regulation of about 20%. This figure and Table 5.20 illustrate the increase both in active life and volumetric energy density as battery size increases. They also show the loss in energy density which occurs in small batteries when their discharges are ended by electrolyte freezing, not by chemical exhaustion.

Thermal batteries are heavy, and a relative density of 2.5 may be used to calculate the approximate weights of these and other examples.

The on-load voltage of an individual cell is between 1.8 and 1.5 V, depending on the rate of, and time into, discharge. The current available from a cell will depend on the surface area. Usually, discharge current densities are between 1 A and 100 mA $cm^{-2}$, but they can be several amperes during pulsed loads. The Wh capacity of a cell is determined by its chemical contents, so a design balance is struck between cell area — for current output, and cell thickness — for discharge duration.

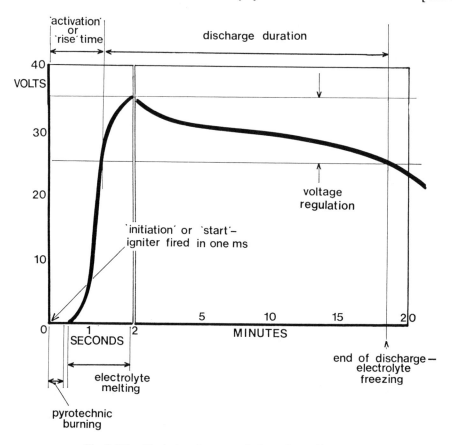

Fig. 5.112 — Typical performance of a large thermal battery.

Battery height (number of cells) is thus determined largely by voltage and duration requirements, whilst output current dictates the diameter. Series-connected stacks of cells may be connected in parallel within a battery to give high currents. It is good practice to aim for a height to diameter ratio of not more than three to one. Long, thin batteries present manufacturing difficulties, and their heat balancing is more difficult. They will have shorter durations, because of more rapid cooling, than a shorter, fatter, battery of the same volume.

Much smaller batteries than those discussed so far are available. Fig. 5.114 shows the performance of one which is 18 mm diameter by 25 mm long and weighs 22 g. Note that general performance is affected little over a temperature range of $-40°C$ to $+70°C$, and that the 'heat balance' of this battery has degraded the $+70°C$ performance compared to that at $-40°C$. The activation time is typical and not the fastest possible for this size. Understandably, it is rather longer at the lower operating temperature.

If a very fast activation time, say 100 ms, is required from a large thermal battery, then a 'starter' battery of the type described in section 5.9.2 can be connected in parallel just to bridge the time from initiation until the large battery has activated.

# Sec. 5.9] Lithium anode thermal batteries

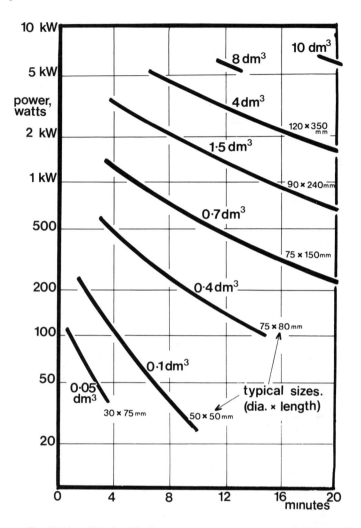

Fig. 5.113 — Relationships between size, power output, and discharge life.

The constructional techniques used in the manufacture of thermal batteries mean that they are inherently rugged and able to withstand many of the shock, vibration, and acceleration requirements of modern weapons and aircraft. They can be designed to withstand gun-launch shock, but performance is degraded at high axial spin rates because electrolyte can be flung out of the cells. The ability to function at spin rates up to 275 rev/s has been reported [119].

## 5.9.7 The use of thermal batteries

As thermal batteries are complex devices, changing their state very rapidly during the early part of their active lives, there has to be close collaboration between equipment designers and the battery makers.

**Table 5.20** — Performance figures of lithium thermal batteries according to size

| Battery size (dia. × length) (mm) | Output (W) | Duration (min) | Wh dm$^{-3}$ |
|---|---|---|---|
| 30 × 75 | 100 | 1 | 35 |
| 30 × 75 | 33 | 3 | 33 |
| 75 × 150 | 1200 | 4 | 110 |
| 75 × 150 | 220 | 20 | 85 |
| 120 × 350 | 8000 | 4 | 135 |
| 120 × 350 | 1800 | 20 | 130 |
| rectangular (10 dm$^3$) | 5200 | 20 | 175 |

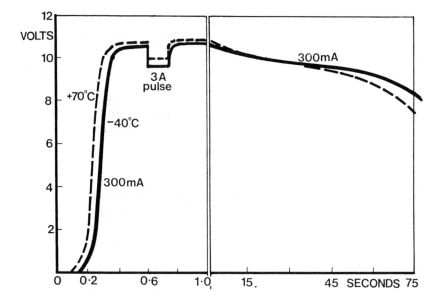

Fig. 5.114 — Performance of a small thermal battery.

Owing to the internal complexity and the high operating temperature, the number of electrical outputs should be minimized. Usually, discussions with the user can reduce these to two or three. As with any type of battery, it is good practice not to draw substantial currents from voltage tappings.

High-voltage outputs (over 100 V), although possible, should be avoided, because of the large number of cells required. Most thermal batteries have voltage outputs between 6 and 60 V.

Although lithium thermal batteries are almost completely inert during storage, very small currents (picoamps) can be drawn from them, and these are often large enough to charge and maintain a capacitor for long periods. If necessary, a very high value resistor across the output terminals will eliminate this ability.

During storage and activation, loads or external power supplies may be connected across the battery without adverse effect. However, the battery may develop a low internal resistance during activation, and, until the rising output voltage matches the value of external supplies, a current will flow into the battery. It may be necessary to use blocking diodes to prevent excessive drain on the external supply. If two different types of battery are connected in parallel, for example a 'starter' and a main battery, it is good design to diode-isolate them to prevent high circulating currents from depressing the output voltage of the combination.

While flow of current into an activating or live battery may affect the regulation of an external power supply, it does not appear to have any adverse effect on the battery [138].

The electrical insulation between the output, the case, and an electric igniter will be very high in the inactive state. However, low values can be encountered during activation, but they will rise rapidly during the early part of active life. It is therefore essential that battery designers should be given full information about external circuitry so that cross-talk brought about by these low insulation values may be considered and appropriate design action be taken to avoid them. Changes to the use of existing batteries should be discussed, as different external wiring may expose internal insulation to potentials higher than those for which they were designed.

Thermal batteries generally have low background electrical noise levels, but it is recommended that noise limits of 500 mV peak to peak be considered, to cater for the very rare appearances of short duration noise spikes.

The way in which the battery is mounted in the equipment can have a marked effect on its performance. If discharged in free air, the case temperatures of some designs can reach at least 300°C before the end of useful life, as there will be little thermal insulation between the cell stacks and the cases. Clamping such designs to a heat sink will remove heat at a rate that will reduce the useful duration. If a high case temperature is a problem, then a thin layer of high-performance external insulation or a slightly larger battery, accommodating more material internally, should be considered.

The mounting of batteries is best done through studs on the battery lid or by clamps around the case. Brackets or other attachments to the case are feasible, but add to the complexity of the design.

Usually, soldering is used to connect to the terminals of small batteries. Spot-welded or wrapped joints are feasible. For high-current or long-duration applications crimped joints can be used. Whatever technique is used, care must be taken to avoid damaging the terminal seals and to leave no residual stress upon them. The use of multi-pin plug and socket connectors should be treated cautiously because of the very high degree of hermeticity required throughout the storage life of the battery.

As with any 'one-shot' device, it is not possible to carry out a functioning test on a thermal battery. During manufacture, non-destructive tests for sealing of the case, for correct output polarities, and for insulation and igniter resistances, are done on every battery. The position of components can be checked by radiography, and samples from each production batch are discharged. Such checks, together with rigorous quality control procedures, ensure the very high reliability of thermal batteries.

Once installed, few checks are possible. The methods used during manufacture to ensure that the output is of the right polarity are not applicable to installed batteries. To detect if a battery has been fired, the resistance of electric igniter circuits can be measured, but because of the harsh thermal environment in a battery, some designs of igniter cannot be guaranteed to go open-circuit after firing. Special electrical indicators, which reliably change their state, are available for remote indication of a fired battery. Various types of temperature-sensitive paints and labels have been used to provide a visual indication, but these must be evaluated to ensure that they do not revert to their original colour or state during exposure to subsequent ambient temperature and humidity extremes or to mechanical pressure or scratching.

### 5.9.8 Alternatives to thermal batteries

The thermal battery is so well-suited to its applications that, in the smaller sizes, it is difficult to postulate an alternative. Thermals are reliable, with 99.97% reliability at a 90% confidence level having been declared [139]. They have a 'zero-cost-of-ownership' because they do not need replacement during the life of a weapon. They present few hazards, being regarded in the UK as non-explosive items when stored and transported.

Larger batteries, up to several kW for tens of minutes (an upper limit determined more by mechanical constraints of pellet manufacture than any fundamental design limitation) have supplanted the equivalent reserve type silver oxide–zinc battery. Large thermals do face competition from reserve type lithium–oxyhalide systems which can be both lighter and smaller. For example, a 2 kW, 20 min Li–$SOCl_2$ reserve battery would give 250 Wh $kg^{-1}$, 340 Wh $cm^{-3}$, showing it to be much lighter and about twice the volumetric energy density of a comparable thermal. Against this must be put the mechanical complexity and cost of an electrolyte transfer system compared to the thermal concept, and the design-for-safety implications of using this system. Nevertheless, development of lithium reserve batteries is active in Europe and the USA.

### 5.9.9 Major suppliers of thermal batteries

Eagle Picher Industries, Couples Division, Joplin, Missouri, USA. MSA (Britain) Ltd, East Shawhead, Coatbridge, Scotland. Power Conversion Inc., 280 Midland Avenue, Saddle Brook, New Jersey, USA. SAFT America Inc., 107 Beaver Court, Cockeysville, Maryland, USA. SAFT, 156 Avenue de Metz, Romainville, France. SAFT (UK) Ltd, Station Road, Hampton, Middx., England. Leclanché S.A. 48, Avenue de Grandson, Yverdon, Switzerland.

### 5.9.10 The future

Thermal batteries will remain as first choice for several defence applications. Other uses will be found for them, as larger sizes become available. However, despite a good volumetric energy and power density, any reduction in size and weight will be an advantage.

No 'quantum leap' in performance improvement can be expected, such as occurred when calcium was replaced by lithium, unless systems based upon chlorine as a cathode material can be evolved. Otherwise, size and weight reductions in the order of 10–20% may result from design improvements and the uses of new cathode or pyrotechnical chemistries.

## 5.10 LITHIUM–MOLYBDENUM DISULPHIDE RECHARGEABLE BATTERIES
(F. C. Laman)

### 5.10.1 Introduction

Research on rechargeable lithium batteries has been done for the last 15 years. A major boost in the effort of producing practical rechargeable lithium batteries came from the studies on layered dichalcogenides [140]. In these materials $Li^+$ ions can be reversibly intercalated, and they can therefore serve as rechargeable cathodes in lithium batteries. The development of molybdenum sulphide as a cathode material for a rechargeable lithium battery was greatly assisted by the discovery in 1977 at the University of British Columbia of a new crystalline phase of $MoS_2$ [141]. This crystalline phase is stabilized by the presence of a small quantity of lithium within the crystal lattice, and differs from the naturally occurring $MoS_2$ in that the atomic coordination of the molybdenum atoms has octahedral symmetry (1T) rather than trigonal prismatic (2H) [142]. This change in coordination shifts the equilibrium potential of the material so that it becomes considerably more electropositive for a wide range of lithium concentrations in $Li_xMoS_2$.

A reversible lithium electrode technology has also been developed. The problem associated with the replating of lithium on a passivated lithium surface has been overcome to the extent that the number of turnovers of the lithium in the anode that can be achieved is greater than 200. One important advantage of molybdenum disulphide over other intercalation compounds is that it is abundantly available in nature and can be used as cathode active material with a minimum amount of processing. Production in the manufacturing plant in early 1989 consisted of 'AA' size cells of two different types referred to as A and B at a rate of 2500 cells/day. The B type cell was introduced in order to deliver more capacity than the standard A type cell within the same voltage range, whilst sacrificing part of the number of charge/discharge cycles achievable. The A and B type cells have slightly different cathodes. These differences are with respect to the density of the $MoS_2$ matrix deposited onto the cathode current collector and the crystal structure of the $MoS_2$ particles.

### 5.10.2 Cell chemistry and operation

The main reaction in the Li–$MoS_2$ cell is the electrochemical oxidation of lithium metal and reduction of molybdenum disulphide via the intercalation reaction:

$$xLi + MoS_2 \longleftrightarrow Li_xMoS_2 \;. \tag{2.23}$$

In this reaction, lithium ions are generated at the anode during discharge and then transported through the liquid electrolyte and inserted into the $MoS_2$ crystals. Intercalation refers to the process of insertions of $Li^+$ ions into the $MoS_2$ crystals without substantially changing the molecular structure. The lack of structural change in the molybdenum disulphide ensures a high degree of reversibility of this process. Electrons are transported through the external circuit and are recombined with lithium ions in the $MoS_2$ host. The open circuit voltage of the $Li-MoS_2$ cell arises from the difference in equilibrium potential between lithium atoms in the metal and in the $MoS_2$ host. The equilibrium potential of lithium in $Li_xMoS_2$ decreases with an increase in the lithium concentration (see section 2.1.2). Therefore, the open circuit voltage decreases as the cell is being discharged. The voltage of the cell varies continuously from about 2.4 V for a value of $x$, in equation (2.23), of near zero, tó about 1.0 V for $x$ near 1. The active cathode material of $MoS_2$ with 1T structure, referred to as β-phase, has a tendency to convert to the naturally occurring form of $MoS_2$ with 2H structure, referred to as α-phase, when the cell is operated at elevated temperatures near its maximum voltage of 2.4 V for some time. This problem is further aggravated when the ratio of discharge to charge rate is high [143]. The α-phase of $MoS_2$ can only reversibly intercalate an amount of lithium corresponding to an $x$-value of about 0.15, see Fig. 5.115. Therefore, the formation of this phase results in a loss in cathode capacity. Phase conversion occurs only when the cell is repeatedly cycled at high voltage, and it can be completely avoided by charging the cell to only 2.0 V at any temperature or to 2.2 V at temperatures of 21°C and lower. The phase conversion reaction can be reversed, i.e. the lost cathode capacity can be restored, by slowly discharging the cell to a voltage of 1.3 V or less. Fig. 5.115 also shows the voltage characteristic of the reaction α-$MoS_2$→ β-$MoS_2$.

Degradation of active cell components leading to a loss in performance is linked to a variety of processes. The two main processes are change in the $MoS_2$ crystal structure, which dominates capacity losses in the early stages of cell life, and reaction between lithium and electrolyte, both at the anode and the cathode, which dominates capacity losses at the final stages of cell life [143]. The electrolyte used in the $Li-MoS_2$ cell consists of a one molar solution of lithium hexafluoroarsenate in a 50/50 mixture by volume per cent of propylene carbonate and ethylene carbonate. Both the solvent mixture and the electrolyte salt decompose as a result of the reaction with lithium. Salt decomposition is mostly responsible for the passivating film on the lithium anode [144], while solvent decomposition produces a gradual pressure build-up inside the cell from gaseous products (propene and ethene) and accumulation of solid products (lithium carbonate) [145] in the porous cell components.

Cell failure, resulting from these processes, occurs only in two modes, either through a high impedance or a low impedance. No cell failures due to an open circuit have been found when cells are used without abusive conditions. The high impedance failure is associated with consumption of electrolyte and accumulation of solid breakdown products, and formation of partially passivated lithium deposits on the anode. All mechanisms lead to a gradual increase in cell impedance. A model using percolation theory proposed by Dahn *et al.* [146], as well as loss of electrochemically active lithium, can explain an accelerated loss in capacity towards the end of

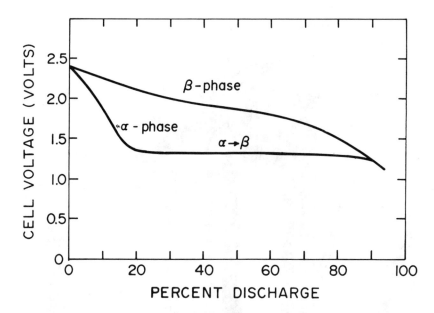

Fig. 5.115 — Discharge characteristics of α-phase and β-phase $MoS_2$, including the phase conversion voltage.

life. The low impedance failure, caused mainly by a deterioration of the current uniformity (producing areas of high current density on charging), leads initially to an intermittent or soft short. This soft short causes a substantial decrease in the charge efficiency, which is normally close to 100%. Eventually, a permanent or hard short will develop, and the cell fails to charge. The frequency and mode of failure is strongly affected by operating conditions [143].

### 5.10.3 Construction

Owing to the consumption of lithium during cell operation, Li–$MoS_2$ cells are constructed in a cathode limited configuration, using a 2.5-fold excess of lithium. The lithium metal anode, consisting of a strip of lithium foil (Fig. 5.116) without current collector, is placed between two $MoS_2$ cathode strips, consisting of aluminium foil current collectors with a mixture of molybdenum sulphide and a polymeric binder deposited on it. To ensure good electronic conductivity of the cathode active material, the molybdenum disulphide particles, which have semiconducting properties in the unintercalated state, are coated with a thin layer of molybdenum dioxide, which is metallic. The complete electrode stack is spirally wound around a central mandrel [147] and placed in a nickel-plated mild steel can. The anode or negative electrode is connected to the can, while the cathode or positive electrode is connected to the centre pin. The centre pin is insulated from the rest of the can via a glass-to-metal seal. The cell is hermetically sealed by laser welding the cell top (containing the centre pin) to the can. Electrolyte is introduced through a filling hole

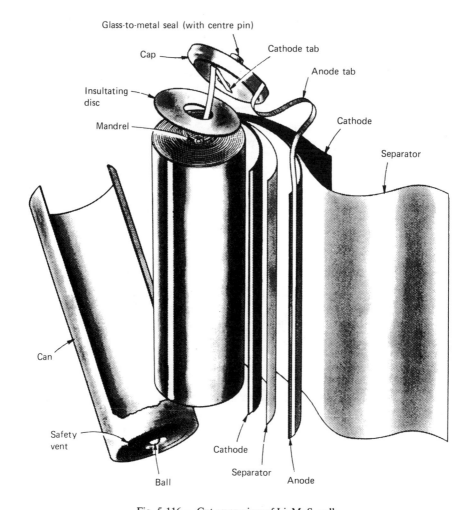

Fig. 5.116 — Cut-away view of Li–MoS$_2$ cell.

in the cell top, using a pressure wetting technique. This hole is then closed with a ball weld. A coined vent is provided in the bottom of the cell to release pressure in a controlled manner, should the cell be subject to extreme abuse.

### 5.10.4 Electrical characteristics

#### 5.10.4.1 Cell voltage

The open circuit voltage of Li–MoS$_2$ cells varies substantially as the battery is discharged. By contrast, the difference between the cell voltage under load and on open circuit is small and does not depend substantially on the state of charge of the cell. These properties provide MOLICEL™ batteries with a state of charge indication capability with the battery on open circuit, both during discharge and

Sec. 5.10]     **Lithium–molybdenum disulphide rechargeable batteries**     427

during charge. However, the voltage variation also means that batteries containing Li–MoS$_2$ cells may not be a direct replacement for other types of battery for use in existing portable devices. Voltage regulation such as that provided by a D.C./D.C. converter is often required to condition the battery voltage for use with many devices.

The change in cell voltage with state-of-charge (i.e. voltage profile) is slightly different for A and B type cells. This is a consequence of the difference in crystal structure of the MoS$_2$ used as cathode material. As is shown in Fig. 5.117, the voltage profile of the B type cell is nearly linear, while the A type cell has a steeper voltage decrease near the beginning and to a lesser degree near the end of discharge. This

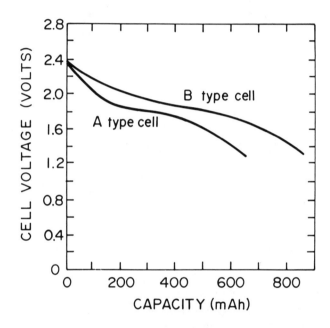

Fig. 5.117 — Discharge characteristics of A and B type Li–MoS$_2$ cells.

difference in discharge profile between A type cell (S-shape) and B type cell (linear) means that the amount of extra capacity deliverable by the B type cell depends very much on the discharge voltage range. The B type cell delivers most of its extra capacity at high voltage and some at low voltage. The difference in deliverable capacity as a function of voltage range for cells discharged at a rate which corresponds to a three-hour discharge at 100% depth, is shown in Fig. 5.118. As was mentioned in the previous section, Li–MoS$_2$ cells suffer loss in performance when cycled continuously at high voltages. The crystal structure of MoS$_2$ in the B type cell is more stable at high voltage than the pure β-phase MoS$_2$ (active material in the A type cell) and the B type cell is therefore less susceptible to this problem.

The Li–MoS$_2$ cell should be used with both charge and discharge control circuitry. The termination voltages on charge and discharge can be set by the

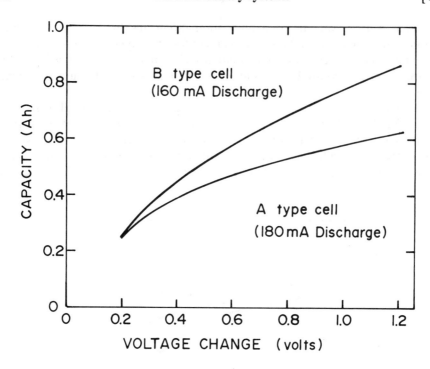

Fig. 5.118 — Deliverable capacity for A and B type cells as a function of voltage change. The voltage change is centred around a cell voltage of 1.8 V.

designer of the power pack for a specific application, but should not exceed 2.4 V per cell for charge and 1.1 V per cell for discharge. The recommended voltage range depends on discharge rate. For rates of 2 h or higher, a voltage range of 2.2–1.1 V/cell is recommended, and for rates of 3 hours and lower, a voltage range of 2.4–1.3 V/cell is advised. To optimize the combination of cycle life and deliverable capacity under a wide variety of discharge conditions, 2.2 V is recommended as the charge termination voltage, and 1.3 V is recommended as the discharge termination voltage. Li–$MoS_2$ cells can be operated even outside the wider voltage limits of 2.4 V and 1.1 V, but performance will degrade much more rapidly, as will the safety characteristics under abusive conditions.

### 5.10.4.2 Internal resistance

Rechargeable lithium cells use a non-aqueous electrolyte. Non-aqueous electrolytes have a poorer conductivity than aqueous electrolytes and therefore contribute more to the internal resistance of the cell. The cell configuration with thin spirally wound electrodes is used to minimize losses associated with the electrolyte resistance. The loss in power capability resulting from the reduced rate capability of lithium cells relative to cells with an aqueous electrolyte, is partly compensated by the higher cell voltage typical for most lithium cells. The internal resistance of a Li–$MoS_2$ cell varies on discharge and shows a minimum at the mid state of charge. The fact that the internal resistance does not continuously increase on discharge, indicates that the change is not caused solely by cell polarisation. Instead, a large part of it is attributable to the properties of the $MoS_2$. The variation of cell impedance with cell voltage, measured from a 4 s pulse for an A type cell, is shown in Fig. 5.119. The internal resistance of the cell can increase dramatically with decreasing cell temperature. Besides temperature and state of charge, the impedance of the Li–$MoS_2$ cell also depends on prior cycling history. Typical values of the cell impedance for the Li–$MoS_2$ cell, both A and B type, at room temperature are in the range of 100–200 m$\Omega$.

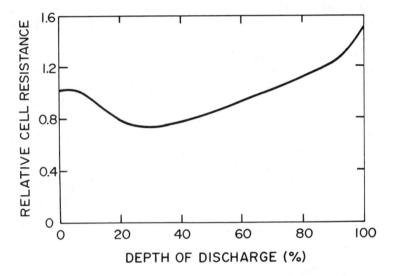

Fig. 5.119 — Relative cell resistance as a function of state of charge.

### 5.10.4.3 Energy efficiency
The 100% charge efficiency and excellent charge retention combined with the low polarization characteristics give the Li–$MoS_2$ cells a 'round trip' energy efficiency in many circumstances of 90% or greater. Round trip efficiency is defined as the ratio of deliverable energy from a fully charged battery to the energy required to charge the battery from a fully discharged state to a fully charged state.

### 5.10.4.4 Charge requirement
The recommended charge procedure for Li–$MoS_2$ cells is charging at a constant current with a ten-hour rate to the termination voltage. The Li–$MoS_2$ cells can also be charged with a constant voltage source; in this case the maximum charge current should not exceed the ten hour rate and the termination voltage should not exceed 2.2 V for an A type cell and 2.3 V for a B type cell. For float charging over extended periods, the maximum charge voltage should be further reduced to 2.0 V. Charge rates should also be adapted to the ambient temperature. Charging of Li–$MoS_2$ cells below $-10°C$ is not recommended, while for charging below 0°C the charge rate should be reduced to 20 hours.

### 5.10.5 Performance characteristics

#### 5.10.5.1 Deliverable capacity

The nominal capacity of the 'AA' size Li–MoS$_2$ cell is 600 mAh for the A type and 800 mAh for the B type. The deliverable capacity depends on the charge and discharge voltage cut-offs used. The sloping voltage characteristic will give an increase in deliverable capacity when operated outside the suggested voltage range. However, extension of voltage range results in penalties in other areas of performance, such as cycle life and safe operation. Alternatively Li–MoS$_2$ cells will deliver less capacity when operated in a narrower voltage range and they will gain in cycle life. A maximum in the combination of deliverable capacity and cycle life is achieved when Li–MoS$_2$ cells are cycled around the mid state of charge or a cell voltage of 1.8 V.

The deliverable capacity between fixed voltage limits also depends on the discharge rate. Because of the internal resistance of the cell, so-called $iR$ losses increase as the discharge current increases and the deliverable capacity between fixed voltage limits decreases. The degree of loss in deliverable capacity with increasing discharge rate is known as the rate capability characteristic of the cell. The lower the losses are, the higher the rate capability of the cell. The internal resistance of the cell increases dramatically with decreasing cell temperature, and therefore losses in deliverable capacity with increasing discharge rate become more severe as the cell temperature is lowered. Discharge curves of A and B type Li–MoS$_2$ cells discharged at different currents and temperatures are shown in Figs 5.120 to 5.123.

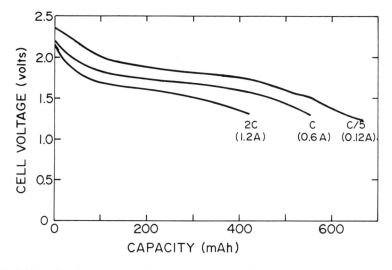

Fig. 5.120 — Discharge curves of A type Li–MoS$_2$ cells discharged from 2.4 V to 1.3 V, with constant currents of 1200 mA, 600 mA, and 120 mA at 21°C.

When cells are in poor thermal contact with the environment, which is the case in enclosed multi-cell battery packs, the performance improves significantly. This is a result of heat produced by resistive heating of cell components during discharge. This heat remains trapped inside the battery and raises its internal temperature.

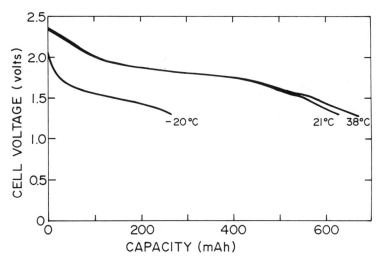

Fig. 5.121 — Discharge curves of A type Li–MoS$_2$ cells discharged from 2.4 V to 1.3 V, at a constant current of 120 mA and ambient temperatures of −20°C, 21°C and 38°C.

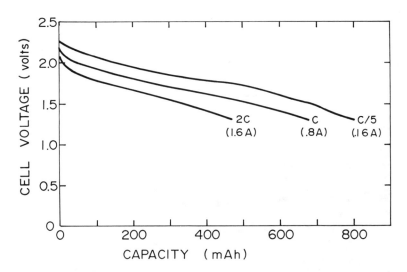

Fig. 5.122 — Discharge curves of B type Li–MoS$_2$ cells discharged from 2.4 V to 1.3 V, at constant currents of 1600 mA, 800 mA and 160 mA, at 21°C.

An example of a discharge showing the effect of self-heating of a single cell is given in Fig. 5.124. Improvements in performance will strongly depend on the thermal environment of the individual cells, therefore general statements about low temperature performance of Li–MoS$_2$ cells can be misleading. However, tests have shown

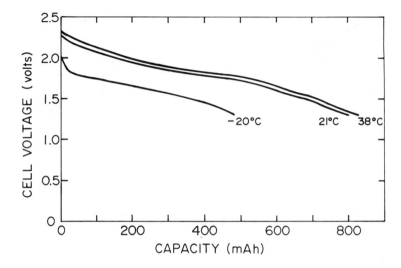

Fig. 5.123 — Discharge curves of B type Li–MoS$_2$ cells discharged from 2.4 V to 1.3 V, at a constant current of 160 mA, and ambient temperatures of −20°C, 21°C, and 38°C.

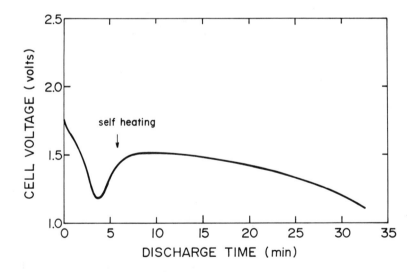

Fig. 5.124 — Discharge curve of a B type Li–MoS$_2$ cell, discharged at a constant current of 700 mA and an ambient temperature of −18°C, showing a self-heating effect after three minutes of discharge.

## Sec. 5.10]  Lithium–molybdenum disulphide rechargeable batteries

that fully engineered Li–MoS$_2$ battery packs can operate a cellular phone at temperatures down to $-20°C$, depending on the configuration of the pack.

Often, the battery load will draw constant power rather than constant current from a battery. Discharge curves of A and B type cells discharged at different constant power levels are shown in Figs 5.125 and 5.126. Since the voltage profile of a

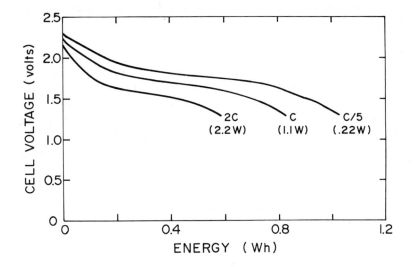

Fig. 5.125 — Discharge curves of A type Li–MoS$_2$ cells, discharged from 2.4 V to 1.3 V, at constant powers of 2.2 W, 1.1 W, and 0.22 W, at 21°C.

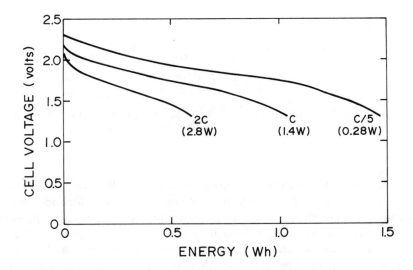

Fig. 5.126 — Discharge curves of B-type Li–MoS$_2$ cells, discharged from 2.4 to 1.3 V, at constant powers of 2.8 W, 1.4 W and 0.28 W, at 21°C.

battery is an important characteristic, it is of interest to note that the voltage profile does not change significantly by changing the discharge mode from constant current to constant power. The magnitude of the discharge power or current is often referred to as the discharge rate and is expressed in number of hours required to completely discharge the battery. For example, a discharge rate of C/2 means a discharge time of two hours, while a 2C rate means a discharge time of half an hour. Values of discharge rates presented here will always be associated with a discharge of the full nominal capacity.

### 5.10.5.2 *Cycle life*

Under most conditions, several hundred full charge/discharge cycles can be achieved with Li–MoS$_2$ cells. The exact number of cycles strongly depends on discharge conditions, such as depth of discharge, voltage range, discharge and charge rate, and operating temperature. Examples of cycle life curves for A and B type cells, repeatedly charged at 60 mA to 2.4 V and discharged at 200 mA to 1.2 V at 21°C, are given in Fig. 5.127. The effect of depth of discharge and voltage range has been

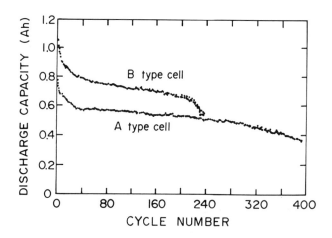

Fig. 5.127 — Cycle life curves of A and B type Li–MoS$_2$ cells, charged to 2.4 V at 60 mA and discharged to 1.2 V at 200 mA, temperature 21°C.

discussed earlier in the section on deliverable capacity. Provided that cells are discharged and charged around their mid state of charge, cycle life increases with decrease in depth of discharge. Cycle life in these circumstances increases more than one would expect on the basis of the deliverable capacity per cycle, i.e. the total delivered capacity accumulated over the life of the cell increases as the depth of discharge decreases [148]. The relationship between depth of discharge and cycle life for A and B type cells when cycled around the mid state-of-charge with discharge currents corresponding to a three-hour rate to 100% depth, is given in Fig. 5.128.

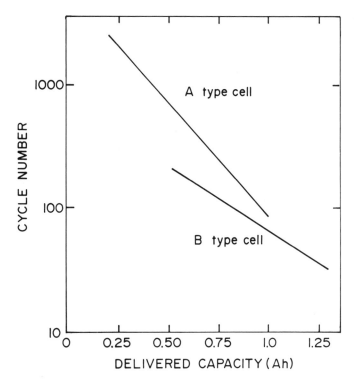

Fig. 5.128 — Cycle life as a function of delivered capacity for A and B type Li/MoS$_2$ cells, cycled at 21°C with a three-hour discharge rate and a ten-hour charge rate, around a cell voltage of 1.8 V.

With respect to voltage range, the following was noted. Owing to the properties of β-phase MoS$_2$, the cells should not be repeatedly discharged and charged at high voltages, otherwise a high rate of capacity fade will occur, resulting in a loss in cycle life. As discussed earlier, the B type cells are more stable at high voltages than the A type cells. This is illustrated in Fig. 5.129, showing cycling data for A and B type cells discharged between 2.4 and 1.5 V. For cycling with very frequent shallow discharges, an upper cut-off voltage of 2.2 V for A type cells and 2.3 V for B type cells is recommended.

The discharge current affects the rates of the various cell degradation mechanisms, hence also cycle life [143]. Cycling results with a sustained constant current discharge show that a maximum in cycle life is obtained with intermediate discharge rates in the order of two to three hours, see Fig. 5.130. In addition to the discharge rate, the duty cycle itself also affects cycle life. In general, pulsed discharges prolong the life over sustained discharges. Cycle life also depends on operating temperature. Owing to the increase in the rate of the degradation processes in general, operating the cell at high temperature reduces cycle life. For an A type cell charged and discharged at 38°C, at intermediate rates, cycle life is reduced by about 50%. The

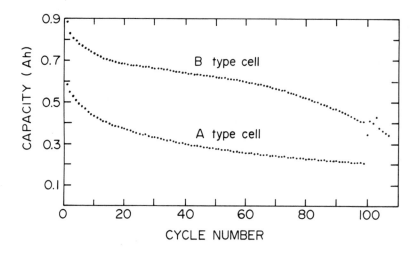

Fig. 5.129 — Capacity fade curves for A and B type Li–MoS$_2$ cells, cycled between 2.4 V and 1.5 V, at 21°C. The A type cell is discharged at 180 mA and charged at 60 mA, and the B type cell is discharged at 160 mA and charged at 80 mA.

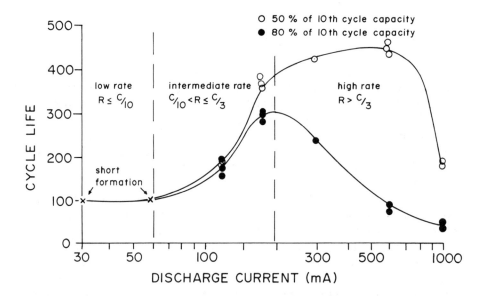

Fig. 5.130 — Cycle life as a function of discharge current of A type Li–MoS$_2$ cells, cycled in the voltage range 2.4 V to 1.1 V, with a charge current of 60 mA, temperature 21°C.

Sec. 5.10]    **Lithium–molybdenum disulphide rechargeable batteries**    437

effect of temperature on charging is mostly observable at low temperature. Decreasing the temperature on charging affects the morphology of the deposited lithium adversely, giving loss in cycle life from early short failures. This problem occurs only when cells are charged at temperatures below 0°C.

### 5.10.5.3  Charge retention
All lithium batteries are characterized by a good charge retention, which results from the passivation of the lithium electrode in contact with the liquid electrolyte. For some lithium primary batteries, this passivation leads to the negative characteristic of voltage delay (see section 2.2.5.2). The Li–$MoS_2$ cell shows a good intermediate behaviour with no voltage delay and a charge retention capability which is significantly better than that for rechargeable cells based on an aqueous electrolyte, but not as good as some lithium primary batteries. When Li–$MoS_2$ cells are stored at high voltage, some of the loss in capacity is associated with a change in the $MoS_2$ crystal structure to a more stable form, and only the remaining loss is the result of a true self-discharge reaction. Cells stored at a voltage of 1,8–1.9 V, where the cathode is stable, do give a measure of true self-discharge. Based on the change in voltage for cells stored at 1.85 V, a self-discharge rate of 5% per year has been found [149]. The rate of self-discharge increases as the temperature increases. When Li–$MoS_2$ cells are cycled before storage, capacity losses associated with a change in the cathode structure are minimal. This is due to the stabilization of the cathode resulting from cycling. In this case, cathode stabilization is also accompanied by some capacity loss. The loss of cell capacity of fully charged non-stabilized (one pre-cycle) and stabilized (10 pre-cycles) A type cells as a function of storage time at 21°C and 38°C is shown in Fig. 5.131. The data show a loss of only 8% in capacity of stabilized cells when stored for one year at room temperature and at 2.4 V. Furthermore, the data show that the charge retention capability of stabilized cells improves more than two-fold when compared to non-stabilized cells. The charge retention capability for A type and B type cells is very similar.

### 5.10.6  Safety
Li–$MoS_2$ cells have many safety design features which result in tolerance of a wide range of abuse without creating a hazard. Even under extreme abuse, such as the cell being punctured, sawn, crushed, or incinerated, the cells will not explode. A safety vent in the bottom of the cells will reliably open to release internal pressure. The cells are normally hermetically sealed. This seal is permanently lost after vent activation, and the cell should not be further used. Li–$MoS_s$ cells use a solid cathode of molybdenum disulphide which is a very innocuous material. The liquid electrolyte has a high flash point of 135°C and is immobilized by absorption in the separator and porous electrode materials. The electrolyte is not dangerous on skin contact although it does cause surface corrosion of mild steel. Vent activation produces ethene and propene gas; about 12 ml of gas are released on venting of an 'AA' size cell.

#### 5.10.6.1  Electrical abuse test results
The response of Li–$MoS_2$ cells to electrical abuse is very similar for A and B type cells and can be summarized as follows. On external short circuit of the 'A' size cells, peak

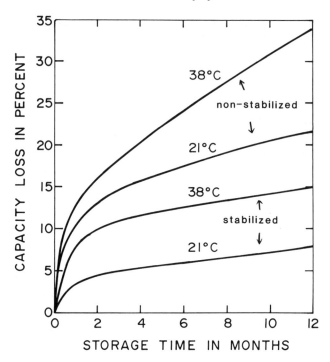

Fig. 5.131 — Capacity loss as a function of storage time of A type Li–MoS$_2$ (stabilized and non-stabilized) cells, after charging to 2.4 V, and at storage temperatures of 21°C and 38°C.

currents can reach 20 A, with an average current of about 10 A. An automatic thermal shut-down mechanism due to fusion of the polypropylene separator will occur if the cell is shorted. The internal cell resistance will rise and the current will fall. A sustained short circuit will always render the cell useless, but momentary short circuits can be tolerated without a performance penalty.

Li–MoS$_2$ cells can be overcharged indefinitely at a one hour rate at room temperature. Sustained higher charge currents can cause cell overheating. After 2.5 h at a 1 A charge current, the cell will have a high probability of burning vigorously. Extended overcharge above 2.6 V will render the cell useless. Li–MoS$_2$ cells are tolerant to overdischarge to a large extent, since there is a considerable amount of capacity available below their normal end-of-discharge voltage of 1.1 V. Most of this capacity occurs at about 0.5 V, and is therefore of limited value. The cell can still be recharged if the cell has not reached zero volts. When the cell is driven below zero volts either deliberately with a power supply or inadvertently by discharging a very imbalanced series string of cells and the current exceeds 3 A, there is a finite probability that the cell will overheat, and it could burn vigorously. When cells are combined in multi-cell batteries, two types of protective device can be used to guard against abuse: electrical fuses and thermal fuses.

### 5.10.7 Shipping and handling

The US Department of Transportation (DOT) has awarded Moli Energy Limited several exemptions regarding the shipment of Li–MoS$_2$ cells and batteries. The most

### Sec. 5.10] Lithium–molybdenum disulphide rechargeable batteries

important ones are: Li–MoS$_2$ cells can be shipped in sizes containing up to 12 g of lithium, provided that there is compliance with the mechanical and thermal abuse tests and packaging and labelling requirements. All modes of transportation are allowed, except passenger aircraft. For passenger aircraft, batteries containing up to eight 'AA' size Li–MoS$_2$ cells are allowed in carry-on or checked baggage, with three of such batteries per passenger.

#### 5.10.8 Reproducibility and reliability

##### 5.10.8.1 Single cells

In 1986 and 1987 Moli Energy Limited produced, in its pilot plant facilities, approximately 30 000 'AA' size cells, mostly of the A type. In the period from August 1987 to December 1988, about 650 000 'AA' size cells were produced in the manufacturing plant, with an approximately equal amount of A and B type cells.

To assess the reproducibility and reliability of the cells, random samples have been subjected to charge/discharge cycle tests and charge retention tests. During the evaluation period 200 000 cell cycles were performed on pilot plant cells and 300 000 cell cycles on manufacturing plant cells. The distribution in deliverable capacities determined at various stages of cycling for the different types of cell produced in the manufacturing plant are given in the histograms in Figs 5.132 and 5.133. A histogram of the self-discharge rates measured for pilot plant cells is shown in Fig. 5.134. While

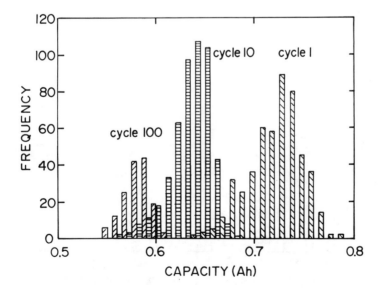

Fig. 5.132 — The distribution of discharge capacities of A type Li–MoS$_2$ cells at cycle numbers 1, 10, and 100. Capacities are determined from a discharge at 120 mA between 2.4 V and 1.3 V.

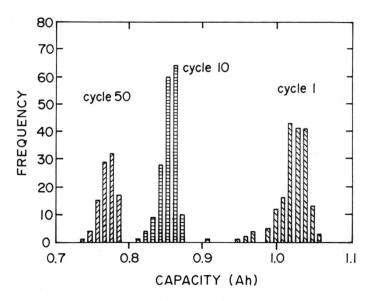

Fig. 5.133 — The distribution of discharge capacities of B type Li–MoS$_2$ cells at cycle numbers 1, 10, and 50. Capacities are determined from a discharge at 160 mA between 2.4 V and 1.1 V.

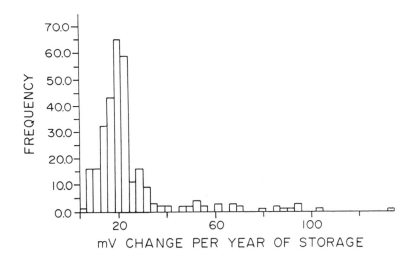

Fig. 5.134 — The distribution of voltage losses associated with self-discharge of A type Li–MoS$_2$ cells, stored at 1.85 V and 21°C.

the distribution of the discharge capacities is close to normal, the distribution of the self-discharge rates deviates from normal. The latter is caused by a low frequency of faulty cells, associated with damaged separators.

The width of the distribution of discharge capacities of fresh cells, measured at one standard deviation and as a percentage of the average discharge capacity, varies from 2 to 4%. Upon cycling the width is further reduced to about 1.5%. The latter is caused by changes occurring in active cell components as a result of repeated charge/discharge cycles. This narrow distribution observed for fresh cells is a direct consequence of the tightness of the control of the cell manufacturing parameters. Fig. 5.135 shows the frequency of cell failure as a function of cycle number for a group of cells repeatedly charged at 60 mA to 2.4 V and discharged at 180 mA to 1.3 V. The data show no early failure.

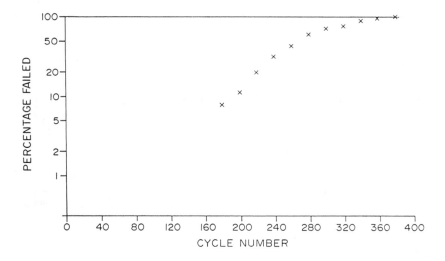

Fig. 5.135 — Frequency of failures as a function of cycle number, for A type Li–MoS$_2$ cells, cycled between 2.4 V and 1.3 V, with a discharge current of 180 mA and a charge current of 60 mA, at 21°C.

### 5.10.8.2 *Multi-cell batteries*

In general, it can be expected that multi-cell batteries do not perform so well as predicted on the basis of the single cell performance specifications. This is because in a multi-cell battery, charge and discharge conditions are controlled for the complete battery and not for individual cells. This lack of stringent control will cause single cells to be charged and discharged outside the range of recommended charge current and voltage. The degree to which cells will be cycled outside the accepted range of conditions, depends on the cell imbalances existing within the battery. It can therefore be minimized by matching of capacity and equilibration of charge of individual cells before battery assembly. This procedure, however, is labour-intensive and adds significant cost to the battery manufacturing process. Overcharge protection for individual cells reduces the development of imbalances between cells assembled within one battery, during operation of the battery. These imbalances will build up as a result of multiple charge and discharge cycles of the battery. Data show that multi-cell batteries, assembled with Li–MoS$_2$ cells, exhibit a performance which is close to what could be expected, based on the properties of single cells, without preselection or charge equilibration of individual cells, see Table 5.21. This excellent behaviour, which is achieved without any overcharge protection of individual cells, is attributable to the following characteristics of Li–MoS$_2$ cells [149].

(a) As was shown in the previous section, the distribution of discharge capacities of uncycled cells is quite narrow and is further reduced by repeated charge/discharge cycling.
(b) The sloping voltage characteristics of Li–MoS$_2$ cells, combined with reserve capacity beyond the recommended charge and discharge voltage cut-off points, improves reliability of the battery in terms of deliverable capacity as well as cycle life.
(c) The near 100% charge efficiency of cells means that all cells in the battery are charged by the same amount, during recharge of the battery in each cycle.

**Table 5.21** — Comparison of capacity in mAh per cell and cycle life to 50% of the 10th cycle capacity for single cells and 4-cell series-connected batteries. Cycling conditions: voltage range per cell 2.4–1.3 V; charge current per cell 60 mA; discharge current per cell 120 mA

| Configuration | Sample size | 10th cycle capacity | | | Cycle life | | |
|---|---|---|---|---|---|---|---|
| | | Mean | Width | Range | Mean | Width | Range |
| Single cell | 19 | 626 | 20.9 | 587–647 | 233 | 18.5 | 187–263 |
| 4-cell battery | 20 | 626.8 | 7.5 | 639–610 | 219 | 11.8 | 205–253 |

Comparison of capacity (in mAh) per cell and cycle life to 80% of the 10th cycle capacity for single cells and 8-cell batteries. 8-cell batteries consist of a parallel/series combination with 4 cells in series. Cycling conditions: voltage range per cell 2.4 to 1.1 V; charge current per cell 60 mA; discharge current per cell 180 mA

| Configuration | Sample size | 10 cycle capacity | | Cycle life | |
|---|---|---|---|---|---|
| | | Mean | Range | Mean | Range |
| Single cell | 4 | 673 | 652–683 | 309 | 265–336 |
| 8-cell battery | 4 | 663 | 646–690 | 293 | 255–350 |

The effect of the sloping voltage and reserve capacity beyond the voltage cut-off points needs further clarification. This characteristic of the Li–MoS$_2$ cell creates an

Sec. 5.10]     **Lithium–molybdenum disulphide rechargeable batteries**     443

averaging effect, such that the discharge capacity of the battery is dictated by the average capacity and not the lowest capacity of the single cells. In addition, this averaging reduces the width of the distribution of discharge capacities of multi-cell batteries by a factor of $Vn$, in which $n$ is the number of series-connected cells in the battery. It furthermore allows cells to be cycled outside the recommended voltage range with minimum penalty in performance and provides a regulating mechanism for the branch currents in parallel-connected strings in the batteries. Both features will result in a cycle life of the battery which is very similar to that of single cells. Failure of a single cell in a multi-cell battery leads to a gradual reduction in battery performance and not to a catastrophic failure of the battery. Again, this can be attributed to the characteristics of the $Li-MoS_2$ cell described in this section [150].

### 5.10.9 Future developments
Work to further improve the lithium rechargeable battery manufactured at Moli Energy Limited focuses on the areas of energy density, rate capability, and safety. Since increasing the open circuit voltage of the battery can lead to a significantly higher gain in energy density than increasing the deliverable capacity of the battery, further work will examine cathode materials other than $MoS_2$. To improve the discharge rate capability, especially at low temperatures, new electrolytes will be evaluated. In addition, the search for a non-toxic but safe electrolyte salt will continue. The work on chemical shuttles to provide an intrinsic overcharge protection mechanism and further improve battery performance relative to single cell performance, will also continue. Work on the anode will mainly be concerned with the replacement of the pure lithium metal anode with a lithium-containing alloy, or low voltage lithium intercalation compound. This will benefit both the cycle life and the safety characteristics of the cell.

### 5.10.10  Summary
Moli Energy Limited, at present, is the only company producing 'AA' size rechargeable lithium cells for a mass market. The technology is based on a molybdenum disulphide cathode. This material was primarily chosen because it is abundant and cheap. The cells are cathode limited and constructed in a spirally wound format with thin electrodes and electrolyte immobilized in a microporous polypropylene separator. For safety purposes, cells are provided with a coined pressure vent in the bottom of the can. Cells of two types, A and B, are being produced in a manufacturing plant. The 1989 rate of production was 50 000 cells per month. The two cell types differ in deliverable capacity, with a nominal capacity of 600 mAh for the A type and 800 mAh for the B type cell and in cycle life, which in most cases is higher for the A type cells. The relative performance of A and B type cells depends strongly on the voltage range and discharge currents used.

Most of the small battery market for portable devices is at present served with the Ni–Cd technology. Batteries made with $Li-MoS_2$ cells have several advantages over small Ni–Cd batteries. The average voltage per cell is about 40 to 50% higher, depending on load conditions. The energy density can be up to 100% higher than conventional, and up to 30% higher than advanced Ni–Cd batteries. The charge retention corresponding to less than 10% loss per year at room temperature is at least one order of magnitude better than that of Ni–Cd cells. In contrast to some primary lithium technologies, the good charge retention capability is not accompanied by voltage-delay effects. The change in open circuit voltage upon discharge gives a state-of-charge indication which is not provided by the constant voltage output of Ni–Cd cells.

The excellent charge retention, combined with state of charge indication, makes Li–MoS$_2$ batteries very suitable for infrequent and irregular use, conditions where Ni–Cd batteries are much less attractive. Li–MoS$_2$ cells show optimum performance with respect to the combination of deliverable capacity and cycle life when discharged at intermediate rates of one to three hours. Li–MoS$_2$ cells are not suitable for applications requiring very high discharge rates, such as cordless drills. Li–MoS$_2$ cells operate safely over a wide range of abusive conditions, and batteries containing up to eight 'AA' size cells are allowed to be carried onto passenger aircraft. Based on a good technical fit, Li–MoS$_2$ batteries are at present being used by consumers in portable cellular phones, lap-top computers, and portable data terminals.

Further development of rechargeable lithium batteries at Moli Energy Limited will focus on increasing energy density, cycle life, low temperature performance, and safe operation.


**REFERENCES**

[1] Reddy, T., Storage data on low-rate TCL cells stored for five years at room temperature (unpublished) (1987).

[2] Berger, C., Advanced lithium batteries for command, control and communications, In: *Proc. Second Annual Battery Conference on Applications and Advances*, Long Beach, California, 13–15 January (1987).

[3] Chang, V. D. A., Wilson, J. P., Bruckner, J., Inenaga, B. & Hall, J. C., Lithium/thionyl chloride batteries for the small intercontinental ballistic missile, In: *Proc. 33rd International Power Sources Symposium*, Cherry Hill, New Jersey, 13–16 June (1988) pp. 322–333.

[4] Blomgren, G. E., Leger, V. Z., Kalnoki-Kis, T., Kronenberg, M. L. & Brodd, R. J. *Power Sources*, **7** 583 (1979).

[5] Bowden, W. L. & Dey, A. N. *J. Electrochem. Soc.*, **127** 1419 (1980).

[6] Carter, B. J., Williams, R. M., Rodrigues, A., Tsay, F. D. & Frank, H, In: *Extended Abstracts of the 162nd Meeting of the Electrochemical Society*, Detroit, Michigan, (October 1982).

[7] Bailey, J. C. & Kohut, J. P., *Power Sources,* **8** 17 (1981).

[8] Dey, A. N. & Bro, P. *J. Electrochem. Soc.*, **125** 1574 (1978).

[9] Tsaur, K. C. & Pollard, R. J., *J. Electrochem. Soc.*, **133** 2296 (1986).

[10] Clark, W., Dampier, F., Lombardi, A. & Cole, T. *Lithium cell reactions*, Report A Final-TR-83-2083 (December 1983).

[11] Subbarao, S. & Frank, H, In: *Proc. 1983 Goddard Space Flight Center Battery Workshop*, NASA Conference Publication NASA CP-2331, 97 (1983).

[12] Schlaiker, C. R, In: *Lithium batteries*, J. P. Gabano, Ed., Academic Press, New York, (1984), pp. 303–370.

[13] Boyd, J. W. *J. Electrochem. Soc.*, **134** 18 (1987).
[14] Ayers, A. D., Horvath, R. E. & Biegger, D. W., Design of 'F' size Li/SOCl$_2$ cells for the BA-6598 battery, In: *Proc. 33rd International Power Sources Symposium,* Cherry Hill, New Jersey, 13–16 June (1988) pp. 212–216.
[15] Catanzarite, V. O., US Patent No. 4,366,616, (1983).
[16] Gibbard, H. F. & Nadkarni, A. J. Advanced Li–SOCl$_2$ C-cell Battery, In: *Proc. 34th International Power Sources Symposium,* Cherry Hill, New Jersey, 25–28 June (1990).
[17] Cieslak, W. R., Compatibility and performance of separators in Li–SOCl$_2$ cells, In: *Proc. 33rd International Power Sources Symposium,* Cherry Hill, New Jersey, 13–16 June (1988) pp. 233–239.
[18] Blomgren, G. E., Bailey, J. C., Bailey, J. W. & Kalisz, D. W., Very high rate, high energy lithium thionyl chloride cells for coupled systems, In: *Extended Abstracts of the national meeting of The Electrochemical Society,* Seattle, Washington, 14–19 (October, 1990) pp. 203–204.
[19] Gibbard, H. F., Heat generation in lithium/thionyl chloride batteries, In: *Proc.* Vol. 80–4, The Electrochemical Society, Princeton, NJ (1980), pp. 510–525.
[20] Phillips, J. & Gibbard, H. F., Microcalorimetric and A.C. impedance measurements on lithium-thionyl chloride cells, In: *Proc.* Vol. 81–4, The Electrochemical Society, Princeton, NJ (1981) pp. 54–63.
[21] Phillips, J., Hall, J. C. & Gibbard, H. F., An investigation of the lithium and carbon electrodes in thionyl chloride solutions, In: *Proc.* Vol. 81–4, The Electrochemical Society, Princeton, NJ (1981) pp. 41–53.
[22] Boyle, G. H. & Goebel, F., High rate lithium thionyl chloride batteries, In: *Proc. 33rd International Power Sources Symposium,* Cherry Hill, New Jersey, June (1988) p. 215.
[23] Planchat, J. P., Decroix, J. P. & Sarre, G. Reserve lithium-thionyl chloride battery for missile applications, In: *Proc. 33rd International Power Sources Symposium,* Cherry Hill, New Jersey, 13–16 June (1988) pp. 300–311.
[24] Peabody, M. & Brown, R. A. Reserve lithium-thionyl chloride battery for high rate extended mission applications, In: *Proc. the 33rd International Power Sources Symposium,* Cherry Hill, New Jersey, 13–16 June (1988) pp. 312–321.
[25] Eichinger, G., Semran, G., Jacobi, W. & Heydecke, J., Development of high power high energy lithium thionyl chloride batteries, In: *Proc. 34th International Power Sources Symposium,* Cherry Hill, New Jersey, 25–28 June (1990).
[26] Doddapaneni, M., The effect of cathode additives on Li–SOCl$_2$ cell performance after high temperature storage, In: *Proc. 32nd International Power Sources Symposium,* Cherry Hill, New Jersey, 9–12 June (1986).
[27] Barrella, J., US Patent No. 4,871,946 (3 October 1989).
[28] Swiss Patent No. 202, 941 (16 May, 1939).
[29] Maricle, D. L. & Mohns, J. P.. US Patent No. 3,567,515 (2 March, 1971).
[30] Hoffmann, A. K., US Patent No. 3,953,234 (27 April, 1976).
[31] Reddy, T. B., US Patent No. 3,580,828 (25 May, 1971).
[32] Bowden, W. L., Chow, L., DeMoth, D. & Holmes, R., *J. Electrochem. Soc.*, **131**, 229 (1984).
[33] Reddy, T. B., unpublished results (1968).

[34] US Military Specification MIL-B-49430B/ER (25 April 1988).
[35] Shah, P. M., *Proc. Int'l. Power Sources Symp.*, **27** 59 (1976).
[36] Dimasi, G. J., *Proc. Int'l. Power Sources Symp.*, **27** 75 (1976).
[37] DiMasi, G. J. & Christopulos, J. A., *Proc. Int'l. Power Sources Symp.*, **28** 179 (1978).
[38] Bro, P. & Levy, S. C., Chapter 4 in *Lithium battery technology*, H. V. Venkatasetty, Ed., John Wiley, New York (1984).
[39] Barrella, J. N., US Patent No. 4,184,012 (15 January 1980).
[40] Jagid, B., Watson, T. & Chodosh, S. M., *Proc. Symp. on Power Sources for Biomedical and Implantable Applications and Ambient Temperature Lithium Batteries*, The Electrochemical Society, Vol. 80–4, (1980), p. 615.
[41] Chireau, R. & Andruk, W., *Proc. Int'l. Power Sources Symp.*, **30** 137 (1982).
[42] Reddy, T. B. & Bittner, H. F., *Proc. Int'l. Power Sources Symp.*, **33** 639 (1988).
[43] Barrella, J. N., US Patent No. 4,871,956 (3 October 1989).
[44] Dey, A. N., Kuo, H. C., Piliero, P. & Kallianidis, M., *J. Electrochem. Soc.*, **135** 2115 (1988).
[45] *Lithium batteries technical handbook*, Panasonic Industrial Company, Japan (1990).
[46] Fulenda, M. & Lijima, T., Lithium–Carbon monofluoride cells. In: *Lithium batteries*, Gabano, J. P., Ed., Academic Press, New York (1983).
[47] *Batteries for electronics*, Crompton Parkinson, UK (1990).
[48] Ikeda, H., Saito, T. & Tamura, H., *Manganese Dioxide Symposium*, Vol. 1, Cleveland Section of the Electrochemical Soc. Inc. (1975), p. 384.
[49] Ikeda, H., *Lithium batteries*, Gabano, J. P., Ed., Academic Press, New York (1983), Chapter 8.
[50] Furukawa, N., Moriwaki, K., Narukawa, S. & Hara, M., *14th International Power Sources Symposium* (1984), p. 63.
[51] Ikeda, H. & Furukawa, N., *Electrochem. Soc. Fall Meeting*, San Diego, October (1986), p. 19.
[52] Ikeda, H. & Furukawa, N., *Practical lithium batteries*, JEC Press Inc. (1988), p. 57.
[53] Furukawa, N. & Narukawa, S., *et al.*, *Sanyo Technical Review*, **20** 60 (1988).
[54] Owens, B. B., Oxley, J. E. & Sammells, A. F., Applications of halogenide solid electrolytes, In: Geller, S., Ed. *Topics in Applied Physics*, Vol. 21, *Solid Electrolytes*, Springer-Verlag, Berlin (1977), pp. 67–104.
[55] Gutmann, F., Hermann, A. M. & Rembaum, A., Solid-state electrochemical cells based on charge transfer complexes, *J. Electrochem. Soc.*, **114** (1967), 323–329.
[56] Gutmann, F., Hermann, A. M. & Rembaum, A., Environmental and reaction studies on electrochemical cells based on solid charge-transfer complexes, *J. Electrochem. Soc.*, **115** (1968), 359–362.
[57] Schneider, A. A., Moser, J. R., Webb, T. H. & Desmond, J. E., A new high energy density solid electrolyte cell with a lithium anode, In: *Proc. 24th Power Sources Symposium*, Atlantic City, N.J., (1970). pp. 27–30.
[58] Schneider, A. A. & Moser, J. R., Primary cells and iodine containing cathodes. US Patent No. 3,674,562 (1972).

[59] Schneider, A. A., Harney, D. E. & Harney, M. J., The lithium–iodine cell for medical and commercial applications, *J. of Power Sources*, **5** (1980), 15–23.

[60] Schneider, A. A., Bowser, G. C. & Foxwell, L. H., Lithium iodine primary cells having novel pelletized depolarizer, US Patent No. 4,148,975 (1979).

[61] Schneider, A. A. & Tepper, F., The lithium–iodine cell, In: Thalen, H. J. & Harthorne, J. W., Eds, *To pace or not to pace*. Martinus Nijhoff Medical Division, The Hague (1978), pp. 116–121.

[62] Phillips, L. C., Kelly, R. G., Wagner, J. W. & Moran, P. J. An investigation of the volume change associated with discharge of lithium–iodine batteries via holographic interferometric techniques, *J. Electrochem. Soc.*, **133** (1986), 1–5.

[63] Jolson, J. D. & Schneider, A. A., High rate lithium–iodine cells for military and commercial applications, In: *Proc. 30th Power Sources Symposium*. Atlantic City, NJ (1982), pp. 185–187.

[64] Antonioli, G., Baggioni, G., Consiglio, F., Grassi, G., Lebrun, R. & Zanardi, F., Stimulatore cardiaco implantabile con nuova battaria a stato solido al litio. *Minerva Med.*, **64** (1973), 2298.

[65] Mead, R. T., Greatbatch, W. & Rudolph, F. W., Lithium–iodine battery having coated anode, US Patent No. 3,957,533 (1976).

[66] Mead, R. T., Greatbatch, W., Rudolph, F. & Frenz, N. W., Lithium–iodine cell, US Patent No. 4,210,708 (1980).

[67] Holmes, C. F. & Brown, W. R., In: Owens, B. B. & Margalit, N., Eds, *Power sources for biomedical implantable applications and ambient temperature lithium batteries*, Vol. 80-4, The Electrochemical Society, Pennington, New Jersey (1980), pp. 187–194.

[68] Kelly, R. G. & Moran, P. J., The rate limiting mechanism in Li–$I_2$ (P2VP) batteries, *J. Electrochem. Soc.*, **134** (1987), 25–30.

[69] Holmes, C. F., Lithium halogen batteries, In: Owens, B. B., Ed., *Batteries for implantable biomedical devices*, Plenum Press, New York (1986), pp. 134–136.

[70] Catalyst Research, Lithium cells designed for CMOS RAM data retention and CMOS clock/calendar backup, In: Matsuda, Y. & Schlaiker, C. R., Eds, *Practical lithium batteries*, JEC Press, Cleveland, OH (1988).

[71] Sengupta, R. N., Power supply for CMOS ICs from lithium–iodine battery, *CSIO Commun.*, **11** (1984), 124–125.

[72] Jones, K. J. & Hatch, E. S., Lithium batteries make non-volatile CMOS–RAM possible. *Industrial Research & Development*, Pittsburgh Conference Issue (1982).

[73] Jumel, Y., Lambin, O. & Broussely, M., *Proc. Symposium on Lithium Batteries*, Fall Meeting of the Electrochemical Society, Washington (1983).

[74] Takashi, I., Yoshinoto, T., Joji, N. & Hiromichi, O., *J. Power Sources*, **5** (1980), 99–109.

[75] Furukawa, N., Moriwaki, K. & Ishibashi, C., *Proc. 32nd International Power Sources Symposium* (1986).

[76] Broussely, M., Jumel, Y. & Gabano, J. P., *152nd Electrochemistry Meeting*, Atlanta (October 1977).

[77] Bro, P., Holmes, R., Marincic, N. & Taylor, H., In *Power sources* Collins, D. H., Ed., Academic Press, London (1975), pp. 703–712.

[78] Jumel, Y. & Thunder, A. *et al.* (1980). *Proc. 29th Power Sources Conference*, Atlantic City (1980), p. 222.
[79] Broussely, M., Jumel, Y. & Gabano, J. P., *Proc. Second International Conference on Lithium Batteries* (1984).
[80] Bates, R. E. & Jumel, Y., In: *Lithium batteries* Gabano, J. P., Ed. (1983).
[81] Bates, R. E., Brown, A. E., Baldwin, K. R., Knight, J. & Swan, D. N., In: *Power sources* Pearce, L., Ed., No. 10, pp. 111–125.
[82] Bates, R. E., Murphy, B. P. & White, G. D., *Proc. 30th Power Sources Symposium* (1982).
[83] Murphy, D. W., Christian, P. A., DiSalvo, F. S. & Waszcak, J. V., *Inorg. Chem.*, **18** (1979), 2800.
[84] Murphy, D. W., Christian, P. A., DiSalvo, F. S., Carides, J. N. & Waszcak, J. V., *J. Electrochem. Soc.*, **128** (1981), 2053.
[85] Kahara, T., Horiba, T., Tamura, K. & Fujita, M., *Proc. Symposium on Ambient Temperature Lithium Batteries*, Owens, B. B. & Margalit, N., Eds, The Electrochemical Soc. Inc., NJ (1980).
[86] Dey, A. N., *The Electrochemical Society Extended Abstracts, Fall Meeting, Boston, MA*, 7–11 October (1973) Abstract 54, p. 132.
[87] Horning, R. J., *Record 10th Intersoc. Energy Conv. Eng. Conf.* (1975), pp. 424.
[88] Gibson, J. G. & Sudworth, J. L., in *Specific energies of galvanic reactions and related thermodynamic data*, Chapman & Hall, London (1973).
[89] Auborn, J. J., French, K. W., Lieberman, S. I., Shah, V. K. & Heller, A., *J. Electrochem. Soc.*, **120** (1973), 1613.
[90] Szpak, S. & Venkatasetty, H. V. *J. Electrochem. Soc.*, **131** (1984), 961.
[91] Binder, M., Gilman, S. & Wade, W. Jr., *J. Electrochem. Soc.*, **131** (1984), 1985.
[92] Venkatasetty, H. V., Ed., *Lithium battery technology*, Wiley–Interscience, New York (1984).
[93] Whittingham, M. S., *J. Electrochem. Soc.*, **123** (1976), 315.
[94] Lang, M., Backlund, J. R. & Weidner, E. C., *Proc. of 26th Power Sources Symp.* (1974), pp. 37.
[95] Ebner, W. B. & Walk, C. R., *Proc. of 27th Power Sources Symp.* (1976), p. 48.
[96] Venkatasetty, H. V., *J. Electrochem. Soc.*, **122** (1975), 245.
[97] Venkatasetty, H. V., Ebner, W. B. & Lin, W. H., *Quarterly Technical Report, US Army Laboratory Command Contract #DAA LOI-86-C-0004* (July–September 1986 and June 1987).
[98] Ebner, W. B. & Merz, W. C., *Proc. of 28th Power Sources Symp.* (1978) p. 214.
[99] Levy, S. C., *Proc. of 27th Power Sources Symp.* (1976), p. 52.
[100] Honeywell Power Sources Center, *Final Report, ECOM* 71-0191-*F IV*-1-*IV*-47 (1974).
[101] Horning, R. J. & Viswanathan, S., *Proc. of 29th Power Sources Symp.* (1980), 64.
[102] Horning, R. J. & Rhoback, F. W., *Progress in Batteries and Solar Cells*, **4** (1982), 97.
[103] Keister, P., Mead, R. T., Ebel, S. J. & Fairchild, W. R., *Proc. of 31st Power*

*Sources Symp.* (1984), 331.

[104] Takeuchi, E. S., Zelinsky, M. A. & Keister, P., *Proc. of 32nd Power Sources Symp.* (1986), 268.
[105] Thiebolt, W. C. & Takeuchi, E. S., Abstract No. 20, *Proc. Electrochem. Soc. Meeting, Hawaii, Honolulu* (October 1987).
[106] Keister, P. P., Mead, R. T., Muffoletto, B. C., Takeuchi, E. S., Ebel, S. J., Zelinsky, M. A. & Greenwood, J. M., US Patent 4,830,940 (16 May, 1989).
[107] UK Patent Specification 1,537,323 (1987).
[108] European Patent Application 0,068,230A1, (1983).
[109] European Patent Application 0,049,139A1, (1982).
[110] Jacobsen, A. J. & McCandlish, L. E., *J. Solid State Chemistry*, **29** (1979), 355.
[111] Eisenberg, M. & Willis, M., *Electrochemical Society Extended Abstracts*, Vol. 77-2, Atlanta Georgia (October 1977), Abstract 11, p. 38.
[112] Bennett, O. G. & Woolley, J. P., US patent 3,575,714 (filed August 1953).
[113] Tepper, F., A survey of thermal battery designs *Intersoc. Energy Convers. Conf.*, Amer. Soc. Mech. Engineers, USA (1974), p. 671.
[114] Van Domelen, B. & Wehrli, R. D., A review of thermal battery technology, *Interxoc. Energy Convers. Conf.* Amer. Soc. Mech. Engineers, USA (1974), p. 665.
[115] Attewell, A. & Clark, A. J., Recent developments in thermal batteries, *Power Sources*, Vol. 8, Academic Press, London (1981), pp. 286.
[116] Tepper, F. & Yalom, D., *Handbook of batteries and fuel cells*, Ed., Linden, D. McGraw-Hill, New York (1984), pp. 40–41.
[117] Clark, A. J., Recent advances in thermal battery technology, *Chemistry & Industry*, **17** (1986), 205.
[118] Singh, S. S. & Sachan, D. P., Thermal battery as power source for guided missiles, *Def. Science J. (India)*, **23** (1973), 163.
[119] Briscoe, J. D., Chagnon, G., Gessler, J. L. & Mattson, D., Advances in thermal battery applications, *Proc. 33rd Int. Power Sources Symp.*, Electrochem. Soc. Inc. USA (1988), p. 393.
[120] Bush, D., Advances in pellet-type thermal battery technology, Sandia Labs, USA, *Report SC-RR-69-497A* (October 1972).
[121] Nelson, P. A., *et al.*, Argonne Nat. Lab., *Report ANL-76-45* (1976).
[122] Bowser, G. C., Harney, D. & Tepper, F., High energy density molten anode thermal battery, *Power Sources*, Vol. 6, Academic Press, London (1977), pp. 537.
[123] Searcy, J. Q. & Armijo, J. C., Improvements in thermal battery technology, Sandia Nat. Labs, *USA, Rpt. SAND-82-0565* (June 1982).
[124] James, S. D. & DeVries, L. E., Structure and behaviour of Li–B alloys in KCl–LiCl melt, *J. Electrochem. Soc.*, **123** (1976), 321.
[125] Baird, M. D., Clark, A. J., Feltham, C. R. & Pearce, L. J., Recent developments in high temperature primary lithium batteries, *Power Sources*, Vol. 7, Academic Press, London, (1979), pp. 701.
[126] Street, H. K., Characteristics of MC3573 thermal battery, Sandia National Laboratories, USA, *Report SAND-82-0695* (February 1983).
[127] Frazer, R. T. M., Pearce, L. J. & Birt, O. C. P., High rate lithium-iron

disulphide batteries, *Proc. 29th Power Sources Symp.*, Electrochem. Soc. Inc., USA, (1980), p. 43.

[128] Embrey, J. & Staniewicz, R. J., Alternative electrolytes for thermal batteries, *Proc. 33rd Int. Power Sources Symp.*, Electrochem. Soc. Inc., USA, (1988), p. 351.

[129] Guidotti, R. & Reinhardt, F. W., Alternative electrolytes for use in thermal batteries, *Proc. 33rd Int. Power Sources Symp.*, (1988), p. 369.

[130] Hunt, J. B. & Root, C. B., US Patent 3,498,843 (3 March 1970).

[131] Miles, M. H., McManis, G. E. & Fletcher, A. N., Passivating films on solid and liquid lithium anodes, *Electrochimica Acta*, **30** (1985), 889.

[132] Lin, Y-H., Yu, K-T., Yao, P-G. & Hsu, S-E., LiAl/LiI(Al$_2$O$_3$)/FeS$_2$ thermal batteries, *Proc. 32nd Int. Power Sources Symp.*, Electrochem. Soc. Inc., USA, (1986), p. 664.

[133] Searcy, J. Q., *et al.*, Li$_2$O in the Li/FeS thermal battery system, Sandia National Laboratories USA, *Report SAND-81-1705* (November 1981).

[134] Faul, I., A new high-power thermal battery cathode material, *Proc. 32nd Int. Power Sources Symp.*, Electrochem. Soc. Inc., USA, (1986), p. 636.

[135] Lamb, C. M. *et al.*, Primary lithium-chlorine cells *Proc. 33rd Int. Power Sources Symp.*, Electrochem. Soc. Inc., USA, (1988), p. 404.

[136] Attewell, A. & Grimes, J. H., A novel type of iron powder for thermal battery pyrotechnics, *Proc. 31st Int. Power Sources Symp.*, Electrochem. Soc. Inc., USA (1984), p. 373.

[137] Knight, J. & McKirdy, I., Prediction of thermal battery internal temperatures, *Power Sources*, Vol. 11, Int. Power Sources Symp. Committee, Leatherhead, England (1987), p. 491.

[138] Wells, J., Li(alloy)/FeS$_2$ thermal batteries: safety under 'pump-back' recharging conditions, *Proc. 32nd Int. Power Sources Symp.*, Electrochem. Soc. Inc., USA (1986), p. 648.

[139] Winchester, C. S., The LAN/FeS$_2$ thermal battery system, *Proc. 30th Power Sources Symp.*, Electrochem. Soc. Inc., USA (1982), p. 23.

[140] Whittingham, M. S., Chemistry of intercalation compounds: metal guest in chalcogenides hosts, *Progr. Solid St. Chem.*, **12** (1978), 41–99.

[141] Haering, R. R., Stiles, J. A. R. & Brandt, K., Lithium molybdenum disulfide battery cathode, US Patent 4,224,390 (1980).

[142] Py, M. A. & Haering, R. R., Structural destabilization induced by lithium intercalation in MoS$_2$ and related compounds, *Can. J. Phys.*, **61** (1983), 76–84.

[143] Laman, F. C. & Brandt, K., Effect of discharge current on cycle life of a rechargeable lithium battery, *J. Power Sources*, **24** (1988), 195–206.

[144] Koch, V. R., Reactions of tetrahydofuran and lithium hexafluoroarsenate with lithium, *J. Electrochem. Soc.*, **126** (1979), 181–187.

[145] Dey, A. N., Lithium anode film and organic and inorganic electrolyte batteries, *Thin Solid Films*, **43** (1977), 131–171.

[146] Dahn, J. R., Stiles, J. A. R., Murray, J. J. & Alderson, J. E. A., Percolation theory of capacity fade in lithium molybdenum disulfide secondary cells, *Proc. Symposium on Electrode Materials and Processes for Energy Conversion and Storage*, Electrochemical Society Meeting, Philadelphia, PA (May 1978).

[147] Laman, F. C., Matsen, M. W. & Stiles, J. A. R., (1986) Inductive impedance of a spirally wound Li–MoS$_2$ cell, *J. Electrochem. Soc.*, **133** (1986) 2441–2446.

[148] Johnson, C. J. & Laman, F. C., A rechargeable lithium cell with long cycle life, for aerospace applications, In: Dey, A. N., Ed., *Lithium Batteries*, Vol. 87-1, The Electrochemical Society Inc., Pennington, NJ, (1987) 310–313.

[149] Brandt, K. & Laman, F. C., Reproducibility and reliability of rechargeable lithium molybdenum disulfide batteries, *J. Power Sources*, **25** (1989) 265–276.

[150] Fouchard, D. T. & Taylor, J. B., The Molicel rechargeable lithium system: multicell battery aspects, *J. Power Sources*, **21** (1987), 195–205.

# 6
# Systems under development

## 6.1 NICKEL–ZINC BATTERIES
### (A. Duffield)

Of all rechargeable systems under development, nickel–zinc probably has one of the longest histories. However, this system has never been successful when it has entered the commercial world.

The first work on Ni–Zn appeared in 1887 when Dun & Haslacher [1] patented a number of systems based on nickel–potassium hydroxide secondary batteries. The next recorded work was by Michalowski [2] around 1899 who studied this system, but as with the previous work, no commercial products emerged. The first successful commercial alkaline storage batteries were developed by two independent entrepreneurs: the Swedish Waldemar Jungner and the American Thomas Alva Edison. Toward the end of the 19th century both of these workers investigated the secondary Ni–Zn system. In 1897, after the commercial failure of a zinc project, Jungner gave up the idea of zinc anodes. Edison endeavoured further, investing a large amount of money into Ni–Zn development, but by September 1900, owing to zinc's high solubility which caused replating problems, he wrote in his diary, 'I have become totally disenchanted with zinc' [3], and hence stopped his research on zinc systems.

Since that time research into Ni–Zn batteries has been cyclic, with either new developments in materials or power crises spurring renewed interest. During this period, some success in overcoming the inherent problems of the system have been achieved. In the USSR, large Ni–Zn batteries are still used in industrial traction applications, but the technology behind them is old. In more recent years, Yuasa in Japan have been developing prototype batteries for electric vehicle applications. They have claimed 200 maintenance free cycles at 100% depth of discharge.

Overall, the Ni–Zn system is plagued by a poor cycle life, and it is being surpassed in performance by new systems as well as being disfavoured by the soaring cost of nickel.

### 6.1.1 The nickel–zinc system
The simplest form of the overall reaction for this couple may be written as:

$$2\text{NiOOH} + \text{Zn} + 2\text{H}_2\text{O} \rightleftharpoons 2\text{Ni(OH)}_2 + \text{Zn(OH)}_2 \, . \tag{6.1}$$

The environment of this couple is an alkaline one as with Ni–Cd cells. The electrolyte used is typically potassium hydroxide dissolved in water, with a concentration of 6–7 M (35–40 w/v%), the zinc electrode combined with this electrolyte giving a conductivity maxima at this concentration [4]. In some respects this has been found to be at the extreme for satisfactory operation of the nickel electrode.

Rechargeable zinc electrode battery systems, especially when combined with nickel oxide, have always seemed attractive for several reasons.

(i) Energy densities of 60–70 Wh.kg$^{-1}$ are practical.
(ii) Similarly power densities of 140–200 W.kg$^{-1}$ are available.
(iii) Good low temperature performances of this alkaline system.
(iv) Zinc is highly reactive in terms of oxygen recombination, making it an ideal candidate for sealed cells.
(v) A high open circuit cell voltage of 1.75 V.

There are three main problems which limit the cycle life of the zinc electrode. For the Ni–Zn system a fourth complication arises which also causes a strong reduction in cell capacity with cycling. These four impediments are:

### (i) Dendrite growth
During the charging process dendrites grow at points of high current density. These can pierce the separator materials and touch the other electrode active surface, resulting in an internal short circuit. This can cause premature cell failure. Much research has been done both on separators to prevent this short circuiting, and on additives to the electrode and electrolyte to mitigate zinc dendrite growth. The problem is avoided in flowing electrolyte zinc–halide cells (see section 6.5) as, in this case, dendrites break off and redissolve in the electrolyte or collect in filters.

### (ii) Zinc corrosion
Zinc corrodes to form zinc hydroxide and zincate in alkaline electrolyte. The consequence is that zinc electrodes have relatively poor charge retention on standing and subsequently a low shelf life. In a typical nickel–zinc cell only 70% of the capacity is available after one month of rest at room temperature.

### (iii) Shape change
On continued cycling there is a relocation of zinc from the outside edges of the battery plate to the centre of the electrode. The tendency to agglomerate at the plate centre is also often called densification. This process causes a reduction in the surface area of the porous electrode, in the direction perpendicular to the surface, and occurs during the replating of zinc from solution. Once initiated, this displacement of active zinc progresses with cycling and results in a loss of cycle life due to the reduction of useful capacity. This problem is the most important factor leading to limited Ni–Zn and AgO–Zn battery cycle life [5,6].

Two mechanisms have been suggested for the shape change phenomena. McBreen [6] proposed one which was based on observed experimental data. He suggested that differences in current distribution during charging and discharging result in concentration cells, such that zinc dissolves away from the plate edge to be re-deposited at the centre. Choi *et al.* [7] presented a mathematical model, which is based on convective flows induced by electro-osmotic forces across the membrane separator present in an actual battery. At present, doubts are still attached to both theories by a number of workers [8], including James McBreen himself [2].

*(iv) Zinc poisoning of the nickel electrode*
On continued cycling, particularly in cases of deep discharge, zincate ions reprecipitate as ZnO onto the nickel oxide electrode, resulting in pore blocking and hence loss of capacity [9].

Of all these four problems, the last is perhaps the least studied. One reason for this is that most effort is devoted to controlling the first and third problems, and that poisoning becomes significant only when these are sufficiently offset. With sintered nickel electrodes, which have been widely used, the effect is slight compared to pocket plate [5] and plastic bonded nickel oxide electrodes, where the effect can be severe.

Control of the initial first three problems has been attempted by the addition of additives to the electrode and electrolyte, different separator materials, and changes in the conventional mechanics of cell design. The last of these categories entails flowing electrolytes and vibrating electrodes, both of which minimize the above problems. However, at the same time substantial sacrifices are made in terms of energy density. Corrosion has commonly been impeded by the addition of heavy metals or their oxides to raise the hydrogen overpotential. Of these, mercury has proved the most effective [10], but there is evidence to suggest that this and some other heavy metals increase shape change [11].

Ways of controlling the effect of growth of dendrites have concentrated on separator development. Some success has been obtained with additives used in commercial electroplating processes, as these ensure a smooth redeposition. Separators must be resistant to dendrite penetration yet have a very low electrical resistance. Microporous films of polypropylene or mixed copolymers [12] which have a critical pore size lower than that which allows for dendrite growth, have been popular. For some applications, thin ceramic separators have proved fairly successful. Radiation grafted polyolefin films have shown superior qualities, especially with respect to their relatively low cost [13]. In Japan, this technique has been extended to non-woven fabrics which have been used in experiments with large sealed recombinant cells [14]. So far this has been more successful with Ni–Cd than Ni–Zn systems.

Shape change has been actively attacked from the additives standpoint. Since this is related to zinc solubility in potassium hydroxide (KOH) solutions, materials such as $Ca(OH)_2$ [15] or oxalates have been added. While these reduce shape change effectively they have an adverse effect upon the high rate discharge capability of the zinc electrode. The reason for this is passivation due to the insoluble discharge species blocking the porous electrode surface. Other materials such as $BiO_3$ have a beneficial effect by influencing crystal growth at the electrode surface [16].

Polymeric additives are often used as binders both to give the electrodes some mechanical integrity and to help control the formerly mentioned problems. A common one used in the zinc electrode (as well as other electrode systems) is PTFE.

A series of papers on this subject have been published by McNeil & Hampson [17]. These authors studied polymer bonded secondary zinc electrodes. The overall conclusions which can be drawn from this work are that polymeric binders which show positive effects are those which maintain an open porous structure conjoint with the electrode physical integrity. All polymers in high concentrations ( > 10%) tend to show adverse effects by shielding the electrode active material. This is particularly so if they are hydrophobic in nature and therefore act to suppress the electrodic reactions. PTFE has beneficial effects at low concentrations even though it is highly hydrophobic, because of its unique mechanical properties. PTFE has been seen to form into a fibrous matrix in pasted plate electrodes, binding the electrode together, while maintaining an open network. At higher concentrations, however, PTFE behaves as other polymers, suppressing both electrode dissolution and the active area.

In 1983 a patent by the Lucas Research Centre, England [18], detailed the use of graphite in the formation of porous zinc electrodes for secondary applications in silver oxide–zinc and nickel–zinc cells. The patent covered a pressed plate secondary zinc electrode containing 16 to 50% graphite, 0.5 to 10% PTFE, and small amounts of heavy metal oxides (e.g. HgO, PbO, and CdO). It was claimed that the addition of graphite results in an electrode much less susceptible to shape change and dendritic growth than graphite free electrodes. Consequently, these electrodes show a much better cycle life. Further studies both by Lucas and at Loughborough University, England [19,20] revealed several factors concerning the electrodes.

(i) Graphite particle size. The lower the particle size, the greater the benefit. The size range 150–2.5 $\mu$m was studied. This factor was closely related to the graphite content.
(ii) Graphite content. The higher the content the better, although a marked improvement was seen above 24% w/w graphite. The range 8–47% was studied. However, above 32% w/w the gains were small compared to losses in energy density/cost advantage.
(iii) Electrode expansion. Graphite-containing zinc electrodes require sufficient expansion space on the forming charge to allow approximately 72% porosity.
(iv) Paste pressing. The graphite-containing sheets formed better electrodes if pressed at $7.5 \times 10^7$ Pa.

   The optimum paste mix finally agreed upon contained 32% graphite (2.5 $\mu$m), 5% PTFE, 62% ZnO, 1% metallic oxides.

The particle size effect at 32% concentration was quite marked. At 150 $\mu$m, graphite gave little benefit. On cells of 25 Ah capacity at 100% depth of discharge (DOD) at the 0.2C rate, the 2.5 $\mu$m graphite zinc electrodes gave some 30% extra life than 15 $\mu$m graphite-containing cells. Fig. 6.1 shows the effect on cycle life of 25 Ah Ni–Zn cells caused by different concentrations of graphite under the cycling regime mentioned above, the cells having sintered nickel electrodes.

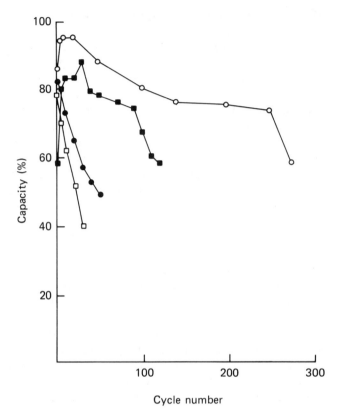

Fig. 6.1 — Capacity vs cycle number for 25 Ah cells at the 0.2 C rate to 100% DOD. Graphite contents of zinc electrodes: □ 0%, ● 8%, ■ 16%, ○ 32%.

The result of detailed studies revealed that the whole structure was modified in a manner which trapped dissolved zinc species within a 3-dimensional matrix until a dissolution–precipitation mechanism occurred. Thus the whole system reformed in a pseudo-solid state manner.

### 6.1.2 Cell and battery construction and performance

The development of Ni–Zn was initially carried out in the same manner as Ni–Cd, the pocket plate route being used initially. Drumm [3] used this design for railway propulsion in the 1930s, and some of the USSR systems are of this form. On the whole, pocket plate Ni–Zn cells exhibit all the disadvantages of this couple with little gain over other systems. Systems using flowing electrolyte through special beds of electrodes are still investigated from time to time. The aim of the latter systems is to have a completely soluble zinc electrode and to use a flowing system to control zinc deposition. Some success has been found this way, but needless to say the complexity, cost, and size of these systems have proved prohibitive in comparison to their benefits.

The more successful Ni–Zn batteries have been based on plastic bonded (usually PTFE at low concentration) pressed plate zinc electrodes together with one of three types of nickel electrode:

(i) Sintered.
(ii) Plastic bonded or pressed plate.
(iii) Fibre or expanded foam.

These nickel electrodes are of the same general types that are found in Ni–Cd systems. Fibre and expanded foam nickel electrodes are the latest developments in the Ni–Zn sphere, and have been studied mostly in Poland [21], Germany, and Japan. Development in this area is still in its early stages, but some improvements have apparently been achieved. If the performance of the fibre nickel cells can match or exceed that achieved by sintered electrodes, conjoint with a cost reduction, then this route could prove promising.

Sintered nickel electroded cells offer the best performance at the highest cost. This concept has been tried extensively, from small sealed cells to electric vehicle size. These cells are normally nickel limiting so as to reduce the depth of discharge required on the zinc electrode. Such cells can be capable of providing many 25 C discharges. However, these cells are still ultimately prone to the problems of the zinc electrode.

One of the most economic routes for the production of the nickel electrodes is the plastic bonded route. A viable technique for producing these was developed at the Lucas Research Centre (Solihull, England) in the mid 1980s. The basic technique was as follows. Dry powders of 65% nickel hydroxide and 30% graphite (LONZA KS44) were first mixed in a plough shear type mixer. Propan-2-ol was then added as a wetting agent, and mixing continued. The remaining 5% constituent was PTFE in suspension which was then added and mixed until an even dough-like paste was obtained. This could then be rolled into a continuous sheet and oven dried, using equipment similar to that found in the baking industry. The sheets were then pressed onto nickel foil collectors and the electrodes were coated with a non-woven polyamide separator material.

Plastic bonded zinc electrodes from the Lucas Research Centre were constructed in the same manner. The principal constituents were zinc oxide, graphite, mercury oxide (1% max), and PTFE (5% max). As discussed above, the most effective combination was 62% ZnO and 32% graphite. The rolled plastic bonded sheet was pressed onto a copper foil collector, wrapped with a layer of non-woven polyamide absorber/separator, and then finally wrapped with three layers of microporous polypropylene separator material (Celgard™ K306). The final form of the zinc electrode can be seen in Fig. 6.2. A layer of PVC tape was frequently included on the edge of the electrode to reduce edge dendrites.

Whilst the performance of this zinc electrode was quite good, two limiting factors existed. One was the high cost of the Celgard™ separator, particularly as three layers of this were required, although some other manufacturers have reported using up to five layers of the same material. At the time of the work in question, no other separator could offer the same performance. However, recent developments in

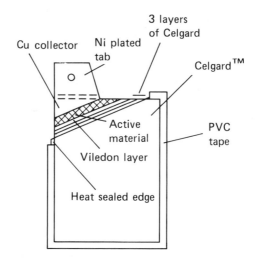

Fig. 6.2 — Zinc electrode construction for Ni–Zn cells.

separator technology may give rise to a viable alternative. The other limiting factor was related to the plastic bonded (PB) nickel electrode. It is found that the PB nickel electrode requires restricted expansion (about 5% from dry uncharged state) for good performance. As mentioned above, the zinc electrode requires quite a large expansion gap (about 70% over dry unformed state) for successful use. Hence the two electrodes were not completely compatible. It was found possible to obtain 300 usable cycles, but the Ni–Zn superior high discharge rate advantage was significantly reduced. It was for this reason that sintered electrodes always gave rise to superior cells.

### 6.1.3 The future for nickel–zinc
The cost of nickel has always been a prohibitive factor of nickel hydroxide cathode cells, but this system has a less hazardous environmental impact than Ni–Cd. World concern over oil in the past has always stimulated electric vehicle interest, and in the latter half of the 20th century this has always caused renewed interest in Ni–Zn.

The most recent advances in nickel electrode technology using fibre and expanded foam substrates may increase the likelihood of successful Ni–Zn commercial use, but ever since the time of Edison it has had to compete with new developments in other cell systems.

## 6.2 SODIUM–SULPHUR BATTERIES
(M. McNamee)

### 6.2.1 Introduction
The many useful characteristics of aqueous battery couples have resulted in their dominance in the existing battery market, despite their relatively low energy density

and employment of costly materials such as lead, cadmium, and nickel. They are also limited by maximum specific energies of approximately 30 Wh kg$^{-1}$ and 50 Wh kg$^{-1}$ for lead–acid and nickel–cadmium respectively. These deficiencies led to a search for electrochemical couples capable of delivering higher specific energy whilst using readily available, non-strategic starting materials. Calculations show that battery specific energies of around 100 Wh kg$^{-1}$ and peak power capabilities of 120 W kg$^{-1}$ are needed for traffic compatible electric vehicle performance with worthwhile ranges. These values are for delivery vans; the requirements in smaller vehicles are more demanding [22].

In the 1960s, J. T. Kummer conceived the idea of employing sulphur and sodium as the electrodes of a secondary cell, but lacked a suitable high temperature compatible, ionically-conducting membrane to use as an electrolyte. At that time, both J. T. Kummer and N. Weber [144] began to search for a suitable candidate electrolyte at the Ford Motor Company. Initially, they concentrated on conducting glasses in the soda aluminosilicate or aluminoborate systems, and they made some primitive electrochemical cells to test Kummer's concept. The conductivity of the glasses was found to be higher with an increased content of sodium meta-aluminate, and they attempted to fabricate samples of this crystalline compound. One of the higher conductivity samples, which was sintered without encapsulation, was found to contain a material with a layered structure known as beta alumina. Kummer and Weber discovered that, whilst many of the properties of beta alumina (and one of its associated formula beta″ alumina) were reported, their large alkali ion mobilities were a notable exception. This formative work was to develop into a multi-million dollar effort, and by the end of the 1980s it is estimated that more than $100 m and approximately 2000 man years of effort have been expended worldwide on sodium–sulphur battery development.

Independently, Levine and Brown [145] of the Dow Chemical Company were also experimenting with the use of hollow glass fibres as a means of providing ionic conduction of sodium. By January 1965 they had shown that sodium in a glass/sulphur sulphide cell could be transported reversibly. However, the Ford Motor Company announced its sodium–sulphur battery based on beta alumina in 1966, and, in the intervening period, this has become the electrolyte of choice for the major developers of this battery.

The basic reaction of the sodium/sulphur cell may be written as

$$2\,Na + 3\,S = Na_2S_3 \ . \tag{6.2}$$

The generally accepted value for the specific energy of this reaction is 760 Wh kg$^{-1}$. The schematic operation of the cell is illustrated in Fig. 6.3. The cells are operated at approximately 350°C to keep the reaction products molten. Electrons are stripped out of the molten sodium and return to the sulphur–electrolyte interface via the load and conducting carbon fibres in the sulphur electrode. The carbon is required to overcome the insulating nature of sulphur which would not transport electrons. The emerging sodium ion reacts with the sulphur to form sodium sulphide. Initially during discharge, the electrode melt consists of two immiscible phases, almost pure

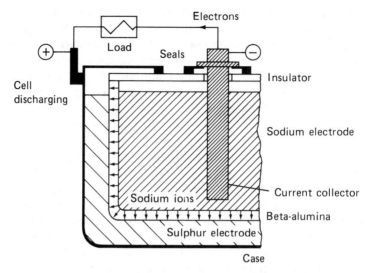

Fig. 6.3 — Schematic of sodium–sulphur cell.

sulphur and $Na_2S_{5.2}$, and the cell e.m.f. remains stable at 2.076 V (at 350°C). The e.m.f. begins to decline when all the pure sulphur has been converted, and at 1.74 V a mixture of $Na_2S_{2.7}$ and solid $Na_2S_2$ begins to form. Whilst cells containing $Na_2S_2$ in the sulphur electrode have been satisfactorily recharged, this condition is considered abnormal and is usually prevented by automatic limitation of the discharge or arrangement of the electrode reactant loadings.

During the development of the sodium–sulphur system a wide variety of individual cell shapes and sizes has been constructed and tested. As an example, Chloride Silent Power Ltd (CSPL) has constructed numbers of cells with energies from 20 Wh to 1300 Wh, probably the widest span researched. Both tubular and flat plate bipolar electrolyte configurations have been studied, and electrolytes have been fabricated up to 50 mm diameter and 600 mm in length. The largest tubes fabricated in any quantity were those made by the US General Electric Company which were 600 mm long and 40 mm in diameter. The bipolar cell design was researched during the early years, but was gradually superseded by the more successful tubular configuration. In the latter design, the sodium may be placed inside the closed-end tube or outside it. Both configurations have been researched, although the central sodium has found favour because of its better energy density. The majority of developers have recently concentrated their efforts on cell designs with capacities in the range of 10 to 100 Ah for electric vehicle applications and 200 to 400 Ah for utility load-levelling applications.

The choice of cell size is a trade-off between battery reliability, voltage, power, space constraints, and cost. As with all electrochemical cells, the power is derived from the electrode surface areas and the energy from the mass of reactants available. Provided that the dead space for seals, etc., is minimized, smaller sizes of electrolyte create more powerful cells because the surface area to volume ratio is increased. The ultimate extension of this relationship is the flat plate bipolar cell. However, the sodium–sulphur cell has a ceramic electrolyte which is, by its nature, relatively brittle. The general theories of fracture predict lower probability of failure for smaller volumes of ceramics. Beta alumina conforms to this relationship, and, all other factors being equal, a large cell would not be as reliable as a smaller one. Tubular structures are more robust than flat plates of the same surface area, and for this reason the closed-end tube has become the norm in the majority of sodium–sulphur cells. A requirement for low electrical resistance of cells has meant that tube wall thicknesses of around 1.5 mm have become the most prevalent in spite of the skill required from the ceramists in the fabrication of such items.

### 6.2.2 Performance characteristics

The within cycle performance of a typical sodium–sulphur cell is shown in Fig. 6.4. The open circuit voltage remains at 2.076 V for approximately 70% of the discharge, after which it begins to decline as the higher sulphur species of sodium polysulphide are formed. Discharge is normally terminated between open circuit voltages of 1.9 V and 1.76 V. The recharge characteristics of the better designs of cell are also stable with no significant polarization until the 2 h rate of charge is exceeded. Further increases in recharge rate are possible, but would require a reduction in the sulphur electrode thickness and an attendant sacrifice in cell capacity.

In independent trials conducted by De Luca at Argonne National Laboratories [23], sodium–sulphur cells and small batteries were tested and compared with other test results from that laboratory. The specific power to specific energy of sodium–sulphur and other electrochemical couples are displayed as a Ragonne plot in Fig. 6.5.

Sodium–sulphur cells have been tested for more than 7000 deep discharge–recharge cycles during more than 7 years of operation. Several developers have tested batteries for more than 500 cycles. Earlier worries about the intrinsic stability of beta″ alumina in the cell environment have largely been allayed, and a large body of data is available worldwide to support this contention. The cycle to cycle reproducibility or cell performance is also excellent. Fig. 6.6 shows the resistance and capacity of a group of 18 cells to 1000 cycles. Rising resistance with increasing cycle life has been eliminated as developers have understood and controlled the deleterious effects of impurities in cells. Sodium–sulphur cells also have good tolerance to varying charge–discharge profiles, and no memory effects have been observed.

The sodium–sulphur system has two unique features. Firstly, the Faradaic efficiency of the cell is 100% both in theory and practice, the amp hour capacity is invariant with discharge rate, and, provided that the recharge rate is not so high as to create gross polarization and premature charger trip out, the amp hours delivered on discharge will exactly match those of the previous recharge. No excess charge is required as no secondary reactions occur. Secondly, the beta alumina is an ionic conductor and an electronic insulator. The self-discharge electron path available in non-ceramic electrolytes is thus absent, and a sodium–sulphur cell will remain at the top of charge for extended periods. Tests with open circuit holds have demonstrated this characteristic to 35 000 hours [24]. High shelf-life is critical for certain space applications.

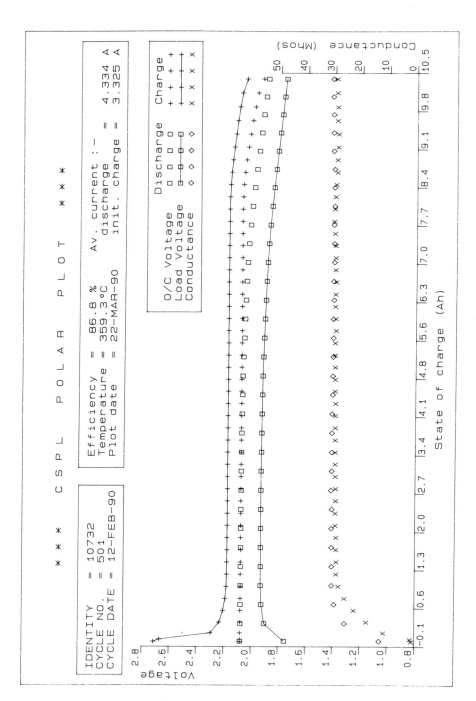

Fig. 6.4 — Typical in-cycle performance of sodium–sulphur cell.

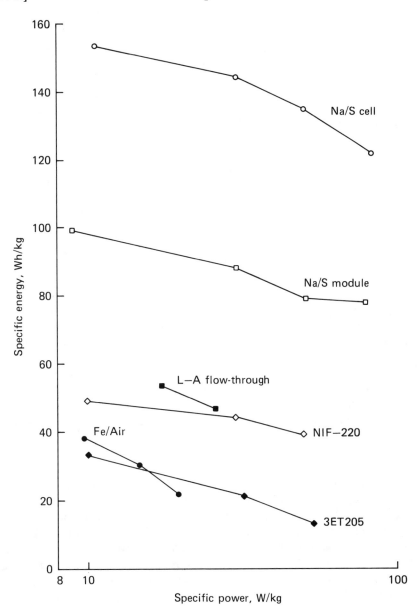

Fig. 6.5 — Ragonne plot of sodium–sulphur systems. Comparisons are made with lead–acid (L-A and 3ET205, nickel–iron (NiF-220) and iron–air (Fe/air).

### 6.2.3 Cell construction

The cell configuration and size have already been discussed. Most tubular sodium–sulphur cells consist of the same basic features regardless of size, especially if the configuration is of the central sodium type, at present the most common. However, a schematic of the three types of cell configuration are shown in Fig. 6.7.

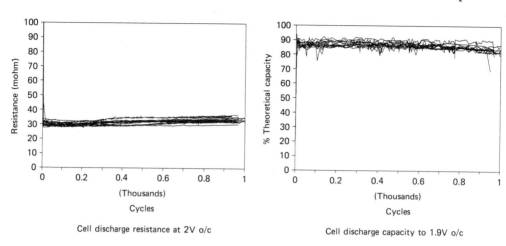

Fig. 6.6 — Resistance and capacities of sodium–sulphur cells to 1000 cycles.

An important component of any sodium–sulphur cell is the sealing arrangement. This is required to prevent the ingress of contaminants to the cells and loss of reactants to the atmosphere. The two electrode seals must be electrically insulated from each other, thus an alpha alumina header is usually attached to the beta alumina electrolyte. The seal at the top of the ceramic usually consists of a glass joint between the beta alumina and an alpha alumina header. Metal components can be sealed by either mechanical means or by a thermocompression bond to the alpha alumina. Aluminium or high aluminium alloys are the most frequently used sealing interstates to provide the required high level of hermeticity. Choice of material for the sodium electrode seal is relatively straightforward; that for the sulphur electrode is dictated by a need for corrosion resistance to sodium polysulphides and sulphur.

The current collector of the central sulphur cell and the container of the central sodium cell perform the same electrochemical function of injecting electrons into the sulphur electrode during discharge. They both have a requirement for chemical stability in the electrode environment. The materials chosen by most developers are based upon aluminium (and its alloys) and/or high chromium alloys (or substrates coated with high chromium compounds). Both the chromium and aluminium based systems have demonstrated stable characteristics over useful cell lives, although the development of inexpensive, production-engineered solutions remains a serious challenge. The success of the sodium–sulphur system in commercial terms depends largely upon the realization of low costs in the highly cost-conscious transport and utility load-levelling markets.

### 6.2.4 The battery system

Battery systems require the interconnection of cells in series and parallel to build up the terminal voltage and capacity. This presents no problems to the aqueous system

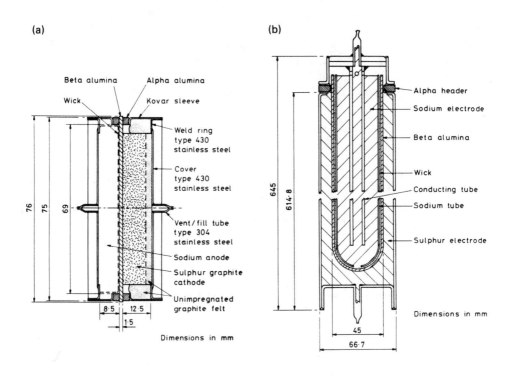

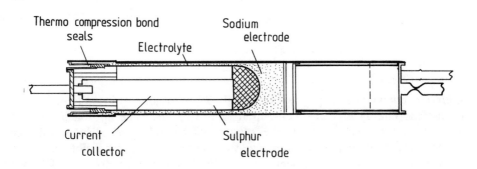

Fig. 6.7 — Configurations of sodium–sulphur cells. (Copyrights (a) 1976 and (b) 1982, EPRI. Reproduced with permission.)

as there are no modes of failure of the liquid electrolytes, and the electrodes are generally solid. In contrast, fracture of the ceramic electrolyte of the sodium–sulphur cell causes mixing of the electrode materials, so some means of controlling the effect of occasional cell failure has to be provided. Individual cells lose voltage and have variable post-failure resistances, and this could open circuit a simple series arrangement or short out a parallel arrangement. In practice, a mixture of series and parallel connection is needed to meet both capacity and voltage requirements. One solution, proposed by Bindin [25], now forms the basis of CSPL's network strategy which exploits the high top of charge. This high resistance, which rises exponentially at the end of charge, acts like a switch. Cells are interconnected on a parallel patch between every fourth series connection. This is illustrated in Fig. 6.8. The choice of 4 cell

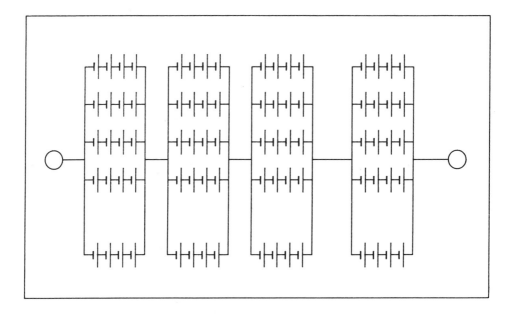

Fig. 6.8 — 4 cell interconnection network in battery.

interconnection is based upon assessment of the probability of all 4 failing, balanced against the energy loss of 4 cells upon failure of one. The approach has been successfully demonstrated in batteries of up to 3500 × 20 Wh cells. With suitable cell life statistics, a fully sealed maintenance-free battery is a highly credible target.

A different approach has been adopted by ASEA Brown Boveri (ABB) who have used a short circuiting looping device around each cell to ensure a short circuit failure. A disadvantage of this approach is that the reliability of the device has to be of a greater order than that of the cell. Cost also has a bearing upon this approach, although the device can be three quarters of the cost of a cell and still compare favourably with the 4 cell string approach. The devices being used by various developers are key components, and their detailed constructions remain closely guarded secrets.

The sodium–sulphur system is operated at between 300°C and 369°C, which requires the use of high thermal performance battery containment. This is a requirement shared by other high temperature systems such as lithium–metal

Sec. 6.2]  **Sodium–sulphur batteries**  467

sulphide. Aqueous battery systems may also require an insulating container for electric vehicle applications to keep the battery warm at low ambient temperatures, although this would have a reduced temperature range. One of the important uses of battery-powered forklift trucks, in cold storage warehousing, also requires an insulating container. A sodium–sulphur battery would not require any modification for operation across the 40°C shift in temperature resulting from cold storage materials warehousing.

Most developers have concentrated on the design of high performance insulated containers with a wall thickness of less than 30 mm. A double vacuum skin device with a radiation barrier within the vacuum is shown in Fig. 6.9. The system is

Fig. 6.9 — Prototype NaS battery for Bedford CF Electric vehicle.

described in more detail by Leadbetter [26]. The 60 kWh battery shown was the first to be located totally outside the cargo space in an electric vehicle. It demonstrated a 100 mile range on the open road on a single charge.

A detailed schematic of a typical EV battery is shown in Fig. 6.10. This design was developed for the Ford ETX$^{TM}$ II vehicle and illustrates the 4-cell interconnections arranged into 30 string monoblocs or 120 cells. Each of these monoblocs is a 300 Ah, 8 V unit, and they are connected in series to produce the required battery voltage (typically around 200 V for these applications). Also shown are the cooling plenums. Although these were intended to cool the battery at high discharge profiles, subsequent battery testing has shown that they are not required, enabling further weight savings to be made in subsequent designs. A battery such as that illustrated

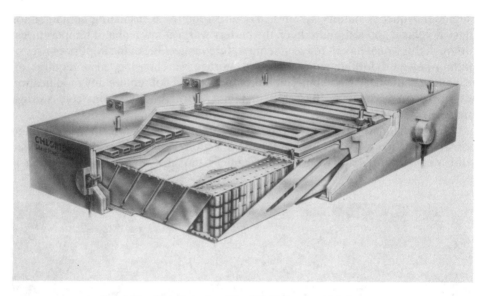

Fig. 6.10 — Detailed schematic of NaS electric vehicle battery.

was delivered to the USA for independent testing under a US Department of Energy contract.

The market for advanced batteries is not limited to electric vehicles. Designs are being developed for use in stationary energy storage (S.E.S) systems, standby power systems, electric fork trucks, and for a variety of space and military applications. A typical S.E.S. design is shown in Fig. 6.11. This is of a 100 MWh unit consisting of

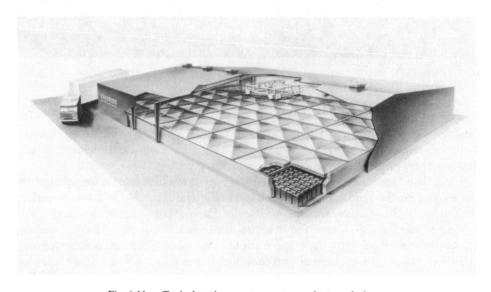

Fig. 6.11 — Typical stationary energy storage battery design.

1.2 m sodium–sulphur cells [27]. In such designs the saving on footprint over lead–acid batteries is substantial.

The weight saving of the sodium–sulphur system can also be exploited in circumstances where the floor loading is critical.

### 6.2.5 Development status

During more than two decades of development, the number of organizations working on sodium–sulphur has fluctuated. A feature of the research effort in the middle to late 1980s has been an increasing number of collaborative ventures based upon pooling of technologies to create strengthened organizations. This has occurred across three continents, in some instances involving Europe, Japan, and North America. A further feature of this period was the increased testing of batteries by third parties. This type of independent testing provides an unbiased assessment of development status and also helps the developer to refine both his design and testing procedures. In the USA, independent battery testing has been led by the Argonne National Laboratories (ANL) [28]. This organization has tested cells, modules, and batteries for the majority of the US-based development efforts and also for those efforts sponsored by the US Department of Energy. One such test was carried out for Chloride Silent Power on a sub-battery of 960 NaS cells representing a third of a subsequent battery delivered for testing at the Ford Motor Company. This battery was tested for 240 electrical cycles at different rates and also to various driving profiles, being cycled to 80% depth of discharge (DOD) (8 V per monobloc, 2 V per cell), with the occasional cycle to 100% DOD (7.6 V per monobloc, 1.9 V per cell). During normal operation, 3 h discharge/9 h recharge, the battery consistently gave approximately 240 Ah when discharged to 80% DOD, as shown in Fig. 6.12.

The testing at ANL was recently followed by independent testing of a delivery van at the Chattanooga test facility of Electrotek Inc. The battery was funded by Southern California Edison and the Tennessee Valley Authority and constructed by Chloride Silent Power, and it is shown in Fig. 6.13. This testing confirmed the findings of the ANL tests, and the vehicle is shown at full speed on the test track in Fig. 6.14. Ranges of up to 163 miles were recorded at 35 mph, and the 120 mile mark was exceeded at 50 mph. At the time of writing the battery had completed almost 3000 miles on the track in addition to a variety of static load and high power testing.

An assessment of the relative technology status of the principal developers is difficult to make in an environment which is moving rapidly toward commercialization. In such circumstances the sponsoring organizations are understandably reticent about their future plans. Both CSPL and ABB have had prototype batteries tested by independent organizations. These have demonstrated that the performance of the batteries matches the predictions that have been made for the NaS system. CSPL is constructing a production module capable of manufacturing 5000 NaS cells per week for use in extensive customer field trials. ABB have been operating a similar capacity manufacturing plant for some years, and have completed an estimated 200 000 km in NaS powered vehicle trials. Yuasa have constructed the largest battery for stationary applications testing (a 400 kWh, 50 kW unit) and have well advanced plans for the construction of a 1 MW, 8MWh unit for stationary testing. Overall, sodium–sulphur is perceived as one of the leading contenders for the developing advanced battery

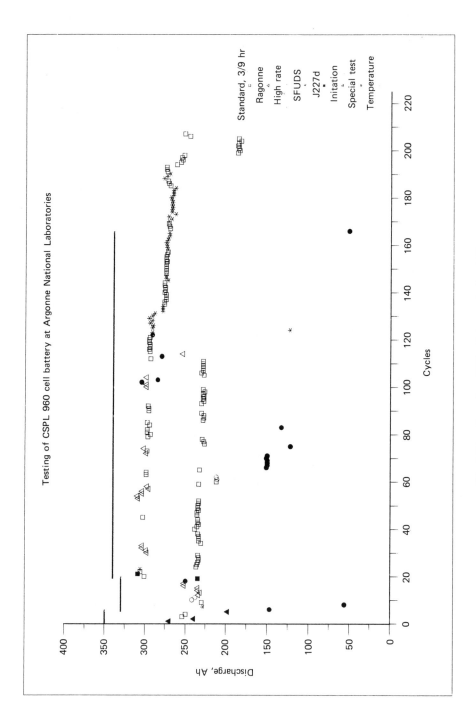

Fig. 6.12 — Capacity of 960 cell demonstration battery.

## Sodium–sulphur batteries

Fig. 6.13 — Battery for GM Griffon Van for Southern California Edison.

Fig. 6.14 — GM Griffon Van at speed on Electrotek Test Track, Chattanooga, TN.

market, particularly if the demonstrated technical benefits can be incorporated into a reliable low cost product.

Based upon the performance of the battery, ANL constructed a table of projected ranges for different simulated driving profiles, comparing the range obtained from sodium–sulphur with other battery systems. Three types of vehicle were considered, a 2.5 ton van (IDSEP), a passenger car (IETV-1), and a 4.5 ton van G-Van. The considerable advantages of the sodium–sulphur system are clearly evident in Table 6.1. Ranges of 148 miles and 182 miles were projected for electric vehicles with CSPL batteries, which exceed lead–acid battery ranges by a factor of 3.

**Table 6.1** — Projected ranges for simulated driving profiles

| Driving profile | | | | |
|---|---|---|---|---|
| Schedule | | SFUDS79 | J227aD | J227aC |
| Vehicle | | IDSEP | IETV-1 | G-Van |
| Battery weight | (kg) | 695 | 488 | 1180 |
| Average speed | (mph) | 19.0 | 28.3 | 15.1 |
| Peak Power | (W $kg^{-1}$) | 79 | 48 | 36 |
| Average power | (W $kg^{-1}$) | 9.9 | 12.0 | 7.3 |
| Battery | | \multicolumn{3}{c}{Battery range in miles} | | |
| Na–S 1/3 Battery | | 148 | 182 | TBD |
| Ni–Fe NIF220 | | 75 | 98 | 93 |
| Lead–acid 3ET205 | | 47 | 54 | 65 |

## 6.3 NICKEL–HYDROGEN BATTERIES

(J. J. Smithrick)

### 6.3.1 General characteristics

The state of development of individual pressure vessel (IPV) nickel–hydrogen battery cells is such that they are acceptable for geosynchronous orbit (GEO) applications where not many cycles are required over the life of the system (1000 cycles, 10 years). For the demanding low earth orbit (LEO) applications (30 000 cycles, 5 years), however, the current cycle life of 6000 to 10 000 cycles at the deep depth of discharge (DOD) of 80% is not acceptable [29,30]. At shallower depths of discharge (10 to 40%) the cycle life is projected to be acceptable; however, the database required for validation has not been fully generated. It is in the process of evolving.

The nickel–hydrogen battery cell is characterized by solid nickel and gaseous hydrogen electrodes housed in a pressure vessel which generally is a cylinder with hemispherical end caps. The nickel–hydrogen system combines the best electrode of the nickel–cadmium cell and the alkaline hydrogen–oxygen fuel cell. IPV cells when connected electrically in series constitute a battery.

Some of the advantages of the nickel–hydrogen compared to nickel–cadmium battery are (a) longer cycle life, (b) higher specific energy, (c) inherent protection against overcharge and overdischarge (reversal), and (d) cell pressure can be used as an indication of state of charge. Some of the disadvantages are (a) relatively high initial cost due to limited production, which could be offset by cycle life costs, and (b) self discharge which is proportional to hydrogen pressure.

Nickel–hydrogen batteries have captured a large share of the space battery market in recent years [31]. They are rapidly replacing nickel–cadmium as the energy storage system of choice.

### 6.3.2 Manufacturers and experience

There are four major manufacturers of nickel–hydrogen battery cells in the United States. They are Eagle–Picher Industries, Whittaker–Yardney Corporation, Gates Aerospace Batteries (formerly General Electric), and Hughes Aerospace Corporation. In Europe and Japan several companies are developing nickel–hydrogen cells and could become manufacturers. They are Harwell/Marconi (ESA contract) — England, SAFT (ESA contract) — France, Daug (ESA contract) — Germany, Furukawa Battery Company, Toshiba (NASDA contract) — Japan. The United States manufacturers and their experience are summarized in Table 6.2.

Table 6.2 — Summary of nickel–hydrogen cell manufacturers and experience

| | |
|---|---|
| Eagle–Picher (Joplin, MO) | Manufacture Air Force/Hughes, Comsat/Intelsat, and Advanced design cells. Manufactured over 10 000 cells as of February 1989. Manufactured battery cells for 16 spacecraft which are on GEO. Eagle–Picher Colorado Springs has manufactured a limited number of 150 ampere hour $4\frac{1}{2}$ inch diameter cells |
| Whittaker–Yardney Pawcatuck, CT) | Manufacture Air Force/Hughes, Comsat/Intelsat, and Advanced design cells. Selected as one of two vendors to supply cells for evaluation as to their suitability for use on Space Station Freedom. |
| Gates Aerospace Batteries (Formerly GE, Gainesville, FL) | Manufacture Air Force/Hughes, Comsat/Intelsat, and Advanced design cells. Selected as one of two vendors to supply cells for evaluation as to their suitability for use on Space Station Freedom. They are relatively new manufacturers of nickel–hydrogen cells; however, they have supplied cells for Superbird and Intelsat-VII communication satellites. |
| Hughes Aerospace Corp. (El Sefundo, CA) | Manufacture Air Force/Hughes, and Advanced design cells. They are under contract to supply nickel–hydrogen batteries for the Intelsat VI series of communication satellites. |

### 6.3.3 Chemistry

The electrochemical reactions for the normal, overcharge, and overdischarge (cell reversal) of a sealed nickel–hydrogen reversible battery cell or summarized in Table 6.3.

**Table 6.3** — Nickel–hydrogen cell electrochemistry

| Operation mode | Electrochemistry |
|---|---|
| **Normal** | |
| Nickel electrode | $Ni(OH)_2 + OH^- \underset{Discharge}{\overset{Charge}{\rightleftharpoons}} NiOOH + H_2O + e^-$ |
| Hydrogen | $H_2O + e^- \underset{Discharge}{\overset{Charge}{\rightleftharpoons}} \tfrac{1}{2} H_2 + OH^-$ |
| Net reaction | $Ni(OH)_2 \underset{Discharge}{\overset{Charge}{\rightleftharpoons}} NiOOH + \tfrac{1}{2} H_2$ |
| **Overcharge** | |
| Nickel electrode | $2OH^- \rightarrow \tfrac{1}{2} O_2 + H_2O + 2e^-$ |
| Hydrogen electrode | $2 H_2O + 2e^- \rightarrow H_2 + 2OH^-$ |
| Net reaction | $H_2 + \tfrac{1}{2} O_2 \rightarrow H_2O + HEAT$ |
| **Overdischarge (cell reversal)** | |
| Nickel electrode | $H_2O + e^- \rightarrow \tfrac{1}{2} H_2 + OH^-$ |
| Hydrogen electrode | $OH^- + \tfrac{1}{2} H_2 \rightarrow H_2O + e^-$ |

#### 6.3.3.1 Normal operation

During normal operation of a sealed rechargeable nickel–hydrogen cell, hydrogen is produced during charge and consumed during discharge at the catalyzed hydrogen electrodes. The pressure is proportional to ampere-hours into or out of the cell, and it can be used as an indicator of state of charge. The hydrogen is not segregated, and it reacts with the nickel oxyhydroxide resulting in a relatively slow self-discharge. As indicated by the net electrochemical reaction, no net water is produced or consumed. Hence there is no net change in the overall potassium hydroxide (KOH) electrolyte concentration.

#### 6.3.3.2 Overcharge

During overcharge, oxygen is generated at the nickel electrodes and hydrogen is generated at the catalyzed hydrogen electrodes. The oxygen level is limited since it recombines rapidly with hydrogen on the catalyzed hydrogen electrodes (Air Force/Hughes and Comsat/Intelsat design cells). For the NASA advanced cell the recombination occurs on the catalyzed wall wick.

### 6.3.3.3 Reversal
The nickel–hydrogen battery is tolerant of reversal since there is no pressure build-up in the sealed battery cells and no heat of recombination. The hydrogen generated at the nickel electrode recombines with the hydroxyl ions at the catalyzed hydrogen electrodes.

### 6.3.4 Cell construction
There are three similarly designed IPV nickel–hydrogen battery cells commercially available at present. They are the Air Force/Hughes, Comsat/Intelsat, and recently the NASA advanced cell. The Air Force/Hughes cell was designed for primarily low earth orbit applications, but is considered an all orbit cell. The Comsat/Intelsat cell was designed for primarily geosynchronous orbit applications. The NASA advanced cell was designed for primarily long life low earth orbit applications.

#### 6.3.4.1 Air Force/Hughes cell
The Air Force/Hughes cell is illustrated in Fig. 6.15. It consists of a stack of nickel electrodes, separators, hydrogen electrodes, and gas screens assembled in a recirculation configuration. In this configuration electrodes of different types directly face each other. The stack is packaged in a cylindrical pressure vessel, with hemispherical end caps. The vessel is made of Inconel 718. It is lined with a zirconium oxide wall wick. The components are shaped in a 'pineapple' slice pattern and the electrodes are connected in parallel.

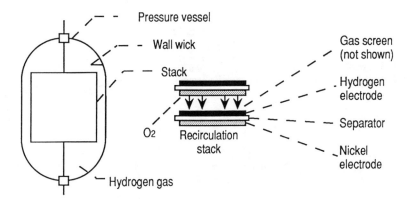

Fig. 6.15 — Air Force/Hughes individual pressure vessel nickel–hydrogen cell.

If a high bubble pressure separator, such as asbestos, is used, the oxygen generated at the nickel electrode on charge is directed to the hydrogen electrode of the next unit cell, where it recombines chemically with hydrogen to form water. If a low bubble pressure separator, such as Zircar™ (ceramic fabric), is used, the oxygen can pass through the separator and recombine on the hydrogen electrode of the same unit cell. This reaction is quite exothermic and care must be taken to limit damage to the hydrogen electrode surface due to the heat of recombination. The separators extend beyond the electrodes to contact the wall wick. Hence, electrolyte which leaves the stack during cycling will be wicked back. The nickel electrode consists of a sintered nickel powder plaque containing a nickel screen substrate, which is electrochemically impregnated with nickel hydroxide active material by the Pickett process [32]. The gas screens are polypropylene. The electrolyte is a 31% aqueous solution of potassium hydroxide.

### 6.3.4.2 Comsat/Intelsat cell

The Comset/Intelsat design is illustrated in Fig. 6.16 [33]. It consists of a stack of nickel electrodes, separators, hydrogen electrodes, and gas screens. The electrodes are connected electrically in parallel and are assembled in a back-to-back configuration. In this configuration the same type of electrodes face each other. Hence, the oxygen generated on charge leaves the stack between the nickel electrodes and re-enters between the hydrogen electrodes to combine chemically with hydrogen at the catalyzed hydrogen electrodes. This cell does not use a wall wick for electrolyte management. The nickel electrode plaques are fabricated by the slurry process and are electrochemically impregnated with active material by a modified Bell process. The separators are fuel cell grade asbestos, and the gas screens are Vexar™.

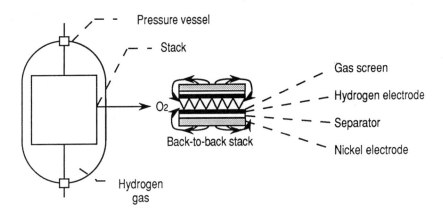

Fig. 6.16 — COMSAT individual pressure vessel nickel–hydrogen cell.

### 6.3.4.3 NASA advanced cell

The NASA Lewis advanced cell is illustrated in Fig. 6.17 [34]. Features of the advanced cell which are new and not incorporated in either the Air Force/Hughes or Comsat/Intelsat cells are (1) use of 26% rather than 31% potassium hydroxide electrolyte to improve cycle life, (2) use of a catalyzed wall wick located on the inside

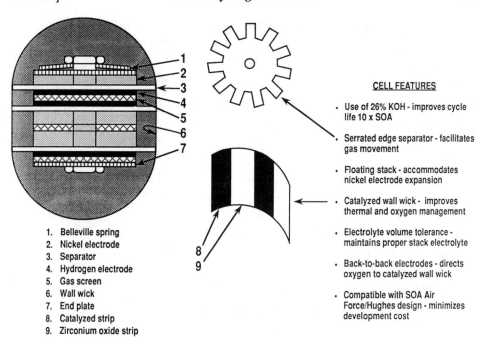

Fig. 6.17 — NASA advanced individual pressure vessel nickel–hydrogen cell-catalyzed wall wick. Comparison is made with state-of-the art (SOA) design.

surface of the pressure vessel wall to chemically recombine the oxygen generated at the end of charge with hydrogen to improve oxygen and thermal management, (3) use of serrated edge separators to facilitate hydrogen and oxygen gas movement within the cell while still maintaining physical contact with the wall wick, and (4) use of a floating rather than a fixed stack (state-of-the-art) to accommodate nickel electrode expansion.

Nickel–hydrogen battery cell information such as capacity, weight, average discharge voltage, and specific energy for three communication satellite batteries is shown in Table 6.4 [35,36,37]. The Spacenet and Intelsat V battery cells are the Comsat design. The Superbird cells are similar to the Air Force/Hughes cell design except the electrodes are in the back-to-back configuration.

### 6.3.5 Battery information

Specific energy and weight for three nickel–hydrogen batteries is shown in Table 6.5 [36,37,38]. They are the Superbird battery, launched June 1989, the Spacenet battery, launched May 1984, and the Intelsat-V battery, launched May 1983. The specific energy for the Superbird battery is 31.5% higher than the Spacenet and 37.2% higher than the Intelsat-V. This increase is due mainly to using a higher capacity cell (83 Ah for Superbird, 40 Ah for Spacenet, and 30 Ah for Intelsat-V). The battery cells account for about 80% of the battery weight. A photograph of an Intelsat-V battery is shown in Fig. 6.18 [39].

The energy per unit volume is shown in Table 6.6.

**Table 6.4** — Nickel–hydrogen battery cell information

| Parameter | Superbird | Spacenet | Intelsat-V |
|---|---|---|---|
| Rated capacity (Ah) | 83 | 40 | 30 |
| Cell weight (kg) | 1.874 | 1.168 | 0.890 |
| Capacity (Ah) | | | |
| 20°C | 92 | 41.67 | 31.91 |
| 10°C | 103 | 48.31 | 34.80 |
| 0°C | — | 49.90 | 35.31 |
| −10°C | 104 | | |
| Average discharge voltage (V) | 1.24 | 1.25 | 1.25 |
| Specific energy (Wh/kg) | | | |
| 20°C | 60.9 | 44.6 | 44.8 |
| 10°C | 68.2 | 51.7 | 48.9 |

**Table 6.5** — Nickel–hydrogen battery specific energy and weight

| Parameter | Superbird | | Spacenet | | Intelsat-V | |
|---|---|---|---|---|---|---|
| | Wt. (kg) | (%) | Wt. (kg) | (%) | Wt. (kg) | (%) |
| Cell weight | 1.874 | — | 1.168 | — | 0.890 | — |
| Number of cells | 27 | | 27 | | 27 | |
| Total cell weight | 50.60 | 78.5 | 25.70 | 78.7 | 24.03 | 79.8 |
| Battery weight other than cell wt. | 13.86 | 21.5 | 6.94 | 21.3 | 6.09 | 20.2 |
| Total weight | 64.46 | 100 | 32.64 | 100 | 30.12 | 100 |
| Energy (Wh) at 10°C | 3448.4 | | 1328.5 | | 1174.5 | |
| Specific energy (Wh/kg) at 10°C | ≅53.5 | | 40.7 | | 39 | |

### 6.3.6 Cell performance

#### 6.3.6.1 Discharge

The effect of temperature on the capacity for a 27 cell Superbird Ni–$H_2$ battery [38] and for a Spacenet battery cell lot of 54 cells is shown in Fig. 6.19 [36].

The capacity decreases, as expected, with increasing temperature. For the Superbird battery the capacity at 35°C is only about 57% of the capacity of 0°C, at

Fig. 6.18 — Intelsat-V nickel–hydrogen battery.

**Table 6.6** —Nickel–hydrogen battery energy per unit volume

| Parameter | Superbird | Spacenet | Intelsat-V |
|---|---|---|---|
| Number of cells | 27 | 22 | 27 |
| Length (cm) | 51.0 | 58.5 | 51.8 |
| Width (cm) | 89.0 | 53.3 | 52.1 |
| Height (cm) | 31.7 | 19.7 | 22.2 |
| Volume (L) | 143.9 | 61.5 | 59.9 |
| Energy (Wh) at 10°C | 3448.4 | 1328.5 | 1174.5 |
| Energy/volume (WH/L) | 24.0 | 21.6 | 19.6 |

30°C about 67%, at 20°C about 81%, and at 10°C about 94%. For the Spacenet cells, on average the capacity at 20°C is about 84% of the capacity at 0°C, and at 10°C about 97%.

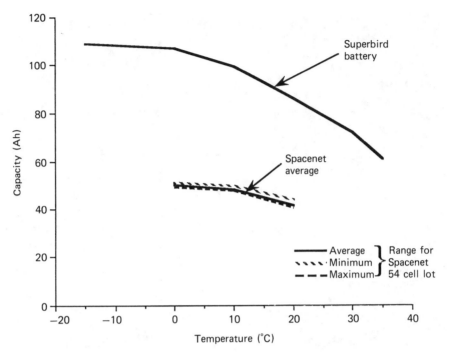

Fig. 6.19 — Effect of temperature on capacity for a Superbird battery and for a Spacenet cell lot.

For a representative Spacenet battery cell the discharge voltage as a function of ampere-hours removed at 0, 10, and 20°C is shown in Fig. 6.20 [36]. The discharge voltage plateau regions are relatively independent of temperature over the range tested.

The effect of discharge rate on capacity at 10°C for a representative 48 ampere-hour Hughes cell is shown in Fig. 6.21. The capacity is relatively independent of the discharge rate over the range tested.

### 6.3.6.2 Charge

The charge voltage and pressure as a function of ampere-hour input at 0, 10, and 20°C for a representative Spacenet battery cell are shown in Fig. 6.22 [36]. In general, the charge voltage increases with decreasing temperature. The pressure increases with state of charge. The temperature effect on cell pressure is minor over the range tested.

Cell voltage and pressure for a representative Spacenet cell during a 72 hour overcharge at the C/20 rate, 10°C is shown in Fig. 6.23 [36]. After about 64 ampere-hours the cell voltage remained constant at 1.5 V. The pressure increase after this was relatively minor (590 to 660 psig in 40 hours).

### 6.3.6.3 State-of-charge

During normal operation of a sealed rechargeable nickel–hydrogen battery cell, hydrogen is produced during charge and consumed during discharge. The hydrogen

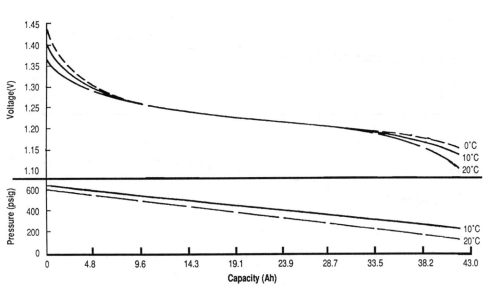

Fig. 6.20 — Effect of temperature on discharge voltage and pressure for a representative Spacenet battery cell.

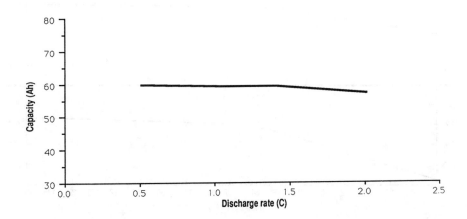

Fig. 6.21 — Effect of discharge rate on capacity for representative Hughes flight IPV Ni–H2 cell at 10°C.

pressure is proportional to the ampere-hours into or out of the cell and can be used as an indicator of state of charge. The data in Figs 6.20 and 6.22 show a linear variation of pressure with state of charge. It should be noted, however, that the cell pressure can increase with cycling; this can cause a shift in the state of charge versus pressure curve.

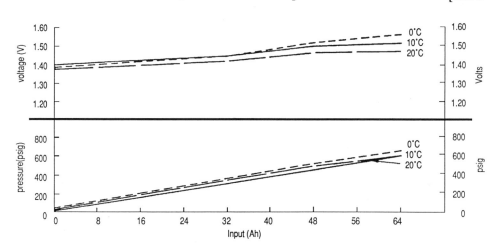

Fig. 6.22 — Effect of temperature on charge voltage and pressure for a representative Spacenet battery cell.

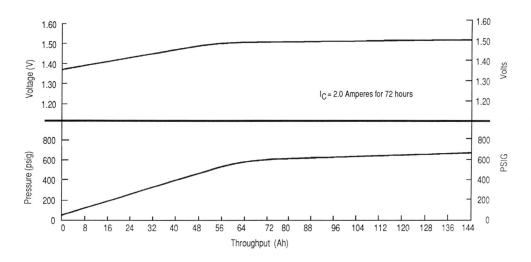

Fig. 6.23 — Effect of overcharge at 10°C on cell voltage and pressure for a representative Spacenet battery cell.

### 6.3.6.4 Capacity retention

In a nickel–hydrogen cell the hydrogen gas is not isolated. It comes into direct contact with the nickel electrodes, resulting in a relatively slow discharge on the NiOOH active electrode mass. The rate of this reaction is dependent on temperature and hydrogen pressure. The Spacenet battery cells on average lose about 15.5% capacity during a 72 hour open circuit stand at 10°C [36]. For a low earth orbit cycle regime the self-discharge is acceptable. For extended storage, however, the battery should be trickle charged if a full charge is required.

### 6.3.6.5 Life test

For geosynchronous orbit applications the cycle life for state-of-the-art nickel–hydrogen batteries is acceptable. An on-orbit database is rapidly being accumulated (16 satellites using Ni–$H_2$ batteries).

For low earth orbit applications such as Space Station Freedom and Hubble Space Telescope, the database is limited. There are virtually no on-orbit data as the terrestrial database is in the process of being developed. Low earth orbit (LEO) cycle life test results from GE Astro Space Division, Martin Marietta Astronautics, and NWSC Crane, Indiana are summarized in Table 6.7 [40,41,42]. The Hughes cells have been cycled at GE Astro Space for over 19 000 cycles at 40% DOD, 10°C (no failures, test continuing). Yardney cells have been cycled at Martin Marietta for over 18 000 cycles at 40% DOD, 10°C (no failures, tests continuing). Yardney cells have been cycled also at NWSC, Crane for over 12 224 cycles at 40% DOD, −5°C and over 14 483 cycles, 10°C (no failures, tests continuing). General Electric (GE, now Gates) cells have been cycled at Martin Marietta for over 15 968 cycles, 10°C and over 16 970 cycles, 20°C (no failures, tests continuing).

The above data were for a 40% DOD. At a deeper depth of discharge (80%) the LEO cycle life for state-of-the-art cells range from 6000 to 10 000 cycles [1,2]. The advantage of operating at a deep depth of discharge is that the battery specific energy is improved (increasing the DOD from 40% to 80% will double the specific energy). This is particularly important at high power levels to limit the battery weight. It is desirable, therefore, to increase the cycle life at deep depths of discharge. One way of doing this, suggested by results of a recent boiler plate LEO cycle life stress test, is to decrease the potassium hydroxide (KOH) electrolyte concentration. For this test failure was defined to occur when the discharge voltage degraded to 0.9 V during the course of the discharge. The effect of KOH concentration on cycle life is shown in Fig. 6.24 [43,44] and shows there to be a strong relationship between the two. The cycle life was improved by greater than a factor of 10 over state-of-the-art cells. Boiler plate cells containing 26% KOH were cycled for 39 600 accelerated LEO cycles at 80% DOD, 23°C, compared to 3500 cycles for cells containing 31% KOH. This was a breakthrough in cycle life. The next step is to validate these results, using flight hardware and a real time test.

The validation test is in progress at the national battery test laboratory, NWSC, Crane, Indiana. Six 48 ampere-hour Hughes flight cells are being cycle life tested under a LEO cycle regime at 80% DOD, 10°C. Three of the cells contain 26% KOH (test cells) and three contain 31% KOH (control cells). The test results are shown in Fig. 6.25. Cells containing 26% KOH have been cycled for over 6288 cycles with no failures during the continuing test. Two of the cells containing 31% KOH have failed (cycle 3729 and 4165).

A validation cycle life test is also being conducted at NWSC on the 125 Ah NASA advanced design catalyzed wall wick cells manufactured by Eagle–Picher. The cycle

**Table 6.7** — Summary of LEO cycle life test results for IPV Ni–H$_2$ battery cells

| Manufacturer | No. cells | Capacity (Ah) | Cycle regime | DOD (%) | T (°C) | Cycles | Status | Test laboratory |
|---|---|---|---|---|---|---|---|---|
| Hughes | 3 | 50 | LEO | 40 | 10 | 19000 | No failures test continuing | GE Astro Space |
| Yardney | 5 | 50 | LEO | 40 | 10 | 18000 | No failures test continuing | Martin Marietta |
| GE (now Gates) | 16 | 50 | LEO | 40 | 10 | 15968 | No failures test continuing | Martin Marietta |
| GE (now Gates) | 7 | 50 | LEO | 40 | 20 | 16970 | No failures test continuing | Martin Marietta |
| Eagle–Picher | 16 | 50 | LEO | 40 | 10 | 16912 | No failures test continuing | Martin Marietta |
| Eagle–Picher | 8 | 100 | LEO | 40 | 10 | 12787 | No failures test continuing | Martin Marietta |
| Yardney | 4 | 100 | LEO | 40 | 10 | 10111 | No failures test continuing | Martin Marietta |
| GE (now Gates) | 5 | 100 | LEO | 40 | 10 | 11961 | No failures test continuing | Martin Marietta |
| Yardney | 12 | 50 | LEO | 40 | −5 | 12224 | No failures test continuing | NWSC, Crane Indiana |
| Yardney | 10 | 50 | LEO | 40 | 10 | 14483 | No failures test continuing | NWSC, Crane Indiana |

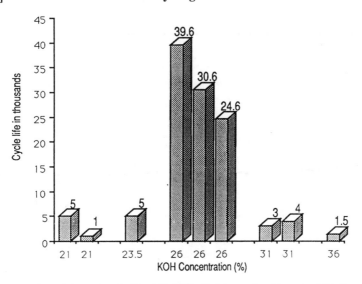

Fig. 6.24 — Comparison of cycle life of Ni–H2 boiler plate cells with various KOH concentrations, 80% DOD, 23°C.

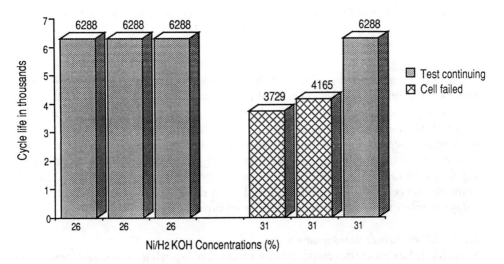

Fig. 6.25 — Comparison of LEO cycle life for Hughes IPV Ni–H2 flight cells containing 26% and 31% KOH electrolyte, 80% DOD, 10°C.

regime is a LEO at 60% DOD, 10°C. The average end of discharge voltage (3 cells) during the continuing test is shown in Fig. 2.26. After 2800 cycles no cells failed, and the discharge voltage was stable.

### 6.3.6.6 Storage
Nickel–hydrogen batteries could undergo a planned or unplanned storage due to delays before launch. It has been reported that the nickel–hydrogen battery could

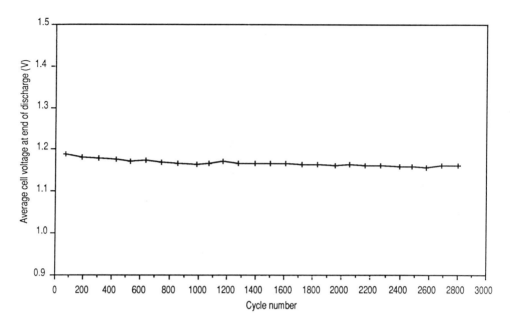

Fig. 6.26 — Effects of LEO cycling at 60% DOD on 125 Ah Eagle–Picher flight cells — 26% electrolyte, catalyzed wall wick, 10°C.

experience a capacity fading during such a storage [45]. The magnitude of the fading and its reversibility depends on several factors such as storage temperature, time, hydrogen pressure, and nickel electrode design.

Intelsat-VI nickel–hydrogen battery cells manufactured by Hughes, stored discharged open circuit at room temperature, undergo varying degrees of capacity fading with storage time [46]. Cell capacity can be restored by high rate reconditioning. Capacity fading does not occur, however, when cells are stored at −20°C or 0°C in the discharged, open circuit condition. Trickle charge, or top charge of cells every 7 days at room temperature also produces no capacity fading.

### 6.3.7 Likely future developments

Some likely key future developments for nickel–hydrogen batteries could include (1) increasing the cycle life for low earth orbit applications to 40 000 cycles (6.9 years) at modest to deep depths of discharge (40% to 80%), (2) increasing the specific energy from the state-of-the-art (SOA) 50 Wh kg$^{-1}$ to 100 Wh kg$^{-1}$ for geosynchronous orbit applications, and (3) developing a bipolar nickel–hydrogen battery for high pulse power applications.

Improving the cycle life will reduce satellite cycle life costs by reducing the frequency of battery replacement. This may be accomplished by modifying the state-of-the-art design to eliminate identified failure mechanisms and by using 26% rather than 31% (SOA) KOH electrolyte.

As the power required for satellites increases, the fraction of satellite mass occupied by the SOA power system will also increase, reducing payload mass. This situation can be altered by increasing the battery specific energy. For a given fixed power requirement, increasing the specific energy will increase the payload mass or decrease the launch mass. Increased specific energy could be accomplished by operating at a deeper depth of discharge and using a thick light-weight nickel electrode substrate which has a high porosity and is heavily loaded.

Bipolar batteries have been under development, using boiler plate pressure vessels. Design feasibility has been demonstrated, and the program is now being directed to design and demonstrate bipolar flight batteries for high power pulse applications.

## 6.4 ALUMINIUM–AIR BATTERIES
(C. D. S. Tuck)

### 6.4.1 Historical introduction

Aluminium is a high energy density fuel, possessing a theoretical energy content of 8.1 Wh $g^{-1}$, thus it is potentially a prime battery anode material. Its use in this manner was first suggested for a voltaic cell in 1855 by Hulot [47], and patents were filed at this time for aluminium–carbon and aluminium–$MnO_2$ cells [48,49]. However, attempts to substitute aluminium for zinc in commercial cylindrical dry cells have failed, largely owing to the unpredictable corrosion behaviour of aluminium, which resulted in a poor shelf life [50]. The coupling of an aluminium anode to an air electrode was first suggested by Zaromb [51]. After some development work on anodes carried out at General Electric [52], and the US Naval Underwater Systems Command [53], the system was evaluated by the US Department of Energy for traction application [54]. A scenario by Cooper [55] indicated that an Al–air battery could power a compact car, operating with traffic-compatible performance, provided that 14 kg of fresh aluminium and 23 l of water could be installed every 400 km and the 41 kg of hydroxide sludge disposed of. The fuel cost was calculated to be somewhere around three times the cost of the internal combustion engine drive system, the aluminium being used as a portable energy storage medium. Cooper estimated that the consumption of aluminium used in such a way, for 1% of cars in the USA, would be 600 000 tonnes of aluminium per year.

As well as the previously mentioned work funded by the US Department of Energy in the late 1970s, which was carried out at the Lawrence Livermore Laboratory, and the work of Littauer at the Lockheed Missiles and Space Company [56] during the same period, the system was also being developed by groups in Europe. Valand [57] at the Norwegian Defence Research Establishment built a portable Al–air battery operating with a non-alkaline electrolyte which used super-purity aluminium and produced 120 W. The battery is shown in Fig. 6.27; it was about the size of a small suitcase. It was developed for powering military communications equipment in the field, and it operated for several days without maintenance. Ruch & Katriniok [58], at the Hoppecke battery company, also developed a battery for military use, employing high-purity aluminium with a potassium hydroxide electrolyte.

The idea of applying saline rather than alkaline electrolyte to the Al–air battery for vehicles is attributable to Despic [59], although Vielstich, of the University of Bonn, also developed small devices employing this electrolyte [60]. Despic

Fig. 6.27 — The aluminium–air battery developed by the Norwegian Defence Research Establishment [57]. (Courtesy of the International Power Sources Symposium Committee.)

devised the construction of a wedge shaped anode which was designed to move down the cell as it was consumed, as shown in Fig. 6.28 [61]. This design was later taken up by the Lawrence Livermore Laboratory as a means of rapidly recharging cells [62].

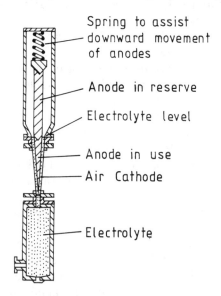

Fig. 6.28 — The wedge-shaped Al–air battery cell designed by Despic [61]. (Reproduced with permission.)

Alcan Aluminium Limited began their Al–air battery development largely through the US Department of Energy programme, working on the development of a crystallizer/separator unit [63]. A range of salt water-based Al–air demonstration battery products were initially developed by Alcan's battery company Alupower, and the potential use of a larger, saline battery of this type as a range-extender for off-road vehicles was demonstrated [64], this battery being a 230 W unit with a specific energy of 106 Wh.Kg$^{-1}$ including the electrolyte changes. In 1986, with the collaboration of the Power and Building Services Division of British Telecom, Alcan undertook to develop an Al–air battery of higher energy for use as a standby/reserve unit, replacing small diesel-powered motor generators in prolonged electricity shutdowns. The result was a 600 W unit capable of continuous operation for 60 hours which had an energy density of 355 Wh.L$^{-1}$, and was first demonstrated in 1987. The development of this system is now the subject of a joint venture between Alupower and Chloride.

### 6.4.2 Chemistry of the system

The chemistry of the Al–air battery is dependent on the electrolyte used, as aluminium displays amphoteric properties. In alkaline electrolytes a soluble aluminate complex Al $(OH)_4^-$ is formed according to the equation

$$Al + 4OH^- \rightarrow Al(OH)_4^- + 3e^- \quad E° = -2.35 \text{ V} . \tag{6.3}$$

The air electrode enables the following oxygen reduction reaction to take place:

$$O_2 + 2H_2O + 4e^- \rightarrow 4OH^- \quad E° = +0.40 \text{ V} . \tag{6.4}$$

Thus, the overall reaction is

$$4Al + 3O_2 + 6H_2O + 4OH^- \rightarrow 4Al(OH)_4^- , \quad E° = 2.75 \text{ V} . \tag{6.5}$$

The actual voltage obtained depends on several factors: the particular alloy used, the current density during operation, the anode cathode gap, and electrolyte conductivity. Typically, it falls in the range 1.0 to 1.5 V, which results an an electrical output of 3.0–4.5 Wh. g$^{-1}$ of aluminium.

As the aluminium dissolves, the aluminate ion concentration increases, which causes the solution viscosity and resistivity to rise. Eventually, the solution becomes supersaturated, and, after this point, aluminium hydroxide precipitates out as the hydrargillite phase. When this occurs, the solution viscosity and resistivity decrease markedly owing to the regeneration of hydroxyl ions according to

$$Al(OH)_4^- \rightarrow Al(OH)_3 + OH^- . \tag{6.6}$$

As well as the above current producing reactions (6.3)–(6.6), corrosion of the aluminium also occurs, accompanied by the production of hydrogen. This process is described by the equation:

$$2\text{Al} + 6\text{H}_2\text{O} + 2\text{OH}^- \rightarrow 2\text{Al(OH)}_4^- + 3\text{H}_2 \ . \tag{6.7}$$

Since the cell voltage is typically about half the theoretical value of 2.75 V, approximately half of the energy content of the aluminium is evolved as heat.

When saline electrolytes or electrolytes of pH <9 are used in Al–air batteries, the aluminium hydroxide products precipitate out very readily and initially cause the electrolyte to gel, before forming as a more crystalline species. Two hydroxide phases are generated, the initial one being amorphous pseudoboemite, usually given the formula AlOOH. During the later stages of the reaction, the more crystalline phase, bayerite develops, which is given the formula $\text{Al(OH)}_3$. Thus the overall cell equation is

$$4\text{Al} + 6\,\text{H}_2\text{O} + 3\text{O}_2 \rightarrow 4\text{Al(OH)}_3 \ , \qquad E^\circ = 2.06 \text{ V} \tag{6.8}$$

with a parasitic corrosion reaction

$$2\text{Al} + 6\text{H}_2\text{O} \rightarrow 2\text{Al(OH)}_3 + 3\text{H}_2 \tag{6.9}$$

A comparison of the energy and power capabilities of practical alkaline and saline Al–air systems gives energy densities of 400 Wh.kg$^{-1}$ and 220 Wh.kg$^{-1}$ and possible peak power densities of 175 W.kg$^{-1}$ and 30 W.kg$^{-1}$ for the two systems respectively.

### 6.4.3 Anode development

Aluminium is well known for its excellent corrosion resistance, and its use as an anode has meant the development of alloys which overcome its passive nature. The first extensive study for elements which, when added to aluminium, cause depassivation and allow the aluminium to assume a more active electrode potential, was carried out by Reding & Newport [65], who surveyed over 2500 experimental alloys. The elements which they found produced this effect tended to be those with low melting points, namely mercury, gallium, indium, tin, and thallium, and combinations of these were added to produce viable battery materials [66]. The mechanism by which these elements produce so-called superactivation of the aluminium, i.e. activation above that of a normal aluminium electrode, was discovered to be related to both their electrochemical and physical properties. The activator species are observed to accumulate on the surface of the dissolving alloy at specific active sites which then possess very little passive nature [67]. Thus they tend to work much more effectively if they are released by the dissolution reaction from a solid solution in the aluminium. Fig. 6.29 shows typical current/voltage curves for 200 ppm aluminium binary alloys of gallium, tin, and indium (mercury alloys being disregarded as possible battery materials on environmental grounds). The curves were produced in an unoptimized laboratory cell with 4 M NaOH at 60°C, and they show that gallium has the most active electrochemical performance. However, in practice, gallium alloys display extremely high open circuit corrosion rates, and are not suitable for battery use.

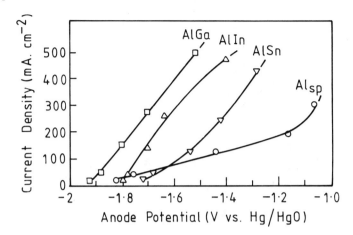

Fig. 6.29 — Polarization curves obtained in 4 M NaOH at 60°C during microcell discharge of super-purity (99.999%) aluminium and the superpurity-based binary alloys with 200 ppm gallium, indium and tin [79]. (Reproduced with permission.)

In alloys with more than one activating element present, it has been found that superactivation is controlled by just one of the additions, termed the dominant activator [68]. For the elements tin, indium, and gallium the order of dominance is tin > indium > gallium, and in a quaternary alloy of the type AlSnInGa, the superactivation is controlled by tin, even if its concentration in the alloy is lower than that of the indium and gallium.

The anodic behaviour of some indium-activated alloys is given in Table 6.8. The particular characteristics of these alloys are a relatively low open circuit corrosion rate before activation and a progressive increase in corrosion rate which is proportional to the current drawn. Various additions have been made to the alloys to try to mitigate this effect, the most successful being magnesium and manganese in combination [69]. Tin superactivated alloys, some of whose properties are given in Table 6.9, are characterized by higher open circuit corrosion rates compared to the indium alloys. However, as the current density is increased, the corrosion rate progressively decreases until coulombic efficiencies of near 100% are produced at 600 mA.cm$^{-2}$. Recently, a ternary alloy displaying excellent coulombic efficiencies over a much wider range of current density (15 mA.cm$^{-2}$ to 1000 mA.cm$^{-2}$) as well as possessing very good polarization characteristics, has been developed [70].

The AlSn alloys readily display the phenomenon which has become known as 'hyperactivation' [70]. This is a display of extremely negative potential behaviour in alkali electrolytes, as shown in Fig. 6.30. Other alloy additions have also been found to cause this effect, and it has been suggested that the phenomenon is due to the formation of a hydride phase; either aluminium hydride by catalytic interaction of the alloying element, or of the alloy metal hydrides themselves.

The effects of the alloying elements mentioned above on aluminium electrochemical activity can also be achieved by adding the activator elements in ionic form to an alkaline solution [71]. Gallium has been used in solution as

Table 6.8 — Comparative performance of indium activated anodes in 4 M NaOH at 60°C [68]

| Alloy system | $I_{oc}$ (mA cm$^{-2}$) | Coulombic efficiency | Anode potential (V) | | | Power density (W cm$^{-2}$) | | | Energy yield (kWh kg$^{-1}$) | | |
|---|---|---|---|---|---|---|---|---|---|---|---|
| | | | $E_{200}$ | $E_{400}$ | $E_{600}$ | $P_{200}$ | $P_{400}$ | $P_{600}$ | $EY_{200}$ | $EY_{400}$ | $EY_{600}$ |
| AlIn | 36 | 52% | 1.69 | 1.66 | 1.65 | 0.296 | 0.524 | 0.708 | 2.30 | 2.03 | 1.83 |
| AlInMn | 63 | 83% | 1.70 | 1.66 | 1.65 | 0.298 | 0.524 | 0.714 | 3.68 | 3.24 | 2.94 |
| AlInMg | 8 | 78% | 1.69 | 1.66 | 1.65 | 0.296 | 0.524 | 0.708 | 3.45 | 3.04 | 2.75 |
| AlInMnMg | 14 | 88% | 1.73 | 1.69 | 1.66 | 0.304 | 0.536 | 0.732 | 3.99 | 3.51 | 3.20 |
| AlInSb | 46 | 90% | 1.67 | 1.66 | 1.64 | 0.292 | 0.524 | 0.696 | 3.92 | 3.52 | 3.11 |
| AlInGa | 26 | 95% | 1.48 | 1.45 | 1.44 | 0.254 | 0.440 | 0.582 | 3.59 | 3.11 | 2.74 |
| AlInGaMn | 16 | 86% | 1.68 | 1.66 | 1.64 | 0.294 | 0.524 | 0.702 | 3.76 | 3.35 | 2.99 |
| AlInGaSb | 75 | 86% | 1.69 | 1.67 | 1.66 | 0.296 | 0.528 | 0.708 | 3.79 | 3.38 | 3.02 |

Notes: — (1) Open circuit corrosion rate measured after activated discharge.
(2) Power density and energy yields calculated for 2 mm gap cell with standard air electrode.

**Table 6.9** — Performance characteristics of tin binary alloys in 4 M NaOH at 60°C [68]

| Alloy composition (wt %) | Activation time (s) | Test time (min) | Current density (mA cm$^{-2}$) | Corrosion rate mg cm$^{-2}$ min$^{-1}$ | Anode potential (V) | Power density (W cm$^{-2}$) | Energy yield (kWh kg$^{-1}$) |
|---|---|---|---|---|---|---|---|
| 99.995 Al | — | 40 | 520 | 0.144 | 1.22 | 0.359 | 1.96 |
| 0.02 Sn | 14 | 40 | 620 | 0.096 | 1.43 | 0.502 | 2.35 |
| 0.05 Sn | 16 | 40 | 630 | 0.138 | 1.46 | 0.518 | 2.36 |
| 0.09 Sn | 14 | 40 | 650 | 0.039 | 1.48 | 0.540 | 2.45 |
| 0.2 Sn | 20 | 40 | 625 | 0.063 | 1.45 | 0.513 | 2.40 |
| 0.4 Sn | 10 | 40 | 630 | 0.19 | 1.47 | 0.525 | 2.36 |

*Notes:* – (1) The 0.02% Sn alloy was pre-activated at open circuit (900 kΩ resistance) prior to high current discharge.
(2) Power Density and Energy Yield calculated for 2 mm gap air cell.

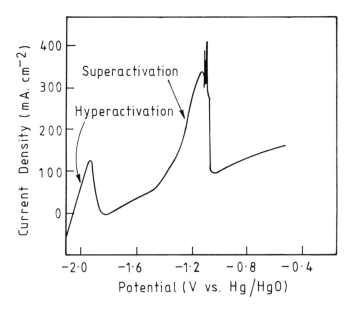

Fig. 6.30 — A rotating disk electrode potential sweep polarization curve obtained from an AlSn alloy in 4 M NaOH at 25°C [70].

gallium nitrate or as the molten salt $Ga(AlCl_4)_3$ to produce an active aluminium surface by dipping [72]. The most widely used electrolyte addition has been sodium stannate [73] which produces a tin superactivation behaviour and acts as a dominant activator when AlGa alloys are used as anodes. The typical concentration of sodium stannate added has been 0.06 M [55].

### 6.4.4 Air cathode development

Air cathode development has entailed optimization of both the oxygen reduction catalyst and carbon substrate as well as considerations of methods and costs of manufacture. Attention has been focused on macrocyclic compounds as efficient catalysts, and cobalt tetramethoxyphenyl porphyrin (Co-TMPP) has been found to perform particularly well [74]. The effectiveness of this catalyst was shown to be good even on carbon blacks with moderate or low surface area. The polarization behaviour in 5 M KOH of two commercial air cathodes is shown in Fig. 6.31, and effective voltages can now be achieved at current densities above 400 mA/cm².

During operation, the carbon undergoes structural changes, and it is electrochemically oxidized. This corrosion reaction has been found to be preceded by oxidation of amorphous carbon sites within the carbon particles, whilst the outer surface, which is more crystalline, remains intact [75]. Heat treatment overcomes this problem to some extent [76], and the use of Shawingan carbon black rather than active carbon has been suggested as a method of lessening carbon corrosion [77].

The continuous production of low-cost air cathodes is necessary for commercial development of any air battery. Fig. 6.32 shows a cross-section through the cathode

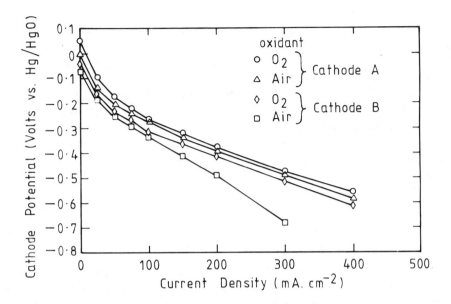

Fig. 6.31 — The polarization characteristics of two commercial air cathodes in 5 M KOH at 60°C, using air and oxygen as oxidants. Cathode A was supplied by Electromedia and cathode B supplied by Alupower [70]. (Courtesy of the International Power Sources Symposium Committee).

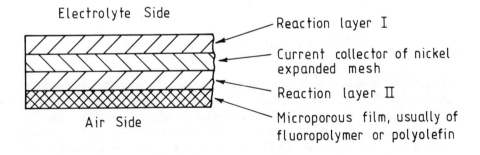

Fig. 6.32 — A cross-section through an Alupower AC45 air cathode showing its multilayer construction. The total thickness is approximately 0.5 mm.

manufactured by Alupower Inc [78], which is produced as a continuous web. The various layers are fed into a roller which binds them together, and the process has successfully produced a number of types of cathode, as each of the layers can be modified.

### 6.4.5 Electrolyte management

The method of dealing with electrolyte during the Al–air battery operation differs, depending on whether caustic electrolytes or neutral electrolytes are used.

With caustic electrolytes it is often necessary to control the precipitation of the hydrargillite from the supersaturated aluminate solution by means of a crystallizer unit, and to separate out the resulting solids so that the remaining liquor can be further used [79]. The crystallizer consists of a vessel containing a suspended mass of seed particles of an appropriate total surface area. The value of surface area must be critically controlled, otherwise an extremely fine precipitate will result which is not easily separated out. In the temperature range 60°C to 90°C the kinetics of hydrargillite crystal growth are related to the caustic and aluminate concentrations by the equation:

$$G = k[(R - R_{eq})/(1 - 1.04R)]^2 \ . \tag{6.10}$$

This results from a model proposed by King [80], where $G$ is the crystal growth rate ($\mu m. h^{-1}$), $k$ is the rate constant ($\mu m. h^{-1}$), $R$ is the ratio of concentration ($g.l^{-1}$) of aluminate ($Al_2O_3$) to caustic (as $Na_2CO_3$), and $R_{eq}$ is the equilibrium alumina/caustic concentration ratio. From Misra & White [81], the equilibrium aluminate–caustic ratio is given by

$$R_{eq} = \exp[5.6742 + (0.63598 - 2486.7/T)] \ , \tag{6.11}$$

where $C$ is the concentration of caustic ($g.l^{-1}$). Between 60°C and 90°C the rate constant $k$ has been found [63] to fit the Arrhenius equation

$$k = 3.91 \times 10^9 \exp(-14155/RT) \mu m. h^{-1} \ . \tag{6.12}$$

Fig. 6.33 shows values of the crystal growth rate, $G$, calculated from the above equations as a function of temperature for different aluminate concentrations in 4 M NaOH, this being the concentration possessing the highest specific conductivity. It can be seen that crystal growth rates are fastest between 80°C and 90°C. Agglomeration is found to be favoured by high crystal growth rates [63], and is thus more efficient at 80°C, leading to a high proportion of particles larger than 40 $\mu m$. The efficiency of agglomeration has also been observed to be lower in KOH than in NaOH because of the more elongated, less spherical crystal morphology developed in KOH. It is also seriously inhibited by any high turbulence and shearing forces present as a result of forced electrolyte pumping.

In the saline Al–air battery, to prevent the clogging of cells by the hydroxide precipitate it has been found necessary to pump the electrolyte continuously. Despic has shown [82] that the capacity of the cells can be more than doubled by reciprocating the electrolyte, that is, pumping the cell electrolyte up and down inside the cells. It has also been demonstrated that stirring the electrolyte by gas injection

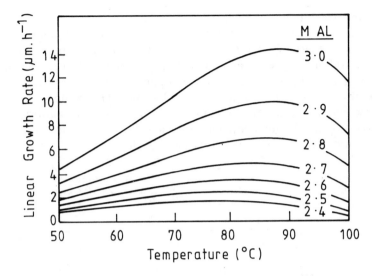

Fig. 6.33 — Crystal growth rates of hydrargillite as a function of temperature from 4 M NaOH containing the dissolved aluminium concentration indicated [63]. (Reproduced with permission.)

has a similar effect [64], and that the capacity of cells in which electrolyte is reciprocated is proportional to the solution conductivity, with 20% KCl achieving 0.42 Ah. cm$^{-3}$ of electrolyte, as shown by Fig. 6.34.

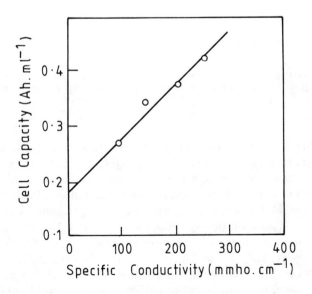

Fig. 6.34 — Electrolyte capacity vs cell electrolyte conductivity for a saline Al–air cell with electrolyte agitation [64].

### 6.4.6 Commercial battery development

The value of Al–air batteries lies in their high energy density, and commercial developments have largely used that aspect. Fig. 6.35 shows the probable commercial niches for different types of aluminium battery, showing a very wide spectrum depending largely on the power requirement. Aluminium battery systems such as the aluminium–silver oxide, developed recently by the SAFT battery company for torpedo power packs [83], or the aluminium–hydrogen peroxide battery for undersea use [84], rely on oxygen-containing substances as a cathode rather than the oxygen content of air, which is used in Al–air batteries.

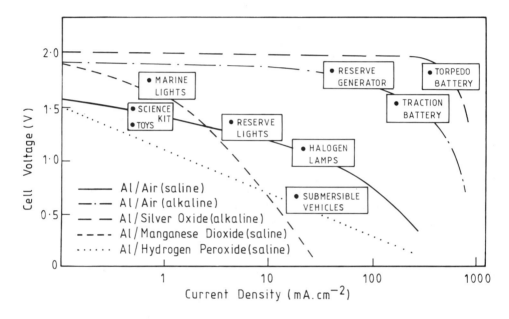

Fig. 6.35 — Polarization characteristics of various types of aluminium cells together with their commercial uses [68]. (Courtesy of the International Power Sources Symposium Committee).

Commercial development of Al–air batteries has initially been in the reserve light area, with products such as the Barge Mooring Light (Fig. 6.36), a saline battery which is used on barges to power signal lights at night, operating continuously for four weeks. A more powerful reserve battery which is designed to back-up telecommunications equipment has been developed by Alcan [85,86], and is rapidly approaching commercialization through Alupower–Chloride Limited, a joint venture between Alcan Aluminium's battery company and Chloride. A prototype version of this system is shown in Fig. 6.37, and it is designed to operate in conjunction with a controller and small rechargeable battery such as lead–acid. The latter supplies the power for all interruptions of less than about two hours' duration. After that time the load is shared between the two systems for half an hour before the

Fig. 6.36 — The Nightstar™ Barge Mooring Light commercialized by Alupower Inc.

Al–air battery completely takes over. The duty requirements of the Al–air battery in this case is to deliver 30 kWh over 50 h at a nominal voltage of 25 V. At an average power output of 600 W, the current is approximately 24 A. Fig. 6.38 shows the battery with the power block housing and electrolyte reservoir wall cut away. There are two sets of twenty cells connected in series, each set arranged in two banks of 10 cells each, both being situated above the reservoir by means of an electrolyte and attached to an electrolyte feed manifold by means of a plug-in O-ring sealed feed tube. Electrolyte is pumped upwards, through the cells, overflows a weir at the top, and passes downward under gravity through a drain tube to the reservoir. Air is blown through the stack from a stuffer box and passes between the cells, an even distribution being achieved by an array of pinholes in the stuffer-box wall. On its exit, the air enters a condenser to recover evaporated water. A heater exchanger is also present through which the electrolyte is pumped. This controls the electrolyte temperature at 60°C. Total parasitic power consumption by the battery is about 66 W, this being about 10% of the output. The volume of the battery is 106 L, giving a volumetric energy density of 356 Wh.L$^{-1}$ and a gravimetric energy density of 365 Wh.kg$^{-1}$.

Fig. 6.37 — The Alupower-Chloride Reserve Power System prototype designed for back-up of telecommunications exchange equipment.

### 6.4.7 Future developments

A great deal of development work is still taking place toward an Al–air traction battery. Eltech Corporation took part in the initial US Department of Energy programme toward this goal, and have continued the development to the point of producing a cell shown in Fig. 6.39. This cell, with a 300 cm$^2$ area of active electrode, operates at 200 mA/cm$^2$ and a voltage of 1.1 V. The weight of a 100 V battery stack of this design would be 211 kg and volume 159 L [87]. Optimization of the design is taking place.

Sec. 6.4] **Aluminium–air batteries** 501

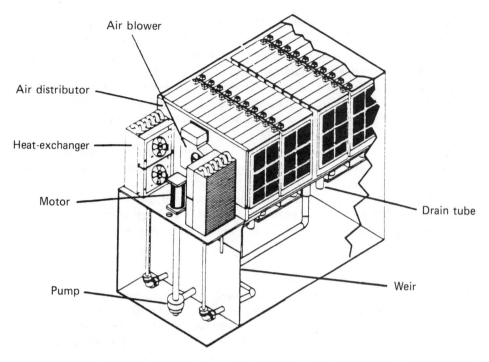

Fig. 6.38 — A cut-away view of the Alupower 40-cell aluminium–air battery designed as a reserve unit for telecommunications equipment [86]. (Reproduced with permission.)

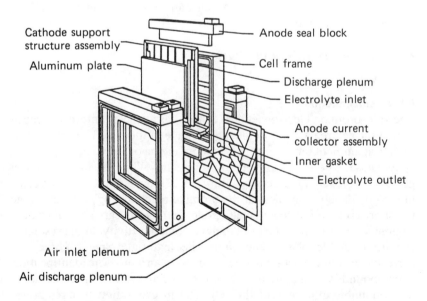

Fig. 6.39 — An exploded view of the aluminium–air bipolar battery developed by the Eltech Corporation [87]. (Reproduced by permission of the publisher, The Electrochemical Society, Inc.)

Alcan has also worked towards Al–air as an electric vehicle battery, but has concentrated on initially developing a hybrid system between Al–air and a secondary lead–acid battery. With a series of 8 lead–acid batteries operating in parallel with a bank of Al–air batteries, a distance of 105 miles has been travelled at 30 mph in a small hatchback vehicle [88]. With 16 lead–acid batteries alone, only 57 miles were achieved. Certain government agencies are realizing the possible benefits of Al–air as a traction battery and are assisting its development both directly and indirectly, as a result of environmental legislation [89].

Another area of development will be aluminium batteries in seawater systems which operate on liquid oxygen, hydrogen peroxide, or oxygen extracted from the seawater [90]. Envisaged applications will be both manned or unmanned submersible vehicles of various types, and it has been proposed that an aluminium battery would enable 100 MWh of energy to be packaged in a 10 m long plug slotted into a conventional diesel electric submarine of 2000 to 2400 tonne displacement [84].

**Summary**

The high energy density of aluminium has begun to be used in commercial battery systems both as reserve power units and as range extenders for electric vehicles. The favoured electrolyte system is based on aqueous alkali hydroxide owing to the ease of electrolyte management, as use of saline-electrolytes produces an insoluble hydroxide product and reduces the power density. Optimization of the aluminium alloy to produce high coulombic and voltage efficiencies has been achieved, and a low-cost air cathode has been developed with rugged character and good electrochemical properties. The success of the Alupower–Chloride telecommunications battery has alerted manufacturers to possible future application and markets for the battery, and, with further development, Al–air batteries will be increasingly used both on land and in the ocean.

## 6.5 ZINC–BROMINE BATTERIES

(A. Leo)

### 6.5.1 Introduction

The zinc–bromine electrochemical couple has attracted interest because of the low cost of the cell reactants and the high voltage of the cell. The cell reactions proceed readily at ambient temperature in aqueous media on inert reaction substrates, so the system has the potential for very long cycle life. While the first references to the cell go back to the late nineteenth century [91,92], development activities were initially limited by the corrosiveness of the bromine reactant, high self-discharge rates, and the short cycle life of the zinc electrode. In recent years a number of factors have combined to accelerate the level of development activity on the system. As low-cost chemically stable plastic and composite materials have become available, the corrosive bromine reactant has become less of an issue. The identification of a family of compounds which absorb evolved bromine and release it on discharge has raised the coulombic efficiency of the cell. The incorporation of circulating electrolyte systems into the battery configuration has eliminated the life-limiting problems of the zinc electrode: shape change, and dendrite shorting. Finally, interest in the system has been renewed because of the anticipated need for low cost, long life energy storage systems for peak shaving (load levelling) and electric vehicle propulsion.

The battery stores energy by the electrolysis of an aqueous zinc bromide salt solution to zinc metal and dissolved bromine liquid. The cell reactions are as follows:

$$\text{Discharged} \rightleftarrows \text{Charged}$$

Zinc electrode: $\quad Zn^{+2}(aq) + 2e^- \rightleftarrows Zn(s) \quad (6.13)$

Bromine electrode: $\quad 2Br^-(aq) \rightleftarrows Br_2(aq) + 2e^- \quad (6.14)$

The theoretical reversible voltage of the couple is 1.83 V, giving a theoretical energy density of 430 Wh.kg$^{-1}$. A number of configurations have been used in battery designs over the period the system has been studied. These have all been approaches to the major performance limiting problems encountered in early cells.

Self-discharge rates were initially high because of the non-Faradaic reaction between dissolved bromine and deposited zinc metal in the cell. One approach to limiting self-discharge has been the use of solid or liquid absorbent materials to extract bromine from the cell electrolyte. Activated carbon was used in some systems [93,94]; however, the carbon-to-bromine weight ratio was high, and the high surface area carbon tended to be unstable in the bromine solution. Organic solvents have also been used as liquid extraction agents [95], and, in one approach, a propionitrile solvent has been studied which forms charge transfer complexes and can be used as the electrolyte in the positive cell channels [96]. The most commonly used bromine extracting agents in the present state-of-the-art systems are alkyl ammonium bromide materials, which form polybromide complexes (R–Br$_3$, R–Br$_5$, etc.), with evolved bromine [97,98].

Another approach to limiting the self-discharge of zinc–bromine cells has been to use a separator material between the zinc and bromine compartments in the cell to limit diffusion of bromine to the zinc metal deposit. Tetrafluoroethylene-based ion exchange membranes have shown excellent chemical stability and very low self-discharge diffusion rates [99]; however, their high cost precludes their use in systems targeted for commercial applications. The most commonly used separator membranes in systems currently under evaluation are low cost polyolefin microporous materials such as Daramic™, from WR Grace. The self-discharge rate of cells using these separators is higher than that of cells using ion exchange membranes; however, the lower coulombic efficiency is acceptable in the light of the much lower cost of the microporous membranes.

Another problem faced in early investigations of the zinc–bromine system was the poor cycle life of the zinc electrode. The performance of the zinc electrode in secondary batteries has always been limited by shape change and dendrite short circuiting. The approach taken in the advanced zinc halogen battery systems has been to circulate the battery electrolyte over the zinc deposit. This eliminates maldistribution of dissolved discharge product, and provides uniform availability of Zn$^{++}$ for deposition on charge.

While many organizations have conducted experimental studies on the zinc–bromine system, the two major industrial developers who emerged in the 1970s were

Exxon Research and Engineering Company and Gould Inc., both in the USA. Extensive research, development, and engineering work was done in programmes funded by the developers, the US Department of Energy (DOE), and the Electric Power Research Institute (EPRI). The Exxon program focused on zinc–bromine batteries for electric vehicle use, but stationary energy storage applications were also considered. The Gould programme considered only stationary energy storage applications, such as utility load levelling. Both developers used bipolar stack configurations, with electrolyte storage and recirculation systems feeding both the positive and negative electrodes. Both technologies used liquid quaternary ammonium bromide materials as bromine storage agents and microporous polyethylene Daramic™ as the separator membrane.

In the Exxon system the organic bromine complex was mixed with the positive electrolyte to form an emulsion, which was circulated through the battery stack. The bipolar electrode used in the Exxon system was an extruded polyolefin/carbon black composite. One side of the electrode served as the zinc deposition substrate. A thin high surface area layer was laminated to the opposite side of the electrode to serve as the bromine reaction substrate. The positive electrolyte and the polybromide emulsion was circulated over the bromine layer, while a separate flow system circulated electrolyte over the zinc electrodes [100,101].

In the Gould zinc–bromine battery the polybromide complex was not circulated through the battery stacks, but was mixed with the aqueous electrolyte in a third recirculation system. Since the polybromide was not circulated through the bromine electrode compartments, the bromine activity in the electrodes was relatively low. While this enhanced the coulombic efficiency of the battery it also impaired the bromine reaction kinetics, so a high surface area flow-through electrode was used to support the bromine reaction. A carbon felt material which filled the bromine electrode compartment was used as the reaction substrate. The bipolar electrodes in the Gould stacks were thin vitreous carbon sheets [102,103].

In the 1980s both Gould and Exxon curtailed their zinc–bromine activities and transferred the battery technology to other developers. The Exxon technology was licensed to Johnson Controls Inc. (USA), Meidensha Electric Mfg. Co. (Japan), and S.E.A. GmbH (Austria). Johnson Controls and S.E.A. have maintained the focus on the electric vehicle application, but both developers are also evaluating stationary applications. These developers have concentrated on reducing the weight of the battery hardware and improving the reliability of stack seals. Meidensha has developed the system for stationary energy storage under the Moonlight Project of the Ministry of International Trade and Industry (MITI) of Japan. The Gould technology was acquired by Energy Research Corporation (USA), which has continued the development of the battery for stationary energy storage applications in cost-shared programmes with EPRI and DOE. ERC has also taken part in a technology exchange programme with Murdoch University (Australia) to advance the battery technology. Because of the long life requirement of the utility load-levelling application, the ERC programme has focused on developing stable cell and stack components. Since some of the high stability materials are relatively expensive, a major ERC focus has been on increasing the amount of capacity which can be stored per unit of cell active area [104,105].

## 6.5.2 Battery configuration

The developers who are at present working on zinc–bromine systems are using flowing electrolyte configurations. Fig. 6.40. shows a simplified schematic of the ERC battery system. The battery stack is fed with two recirculation systems. The electrolyte used in both systems is an aqueous solution of zinc bromide reactants with supporting salts to enhance conductivity, and additives to improve the zinc electrode performance. The positive electrolyte contains from 5 to 15 g.l$^{-1}$ of dissolved bromine, while the level of bromine in the negative electrolyte is less than 1 g.l$^{-1}$. Most of the bromine produced on charge is absorbed into the polybromide complex, which can have a bromine concentration as high as 1800 g.l$^{-1}$.

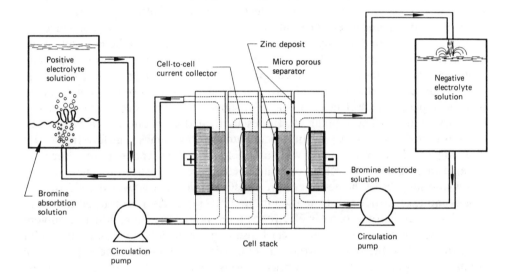

Fig. 6.40 — Zinc–bromine battery system.

The positive electrolyte tank used in the ERC system is specially designed for mixing the separation of the aqueous electrolyte and organic polybromide fluids. As positive electrolyte returns to the tank from the stacks it is sparged into the polybromide, which is kept in the section in the lower area of the tank. An expanded bed of polybromide droplets is formed in the mixing zone, and this provides the contact area needed for bromine transfer between the phases. Settling zones are provided in the tank, in which the droplets collect and are returned to the polybromide storage area in the tank. The electrolyte is clarified by the time it reaches the pump suction port in the tank for return to the stacks. This mixing/settling system has been developed in the ERC programme to eliminate the separate polybromide recirculation system used in the Gould design. In addition to reducing the system cost and complexity, the design confines the polybromide to a small area in the system. This enhances the overall safety of the battery, and limits exposure of critical components (e.g. pumps and instrumentation) to the corrosive polybromide.

The system configuration used by Exxon and its licensees is similar except that the polybromide is circulated through the cell stack to enhance the power capabilities of the system. The complex is collected in the positive electrolyte tank on charge and metered into the electrolyte stream on discharge.

The Gould–ERC system approach of circulating only aqueous catholyte into the stack is especially applicable to stationary energy storage systems which operate over a narrow power range. The power capabilities of this system are limited by the rate of bromine delivery to the stack in the electrolyte stream. If sustained power above the rated level is demanded by an application, the positive electrolyte flow rate needs to be increased, with a resulting increase in parasitic power. The Exxon system design delivers much more bromine to the stack, providing the capability of meeting the peak requirements in missions such as electric vehicle propulsion.

The electrochemical cell stacks at present used by developers are configured in bipolar arrangements, in which each cell is conected in series to the next cell in the stack. The predominance of the bipolar arrangement is due to the lower weight and cost of composite bipolar materials compared to the metal current collectors used in monopolar cells. Metal current collectors would also be prone to corrosion from the bromine reactant in the electrolyte. Fig. 6.41 shows the cell stack components used in the ERC battery. The bipolar electrodes used in the system are carbon/plastic composite sheets which are solvent bonded into the flow frames. The flow frames are injection moulded components with manifolds and flow channels to direct the positive and negative electrolytes over the electrodes. Flow distribution plenums just below and just above the active cell area ensure uniform flow over the electrodes. The bromine electrode substrate used in the ERC system is a carbon felt material, which is bonded to the bipolar electrodes with an electrically conductive solvent bond. A specially designed grid spacer is used on the negative side of each bipolar plate to maintain the zinc electrode gap dimension. The positive and negative cell compartments are separated by a microporous membrane, which limits bromine diffusion to the zinc electrodes and prevents bulk mixing of the electrolytes.

The electrode/frame sub-assemblies are stacked to the desired voltage level and held together with a clamping assembly. The critical stack design issues that have been addressed by the developers have been materials of construction, stack fabrication and sealing methods, and shunt currents.

### 6.5.3 Bipolar stack shunt current management

Shunt currents result from the fact that the cells in the bipolar stack are fed electrolyte from a common manifold system, the large ports at the top and bottom of the frame in Fig. 6.41. The voltage difference between cells in the stack can drive parasitic currents through the conductive electrolyte in the flow channels and manifolds. The flow channels leading from the manifold to the cell active area are made long and thin, to increase the electrical resistance in the shunt circuit and limit the current drain. This also increases the pumping power needed to circulate electrolyte, so the design of these flow channels must minimize the sum of the shunt current and pumping power losses.

In addition to reducing the coulombic efficiency of the battery, shunt currents present two additional problems. If the currents are high enough, zinc deposition at

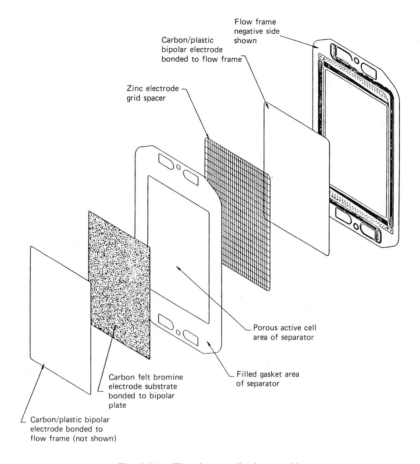

Fig. 6.41 — Flow frame cell sub-assembly.

the negative end of the circuit can result in zinc growing into the flow channels. This was observed in early stack tests, but has been eliminated in subsequent systems with optimized flow channel design. The second problem caused by shunt currents is stack capacity redistribution. Because of the distribution of the currents in the battery stack the cells near the centre of the stack lose more capacity than the cells near the ends of the stack. Left uncorrected, this capacity redistribution would eventually lead to short circuiting of cells at the ends of the stack, which have a Zn capacity excess. Equalization cycles — deep discharges in which all of the zinc is removed from the stack — are periodically required to maintain stack performance.

A method of reducing the capacity redistribution associated with shunt currents was developed at Exxon and is used by its licensees. In the shunt current protection system a bias current runs through the manifold or a parallel tunnel port, creating a voltage gradient which matches the voltage gradient in the active cells. There is still a coulombic loss (the bias current), but since the ionic bridge portion of the current is not through the cell feed channels the capacity redistribution is eliminated. Periodic equalization cycles are still required, since small differences in zinc deposit quality within each cell are magnified with repeated partial depth of discharge cycling.

### 6.5.4 Stack fabrication and sealing methods

Stack sealing is a design issue in which the approaches taken by the developers vary significantly. The ERC battery uses the approach developed at Gould, in which shallow ridges are designed around the flow channels and manifolds in the frames. These ridges press into the separator, which is filled (outside of the cell active area) with an elastomer so that it serves as a gasket. The stack is held together with a tie rod and steel strong back assembly, shown in Fig. 6.42. Packs of belleville spring washers are used with the tie rods to maintain compressive load as the stack seals relax.

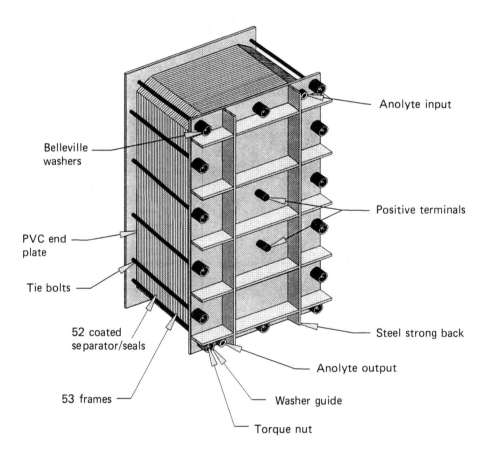

Fig. 6.42 — Zinc–bromine battery stack.

The stack sealing method used by Exxon also used compression seals, but in a slightly different configuration. Instead of bonding the bipolar electrode into the flow frame, the separator was bonded in the frame. This was done by insert-moulding the frame around the separator. The bipolar electrodes were extruded carbon plastic strips with unfilled, non-conducting strips of plastic along the sides of the sheet. The flow channels in the frames sealed against these unfilled strips. Adhesive material was sometimes used around the flow channel seals to improve the seal reliability. A strong back assembly was used, but the end blocks were reinforced plastic instead of steel.

Meidensha has also developed a steel strong back clamping system which uses spring washers to maintain stack compression. Since their target application is stationary energy storage, the extra weight of the steel strong back is not an issue. Johnson Controls and S.E.A. have both been developing methods of sealing the stack with heat welding instead of gasket seals. In the Johnson Controls design all of the manifold and flow channel seals are made with vibration welding. Both the bipolar electrode and the separator membrane are bonded to individual flow frames. A stack is formed by alternately welding separator frames to electrode frames until the desired stack size is reached [106].

The process of forming reliable weld seals around all of the manifold and flow channel perimeters is difficult, and the S.E.A. design has addressed this with a more simple frame design. The S.E.A. flow frame has no flow channels. Entrance and exit ports are provided on the side of the flow frame, and flow distribution plenums are provided at the inlet and outlet of the cell active area, but there are no manifold holes or long channels in the flow frame. Manifolding is done externally, with long thin tubes which run from a manifold header to the entrance or exit ports in each frame. The long tubes provide the shunt current resistance drop. This design greatly simplifies the task of heat sealing frames together to form a stack, since only the perimeter of the active area and plenum needs to be sealed [107].

The main objective behind both the Johnson Controls and S.E.A. designs is to reduce the weight of the system by eliminating the stack clamping assembly. The developers project energy densities in the range of 70 to 85 $Wh.kg^{-1}$, compared to the 50 $Wh.kg^{-1}$ demonstrated with the Exxon technology. The energy densities of battery systems built by Meidensha and ERC are much lower (15 to 25 $Wh.kg^{-1}$), since these systems are not optimized for low weight.

### 6.5.5 Materials of construction

The selection of materials for use in the battery system is primarily driven by the need for long term chemical stability in the bromine–electrolyte solution. The materials must not only maintain sufficient mechanical properties over the life of the application, but they must also not leach impurities into the electrolyte which would impair the system performance. For electric vehicle batteries, low specific weight materials are preferred. For many stationary energy storage applications very long life — of the order of ten years for the stack components — is required to provide sufficiently low life cycle cost. Regardless of the application, the materials chosen must be relatively low in cost. Very stable materials, such as Teflon™, cannot be used extensively in the system if the target mission is a consumer or commercial application.

All of the developers are using plastic and composite materials for most of the components of their battery systems. Polyolefin materials were used extensively in the Exxon design, and Johnson Controls and S.E.A. use high molecular weight polyethylene for the flow frames, tanks, and flow system plumbing. These materials are lightweight and appear to provide sufficient stability for the EV mission.

Polypropylene was initially used as the flow frame material in the Gould programme. Toward the end of the programmes, stability studies began to indicate that polypropylene would not be stable enough to last the 10 year life goal of the load levelling mission. Gould began using PVC as the flow frame material, and this is the material at present used by ERC. Stability tests have been done on a variety of moulding grades to select the most stable material among grades suitable for moulding the frame. The selection of a PVC material is complicated by the fact that moulding grades are not pure PVC, but include additives to improve mould flow and provide thermal stability during moulding. An acceptable material must include additives that are stable in the electrolyte and which do not leach out and poison the electrolyte.

Tanks and plumbing components used in ERC batteries are made of polypropylene because piping components are commonly available and the material can easily be fabricated into custom tank configurations. While the long term stability of the material may not be acceptable, it is sufficient for use in prototype systems.

Investigators at Meidensha have been using glass-filled polyethylene as the flow frame material until recently. A crosslinked polyethylene elastomer is now used as the frame material, and this is reported to make a better seal against the unfilled strips in the bipolar electrode sheet. The electrolyte tanks in Meidensha batteries have been made from polypropylene or PVC. The flow system piping components are standard PVC materials [108].

The bipolar electrodes used in the battery are subjected to one of the most corrosive environments in the system. Local bromine concentrations can be very high, and potential gradients exist which could accelerate corrosion. A major problem encountered in the Exxon work was electrode warping. After exposure to the battery electrolyte, the electrode material would swell, which caused a warping of the confined plate. The problem could be reduced to some extent by using thicker electrodes (2.5 mm instead of the standard 0.6 to 0.8 mm), but at the expense of higher cost and weight. In the Johnson Controls design, electrode expansion has been reduced four-fold by using high density polyethylene with glass-fibre reinforcement. Studies of Meidensha have focused on reducing the carbon content in the plate as a method of improving dimensional stability.

In the Gould zinc–bromine system the bipolar electrodes were sheets of vitreous carbon about 1 mm thick. The electrodes were made by compression moulding a composite of graphite and phenolic thermoset resin. The plate was then heat-treated to convert the resin to a vitreous carbon. While the chemical stability of the material was good, the plates were brittle and difficult to seal into the flow frame. The sealing problem was made more difficult by the fact that the coefficient of thermal expansion of the plate was much lower than that of the plastic frame. Late in the programme, ERC began working with Gould on a subcontract basis to develop a stable plastic composite electrode. The effort continued after ERC acquired the Gould technology, and a material was developed which appears to meet the stability requirements of the load-levelling application.

Sec. 6.5]  **Zinc–bromine batteries**  511

The first carbon/plastic formulations evaluated in the ERC studies were composites of carbon black and polyethylene or polyvinyledene fluoride (PVDF, trade name Kynar™ from Pennwalt corporation). The PVDF composite exhibited excellent stability in electrolyte–bromine solutions, but swelled and warped when exposed to the polybromide complexing agent. Stable composite materials were made by replacing the carbon black with pretreated graphite materials. Electrodes made with these materials exhibit greater dimensional stability than even the vitreous electrodes, and this composite is at present the baseline bipolar electrode in the ERC battery. Accelerated chemical and electrochemical stability tests have indicated that the electrode will maintain sufficient dimensional and strength stability to operate over a ten year life in the battery.

### 6.5.6 System design

The system design approach used for the battery is highly dependent on the application. For vehicle applications, the electrolyte storage reservoirs can be specially shaped to fit into a variety of spaces in the vehicle. Fig. 6.43 shows a 30 kWh

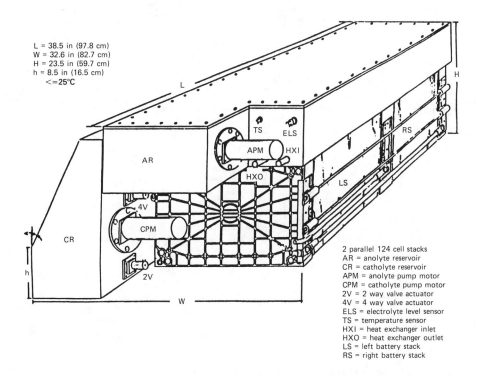

Fig. 6.43 — 30 kWh Exxon electrical vehicle battery.

Exxon battery custom designed to fit into available space in the Ford ETX™ electric vehicle. Because of this unique ability to be configured for specific applications, the system can be highly volume efficient. Similar custom configurations with up to 45 kWh capacity have been fabricated by S.E.A. and operated successfully in electric vehicles in Austria.

For the stationary storage applications the system configuration is somewhat more straightforward. For large utility batteries long strings of stacks would be fed from large cylindrical tanks using common chemical process system design practices. Fig. 6.44 shows a conceptual design for a 50 MWh utility battery. By feeding as many stacks as possible with a common circulation system the cost of auxiliaries in a large stationary system is significantly reduced.

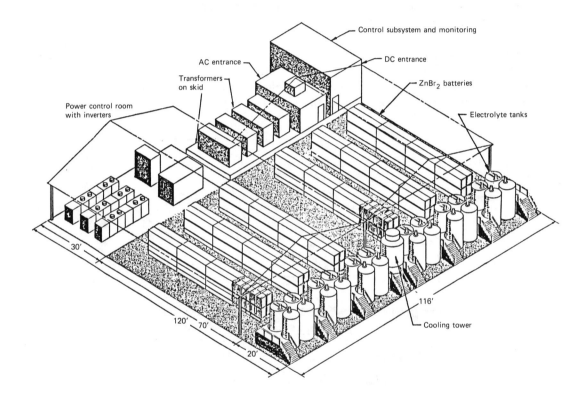

Fig. 6.44 — Conceptual design of 50 MWh zinc–bromide battery system.

### 6.5.7 Battery performance characteristics

The reversible potential of the couple is 1.83 V, and operating cells typically exhibit open circuit voltages of 1.75 to 1.80 V, depending on state of charge. The cells charge at 1.9 to 2.1 V, and discharge plateau voltages are typically 1.5 to 1.6 V. Fig. 6.45 shows voltage data for charge/discharge cycle tests of single cells with 872 cm$^2$ active cell area. These cells were built with carbon felt flow-through electrodes of different designs, and the impact of bromine electrode design on the voltaic efficiency of the

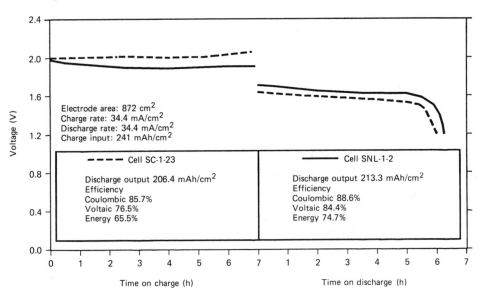

Fig. 6.45 — Performance of 872 cm² zinc–bromine single cells.

system can be seen in the test data. The cells were operated with capacity loadings in excess of 200 mAh.cm$^{-2}$, a level of performance that has been a major target of the ERC work. Long term cycle testing of multi-cell stacks has not yet been done at that level because the required cell-to-cell flow uniformity has not yet been achieved. The baseline cycle test used for multi-cell stack tests at ERC uses a charge input of 172 mAh.cm$^{-2}$, with a delivered capacity density of about 150 mAh.cm$^{-2}$. The charge and discharge rate used is 34 mA.cm$^{-2}$. Fig. 6.46 shows the charge/discharge performance of the 30-cell stack on the baseline stack cycle test. Fig. 6.47 shows polarization data for a 30-cell stack tested at Murdoch University.

The ERC approach toward the battery development has been to evolve very stable component designs which can survive the long target life of the load-levelling application. Some of these components — the PVDF–carbon bipolar plate in particular — are more expensive than those in the lightweight Exxon technology. The high capacity loading targets of the design reduce the cost per kWh of the stacks and associated feed plumbing by requiring fewer stacks per system.

In the battery tests conducted by Exxon and its licensees, lower capacity and current densities are generally used. Constant current battery testing at Exxon was typically done at 17 to 25 mA.cm$^{-2}$ and zinc loadings up to 90 mAh.cm$^{-2}$. Energy efficiencies for stacks were typically 65% for small test batteries (500 Wh) and 60% for larger (20–30 kWh) systems. A typical constant current cycle used in the Johnson Controls programme uses a 18.1 mA.cm$^{-2}$, 5 hour charge and a 28.3 mA.cm$^{-2}$ discharge. Energy efficiencies of 60 to 70% have been reported, depending on the level of supporting salts in the electrolyte. Exxon, Johnson Controls, and S.E.A. have all done additional testing on simulated electric vehicle driving cycles. On the Simplified Federal Urban Driving Schedule, Johnson Controls has reported test

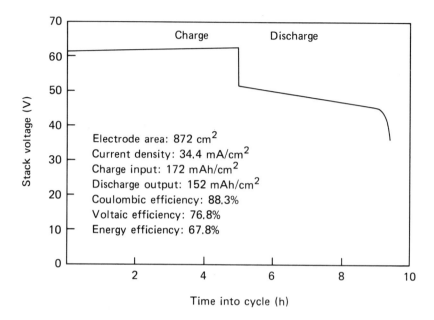

Fig. 6.46 — Cycle performance of 30-cell, 1.4 kW stack.

results which are projected to full system energy densities of 67 to 72 Wh.kg$^{-1}$, with vehicle ranges of 125 to 134 miles.

In the Meidensha programme the target application is utility load-levelling, but in the Japanese energy mix the primary competitive technology is pumped hydro storage. The system therefore has a much higher target efficiency than the ERC system. The competitive technologies for load-levelling batteries in the US are gas-fired turbine generators and additional base load capacity. Against these technologies the primary cost driver is capital cost. If pumped hydro storage is an alternative, the battery system must compete on an efficiency basis. For this reason the Meidensha systems are run at low current and capacity densities with very high efficiencies. Typical charge and discharge current densities are 12 to 13 mA.cm$^{-2}$, with capacity loadings of only 100 mAh.cm$^{-2}$. On this cycle, battery D.C. energy efficiencies are typically 79 to 82%. A.C. energy efficiencies are 76 to 79%.

### 6.5.8 Technology status

All of the developers have reported very long cycle life on small test systems. Small cells have been cycled for up to 3000 cycles at ERC, and a 500 Wh Exxon stack was tested for more than 2000 cycles at Sandia National Laboratories. Because of the unique design of the system, where the electrodes serve only as inert reaction substrates, the life cycle capabilities of the battery are very good. Large battery stack cycle life has been limited to the 400 to 600 cycle range primarily by hardware failures, as opposed to fundamental electrochemical problems. As additional engineering optimization work is done on large battery designs, the excellent cycle capabilities of the battery should be reflected in large system test results.

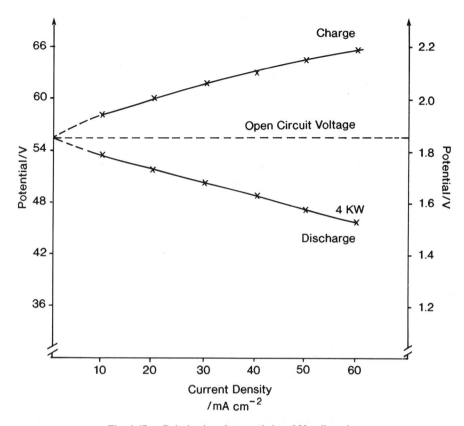

Fig. 6.47 — Polarization characteristics of 30-cell stack.

Among the electric vehicle technologies, batteries as large as 40 kWh have been operated in vehicles by S.E.A. The Exxon technology has been tested at the 27 kWh level. Johnson Controls is planning to test the new vibration-welded stack design in a 50 KWh system for a Ford Aerostar™ van in the near future. With continued successful demonstrations of the technology at these levels, the system should prove to be a leading candidate for EV propulsion as environmental and energy policies accelerate the need for EV systems. The key technological hurdles to be overcome are: demonstrating that the system hardware can operate reliably for long life times, and demonstrating that the system can be configured in ways that ensure that the battery is safe enough for vehicle use.

For the stationary energy storage applications, systems as large as 400 kWh have been built by Meidensha. Installation of a 4 MWh demonstration system is scheduled for completion in 1991. The ERC programme has focused more on core technology studies, such as improved zinc deposition at high capacity densities, and has been demonstrated at the 18 kWh level. Large scale demonstrations of the technology are essential for the system to be accepted for utility use. As with the EV systems, a key area for ongoing work is the demonstration of reliable battery hardware in large systems over long term cycle tests.

Given the fundamental advantages of the system, it is likely that the battery will be applied to some of the energy problems that face the industrialized world in the 1990s. The developers have demonstrated the viability of the basic system chemistry as well as a variety of system configuration approaches.

## 6.6 RECHARGEABLE LITHIUM–IRON SULPHIDE BATTERIES
### (H. F. Gibbard)

### 6.6.1 General characteristics

The rechargeable lithium(alloy)–iron sulphide battery is a descendant of the high-temperature lithium–sulphur system investigated at Argonne National Laboratory in the late 1960s. Despite its attractive theoretical specific energy density of approximately 2600 Wh kg$^{-1}$, the lithium–sulphur battery with two liquid active materials and a liquid molten-salt electrolyte was unpractical. Workable batteries were developed by alloying lithium with aluminium or silicon, and by the substitution of metal sulphides of high sulphur activity for elemental sulphur, to yield solid active materials at the operating temperature of the battery, typically 450°C to 500°C. The lithium(aluminium)–iron monosulphide (LiAl–FeS) secondary system, to which most attention is given in this chapter, has been under development since 1976 and is the closest to commercial production.

Although some consideration was given to utility off-peak energy storage early in the development of lithium–iron sulphide batteries, nearly all effort is now centred on traction applications for electric vehicles. Fleets of delivery vans, fork lift trucks, aircraft towing trucks, and underground mining vehicles are considered likely early uses.

Operating on a single voltage plateau with an open-circuit voltage of 1.33 V, the LiAl–FeS system yields a theoretical specific energy of 458 Wh kg$^{-1}$. Engineering development cells with capacities of 200 Ah have yielded a specific energy of 105 Wh kg$^{-1}$ at the three-hour discharge rate, and their energy density of 235 Wh dm$^{-3}$ is among the highest of any near-term rechargeable system. The ability to use inexpensive steel or nickel current-collector hardware constitutes a decided practical advantage over the higher-voltage iron disulphide system, which requires the use of expensive refractory metal current collectors. LiAl–FeS cells can be thermally cycled repeatedly between the operating temperature of 450–500°C and room temperature with no deleterious effects.

The reliability of the system is also high because, if cell failure does occur, the cells fail in a fully shorted condition. Batteries containing failed cells have undergone many cycles without developing internal pressure due to gas generation, and without catastrophic failure. Laboratory tests indicate that a battery containing up to 10% of failed cells could continue to function without serious deterioration in performance in an electric vehicle application.

The high-temperature LiAl–FeS battery requires three major subsystems for its operation: the cells, a thermal management system, and a specialized charger. Most of the development work on the system has concentrated on cell development, and it appears that the only major problem which remains in achieving the performance goals of electric vehicle propulsion is that of ensuring adequate cycle life and life at temperature. Cells of the design now made by Westinghouse — Oceanic Division achieve a cycle life of approximately 350 cycles in cell tests and have given about 120 cycles in battery tests. The predominant failure mechanism is the deposition of conductive, metallic particles in the separator during cycling. The present approach to this problem is careful control of the voltage during charging, to prevent the positive electrode from reaching a potential high enough to produce soluble iron species. The existing charger, developed at Argonne National Laboratory [109], can be operated in a taper-charge, constant-voltage mode during the final charge-equalization phase of the charging process. This rather complicated charger is required because the present LiAl–FeS system has little tolerance to overcharge.

The thermal management system, also developed by Argonne National Laboratory [110], contains both active and passive elements to achieve and maintain the proper operating temperature for the battery. Electric resistance heaters are provided in the trays containing the cell modules (typically groups of nine or more cells); these are operated on external power during charging and as a parasitic load on the battery when it is free from external power. Operating under thermostatic control, the heaters serve to raise the battery from ambient to operating temperature and to maintain it there. The trays of the battery modules also contain an air heat exchanger which can extract heat from the battery during high-rate discharge to keep it within the desired operating range. The amount of heat produced by the discharge reaction has been quantified by calorimetric measurements [111,112], and the performance of a thermal management system has been described in detail [113].

The passive elements of the thermal management system are an enclosure equipped with walls containing high-performance, vacuum-multifoil thermal insulation, and an end cover for the container. The end cover contains a plug of thermal insulation and is penetrated by the electrical leads for the battery and the tubes for the heat exchanger. Various insulating materials have been considered for the side walls of the container, including evacuated powder beds and evacuated, metallized glass spheres; but the best performance has been obtained when using alternating layers of aluminium foil and glass fibre paper filling an evacuated annular space [110]. This structure yields a thermal conductivity of approximately $0.001$ $W (m K)^{-1}$.

Figs 6.48 and 6.49 show a 36 V, 8 kWh LiAl–FeS battery made by Westinghouse — Oceanic Division, Chardon, Ohio. The end cover (Fig. 6.48) and the thermal enclosure (Fig. 6.49) assembly were fabricated by F. Meyer Tool and Manufacturing Company of Oak Lawn, Illinois. Two of these batteries and two of the component nine-cell modules have been tested at Argonne National Laboratory at constant current and under simulated electric vehicle driving profiles. The best performance corresponded to an estimated vehicle range of more than 160 km.

The safety of lithium alloy–iron sulphide batteries is evidently superior to that of several other high energy rechargeable systems. Cells exposed to air have been crushed with a drop hammer at operating temperature, simulating a 30 mph vehicle crash, without explosion, fire, or the evolution of harmful gases. When the Mark IA 40 kWh battery experienced massive internal shorting [114], essentially all of the battery's energy was converted into heat. Some metallic cell components fused, but the exterior temperature of the thermal enclosure rose to only about 100°C, with no evident hazard to personnel or property. When lithium/metal sulphide batteries and

Fig. 6.48 — 8 kWh LiAl–FeS battery made by Westinghouse — Oceanic Division, Chardon, Ohio. Left: end cover.

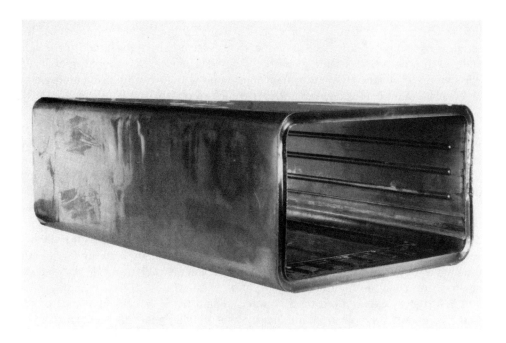

Fig. 6.49 — Thermal enclosure for 8 kWh LiAl–FeS battery.

Sec. 6.6]     **Rechargeable lithium–iron sulphide batteries**     519

modules are tested at the Advanced Battery Test Laboratory at Argonne National Laboratory, no special precautions are required to remove explosive or toxic gases, as is the case with lead–acid and zinc–bromine batteries.

### 6.6.2 Manufacturers

The lithium alloy–iron sulphide battery was invented at Argonne National Laboratory in the early 1970s [115,116]. By 1976 the US Department of Energy (DOE) was supporting development programs at Rockwell International, Eagle Picher Industries, and Gould, Inc. In 1981 development programmes were underway in the US, the UK, Canada, Japan, Korea, West Germany, and the USSR. Until late in 1989 the most advanced development effort was being conducted at Westinghouse — Oceanic Division in Chardon, Ohio. This successor to the Gould programme was supported by the Electric Power Research Institute (EPRI), Palo Alto, California.

The Westinghouse programme was at an advanced engineering stage of development, with several 12 V and 36 V battery modules having been built and tested for performance and cycle life. It was complemented by battery design studies carried out at Argonne National Laboratory under support of the DOE. At the time of writing of this section, 200 Ah cells are available for purchase from Westinghouse in limited quantity for evaluation purposes.

In December 1989 SAFT America was awarded a $3.4 million contract from the US Department of Energy to develop a lithium–iron sulphide battery for electric vehicle propulsion. SAFT's programme will be conducted at the new research and development centre in Cockeysville, Maryland. The present goal of this programme is the delivery of two 50 kWh batteries for in-vehicle tests within three years. The performance objectives for these batteries are a specific energy of 100 Wh $kg^{-1}$, specific power of 110 W $kg^{-1}$ at 80% depth of discharge, and life of 600 cycles [117].

The Electro Fuel company in Toronto, Canada, is also developing lithium–alloy–iron sulphide batteries for electric vehicle propulsion; their plans have included construction of a 10 kWh module. Electro Fuels's cell employs a separator of fibrous boron nitride, in contrast to the magnesium oxide powder separator used by Westinghouse. Early Eagle Picher and Gould cells employing BN separators achieved long cycle life and long life at operating temperature [118], and its use would appear to be desirable if the material can be obtained at a competitive cost.

### 6.6.3 Chemistry

#### 6.6.3.1 Negative electrode

The discharge reaction of the lithium alloy negative electrode typically occurs through a series of voltage plateaux, during each of which two solid phases are present in equilibrium. If the compositions of the two phases are denoted by $Li_aM$ and $Li_bM$, where M = Al or Si, then the discharge reaction on the plateau corresponding to $Li_aM \rightarrow Li_bM$ can be written as

$$1/(a-b)\ Li_aM \rightarrow 1/(a-b)\ Li_bM + Li^+ + e^- \ . \tag{6.15}$$

Lithium–silicon alloys were considered for the negative electrode in iron sulphide cells by the Atomics International division of Rockwell International and General

Motors Corporation. Sharma and Seefurth [119] determined the phase diagram of the lithium–silicon system and found a series of solid phases with the compositions Si, $Li_2Si$, $Li_{21}Si_8$, $Li_{15}Si_4$, and $Li_{22}Si_5$. In fair agreement with the work of Sharma & Seefurth, Lai [120] found a series of voltage plateaux with potentials ranging from 41 mV to 329 mV positive of elemental lithium at 45°C, as shown in Table 6.10. The average potential over the useful composition range from about 80 atom per cent lithium to 50 atom per cent is 195 mV positive of lithium.

Table 6.10 — Potential of lithium–silicon alloy voltage plateaux with respect to elemental lithium at 450°C

| Composition of coexisting phases, $Li_aSi$ and $Li_bSi$ | | Potential, mV vs $Li/Li^+$ |
|---|---|---|
| a | b | |
| 5.0 | 4.1 | 41 |
| 4.1 | 2.8 | 149 |
| 2.8 | 2.0 | 276 |
| 2.0 | 1.5 | 329 |

The LiAl system at 450°C exhibits only two solid phases, thus a constant electrode potential at equilibrium, between the useful operating ranges of 10 and 47 atom percent lithium. Over the range of temperatures of interest for molten salt batteries, 350°C to 50°C, the electrode potential is more positive than that of lithium electrode by 314 to 270 mV. Its specific capacity when cycled between 10 and 47 atom percent lithium is 0.610 Ah $g^{-1}$.

The specific capacity of a LiSi electrode cycled between 50 and 80 atom per cent lithium is 1.440 Ah $g^{-1}$. Both the specific capacity and the volumetric capacity density of LiSi are considerably higher than those of LiAl, and the potentials of LiSi cells are higher than that of LiAl cells by an average of about 0.1 V. Nevertheless, much more development work has been carried out with cells containing the aluminium alloy. The main reasons for this are the corrosiveness of the silicon alloy in contact with steel cell components [121] and the tendency of LiAl to form a self-supporting, stable structure capable of more than 1000 deep discharge cycles [110].

Interestingly, the negative electrodes of recent Gould/Westinghouse cells contain both the LiSi and LiAl alloys, with the ratio by weight in the range of 3.5:1 to 4.0:1 [110]. Apparently the LiSi additive yields higher specific and volumetric capacity with neither the requirement for a supporting structure nor corrosion of the steel hardware.

### 6.6.3.2 *Positive electrode*

The chemistry and electrochemistry of the FeS electrode are quite complicated, as several detailed investigations [122,123] have shown. Much of the complexity in cells containing the commonly used LiCl–KCl electrolyte is due to the presence of potassium ions, which become incorporated in intermediate phases during the discharge. The cell reaction becomes much simpler, however, when an all-lithium-cation electrolyte such as LiF–LiCl–LiBr is employed. Then the discharge reaction of the FeS electrode occurs in two steps, as predicted from the Li–Fe–S phase diagram at 450°C [124]:

$$2FeS + 2Li^+ + 2e^- \rightarrow Li_2FeS_2 + Fe \qquad (6.16)$$

$$Li_2FeS_2 + 2Li^+ + 2e^- \rightarrow 2Li_2S + Fe \ . \qquad (6.17)$$

The discharge reaction occurs on a single voltage plateau, because the equilbrium potentials of the two reactions differ by only about 20 mV. Thus the overall cell reaction for a LiAl–FeS cell can be written as

$$1/(a-b) \ Li_aAl + FeS \rightarrow 1/(a-b) \ Li_bAl + Li_2S + Fe \qquad (6.18)$$

For a cell in which the negative electrode is cycled between 47 atom per cent and 10 atom per cent lithium, the cell reaction is

$$2.577 \ Li_{0.887}Al + FeS \rightarrow 2.577 \ Li_{0.111}Al + Li_2S + Fe \ . \qquad (6.19)$$

Writing the cell reaction in this way has the advantage of showing explicitly the composition of the negative electrode for the discharge between defined initial and final states. It avoids the common, but misleading, practice of writing elemental aluminium as a reaction product.

### 6.6.3.3 Negative-to-positive capacity ratio

A major recent accomplishment in the development of Li alloy–FeS batteries is a very substantial increase in their specific power, which has been found to be strongly dependent on the relative capacities of the negative and positive electrodes. Although initial development of the LiAl–FeS system concentrated on specific energy and life, recent efforts have focused on specific power, especially at deep discharge. High specific power provides the acceleration capability required for an electric vehicle battery. The cells built several years ago were positive limited, with a negative-to-positive capacity ratio of about 1.3:1. The internal resistance of these cells rose rapidly when they were discharged to more than 60% of their capacity, and the peak power dropped from 175 W kg$^{-1}$ in the fully charged state to only about 55 W kg$^{-1}$ at 80% depth of discharge.

Studies at Argonne National Laboratory and at Gould showed that changes in the positive electrode were responsible for most of the rise in internal resistance [110]. Rather than try to increase the conductivity of the positive electrode through the addition of conductive powders, Gould decreased the negative-to-positive capacity ratio to 0.95:1 to produce a negative limited cell. This change more than doubled the peak specific power at 80% depth of discharge, to 130 W kg$^{-1}$.

### 6.6.3.4 Separator

Only a few materials, most notably boron nitride and magnesium oxide, can withstand long term contact with both a high lithium activity alloy and iron sulphide at elevated temperature. Boron nitride can be made as fibres and was used in early cells in the form of felt or cloth. Dissatisfied with the cost projections for this material, Gould chose to develop a separator based on magnesium oxide powder. This separator is used in the present generation of Westinghouse cells.

The Gould/Westinghouse separator is a ceramic plate about 1.5 mm thick, cold pressed from a powdered mixture of magnesium oxide and electrolyte. The typical composition is 35% by weight MgO and 65% electrolyte. This is the same as the composition of the separator/electrolyte component in primary thermal batteries based on the lithium alloy–iron disulphide system [125]. The void volume of the separator at room temperature is approximately 25% of the total volume; at 500°C the molten electrolyte expands to yield a void volume of less than 10% of the total. This 'electrolyte starved' design is necessary for a cell with a ceramic powder separator, which would become fluid if exposed to an excess of electrolyte. The ceramic plate separator considerably simplifies the process of cell assembly. At room temperature the electrolyte is present only in the solid plates, which are easily stacked to form cells. The flooded cells using BN separators contained an excess of electrolyte and were filled with molten salt in an evacuated furnace.

### 6.6.4 Construction

Fig. 6.50 shows a 7 plate, 200 Ah cell of the Gould/Westinghouse design. The cell consists of three positive plates interleaved with four negative plates, the outside two of which are of half thickness. The positive terminal exits the cell case through an insulated feedthrough; the negative terminal is welded to the stainless steel case. Active electrode components are fabricated by cold pressing powders of the selected composition into half electrodes. The height and width of the active electrode components are approximately 12.5 cm by 17.0 cm. Each electrode, except the two end negatives, is assembled by two half electrodes with a central current collector made of 0.025 cm nickel or stainless steel sheet. At the top of each central current collector is a thickened bus bar terminating in an integral electrode tab, as shown in Fig. 6.51.

It is interesting to note that a few years ago Gould constructed cells using electrodes of approximately the same dimensions as those now used, but with the terminals at the short ends. An analysis of the 'terminal effect,' which describes how the ohmic resistance of the current collector plates affects current distribution over the electrode surface, showed that a cell of low aspect ratio would provide improved capacity and power, particularly at high discharge rates [126]. Experimental confirmation led to immediate adoption of the low profile cell.

Each electrode is surrounded by a particle retainer basket made from stainless steel or nickel foil photo-etched with a pattern of 0.03 cm holes. The particle retainer basket restricts the movement of particles of the active materials during cycling, and thus extends cycle life. The baskets are made of two halves which fit snugly over the electrodes and are notched at the top for the electrode tab. The two halves of the baskets are spot welded at their overlapping, non-photo-etched edges to prevent loss of particles of active material at the electrode edge.

The construction of the copper cored, nickel clad cell terminals and the positive seal is shown in Fig. 6.52. The seal is formed by a bed of boron nitride powder

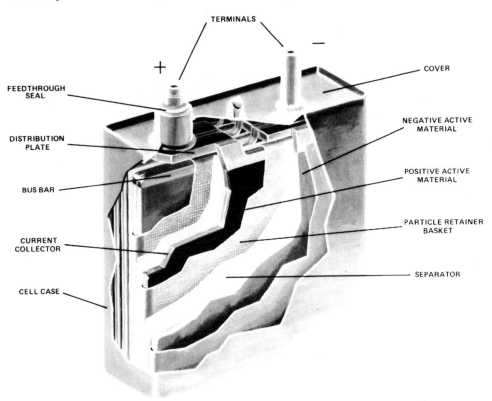

Fig. 6.50 — 7 Plate, 200 Ah cell of Gould/Westinghouse design.

compressed between beryllia bushings. The integrity of the seal depends on the lack of wetting of boron nitride by the electrolyte. The seal is not hermetic and tends to leak electrolyte if the surface of the boron nitride particles is converted to wettable boron oxide through reaction with oxygen from the air.

During cell fabrication the positive and negative electrodes are interleaved with separators to form the cell assembly. Pressed plaques of MgO and electrolyte, similar to the electrolyte but thicker, are added to the bottom and edges of the assembly to insulate the cell edges; and the entire assembly is slipped into the can. The can cover assembly containing the positive feedthrough is positioned so that the positive current distribution plate can be welded to the positive electrode tabs. The negative current distribution plate is welded onto the negative electrode tabs, the can cover is welded to the can, and the negative terminal is welded to the cover. The cell is evacuated through the tube located between the terminals and heated to de-gas the components. The tube is pinched off and welded shut to complete the assembly process.

The entire cell fabrication procedure, from the processing of the electrode and separator powders to the final assembly, is carried out in a dry room with a typical relative humidity of 1–3%. The cell is fabricated in the fully charged state. This has

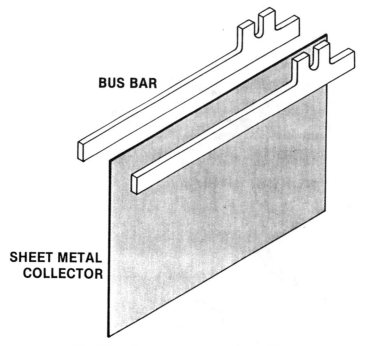

Fig. 6.51 — Current collector assembly for 200 Ah cell.

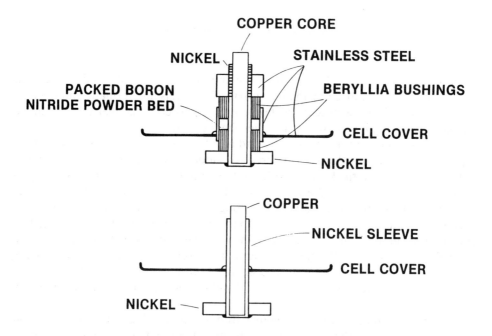

Fig. 6.52 — Cross-sections of positive and negative terminals for 200 Ah cell.

Sec. 6.6]     **Rechargeable lithium–iron sulphide batteries**     525

several advantages over fabrication in the discharged or partly charged state. The most important of these is avoiding the handling of lithium sulphide, a discharge product of the cell reaction which is difficult to obtain in pure form and readily reacts with even traces of water to yield hydrogen sulphide. Cells which are used in a battery module are electrically insulated from each other, either by coating them with ceramic paint or by placing mica sheets between them. Typically, nine cells are connected in series and bound tightly together, with two strong but lightweight end plates (Fig. 6.53), to prevent deformation of the cell case because of volume changes in the positive electrode during cycling. The completed modules are placed in the trays containing the active thermal management elements and installed in the thermal enclosure (Figs 6.48 and 6.49).

Fig. 6.53 — 2.5 kWh Li–Al/FeS battery module.

### 6.6.5 Performance
Performance data are available for cells built by Eagle Picher, Gould, and Argonne National Laboratory. Only results on the latest Westinghouse cells and batteries are considered here, since none of the others is close to commercial production. The

performance on a cell basis of four batteries ranging from 2.5 kWh to 7.5 kWh [127] is summarized in Fig. 6.54, which depicts the specific energy and specific power for constant current discharge at various discharge times. For times less than four hours the performance curve has been extrapolated by using earlier performance data for similar cells [126]. The mass of the seven-plate cells in these batteries was 2.8 kg; that of nine-cell modules including restraint hardware was 28.8 kg. The nominal capacity at the three hour rate was 200 Ah.

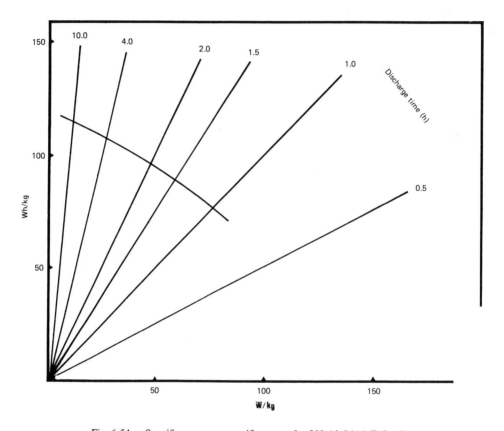

Fig. 6.54 — Specific energy vs specific power for 200 Ah LiAl–FeS cell.

In addition to constant current tests, the Westinghouse batteries were tested under loads which simulated electric vehicle driving profiles. One of the 36 V batteries yielded a specific energy of 112 Wh kg$^{-1}$ on the basis of the cell mass and 98 Wh kg$^{-1}$ for the module, when discharged on the SAE J227a/C test. The mass of the thermal enclosure was not included in the computation of specific energy. This performance was estimated to give a range of over 100 miles in an electric vehicle with a full-scale battery.

In tests at Argonne National Laboratories the Westinghouse batteries achieved a cycle life of about 120 cycles, only a third the cycle life of individual cells. This was attributed to overcharging, during which soluble iron species were believed to be produced in the positive electrode. Migration of these species into the separator, followed by reduction to elemental iron, caused shorting and failure of the cells. Additionally, testing in a thermal enclosure specifically designed for the 36 V battery showed some deficiencies of the thermal management system. During operation at high power, the heat exchanger overcooled some cells; the resulting part freezing of the electrolyte in these cells caused poor performance.

In summary of their performance, LiAl–FeS cells achieve levels of specific energy and specific power which are attractive for use in near term electric vehicle applications. Essentially the same levels of performance are attained in batteries. Major improvements are necessary in the cycle life of cells and batteries, and the charging and thermal management systems must be further developed before these batteries can find use in commercial electric vehicles. Nevertheless, construction of a full scale battery for in-vehicle tests is planned under the SAFT/DOE programme within the next three years.

### 6.6.6 Future developments

The technology of the monopolar LiAl–FeS system, as described above, can be regarded as well established. It apparently requires a series of evolutionary improvements to become a commercially viable system for traction applications. This prospect appears realistic, provided that adequate financial support is available over the next few years. In the longer term, a programme of advanced development could very substantially improve the specific energy and specific power of lithium alloy–iron sulphide systems. Specific improvements which are now under study include bipolar design and the use of iron disulphide as the positive active material.

Work on the LiAl–FeS$_2$ system has been underway at General Motors Corporation [128] and at Argonne National Laboratory [129] for several years. The discharge reaction for this system occurs on two plateaux at potentials of 1.76 V and 1.33 V, with a theoretical energy density of 464 Wh/kg for the utilization of both plateaux.

Bipolar design has been used for many years in the primary thermal battery based on the lithium–iron disulphide system. By eliminating intercell connectors, bipolar design not only saves mass and volume, but yields a much more uniform current distribution and higher utilization of the active material at high discharge rates. Calculations indicate that the specific energy of the 1000 cell bipolar battery stack could attain 150 Wh kg$^{-1}$. The problem of obtaining an electrolyte-proof seal at the edge of the electrodes is usually regarded as a major difficulty in developing a high temperature bipolar battery. In the absence of an edge seal, leakage currents between cells cause self-discharge at a rate which varies from cell to cell. Another major problem is that of charge equalization when the cells become unbalanced after many cycles. In the early 1980s Gould constructed a four-cell bipolar LiAl–FeS battery, which operated for 40 cycles with no edge seal before the problem of cell imbalance terminated its life [126].

Another major problem with batteries based on the FeS$_2$ positive electrode has been the loss in capacity with cycling and on stand at the operating temperature, due to decomposition of the positive active material. Recently, workers at Argonne National Laboratory [130,131] have reported major progress in stabilizing the capacity through the use of a LiCl–LiBr–KBr electrolyte in cells which operate only on the upper voltage plateau. Within the past few months, bipolar four-cell batteries of 0.5 Ah capacity have achieved 100 deep discharge cycles at ANL [132]. Charge equalization is accomplished by trickle charging at the end of the charging process, which is possible because the cells tolerate low level overcharge without damage. This promising work is supported by the DOE through Lawrence Berkeley Laboratory and through a technology development programme of the state of Illinois.

## 6.7 LITHIUM SOLID STATE BATTERIES

(R. J. Neat)

### 6.7.1 Introduction

The concept of an all solid state battery is very appealing. Solid state batteries are intrinsically spill proof and rugged, have a wide operating temperature range, long shelf-life, and are capable of being engineered into any geometry. In addition, since they can be constructed with an excellent packing efficiency of active components, energy densities are maximized. Practical solid state batteries became a possibility when a range of materials known as *fast ion conductors* or *solid electrolytes* were discovered. These materials (e.g. β-alumina and $Li_3N$) were capable of conducting alkali metal ions (e.g. $Na^+$ and $Li^+$) to levels greater than $10^{-4}$ S/cm. Although this level of ionic conductivity is substantially lower than that of liquid battery electrolytes, the ability to engineer the solid electrolyte in thin ($< 100$ $\mu$m), high area sheets allows reasonably low electrolyte resistance to be attained.

Although many highly conducting solid electrolytes have been known for a number of years, only one type of solid state battery, the $Li-I_2$ heart pacemaker, is commercially available. The problem lies in the nature of the contact between two solids; although the electrode surfaces can be machined flat, the volumetric changes which occur during discharge, and especially during recharge, completely disrupt the cell interfaces. This is not a problem in liquid cells since the electrolyte can easily flow into any voids, but it results in solid state cells which have very limited current densities and an inability to recharge. Both these features are acceptable to the pacemaker application.

The discovery of a new class of solid electrolytes, the *solid polymer electrolytes*, by Armand in 1978 [133,134] allowed many of the problems discussed above to be solved. Armand reported that certain heteroatom containing high molecular weight polymers could conduct alkali metal ions to levels $> 10^{-4}$ S/cm. The physical properties of polymer electrolytes are such that they can easily be produced in thin (20–100 $\mu$m), high area sheets using standard thick film processing techniques. Although physically solids, polymer electrolytes behave as viscous liquids at the microscopic level and hence form excellent interfaces with solid electrodes. Moreover, they can easily accommodate electrode volume changes during cycling by microscopic polymer flow. Thus these newly discovered ion-conducting polymers were seen as the ideal electrolytes for all-solid-state cells and batteries.

In 1983, Hooper & North [135] reported an all-solid-state cell based on a lithium ion conducting polymer electrolyte and a solid intercalation cathode material. The cell configuration, shown in Fig. 6.55, is a four layer laminate consisting of a lithium

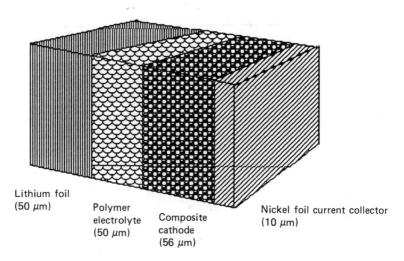

Fig. 6.55 — Lithium solid state cell.

foil anode, a polymer electrolyte, a composite cathode containing an active intercalation material, and a metal foil current collector. Cells are typically in the range of 150–200 $\mu$m in thickness and can have any area from a few square centimetres to many square metres. It is this particular example of a solid state lithium battery which is the subject of this section. Many of the results presented here are taken from the Harwell-based programme, and the section is not intended to be a comprehensive review of the worldwide effort in polymer electrolyte batteries. If such a review is required, the reader is referred to two recent papers [136,137].

### 6.7.2 The cell components

#### 6.7.2.1 Lithium anodes
Lithium metal in foil form is commercially available in thicknesses above 100 $\mu$m. However, even at 100 $\mu$m there is a large excess of lithium, and efforts must be made to reduce the anode thickness to 30–40 $\mu$m per cell.

#### 6.7.2.2 Polymer electrolytes
The key component in the all-solid-state battery is the polymer electrolyte. A solid polymer electrolyte is formed when a metal salt is dissolved in a high molecular weight polymer. The best and most common example is poly(ethylene oxide) [PEO] with dissolved lithium triflate. There are, however, many other polymers that are capable of dissolving a vast array of cations and anions from all corners of the periodic table. If more detail of this fascinating branch of material science is required, the reader is referred to some excellent reviews [138,139,140]. This section is thus confined to polymer electrolytes based on PEO and lithium salts.

The ionic conductivity as a function of temperature for several polymer electrolytes used during development work at Harwell is shown in Fig. 6.56. The original

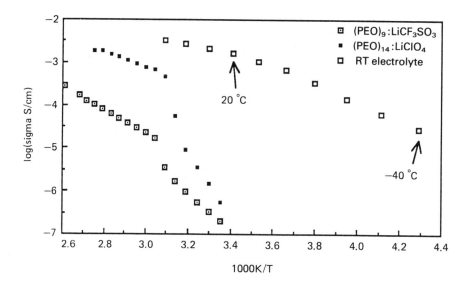

Fig. 6.56 — Ionic conductivity as a function of reciprocal temperature for some polymer electrolytes used in the Harwell development programme.

polymer electrolyte used to construct cells and batteries was $PEO_9:LiCF_3SO_3$ (the subscript denotes there are nine ethylene oxide units per cation). The ionic conductivity of this material is very low at room temperature, but gradually increases as the temperature is raised; a larger rate of increase is observed between 55°C and 70°C. The origin of this increase is a melting transition within the polymer film, and illustrates that it is the amorphous phase of the polymer electrolyte which conducts the ions; the crystalline regions are, in fact, insulating. Thus, since a conductivity level of $> 10^{-4}$ S cm$^{-1}$ is required for medium to high rate applications, an operating temperature of 100–140°C is necessary for cells based on this electrolyte.† Further development work has shown the electrolyte $PEO_{14}:LiClO_4$ to be the best conductor of temperatures $> 100°C$, and this has become the standard for operating in that temperature range.

### 6.7.2.3 Intercalation composite cathodes

The cathode compartment of the lithium polymer solid state battery is a composite of an intercalation or insertion compound, a suitable electronic conductor, and the polymer material used in the electrolyte. Work at Harwell has concentrated on $V_6O_{13}$-based composite cathodes, although elsewhere $TiS_2$ [141], $MoO_2$ [141,142], $Li_{1+x}V_3O_8$ [143] and other $VO_x$ [142] materials have been studied within the polymer battery configuration. All the above compounds have in common the ability to reversibly intercalate lithium ions with an associated reduction/oxidation process taking place. In short, these compounds confer rechargeability on the system. It is,

---

† An ionic conductivity of $10^{-4}$ S cm$^{-1}$ is often seen as a minimum for 100 $\mu$m solid electrolytes intended for medium to high rate applications.

however, possible to construct a primary version by incorporating a material such as $MnO_2$. In a typical composite cathode 45% of the volume is taken up by $V_6O_{13}$, 5% by a carbon black, and 50% by the polymer electrolyte material. Resultant cathode capacities are in the range 0.5–3.0 mA h cm$^{-2}$.

### 6.7.2.4 Component fabrication

Perhaps the most important feature of this battery is the ease with which the components can be fabricated on a large scale. There are numerous industrial methods of producing thin (20–100 μm) polymer films, and one such method, doctor blade solvent casting, is illustrated Fig. 6.57. The polymer electrolyte is formed by dissolving the PEO and lithium salt in a suitable solvent (e.g. methanol) and casting the resultant viscous solution on a moving releasing substrate using an adjustable blade gap. Solvent evaporation results in a thin film which can be many metres wide and many kilometres long. The composite cathode is fabricated in a similar manner, but is cast onto a metal foil (e.g. nickel) to act as a current collector.

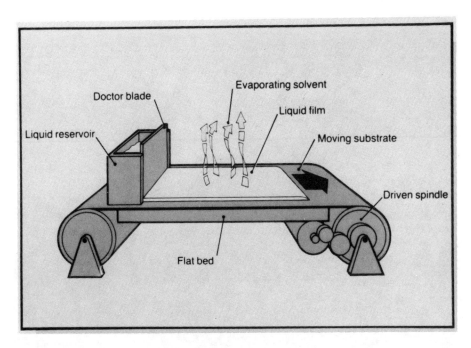

Fig. 6.57 — Doctorblade casting.

Thus both the polymer electrolyte and the composite cathode can be produced on a continuous basis in very large quantities (processing speeds found in the casting/coating industry are typically many metres per second). This is a distinct advantage over advanced batteries (e.g. sodium–sulphur) where the transition from a batch to a continuous process is very difficult.

532                    **Systems under development**                    [Ch. 6

### *6.7.2.5 Cell fabrication*
Constructing a lithium polymer electrolyte cell is a simple matter of laminating together the three components described above. Again, this process can be easily automated and performed on a continuous large scale basis. Cells of any active ares, and therefore any capacity, can be fabricated, and the resultant laminate rolled, folded, or stacked to provide a box of energy which itself can be almost any shape. This design freedom is another attractive feature of this battery, and will be illustrated later.

### 6.7.3 Cell performance
Fig. 6.58 shows a typical test cell used for performance evaluation at Harwell. The three cell components described above are clearly visible, as considerable overlap of the layers is employed to avoid the possibility of edged shorting. The lithium active area is 40 cm$^2$ which gives a cell capacity of 80 mA h; the energy density of the cell is 211 W h kg$^{-1}$, which includes the obvious redundancy associated with the overlap. Test cells are packaged in a suitable hermetically sealable pouch to allow evaluation in the normal laboratory atmosphere.

Fig. 6.58 — Typical test cell.

Fig. 6.59 shows the first discharge curve of a typical test cell operating at 120°C. The distinctive plateaux in the voltage–time curve are a characteristic of lithium ion insertion into $V_6O_{13}$. Fig. 6.59 illustrates that, when discharged to 1.7 V (the theoretical potential of $Li_8V_6O_{13}$ *vs.* lithium metal) at the C/10 rate (0.2 mA cm$^{-2}$),

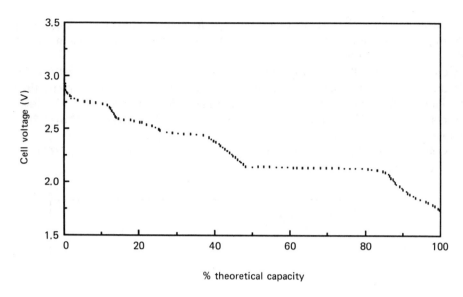

Fig. 6.59 — First discharge at the C/10 rate of a typical test at 120°C.

100% of the theoretical cell capacity is obtained. This fact, combined with the overall sharpness of the voltage–time curve, indicates that the discharge kinetics are very good. The performance of the test cell at different discharge rates is shown in Fig. 6.60. Thus, more than 50% of the cell capacity is available at the C rate, 80% at the

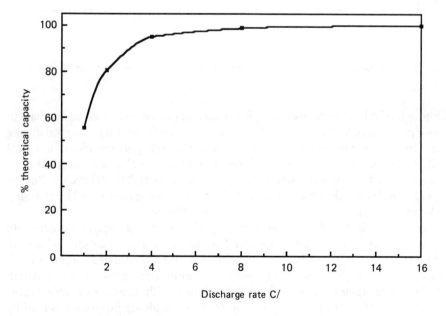

Fig. 6.60 — Rate performance of a typical test cell at 120°C.

C/2 rate, and 100% at the C/4 rate or less. This performance at the higher rates is very encouraging, and illustrates the significant potential this battery system has in traction type applications.

Considering the all-solid nature of the cell, the performance of test cells under pulsed discharge conditions is remarkably good. Fig. 6.61 shows the current–time data obtained from a 40 cm² cell when the terminal voltage was held at 1.7 V. Peak currents of 1.8 A (40C rate) were observed, declining to 0.5 A (10C rate) over 20 s. Average power densities of over 1.85 kW kg$^{-1}$ have been achieved over twenty 20 s pulses, and 80% of the theoretical capacity can be removed in 1800 s (i.e. 3C rate) under pulse discharge conditions.

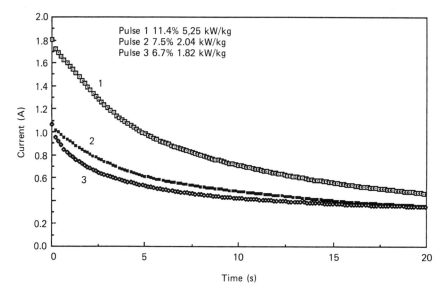

Fig. 6.61 — Potential pulse performance of a typical test cell at 120°C. The first three 20s pulses are shown.

Although cells have demonstrated good rate and pulsed discharge performance, the major problem of $V_6O_{13}$ based cells remains the capacity decline observed during cycling. This is illustrated in Fig. 6.62 which shows the cycle performance of a typical test cell. The capacity obtained on each discharge falls to 65% of theoretical after 20–30 cycles, but then begins a steady decline over the next 200–300 cycles. Recent results at Harwell have shown that much of this decline can be eliminated by charging the cell correctly; long term cycle evaluation is continuing.

The results presented above show that the lithium solid state polymer electrolyte battery should have significant commercial possibilities in applications where an operating temperature of 120°C can easily be tolerated. The major application is therefore seen as vehicle and satellite power. The demonstrated high power and high energy density characteristics of the battery, coupled with its ease of manufacture, low cost, and safety, make it an ideal traction battery. Scale-up programmes leading to demonstration traction units are underway at Harwell.

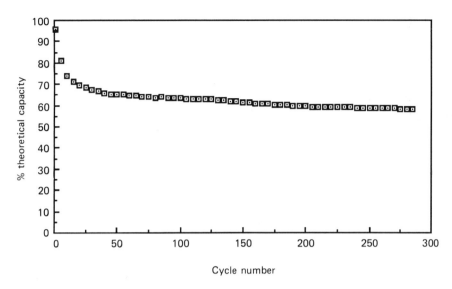

Fig. 6.62 — Cycle performance of a typical test cell at 120°C.

### 6.7.4 'Room temperature' performance

Many of the attractive features of the polymer electrolyte battery, e.g. high energy density, good safety, and especially the variable geometry, are very desirable for small portable equipment applications. Thus the main driving force behind research work in this area over the past few years has been to translate the performance obtained at 120°C to room temperature (10–30°C). The major thrust of this research has been to find a polymer electrolyte with a suitable conductivity at the reduced temperatures. Fig. 6.56 shows the conductivity of such a polymer electrolyte developed at Harwell. This material, which is a modified PEO, exhibits levels of conductivity which are higher than the unmodified PEOs at all temperatures. The new material can be handled in the same manner as $PEO_{14}$: $LiClO_4$, and thus cells and batteries can readily be fabricated.

Fig. 6.63 shows a first discharge curve of an 18 cm$^2$ test cell at the C/6 rate. The voltage–time curve again exhibits the steps and plateaux associated with $V_6O_{13}$, but the last plateau is reduced, limiting the cell capacity to 90% of the theoretical figure. The high rate performance of the room temperature version is also reduced, as is the rechargability. Although the performance at room temperature is, as yet, poorer than the 120°C operating temperature version, the possible advantage of this system in portable equipment, illustrated in Fig. 6.64, is greater. Fig. 6.64 shows a polymer electrolyte battery shaped in the word HARWELL, driving a small motor connected to the UKAEA logo. It is this ability to shape a cell or battery that will provide equipment design engineers with a much greater degree of flexibility.

Room temperature polymer electrolyte batteries have also been reported by several other groups. A collaboration between the Mead paper company (USA) and the Energy Research Laboratory (Denmark) has reported, and has demonstrated

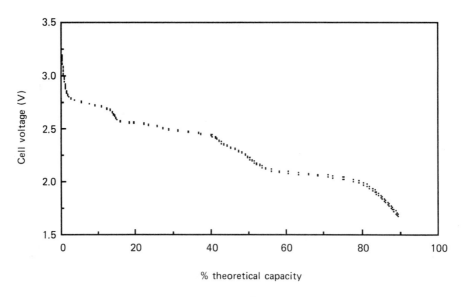

Fig. 6.63 — First discharge (C/6 rate) of an 18 cm$^2$ test cell incorporating a modified polymer electrolyte 15°C.

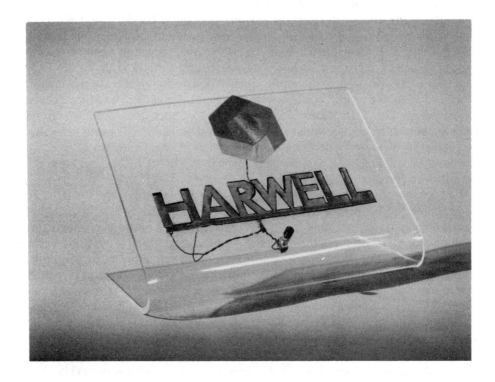

Fig. 6.64 — Room temperature cell in the shape HARWELL driving a small motor..

test cells powering small motors at room temperature. A group at IREQ (Insitut de Recherche d'Hydro-Quebec; Canada) have also reported room temperature operation, although at lower discharge rates.

### 6.7.5 Summary

The lithium solid state polymer electrolyte battery can be viewed as a system suitable for all applications, from small area miniature batteries for 'on-chip' power supplies to large load levelling batteries. At present there are two general areas of ongoing development work. The first concerns the original technology, and is essentially a scale-up programme with continued effort on extending cycle life. The goal of this work is to demonstrate the polymer electrolyte battery as an improved traction power source. The second development area concerns room temperature technology, where improvements are required in the areas of cyclability, sub-ambient, and high power performance. It must, however, be recognized that considerable progress has been made in a relatively short time, and that it may not be long before the 'room temperature' technology is as good as, if not better than, the original system.

### ACKNOWLEDGEMENTS

C. D. S. Tuck would like to acknowledge with thanks the advice he received on the section concerning aluminium–air batteries from his colleagues at Alcan. Particularly useful suggestions were made by Dr W. B. O'Callaghan, Dr N. P. Fitzpatrick, Dr R. P. Hamlen, Dr W. Hoge, Dr J. A. Hunter, and Dr G. M. Scamans.

H. F. Gibbard is indebted to Mr Geoffrey Barlow of Westinghouse and Mr Albert Chilenskas of Argonne National Laboratories for providing information and illustrations for his section on rechargeable lithium–iron sulphide batteries.

R. J. Neat wishes to acknowledge the financial support of the Harwell based Solid State Battery Working Party, the Office of Naval Research (USA), the Commission of the European Communities, and the Dowty Group plc.

### REFERENCES

[1] German Patent 38,383 (1887).
[2] McBreen, J. & Cairns, E. J., *Adv. Electrochem. and Electrochem. Eng.*, **11** (1978), 273.
[3] Shallenberg, R. H., *Bottled energy*, American Philosophical Society (1982).
[4] Dirkse, T. P. & Hampson, N. A., *Electrochemica Acta*, **17** (1972), 135.
[5] Falk, U. & Salkind, A. J., *Alkaline storage batteries*, John Wiley, New York (1969).
[6] McBreen, J., *J. Electrochem. Soc.*, **119** (1972), 1620.
[7] Choi, K. W., Bennion, D. N. & Newman, J., *J. Electrochem Soc.*, **123** (1976), 1616.
[8] Hendrix, J., PhD thesis, Eindhoven University, Netherlands (1984).
[9] Michalenko, M. G., *Zh. Prikladnoi Khimii*, **45** (1972), 1953.
[10] Gregory, D. P., Jones, P. C. & Redfern, D. P., *J. Electrochem. Soc.*, **119** (1972), 1288.

[11] Himy, A. & Wagner, O., *Proc. 27th Power Sources Conference,* Cherry Hill, NJ, **27** (1976), 135.
[12] Poa, D. S., US Patent application 6,580,982 (1984).
[13] Tanso, S. & Okamoto, J., *Yuasa-Jiho,* **54** (1983), 57.
[14] Tanso, S., Yoshida, S. & Senoo, K., *Yuasa–Jiho,* **59** (1985), 35.
[15] Gagnon, E. G. & Hill, B. S., *J. Electrochem. Soc.,* **137** (1990), 377.
[16] McBreen, J. & Gannon, E., *J. Power Sources,* **15** (1985), 169.
[17] McNeil, A. J. S. & Hampson, N. A., *J. Power Sources,* **15** (1985), 261.
[18] Lucas Industries, UK Patent 2054252B (1983).
[19] Duffield, A., Mitchell, P. J., Shield, D. W. & Kumar, N., *Power Sources,* Vol. 11, Academic Press, London (1987), p. 253.
[20] Duffield, A., *Chemistry and Industry,* 1 February (1988), 88.
[21] Skowronski, J. M., Reksc, W. & Jurewecz, K., *J. Power Sources,* **23** (1988), 351.
[22] US Dept. of Energy, Task Force on Electric Vehicle Goals, *Recommended mission directed goals for electric vehicle battery research and development,* DoE/CE-0148 (March 1986).
[23] DeLuca, W. H., Kulaga, J. E., Hogrefe, R. L., Tummillo, A. F. & Webster, C. E., *Laboratory evaluation of advanced battery technologies for electric vehicle applications,* SAE Technical Paper Series 890820, International Congress and Exposition, Detroit, MI (February 1989).
[24] Chloride Silent Power Limited, *Theoretical and experimental investigations of factors affecting the performance and reliability of sodium sulphur cells and batteries,* Electric Power Research Institute, Palo Alto, CA. Contract 128–13 (to be published).
[25] British Patent GB 2102 622B. P. J. Bindin (2 February 1983).
[26] Leadbetter, A., *Heat transfer in electrochemical systems — Thermal enclosures for advanced electric vehicle batteries,* AICHE Conf. (November 1987).
[27] Auxer, W., Sodium sulphur battery commercialization, *Ninth Battery and Electrochemical Contractors Conference,* Alexandria, VA, (November 1989, US Department of Energy, CONF-891132-SUMMS (DE 89016837).
[28] DeLuca, W. H. & Tumillo, A. F., Sodium sulphur battery testing at Argonne National Laboratory, *Ninth Battery and Electrochemical Contractors Conference, Alexandria, VA* (November 1989), US Department of Energy, CONF-891132-SUMMS (DE 89016837).
[29] Adler, E. et al., *Nickel–hydrogen battery advanced development program,* AFWAL TR-80-2044 (1980).
[30] Warnock, D., *The 1981 Goddard Space Flight Center Battery Workshop,* G. Halpert, Ed., NASA CP-2217 (1982), p. 487.
[31] Miller, L., *Intersociety Energy Conversion Engineering Conference,* Vol. 2, American Society of Mechanical Engineers (1988), p. 489.
[32] Pickett, D. F., *Preparation of nickel electrodes,* US Patent 3,827,911 (August 1974).
[33] Dunlop, J. D. et al., *Power Sources,* **5** (1974), 315.
[34] Smithrick, J. J. et al., *Advanced energy systems — Their role in our future,* Vol. 1, American Nuclear Society (1984), p. 631.

[35] Yang, T. M., Ford Aerospace, telephone communication (1989).
[36] Gaston, S. J., *Intersociety Energy Conversion Engineering Conference*, Vol. 1, American Nuclear Society (1984), p. 258.
[37] Dunlop, J. D. & Stockel, J. F., *Intersociety Energy Conversion Engineering Conference*, Vol. 3, The American Institute of Aeronautics and Astronautics (1980), p. 1878.
[38] Yang, T. M. et al., *Intersociety Energy Conversion Engineering Conference*, Vol. 1, Institute of Electrical and Electronic Engineers (1989), p. 1375.
[39] Hudak, R. E., *Intersociety Energy Conversion Engineering Conference*, Vol. 2, American Institute of Aeronautics and Astronautics (1987), p. 897.
[40] Schiffer, S., G.E. Astro Space, telephone communication (1989).
[41] Fuhr, K., Martin Marietta, telephone communication (1989).
[42] Hall, S., Naval Weapons Surface Center (NWSC) Crane, Indiana, telephone communication (1989).
[43] Lim, H. S. & Verzwyvelt, S. A., *J. Power Sources,* **22** (1988), 312.
[44] Lim, H. S. & Verzwyvelt, S. A., *J. Power Sources* (1991) (in press).
[45] Manzo, M. A. et al., *J. Power Sources* (1991) (in press).
[46] Vaidyanathan, H. & Dunlop, J., *Intersociety Energy Conversion Engineering Conference,* Vol. 3, American Chemical Society (1986), p. 1560.
[47] Hulot, M., *Compt. Rend.,* **40** (1855), 1148.
[48] Mennons, M. A. F., British Patent Application, 296 (1858).
[49] Stokes, J. J., The aluminium dry cell, *Electrochemical Technology,* **6** (1968), 36, and US Patent 2,838,591 (1958).
[50] Stokes, J. J. & Belistkus, D., *The primary battery*, Cahoon & Heise, Eds, John Wiley, New York (1976), p. 17.
[51] Zaromb, S., *J. Electrochem. Soc.,* **109** (1962), 1125.
[52] Hamlen, R. P., Jerabek, E. C., Ruzzo, J. C. & Siwek, E. G., *J. Electrochem. Soc.,* **116** (1969), 1588.
[53] Moden, J. R. & Perkons, G., US Patents 4,107,406 (1978) and 4,150,204 (1979).
[54] Cooper, J. F., *International Journal Engineering Science*, No. 6 (1983), 20.
[55] Cooper, J. F., Estimates of the cost and energy consumption of Al–Air electric vehicles. Paper presented at *Fall Meeting of the Electrochemical Society, Hollywood, Florida* (5–10 October 1970).
[56] Cooper, J. F. & Littauer, E. L., Mechanically rechargeable metal–air batteries for auto propulsion, *13th Intersoc. Energy Conversion Engineering Conference*, Vol. 1, San Diego, CA (20–25 August 1978).
[57] Valand, T., Mollestad, O. & Nilsson, G., Al–Air cells — potential small electric generators for field use, *Power Sources*, Vol. 8, Thompson, J., Ed., Academic Press, London (1981).
[58] Katryniok, D., Ruch, J. & Schmode, P., *Elektrotechnische Zeitschrift,* **102** (1981), 1094.
[59] Despic, A. R. & Milanovic, P. D., Aluminium–Air battery for electric vehicles, *Recueil des Travaux de l'Institut des Sciences Techniques de l'Academio Serbe des Science et Arts,* **12** (1979), 1.
[60] Ritschel, M. & Vielstich, W., *Electrochimica Acta,* **24** (1979), 885.

[61] Despic, A. R., Drazic, D. M., Purenovic, M. M. & N. Cikovic, *J. Appl. Electrochem.*, **6** (1976), 527.
[62] Cooper, J. F., Experimentation with wedge-shaped cells, paper presented at *2nd Intern, Reactive Metal–Air Battery Workshop, Palo Alto* (May 1982).
[63] O'Callaghan, W. B., *7th D.O.E. Contractors Conference, Washington*, (1985).
[64] Fitzpatrick, N. P. & Strong, D. S., *9th International Electric Vehicle Symposium, Toronto* (November 1988).
[65] Reding, J. T. & Newport, J. J., *Materials Protection*, **5** (1966), 15.
[66] Pryor, M. J., US Patent 3,250,649 (1966).
[67] Tuck C. D. S., Hunter, J. A. & Scamans, G. M., *J. Electrochem. Soc.*, **134** (1987), 2070.
[68] Scamans, G. M., Hunter, J. A., Tuck, C. D. S., Hamlen, R. P. & Fitzpatrick, N. P., *Power Sources*, Vol. 12, 363. Keily, T. & Baxter, B. W., Eds, Academic Press, London (1988), p. 363.
[69] Jeffrey, P. W. & Halliop, W., Extended Abstract, *Electrochemical Society Fall Meeting, Hawaii* (1987).
[70] Hunter, J. A., Scamans, G. M. & Sykes, J., *Power Sources*, Vol. 13, Keily, T. & Baxter, B. W., Eds, Academic Press, London (1991), p. 193.
[71] Bohnstedt, W. J., *J. Power Sources*, **5** (1980), 245.
[72] Despic, A. R., Stevanovic, R. M. & Vorkapic, A. M., Extended Abstract A2019 of *35th ISE Meeting, Berkeley, California* (1984).
[73] Katoh, M., US patent 3,563,803 (1967).
[74] Yeager, E. B., *Electrochim. Acta*, **29**, 1527 (1984).
[75] Gruver, G. A., *J. Electrochem. Soc.*, **125** (1978), 1719.
[76] MacDonald, J. P. & Stonehart, P., Ext. Abs. Program, *Biennial Conference on Carbon*, Vol. 14 (1979), p. 183.
[77] Solomon, F., Knerr, L. A. & Wheeler, D. J., Ext. Abs. No. 304, *Electrochemical Society Spring Meeting, Toronto* (1985).
[78 Hoge, W., US Patents 4,885,217 (1989) and 4,906,535 (1990).
[79] Scamans, G. M., O'Callaghan, W. B., Fitzpatrick, N. P. & Hamlen, R. P., *21st Interscience Energy Conversion Conference, San Diego* (1986). American Chemical Society.).
[80] King, W. R, *Light Metals*, Vol. 2, published by AIME, New York (1973).
[81] Misra, C. & White C. T., *Chem. Eng. Prog. Symposium*, Vol. 67 (1971), p. 53.
[82] Despic, A. R. & Drazic, D. M., *Bull-Soc. Chim., Beograd.*, Suppl., **48** (1983), 299.
[83] Danel, V., *Proc. 32nd International Power Sources Symposium, Cherry Hill, 1986*, Electrochemical Society (1986).
[84] Scamans, G. M., Stannard, J. H., Fitzpatrick, N. P., Rao, B. & Hamlen, R. P., *Proc. Conf. on Defence Oceanography, Brighton* (1990).
[85] O'Callaghan, W. B., Fitzpatrick, N. P. & Peters, K., *Proc. INTERLEC '89 Conference, Florence* (October 1989).
[86] Scamans, G. M., Warner, S. M., *Proc. 6th Battery Conf. and Exhibition*, ERA Technology, London (1990).
[87] Rudd, E. J., *33rd International Power Sources Symposium, Cherry Hill, NJ*

(1988), p. 427.
[88] Parish, D. W., Fitzpatrick, N. P., O'Callaghan, W. B. & Anderson, W. M., *Proc. Conf. Future Transportation Technology, Vancouver*, (August 1989).
[89] Fitzpatrick, N. P., *Aluminium*, **66** (1990), 355.
[90] Niksa, M. J., Coin, R. J. & Turley, H. L., *33rd International Power Sources Symposium, Cherry Hill, NJ* (1988), p. 421.
[91] Bradley, C. S., US Patent 312,802 (1885).
[92] Bradley, C. S., US Patent 409,448 (1889).
[93] Zito, R., US Patent 3,382,102 (1968).
[94] Barnartt, S. & Forejt, D., *J. Electrochemical. Soc.*, **111**, 1201–1204.
[95] Blue, R. & Leddy, J., U.S. Patent 3,408,232 (1968).
[96] Singh, P., White, K. & Parker, A., *J. Power Sources,* **10** (1983), 309.
[97] Bloch, M., US Patent 2,566,114 (1951).
[98] Walsh, M., US Patent 3,816,177 (1974).
[99] Will, F., *Proceedings of the 12th IECEC Conference* (1977), pp. 250–255.
[100] Venero, A., US Patent 4,105,829 (1978).
[101] Bellows, R., Einstein, H., Grimes, P., Kantner, E., Malachesky, P. & Newby, K., *Development of a circulating zinc–bromine battery — Final Report* (Final Report to US DOE, Sandia National Laboratories (1987).
[102] Putt, R. & Montgomery, M., U.S. Patent 4,162,351 (1979).
[103] Putt, R., *Development of zinc–bromide batteries for stationary energy storage*, EPRI EM-2497, Project 635-2 Final Report (1982).
[104] Singh, P. & Leo, A., Development of Zinc–Bromine Battery Systems For Stationary Energy Storage, presented at the *2nd Annual International Conference on Batteries For Utility Energy Storage*, Newport Beach, CA (1989).
[105] Leo, A., Status of Zinc-Bromine Battery Development at Energy Research Corporation, *Proc. 24th IECEC Conference* (1989).
[106] Bolstad, J. & Miles, C., *Development of zinc–bromine IECEC Conference* (1989).
[107] Kordesh, K., Fabjan, Ch. & Tomazic, G., Zinc–bromine battery: advanced technology, *Electrochemical Society Extended Abstracts, Fall Meeting*, Honolulu, Hawaii (1987).
[108] Fujii, T., Igarashi, M., Fushimi, K., Hashimoto, T., Hirota, A., Itoh, H., Jinnai, K., Tagami, Y., Kouzuma, I., Sera, Y. & Jakayama, T., 4 MW zinc–bromine battery for electric power storage, *Proc. 24th IECEC Conference* (1989)..
[109] De Luca, W., Chilenskas, A. & Hornstra, F., A charging system for the LiAl–FeS electric vehicle battery, In: *Proc. 14th Intersociety Energy Conversion Engineering Conference, Boston MA*, (5–10 August 1979), pp. 665–670.
[110] Gay, E. C., Steunenberg, R. K., Miller, W. E., Battles, J. E., Kaun, T. D., Martino, F. J., Smaga, J. A. & Chilenskas, A. A., Li–Alloy/FeS cell design and analysis report, In: *Argonne National Laboratory Report ANL-84-93* (July 1985).
[111] Hansen, L. D., Hart, R. H., Chen, D. M. & Gibbard, H. F., High-temperature battery calorimeter, *Rev. Sci. Instrum.*, **53** (1982), 45.
[112] Chen, C. C. & Gibbard, H. F., Thermal energy generation of LiAl/FeS cells,

*J. Electrochem. Soc.*, **130** (1983), 1975–1979.

[113] Gibbard, H. F., Chen, D. M., Chen, C. C. & Olszanski, T. W. (1981) Thermal Properties of LiAl–FeS Batteries, In: *16th Intersociety Energy Conversion Engineering Conference, Atlanta, GA*, (9–14 August, 1981), paper number 819362.

[114] Kolba, V. M., Failure analysis of the Mark IA lithium–iron sulfide battery, In: *Argonne national Laboratory Report ANL-80-44* (October, 1980).

[115] Yao, N. P., Heredy, L. A. & Saunders, R. C., *J. Electrochem. Soc.*, **118** (1971), 1039.

[116] Vissers, D. R., Tomczuk, Z. & Steunenberg, R. K., *J. Electrochem. Soc.*, **121** (1982), 665.

[117] Embry, J., personal communication (1990).

[118] Martino, F. J., Gay, E. C. & Shimotake, H., Cycle life studies of LiAl–FeS cells using BN felt separators, In: *Proc. 15th Intersociety Energy Conversion Engineering Conference, Seattle, Washington* (August 1980), pp. 205–210.

[119] Sharma, R. A. & Seefurth, R. N., Thermodynamic properties of the lithium–silicon system, *J. Electrochem. Soc.*, **123** (1977), 1763–1768.

[120] Lai, S. C., Solid lithium–silicon electrode, *J. Electrochem. Soc.*, **123** (1976), 1196–1197.

[121] Sudar, S., Hall, J. C., Heredy, L. A. & McCoy, L. R., Development status of lithium–silicon/iron sulfide load leveling batteries, In: *Proc. 12th Intersociety Energy Conversion Engineering Conference, Washington, DC* (28 August–2 September 1977).

[122] Tomczuk, Z., Preto, S. K. & Roche, M. F., Reactions of FeS electrodes in LiCl–KCl electrolyte, *J. Electrochem. Soc.*, **128** (1981), 760–772.

[123] Tomczuk, Z., Tani, B., Otto, N. C., Roche, M. F. & Vissers, D. R., Phase relationships in positive electrodes of high temperature Li–Al/LiCl–KCl/FeS$_2$ cells, *J. Electrochem. Soc.*, **129** (1982), 925–931.

[124] Martin, A. E., In: High performance batteries for electric-vehicle propulsion and stationary energy storage, *Argonne National Laboratory Report ANL-78-94* (1980), p. 167.

[125] Quinn, R. K., Baldwin, A. R., Armijo, J. R., Neiswander, P. G. & Zurawski, D. E., Development of a lithium alloy–iron disulfide, 60-minute primary thermal battery, In: *Sandia National Laboratory Report SAND79-0814* (1979).

[126] Barlow, G., High energy density rechargeable battery phase II (lithium–metal sulfide development), *Final Report to Wright–Patterson Air Force Base under contract #F33615-82-C2223* (December 1984).

[127] Barlow, G., The development and performance of lithium alloy/iron sulfide propulsion batteries, In: *Proc. of the 33rd International Power Sources Symposium, Cherry Hill, NJ* (13–16 June 1988), pp. 539–546.

[128] Dunning, J. S. & Seefurth, R. N., In: *Proc. Fourth US Department of Energy Battery and Electrochemical Contractors' Conference, Washington DC*, (2–4 June 1981).

[129] Martino, F. J., Moore, W. E. & Gay, E. C., In: *Argonne National Laboratory Report ANL-83-62* (1983), pp. 29–38.

[130] Kaun, T. D., An advanced lithium–aluminium/iron disulfide secondary cell, Inn: *Proc. 32nd International Power Sources Symposium, Cherry Hill NJ*, (9–12 June 1986), pp. 16–22.
[131] Nelson, P. A. & Chilenskas, A. A., Summary of the status of lithium–metal sulfide batteries, In: Sen, R. K., Ed., *Assessment of high-temperature battery systems. Summary and recommendations of the high-temperature battery systems workshop, Washington, DC* (16–17 August 1988), pp. 3.1–3.36.
[132] Kaun, T. D., personal communication (1990).
[133] Armand, M. B., Chabagno, J. M. & Duclot, M. J., *2nd International Meeting on Solid Electrolytes, St. Andrews, Scotland*, Extended Abstract (September 1978).
[134] Armand, M. B., Chabagno, J. M. & Duclot, M. J., In: *Fast ion transport in solids*, Vashista, P., *et al.*, Eds, North Holland, New York (1979), p. 131.
[135] Hooper, A. & North, J. M., *Solid state ionic* **9** & **10** (1983), 1161.
[136] Hooper, A., Gauthier, M. & Belanger, A., Polymer electrolyte lithium batteries, In: *Electrochemical Science and Technology of Polymers*, Vol. 2, Linford, R. G., Ed., Elsevier Applied Science, Barking, Essex (1990).
[137] Gauthier, M., Belanger, A., Kapfer, B., Vassort, G. & Armand, M. B., Solid polymer electrolyte lithium batteries, In: *Polymer Electrolyte Reviews*, Vol. 2, MacCallum, J. R., & Vincent, C. A., Eds, Elsevier Applied Science, Barking, Essex (1989).
[138] Vincent, C. A., *Progress in Solid State Chemistry*, **17** (1987), 145.
[139] Various articles in *Polymer Electrolyte Reviews*, Vol. 1, MacCallum, J. R. & Vincent, C. A., Eds, Elsevier Applied Science, Barking, Essex (1987).
[140] Various articles in *Polymer Electrolyte Reviews*, Vol. 2, MacCallum, J. R. & Vincent, C. A., Eds, Elsevier Applied Science, Barking, Essex (1989).
[141] Gauthier, M., Fauteux, D., Vassort, G., Belanger, A., Duval, M., Ricoux, P., Chabagno, J-M., Muller, D., Rigaud., Armand, M. B. & Deroo, D., *Proc. 2nd International Meeting on Lithium Batteries, Paris* (April 1984); *J. Power Sources*, **14** (1985), 23.
[142] Gauthier, M. *et al.*, *Proc. 3rd International Meeting on Lithium Battreries, Kyoto* (May 1986). Extended Abstracts ST11 and ST12.
[143] Bonino, F., Ottaviani, M., Scrosati, B. & Pistoia, G. *J. Electrochem. Soc.*, **135** (1988), 12.
[144] Kummer, J. T. & Weber, N., *Automot. Eng. Cong.*, Detroit, Jan 1–13, 1967.
[145] Levine, C. A., *Proc. 10th IECEC*, Inst. of Electrical & Electronic Engineers, New York (1975), p. 621.

# Appendix A
# Performance characteristics of battery systems included in this volume (in alphabetical order of common name)

| | |
|---|---|
| **Cell type** | **Alkaline–manganese (cylindrical)** |
| Anode | Zinc |
| Cathode | $MnO_2$ |
| Electrolyte | KOH |
| Open circuit voltage (V) | 1.5 |
| Voltage under load (V) | Variable. Can start near 1.5 — lower limit depends on application |
| Discharge voltage profile | Sloping |
| Gravimetric energy density (Wh.kg$^{-1}$) | 9 to 20 |
| Volumetric energy density (Wh.dm$^{-3}$) | 150 to 270 |
| Gravimetric power density (W.kg$^{-1}$) | Up to 23 |
| Volumetric power density (W.dm$^{-3}$) | Up to 57 |
| Temperature range for storage (°C) | −40 to +50 |
| Temperature range for operation (°C) | −30 to +55 |
| Self-discharge rate | Less than 3% per year (20°C) |
| Cell types available | D, C, AA, 3A, N 9-volt, lantern battery (6 V) |

| | |
|---|---|
| **Cell type** | **Aluminium–air** |
| Anode | Aluminium alloy |
| Cathode | Oxygen |
| Electrolyte | (a) KOH or NaOH |
| | (b) NaCl |

| | |
|---|---|
| Open circuit voltage (V) | (a) 1.9 |
| | (b) 1.5 |
| Voltage under load (V) | (a) 1.2 to 1.6 |
| | (b) 0.8 to 1.2 |
| Discharge voltage profile | (a) Flat |
| | (b) Flat |
| Gravimetric energy density (Wh.kg$^{-1}$) | (a) up to 500 |
| | (b) up to 100 |
| Volumetric energy density (Wh.dm$^{-3}$) | (a) up to 500 |
| | (b) up to 110 |
| Gravimetric power density (W.kg$^{-1}$) | (a) 7 [90 has been achieved] |
| | (b) 0.5 |
| Volumetric power density (W.dm$^{-3}$) | (a) 7 [50 has been achieved] |
| | (b) 0.6 |
| Temperature range for storage (°C) | (a) −40 to +60 |
| | (b) −10 to +40 |
| Temperature range for operation (°C) | (a) −40 to +60 [special start-up procedure for −40°C] |
| | (b) +10 to +70 |
| Self-discharge rate | Not applicable (reserve applications) |
| Cell types available | (a) Telecommunications reserve battery (prismatic) |
| | (b) Demonstration units only (prismatic) |

| | |
|---|---|
| **Cell type** | **Lead–acid (motive power)** |
| Anode | Lead |
| Cathode | Lead dioxide |
| Electrolyte | $H_2SO_4$ |
| Open circuit voltage (v) | 2.10 |
| Voltage under load (V) | 1.5 to 2.0 per cell |
| Discharge voltage profile | Flat |
| Gravimetric energy density (Wh.kg$^{-1}$) | 25 to 35 |
| Volumetric energy density (Wh.dm$^{-3}$) | 80 |
| Gravimetric power density (W.kg$^{-1}$) | 3.5 to 6.5 Depends on discharge rate |
| Volumetric power density (W.dm$^{-3}$) | 7 to 13 Depends on discharge rate |
| Temperature range for storage (°C) | Ambient |
| Temperature range for operation (°C) | −18 to +50 |
| Self-discharge rate | 5% per month |
| Cell types available | Tubular/flat plate |
| Cycle life | 1500 (80% depth of discharge) |
| Calendar life | 5/6 years |

| Cell type | Lead–acid (portable) |
|---|---|
| Anode | Lead |
| Cathode | Lead dioxide |
| Electrolyte | $H_2SO_4$ |
| Open circuit voltage (V) | 2.10 |
| Voltage under load (V) | 1.8 to 2.0 |
| Discharge voltage profile | Flat |
| Gravimetric energy density (Wh.kg$^{-1}$) | 30 |
| Volumetric energy density (Wh.dm$^{-3}$) | 80 |
| Gravimetric power density (W.kg$^{-1}$) | Up to 50 |
| | Depends on discharge rate |
| Volumetric power density (W.dm$^{-3}$) | Up to 130 |
| | Depends on discharge rate |
| Temperature range for storage (°C) | Ambient |
| Temperature range for operation (°C) | 0 to 40 |
| Self-discharge rate | 2% per month |
| Cell types available | Sealed (gelled/absorbed electrolyte) |
| Cycle life | 500 (50% DOD) |
| Calendar life | 5 years |

| Cell type | Lead–acid (SLI) |
|---|---|
| Anode | Lead |
| Cathode | Lead dioxide |
| Electrolyte | $H_2SO_4$ |
| Open circuit voltage (V) | 2.10 |
| Voltage under load (V) | 1.5 to 2.0 per cell |
| Discharge voltage profile | Flat |
| Gravimetric energy density (Wh.kg$^{-1}$) | 35 |
| Volumetric energy density (Wh.dm$^{-3}$) | 70 |
| Gravimetric power density (W.kg$^{-1}$) | 2 to 260 |
| | Depends on discharge rate |
| Volumetric power density (W.dm$^{-3}$) | 4 to 500 |
| | Depends on discharge rate |
| Temperature range for storage (°C) | Ambient |
| Temperature range for operation (°C) | $-18$ to $+70$ |
| Self-discharge rate | 2% to 3% per month |
| Cell types available | Prismatic |
| Cycle life | 200 |
| Calendar life | 3 to 5 years |

| | |
|---|---|
| **Cell type** | **Lead–acid (stationary)** |
| Anode | Lead |
| Cathode | Lead dioxide |
| Electrolyte | $H_2SO_4$ |
| Open circuit voltage (V) | 2.10 |
| Voltage under load (V) | 1.80 to 2.00 per cell |
| Discharge voltage profile | Flat |
| Gravimetric energy density (Wh.kg$^{-1}$) | 10 to 25 |
| Volumetric energy density (Wh.dm$^{-3}$) | 50 to 90 |
| Gravimetric power density (W.kg$^{-1}$) | 3 to 90 |
| | Depends on discharge rate |
| Volumetric power density (W.dm$^{-3}$) | 15 to 450 |
| | Depends on discharge rate |
| Temperature range for storage (°C) | Ambient |
| Temperature range for operation (°C) | 0 to 40 |
| Self-discharge rate | 2% per month |
| Cell types available | Flat plate, tubular, round cell, sealed |
| Cycle life | 400 |
| Calendar life | 10 to 15 years |

| | |
|---|---|
| **Cell type** | **Lithium–carbon monofluoride (BR 2330 coin cell)** |
| Anode | Lithium |
| Cathode | $CF_x$, carbon, binder |
| Electrolyte | 50:50 DME:PC with $LiBF_4$ |
| Open circuit voltage (V) | 3.3 to 3.5 |
| Voltage under load (V) | 2.5 to 2.8 |
| Discharge voltage profile | Flat |
| Gravimetric energy density (Wh.kg$^{-1}$) | 235 maxm |
| Volumetric energy density (Wh.dm$^{-3}$) | 587 maxm |
| Gravimetric power density (W.kg$^{-1}$) | 23 maxm |
| Volumetric power density (W.dm$^{-3}$) | 57 maxm |
| Temperature range for storage (°C) | −45 to +85 |
| Temperature range for operation (°C) | −45 to +85 |
| Self-discharge rate | 1% per year |
| Cell types available | Cylindrical, coin and pin as per Tables 5.10, 5.11 and 5.12 |

| | |
|---|---|
| **Cell type** | **Lithium–copper oxide** |
| Anode | Lithium |
| Cathode | Copper oxide |
| Electrolyte | Lithium perchlorate/1,3-dioxolane |
| Open circuit voltage (V) | 2.4 |
| Voltage under load (V) | 1.5 |
| Discharge voltage profile | Depends on current |

| | |
|---|---|
| Gravimetric energy density (Wh.kg$^{-1}$) | 300 maxm |
| Volumetric energy density (Wh.dm$^{-3}$) | 630 maxm |
| Gravimetric power density (W.kg$^{-1}$) | Information not available (not designed as power cells) |
| Volumetric power density (W.dm$^{-3}$) | Information not available (not designed as power cells) |
| Temperature range for storage (°C) | −40 to +55 |
| Temperature range for operation (°C) | −40 to +55 |
| Self-discharge rate | 1% per year at room temperature |
| Cell types available | AA and ½AA |

| **Cell types** | **Lithium–iodine** |
|---|---|
| Anode | Lithium metal |
| Cathode | Iodine |
| Electrolyte | Lithium iodide |
| Open circuit voltage (V) | 2.8 |
| Voltage under load (V) | 2.5 to 2.8 |
| Discharge voltage profile | Linear decline through most of life |
| Gravimetric energy density (Wh.kg$^{-1}$) | 330 (maxm) [packaged] |
| Volumetric energy density (Wh.dm$^{-3}$) | 1300 (maxm) [packaged] |
| Gravimetric power density (W.kg$^{-1}$) | 0.3 |
| Volumetric power density (W.dm$^{-3}$) | 1 |
| Temperature range for storage (°C) | Maxm range −55 to +125 |
| Temperature range for operation (°C) | Maxm range −55 to +125 |
| Self-discharge rate | ≤5% in 10 years |
| Cell types available | Medical grade (custom sizes and shapes) Industrial grade (button, rectilinear, cylindrical and custom) |

| **Cell type** | **Lithium–iron sulphide (rechargeable)** |
|---|---|
| Anode | Li–Al or Li–Si |
| Cathode | FeS |
| Electrolyte | Mixture of alkali or halide salts, e.g. LiCl–KCl or LiBr–LiCl–LiF |
| Open circuit voltage (V) | 1.35 (Li–Al) to 1.48 (Li–Si) |
| Voltage under load (V) | 1.25 (typical) |
| Discharge voltage profile | Fairly flat |
| Gravimetric energy density (Wh.kg$^{-1}$) | 105 (cell basis) |
| Volumetric energy density (Wh.dm$^{-3}$) | 200 (cell basis) |
| Gravimetric power density (W.kg$^{-1}$) | Up to 200 (cell basis) |
| Volumetric power density (W.dm$^{-3}$) | Up to 400 (cell basis) |
| Temperature range for storage (°C) | high-temperature cell |
| Temperature range for operation (°C) | 400 to 500 |
| Self-discharge rate | 0.3% to 3% per month |
| Cell types available | Prismatic development cells (100–400 Ah) |

| | |
|---|---|
| Cycle life | 250–1500 |
| Calendar life | Not yet known |

| | |
|---|---|
| **Cell type** | **Lithium–iron disulphide (AA size)** |
| Anode | Lithium metal foil |
| Cathode | $FeS_2$+graphite+binder |
| Electrolyte | $LiCF_3SO_3$ in mixed organic solvents |
| Open circuit voltage (V) | 1.9 |
| Voltage under load (V) | 1.5 (at C/100 rate) [variable with load] |
| Discharge voltage profile | Flat at medium to low drains |
| Gravimetric energy density (Wh.kg$^{-1}$) | 240 at C/100 rate |
| Volumetric energy density (Wh.dm$^{-3}$) | 550 at C/100 rate |
| Gravimetric power density (W.kg$^{-1}$) | 32 |
| Volumetric power density (W.dm$^{-3}$) | 73 |
| Temperature range for storage (°C) | −40 to +60 |
| Temperature range for operation (°C) | −20 to +50 |
| Self-discharge rate | <2% per year |
| Cell types available | AA at present |

| | |
|---|---|
| **Cell type** | **Lithium–manganese dioxide** |
| Anode | Lithium metal |
| Cathode | $MnO_2$ |
| Electrolyte | Lithium salt in organic solvent |
| Open circuit voltage (V) | 3.0 to 3.6 |
| Voltage under load (V) | 2.7 to 3.0 |
| Discharge voltage profile | Relatively flat |
| Gravimetric energy density (Wh.kg$^{-1}$) | 150 to 350 |
| Volumetric energy density (Wh.dm$^{-3}$) | 450 to 750 |
| Gravimetric power density (W.kg$^{-1}$) | Up to 40 |
| Volumetric power density (W.dm$^{-3}$) | Up to 80 |
| Temperature range for storage (°C) | −20 to +70 (crimp-sealed type) <br> −40 to +85 (laser-sealed type) |
| Temperature range for operation (°C) | −20 to +70 (crimp-sealed type) <br> −40 to +85 (laser-sealed type) |
| Self-discharge rate | 1% per year (crimp-sealed type) <br> 0.5% per year (laser-sealed type) |
| Cell types available | Coin, cylindrical, flat, prismatic |

| | |
|---|---|
| **Cell type** | **Lithium–molybdenum disulphide (AA size)** |
| Anode | Lithium metal |
| Cathode | $MoS_2$ |
| Electrolyte | 1M $LiAsF_6$ in PC/EC 50% vol. |
| Open circuit voltage (V) | 2.4 to 1.1 |

| | |
|---|---|
| Voltage under load (V) | 1.85 at C/10 (plateau) |
| | 1.80 at C/10 (plateau) |
| Discharge voltage profile | Sloping (see section 5.10) |
| Gravimetric energy density (Wh.kg$^{-1}$) | 61 |
| Volumetric energy density (Wh.dm$^{-3}$) | 175 |
| Gravimetric power density (W.kg$^{-1}$) | 130 |
| Volumetric power density (W.dm$^{-3}$) | 375 |
| Temperature range for storage (°C) | −54 to +55 |
| Temperature range for operation (°C) | −30 to +55 (discharge) |
| | −10 to +45 (charge) |
| Self-discharge rate | 5% per year at 21°C |
| Cell types available | AA |
| Cycle life | 200 cycles typical—dependent on discharge rate and DOD |
| Calendar life | 10 years (estimate) |

| | |
|---|---|
| **Cell type** | **Lithium solid state** |
| Anode | Lithium foil |
| Cathode | V$_6$O$_{13}$-based composite |
| Electrolyte | Polyethylene oxide-based polymer with LiClO$_4$ |
| Open circuit voltage (V) | 3.20 |
| Voltage under load (V) | 3.25 to 1.70 (variable) |
| Discharge voltage profile | Stepped |
| Gravimetric energy density (Wh.kg$^{-1}$) | 200 |
| Volumetric energy density (Wh.dm$^{-3}$) | 250 |
| Gravimetric power density (W.kg$^{-1}$) | 150 (1 h rate), 2000 (20 s rate) |
| Volumetric power density (W.dm$^{-3}$) | 175 (1 h rate), 2300 (20 s rate) |
| Temperature range for storage (°C) | −40 to +60 |
| Temperature range for operation (°C) | (a) +100 to +140 |
| | (b) +5 to +40 |
| Self-discharge rate | 1% |
| Cell types available | Development cells only |
| Cycle life | (a) 300 to 500 (100°C to 140°C) |
| | (b) 50 to 100 (20°C) |
| Calendar life | Not known |

| | |
|---|---|
| **Cell type** | **Lithium–sulphur dioxide** |
| Anode | Lithium metal foil |
| Cathode | Acetylene black/PTFE binder/Al grid |
| Electrolyte | Sulphur dioxide, acetonitrile, LiBr |
| Open circuit voltage (V) | 2.95 |

| | |
|---|---|
| Voltage under load (V) | 2.7–2.9 |
| Discharge voltage profile | Very flat |
| Gravimetric energy density (Wh.kg$^{-1}$) | 275 |
| Volumetric energy density (Wh.dm$^{-3}$) | 520 |
| Gravimetric power density (W.kg$^{-1}$) | 220 |
| Volumetric power density (W.dm$^{-3}$) | 350 |
| Temperature range for storage (°C) | −54 to +71 |
| Temperature range for operation (°C) | −54 to +54 |
| Self-discharge rate | 1% per year at room temperature |
| Cell types available | 0.50–35.0 Ah in $\frac{1}{2}$AA, $\frac{1}{2}$A, AA, $\frac{2}{3}$A, $\frac{3}{4}$C, C, $1\frac{1}{4}$C, D, squat D, DD and non-standard sizes |

| | |
|---|---|
| **Cell type** | **Lithium thermal** |
| Anode | Lithium metal |
| Cathode | FeS$_2$ |
| Electrolyte | LiCl:KCl (molten) |
| Open circuit voltage (V) | 1.75 |
| Voltage under load (V) | 1.6 |
| Discharge voltage profile | Sloping |
| Gravimetric energy density (Wh.kg$^{-1}$) | 150 (50 with pyrotechnic included) at 5 min rate of discharge |
| Volumetric energy density (Wh.dm$^{-3}$) | 400 (130 with pyrotechnic included) at 5 min rate of discharge |
| Gravimetric power density (W.kg$^{-1}$) | 2500 (800 with pyrotechnic included) at 5 min rate of discharge |
| Volumetric power density (W.dm$^{-3}$) | 6000 (2000 with pyrotechnic included) at 5 min rate of discharge |
| Temperature range for storage (°C) | −60 to +120 |
| Temperature range for operation (°C) | −40 to +70 (typical) |
| Self-discharge rate | Not applicable |
| Cell types available | Not applicable |

| | |
|---|---|
| **Cell type** | **Lithium–thionyl chloride** |
| Anode | Lithium |
| Cathode | SOCl$_2$ |
| Electrolyte | Lithium salt in thionyl chloride (usually LiAlCl$_4$) |
| Open circuit voltage (V) | 3.66 |
| Voltage under load (V) | 3.0 to 3.5 |
| Discharge voltage profile | Flat |
| Gravimetric energy density (Wh.kg$^{-1}$) | 500 to 1500 |
| Volumetric energy density (Wh.dm$^{-3}$) | 1000 to 3000 |
| Gravimetric power density (W.kg$^{-1}$) | Up to 100 |
| Volumetric power density (W.dm$^{-3}$) | Up to 200 |
| Temperature range for storage (°C) | −55 to +50 |
| Temperature range for operation (°C) | −55 to +150 |

| | |
|---|---|
| Self-discharge rate | <1% per year (room temperature) |
| Cell types available | Button |
| | Flat, cylindrical, and prismatic 1 Ah to 20 000 Ah |

| | |
|---|---|
| **Cell type** | **Lithium–vanadium pentoxide** |
| Anode | Lithium metal |
| Cathode | $V_2O_5$ mixed with carbon |
| Electrolyte | Methyl formate with $LiAsF_6+LiBF_4$ |
| Open circuit voltage (V) | 3.4 |
| Voltage under load (V) | (a) Reserve cell: 3.35, 3.2, and 2.5 |
| | (b) Active cell: 3.2, 2.9, and 2.7 |
| Discharge voltage profile | (a) At low discharge rate (0.4 mA.cm$^{-2}$) three voltage profiles |
| | (b) At high discharge rates (1 mA.cm$^{-2}$) two voltage profiles |
| Gravimetric energy density (Wh.kg$^{-1}$) | 264 |
| Volumetric energy density (Wh.dm$^{-3}$) | 670 |
| Gravimetric power density (W.kg$^{-1}$) | 220 |
| Volumetric power density (W.dm$^{-3}$) | 610 |
| Temperature range for storage (°C) | 21 to 54 |
| Temperature range for operation (°C) | −54 to +54 |
| Self-discharge rate | For active cells, <1% per year |
| Cell types available | Cylindrical, prismatic, and button |

| | |
|---|---|
| **Cell type** | **Nickel–cadmium (pocket plate)** |
| Anode | Cadmium |
| Cathode | $Ni(OH)_2$/graphite |
| Electrolyte | KOH |
| Open circuit voltage (V) | 1.28 |
| Voltage under load (V) | 0.65 to 1.14 per cell (dependent on load) |
| Discharge voltage profile | Relatively flat |
| Gravimetric energy density (Wh.kg$^{-1}$) | 15 to 35 |
| Volumetric energy density (Wh.dm$^{-3}$) | 25 to 50 |
| Gravimetric power density (W.kg$^{-1}$) | 30 to 150 ⎫ (lower values include |
| Volumetric power density (W.dm$^{-3}$) | 50 to 300 ⎭ electrolyte reserve for low maintenance) |
| Temperature range for storage (°C) | −60 to +60 |
| Temperature range for operation (°C) | −40 to +60 |
| Self-discharge rate | Less than 5% per year (between 0% and 80% charged) |
| Cell types available | Low, medium, and high rate discharge types; also deep cycling range; sizes 10 to 1500 Ah |
| Cycle life | 2500 (100% DOD) [Depends on design] |
| Calendar life | 25 years |

| | |
|---|---|
| **Cell type** | **Nickel–cadmium (sealed cylindrical)** |
| Anode | Cadmium |
| Cathode | Nickel oxyhydroxide |
| Electrolyte | KOH |
| Open circuit voltage | 1.3 |
| Voltage under load (V) | 1.2 Nominal |
| | 1.3 to 1.0 |
| | (according to conditions) |
| Discharge voltage profile | Flat |
| Gravimetric energy density (Wh.kg$^{-1}$) | 13 to 35 |
| Volumetric energy density (Wh.dm$^{-3}$) | 50 to 120 (according to cell size) |
| Gravimetric power density (W.kg$^{-1}$) | 750 to 1000 (according to cell size) |
| Volumetric power density (W.dm$^{-3}$) | 2500 to 3500 (according to cell size) |
| Temperature range for storage (°C) | −40 to +70 |
| | (+5 to +25 optimum) |
| Temperature range for operation (°C) | −40 to +70 |
| | (+5 to +25 optimum) |
| Self-discharge rate | 10% to 25% per month |
| | (according to type) |
| Cell types available | Cylindrical (0.1 to 10 Ah) |
| Cycle life | More than 500 |
| | (according to conditions) |
| Calendar life | 4 to 8 years operational |
| | 5 to 10 years storage |
| | (according to conditions) |

| | |
|---|---|
| **Cell type** | **Nickel–hydrogen** |
| Anode | Hydrogen electrode |
| Cathode | Nickel electrode |
| Electrolyte | KOH |
| Open circuit voltage (V) | 1.4 |
| Voltage under load (V) | 1.2–1.3 per cell |
| Discharge voltage profile | Relatively flat |
| Gravimetric energy density (Wh.kg$^{-1}$) | 49 |
| Volumetric energy density (Wh.dm$^{-3}$) | 57 |
| Gravimetric power density (W.kg$^{-1}$) | 135 (6.7C rate) |
| Volumetric power density (W.dm$^{-3}$) | 157 (6.7C rate) |
| Temperature range for storage (°C) | −20 to 0 (open circuit discharge) |
| Temperature range for operation (°C) | −5 to +20 |
| Self-discharge rate | 15% at 10°C in 72 h |
| Cell types available | Individual pressure vessel |

| | |
|---|---|
| Cycle life | Greater than 20 000 LEO cycles at 40% DOD, 10°C |
| Calendar life | Greater than 10 years |

| | |
|---|---|
| **Cell type** | **Nickel–zinc (alkaline)** |
| Anode | Zinc oxide/zinc |
| Cathode | Nickel hydroxide/oxyhydroxide |
| Electrolyte | 30–35% KOH |
| Open circuit voltage (V) | 1.75 |
| Voltage under load (V) | 1.4 to 1.6 |
| Discharge voltage profile | Flat (sintered nickel positives) |
| | Slight taper (pressed nickel positives) |
| Gravimetric energy density (Wh.kg$^{-1}$) | 60 to 70 |
| Volumetric energy density (Wh.dm$^{-3}$) | 90 to 100 |
| Gravimetric power density (W.kg$^{-1}$) | 140 to 200 |
| Volumetric power density (W.dm$^{-3}$) | 160 to 240 |
| Temperature range for storage (°C) | $-40$ to $+20$ |
| Temperature range for operation (°C) | $-40$ to $+50$ |
| Self-discharge rate | Up to 25% in the first month; 5% in later months |
| Cell types available | Large and small sealed cells have been developed, but mostly large vented cells |
| Cycle life | 300 cycles at 100% depth of discharge (0.2C rate) |
| | 3000 cycles at 20% depth of discharge (0.2C rate) |
| Calendar life | Not known as yet |

| | |
|---|---|
| **Cell type** | **Sodium–sulphur** |
| Anode | Sodium |
| Cathode | Sulphur |
| Electrolyte | Beta″ alumina |
| Open circuit voltage (V) | 2.076 at 350°C |
| Voltage under load (V) | 1.95 (C/3 rate) |
| Discharge voltage profile | 2.076 V for 57% of discharge |
| | Linear reduction to 1.76 V open circuit (voc) with respect to $Na_2S_3$ |
| Gravimetric energy density (Wh.kg$^{-1}$) | 188 to 1.76 voc |
| Volumetric energy density (Wh.dm$^{-3}$) | 370 to 1.76 voc |
| Gravimetric power density (W.kg$^{-1}$) | 210 at 66% V of open circuit voltage |
| Volumetric power density (W.dm$^{-3}$) | 440 at 66% V of open circuit voltage |
| Temperature range for storage (°C) | Insensitive |

| | |
|---|---|
| Temperature range for operation (°C) | Insensitive |
| Self-discharge rate | Zero |
| Cell types available | 20 Ah PB |
| Cycle life | 8000 (maxm cell to date) |
| | 500 (maxm battery to date) |
| Calendar life | 8 years (maxm cell to date) |
| | 1 year (maxm battery to date) |

| | |
|---|---|
| **Cell type** | **Premium zinc–air (button cell)** |
| Anode | Zinc |
| Cathode | Oxygen |
| Electrolyte | 30% KOH +2% ZnO |
| Open circuit voltage (V) | 1.40 |
| Voltage under load (V) | 1.23 |
| Discharge voltage profile | Flat |
| Gravimetric energy density (Wh.kg$^{-1}$) | 400 |
| Volumetric energy density (Wh.dm$^{-3}$) | 1200 |
| Gravimetric power density (W.kg$^{-1}$) | 1.47 |
| Volumetric power density (W.dm$^{-3}$) | 4.39 |
| Temperature range for storage (°C) | −40 to +50 |
| Temperature range for operation (°C) | −10 to +55 |
| Self-discharge rate | 3% per year (21°C) |
| Cell types available | Button |

| | |
|---|---|
| **Cell type** | **Large zinc–air** |
| Anode | Zinc |
| Cathode | Oxygen |
| Electrolyte | 5M KOH |
| Open circuit voltage (V) | 1.45 to 1.52 |
| Voltage under load (V) | 1.25 nominal at 0.5 A drain rate for 1000 Ah cell |
| Discharge voltage profile | Flat over 85% of rated capacity |
| Gravimetric energy density (Wh.kg$^{-1}$) | 165 (wet cell) |
| | 265 (dry cell) |
| Volumetric energy density (Wh.dm$^{-3}$) | 215 (wet cell) |
| | 340 (dry cell) |
| Gravimetric power density (W.kg$^{-1}$) | 0.17 (wet cell) |
| | 0.24 (dry cell) |
| Volumetric power density (W.dm$^{-3}$) | 0.20 (wet cell) |
| | 0.27 (dry cell) |
| Temperature range for storage (°C) | +10 to +30 possible |
| | Ambient 25°C, 50% RH for most cases |
| Temperature range for operation (°C) | −20 to +45 (wet cell) |
| | −10 to +40 (dry cell) |

| | |
|---|---|
| Self-discharge rate | 1–3% per year (wet cell)<br>5–15% per year (dry cell)<br>Open circuit conditions |
| Cell types available | Wet type (1000–1100 Ah at 2.5–3.75 V and 2000–3300 Ah at 1.25 V)<br>Dry type (1200–3600 Ah at 1.25 V) |

| | |
|---|---|
| **Cell type** | **Zinc–bromine aqueous rechargeable** |
| Anode | Zinc metal<br>Deposited on an inert substrate |
| Cathode | Bromine<br>Evolved and consumed on an inert substrate |
| Electrolyte | Aqueous $ZnBr_2$ solution |
| Open circuit voltage (V) | 1.75 to 1.80 |
| Voltage under load (V) | 1.5 to 1.6 |
| Discharge voltage profile | Flat: one-step discharge |
| Gravimetric energy density (Wh.kg$^{-1}$) | 15 to 85 depending on design |
| Volumetric energy density (Wh.dm$^{-3}$) | 10 to 50 depending on design |
| Gravimetric power density (W.kg$^{-1}$) | 50 to 150 |
| Volumetric power density (W.dm$^{-3}$) | 40 to 90 |
| Temperature range for storage (°C) | 0 to 50 |
| Temperature range for operation (°C) | 15 to 45 |
| Self-discharge rate | Initially 5% in 2 days, then 1% per month |
| Cell types available | None commercially available<br>Two designs are under development for stationary energy storage and vehicle propulsion; both flowing electrolyte systems |
| Cycle life | 3000 cycles projected and demonstrated cells |
| Calendar life | 10 years |

| | |
|---|---|
| **Cell type** | **Premium zinc–carbon** |
| Anode | Zinc |
| Cathode | $MnO_2$ |
| Electrolyte | $ZnCl_2$ + 2–4% $NH_4Cl$ |
| Open circuit voltage (V) | 1.70 |
| Voltage under load (V) | 1.40 to 0.75 |
| Discharge voltage profile | Sloping |
| Gravimetric energy density (Wh.kg$^{-1}$) | 60 to 110 |
| Volumetric energy density (Wh.dm$^{-3}$) | 120 to 250 |
| Gravimetric power density (W.kg$^{-1}$) | 10 to 20 |
| Volumetric power density (W.dm$^{-3}$) | 20 to 40 |
| Temperature range for storage (°C) | −40 to +54 |

| | |
|---|---|
| Temperature range for operation (°C) | −20 to +60 |
| Self-discharge rate | 10% per year |
| Cell types available | General-purpose Leclanché |
| | Premium Leclanché and zinc chloride |
| | (cylindrical or flat geometry) |

| | |
|---|---|
| **Cell type** | **Premium zinc–mercury oxide** |
| Anode | Zinc |
| Cathode | HgO |
| Electrolyte | 40% KOH+1% ZnO |
| Open circuit voltage (V) | 1.35 |
| Voltage under load (V) | 1.25 |
| Discharge voltage profile | Flat |
| Gravimetric energy density (Wh.kg$^{-1}$) | 110 |
| Volumetric energy density (Wh.dm$^{-3}$) | 600 |
| Gravimetric power density (W.kg$^{-1}$) | 0.96 |
| Volumetric power density (W.dm$^{-3}$) | 4.39 |
| Temperature range for storage (°C) | −40 to +60 |
| Temperature range for operation (°C) | −10 to +55 |
| Self-discharge rate | 4% per year (21°C) |
| Cell types available | Buttons, small cylindricals |

| | |
|---|---|
| **Cell type** | **Zinc–plumbate** |
| Anode | Zinc |
| Cathode | Silver 'plumbate' |
| Electrolyte | 40% KOH+1% ZnO |
| Open circuit voltage (V) | 1.70 |
| Voltage under load (V) | 1.50 |
| Discharge voltage profile | Flat |
| Gravimetric energy density (Wh.kg$^{-1}$) | 155 |
| Volumetric energy density (Wh.dm$^{-3}$) | 660 |
| Gravimetric power density (W.kg$^{-1}$) | 1.60 |
| Volumetric power density (W.dm$^{-3}$) | 7.25 |
| Temperature range for storage (°C) | −40 to +60 |
| Temperature range for operation (°C) | −10 to +55 |
| Self-discharge rate | 3% per year |
| Cell types available | Buttons |

| | |
|---|---|
| **Cell type** | **Zinc–silver nickel oxide** |
| Anode | Zinc |
| Cathode | AgNiO$_2$ |
| Electrolyte | 40% KOH+1% ZnO |
| Open circuit voltage (V) | 1.60 |
| Voltage under load (V) | 1.45 |
| Discharge voltage profile | Flat—two steps |
| Gravimetric energy density (Wh.kg$^{-1}$) | 145 |

| | |
|---|---|
| Volumetric energy density (Wh.dm$^{-3}$) | 600 |
| Gravimetric power density (W.kg$^{-1}$) | 1.49 |
| Volumetric power density (W.dm$^{-3}$) | 6.78 |
| Temperature range for storage (°C) | −40 to +60 |
| Temperature range for operation (°C) | −10 to +55 |
| Self-discharge rate | 5% |
| Cell types available | Buttons |

| | |
|---|---|
| **Cell type** | **Zinc–silver oxide (AgO)** |
| Anode | Zinc |
| Cathode | AgO |
| Electrolyte | 40% KOH+1% ZnO |
| Open circuit voltage (V) | 1.70 |
| Voltage under load (V) | 1.50 |
| | 6500 Ω cont., 21°C |
| Discharge voltage profile | Flat |
| Gravimetric energy density (Wh.kg$^{-1}$) | 160 |
| Volumetric energy density (Wh.dm$^{-3}$) | 680 |
| Gravimetric power density (W.kg$^{-1}$) | 1.60 |
| Volumetric power density (W.dm$^{-3}$) | 7.25 |
| Temperature range for storage (°C) | −40 to +60 |
| Temperature range for operation (°C) | −10 to +55 |
| Self-discharge rate | 4% per year (21°C) |
| Cell types available | Buttons |

| | |
|---|---|
| **Cell type** | **Zinc–silver oxide (Ag$_2$O)** |
| Anode | Zinc |
| Cathode | Ag$_2$O |
| Electrolyte | 40% KOH+1% ZnO |
| Open circuit voltage (V) | 1.60 |
| Voltage under load (V) | 1.50 |
| | 6500 Ω cont., 21°C |
| Discharge voltage profile | Flat |
| Gravimetric energy density (Wh.kg$^{-1}$) | 120 |
| Volumetric energy density (Wh.dm$^{-3}$) | 500 |
| Gravimetric power density (W.kg$^{-1}$) | 1.60 |
| Volumetric power density (W.dm$^{-3}$) | 7.25 |
| Temperature range for storage (°C) | −40 to +60 |
| Temperature range for operation (°C) | −10 to +55 |
| Self-discharge rate | 3% per year (21°C) |
| Cell types available | Buttons |

# Appendix B
# Standard potentials of various battery electrode reactions

Reactions at 25°C (in alphabetical order of reacting component).

| Electrode reaction | $E°(V)$ |
| --- | --- |
| $AgCl + e^- \rightarrow Ag + Cl^-$ | $+0.222$ |
| $Ag_2O + H_2O + 2e^- \rightarrow 2Ag + 2OH^-$ | $+0.344$ |
| $2AgO + H_2O + 2e^- \rightarrow Ag_2O + 2OH^-$ | $+0.57$ |
| $Al^{3+} + 3e^- \rightarrow Al$ | $-1.66$ |
| $Al(OH)_4^- + 3e^- \rightarrow Al + 4OH^-$ | $-2.35$ |
| $Br_2(l.) + 2e^- \rightarrow 2Br^-$ | $+1.065$ |
| $Ca^{2+} + 2e^- \rightarrow Ca$ | $-2.84$ |
| $Cd(OH)_2 + 2e^- \rightarrow Cd + 2OH^-$ | $-0.809$ |
| $Cl_2 + 2e^- \rightarrow 2Cl^-$ | $+1.360$ |
| $Cu^{2+} + 2e^- \rightarrow Cu$ | $+0.337$ |
| $CuCl + e^- \rightarrow Cu + Cl^-$ | $+0.137$ |
| $CuCNS + e^- \rightarrow Cu + CNS^-$ | $-0.27$ |
| $CuI + e^- \rightarrow Cu + I^-$ | $-0.185$ |
| $Cu_2O + H_2O + 2e^- \rightarrow 2Cu + 2OH^-$ | $-0.358$ |
| $Fe^{2+} + 2e^- \rightarrow Fe$ | $-0.440$ |
| $Fe(CN)_6^{3-} + e^- \rightarrow Fe(CN)_6^{4-}$ | $+0.36$ |
| $2H^+ + 2e^- \rightarrow H_2$ | $0.000$ |
| $2H_2O + 2e^- \rightarrow H_2 + 2OH^-$ | $-0.828$ |
| $H_2O_2 + 2H^+ + 2e^- \rightarrow 2H_2O$ | $+1.77$ |
| $HgO(red) + H_2O + 2e^- \rightarrow Hg + 2OH^-$ | $+0.098$ |
| $I_2 + 2e^- \rightarrow 2I^-$ | $+0.536$ |
| $Li^+ + e^- \rightarrow Li$ | $-3.045$ |
| $Mg^{2+} + 2e^- \rightarrow Mg$ | $-2.37$ |
| $MnO_2 + 4H^+ + 2e^- \rightarrow Mn^{2+} + 2H_2O$ | $+1.23$ |
| $\gamma\text{-}MnO_2 + H_2O + e^- \rightarrow \alpha\text{-}MnOOH + OH^-$ | $+0.30$ |

| Electrode reaction | $E°(V)$ |
|---|---|
| $\gamma\text{-}MnO_2 + H_2O + e^- \rightarrow \gamma\text{-}MnOOH + OH^-$ | $+0.36$ |
| $Na^+ + e^- \rightarrow Na$ | $-2.74$ |
| $NiOOH + H_2O + e^- \rightarrow Ni(OH)_2 + OH^-$ | $+0.45$ |
| $O_2 + 4H^+ + 4e^- \rightarrow 2H_2O$ | $+1.229$ |
| $O_2 + H_2O + 2e^- \rightarrow HO_2^- + OH^-$ | $-0.076$ |
| $O_2 + 2H_2O + 4e^- \rightarrow 4OH^-$ | $+0.401$ |
| $PbCl_2 + 2e^- \rightarrow Pb + 2Cl^-$ | $-0.268$ |
| $PbI_2 + 2e^- \rightarrow Pb + 2I^-$ | $-0.365$ |
| $PbO_2 + 4H^+ + 2e^- \rightarrow Pb^{2+} + 2H_2O$ | $+1.455$ |
| $PbO_2 + SO_4^{2-} + 4H^+ + 2e^- \rightarrow PbSO_4 + 2H_2O$ | $+1.685$ |
| $PbSO_4 + 2e^- \rightarrow Pb + SO_4^{2-}$ | $-0.356$ |
| $S + 2e^- \rightarrow S^{2-}$ | $-0.48$ |
| $Zn^{2+} + 2e^- \rightarrow Zn$ | $-0.763$ |
| $Zn(OH)_2 + 2e^- \rightarrow Zn + 2OH^-$ | $-1.245$ |

# Appendix C

## Electrochemical equivalent per mass and volume of possible battery electrode elements and compounds

| Element or compound | Symbol or formula | No. of redox electrons | Ah.g$^{-1}$ | g.(Ah)$^{-1}$ | Ah.cm$^{-3}$ | cm$^3$(Ah)$^{-1}$ |
|---|---|---|---|---|---|---|
| Aluminium | Al | 3 | 2.98 | 0.34 | 8.05 | 0.12 |
| Bromine | Br | 1 | 0.34 | 2.98 | 0.98 | 1.02 |
| Cadmium | Cd | 2 | 0.47 | 2.10 | 4.12 | 0.24 |
| Calcium | Ca | 2 | 1.33 | 0.75 | 2.06 | 0.49 |
| Carbon (graphite) | C | 4 | 8.92 | 0.11 | 20.08 | 0.05 |
| Carbon monofluoride | CF$_x$ | 1 | 0.86 | 1.16 | 2.33 | 0.42 |
| Chlorine | Cl | 1 | 0.75 | 1.32 | — | — |
| Copper | Cu | 2 | 0.84 | 1.19 | 7.49 | 0.13 |
|  |  | 1 | 0.42 | 2.37 | 3.75 | 0.27 |
| Copper (I) chloride | CuCl | 1 | 0.27 | 3.69 | 1.12 | 0.89 |
| Copper (II) oxide | CuO | 2 | 0.67 | 1.48 | 4.25 | 0.24 |
| Copper (I) thiocyanate | CuSCN | 1 | 0.22 | 4.53 | 0.63 | 1.60 |
| Fluorine | F | 1 | 1.41 | 0.71 | — | — |
| Hydrogen | H | 1 | 26.59 | 0.038 | 2.04 (l) | 0.49 (l) |
| Hydrogen peroxide (80$^w$/$_o$) | H$_2$O$_2$ | 2 | 1.26 | 0.79 | 1.68 | 0.60 |

| Element or compound | Symbol or formula | No. of redox electrons | Ah.g$^{-1}$ | g.(Ah)$^{-1}$ | Ah.cm$^{-3}$ | cm$^3$(Ah)$^{-1}$ |
|---|---|---|---|---|---|---|
| Hydrogen peroxide (60$^w$/$_o$) | H$_2$O$_2$ | 2 | 0.95 | 1.05 | 1.17 | 0.85 |
| Iodine | I | 1 | 0.211 | 4.74 | 1.04 | 0.96 |
| Iron | Fe | 2 | 0.96 | 1.04 | 7.54 | 0.13 |
| Iron (I) sulphide | FeS | 2 | 0.61 | 1.64 | 2.95 | 0.39 |
| Iron (II) sulphide | FeS$_2$ | 4 | 0.89 | 1.12 | 4.35 | 0.23 |
| Lead | Pb | 2 | 0.26 | 3.87 | 2.93 | 0.34 |
| Lead chloride | PbCl$_2$ | 2 | 0.19 | 5.19 | 1.12 | 0.89 |
| Lead dioxide | PbO$_2$ | 2 | 0.22 | 4.46 | 2.10 | 0.48 |
| Lead iodide | PbI$_2$ | 2 | 0.12 | 8.60 | 0.72 | 1.40 |
| Lithium | Li | 1 | 3.86 | 0.26 | 2.06 | 0.48 |
| Magnesium | Mg | 2 | 2.2 | 0.45 | 3.83 | 0.26 |
| Manganese dioxide | MnO$_2$ | 1 | 0.31 | 3.24 | 1.55 | 0.65 |
| Mercuric oxide | HgO | 2 | 0.25 | 4.04 | 2.75 | 0.36 |
| Nickel | Ni | 2 | 0.91 | 1.09 | 8.10 | 0.12 |
| Nickel oxyhydroxide | NiOOH | 1 | 0.29 | 3.42 | 2.16 | 0.46 |
| Oxygen | O | 2 | 3.35 | 0.30 | 3.82 (l) | 0.26 (l) |
| Silver chloride | AgCl | 1 | 0.19 | 5.35 | 1.04 | 0.96 |
| Silver (I) oxide | Ag$_2$O | 2 | 0.23 | 4.32 | 1.65 | 0.61 |
| Silver (II) oxide | AgO | 2 | 0.43 | 2.31 | 3.20 | 0.31 |
| Sodium | Na | 1 | 1.17 | 0.86 | 1.13 | 0.88 |
| Sulphur | S | 2 | 1.67 | 0.60 | 3.46 | 0.29 |
| Sulphur dioxide | SO$_2$ | 1 | 0.41 | 2.39 | 0.60 (l) | 1.67 (l) |
| Thionyl chloride | SOCl$_2$ | 2 | 0.45 | 2.22 | 0.75 | 1.34 |
| Vanadium pentoxide | V$_2$O$_5$ | 1 | 0.15 | 6.79 | 0.49 | 2.02 |
| Zinc | Zn | 2 | 0.82 | 1.22 | 5.85 | 0.17 |

# Appendix D
# Conversion tables and physical constants

### MASS

|  | g | kg | lb |
|---|---|---|---|
| 1 gram = | 1 | 0.001 | $2.205 \times 10^{-3}$ |
| 1 kilogram = | 1000 | 1 | 2.205 |
| 1 pound (avoirdupois) = | 453.6 | 0.4536 | 1 |

1 tonne = 1000 kg    1 ton = 1016 kg

### LENGTH

|  | cm | metre | km | in | ft |
|---|---|---|---|---|---|
| 1 centimetre = | 1 | $10^{-2}$ | $10^{-5}$ | 0.3937 | $3.281 \times 10^{-2}$ |
| 1 metre = | 100 | 1 | $10^{-3}$ | 39.37 | 3.281 |
| 1 kilometre = | $10^5$ | 1000 | 1 | $3.937 \times 10^4$ | 3281 |
| 1 inch = | 2.540 | $2.540 \times 10^{-2}$ | $2.540 \times 10^{-5}$ | 1 | $8.333 \times 10^{-2}$ |
| 1 foot = | 30.48 | 0.3048 | $3.048 \times 10^{-4}$ | 12 | 1 |

1 angstrom (Å) = $10^{-10}$ m    1 micron (micrometer, $\mu$m) = $10^{-6}$ m    1 mil = $10^{-3}$ in = 25.4 $\mu$m

### AREA

|  | m² | cm² | ft² | in² |
|---|---|---|---|---|
| 1 square metre = | 1 | $10^4$ | 10.76 | 1550 |
| 1 square centimetre = | $10^{-4}$ | 1 | $1.076 \times 10^{-3}$ | 0.1550 |
| 1 square foot = | $9.290 \times 10^{-2}$ | 929.0 | 1 | 144 |
| 1 square inch = | $6.452 \times 10^{-4}$ | 6.452 | $6.944 \times 10^{-3}$ | 1 |

## VOLUME

|  | m³ | cm³ | l | ft³ | in³ |
|---|---|---|---|---|---|
| 1 cubic metre = | 1 | $10^6$ | 1000 | 35.31 | $6.102 \times 10^4$ |
| 1 cubic centimetre = | $10^{-6}$ | 1 | $1.000 \times 10^{-3}$ | $3.531 \times 10^{-5}$ | $6.102 \times 10^{-2}$ |
| 1 litre = | $1.000 \times 10^{-3}$ | 1000 | 1 | $3.531 \times 10^{-2}$ | 61.02 |
| 1 cubic foot = | $2.832 \times 10^{-2}$ | $2.832 \times 10^4$ | 28.32 | 1 | 1728 |
| 1 cubic inch = | $1.639 \times 10^{-5}$ | 16.39 | $1.639 \times 10^{-2}$ | $5.787 \times 10^{-4}$ | 1 |

1 US fluid gallon = 4 US fluid quarts = 8 US pints = 128 US fluid ounces = 231 in³. = 3.786 litre.
1 British imperial gallon = the volume of 10 lb of water at 62°F = 277.42 in³. = 4.55 litre.
1 litre = the volume of 1 kg of water at its maximum density = 1000.028 cm³.

## POWER

|  | Btu/h | hp | cal/s | kW | W |
|---|---|---|---|---|---|
| 1 British thermal unit per hour = | 1 | $3.929 \times 10^{-4}$ | $7.000 \times 10^{-2}$ | $2.930 \times 10^{-4}$ | 0.2930 |
| 1 horsepower = | 2545 | 1 | 178.2 | 0.7457 | 745.7 |
| 1 calorie per second = | 14.29 | $5.613 \times 10^{-3}$ | 1 | $4.186 \times 10^{-3}$ | 4.186 |
| 1 kilowatt = | 3413 | 1.341 | 238.9 | 1 | 1000 |
| 1 watt = | 3.413 | $1.341 \times 10^{-3}$ | 0.2389 | 0.001 | 1 |

## ENERGY

|  | Btu | ft-lb | hp-hr | J | cal | kWh | eV |
|---|---|---|---|---|---|---|---|
| 1 British thermal unit = | 1 | 777.9 | $3.929 \times 10^{-4}$ | 1055 | 252.0 | $2.930 \times 10^{-4}$ | $6.585 \times 10^{21}$ |
| 1 foot-pound = | $1.285 \times 10^{-3}$ | 1 | $5.051 \times 10^{-7}$ | 1.356 | 0.3239 | $3.766 \times 10^{-7}$ | $8.464 \times 10^{18}$ |
| 1 horsepower-hour = | 2545 | $1.980 \times 10^6$ | 1 | $2.685 \times 10^6$ | $6.414 \times 10^5$ | 0.7457 | $1.676 \times 10^{25}$ |
| 1 joule = | $9.481 \times 10^{-4}$ | 0.7376 | $3.725 \times 10^{-7}$ | 1 | 0.2389 | $2.778 \times 10^{-7}$ | $6.242 \times 10^{18}$ |
| 1 calorie = | $3.968 \times 10^{-3}$ | 3.087 | $1.559 \times 10^{-6}$ | 4.186 | 1 | $1.163 \times 10^{-6}$ | $2.613 \times 10^{19}$ |
| 1 kilowatt-hour = | 3413 | $2.655 \times 10^6$ | 1.341 | $3.6 \times 10^6$ | $8.601 \times 10^5$ | 1 | $2.247 \times 10^{25}$ |
| 1 electronvolt | $1.519 \times 10^{-22}$ | $1.182 \times 10^{-19}$ | $5.967 \times 10^{-26}$ | $1.602 \times 10^{-19}$ | $3.827 \times 10^{-20}$ | $4.450 \times 10^{-26}$ | 1 |

## ELECTRIC CHARGE

|  | Ah | C | faraday |
|---|---|---|---|
| 1 ampere-hour = | 1 | 3600 | $3.731 \times 10^{-2}$ |
| 1 coulomb = | $2.778 \times 10^{-4}$ | 1 | $1.036 \times 10^{-5}$ |
| 1 faraday = | 26.80 | $9.6487 \times 10^4$ | 1 |

## GRAVIMETRIC AND VOLUMETRIC ENERGY DENSITIES

|  | W(h).in$^{-3}$ | W(h).dm$^{-3}$ | W(h).ft$^{-3}$ | W(h).lb$^{-1}$ | W(h).kg$^{-1}$ |
|---|---|---|---|---|---|
| 1 W(h).in$^{-3}$ = | 1 | 61.023 | 1728 | — | — |
| 1 W(h).dm$^{-3}$ = | 1.639 × 10$^{-2}$ | 1 | 28.32 | — | — |
| 1 W(h).ft$^{-3}$ = | 5.787 × 10$^{-4}$ | 3.531 × 10$^{-2}$ | 1 | — | — |
| 1 W(h).lb$^{-1}$ = | — | — | — | 1 | 2.20 |
| 1 W(h).kg$^{-1}$ = | — | — | — | 0.4536 | 1 |

## CURRENT DENSITY

|  | mA.cm$^{-2}$ | A.m$^{-2}$ | mA.in$^{-2}$ | A.ft$^{-2}$ |
|---|---|---|---|---|
| 1 mA.cm$^{-2}$ = | 1 | 10 | 6.452 | 0.929 |
| 1 A.m$^{-2}$ = | 0.1 | 1 | 0.645 | 0.29 × 10$^{-2}$ |
| 1 mA.in$^{-2}$ = | 0.1550 | 1.550 | 1 | 0.144 |
| 1 A.ft$^{-2}$ = | 1.076 | 10.76 | 6.94 | 1 |

## CONSTANTS

| | |
|---|---|
| Avogadro's number ($L$) | 6.022 × 10$^{23}$ molecules.mol$^{-1}$ |
| Faraday's constant ($F$) | 9.6487 × 10$^4$ C.mol$^{-1}$ |
| Gas constant ($R$) | 8.314 J.K$^{-1}$.mol$^{-1}$ |
|  | 1.980 cal.K$^{-1}$.mol$^{-1}$ |
| 2.303 RT/F | 59.16 mV at 25°C |

## TEMPERATURE

| Degrees celsius (°C) | Degrees fahrenheit (°F) |
|---|---|
| −60 | −76.0 |
| −58 | −72.4 |
| −56 | −68.8 |
| −54 | −65.2 |
| −52 | −61.6 |
| −50 | −58.0 |
| −48 | −54.4 |
| −46 | −50.8 |
| −44 | −47.2 |
| −42 | −43.6 |
| −40 | −40.0 |
| −38 | −36.4 |
| −36 | −32.8 |
| −34 | −29.2 |
| −32 | −25.6 |
| −30 | −22.0 |
| −28 | −18.4 |
| −26 | −14.8 |
| −24 | −11.2 |
| −22 | −7.6 |

| Degrees celsius (°C) | Degrees fahrenheit (°F) |
|---|---|
| −20 | −4.0 |
| −18 | −0.4 |
| −16 | 3.2 |
| −14 | 6.8 |
| −12 | 10.4 |
| −10 | 14.0 |
| −8 | 17.6 |
| −6 | 21.2 |
| −4 | 24.8 |
| −2 | 28.4 |
| 0 | 32.0 |
| 2 | 35.6 |
| 4 | 39.2 |
| 6 | 42.8 |
| 8 | 46.4 |
| 10 | 50.0 |
| 12 | 53.6 |
| 14 | 57.2 |
| 16 | 60.8 |
| 18 | 64.4 |
| 20 | 68.0 |
| 22 | 71.6 |
| 24 | 75.2 |
| 26 | 78.8 |
| 28 | 82.4 |
| 30 | 86.0 |
| 32 | 89.6 |
| 34 | 93.2 |
| 36 | 96.8 |
| 38 | 100.4 |
| 40 | 104.0 |
| 42 | 107.6 |
| 44 | 111.2 |
| 46 | 114.8 |
| 48 | 118.4 |
| 50 | 122.0 |
| 52 | 125.6 |
| 54 | 129.2 |
| 56 | 132.8 |
| 58 | 136.4 |
| 60 | 140.0 |
| 62 | 143.6 |
| 64 | 147.2 |
| 66 | 150.8 |
| 68 | 154.4 |
| 70 | 158.0 |

# Index

acetonitrile, 72, 314
  thermal expansion, 316
acetylene black, 49, 100, 291, 311, 318, 338
  *see also* carbon black
air cathode
  carbonation, 146
  catalyst, 167, 494
  construction, 128, 162–170, 494
  electrochemistry, 127, 168, 489
  fired, 164
  gas flow, 168
  heat treatment, 494
  peroxide reaction, 127, 166, 168–170
  plastic-bonded, 163, 165
  pressure differentials, 166
  undersea use, 42
aircraft applications, 213, 296, 333, 419
aircraft service vehicles, 219
airship, 26
alarm applications, 293, 353
  *see also* security systems
alkaline batteries, 24
  *see also* alkaline–manganese batteries
alkaline batteries (small size), 125–160
alkaline–manganese batteries, 19, 21, 24, 39, 89, 111–125, 313, 328, 348, 402, 405, 544
  applications, 112
  can manufacture, 118
  chemistry, 115–117
  construction, 113–115
  cost, 124
  heavy drain continuous test, 119
  history, 112
  light drain test, 119
  manganese dioxide cathode, 117
  market, 112
  materials used, 117
  performance characteristics, 118–124
    comparison with lithium batteries, 122
    comparison with nickel–cadmium, 120
    comparison with zinc–carbon, 119, 122–123
    effect of temperature, 121
    input capacity, 118
    output capacity, 119
  safety, 115

  sealing, 123
  standard sizes, 112
  storage, 122
  zinc anode, 115
alkaline zinc–manganese dioxide batteries *see* alkaline–manganese batteries
alloy formation effect, 276
aluminium
  corrosion, 490
  sodium–sulphur battery use, 464
  *see also* electrode, aluminium; lithium–aluminium anode
aluminium–air batteries, 20, 38, 487–502, 544
  air cathode development, 494
  anode alloys, 490–494
  *see also* electrode, aluminium alloys
  anode development, 490–494
  chemistry, 489
  commercial battery development, 498–502
  discharge behaviour, 36
  electrolyte management, 496
  history, 487–489
aluminium chloride, 289
aluminium–hydrogen peroxide batteries, 498
aluminium hydroxide, 36
  precipitation, 496
  structural forms, 490
aluminium–manganese dioxide batteries, 498
aluminium–silver oxide batteries, 498
amalgamation
  aluminium, 490
  nickel–zinc batteries, 454
  reduction of use, 125, 186
  zinc, 100, 117, 127, 177
Amein, 26
ammonia, 97, 338
ammonium chloride, 89, 111
Ampère, 23
ampoule, 309, 396, 400
anaglyph, 64
André, 27, 129
anode
  definition, 17
  *for anode types see* electrode

ANSI standard cells, 106
archaeological work, 23
Aron, 24
Arrhenius, 23
arrythmia, 400
asbestos, *(see* separators, asbestos)
asphalt, 101
atomic absorption spectroscopy (AAS), 349
audio devices, 111, 113
auditory canal, temperature and humidity, 145
Auger electron spectroscopy (AES), 69
automatic watering device, 244
automobile industry, 194

Babylonians, 23
back-scattered electrons, 62
back-up power supplies, 241, 302, 306, 333, 336
   *see also* uninterruptible power systems
Bacon, 23
Bailhache, 24
barge, 499
barium sulphate, 196
battery
   archaeology, 23
   charging procedures, 224, 269–272, 281–283
   chemistry and electrochemistry, 31–60
   definition, 17
   discharge behaviour, 58
   disposal, 100, 186, 316, 327, 409
   effect of electrolyte concentration, 35–36
   effect of temperature on performance, 121
   electrochemistry, 36–48
   energy storage schemes, 242, 468, 513
      *see also* load-levelling
   environmental concerns, 155, 186, 243, 349, 409
   generalized cell, 17
   Gibbs free energy, 33
   history, 23–30, 112, 125, 194–196, 312, 366, 452, 458, 487–489
   inefficiencies, 43–48
   manufacturers, *see* manufacturers
   market trends, 28
   markets, 112, 195, 407
   open circuit voltages on discharge, 36
   origin of term, 17
   patents, 18, 338
   performance characteristics tables, 544–558
   production values, 28
   reaction enthalpy, 32
   safety, 23, 35, 100, 105, 110, 115, 287, 289, 316, 319, 321, 327, 334, 336, 348, 368, 378, 384, 392, 399, 437, 516–517, 535
   selection criteria, 18
   standard e.m.f., 35
   standard sizes, *see* standard sizes
   thermodynamics, 31–36
   types, 19–20
   voltage losses, 37
battery charging procedures, 224, 281–283
   current limiting, 233
   differential temperature, 282

tapering current, 224, 517
temperature target, 282
voltage derivatives, 282
voltage target, 282
battery design, 20–21
   construction
      aesthetic aspects, 97
      electrical aspects, 20, 38, 522
      mechanical aspects, 20
      electrode kinetics aspects, 20
      mass transport effects, 20
      thermodynamic aspects, 20
Becquerel, 23
Bedford van, 467
BET equation, 75
biomedical devices, 369
bipolar cells, 26
bipolar electrodes, 309, 460, 486, 503, 527
bismuth–copper oxide cells, 26
blocking, *see* electrode, blocking of
Boettcher, 24
boron, *see* lithium–boron anode
boron nitride, *see* separators, boron nitride
bromine complex, 504
Bunsen, 23, 25
buoys, 88, 160, 316, 334
burn-in, 378
Butler–Volmer equation, 40
button cells, 17, 125–160, 264, 385
$\gamma$-butyrolactone, 339, 398

C-rating, 236
cadmium, *see* electrode, cadmium; copper oxide–cadmium cells; nickel–cadmium batteries
cadmium hydroxide, 67
calcium, *see* electrode, calcium
calcium hydroxide, 162
calcium–potassium dichromate, thermal batteries, 412
calcium thermal batteries, 409, 417
calcium–thionyl chloride cells, 48
calculators, 113
Calot, 97
cameras, 111, 113, 124, 336, 340, 352, 360, 366
can manufacture
   alkaline–manganese batteries, 118
   zinc–carbon batteries, 97
carbon
   fibres, 459
   graphite, *see* graphite
   microfibres, 311
   vitreous, 510
   *see also* electrode, carbon
carbon black, 87, 164, 196, 494, 531
   purity requirements, 100
   surface area measurement, 77
   *see also* acetylene black
carbon monofluoride, 336
   properties, 338
cardiac defibrillator batteries
   technical requirements, 400
   *see also* defibrillator

**Index** 569

cardiac pacemaker batteries,
  technical requirements, 400
cardiac pacers, *see* heart pacers
cassette players, 124
catalyst, *see* air cathode, catalyst; macrocyclics, cobalt
cathode
  charge transfer complex, 367
  definition, 17
  intercalation type, 36, 393, 423, 528, 530
  pore blocking, *see* pores, blocking of
  *for other cathode types see* electrode
cell
  discharge behaviour, 56
  origin of term, 17
  standard sizes, 101
cell voltage
  practical, 37
  theoretical, 33
cereal paste, *see* separators, cereal paste
Chaperon, 24, 112
charge acceptance, 208, 212
charge control circuits, 29
charge retention, 47
charge transfer, 37
  kinetics, 39–42
  resistance, 54
charging of batteries, *see* battery, charging procedures
chemical bonding, 69
chemical shuttle, 48, 272
chlorine, 415
chlorine–zinc batteries, *see* zinc–chlorine batteries
chromates, alkaline earth, 409
chromium, *see* electrode, chromium
chromium oxides, electron microscopy, 65
clock applications, 302, 377
cobalt *see* electrode, cobalt
cobalt oxide, *see* iron–cobalt oxide cells
cold cranking current (CCA), 218
Cole–Cole plot, 54
Commelin, 24
communications equipment, 279, 313, 328, 330, 334
  *see also* telecommunications applications; walkie-talkie
compact disk players, 124
computer applications, 234–236, 292, 296, 333, 336, 340, 356, 377, 384, 444
  *see also* uninterruptible power systems
computer modelling, 417
constants (fundamental), 565
conversion tables, 563–566
  area, 563
  current density, 565
  electric charge, 564
  energy, 564
  energy densities, 565
  length, 563
  mass, 563
  power, 564

  temperature, 565
  volume, 564
copper, *see* electrode, copper
copper oxide, 24, 385, 392
copper oxide–bismuth cells, *see* bismuth–copper oxide cells
copper oxide–cadmium cells, 26
copper oxide–zinc cells, *see* zinc-copper oxide batteries
copper oxyphosphate, *see* lithium–copper oxyphosphate batteries
cordless tools, 263, 340, 444
corrosion, 43, 311, 318–320, 398, 453, 490, 494
  aluminium, 490
  zinc, 146, 175
corrosion cell, 31
corrosion inhibitors, 117
Coulombic efficiency, 45
Coulter counter, 81
counter electrode, 49
Cruikshank, 17, 23
crystal growth kinetics, 496
crystallizer, 496
cyclic voltammetry, 52
cyclic voltammogram, 52
cylindrical alkaline batteries, *see* alkaline–manganese batteries

Daniell, 23
Daniell cells, 17, 25, 31–36
Darrieus, 26
Davy, 17, 23
De Virloy, 24
defibrillator, 400, 401
  *see also* cardiac defibrillator batteries
dendrites, 47, 276, 453, 503
dendritic deposition, 46, 154, 336, 453, 517
Desmazures, 24
differential scanning calorimetry (DSC), 74
differential thermal analysis (DTA), 73
diffusion, 42, 52, 55, 97, 204, 213, 263, 368
diffusion coefficients, 43, 212
diffusion membrane, 128
dimethoxyethane (DME), 72, 339, 351, 393, 403
diode protection, 296, 327
dioxolane (1,3-dioxolane), 47, 383
dirigible, 26
discharge circuit, 328
dislocations, 65
disposal, *see* battery, disposal
doctor blade casting, 531
dominant activator, 491
double layer, 40
double layer capacitance, 54–55
double-sulphate theory, 24
drawing, 99
driving profiles, 469
drug-delivery systems, 401
dry cells, 348
  *see also* zinc–carbon batteries
dry room, 367, 370, 410, 523
dual in-line pin (DIP) connectors, 302

Dun, 24, 452
dynamo, 23

ear, *see* auditory canal
Edison, 23–26, 244, 452
Edison–Lalande cells, 24
EIS, 53
electric car, 18, 224, 487
  *see also* electric vehicles
electric self-starter, 195
electric torch, 19, 88, 113, 366
electric vehicle batteries, comparison, 472
electric vehicles, 24, 26, 213–227, 452, 458, 467, 469, 489, 500, 502, 504, 512, 516–519, 526, 534, 537
  *see also* electric car
electrical double layer, 40
electrochemical equivalents, various battery materials, 561
electrochemical techniques
  single electrode
    A.C. techniques, 53–56
    D.C. techniques, 48–53
  whole cell (two electrode)
    A.C. techniques, 58
    D.C. techniques, 56
electrode
  air, 41
    *see also* air cathode
  aluminium, 41, 44, 489
  aluminium alloys, 50, 492–493
  aluminium–gallium alloy, 494
  aluminium–indium alloy, 64, 490
  aluminium–tin alloy, 62, 490
  bipolar, *see* bipolar electrode
  blocking of, 49, 204, 454
  cadmium, 24, 58, 244, 255, 263, 264
  calcium, 48
  carbon, 48
  chromium, 45
  cobalt, 45
  contaminant, analysis of, 64
  copper, 32
  iron, 41, 45
  lead, 52
  lead dioxide, 197
  lithium, 58, 69, 70, 72
  magnesium, 45
  manganese dioxide, *see* manganese dioxide, electrochemistry
  nickel, 45
  pore, geometry of, 55
  reference, *see* reference electrode
  silver, 31
  vibrating, 454
  working, 49
  zinc, 31, 41, 45, 54, 96, 453, 505
electrode potential, 33
  battery electrodes, 559
electrolyte
  analysis (lithium electrolyte), 290
  gelled, 179
  solid, 528
    *see also* solid electrolytes
  solid polymer, 528
electrolyte properties
  γ-butyrolactone, 398
  dimethoxyethane/propylene carbonate, 350, 403
  ethyl acetate, 398
  gelled electrolyte, 183–186
  lithium tetrachloroaluminate, 290
  methyl formate, 397
  molten salt, 409, 414
  polyethylene oxide, 529–530
  potassium hydroxide, 170, 251, 265, 453
  propyl acetate, 398
  sulphur dioxide/acetonitrile, 314
  sulphuric acid, 202
electrolyte reciprocation, 496
electromigration, 43
electron beam, 61
  penetration depth, 62
electron diffraction, 65
electron microscopy
  scanning, 61–65, 80
  specimen preparation, 64, 80
  transmission, 65, 80
electron optics, 64
electron spectroscopy, 69–70
electron-tunnelling, 46
electronic applications, 19, 24, 287, 313, 327, 334
  *see also* computer applications; portable electronics
electropolishing, 65
emergency lighting, 227, 263, 272, 353, 498
emergency power, 195
energy densities
  calculation, 39
  definition, 34
energy dispersive spectroscopy (EDS), 64–65
energy storage, *see* load-levelling
engine start capability, 217
Entz, 24
environmental concerns, *see* battery, environmental concerns
equilibrium cell potential, 31
equivalent circuit, 54
Estelle, 26
ethyl acetate, 398
exchange current density, 41, 55
expansion, thermal, 316
extended X-ray absorption fine structure (EXAFS), 67
extrusion, 97, 318

fabrics, non-woven, *see* separators, non-woven fabrics
Faradaic reaction, 54
Faraday, 23
Faraday's constant, 33
  derivation, 39
Faraday's laws of electrolysis, 38
Fauré, 194

ferrocene, 48
ferrous/ferric ion exchange, 54
fluorosurfactants, 311
Ford Aerostar van, 515
Ford ETX, 511
fuses, lithium batteries, 327, 412

Galvani, 31
galvanostat, 50
Gassner, 112
gelled electrolyte, 179
gelling agents, 117, 126, 183, 231
General Motors G-van, 223–225, 472
General Motors Griffon van, 471
General Motors Impact electric car, 226
geosynchronous earth orbit (GEO) applications, 472, 483
Gibbs free energy, 33
glass, *see* separators
glass-to-metal seals,
    corrosion, 319
    corrosion protection, 319
Glasstone, 24
golf carts, 219
grain boundaries, 65
graphite, 100, 135, 164, 244, 246, 383, 385, 392, 394, 403, 455, 511
Grove, 23, 25
gum karaya, 89

half-cell reactions, 31, 32
Haslacher, 24, 452
hearing aid batteries, 137-147, 155
hearing aids, 125, 135
heart pacers, 27, 369, 374, 400, 528
    *see also* cardiac pacemaker batteries
heat paper, 417
Heise, 160
Herbert, 112
heterolite, 96
history, *see* battery, history
hybrid battery, general description, 20
hydrargillite (aluminium hydroxide), 497
hydride, *see* metal hydride
hydro energy, 243
hydro energy storage, 514
hydrogen evolution, 43, 48, 117, 154, 176, 208, 233, 255, 271, 283, 406, 474, 490
hydrogen fluoride, 398
hydrogen–oxygen fuel cells, 472
hydrogen peroxide,
    *see* aluminium–hydrogen peroxide batteries;
    air cathode, peroxide reaction
hydrogen sulphide, 525
    *see also* air cathode, peroxide reaction
hyperactivation, 491

IEC standards, 101, 106
image resolution, 61, 64
impedance hump, 144
impedance spectroscopy, 53, 58
impedance-up, 150

implantable devices, 368
impregnation, 264
infrared spectroscopy
    lithium electrolyte analysis, 290
    lithium surface studies, 70
insulin pump, 400
intercalation, *see* cathode, intercalation type
internal resistance
    alkaline–manganese, 119, 122
    general, 20, 41
    lead–acid, 213, 224
    lithium–iodine, 368, 372, 375, 380
    lithium–iron sulphide, 521
    lithium–molybdenum disulphide, 424, 428, 430
    lithium thermal, 421
    nickel–cadmium, 269
    sodium–sulphur, 461
    zinc–air, 141
    zinc–carbon, 104
iodine, 367, 372
ion-beam etching, 65
ion-exchange resins, 89
ion microanalysis, 349
iron
    impurity in lithium battery electrolytes, 290
    *see also* electrode, iron; nickel–iron batteries
iron–cobalt oxide cells, 26
iron disulphide, 402, 415
    lithium incorporation, 407, 413
iron oxide, 409
iron sulphide, *see* lithium–iron sulphide batteries

jelly-roll, 120, 317, 425
Jungner, 23, 24, 244, 452

Kelvin equation, 77
Kettering, 195

lacing, 248
Lalande, 24, 112
laminating, 532
lamp, halogen, 498
land speed record, 18
Lange, 265
Langguth, 245
laser diffraction, 75
laser sealing, 348, 370, 379, 400
lead, *see* electrode, lead
lead–acid batteries, 19, 21, 24, 60, 111, 194–244, 307, 414, 459, 469, 472, 502, 519
    addition of phosphoric acid, 52, 236
    addition of sodium sulphate, 236
    charge acceptance, 208
    charging behaviour, 207–210
    charging methods, 226, 233, 240
    construction, 213–235
    discharge behaviour, 204, 244
    electrochemistry, 203–204
    electrolyte, 200–202
    energy density improvement, 222
    flat pack design, 236
    float voltages (standby batteries), 232

# Index

gas recombination behaviour, 211
grid alloys, 199–200, 211
history, 194–196
improvements, 214, 223-224
lead–antimony alloys, 24, 199
lead–calcium alloys, 199
load-levelling applications, 241–243
markets, 195–196
motive power batteries, 207, 213-227, 472, 545
negative active material
  composition, 196
  porosity changes, 196
portable batteries, 235–241, 546
positive active material
  composition, 197
  porosity changes, 198
sealed batteries, 211–213, 229–235
sealed (horizontal electrode) batteries, 29
self-discharge behaviour, 210-213
separators, 202–203
SLI batteries, 207–208, 213–219, 546
standard testing procedures, 217–219
standby batteries, 227–235, 547
state of charge indication, 235
storage of renewable energy, 243
telecommunications applications, 231
lead–antimony alloys, 24, 199
lead–calcium alloys, 199, 229
lead dioxide, 197
lead sulphate, 52, 204, 210, 229
Leclanché, 23, 112
Leclanché cells, 87
  *see also* zinc–carbon batteries
Lewis acid, 289
ligand gases, 312
lignosulphonates, 196
lime, 171
limiting current, 42
lithium, 58, 385
  battery grade, 290
  surface films, 46, 287, 314
  surface studies, 72
  *see also* electrode, lithium
lithium–aluminium anode, 47, 52, 519
lithium batteries, 18, 46–48, 118–119, 122, 284, 365
  aluminium anode substrate, 47, 52
  cathode catalysts, 289, 291, 306
  chromium oxide cathode, 65
  copper anode substrate, 318
  discharge behaviour, 36
  electrolyte, purity requirements, 289
  electrolyte analysis, 290
  Japanese growth in production, 350
  rechargeable, 335, 353, 423–444
  safety, 287, 296, 316, 319, 321, 327, 334, 336, 348, 368, 378, 384, 392, 399, 437
lithium–boron anode, 414
lithium bromide, 47, 314
lithium–carbon monofluoride batteries, 336–348, 547
  applications, 336, 340

cell components, 337
cell types and performance, 339–345
chemistry, 337
coin cells, 340
comparison with lithium–manganese dioxide, 336
comparison with zinc–carbon, 340
cylindrical cells, 339
manufacturers, 336
pin-type cells, 340
pulse discharge behaviour, 340
sealing, 339, 345
lithium carbonate, 424
lithium–copper oxide batteries, 383–393, 547
  anode assembly, 385
  cathode assembly, 385
  chemistry, 388–392
  construction, 384–385
  discharge curves, 386
  discharge mechanism studies, 392
  electrolyte, 385
  electrolyte studies, 392
  performance, 385
  safety, 392
  sealing, 384–385
  separator, 385
  storage, 387
lithium–copper oxyphosphate batteries, 384
lithium cyanide, 316
lithium dithionite, 314, 318
lithium hexafluoroarsenate, 47, 316
  hydrolysis, 399
lithium iodide, 367
lithium–iodine batteries, 365-383, 528, 548
  chemistry, 367–368
  future developments, 381
  history, 366
  industrial batteries, 377–380
    available types, 378
    construction, 379
    performance characteristics, 380
    sealing, 379
  medical batteries, 368–377
    available types, 369
    construction, 370–374
    performance characteristics, 374
    precoated anodes, 375
    quality control, 373
    sealing, 370
  quality control, 380
lithium–iron disulphide batteries, 402–409, 516, 549
  cell chemistry, 403
  cell designs, 403
  comparison with alkaline–manganese, 405
  comparison with zinc–silver oxide, 405
  discharge mechanisms, 406
  market reaction
    coin cells, 407
    cylindrical cells, 408
  performance characteristics, 404
  pulse discharge, 405

## Index

sealing, 404
lithium–iron sulphide batteries (rechargeable), 516–528, 548
  bipolar design, 527
  chemistry, 519
  construction, 522
  future developments, 527
  general characteristics, 516–519
  iron sulphide cathode, 520
  lithium–aluminium anode, 519
  lithium–silicon anode, 519
  manufacturers, 519
  negative-to-positive capacity ratio, 521
  performance, 525
  sealing, 522
  separators, 521
  thermal management system, 517
lithium-manganese dioxide batteries, 336, 348–365, 384, 393, 549
  applications, 358–359
  camera test results, 365–366
  chemistry, 349
  construction details, 352
  cylindrical type, 352
    inside-out structure, 356
    spiral structure, 358
  flat type, 352
    performance characteristics, 353
  future developments, 365
  general characteristics, 348
  manufacturers and statistics, 349
  performance characteristics, 353–365
  pulse discharge, 353
  safety, 348
  sealing, 348, 352
lithium–metal sulphide batteries, 466
lithium–molybdenum disulphide batteries, 423–444, 549
  applications, 444
  cell chemistry and operation, 423–425
  cell construction, 425
  cell voltage, 426
  charge retention, 437
  charging procedures, 427, 429
  comparison with nickel–cadmium, 443
  cycle life, 434
  deliverable capacity, 430
  electrical abuse test results, 437
  electrical characteristics, 426-429
  energy efficiency, 429
  future developments, 443
  internal resistance, 428
  performance characteristics, 430–437
  reproducibility and reliability, 439
    multi-cell batteries, 441
    single cells, 439
  safety characteristics, 437
  self heating effect, 432
  shipping and handling, 438
lithium nitride, 290
lithium perchlorate, 47, 69, 72
lithium polymer electrolyte batteries, *see* lithium solid-state batteries
lithium–silicon anode, 414, 519–520
lithium–silver vanadium pentoxide batteries, 401
lithium solid-state batteries, 528–537, 550
  cell components, 529
  cell fabrication, 532
  cell performance, 532
  variable geometry, 535
lithium sulphide, 409, 525
lithium–sulphur batteries, 516
lithium–sulphur dioxide batteries, 296, 312–336, 384, 414, 550
  anode construction, 318, 321
  applications, 333–334
  balanced cell concept, 316, 322
  cathode construction, 318
  cell case, 318, 320
  cell construction, 316–320
  cell performance, 323–327
  chemistry, 314–316
  commercially available cells, 320–327
  comparison with alkaline–manganese, 313, 330
  comparison with magnesium batteries, 330
  comparison with magnesium–manganese dioxide, 313
  comparison with zinc–carbon, 313
  design and performance, 327–331
  diode protection, 327
  discharge circuits, 327
  disposal, 327
  electrolyte
    properties, 314
    stability, 315
  fuses, 327
  history, 312
  industrial and electronic applications, 333
  military applications, 328, 334–336
  quality control, 328
  rechargeable, 335
  safety, 319, 321, 327, 334
  sealing, 316, 319, 322
  separators and insulators, 318
  sleeving, 328
  spiral wound construction, 317–320
  standard tests, 328
  technical advantages, 313
  US military battery types, 328
  vent designs, 319
lithium tetrachloroaluminate, 47, 290
lithium thermal batteries, 409–423, 551
  alternatives, 422
  applications, 419–422
  cathodes, 415
  cell components and electrochemistry, 413–416
  computer modelling, 417
  connections, 421
  construction, 410
  electrolytes, 414
  fuses, 412
  lithium anode, 414
  performance characteristics, 417
  quality control, 421

sizes, 417
starter batteries, 418
storage, 421
lithium-thionyl chloride batteries, 46, 287–321, 384, 422, 551
  bobbin cells, 291–294
    applications, 292
    configurations and sizes, 291
    discharge characteristics, 291
  cathode, 291
  cathode catalysts, 289
  chemistry, 288
  electrocatalysts, 310
  electrolyte, 289
  electrolyte additions, 290
  future developments, 310–312
  general characteristics, 287
  improvements in high-rate cells, 312
  improvements in low-rate cells, 311
  improvements in moderate-rate cells, 312
  manufacturers, 288
  materials, 289–291
  prismatic cells, 302–307
    configurations and sizes, 303
    discharge characteristics, 306
    pulse discharge behaviour, 311
  reserve cells, 307-310
    applications, 309
    configurations and sizes, 307
    discharge characteristics, 309
  safety, 289
  separators, 290
  wound cells, 294–301
    characteristics, 295
    high-rate, 296
    low–rate, 294
lithium–vanadium pentoxide batteries, 56, 307, 393–402, 552
  active cells, 398
    discharge characteristics, 398
  cell design and performance, 393
  effect of storage temperature, 399
  medical batteries, 400
  reserve cells, 395
    performance characteristics, 397
  safety, 399
  sealing, 395
  separator, 394
load-levelling, 17, 29, 241–243, 460, 468, 504, 513, 537
low Earth orbit (LEO) applications, 472, 483
Luggin capillary, 50

macrocyclics, cobalt, 291, 311, 494
magnesium, *see* electrode, magnesium
magnesium batteries, 46, 313, 330
magnesium dioxide, 58
  heat treatment, 353
magnesium oxide, 415
  *see also* separators, magnesium oxide
magnesium–silver chloride batteries, 58
magnesium thermal batteries, 409

maintenance free (MF) batteries, *see* lead–acid batteries, sealed
manganese dioxide, 135, 137
  air cathode use, 167
  composite dimensional (CDMO), 353
  electrochemistry, 96, 115, 349
  electrolytically deposited (EMD), 88
  heat treatment, 74, 351
  lithium ion incorporation, 349–350
  original use, 24
  particle size analysis, 82
  purity requirements, 100
  rechargeability, 111
  TGA study, 74
  types, 100, 117
  use in lithium batteries, 348
manganese dioxide-zinc batteries, *see* alkaline–manganese batteries
manufacturers, 88–89, 288, 336, 349, 384, 473, 519
markets, *see* battery, markets
mass transfer, 54
mass transport, kinetics, 42–43
mechanically rechargeable battery, general description, 20
medical applications, 125, 368
  *see also* cardiac defibrillator batteries; hearing aids; heart pacers
medical equipment, 24, 368, 393
memory back-up applications 384
  *see also* computer applications
memory back-up circuits, 272, 378
memory effect, 276
mercuric oxide,
  electrochemistry, 127
  original use, 24
mercury porosimetry, 79
metal–air cells, 41
metal hydride, 491
  electron spectroscopy, 70
methane, 316
methyl formate, 395, 397, 400
methyl oxazolidone, 403
mica, 525
microbalance, 77
microcalorimetry, 380
microdensitometer, 80
micromerograph, 82
microphones, 340
micropore, 78
microtomy, 65
military applications, 313, 327–336, 393, 409
  *see also* aircraft applications; communications equipment; naval applications; submarine applications
missile applications, 302, 308, 334, 409
molten salt electrolyte, 409, 414, 520
molybdenum disulphide
  lithium intercalation, 423
  phases, 424
  structural changes on cycling, 427
  structure, 423

**Index** 575

Mössbauer spectroscopy, 406

naval applications, 331
  *see also* submarine applications
negative difference effect, 45
Nernst, 23
Nernst diffusion layer, 42
neutron diffraction, 65
nickel, *see* electrode, nickel
nickel–cadmium batteries, 19, 26, 28, 48, 58–59, 111, 119–120, 124, 244–284, 443, 459, 472
  electrolyte carbonation, 253, 276
  overcharging reactions, 255
  pocket plate cells, 246–263, 552
    battery ratings, 258
    charge retention, 261
    charging and discharging, 256
    construction, 246–254
    electrolyte, 250
    float voltages, 260
    formation, 250
    life, 261
    manufacture of active materials, 246
    manufacture of perforated strip, 246
    rapid recharge, 258
    separators, 249
    strip filling, 247
    Teflon™-bonded plates, 263
    temperature effects, 260
  reaction mechanisms, 254–257
  sealed cylindrical cells, 263–284, 553
    alloy formation effect, 276–278
    applications and battery design, 277–283
    battery capacity, 278
    cell matching, 283–284
    cell selection, 277–280
    charge characteristics, 269–272
    charge efficiency, 272
    charge retention, 272
    charging methods, 281–283
    constant current charging, 269
    constant potential charging, 271
    construction, 264–266
    discharge capacity, 266–271
    effect of discharge rate on capacity, 267
    effect of temperature on capacity, 268
    electrical characteristics, 266–277
    electrodes, 264
    electrolyte, 265
    failure modes, 275
    improvements, 283
    internal resistance, 269
    mechanical construction, 265
    memory effect, 276–278
    operating voltage, 277
    operational life, 275-277
    overcharging, 281
    peak discharge, 269
    recombination, 265
    safety, 271
    sealing, 265
    self-discharge, 279

    temperature dependence, 279
    use of lithium hydroxide, 272
    self-discharge, 123
    standard rated capacity measurement, 267
    use of lithium hydroxide, general, 245, 253, 262, 265, 272
    X-ray diffraction, 65
nickel fibre, 457
nickel foam, 264, 457
nickel–hydrogen batteries, 27, 472–487, 553
  Air Force/Hughes cell, 475
  bipolar design, 486
  capacity retention, 482
  cell construction, 475
  charging, 474, 478
  chemistry, 474
  comparison with nickel–cadmium, 473
  Comsat/Intelsat cell, 476
  general characteristics, 472
  manufacturers, 473
  NASA Advanced cell, 476
  performance characteristics, 477, 484
  storage, 485
nickel hydroxide, 58
  original use, 24
  structural forms, 254
  X-ray diffraction, 65
nickel–iron batteries, 24, 26, 48, 252, 472
nickel iron hydroxide, DSC study, 74
nickel–metal hydride batteries, 284
nickel oxy-hydroxide, 474
  electrochemistry, 58, 254
  preparation, 244
nickel–zinc batteries, 452–458, 554
  chemistry, 452
  construction and performance, 456–458
  history, 452
  polymer-bonded electrodes, 455, 457
  separators, 454
  shape change, 453
  sintered nickel electrodes, 457
  zinc poisoning, 454
Nyquist plot, 54


ocean buoys, *see* buoys
oceanographic application, 333
Oersted, 23
Ohmic drop, 37, 50
Ohmic resistance, 60
oil-well applications, 293, 384
optical microscopy, 80
overcharge reactions, 48
overvoltage (overpotential), 40
oxygen (liquid), 502
oxygen cycle
  lead–acid batteries, 211
  nickel–cadmium batteries, 48, 211, 262, 265
oxygen evolution, 48, 208, 210, 253, 255, 263, 266, 476

pagers, 113, 125, 405

# Index

paper, *see* separators
particle size, 79, 392
particle sizing and counting, 80–82
particulates, analysis of, 75–82
passivation, 45, 144, 287, 313, 423, 437
patents, *see* battery, patents
perchlorate powders, 412
performance characteristics tables, 544–558
Permion™ membranes, 135, 154
Perovskite structure, 393
peroxide ion, *see* air cathode, peroxide reaction
Peukert equation, 204
Phillips, 24
phonons, 67
phosphoric acid, 52, 236
photovoltaic energy, 243, 254
physical techniques, 60–75
Planté, 23, 194
Planté cells, 24, 228
plaque, 264, 476
plastics, battery containers, 249, 509
platinum, 306
polarization, 37, 40, 102, 119, 375
Pollack, 26
polyamide, *see* separators, polyamide
polyethylene, *see* separators, microporous polyethylene
polyethylene oxide, 529
polyethylene oxide compounds, EXAFS studies, 68
polymer-bonded electrodes, 163, 294, 455, 510
 manufacture, 338, 457
polymer electrolyte, *see* lithium polymer electrolyte batteries
polyolefin, *see* separators
polypropylene, *see* separators
polypropylene seals, 385
polytetrafluoroethylene (PTFE), 129, 135, 263, 291, 318, 338, 393, 403, 455, 495, 503
 reaction with lithium, 319
polyvinylchloride (PVC),
 in lithium–thionyl chloride batteries, 290
 lithium antipassivation agent, 311
polyvinylpyridine (P2VP), 367, 372
pores
 blocking of, 152, 288
 flow models, 168
 impedance, 55
 mass transport, 292
 microporous polypropylene, 319
 non-woven polypropylene, 319
 size distribution, 164
porosity and density measurement, 77–80
portable electronics, 111, 113, 124, 235, 284, 336, 360, 365, 366, 535
 *see also* cameras; computer applications
potassium hydroxide, electrolyte properties, 170–173, 251, 265, 453
potentiostat, 50
Pourbaix diagram, zinc, 175
pressure relief valve, 236
primary battery, general description, 19

process controller circuitry, 296
propyl acetate, 398
propylene carbonate (PC), 47, 52, 70, 72, 339, 351, 393, 403
 reaction with lithium, 424
protective devices, 384
pulse discharge
 hearing aid batteries, 139
 lithium–carbon monofluoride, 340
 lithium–iron disulphide, 405
 lithium–manganese dioxide, 353
 lithium–silver vanadium pentoxide, 401
 lithium solid-state, 533–534
 lithium–sulphur dioxide, 316
 lithium–thionyl chloride, 311
 lithium–vanadium pentoxide, 400
 nickel–cadmium, 279
 watch batteries, 152
pyrotechnic, 412
 thermal battery requirements, 416

radio, advanced military, 334
Ragonne plot, 411, 463
railway propulsion, 456
railway signalling, 24, 160
railway train lighting, 195
Raman spectroscopy
 lithium deposition studies, 70
 lithium surface studies, 70
Randles circuit, 54
range-extender, electric vehicles, 489, 502
recombination efficiency, 213, 265
reference electrode
 cadmium, 203
 calomel, 49
 hydrogen, 48
 lithium, 49
 mercury–mercuric oxide, 49, 256
 mercury–mercury sulphate, 203
 silver–silver chloride, 48
Renard, 26
robotics, 219
rocket propulsion applications, 333
Ruben, 24
Rutherford, 23

safety, *see* battery, safety
safety tests, 517
scanning electron microscopy (SEM), *see* electron microscopy, scanning
Schoop, 24
sealing
 button cells, 136
 lithium–manganese dioxide cells, 355
 *see also under individual battery systems*
seawater batteries, 42, 502
second phase particles, 44
secondary battery, general description, 19
secondary electron imaging (SEI), 62
secondary electrons, 62
secondary ion mass spectrometry (SIMS), 70
security systems, 293, 336, 353

# Index

sedimentation, 82
self-discharge, 43
  alkaline–manganese, 123
  electrochemical explanation, 43
  lead–acid, 210
  lithium–carbon monofluoride, 340
  lithium–iodine, 368, 378
  lithium–iron sulphide, 527
  lithium–manganese dioxide, 348, 356
  lithium–molybdenum disulphide, 437
  lithium–silver vanadium pentoxide, 401
  lithium–sulphur dioxide, 314
  lithium–thionyl chloride, 289
  nickel–cadmium, 268, 274, 279
  nickel–hydrogen, 473
  watch batteries, 152
  zinc, 31, 39, 176
  zinc–air, 146
  zinc–bromine, 503
  zinc–carbon, 105
Sellow, 24
sensors, 293
separators
  absorbent glass mat, 203
  asbestos, 475
  boron nitride, 521
  cellulose, 203
  cereal paste, 91
  coated paper, 91
  glass microfibre, 236
  lead–acid batteries, 202
  magnesium oxide, 521
  microporous polyethylene, 203, 404
  microporous polypropylene, 454
  microporous polyvinylchloride, 203
  non-woven fabrics, 454
  non-woven glass, 203, 288, 290, 383, 385
  non-woven polypropylene, 203, 319
  paper, 114
  phenolic base, 203
  polyamide, 265
  polymeric, 288
  polyolefin cloth, 339
  polyolefin film, 318, 454, 503
  polypropylene, 276, 319, 339, 352
  PTFE, 503
  rubber, 203
  sintered polyvinylchloride, 203
  technical requirements, 202, 385
  Tefzel™, 290
  Zircar™, 476
shelf life, 19, 152, 182, 313–314, 334, 336, 367, 380, 384
  see also self-discharge
shunt currents, 506
shuttle, chemical, see chemical shuttle
Siemens, 23
silicates, 144
silicon, see lithium–silicon anode
silver, see electrode, silver
silver chloride, see magnesium–silver chloride batteries

silver nickel oxide, 132
silver oxide
  electrochemistry, 57, 129
  see also aluminium–silver oxide batteries
silver 'plumbate', 130
silver vanadium oxide, 401
silver–zinc batteries, see zinc–silver oxide batteries
sintered nickel, 245, 264, 457
SLI life test, 219
Sluyters plot, 54
smart cards, 383
smoke detectors, 113
sodium beta″ alumina, 459, 528
sodium–metal chloride batteries, 194
sodium stannate, 494
sodium sulphate, 236
sodium–sulphur batteries, 194, 458–469, 554
  applications, 467
  cell construction, 463
  chemistry, 459
  containers, 467
  development status, 469
  electric vehicle applications, 469
  history, 458
  performance characteristics, 461
  sealing, 464
  system construction, 464–469
sodium–sulphur dioxide batteries, 312
solid electrolyte interphase (SEI), 46
solid electrolytes, 367, 415
solid-state batteries, 365, 528
solution resistance, 50, 54–55
space applications, 309, 472, 534
spinning reserve credit, 242
standard deviation, 441
standard hydrogen electrode (SHE), 33
standard sizes
  cylindrical cells, 106
  flat cells, 107
  hearing aid batteries, 156
  lithium–carbon monofluoride cells, 341–342, 346
  lithium–iodine batteries, 369–371, 381
  lithium–manganese dioxide cells, 357
  lithium–sulphur dioxide cells, 321–322
  lithium thermal batteries, 420
  lithium–thionyl chloride batteries, 295, 305
  nickel–cadmium sealed cylindrical cells, 279
  watch batteries, 157–159
standby applications, 195, 227-235, 245, 275, 281, 468, 489, 547
  see also emergency lighting
starch, 101, 117, 183
starting, lighting and ignition (SLI), 18, 19, 213–219, 245, 546
  life test, 219
state-of-charge (SOC), 59–60, 426, 480
state-of-charge indicator, 60, 426, 481
stationary energy storage (SES), 468, 512
stereo-imaging, 64
Stokes diameters, 82

storage, energy, *see* load-levelling
stratification, 227
strobe light, 363
styrene butadiene, 339
submarine applications, 24, 195, 213, 301, 309, 498
sulphur, 415
   *see also* lithium–sulphur batteries; sodium–sulphur batteries
sulphur dioxide, 312, 314
   vapour pressure, 314
sulphuric acid
   oxygen diffusion, 212
   oxygen solubility, 212
   resistivity, 202
superactivation, 490
surface area, measurement, 75–77
surface morphology, 62
switchgear applications, 333

Tafel plots, 41
tagging (security tagging), 336
tape recorders, 360
Teflon™, *see* polytetrafluoroethylene
Tefzel™, *see* separators, Tefzel™
telecommunications applications, 195, 227, 263, 288, 296, 487, 498
   *see also* communications equipment; telephone, mobile
telegraphy, 194
telephone, mobile, 284, 293, 365, 433, 444
televisions, 235, 360, 366
tetrahydrofuran (THF), 47, 72, 403
thermal batteries, 27
   definition, 409
   *see also* calcium thermal batteries; lithium thermal batteries; magnesium thermal batteries
thermal expansion, acetonitrile, 316
thermal management, 312
thermal protectors, 352
thermogravimetric analysis (TGA), 73
thionyl chloride, 287–291
tidal energy, 243
time constant, 59
titanium, 101
titanium dioxide, 413
titanium disulphide cathode, 532
torch, *see* electric torch
torsion balance, 82
toy applications, 104, 111, 113, 124, 340, 498
transmission electron microscopy (TEM), *see* electron microscopy, transmission
Tribe, 24
trolley car, 24
   *see also* electric vehicles
tungsten oxides, 409

ultra-violet photoelectron spectroscopy (UPS), 69
ultra-violet visible spectroscopy, lithium electrolyte analysis, 290
uninterruptible power systems/supplies (UPS), 227–235

van Marum, 160
Van't Hoff isotherm, 35
vanadium oxide, 409
   lithiated, 415
   $V_6O_{13}$ cathode, 530
vanadium pentoxide, 393, 415
vapour pressure, sulphur dioxide, 314
vent, 137, 177, 229, 265, 294, 316, 319, 321, 327, 339, 400, 426, 437
vibrating electrodes, 454
video recorders, 235, 360
Volta, 23, 31
voltage delay, 46, 287–290, 296, 364, 387, 437
voltage-up, 150
Voltaic pile, 23, 31
volume change, 368, 407

Waddell, 24
walkie-talkie, 125, 334
Warburg impedance, 55
watch batteries, 147–155, 336
watches, 126, 340, 366, 383, 402, 407
wave soldering, 378
weather balloons, 333
wick (wall wick), 476
wind energy, 243
working electrode, 49

X-ray analysis, 64
X-ray diffraction, 65, 81, 254, 339, 349, 393, 406
X-ray examination, 373
X-ray photoelectron spectroscopy (XPS), 69

zinc
   alloying, 100, 117, 126, 186
   corrosion, 146, 175
   electrochemistry, 126, 175
   polymer bonded, 455
   powder, *see* zinc powder
   purity requirements, 100, 117, 176
   shape change, 47, 453, 503
   *see also* electrode, zinc
zinc–air batteries, 20, 116
   air cathode, 127
   environmental effects, 144
zinc–air batteries (large size), 160–190, 555
   applications, 185
   cell chemistry, 161
   cell components, 162–177
   construction
      gelled electroyte type, 179–185
      wet cell type, 177
   current collectors, 177
   electrolyte, 170–175
   general characteristics, 161
   performance characteristics, 185
   zinc anode, 175
zinc–air batteries (small size), 125-160, 555
   cell impedance, 141
   construction, 133

humidity sensitivity, 146
overactivation, 146
performance characteristics, 137–141
sizes and types, 155
storage, 146
underactivation, 146
zinc–bromine batteries, 20, 26, 194, 453, 502–516, 519, 556
- applications, 503
- battery configuration, 505
- chemistry, 503
- electric vehicles, 511, 515
- materials of construction, 509
- performance characteristics, 512
- sealing, 508
- self-discharge, 503
- shunt current management, 506
- stack fabrication, 508
- stationary storage applications, 512
- system design, 511
- technology status, 514

zinc–carbon batteries, 19, 21, 87–111, 119, 313, 340, 348, 487, 556
- can manufacture, 97
- cathode, 100
- chemistry, 89–97
- competition with alkaline–manganese, 89
- construction, 97–102
- discharge tests, 102
- flash current test, 103
- future developments, 110
- manufacturers, 88–90
- performance, 102–110
- production statistics, 91–95
- projected growth in performance, 111
- rechargeable, 111
- safety, 105–110
- sales statistics, 89, 91
- seals, 100
- separators, 100
- shelf life, 105
- types, 97
- zinc anode, 100

zinc chloride cells, *see* zinc–carbon batteries
zinc–chlorine batteries, 26
zinc–copper oxide batteries, 24, 112
zinc–manganese batteries, *see* alkaline–manganese batteries
zinc–mercury oxide batteries, 24, 26, 60, 125–160, 557
- button cell construction, 133
- cell impedance, 141
- environmental effects, 144
- mercuric oxide cathode, 127
- performance characteristics, 137–141
- sizes and types, 155
- use of manganese dioxide, 135

zinc oxide
- solubility in potassium hydroxide, 174
- surface area measurement, 77

zinc oxychloride, 96
zinc–plumbate batteries, 557
zinc powder, 112-117, 182, 186
- surface area measurement, 77

zinc–silver nickel oxide batteries, 557
zinc–silver oxide batteries, 27, 48, 58, 125–160, 307, 402–407, 453, 455, 558
- button cell construction, 133
- double treatment approach, 130
- energy densities, 148
- environmental effects and storage, 152
- performance characteristics, 149–152
- reserve type, 422
- silver–nickel oxide cathode, 132
- silver oxide cathode, 129
- silver 'plumbate' cathode, 130
- sizes and types, 155

Zircar™, *see* separators, Zircar™

# ELLIS HORWOOD SERIES IN
# APPLIED SCIENCE AND INDUSTRIAL TECHNOLOGY

*Series Editor:* Dr D. H. SHARP, OBE, former General Secretary, Society of Chemical Industry; formerly General Secretary, Institution of Chemical Engineers; and former Technical Director, Confederation of British Industry.

**MECHANICS OF WOOL STRUCTURES**
R. POSTLE, University of New South Wales, Sydney, Australia, G. A. CARNABY, Wool Research Organization of New Zealand, Lincoln, New Zealand, and S. de JONG, CSIRO, New South Wales, Australia

**MICROCOMPUTERS IN THE PROCESS INDUSTRY**
E. R. ROBINSON, Head of Chemical Engineering, North East London Polytechnic

**BIOPROTEIN MANUFACTURE: A Critical Assessment**
D. H. SHARP, OBE, former General Secretary, Society of Chemical Industry; formerly General Secretary, Institution of Chemical Engineers; and former Technical Director, Confederation of British Industry

**QUALITY ASSURANCE: The Route to Efficiency and Competitiveness, Second Edition**
L. STEBBING, Quality Management Consultant

**QUALITY MANAGEMENT IN THE SERVICE INDUSTRY**
L. STEBBING, Quality Management Consultant

**INDUSTRIAL CHEMISTRY**
E. STOCCHI, Milan, with additions by K. A. K. LOTT and E. L. SHORT, Brunel

**REFRACTORIES TECHNOLOGY**
C. STOREY, Consultant, Durham; former General Manager, Refractories, British Steel Corporation

**COATINGS AND SURFACE TREATMENT FOR CORROSION AND WEAR RESISTANCE**
K. N. STRAFFORD and P. K. DATTA, School of Material Engineering, Newcastle upon Tyne Polytechnic, and C. G. GOOGAN, Global Corrosion Consultants Limited, Telford

**TEXTILE OBJECTIVE MEASUREMENT AND AUTOMATION IN GARMENT MANUFACTURE**
G. STYLIOS, Department of Industrial Technology, University of Bradford

**INDUSTRIAL PAINT FINISHING TECHNIQUES AND PROCESSES**
G. F. TANK, Educational Services, Graco Robotics Inc., Michigan, USA

**MODERN BATTERY TECHNOLOGY**
Editor: C. D. S. TUCK, Alcan International Ltd, Oxon

**FIRE AND EXPLOSION PROTECTION: A Systems Approach**
D. TUHTAR, Institute of Fire and Explosion Protection, Yugoslavia

**PERFUMERY TECHNOLOGY 2nd Edition**
F. V. WELLS, Consultant Perfumer and former Editor of *Soap, Perfumery and Cosmetics,* and M. BILLOT, former Chief Perfumer to Houbigant-Cheramy, Paris, Président d'Honneur de la Société Technique des Parfumeurs de la France

**THE MANUFACTURE OF SOAPS, OTHER DETERGENTS AND GLYCERINE**
E. WOOLLATT, Consultant, formerly Unilever plc

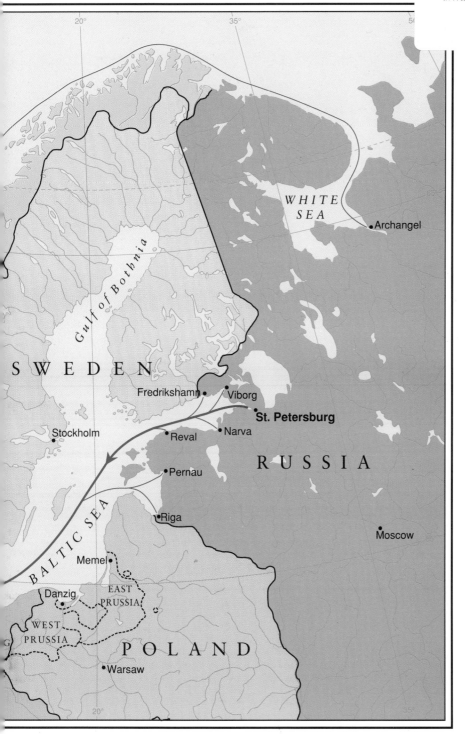

# Russian Overseas Commerce
# with Great Britain
# During the Reign of Catherine II

# Russian Overseas Commerce with Great Britain During the Reign of Catherine II

HERBERT H. KAPLAN

American Philosophical Society
Independence Square • Philadelphia
1995

*Memoirs
of the
American Philosophical Society
Held at Philadelphia
for Promoting Useful Knowledge*
Volume 218

Copyright ©1995 by the American Philosophical Society for its Memoirs series, Volume 218. All rights reserved. Reproduction in whole or in part in any media is restricted. Publication has been subsidized by the Lewis Fund of the American Philosophical Society.

*Endleaf design:* Map of the Russian Empire showing northern overseas trade routes. Design by Swanston Graphics, Derby U.K.

*Dust jacket design:* Map of Russia, 1796. In Thomas Dobson. *Encyclopedia,* Vol. 16, opp. p. 554.

Library of Congress Cataloging in Publication Data:

Kaplan, Herbert H.

Russian Overseas Commerce with Great Britain During the Reign of Catherine II

Index, bibliography, tables, map

1. Russia, eighteenth century history 2. England, eighteenth century history 3. Catherine II 4. Russian-British Commerce 5. Trade, Europe, eighteenth century

ISBN: 0-87169-218-X            94-78517
US ISSN: 0065-9738

*IN MEMORY*

Anna Kaplan
1895–1970

Meyer Polan
1912–1975

Jacob Polan
1902–1991

Michael Gersten
1943–1994

Minna Polan
1913–1995

# Contents

| | |
|---|---|
| Map | Endleaves |
| Abbreviations | ix |
| Weights and Measures | xiii |
| List of Tables | xv |
| List of Figures | xx |
| Acknowledgments | xxiii |
| Preface | xxv |

PART I  The Continuity of Commercial Policy: Elizabeth—Peter III—Catherine II    1

    Chapter I    Elizabeth and Peter III and the Anglo-Russian Trade    3
    Chapter II    Catherine II and the Formulation of Commercial Policy    15
    Chapter III    The Anglo-Russian Commercial Treaty of 1766 and the Russian Tariff of 1766/1767    32

PART II  Russian Commodity Exports and Great Britain's Economic Development: From the Seven Years' War to the Armed Neutrality    49

    Introduction: The Anglo-Russian Overseas Commodity Trade Balance    51
    Chapter IV    Bar Iron    55
    Chapter V    Hemp, Flax and Linen    63
    Chapter VI    Timber    88

PART III  War and Commerce on the High Seas: Conventions, Commodities, Contraband    111

    Chapter VII    Neutrality Unarmed    113
    Chapter VIII    The Expiration of the Anglo-Russian Commercial Treaty of 1766    132

|             |                    |                                                                                                                      |     |
|-------------|--------------------|----------------------------------------------------------------------------------------------------------------------|-----|
|             | Chapter IX         | International Politics and Commerce in Crisis                                                                        | 149 |
| PART IV     | The Balance of Trade and Culture |                                                                                                        | 167 |
|             | Chapter X          | Merchant Nationals                                                                                                   | 169 |
|             | Chapter XI         | Ships and Tonnage                                                                                                    | 198 |
|             | Chapter XII        | Russian Commodity Exports and Great Britain's Economic Development: From the Peace of Paris to the end of Catherine II's Reign | 211 |
|             | Chapter XIII       | The Material of Culture                                                                                              | 231 |
|             | Chapter XIV        | Russia's Foreign Commodity Trade Turnover: The Principal Ports and Great Britain                                     | 245 |
|             | Conclusion         |                                                                                                                      | 269 |
|             | Bibliography       |                                                                                                                      | 275 |
|             | Index              |                                                                                                                      | 303 |

# ABBREVIATIONS

PRO            Public Record Office, London
       BT     Board of Trade
       CO     Colonial Office
       CU     Customs Ledgers
       FO      Foreign Office
       S P     State Papers

SRO           Scottish Record Office, Edinburgh
       CO     Colonial Office

AN            *The Armed Neutralities Of 1780 and 1800. A Collection of Official Documents Preceded by the Views of Representative Publicists.* Edited by James Brown Scott. Carnegie Endowment for International Peace. New York: Oxford University Press, 1918.

BAR IRON    "An Account of the Quantity of Bar Iron imported into England [Scotland] between the 5 January 1764 and the 5th January 1786 [and between 1786 and 1799]; distinguishing each Year and the principal Countries from whence imported." Reports of the Inspector–General, Custom House, London, PRO BT 6/240 (1764–1785), ff. 119–121, and BT 6/230 (1786–1799), ff. 39, 41.

CT            *A Collection Of All The Treaties Of Peace, Alliance, And Commerce, Between Great Britain And Other Powers, From the Treaty signed at Munster in 1648, to the Treaties signed at Paris in 1783. To which is prefixed, A Discourse on the Conduct of the Government of Great-Britain in Respect to Neutral Nations, by the Right Hon. Charles Jenkinson.* London: 1785. Reprint, New York: Augustus M. Kelley, 1969.

CTS           *The Consolidated Treaty Series.* Edited and annotated by Clive Parry. Dobbs Ferry, NY: Oceana Publiations, 1969.

FLAX         "An Account of the Quantity of Flax imported

|  |  |
|---|---|
| | into England [Scotland] between the 5 Jany. 1764 & the 5 Jany. 1786 [and between 1786 and 1799]; distinguishing each Year & the principal Countries from whence imported." Reports of the Inspector–General, Custom House, London, PRO BT 6/240 (1764–1785), ff. 89, 91, and BT 6/230 (1786–1799), ff. 3, 5. |
| HEMP | "An Account of the Quantity of Hemp imported into England [Scotland] between the 5 Jany. 1764 & the 5 Jany. 1786 [and between 1786 and 1799]; distinguishing each Year & the principal Countries from whence imported." Reports of the Inspector–General, Custom House, London, PRO BT 6/240 (1764–1785), ff. 93, 95, and BT 6/230 (1786–1799), ff. 7, 9. |
| JCTP | *Journal of the Commissioners for Trade and Plantations.* London: 1933–1938. |
| MASTS | "An Account of the Quantity of Bowsprits, Masts & Yards imported into England [Scotland] between the 5th Jany. 1764 & the 5 Jany. 1786 [and between 1786 and 1799]; distinguishing each Year, and the principal Countries from whence imported." Reports of the Inspector–General, Custom House, London, BT 6/240 (1764 1785), ff. 109–111 adn BT 6/230 (1786–1799), ff. 33–34. |
| MATHIAS | Peter Mathias, "Russia and the British Industrial Revolution in the Eighteenth Century." A Commentary on a paper by Professor H. H. Kaplan delivered at the 93rd Annual meeting of the American Historical Association, 28 December 1978, San Francisco, 17 pp. |
| P. P. Customs, 1897 | Great Britain, Parliament. Presentation to both Houses. "Customs Tariffs of the United Kingdom, from 1800 to 1897. With some notes upon the history of the more important branches of receipt from the year 1660." *Accounts and Papers.* XXXIV [c. 8706.], London, 1898. |
| P. P. Foreign Trade, 1820 | Great Britain, Parliament. House of Lords. "First Report from The Select Committee of the House of Lords, appointed to inquire into the means of extending and securing the Foreign Trade of the |

Country, and to report to the House; together with the Minutes of Evidence taken before the said Committee—3 July 1820." *Reports From Select Committees Of The House of Lords* (269). Volume III. London, 1820.

P. P. Linen, 1773
Great Britain, Parliament. House of Commons. "Report From The Committee Appointed to Enquire Into The Present State Of The Linen Trade In Great Britain And Ireland. Reported on the Twenty-fifth Day of May 1773." *Reports From Committees Of The House of Commons*. III, London, 1803, pp. 99–133.

P. P. Timber, 1771
Great Britain, Parliament. House of Commons. "A Report From The Committee Appointed To Consider How His Majesty's Navy may be better supplied with TIMBER. The 6th of May 1771." *Reports From Committees Of The House of Commons*. III, London, 1803, pp. 15–53.

P. P. Timber, 1835
Great Britain, Parliament. House of Commons. "Report From Select Committee on Timber Duties Together With Minutes of Evidence, An Appendix, And Index." *Reports From Committees*. XIX (519), London, 1835.

PSZ
*Polnoe Sobranie Zakonov Rossiiskoi Imperii*. Sankt Peterburg, 1830.

RC
"Russia Company. The Court Minute Books." MSS. 11, 741, Vols. 7–9. Guildhall, London.

SAILCLOTH
"An Account of the Quantity of Sail Cloth imported into England [Scotland] between the 5 Jany. 1764 and the 5th Jany. 1786 [and between 1786 and 1799]; distinguishing each Year & the principal Countries from whence imported." Reports of the Inspector–General, Custom House, London, PRO BT 6/240 (1764–1785), ff. 105, 107 and BT 6/230 (1786–1799), ff. 11, 13.

SIRIO
*Sbornik Imperatorskago Russkago Istoricheskago Obshchestva*. St. Peterburg 1867–1916.

SOT
Sir Charles Whitworth. *State of the Trade of Great Britain in its Imports and Exports, Progressively from the Year 1697: also the Trade of each particular Country, during the above Period, distinguishing each Year. In two parts with a Preface and Introduction, setting*

|            | *forth the Articles whereof each Trade consists.* PRO BT 6/185, ff. 81–181. London, 1776, the hand-revised and amended edition to 1801. |
|------------|---|
| TARIFF     | *A Tariff, or Book of Rates, of the Customs to be levied, on Goods imported & exported, at the Port and Frontier Custom-Houses.* Translated from the Russ. PRO SP 91/78. St. Petersburg: Imperial Academy of Sciences, 1767. |
| VEDOMOSTI  | Lists; Registers. |
| VT         | "An Account of the Number of Vessels and the Amount of their Tonnage that have entered Inwards and cleared Outwards in the Several Ports of Great Britain between the 5th January 1771 and the 5th January [1802], distinguishing England from Scotland the British and Foreign Vessels; also distinguishing the Countries from whence such Vessels arrived or to which cleared." PRO BT 6/185, ff. 255–70. |
| VTR        | Indova, E. I., et al, eds. *Vneshniaia torgovlia Rossii cherez Peterburgskii port vo votoroi polovine XVIII—nachale XIX v. Vedomosti o sostave kuptsov i ikh torgovykh oborotakh.* Moskva: Akademii Nauk, 1981. |
| AHR        | *The American Historical Review* |
| CASS/CSS   | *Canadian-American Slavic Studies/Canadian Slavic Studies* |
| EHR        | *The English Historical Review* |
| ESR        | *European Studies Review* |
| FOG        | *Forschungen zur osteuropäischen Geschichte* |
| HJ         | *The Historical Journal* |
| IZ         | *Istoricheskie zapiski* |
| JEEH       | *The Journal of European Economic History* |
| JGO        | *Jahrbücher für Geschichte Osteuropas,* Neue Folge |
| MM         | *The Mariner's Mirror* |
| RR         | *The Russian Review* |
| SEER       | *The Slavonic and East European Review* |
| SR         | *Slavic Review* |

# Weights and Measures

Although not a mathematically exact measurement, British merchants accommodated 63 Russian poods to a British ton = 20 hundredweight = 20 Cwt [1 Cwt = 112 pounds (50.802 kg)] = 2,240 pounds (1016.040 kg); a Russian pood = 40 funty = 16.38 kilograms = 36.113 pounds; therefore, 63 Russian poods = 1 British ton (not 2,240 pounds but 2,275 pounds, a difference of almost 1 pood).

One arshin = 0.77 yard, or 2.33 feet, or 27.96 inches. British merchants accommodated 1 arshin to 28 inches. Forty-five arshins = 35 yards = 1 piece.[1]

---

1. See J. Jepson Oddy, *European Commerce, Shewing New and Secure Channels of Trade with the Continent of Europe: Detailing the Produce, Manufactures and Commerce, of Russia, Prussia, Sweden, Denmark, and Germany; as well as the Trade of the Rivers Elbe, Weser, and Ems; with a General View of the Trade, Navigation, Produce and Manufactures, of the United Kingdom of Great Britain and Ireland; and its Unexplored and Improvable Resources and Interior Wealth* (London, 1805), p. 137, "9 arsheens are 7 yards English, or the arsheen 28 inches"; and B. R. Mitchell with the collaboration of Phyllis Deane, *Abstract of British Historical Statistics* (Cambridge: Cambridge University Press, 1962), p. 201, n. (d), "Prior to 1810. . . . For this table, these latter entries have been converted to yards on the assumption that each piece was 35 yards long. (This assumption is based on the rates of valuation. . . .)"

The author is aware that there is not one universal equivalent of Russian piece in arshins and has used the most appropriate equivalent based upon a thorough examination of the manuscript sources.

The complexity of determining this measure is illustrated in the following studies:

Oddy on p. 119: [1 piece =] "Ravenducks, 50 arsheens long, 28, 31 1/2, 36 inches wide. . . . Flems, 50 and 57 arsheens long, 42 and 45 inches wide. . . . Drillings, bleached and unbleached, 50 arsheens long, 28 inches wide. . . ."

Marten G. Buist, *At Spes Non Fracta. Hope & Co. 1770–1815. Merchant Bankers and Diplomats at Work.* (The Hague: Martinus Nijhoff, 1974), p. 508: "Calamancoes (also known as drillings) . . . (1 piece = approx. 35 arshins)."

S. D. Chapman and S. Chassagne, *European Textile Printers in the Eighteenth Century: A Study of Peel and Oberkampf* (London: Heinemann Educational Books, The Pasold Fund, 1981), p. 213: "Calculations: French equivalents of English measures in brackets. The English piece measured 28 yards long in the lower price brackets (60% of output) and 30 yards in the higher price ranges; averages 28.8 yards. The Continental piece averaged 20 aunes = 24m. = 26.3 yds."

# List of Tables

*Chapter I*

none

*Chapter II*

none

*Chapter III*

| | | |
|---|---|---|
| TABLE MN–I. | Comparison Between the Value of British Merchant Overseas Commodity Trade Turnover in St. Petersburg, Riga, and Narva: Annual Average 1763–1764 | 46 |
| TABLE MN–II. | Comparison Between the Value of the British Merchant Overseas Commodity Trade Turnover in Riga and Narva and the Russian Merchant Overseas Commodity Trade Turnover in St. Petersburg in 1764 | 46 |

*Chapter IV*

| | | |
|---|---|---|
| TABLE BI–I. | Great Britain's Bar Iron Imports from 1764 through 1782 | 58 |
| TABLE BI–II. | Bar Iron Exports from St. Petersburg from 1753 through 1782 | 60 |
| TABLE BI–III. | Value of St. Petersburg's Commodity and Bar Iron Exports from 1768 through 1779 | 62 |

*Chapter V*

| | | |
|---|---|---|
| TABLE H–I. | Great Britain's Hemp Imports from 1764 through 1782 | 67 |

| | | |
|---|---|---|
| TABLE H–II. | Hemp Exports from St. Petersburg from 1753 through 1782 | 69 |
| TABLE H–III. | Value of St. Petersburg's Commodity and Hemp Exports from 1768 through 1779 | 71 |
| TABLE H–IV. | Value of St. Petersburg's Commodity and Clean Hemp Exports from 1768 through 1779 | 72 |
| TABLE F–I. | Great Britain's Flax Imports from 1764 through 1782 | 73 |
| TABLE F–II. | Flax Exports from St. Petersburg from 1753 through 1782 | 74 |
| TABLE F–III. | Value of St. Petersburg's Commodity and Flax Exports from 1768 through 1779 | 76 |
| TABLE L–I. | The Approximate Linen Equivalent of England's Russian Flax Imports: 1765–1785 | 77 |
| TABLE L–II. | Russia's Estimated Share of England's Total Foreign Linen Imports, Re-exports, and Retained Imports from 1752 through 1771 | 82 |
| TABLE L–III. | Minimum Estimates of Selected Linens Exported from St. Petersburg from 1763 through 1782 | 84 |
| TABLE L–IV. | Value of St. Petersburg's Commodity and Selected Linen Exports from 1768 through 1779 | 87 |

*Chapter VI*

| | | |
|---|---|---|
| TABLE T–I. | St. Petersburg's Deal Exports from 1763 through 1782 | 101 |
| TABLE T–II. | Great Britain's Importation of Great Masts from 1764 through 1782 | 108 |
| TABLE T–III. | Great Britain's Importation of Middle Masts from 1764 through 1782 | 108 |
| TABLE T–IV. | Great Britain's Importation of Small Masts from 1764 through 1782 | 109 |

*Chapter VII*

none

*Chapter VIII*

none

*Chapter IX*

none

*Chapter X*

| | | |
|---|---|---|
| TABLE MN–III. | Value of Prima Facie Exports by Merchant Nationals as Percentage of the Total Prima Facie Value of St. Petersburg's Overseas Commodity Exports: 1764–1796 | 177 |
| TABLE MN–IV. | Value of Prima Facie Imports by Merchant Nationals as Percentage of the Total Prima Facie Value of St. Petersburg's Overseas Commodity Imports: 1764–1796 | 180 |
| TABLE MN–V. | Comparison Between the Value of the Total Overseas Commodity Trade Turnover by Russian, British, and Hanse Merchants at St. Petersburg: 1787 | 183 |
| TABLE MN–VI. | Comparison Between the Value of the Total Overseas Commodity Trade Turnover by Russian, British, and Hanse Merchants at St. Petersburg: 1792 | 185 |
| TABLE MN–VII. | Comparison Between the Value of the Total Overseas Commodity Trade Turnover by Russian and British Merchants at St. Petersburg: 1793 | 187 |
| TABLE MN–VIII. | Comparison Between the Value of the Total Overseas Commodity Trade Turnover by Russian and British Merchants at St. Petersburg: 1795 | 188 |
| TABLE MN–IX. | Recomputed Value of Exports by Russian and British Merchants as Percentage of | 192 |

|  |  |  |
|---|---|---|
|  | Total Value of St. Petersburg's Overseas Commodity Exports: 1764–1796 |  |
| TABLE MN–X. | Recomputed Value of Imports by Russian and British Merchants as Percentage of Total Value of St. Petersburg's Overseas Commodity Imports: 1764–1796 | 196 |
| TABLE MN–XI. | Recomputed Value of Trade Turnover by Russian and British Merchants as Percentage of the Total Value of St. Petersburg's Overseas Commodity Trade Turnover: 1764–1796 | 197 |

*Chapter XI*

|  |  |  |
|---|---|---|
| TABLE S–I. | Ships Departing St. Petersburg/Kronstadt from 1753 through 1801 | 200 |
| TABLE VT–I. | Great Britain's Overseas Commerce with "Foreign Europe" and Russia from 1771 through 1801: Vessels and Tonnage | 204 |

*Chapter XII*

|  |  |  |
|---|---|---|
| TABLE BI–IV. | Great Britain's Bar Iron Imports from 1783 through 1796 | 213 |
| TABLE BI–V. | Value of St. Petersburg's Commodity and Bar Iron Exports: 1783–1791, 1793 | 215 |
| TABLE H–V. | Great Britain's Hemp Imports from 1783 through 1796 | 216 |
| TABLE H–VI. | Value of St. Petersburg's Commodity and Hemp Exports: 1783–1791, 1793 | 217 |
| TABLE F–IV. | Great Britain's Flax Imports from 1783 through 1796 | 219 |
| TABLE F–V. | Value of St. Petersburg's Commodity and Flax Exports: 1783–1791, 1793 | 220 |
| TABLE L–V. | Minimum Estimate of Selected Linens Exported from St. Petersburg from 1783 through 1796 | 222 |
| TABLE L–VI. | Minimum Estimate of the Value of St. | 223 |

|  |  |  |
|---|---|---|
|  | Petersburg's Commodity and Selected Linen Exports: 1783–1791, 1793 |  |
| TABLE T–V. | Great Britain's Imports of "Fir Timber Eight Inches Square and Upwards" from 1788 through 1796 | 225 |
| TABLE T–VI. | Approximate Number of Great Masts Imported into Great Britain from 1783 through 1796 | 227 |
| TABLE T–VII. | Great Britain's Imports of "Deals and Deal Ends" and "Battens and Batten Ends" from 1788 through 1796 | 228 |
| TABLE T–VIII. | St. Petersburg's Deal Exports from 1783 through 1796 | 230 |

*Chapter XIII*

| | | |
|---|---|---|
| TABLE CI–I. | Value of Selected Categories of Imports as Percentage of Total Value of Commodity Imports into St. Petersburg: 1764–1796 | 234 |
| TABLE CI–II. | Most Valued Commodities Imported into St. Petersburg: 1783–1796 | 236 |
| TABLE CI–III. | St. Petersburg's Overseas Textile Imports: 1783–1796 | 237 |

*Chapter XIV*

| | | |
|---|---|---|
| TABLE VRT–I. | Approximate Value of Russia's Overseas and Overland Foreign Commodity Trade Turnover: 1775–1801 | 247 |
| TABLE VRT–II. | Approximate Value of Principal Ports' Overseas Commodity Trade Turnover from 1783 through 1801 | 248 |
| TABLE VRT–III. | Approximate Value of Russia's Overseas and Overland Commodity Trade Balance: 1775–1801 | 251 |
| TABLE VRT–IV. | Custom Revenue as a Percent of the Value of Russia's Foreign Commodity Trade Turnover: 1775–1801 | 254 |
| TABLE VRT–V. | Comparison of Custom Revenue as a | 256 |

| | Percent of Russia's and St. Petersburg's Foreign Commodity Trade Turnover: 1768–1801 | |
|---|---|---|
| TABLE MN–XII. | Approximate Value of British Merchants' Overseas Commodity Trade Turnover at St. Petersburg: 1783–1796 | 261 |
| TABLE BT–I. | Approximate Value of Great Britain's Overseas Commodity Trade Balance Deficit with St. Petersburg: 1783–1796 | 262 |

# List of Figures

Fig. 1. Value of St. Petersburg's Overseas Commodity Exports from 1768 through 1779 (all ships)   61

Fig. 2. Value of St. Petersburg's Overseas Commodity Exports from 1768 through 1779 (all ships)   71

Fig. 3. Value of St. Petersburg's Overseas Commodity Exports from 1768 through 1779 (all ships)   75

Fig. 4. Value of St. Petersburg's Overseas Commodity Exports from 1768 through 1779 (all ships)   86

Fig. 5. Great Britain's Importation of Great Masts from 1764 through 1782 (annual averages)   107

Fig. 6. Value of Prima Facie Exports by Merchant Nationals through St. Petersburg, 1768–1779 (annual average)   176

Fig. 7. Value of Prima Facie Exports by Merchant Nationals through St. Petersburg, 1768–1779 (annual average)   179

Fig. 8. Recomputed Value Compared to Prima Facie Value of Exports by Merchant Nationals through St. Petersburg, 1783–1787   189

Fig. 9. Recomputed Value Compared to Prima Facie Value of Exports by Merchant Nationals through St. Petersburg, 1788–1792   190

Fig. 10. Recomputed Value Compared to Prima Facie Value of Exports by Merchant Nationals through St. Petersburg, 1793/1795–96   191

Fig. 11. Recomputed Value Compared to Prima Facie Value of

Imports by Merchant Nationals through St. Petersburg, 1783–1787   193

Fig. 12. Recomputed Value Compared to Prima Facie Value of Imports by Merchant Nationals through St. Petersburg, 1788–1792   194

Fig. 13. Recomputed Value Compared to Prima Facie Value of Imports by Merchant Nationals through St. Petersburg, 1793/1795–96   195

Fig. 14. Value of St. Petersburg's Overseas Commodity Exports from 1783 through 1791   212

Fig. 15. Great Britain's Importation of Great Masts from 1783 through 1796 (annual averages).   226

Fig. 16. Value of Russia's Overseas and Overland Foreign Commodity Trade Turnover, 1775–1796   246

Fig. 17. Value of Principal Ports' Overseas Commodity Exports from 1783 through 1796   249

Fig. 18. Value of Principal Ports' Overseas Commodity Imports from 1783 through 1796   250

# Acknowledgments

I wish to thank the John Simon Guggenheim Memorial Foundation, the American Philosophical Society, the Institute for Advanced Studies in the Humanities of the University of Edinburgh, and Indiana University for fellowship support while researching and writing this book. I am indebted to the staffs of the British Library, the Public Record Office (London and Edinburgh), the Guildhall Library, the University of London Library, the Staatsarchiv Hamburg, and the Indiana University Library for their assistance.

Very early in my research Professor D. C. Coleman, in a chance meeting in London, encouraged me by telling me that I was on the right track. David Spaeder and Professor Emeritus L. L. Waters made useful suggestions to early drafts of the manuscript. The readers for the American Philosophical Society were Professor Jeremy Black of the University of Durham and Dr. Gordon Jackson of the University of Strathclyde, United Kingdom. I am exceedingly grateful for their thorough and constructive critique of every datum and interpretation.

Debbie Chase typed the initial manuscript, and Beth Olson edited and prepared it for submission to the press. Steven T. Duke prepared the figures. Elke and Greg Rogers, and Eric Hender assisted in reading galley and page proof. Susan M. Babbitt, Assistant Editor, and Carole Le Faivre-Rochester, Associate Editor, American Philosophical Society, edited and prepared the manuscript for publication with skill, talent, and tact. However, I am solely responsible for the book and its errors.

I am particularly thankful to my old and dear friends, Joseph Breu and Professor Aron Rodrigue. Aron's patience, wisdom, and faith in me were critical to the book's completion. Joe re-read every word and repeatedly challenged every idea and interpretation in the manuscript. Becaue of him, it is a better book. My beloved wife, Barbara, listened patiently to every word, every idea, and every interpretation. In all this, we were constantly comforted by the companionship of Baby, Fluffie, and Tortie.

# PREFACE

This study grew out of earlier investigations in the Soviet archives into the social and economic management of the landed estates of the very wealthy in eighteenth-century Russia. Those estates had an international commercial component to their economies that led me to delve into archival resources in England and Scotland.

What struck my attention quite early in that research was that the data I kept finding in British archives were Russian customs and commercial data taken from the St. Petersburg Custom House during the reigns of Catherine II and Paul I. In the period following the Seven Years' War, the British Board of Trade and the Russia Company in Great Britain had requested information on Russian overseas commerce from the British Consul General and the British trading representatives in St. Petersburg. As a consequence, commercial data from the St. Petersburg Custom House, including sources from other custom houses in the Empire and from central state agencies, were regularly sent to Britain. For two hundred years this material has rested in the British archives in company with British overseas trade, shipping, and customs documents. The wealth of these materials persuaded me that they would support a comprehensive study on Russia's overseas commerce with Great Britain during the reign of Catherine II (1762–1796).

The significance of foreign trade in Great Britain's economic development has been well-established, but Russia's role in it has not been. Yet Russia's impact on the British economy during Catherine's reign, a time contemporaneous with the British industrial revolution,[1] was both profound and widespread. The quantity, quality, and range of Russian grown and manufactured products, transacted by British merchants at home and resident in Russia, amply demonstrate this. It was the lack of sufficient advanced tech-

---

1. For a recent review of the literature, see Patrick K. O'Brien, "Introduction: Modern Conceptions of the Industrial Revolution," in *The Industrial Revolution and British Society*, edited by Patrick K. O'Brien and Robert Quinault (Cambridge: Cambridge University Press, 1993), pp. 1–30.

nological innovation and diffusion in Britain and the recurrence of international wartime crises that made this so.

In this work, I examine the volume and the monetary value, in both rubles and English pounds sterling, of Russia's exports of natural and manufactured resources to Great Britain, and assess the impact these had on Britain's industrial revolution and on her strategic military capability. Great Britain's exports and re-exports also are examined and evaluated for their significance in contributing to Russia's state building and evolving "westernization," both economically and culturally. To this end, I quantify and analyze the trading activity and the shipping data of the maritime nations in Russia's overseas commerce to resolve the historical question of whether the British dominated Russia's overseas commercial market. I also provide an examination of the political decision-making process at the Russian and English courts to trace the development of their maritime and mercantile policies in times of both peace and war. It is a thesis of this work that Catherine, while following the general commercial policies adopted by her predecessors, recognized the interdependent nature of the British-Russian commercial relationship, and that she sought to exploit this to make Russia an independent commercial and maritime power and to make herself a major player in greater European political affairs.

In addition to the material in the British archives, I have made use of several studies undertaken by Soviet scholars that relate to the role of merchant nationals in Russia's overseas commerce. In a monograph published in 1955, N. L. Rubinshtein used Russian archival materials in an effort to emphasize the role of native Russian merchants in this trade. In the 1970s, B. N. Mironov critically re-examined Rubinshtein's thesis in monographs that also relied on Russian archival material. In 1981, E. I. Indova performed an immensely valuable service by publishing verbatim the most important of the archival sources upon which many of Rubinshtein's and Mironov's arguments depend. These materials, along with the Reval [Tallinn] archive findings by C. F. Menke in his Göttingen dissertation in 1959, are discussed in detail in Chapter X.

The materials used in this study present some important historiographical issues. One relates to the validity of the data to be found in British customs sources and the other to the reliance on material that derives almost exclusively from a single Russian port, St. Petersburg/Kronstadt. Both of these require explanation.

British customs data and commercial statistics are expressed in "official values" of English pounds sterling. Since George N.

Clark published his *Guide to English Commercial Statistics* in 1938,[2] scholars have had a distrust for the veracity of those eighteenth-century sources, in part because of their belief that the designation of "official values" could not be computed in current real values, and thus represented an insurmountable challenge to reliable research results.[3]

With regard to these concerns, I am in agreement with the thinking of G. D. Ramsay, a well-known inquirer into English commercial history:

> Official statistics certainly cannot be accepted as gospel, save for revenue accounting, but nobody in his senses would dismiss them as "virtually useless." The truth lies always somewhere in between these extremes: precisely where is a matter for individual investigation on each occasion. Trends, patterns, and even some idea of volume and value may legitimately be evoked from these tantalizing figures, with the support of other evidence, even if not always with sufficient exactitude to provide a foothold for the bold explorations of the historical economist. *Fortunately, there is no lack of complementary sources to supplement English customs returns,* especially for the eighteenth century.[4]

Concerning the specific challenge presented by the "official values" in the customs sources, I concur with Arthur D. Gayer, Walt W. Rostow and Anna Jacobson Schwartz about the relationship between computed current real values and the "official values" in determining the importance of the Russian trade to Great Britain: ". . . because of the process by which the official value figures are calculated the absolute level of this percentage is not necessarily significant. *Changes in the percentage are, however, a fairly reliable index of the altering importance.*"[5]

In this regard, my study has benefited from the research of Ralph Davis, in particular from his computation of the current real value of British imports during the latter part of the eighteenth

---

2. George N. Clark, *Guide to English Commercial Statistics, 1696–1782* (London: Royal Historical Society, 1938).
3. See Ralph Davis, *The Industrial Revolution and British Overseas Trade* (Leicester, England: Leicester University Press, 1979), pp. 77–86.
4. Quoted from G. D. Ramsay's review of *The Growth of English Overseas Trade in the Seventeenth and Eighteenth Centuries,* ed. W. E. Minchinton in *The Economic History Review,* 2nd series, 23:3 (December 1970): p. 572. Italics mine.
5. Arthur D. Gayer, Walt W. Rostow, and Anna Jacobson Schwartz, *The Growth and Fluctuation of the British Economy 1790–1850: An Historical, Statistical, and Theoretical Study of Britain's Economic Development,* 2 vols. (New York: Barnes and Noble, 1975), vol. 1, p. 70n. Italics mine.

century. His findings not only confirm the increasing magnitude of Great Britain's import trade from Russia found in the "official values" but also show that the "official values" actually understate the increase.[6]

Furthermore, the Russian customs and commercial data in the British archives, expressed in current rubles, provide confirmation of the growth of the Anglo-Russian overseas trade.

The available archival material for the study of Russia's overseas commerce is more abundant for the capital port of St. Petersburg/Kronstadt than for any other port of the Russian Empire. The role of that port in the foreign trade of the Empire is difficult to overstate. The capital port accounted for more than half of Russia's overall foreign trade turnover and more than 70 percent of the overseas trade turnover. It is also true that after the Seven Years' War, British commercial hegemony in Russia's overseas commodity trade was concentrated in Russia's northern ports. This was particularly true with regard to St. Petersburg, where British merchants held a significant share in imports and predominated in exports.

For those reasons this study often focuses on that port's overseas trade with Great Britain to determine the nature, volume, and value of that trade. What these data demonstrate is that the Anglo-Russian overseas commerce during the reign of Catherine II significantly affected Great Britain's economic development during its industrial revolution. Not only did Russia supply Great Britain with enormous amounts of valuable material and manufactured resources vital to Britain during periods of peace and war, which it could not conveniently and economically obtain elsewhere, but also Britain paid for these imports in part by exporting to Russia a considerable amount of British manufactured and foreign re-exported goods. This, as Peter Mathias has stated, "becomes a salutary reminder to all economic historians interested in the Industrial Revolution of the degree of external commitment which was involved in the process of industrialization being experienced by eighteenth-century Britain."[7]

Without the Russian archival materials reposing in the British archives this topic could not have been successfully researched. Ironically, while writing about the British consular service more than fifty years ago, David Bayne Horn had recognized their importance when he refused to dismiss out of hand the historic and

---

6. I discuss this in Chaps. I and XIV.
7. MATHIAS, p. 1.

pragmatic value of eighteenth-century British official statistics, much the way Ramsay would implicitly do later by drawing attention to them as significant complementary sources. Horn stated that the Board of Trade thanked Samuel Swallow, the Consul General in St. Petersburg, for the "accuracy of the accounts transmitted" and further wrote:

> Since these statistics are often expressly stated to be copied from the customhouse books of the foreign ports, they would seem to offer figures independently compiled, which might profitably be compared with the official British statistics to which Professor G. N. Clark has recently drawn attention. Even if, with Professor Clark, we doubt the reliability of all eighteenth-century commercial statistics, the accompanying reports of the consuls and other ministers provide strictly contemporary and well-informed comments on the nature and extent of British foreign trade.[8]

I have made several presentations, and published articles on this topic. In 1974, I delivered my preliminary findings on Russian overseas commerce with Great Britain at the First Conference of Polish and American Historians in Poland. In 1978, at the annual meeting of the American Historical Association, I was invited to present a seminar entitled: "Russia's Impact on the Industrial Revolution in Great Britain During the Second Half of the Eighteenth Century: The Significance of International Commerce," which was published in 1981 under the same title in the *Forschungen zur osteuropäischen Geschichte*, complete with text and tables from the seminar presentation.[9] The two invited commentators of the seminar were Professor Arcadius Kahan of the University of Chicago and Professor Peter Mathias of Oxford University. Kahan later incorporated some of my findings in his book *The Plow The Hammer And The Knout. An Economic History of Eighteenth-Century Russia*.[10] At that meeting, Professor Mathias challenged me to demonstrate further my thesis of Russia's impact on the industrial revolution in terms of monetary value, as I had already done for commodity volumes at the seminar. In 1986, I published some of those findings under

---

8. David Bayne Horn, "The Board of Trade and Consular Reports, 1696–1782," EHR 54 (July 1939): pp. 476–80.
9. Herbert H. Kaplan, "Russia's Impact on the Industrial Revolution in Great Britain During the Second Half of the Eighteenth Century: The Significance of International Commerce," *FOG* 29 (Berlin: Otto Harrassowitz Wiesbaden, 1981): pp. 7–59.
10. Arcadius Kahan, *The Plow The Hammer and The Knout. An Economic History of Eighteenth-Century Russia* (Chicago: University of Chicago Press, 1985), pp. 163–266.
11. Herbert H. Kaplan, "Observations on the Value of Russia's Overseas Commerce with

the title, "Observations on the Value of Russia's Overseas Commerce with Great Britain During the Second Half of the Eighteenth Century," in the *Slavic Review*.[11]

The bibliography includes only those works cited and consulted. There is a list of abbreviations used in the footnotes and a list of tables and figures to be found in the text. All dates used are adjusted to the New Style Gregorian calendar, which during the second half of the eighteenth century was eleven days ahead of the Julian calendar, but in many cases I have used both dates as they appeared on many documents. The transliteration of Russian terms follows the modified Library of Congress system. First names with common English equivalents have been rendered in English (thus: Catherine, rather than Ekaterina, and Paul, rather than Pavel).

---

Great Britain During the Second Half of the Eighteenth Century," *SR* 45:1 (1986): pp. 85–94.
EOT phkapl$$on

The imagination of Catherine could not remain inactive; thus her plans were rather precipitate than well digested; and it was manifest that this precipitation destroyed in the end a part of the creations of her genius.

She wished, at one and the same time, to form a middle class, to admit foreign commerce, to introduce manufactures, to establish credit, to increase paper-money, to raise the exchanges, to lower the interest of money, to build cities, to create acadamies, to people deserts, to cover the Black Sea with numerous squadrons, to annihilate the Tartars, to invade Persia, to continue progressively her conquests from the Turks, to fetter Poland and to extend her influence over the whole of Europe.

These were no common undertakings, either in difficulty or extent; and, although there was undoubtedly much to do in a country so new to civilization, the actual success would have been much greater had fewer objects been attempted at a time, or if at least all projects of conquest had been renounced, and attention had been directed exclusively to internal prosperity, the only true glory of sovereigns.

Count Louis Philippe de Segur

*Memoirs and Recollections of Count Segur, Ambassador from France to the Courts of Russia and Prussia, &c. &c. Written by Himself* (London: Henry Colburn, 1827; reprint, New York: Arno Press and *The New York Times*, 1970), vol. III: p. 84.

# Part I
# The Continuity of Commercial Policy: Elizabeth—Peter III—Catherine II

# Chapter I
# Elizabeth and Peter III and the Anglo-Russian Trade

For most of the governments of Europe, the Seven Years' War was an economic disaster and for their peoples a tragedy. Millions of ill-clad and underfed subjects suffered the consequences of their sovereigns' dreams of speedy victory. Within a few years after its commencement in 1756, Europe's population became physically, economically, and morally exhausted, and the diplomats were compelled to seek an end to this ruinous war.

Nowhere was this truer than in Russia. Originally a supporter of the war, M. V. Lomonosov expressed his moral indignation at its continuation in 1759 in a poem celebrating the nameday of Empress Elizabeth (1741–1762), whom he called "the angel of peace."

> We await the longed for words:
> "Enough of victories—no more,
> No more the ravages of battles."
> O Lord, Lord of peace, arise,
> Let your love for all of us be shown,
> . . . . . . . . . . . . . . . . . . . .
> Let the gate to war be forever closed.
> . . . Have we not endured enough
> Of the woes of this world?
> Behold the tears of the orphaned,
> Behold the blood of thy servants,
> It is to you that Elizabeth calls on this day . . .
> Banish wars from the ends of the earth.[1]

However, the most compelling reason for Elizabeth to withdraw from the war was that, having entered the war with a budgetary deficit, Russia's fiscal condition by November 1760 was critical. By that time the war had cost Russia more than 40 million rubles

---

1. Quoted in Walter J. Gleason, *Moral Idealists, Bureaucracy, and Catherine the Great* (New Brunswick, NJ: Rutgers University Press, 1981), p. 11; and cf. Robert E. Jones, "Opposition to War and Expansion in Late Eighteenth Century Russia," JGO 32:1 (1984): p. 37.

and it became evident to Elizabeth's Grand Chancellor, Mikhail I. Vorontsov, that Russia did not have the financial resources to continue the fighting. He proposed an end to its participation in the war.[2] But getting out of a war is generally more difficult than getting into one, and it was not until 1762 that Emperor Peter III, Elizabeth's nephew and successor, withdrew Russian troops from the field and concluded an armistice with Prussia. "Russians responded to the long awaited peace with relief," for the "war had become a financial and social issue."[3]

The consequences of Russia's involvement in the war made economic reform imperative. Peter III

> spent the next six months cutting expenses.... He slashed inessential military and civilian personnel.... He resorted to a savage confiscation of the wealth of the Church ... he debased the currency.... and created a new state bank to issue paper money, in effect a partial repudiation of debts.[4]

Peter declared his intention to modify tariff regulations, to pursue monetary reform, and to abolish monopolies.[5] He was alert to the potential benefits of industry and commerce: "Crafts and industry must be made more profitable." "All trade should be without restraint."[6] Thus, Peter was intent on continuing several of Elizabeth's policies.

Even before the war had begun, Empress Elizabeth had initiated policies directed toward extricating Russia from its chronic financial crises. Although she sought to achieve a balanced budget through reforms of currency, banking, and taxation rates, she emphasized commerce as the key to Russia's economic stability and development. Russia's most articulate and innovative elites offered proposals—some general, others more detailed and not always in harmony with each other—to facilitate immediate state action. Of great significance were the proposals of the influential senator Count Peter I. Shuvalov, general-master of ordnance and leading member of a powerful family of court favorites, and Dmitrii V. Vol-

---

2. For the financial state of Russian affairs at this time, see S. M. Troitskii, *Finansovaia politika russkogo absoliutizma v XVIII veke* (Moskva: Nauka, 1966), pp. 230–31, 247; and Carol S. Leonard, *Reform and Regicide. The Reign of Peter III of Russia* (Bloomington: Indiana University Press, 1993), pp. 90–93, 183n3, and Chap. IV, passim.
3. Carol S. Leonard, "The Reputation of Peter III," RR 47 (1988): pp. 263–92, quotation on p. 266.
4. Leonard, *Reform and Regicide*, p. 90.
5. Ibid., pp. 104–10, passim.
6. Quotations ibid., p. 107.

kov, secretary to the authoritative Conference at the Imperial Court.

Shuvalov's influence was evident in a variety of policies. In 1754 Elizabeth abolished internal customs duties and established the Bank of the Nobility and the Commercial Bank. In 1756 the Mint converted copper coins into lighter weight for circulation. Although Shuvalov was less successful in advocating canal building to enhance Russia's internal transport infrastructure, in 1757 Elizabeth followed his advice in publishing a new tariff that continued the protectionist policies of her predecessors. In May 1758 Shuvalov proposed the reestablishment of the Commission on Commerce, which had existed previously between 1720 and 1749. The new Commission would recommend measures for the expansion of foreign trade, the organization of trade consulates abroad, the improvement of credit facilities and the future disposition of trade monopolies. The Commission would also investigate and assess the principles underlying a favorable balance of trade and identify those branches of industry and agriculture that could be encouraged to produce more for the export market.[7]

The Senate's codification commission deliberated a forward looking position intended to enhance foreign trade:

> Nothing is so harmful to the merchantry or so damaging to the public as monopolies . . . because prices are fixed. . . . We will not have laws granting privileges or establishing monopolies in the export, shipping, and exchange of any good . . . and we will abolish those that exist, firmly granting merchants freedom in commerce and manufacturing.[8]

---

7. Troitskii, pp. 35–113, 180–84, 248–49; Philip H. Clendenning, "The Economic Awakening of Russia in the Eighteenth Century," JEEH 14:3 (Winter 1985): pp. 443–71, and "Eighteenth Century Russian Translations of Western Economic Works," JEEH 1:3 (1972): pp. 745–53; N. D. Chechulin, *Ocherki po istorii russkikh finansov v tsarstvovanie Ekateriny II* (St. Peterburg: Senatskaia Tipografiia, 1906), pp. 207–8; V. I. Pokrovskii, ed., *Sbornik svedenii po istorii i statistike vneshnei torgovli Rossii* (St. Peterburg: M. N. Frolov, 1902), I, p. xxvi; Konstantin N. Lodyzhenskii, *Istoriia russkago tamozhennago tarifa* (St. Peterburg: V. S. Balashev, 1886), pp. 81–93; Robert E. Jones, "Getting the Goods to St. Petersburg: Water Transport from the Interior 1703–1811," SR 43:3 (Fall 1984): p. 429; M. Ia. Volkov, "Tamozhennaia reforma 1753–1757 gg.," IZ 71 (Moskva, 1962): pp. 134–57; Zenon E. Kohut, *Russian Centralism and Ukrainian Autonomy. Imperial Absorption of the Hetmanate 1760s–1830s* (Cambridge, MA: Harvard Ukrainian Research Institute, 1988), pp. 50–55, 74; N. L. Rubinshtein, "Vneshniaia torgovlia Rossii i russkoe kupechestvo vo vtoroi polovine XVIII v.," IZ 54 (Moskva, 1955): pp. 346–47; John P. LeDonne, *Ruling Russia: Politics and Administration in the Age of Absolutism 1762–1796* (Princeton: Princeton University Press, 1984), pp. 34, 89, 107, 151, 195, 205, 210–11, 284; Ian Blanchard, "Russia. Money Supply," Chap. 4 in *Russia's Age of Silver.' Precious-Metal Production and Economic Growth in the Eighteenth Century* (London: Routledge, 1989), pp. 163–93, in particular, pp. 186–89; and for the interpretation concerning a physiocratic emphasis among some of the elite, Leonard, *Reform and Regicide*, Chap. IV, passim.
8. Quoted in Leonard, *Reform and Regicide*, p. 101.

Volkov was an exponent of a freer trade policy for Russia. His position, originally outlined in a memorandum of December 19/30, 1760, to the Commission on Commerce and subsequently supported by the Conference at the Imperial Court on January 5/16, 1761, called for major tariff reform and liberalization of commercial regulations. He opposed monopolies and other limitations on commerce such as trading companies: "We have been clearly shown that trade dominated by the state tends to decline." To that end a freer export policy that would enhance the exploitation of Russia's trade with Persia, Central Asia, and other eastern lands was recommended. The Persian Trading Company would have to be abolished because it thwarted Russia's trade with Persia. Greater numbers of tradesmen and artisans would have to settle in the region of Astrakhan, whose port would be made duty free. To improve Russia's balance of payments it would not be enough to restrict the importation of luxuries because that would cause Russia's trading partners to reduce their purchase of Russian merchandise. Great Britain was an "already useful market" for Russian goods but it "could be made even more useful." Russian bar iron was cheaper than Sweden's and more of it could be exported and sold on the British market. Restrictions on timber exports should be lifted because exports were not the chief cause of deforestation compared to the lumber used for domestic construction and fuel. The solution to the conservation of forests should be found in improved management of resources. Similarly, restrictions on the export of raw materials should be relaxed: "We should begin freely to export agricultural products, including those which have been prohibited from time to time. We should remove all duties from items or subject them to light duties that cannot be felt."[9] The Senate, however, did not make Volkov's reform package the law of the land and in February 1761 sent it back to the Commission on Commerce, characterizing it as "not useful to the state."[10]

Although Empress Elizabeth did not obtain a consensus for immediate innovations in Russia's trade policy, she nevertheless pursued her general program of economic rehabilitation by con-

---

9. Quoted ibid., pp. 102–3. See also Wallace L. Daniel, *Grigorii Teplov: A Statesman at the Court of Catherine the Great* (Newtonville, MA: Oriental Research Partners, 1991), pp. 34–37. For background documentary information concerning Russia's extensive commercial involvement in Astrakhan, the Persian and Central Asian trade, and even Russia's interests in India, see R. V. Ovchinnikov and M. A. Sidorov, comp., *Russko-Indiiskie otnosheniia v XVIII v. Sbornik dokumentov* (Moskva: Nauka, 1965); Roger P. Bartlett, *Human Capital. The settlement of foreigners in Russia 1762–1804* (Cambridge: Cambridge University Press, 1979), pp. 149–55; and Daniel, pp. 117–23, 132.
10. Leonard, *Reform and Regicide*, p. 102; and Troitskii, pp. 182–85.

centrating on Russia's overseas commerce with Great Britain. Anglo-Russian trade dated from 1555, when Muscovy voluntarily extended privileges to English merchants who, in search of a northeast passage to the spices and gold of the East, formed themselves into the Muscovy Company, commonly known as the Russia Company. This new direct trade with Muscovy allowed English cloths to move eastward and Muscovy's naval stores and furs to move westward. Some silks and spices were included in this trade when the Company in 1557 established a trading link with Persia through Muscovy but it was cut off by the Ottoman Empire in 1580. During the seventeenth century the English trade with Muscovy fluctuated and expanded, including attempts to make Muscovy a market for English colonial tobacco reexports. It was, however, only in 1734 that reciprocal benefits were stipulated in an Anglo-Russian Commercial Treaty.[11] This protected commercial channel to England laid the groundwork for the accelerated pace of Russian exports before the Seven Years' War and an English commodity trade balance deficit with Russia. From 1710–14 to 1750–54 England increased its imports from Russia three and one half times while its exports to Russia actually declined ("Official Values").[12]

---

11. For early English merchant involvement in Muscovy, see T. S. Willan, *The Muscovy Merchants of 1555* (Manchester, 1953); idem, *The Early History of the Russia Company 1553–1603* (Manchester, 1956); Robert Brenner, *Merchants and Revolution. Commercial Change, Political Conflict, and London's Overseas Traders, 1550–1653* (Princeton: Princeton University Press, 1993), p. 13 and passim; Jacob M. Price, *The Tobacco Adventure to Russia. Enterprise, Politics, and Diplomacy in the Quest for a Northern Market for English Colonial Tobacco, 1676–1722. Transactions of the American Philosophical Society*. New Series, Volume 51, part 1 (Philadelphia: American Philosophical Society, 1961); Geraldine M. Phipps, *Sir John Merrick: English Merchant-Diplomat in Seventeenth-Century Russia* (Newtonville, MA, 1983); Paul Bushkovitch, *The Merchants of Moscow 1580–1650* (Cambridge: Cambridge University Press, 1980); and Hermann Kellenbenz, "The Economic Significance of the Archangel Route (from the late 16th to the late 18th century)," JEEH 2:3 (Winter 1973): pp. 541–81. The standard work on the 1734 treaty is Douglas K. Reading, *The Anglo-Russian Commercial Treaty of 1734* (New Haven: Yale University Press, 1938). For additional commentary see Sandys et al. to Bute, May 18, 1762, PRO CO 389/31, pp. 26–61, and PRO SP 91/69, ff. 234–45.

12. SOT, f. 94. The "Official Values" compiled in SOT are from the Custom House, London, the Office of the Inspector-General of Imports and Exports and are not the same as the current real values. There is a considerable history and historiography devoted to the reliability of using the "Official Values" as a measure of the British balance of trade. Ralph Davis has written succinctly about it in *The Industrial Revolution and British Overseas Trade* (Leicester: Leicester University Press, England, 1979), pp. 77–86. For estimated totals of Russian foreign trade turnover between 1742 and 1755, see Troitskii, p. 185ff.

It is not always clear whether British imports from Riga were recorded under Russia or East Country in British customs records. Charles Whitworth's introduction to his compilation of England's and Scotland's imports and exports states:

> THE EAST COUNTRY ... Under this name was formerly included Norway, Sweden, Denmark, Poland, Prussia, and all other parts of the Baltic, exclusive of Narva. For with these countries was granted the exclusive trade to a com-

If Russia viewed its overseas commerce with Great Britain as a means of improving its overall economic and financial condition, Great Britain's reason for continuing its commercial relationship with Russia was its recognition of the scarcity of the resources that it needed to sustain its economic growth. Either Britain's colonies did not possess sufficient quantities of the resources that Britain's accelerating economy demanded, or their resources were so awkwardly or distantly located that the cost of shipping them to the mother country made them uneconomical. Of necessity Great Britain had to look elsewhere for its requirements, in particular, to Russia.

Great Britain's international trade was quite significant during the period following the Seven Years' War. It is estimated that by the time of the Napoleonic Wars it "reached the all-time high of one third of world trade."[13] The contribution of foreign trade to Britain's economic development has been clearly delineated. For the entire eighteenth century Britain's foreign trade

> expanded more rapidly than did the economy as a whole; export industries in particular grew more rapidly than the industrial sector ... and many of the external signs of economic growth during the century were closely associated with the expansion of foreign commerce.... Imports expanded 523 per cent and the exports and re-exports used to pay for the imports 568 per cent and 906 per cent.... Since the population of England increased by only 257 per cent, it is clear that foreign trade became increasingly important on a per capita basis. There is little doubt that foreign commerce over the century became a more important component of national income.

The rate of growth of this expansion from 1700 to 1740 was 0.8 percent annually; for the period 1740–1770 it doubled to 1.7 percent annually; and, for the last thirty years, from 1770 to 1800, the rate of growth was 2.6 percent annually.[14]

---

      pany of merchants, called the Eastland company.
          The first factory settled at Elbing; but ... the factory removed, and settled at Dantzig, Riga, and Königsberg.

SOT, pp. xx–xxi.
13. Carl-Ludwig Holtfrerich, "Introduction: The Evolution of World Trade, 1720 to the Present," in *Interactions in the World Economy. Perspectives from International Economic History*, ed. Carl-Ludwig Holtfrerich (New York: New York University Press, 1989), p. 4.
14. For a concise evaluation of the scholarship, see R. P. Thomas and D. N. McCloskey, "Overseas trade and empire 1700–1860," Chap. 5, in *The Economic History of Britain since 1700. Volume 1: 1700–1860*, ed. Roderick Floud and Donald McCloskey (Cambridge: Cambridge University Press, 1981), pp. 87–102, quotations from pp. 87–90, passim. See also François Crouzet, *Britain ascendant: Comparative studies in Franco-British economic history* (Cam-

Russian and British customs data for the century clearly support Britain's growing dependence upon Russian goods and Russia's dominance of the trade balance. Britain imported huge quantities of cereals, hemp, flax, varieties of linen and timber, bar iron, bristles, hides, isinglass, tallow, pitch and tar, and dozens of other products grown or manufactured in Russia. During the first half of the century Russia's portion of England's total imports in English pounds sterling ("Official Values") amounted to an annual average of 3.3 percent, ranking it seventh, behind Germany, Holland, Italy, Portugal, Spain, and Turkey, but ahead of Sweden, Denmark-Norway, Flanders, Venice, and France, in that order. In the second half of the century Russia's share of England's total imports jumped to an annual average of 8.1 percent, eclipsing all other European countries. During the Seven Years' War, Great Britain's commodity trade balance deficit with Russia deepened. Britain's imports from Russia increased by about 37 percent while its exports to Russia fell by about 40 percent ("Official Values").[15] For the period following Russia's withdrawal from the Seven Years' War, from the reign of Peter III in 1762 to the end of Catherine II's reign in 1796, Russia's percentage of England's and Scotland's total imports was an annual average of 8.5 and 17.3 percent, respectively.[16] However, when compared to the "computed current real values," as detailed in Chapter XIV, the "Official Values" significantly underestimate Russia's contribution to Great Britain's foreign trade.

On September 18/29, 1759, Empress Elizabeth publicly recognized Russia's potential reward from the British trade when she declared a temporary extension of the Commercial Treaty, which by then had expired.[17]

The more aware the Russians became of their potential economic benefits, the more natural was their wish to obtain greater advantages in a new treaty with Britain. Grand Chancellor Vorontsov viewed such a treaty as a way not only to extricate the country from the effects of the war but also to place it on the path toward economic recovery. In the spring of 1760 he informed Robert Keith, the British envoy in St. Petersburg, that he was preparing a

---

bridge: Cambridge University Press, 1990), especially Chap. 6, "Towards an export economy: British exports during the Industrial Revolution."
15. Compiled and computed from SOT, ff. 94, 180.
16. Compiled and computed from SOT, ff. 81–181.
17. "Note pour Monsieur de Keith, Envoyé Extraordinaire de Sa Majesté le Roy de la Grande Bretagne," St. Petersburg, September 18/29, 1759, in Carmarthen to Fitzherbert, November 17, 1786, PRO FO 97/341, No. 15.

revision of the Anglo-Russian Commercial Treaty of 1734.[18] In the summer of 1761 the Russian court firmly reminded Keith in writing that Elizabeth had extended the life of the expired treaty only "temporarily." This prompted Keith to write his ministry: "For the Present, we are at the Mercy of this Court for the few Advantages we enjoy in Commerce."[19]

The Russian court sought to exploit this state of affairs in the terms set forth in the revised treaty draft of July 1761. If the British accepted it, they would be deprived of their former privileges and the Act of Navigation would be undermined.[20] The Earl of Bute, who headed the British ministry at that time, exercised the long-established procedure of calling for the recommendations of the Board of Trade, Customs, the Advocate General and the governing body of the Russia Company. However, he cautioned his colleagues: "As Our Merchants, trading with Russia, have been very Sollicitous to obtain a Renewal of their Old Treaty, Their Privileges seeming, since the Expiration of it, to lye at the Mercy of the Czarina."[21]

The Board of Trade's immediate response was that the Russian "Proposal was so adverse both in principle and Provision, and so inconsistent with that Equity and Moderation which was shewn by that Court in concluding the Treaty of 1734."[22] Yet, despite the apparent urgency, nearly a year elapsed before the Board produced its own counter-project. This kind of procrastination eventually led Edmund Burke to describe the Board as "a sort of

---

18. Keith to Bute, July 12/23, 1761, PRO SP 91/68.
19. "Note pour Monsieur de Keith, Envoyé Extraordinaire de Sa Majesté le Roi de la Grande Bretagne," dated in St. Petersburg, July 3/14, 1761, in Keith to Bute, July 12/23, 1761, PRO SP 91/68, ff. 265–66; and Bute to Board of Trade, Aug. 17, 1761, PRO CO 388/49.
20. Keith to Bute, July 12/23, 1761, PRO SP 91/68, and "Projet d'un nouveau Traité de Commerce avec la Cour d'Angletterre [sic]," PRO SP 91/68, ff. 253–64.
21. Bute to Board of Trade, Aug. 17, 1761, PRO CO 388/49; Bute to Keith, Aug. 18, 1761, PRO SP 91/68; JCTP, 1759–1763, pp. 208–10, 216, 277; "Projet d'un nouveau Traité de Commerce avec la Cour d'Angletterre [sic]," in Keith to Bute, July 12/23, 1761, PRO SP 91/68, ff. 254–62, and enclosures; Sandys et al. to Bute, May 18, 1762, PRO CO 389/31, pp. 26–106, and PRO SP 91/69, ff. 234–61 (which contains a variation on spelling); William Wood to John Pownall, Sept. 19, 1761, and Hay to Pownall, May 13, 1762, PRO CO 388/49; and RC, VII, passim.

The Board of Trade influenced the shaping of British commercial policy and in many ways was the counterpart to the Russian Commission on Commerce. For a general characterization of its functions and a summary of its origins, see Sir Frank Lee, *The Board of Trade* (*The Stamp Memorial Lecture delivered before the University of London on 11 November 1958*) (London, 1958), pp. 7–11; Mary Patterson Clarke, "The Board of Trade at Work," AHR 17 (1912): pp. 17–43; and Anna Lane Lingelbach, "The Inception of the British Board of Trade," AHR 30 (1925): pp. 701–27.
22. Sandys et al. to Bute, May 18, 1762, PRO CO 389/31, p. 68 and PRO SP 91/69, f. 247 (which contains variations in spelling).

temperate bed of influence; a sort of gently ripening hothouse where eight members of Parliament receive salaries of a thousand a year, for a certain given time, in order to mature at a proper season, a claim to two thousand."[23]

Although the Board did not consider all the issues of disagreement between the British and Russians to be of equal significance, it did believe that when the Russians changed the wording of an article even slightly, inequities were created. In particular, the Russian re-wording of Article II alarmed the British. The 1734 treaty had provided for freedom of commerce and navigation in the European territories of both countries, where it had already been permitted, or where it would thereafter be allowed to subjects of other nations. The Russian draft omitted the latter phrase. The very underpinning of the 1734 treaty—the most-favored-nation principle—was being modified. The British feared that the Russians might in the future grant other nations trading privileges that would not be accorded British merchants. The British, therefore, reinserted the original wording of Article II in their counter-project.[24]

In Article III the Russians described the carriers that could enter freely into one country or the other. Vessels belonging to Russian subjects but built in a foreign country and crewed by "two-thirds" Russian nationals were defined as "Russian ships." The British opposed this definition because it contravened the Act of Navigation, which required three-fourths.[25]

The British had several objections to the Russian draft of Article IV, which dealt chiefly with the goods (grown or manufactured) of Russia, Great Britain, and Asia that could be imported into or exported from either of the contracting countries. The Russians had omitted two significant provisos: that Russian subjects had to pay the same duties as British subjects on the goods they exported from Russia to Great Britain and that these goods were not forbidden by some law in force in Great Britain.[26] The Russia

---

23. Quoted in Lingelbach, p. 702.
24. "Projet d'un nouveau Traité de Commerce avec la Cour d'Angletterre [sic]," in Keith to Bute, July 12/23, 1761, PRO SP 91/68, ff. 254, 261–62; Sandys et al. to Bute, May 18, 1762, PRO CO 389/31, pp. 69–73, and PRO SP 91/69, ff. 247–49.
25. "Projet d'un nouveau Traité de Commerce avec la Cour d'Angletterre [sic]," in Keith to Bute, July 12/23, 1761, PRO SP 91/68, f. 255; Sandys et al. to Bute, May 18, 1762, PRO CO 389/31, pp. 73–74, and PRO SP 91/69, f. 249; and William Wood to John Pownall, Sept. 19, 1761, PRO CO 388/49.
26. "Projet d'un nouveau Traité de Commerce avec la Cour d'Angletterre [sic]," in Keith to Bute, July 12/23, 1761, PRO SP 91/68, ff. 255–56; Sandys et al. to Bute, May 18, 1762, PRO CO 389/31, pp. 74–76 and PRO SP 91/69, ff. 249–50; and William Wood to John Pownall, Sept. 19, 1761, PRO CO 388/49.

Company also had a long-standing grievance that it wanted addressed in Article IV. According to Russian regulations, a British merchant who had purchased merchandise in Russia could only export it himself. If he were unable to ship it abroad, he was precluded from selling his cargo locally and would thus have to sustain the loss personally. The Company wanted to amend this article to permit British merchants to sell the goods they had purchased wholesale to any person.[27]

The Russians provoked the British when they addressed the meaning of contraband of war and the procedures employed for the release of ships seized as prizes. The Russians raised these issues because of Great Britain's ruthless behavior toward neutral shipping on the high seas during the Seven Years' War and, in doing so, anticipated the crisis of the Armed Neutrality some twenty years later. Maintaining that warlike materials such as arms and munitions were contraband of war, the Russians stipulated the exclusion of several items from the contraband list; among these were grains, flax, hemp, masts, building timber, sails, and tar (Article IX). While these assertions upset the British and led them to claim that the above-mentioned items had always been considered contraband, they could not have forgotten that Article IV of the "Commercial and Maritime Treaty Between Charles II of Great Britain and The States-General of The United Provinces," concluded in 1674, had provided for just such a "Free List, not contraband." The Russians also wanted ships seized as prizes discharged immediately at the request of the resident envoy (Articles X and XII). The British declared that their law required the admiralty court to conduct hearings before prize ships could be released.[28]

A final point of contention arose from complaints of British merchants in Riga. In their view, the regulations of an Ordinance proclaimed by the Magistracy of Riga in 1756 and confirmed by the Russian Senate in 1760 was contrary to the 1734 treaty, because the status of the British merchants in Riga would not be equal to that of the British merchants in St. Petersburg. The Magistracy of Riga argued that the Ordinance was exempt from the 1734

---

27. Sandys et al. to Bute, May 18, 1762, PRO CO 389/31, pp. 83–85; and PRO SP 91/69, ff. 252–53.
28. "Projet d'un nouveau Traité de Commerce avec la Cour d'Angletterre [sic]," in Keith to Bute, July 12/23, 1761, PRO SP 91/68, ff. 257–60; Sandys et al. to Bute, May 18, 1762, PRO CO 389/31, pp. 77–80, and PRO SP 91/69, ff. 250–51; and Hay to Pownall, May 13, 1762, PRO CO 388/49. See also RC, Oct. 1758, VII, pp. 43–49. For the text of the Anglo-Dutch Treaty of 1674 see Daniel A. Miller, *Sir Joseph Yorke and Anglo-Dutch Relations, 1774–1780* (The Hague: Mouton, 1970), Appendix C, pp. 118–19.

treaty, noting that when Riga had capitulated to Russia, it had reserved to itself commercial privileges set forth by the King of Sweden in an Edict of 1690 and, therefore, the Magistracy could make alterations in the trading operations of Riga as it deemed proper. The British ministry viewed the 1734 treaty as encompassing the entire Russian Empire and wanted this explicitly stated in the new treaty.[29]

By the time the Board of Trade had completed its extensive commercial counter-project on May 18, 1762, Russian history had dramatically changed with the death of Empress Elizabeth on December 25, 1761/January 5, 1762, the ascension of her nephew as Emperor Peter III and the withdrawal of Russia from the war. The British were apprehensive that their political and commercial policies with Russia would be adversely affected.[30]

In almost revolutionary fashion during his brief reign of six months (January–July, 1762), Peter III set out to reform Russia's major policies and institutions. From the standpoint of other European powers, Peter's most startling decisions were his armistice with Frederick II of Prussia, which effectively removed Russia from the Seven Years' War, and his subsequent negotiation for a political alliance with Prussia, which had the potential of another so-called "Diplomatic Revolution." Among Peter's important domestic reforms, he sought to make significant changes in commercial policy. As a sign of his determination he appointed Elizabeth's advisor Dmitrii V. Volkov as his private secretary, who championed freer trade, the favorable balance of trade, the expansion of exports, and the reduction of luxury imports.[31]

In the decree of March 28/April 8, 1762, which the Senate subsequently labeled "The Destruction of Monopolies and the Establishment of complete Freedom of Trade,"[32] Peter relaxed restrictions on grain exports, abolished the Persian Trading

---

29. JCTP 1759–1763 (May 1762): pp. 276–77; "The Humble Petition of His Majesty's trading Subjects residing at Riga [ca. 1761–1762]," "His Majesty's Ordinance of Trade newly augmented and confirmed For The City of Riga. Dated At Stockholm the 10th of October 1690," and "The Ordinance of The Magistrates of The Imperial City of Riga For The foreign Traders arriving and residing here," published in Riga, June 17, 1756, in PRO CO 388/50; Sandys et al. to Bute, May 18, 1762, PRO CO 389/31, pp. 91–98 and PRO SP 91/69, ff. 254–57; and RC, VII, pp. 125, 127, 136–39, 141. See also Bartlett, pp. 85–94; and Edward C. Thaden, Chaps. 1 and 2 in *Russia's Western Borderlands, 1710–1870* (Princeton: Princeton University Press, 1984).
30. PRO CO 388/49; Sandys et al. to Bute, May 18, 1762, PRO CO 389/31, pp. 88–91, and PRO SP 91/69, f. 254; and JCTP 1759–1763, p. 22.
31. Leonard, *Reform and Regicide*, Chap. IV, passim.
32. Ibid., p. 108, and PSZ SV/11,489.

Company, and advocated the export of agricultural commodities, Ukrainian cattle, timber, and forest products.[33]

Peter III reportedly favored the renewal of the Anglo-Russian commercial treaty but he had little time to do anything about it because a *coup* on June 28/July 9 brought Grand Duchess Catherine to the throne.[34] In justifying her claim to the throne, Catherine II publicly warned in a manifesto about the "danger there was to the whole Russian state." Denouncing Peter for withdrawing from the Seven Years' War, she stated: "Russian glory, brought to its height by the victory of arms and by much bloodshed, has now fallen in complete enslavement to the enemy by the conclusion of peace."[35]

Politicians at both the Russian and English courts were unable to predict whether Catherine II would continue the policies of her predecessors. The commentary of C. F. Schwan, a German officer residing in the Russian capital, summed up the situation:

> As soon as Peter III ascended the throne, he became great and wise in all respects. [Everyone] talked about his superb spirit; he was credited, in fact, for virtues he did not possess. . . . After all, was he not the grandson of Peter the Great? No sooner had his misfortune [deposition] occurred, than . . . he was nothing but a traitor, unworthy of occupying his ancestral throne.[36]

---

33. Ibid., pp. 106–7 et seq. For the geo-economic circumstances of the grain trade see Jones, pp. 416–17, 424–25; B. N. Mironov, "Eksport russkogo khleba vo vtoroi polovine XVIII-nachale XIX v.," IZ 93 (Moskva, 1974): pp. 149–88; and Edmund Cieslak, "Aspects of Baltic Sea-borne Trade in the Eighteenth Century: The Trade Relations between Sweden, Poland, Russia and Prussia," JEEH 12:2 (Fall 1983): pp. 239–70.

Until 1758 the Persian trade in silk and precious metals had been profitable for Russia, but after the Persian Trading Company was established in Astrakhan it lost money and precipitated many complaints from local merchants. Peter III believed that such companies are "a refuge for bankrupt merchants," and he sought to improve Russia's trade in that region by abolishing the company and setting up a commercial consulate there. Leonard, *Reform and Regicide*, p. 107.

34. Bute to Keith, May 26, 1762, PRO SP 91/69; Keith to Grenville, June 11/22, 1762, Swallow to Bute, May 27/June 7, 1762, and Grenville to Swallow, July 6, 1762, PRO SP 91/70.

35. Quoted in Leonard, "The Reputation of Peter III," p. 263. Also see p. 266.

36. Quoted ibid., p. 275.

# Chapter II
# Catherine II and the Formulation of Commercial Policy

In the wake of the *coup* that brought Catherine II to the throne, the courtiers who had surrounded her, Peter III, and Elizabeth, vied for influence and power. The process that determined "the political core of the ruling class" who would assist Catherine in governing Russia encouraged rivalry for political turf. Although Catherine, as supreme legislator and supreme judge, "had the authority to appoint all senators, presidents, members and procurators of all colleges and other central agencies. . . . they had to be selected from a pool of candidates submitted by the ruling families."[1]

This was evident among those who came to shape Catherine's domestic and foreign trade policy. For example, the Vorontsov family, which over the decades had consolidated its position in the top stratum of the ruling class and maintained that distinction to the end of Catherine's reign, played a significant role in guiding Russia's commercial policy. Roman I. Vorontsov (1707–1783), the grand chancellor's brother, became director of the College of Audit in 1766. His son, Alexander Romanovich (1741–1805), served as the Russian ambassador to Great Britain and the United Provinces. He also participated in extraordinarily important agencies that developed fiscal and commercial policies: as a president of the College of Commerce (1773–1794); as a member of the Third Department of the Senate (1779–1793), which had administrative and judicial responsibilities for the southern and western borderlands; as a member of a commission in 1783 to reduce Russia's enormous deficit; and as "the architect of the new tariff" that same year. Ironically, the leading member of the family, Grand Chancellor Mikhail I. Vorontsov (1714–1767), was experiencing a rapid

---

1. John P. LeDonne, *Ruling Russia: Politics and Administration in the Age of Absolutism 1762–1796* (Princeton: Princeton University Press, 1984), quotation on p. 24. Also see pp. 25–30, passim.

loss in real authority. As if to punctuate this decline, Catherine rehabilitated his enemy, the man whom Vorontsov had succeeded as grand chancellor, Count Aleksei P. Bestuzhev-Riumin, who at age seventy-nine had not lost his political ambition. By using his past friendship with Catherine and her favorites, the Orlov brothers, Bestuzhev-Riumin soon promoted himself as a power broker in the Anglo-Russian negotiations and competed with Nikita Panin for power at the court.[2]

Through his political acumen, Nikita Ivanovich Panin (1718–1783) soon became supreme in foreign affairs and thereby also became Catherine's spokesperson in commercial policy. However, he was never granted the title of grand chancellor. Panin's family dated its service to the court back to Ivan the Terrible. Nikita Panin himself had long served with merit in the diplomatic service. In 1760 Elizabeth appointed him *Ober-Hofmeister* and tutor to the son of the aspiring rulers of Russia, Peter and Catherine. In particular, he cultivated a rewarding association with Catherine, becoming a conspirator in the *coup* that brought her to the throne.[3]

Catherine ruled these men and autocratic Russia with her swift mind and captivating personality. While still chancellor under Elizabeth, Bestuzhev-Riumin had observed that Catherine exhibited "more steadiness and resolution" than anyone else at court,[4] qualities she would continue to exhibit after becoming empress. But to express her sovereign will Catherine had to cope with councils, commissions and cadres as well as decrees, memoranda and reports produced by an ever-expanding officialdom not free of corruption and inefficiency. It has been argued that Catherine sought to rule Russia within "the prevailing political and administrative notions of continental Europe," which Marc Raeff calls "cameralism and mercantilism."[5] Other historians assert that phys-

---

2. Ibid., quotation on p. 271 and see pp. 4–5, 25, 32, 35n, 60, 65, 105–6, 208, 211, 237, 261, and 359; and David L. Ransel, *The Politics of Catherinian Russia: The Panin Party* (New Haven: Yale University Press, 1975), pp. 105–6.
3. Ransel, "Nikita Panin's Role in Russian Court Politics of the Seventeen Sixties: A Critique of the Gentry Opposition Thesis" (Ph.D. diss., Yale University, 1968), pp. 158–94; idem, "The 'Memoirs' of Count Münnich," SR 30:4 (Dec. 1971): pp. 843–52. For Panin's career and activities in particular, see idem, *The Politics;* and LeDonne, pp. 26, 31, 57, 60, 86, 150–51, 211, 242.
4. Williams to Holderness, Sept. 21/Oct. 2, 1755, PRO SP 91/61.
5. Marc Raeff, "Uniformity, Diversity, and the Imperial Administration," in *Osteuropa in Geschichte und Gegenwart: Festschrift für Günther Stökl zum 60. Geburtstag,* ed. Hans Lemberg, et al. (Köln Wien: Bohlau Verlag, 1977), pp. 98–104; and idem, *The Well-Ordered Police State: Social and Institutional Change Through Law in the Germanies and Russia, 1600–1800* (New Haven and London: Yale University Press, 1983). Cf., Isabel de Madariaga's review of Raeff's book in the *London Times* publication *Times Literary Supplement,* November 25, 1983, p. 1326.

iocratism was influential in Russia at this time.⁶ But Catherine was above all a pragmatist.

In order to increase state revenues and accelerate Russia's recovery from the disastrous consequences of the Seven Years' War, Catherine formulated her early economic policies at the same time that she contemplated governmental reorganization. She had somehow to reform the inefficient financial administration she had inherited.⁷

She recruited the able services of Prince Alexander Alexeevich Viazemskii (1727–1793) as procurator general of the Senate in 1764, entrusting to him the most important fiscal affairs, especially the compilation of a unified state budget—never before achieved. But even under Viazemskii's powerful leadership such an accomplishment would not take place until the 1780s.⁸ In commercial affairs Catherine followed Elizabeth's and Peter III's policies by placing her own signature on acts freeing trade. Not only did she reaffirm the abolition of the Persian Trading Company, but she also terminated the official state trade with China, thereby opening the eastern markets to all Russian merchants. She ended farm leases in seal fisheries and tobacco and returned N. P. Shemiakin's customs farm, which had functioned ineffectually since 1758, to state control.⁹

Nikita Panin best summed up Catherine's program when, in the fall of 1762, he reportedly told the French envoy in St. Petersburg:

> The interests of Russia require that she occupy herself for many years only with the general reestablishment of all parts of her

---

6. Carol S. Leonard, *Reform and Regicide. The Reign of Peter III of Russia* (Bloomington: Indiana University Press, 1993), pp. 23, 25–27, 52, 54–56, 88, 92, 98, 104, 112, 114, 165; P. H. Clendenning, "The Economic Awakening of Russia in the Eighteenth Century," JEEH 14:3 (Winter 1985): pp. 443–71, passim; and Wallace L. Daniel, *Grigorii Teplov: A Statesman at the Court of Catherine the Great* (Newtonville, MA: Oriental Research Partners, 1991), pp. 35, 70, 98.
7. James A. Duran, Jr., "The Reform of Financial Administration in Russia during the Reign of Catherine II," CASS/CSS 4:3 (Fall 1970): p. 485, and see pp. 486–96.
8. James E. Hassell, "Catherine II and Procurator General Vjazemskij," JGO 24:1 (1976): pp. 23–30.
9. Leonard, p. 110; LeDonne, p. 265; Konstantin N. Lodyzhenskii, *Istoriia russkago tamozhennago tarifa* (St. Peterburg: V. S. Balashev, 1886), pp. 108–9; N. D. Chechulin, *Ocherki po istorii russkikh finansov v tsarstvovanie Ekateriny II* (St. Peterburg: Senatskaia Tipografiia, 1906), pp. 208–10; S. M. Troitskii, *Finansovaia politika russkogo absoliutizma v XVIII veke* (Moskva: Nauka, 1966), pp. 32, 85–87, 184; V. I. Pokrovskii, ed., *Sbornik svedenii po istorii i statistike vneshnei torgovli Rossii* (St. Peterburg: M. P. Frolov, 1902), I, pp. xxv-xxvi; N. L. Rubinshtein, "Vneshniaia torgovlia Rossii i russkoe kupechestvo vo vtoroi polovine XVIII v.," IZ 54 (Moskva, 1955): pp. 346, 359; Clifford M. Foust, "Russia's Peking Caravan, 1689–1762," *The South Atlantic Quarterly* 67:1 (Winter 1968): pp. 108–24; and idem, *Muscovite and Manda-*

internal administration which is in such a state of disorder as to demand prompt remedies, and in order to accomplish this he [Panin] proposed to occupy himself particularly with commerce.[10]

It becomes important, therefore, when considering the Anglo-Russian relationship immediately following the Seven Years' War, not to overemphasize the potential for a solely defensive alliance between the two countries. To focus on the relationship simply in terms of political diplomacy, as some British historians have done, is to miss the total picture. In fact the Anglo-Russian relationship operated on at least four concurrent and interdependent levels: the renewal of the political and defensive alliance (concluded in 1742, reaffirmed and extended in 1755, and subsequently allowed to lapse);[11] the commercial treaty of 1734, the privileges of which had already been extended by Elizabeth and the renewal of which was still in negotiation; the commercial intercourse between the two countries in commodities, which had not ceased even during the Seven Years' War when they were on opposite sides; and the continuum of informal English cultural influences in Russia that some scholars would call "Westernization."

The ongoing Anglo-Russian dialogue in the 1760s was as much about the substance of commerce as it was about the language of a political and defensive alliance. It is also evident that the distinction between commercial and political considerations became blurred almost immediately, for Catherine saw the opportunity of fitting trade into her more complex diplomatic strategy—the "Northern System" of European alliances. Catherine had conceived of this system long before she ascended the Russian throne, and it has been misunderstood. Already in 1756 Catherine had confessed to the British envoy that "nothing can save Europe [against the southern alliance of Roman Catholic powers] but an alliance between England, Russia, Prussia, & Holland, to which some of the German Princes might be invited to accede, and declared She will attempt it—the Day She is in a Situation to do it."[12]

---

rin: *Russia's Trade with China and Its Setting, 1727–1805* (North Carolina: The University of North Carolina Press, 1969).
10. Quoted in Ransel, "Nikita Panin's Role," p. 120. See also idem, *The Politics*, p. 145.
11. For texts of these treaties, CT, III, pp. 30–36, and 37–47, respectively.
12. Williams to Holderness, Sept. 17/28, 1756, PRO SP 91/64; and Herbert H. Kaplan, *Russia and the Outbreak of the Seven Years' War* (California: University of California Press, 1968), p. 105. For different discussions of the "Northern System," its origins and meaning, see N. D. Chechulin, *Vneshniaia politika Rossii v nachale tsarstvovanie Ekateriny II, 1762–1774* (St. Peterburg, 1895); K. Rahbek Schmidt, "Wie ist Panins Plan zu einem Nordischen System entstanden?" *Zeitschrift für Slavistik* 2:3 (Berlin, 1957): pp. 406–22; David M. Griffiths, "Rus-

Now, as empress, Catherine was in a position to transform this much-talked-about idea into policy and she requested that Panin execute it.

Catherine intended to use the commercial negotiations to involve Britain in her political machinations in Poland under the "Northern System." Both Peter III and Catherine II anticipated the crisis that would result from a vacancy on the Polish throne and made "the Polish Question" a key feature in their political and military foreign policies as well as in their respective relationships with Frederick II and George III. Such policies would be played out in the Polish-Lithuanian Commonwealth, Denmark, and Sweden. In 1762 Peter III had taken the initiative to ally himself with Frederick II, a negotiation in which the future of the Polish-Lithuanian Commonwealth figured prominently. Catherine continued Peter's policy by concluding that alliance with Frederick in 1764, wherein both parties agreed upon the fate of the Commonwealth. The sources are clear on this matter.[13] But it must be emphasized that the principles underlying the "Northern System" of non-Roman Catholic powers were inconsistently applied and never did work as originally conceived. This has led one scholar to write about it as "Die Fiktion des 'Nordischen Systems.'"[14] When the resolution of the Polish question in 1764 destroyed its applicability and when Russia chose to ally with Roman Catholic Austria to dismember Poland, the "Northern System" came to an end.[15]

---

sian Court Politics and the Question of an Expansionist Foreign Policy under Catherine II, 1762–1783" (Ph.D. diss., Cornell University, 1967), pp. 1–5, 28–53, 317–18 and passim; idem, "The Rise and Fall of the Northern System: Court Politics and Foreign Policy in the First Half of Catherine II's Reign," CASS/CSS 4:3 (Fall 1970): pp. 547, 551–54, 556–57; and Michael Roberts, "Macartney in Russia," EHR Supplement 7 (1974).

13. Kaplan, *The First Partition of Poland* (New York: Columbia University Press, 1962), pp. 1–45. Cf., H. M. Scott, "Great Britain, Poland and the Russian Alliance, 1763–1767," HJ 19:1 (1976): pp. 53–74; idem, "Frederick II, the Ottoman Empire and the Origins of the Russo-Prussian Alliance of April 1764," ESR 7 (1977): pp. 153–75; Herbert Butterfield, "Review Article: British Foreign Policy, 1762–5," HJ 6:1 (1963): pp. 131–40 (where he critiques *The Fourth Earl of Sandwich: Diplomatic Correspondence 1763–1765*, ed. Frank Spencer); M. S. Anderson, "Great Britain and the Russo-Turkish War of 1768–74," EHR 69 (1954): pp. 39–58; idem, "Great Britain and the Russian Fleet, 1769–70," SEER 31 (December 1952): pp. 148–63; Roberts, *Splendid Isolation, 1763–1780* (Reading: University of Reading Press, 1970); and idem, "Macartney in Russia."

14. Andreas Bode, *Die flottenpolitik Katharinas II und die konflikte mit Schweden und der Türkei (1768–1792)* (Wiesbaden: Otto Harrassowitz, 1979), pp. 57–59.

15. Kaplan, *The First Partition of Poland*, pp. 13, 17, 21, 35, 184, 187. Cf., Derek McKay and H. M. Scott, "Partition Diplomacy in Eastern Europe, 1763–1795," Chap. 8 in *The Rise of the Great Powers 1648–1815* (London: Longman, 1983); Derek Beales, "Foreign Policy During the Co-regency: I. The Eastern Question, 1765–1776," Chap. 9 in *Joseph II*, vol. 1 of *In the Shadow of Maria Theresa 1741–1780* (Cambridge: Cambridge University Press, 1987); and Jerzy Topolski, "Reflections on the first partition of Poland (1772)," *Acta Poloniae Historica* 27 (1973): pp. 89–104.

Late in 1762 the Earl of Buckinghamshire, the newly appointed British envoy to Russia, arrived in St. Petersburg with instructions to negotiate not only a treaty of commerce, but also a separate political alliance with Russia. Despite the repeated assurances Buckinghamshire received from Vice Chancellor Alexander M. Golitsyn of Catherine's belief in the "Expediency of setting a new Treaty of Commerce," and from Chancellor Vorontsov that the Empress wanted only "to cultivate the Friendship of England in every Instance," the Englishman sensed that the trade negotiation would "necessarily be spun out to a very great length."[16]

The Englishman's instinct was correct. Catherine was in no hurry to accommodate the British, for she believed Russia could gain more by keeping them waiting and wondering. By January 1763 the British ministry, fearing the possible loss of a trade treaty altogether, was willing to settle for a simple renewal of the 1734 treaty, which meant dropping its additional demands and disregarding the objections of the Russia Company. The Board of Trade had made it clear that the 1734 treaty contained more privileges for British commerce and merchants than any other treaty of its kind with any country. Had it not been for the "Wisdom" and "Moderation" of the Russian court, "the Trade of Great Britain in general and the Interest of every Individual engaged in that Commerce, would have stood in a very precarious Situation."[17]

But before the British could act effectively, their political and commercial positions with the Russians grew much worse. In mid-February Golitsyn told Buckinghamshire that the French had proposed a plan for a trade treaty and that Russia might find it easier to deal with them.[18] The French had attempted for many years to penetrate the lucrative Russian commercial market on a large scale and since 1761 had endeavored to initiate trade discussions with Russia. But for several reasons—the absence of their total commitment, the inadequacy of a French merchant fleet for the Baltic trade, the exigencies of the Seven Years' War, their dependence

---

16. Kaplan, *The First Partition of Poland*, pp. 15–16; Keith to Grenville, July 26/Aug. 6, King George III to Buckinghamshire, Aug. 13, 1762, PRO SP 91/70; and Buckinghamshire to Halifax, Nov. 11/22, 18/29, Dec. 9/20, and to Grenville, Nov. 4/15, 1762, PRO SP 91/70 and Buckinghamshire to Halifax, Jan. 8/19, 1763, PRO SP 91/71.
17. Halifax to Board of Trade, Jan. 19, 1763, PRO SP 91/70; JCTP 1759–1763 (Jan. 27, 1763): p. 329; JCTP 1759–1763 (Feb. 3, 1763): pp. 332–33; and Halifax to Buckinghamshire, Feb. 25, 1763, with enclosures, PRO SP 91/71. See also George Morley to Board of Trade, Nov. [25], 1762, PRO CO 388/50; Halifax to Buckinghamshire, Nov. 26, 1762, PRO SP 91/70, ff. 204–7; and Sandys et al. to Halifax, Feb. 11, 1763, PRO CO 389/31, pp. 136–52.
18. Buckinghamshire to Halifax, Feb. 10/21, 1763, PRO SP 91/71; and Kaplan, *The First Partition of Poland*, p. 17.

upon intermediaries carrying their commerce—they had not been successful. France and Russia had severed diplomatic relations from 1748 to 1756, had become allies during the Seven Years' War, and had a modest amount of trade in Russia's favor. The French now were of the two-fold opinion that if Russia would grant them significant trading privileges, they could not only emancipate Russia from English commercial domination, but also make the Atlantic and Mediterranean into a major thoroughfare for Franco-Russian commerce. French entrepreneurs had proposed establishing sugar refineries and ship-building projects in Russia and Russian Francophiles had tried to sell Ukrainian tobacco on the French market. In 1758 Senator Peter Shuvalov had obtained a twenty-year monopoly to export hundreds of thousands of poods per annum of Ukrainian tobacco to France. Thus, from mutual political and commercial interests Russia and France had the potential for forming compatible foreign policies that might rival Russia's relationship with Great Britain.[19]

French machinations were sufficient to pique Buckinghamshire but at the end of February 1763 Vorontsov told him that the Dutch and the Swedes had also submitted proposals for commercial treaties.[20] Given the historic antagonisms in the Baltic, the British had relatively less to fear from Russia entering into formal commercial arrangements with the Swedes and the Dutch than they did from a Franco-Russian alliance. Except for Sweden's occasional need for Russian grain, the similarity of Sweden's and Russia's natural resources actually made them competitors for the British market. A Swedish-Russian trade agreement would more likely codify whatever exchange normally took place between their respective Baltic ports, than create an obstacle to their overseas commercial interests. Politically, Swedish-Russian relations would always be complicated and uneasy, if not hostile.[21]

Regarding Dutch trade with Russia, despite Peter III's most

---

19. One pood = 36.1 pounds; 63 poods = one ton. Frank Fox, "French-Russian Commercial Relations in the Eighteenth Century and the French-Russian Commercial Treaty of 1787" (Ph.D. diss., University of Delaware, 1968), pp. vi-45, 53–103, 129–30, 187–92, 229, 234, 356–57, 379–80; Walther Kirchner, "Franco-Russian Economic Relations in the Eighteenth Century," and "Ukrainian Tobacco for France," Chaps. 7 and 8 in *Commercial Relations between Russia and Europe 1400 to 1800. Collected Essays* (Bloomington: Indiana University Press, 1966); Paul W. Bamford, *Forests and French Sea Power, 1660–1789* (Toronto, 1956); Jay L. Oliva, *Misalliance: A Study of French Policy in Russia during the Seven Years' War* (New York: New York University Press, 1964); and Kaplan, *Russia and the Outbreak of the Seven Years' War.*
20. Buckinghamshire to Halifax, Feb. 10/21, and 17/28, 1763, PRO SP 91/71.
21. Roberts, *British Diplomacy and Swedish Politics, 1758–1773* (Minnesota: University of Minnesota Press, 1980), pp. xiii-63, passim; and idem, "Macartney in Russia," pp. 40 et seq.

recent overture to set up a commercial consulate in the Hague, which might have led eventually to a commercial agreement,[22] a leading scholar on the subject has concluded:

> Dutch trade with Russia . . . was neither hindered nor protected by political, military or commercial treaties. The direct trade between Russia and Holland . . . was determined as far as Holland was concerned primarily by commercial motives. . . . The decline in Russian exports to Amsterdam must be looked at as part of the decreased demands of Holland herself.[23]

Whatever the outcome of these new negotiations, the British would suffer delays in getting what they wanted from Russia. Vorontsov informed Buckinghamshire that although Catherine would now appoint a commission to deal with all four countries, the British would continue to enjoy their privileges under the old treaty until they agreed upon a new one. By threatening to deal simultaneously with four countries, Catherine enhanced her commercial position and reinforced her political aspirations.[24] This was made clear when the Russians showed Buckinghamshire their draft *Article Séparé* for a defensive political alliance that stated that only after its ratification would both countries enter into negotiations for a treaty of commerce.[25] An astonished British ministry immediately instructed Buckinghamshire "in express Command from the King [that] the Treaty of Commerce and that of the defensive Alliance should go on *pari passu*, & be signed together."[26]

Russia's deliberate procrastination reflected more than just the expected political and diplomatic maneuvering in a treaty negotiation. Catherine had been searching for more efficient ways of overseeing the management of Russia's foreign trade. It was only at the end of 1763 that she appointed a group of officials who had a wide range of governmental experience both at home and abroad to the three central administrative agencies vested with foreign trade authority: the Customs Chancery, the College of Commerce and the newly appointed Commission on Commerce.

On November 20/December 1, 1763, Catherine appointed

---

22. Leonard, p. 97.
23. Jake V. T. Knoppers, *Dutch Trade with Russia from the Time of Peter I to Alexander I. A Quantitative Study in Eighteenth Century Shipping* (Montréal: Interuniversity Centre for European Studies, 1976), I, p. 333.
24. Buckinghamshire to Halifax, Mar. 3/14, and Apr. 14/25, 1763, PRO SP 91/71; and Frank Spencer, ed., *The Fourth Earl of Sandwich: Diplomatic Correspondence, 1763–1765* (Manchester: Manchester University Press, 1961), p. 44.
25. Buckinghamshire to Halifax, July 11/22, and Aug. 11/22, 1763, PRO SP 91/72.
26. Sandwich to Buckinghamshire, Sept. 23, 1763, PRO SP 91/72.

Ernst von Münnich (1707–1788), the son of Field Marshal Count Burchard Christoph von Münnich, as director of Customs (1763–1784) to bring efficiency and honesty to the system. The younger Münnich, well educated in political economy, natural law, languages, history and philosophy, had previously held diplomatic positions in Turin and Paris. He would also serve as president of the College of Commerce (1763–1773) and as president of the Commission on Commerce (1766–1779). Beginning on December 8/19, Catherine appointed, among others, Münnich, Senator Ivan I. Nepliuev, Timofei von Klingstedt and Grigorii N. Teplov to the new Commission on Commerce with Prince Iakov P. Shakhovskoi as president. Like its predecessor, the new Commission was "to undertake a comprehensive study of Russian trade," determine how to exploit more natural resources for state use, and foster the activities of Russian and foreign merchantry to increase Russian state revenues. The Commission was also to formulate a new customs tariff.[27]

Catherine had thus engaged in a full-scale attempt to expand Russia's commerce and increase its agricultural and industrial productive capacities. To this end the Commission on Commerce would serve as an investigative as well as a policy-formulating body. Inevitably, the Commission would have to estimate the quantity and quality of native-grown and manufactured products for the international market, designate what industries should be encouraged or discouraged from entering into that competition and determine the effective demand for foreign goods in Russia. A whole range of related matters would require review, including the balance of commodity trade and payments, merchant fleets and sea transport, the most-favored-nation principle, currency exchanges, banking, credit and investment. The Commission would also have a direct influence on the negotiations of the Anglo-Russian trade

---

27. Daniel, *Teplov : A Statesman*, pp. 46–50, 63–73; idem, "Grigorii Teplov and the Conception of Order: The Commission on Commerce and the Role of the Merchantry in Russia," CASS/CSS 16:3–4 (Fall-Winter 1982): pp. 410–31; and idem, "Russian Attitudes Toward Modernization: The Merchant-Nobility Conflict in the Legislative Commission, 1767–1774" (Ph.D. diss., University of North Carolina, 1973), pp. 60–72. Earlier standard works on the Commission are N. N. Firsov, *Pravitel'stvo i obshchestvo v ikh otnosheniiakh k vneshnei torgovle Rossii vo tsarstvovanie imperatritsy Ekateriny II* (Kazan': Imperatskii Universitet, 1902); and A. S. Lappo-Danilevskii, "Die russische Handelskommission von 1763–1796," in *Beiträge zur russischen Geschichte*, ed. Otto Hötzsch (Berlin, 1907). See also N. Kaidonov, ed., *Sistematicheskii katalog delam kommissii o kommertsii i o poshlinakh, khraniashchimsia v Arkhive Departmenta Tamozhennykh Sborov* (St. Peterburg: V. Kirshbaum, 1887), pp. iii-iv; Lodyzhenskii, pp. 108–13; Pokrovskii, I, pp. xxvi-xxvii; Chechulin, *Ocherki*, pp. 210–12; Ransel, *The Politics*, pp. 145–50; LeDonne, p. 208; and Clendenning, pp. 451 et seq.

treaty because Münnich and Teplov would soon join Golitsyn and Panin as negotiators.[28]

Opinion is mixed as to who was actually responsible for the plan to set up the new Commission at this time, allowing for a range of interpretations identifying those principally responsible for Catherine's commercial policy. David L. Ransel believes Nikita Panin to have been the likely person due to his central position at court and because the principal members of the Commission were his supporters. Catherine had written to Panin in 1763 requesting his opinion about the instructions to be given to the Commission. Panin's reply to Catherine, according to Ransel, is contained in an anonymous document that in style and substance apparently became the basis for State-Secretary Grigorii N. Teplov's final instruction to the Commission:

> What did Your Majesty find upon ascending the throne?
> . . .
> Revenue sources were exhausted and in the greatest disorder; the treasury was plundered and stripped of a large part of its most reliable income; various revenue sources without any reason were handed over to private persons; administration was weak; *there was no direct commerce with foreign nations and no provision for carrying wares in Russian bottoms;* domestic trade everywhere [labored] under restraints; manufactures instead of being improved languished in their former backwardness; currency of a single denomination varied in worth. The Commerce Collegium and Magistracy were staffed by ignoramuses; and finally there was a nobility burdened by useless luxury, and a people suffering from oppression and poverty for lack of trades or the ability to perform crafts.[29]

Consistent with this viewpoint, Panin and Teplov stressed the importance other countries attached to Russia's raw materials, manufactured goods and agricultural products. They wanted a freer domestic and foreign market and recommended the further abolition of restraints on commerce. Panin desired an improvement in Russia's internal waterway and communications system. Teplov sought better conditions for Russian merchant trade—a larger merchant fleet and efficient credit and banking facilities.

---

28. Panin informed Buckinghamshire of this in January 1764, Buckinghamshire to Sandwich, Dec. 30, 1763/Jan. 10, 1764, PRO SP 91/73.
29. Quoted in Ransel, *The Politics*, p. 146. For additional details, see Daniel, *Teplov: A Statesman*, pp. 39–40, 149–50. For the text of the anonymous document that Ransel attributes to Panin, see Firsov, pp. 4–5, 52–56. For Teplov's role, see Daniel, "Grigorii Teplov," pp. 410–31, passim.

"Russian merchants have been nothing but hirelings, or better, carters for foreign merchants."[30] Both Panin and Teplov favored tariffs to protect domestic industry and reduce luxury imports.

Undoubtedly, Panin represented a forceful presence at Catherine's court. But Ransel overemphasizes Panin's influence in this case. For example: "The Panin-Teplov memoirs were, however, the first guidelines of this nature to form the basic instruction of an official government commerce commission, and in this sense their 'liberal' program marked an altogether new departure."[31] And in another place:

> Despite the opposition of some members of the commerce commission, the empress gave the Panin-Teplov program her full support, another example of the Panin party's influence in this period. She repeated the program's fundamental principles a couple of years later in her *Instruction* to the Legislative Commission and followed this modified free trade policy through most of her reign. While the question of whether this policy had appealed to her earlier or had won her approval as a result of the Panin's party's arguments must remain speculative, once again the proposals of Panin and his collaborators had proved determining.[32]

However, this question is not really as speculative as Ransel suggests. Because several Commission papers remained unsigned, attribution of respective authorship and programs could always remain a puzzle.[33] But more important, the so-called "Panin-Teplov program" cannot be considered "an altogether new departure" when it reflects the continuity of tradition and decrees that Elizabeth and Peter III had previously established in Russian commercial affairs.

It was also significant that although the Commission members committed themselves to the improvement of Russia's commerce, there was no consensus as to how that could best be accomplished. Given their varied backgrounds and longtime service careers, it would have been unreasonable to expect such a consensus among the members. According to Wallace L. Daniel, author of a biography on Grigorii Teplov and a scholar of Catherine II's Commission on Commerce: "The Commission's activities and the opinions of

---

30. Ransel, *The Politics*, pp. 147–48, et seq., quotation on p. 148.
31. Ibid.
32. Ibid., p. 150.
33. Wallace Daniel affirms: "These papers are unsigned and their authors are difficult to establish, either by textual criticism or by their handwriting." "Grigorii Teplov," note 4 on p. 411, and see also pp. 410–12.

its members concerning the national economy provide a view of the conflicting economic ideas of the leading Russian officials during the first years of Catherine's reign."[34] A brief review of their backgrounds and views on commerce will illustrate this.

Senator Prince Iakov Petrovich Shakhovskoi (1705–1777) had earned his senior status on the Commission with a long history of state service and personal integrity. In preceding reigns, he was *Ober-Procurator* of the Synod (1742–1753), quartermaster general of the Army (1753–1760), and procurator general of the Senate (1760–1762). Under Catherine he held the position of director of the College of Audit (1762–1766) in addition to presiding over the Commission on Commerce. Although Shakhovskoi applauded the past achievements of Russian foreign trade, viewing Russia's "national wealth primarily in terms of its trade balances" and arguing "that the state's primary need lay in the expansion of foreign commerce," he believed that the merchantry could be encouraged to greater productivity in the export trade if they were provided with additional incentives. For example, they could be awarded social and legal privileges, such as "the 'right' to ride in carriages, to wear swords, and to own household serfs." Because "growth of capital" was the key element in the expansion of trade, merchants should also be given increased banking and credit support.[35]

Like Shakhovskoi, the Pomeranian Timofei von Klingstedt (1710–1786) recognized that Russia's favorable balance of trade confirmed that the economic structure was strong and that no essential changes were necessary. He had entered Russian service in 1740, rising to the rank of state counselor, and later would be appointed vice president of the College of Justice for Lifland, Finland, and Estland Affairs (1765–1770). During the Seven Years' War he had been an administrator of finances in Königsberg, and thereby brought to the Commission specific and practical experience in economic matters. He was not persuaded that new outlets for Russian goods, the establishment of consuls abroad, or the shipping of Russian commodities in Russian vessels would necessarily benefit Russia. Russia really needed improved credit facilities, greater honesty in customs practices and an effective training

---

34. Ibid., pp. 410–13, passim, especially note 10.
35. Robert E. Jones's "Introduction" to *Zapiski Kniazia Iakova Petrovicha Shakhovskago, 1705–1777* (Cambridge, MA: Oriental Research Partners, 1974; originally published in St. Petersburg, 1872), pp. ii–iii; Daniel, *Teplov: A Statesman*, p. 69; idem, "Russian Attitudes," pp. 69–71; and idem, "Grigorii Teplov," p. 425.

program for Russian merchants that would include efficient accounting practices and the study of foreign languages.[36]

> We should try to send young people to foreign countries, especially to England, Holland, France, Italy, and Germany, then to make them serve an apprenticeship in commerce for seven years; if the sending of fifty of them results in ten people who are capable in commercial affairs such policies can be judged successful.[37]

Most striking, however, was von Klingstedt's belief that Russia should form a good commercial relationship with Great Britain, as P. H. Clendenning informs us:

> Closely allied to his agricultural ideology was his belief that Britain, above all other countries, applied itself in thought and money to the improvements of all facets of agriculture. Indeed, he considered that it would be difficult to determine whether commerce or agriculture had contributed more fully to that country's enviable national prosperity. He was firmly convinced that Russia's course was to establish close trade links with Britain, which would lead to greater benefits from the latter's vast pool of techniques and experience.[38]

The perception that Russia's future economic growth would be the result of a well-integrated relationship between foreign and domestic trade was shared by Ernst von Münnich. Through his responsibilities in Customs and his diplomatic experience, he reasoned that there had to be more cooperation among those involved in industry, agriculture, and commerce. Aware that England had to import flax and hemp to sustain its textile industry, von Münnich believed that Russia, with its own abundant resources of flax and hemp, could manufacture enough linen for export to England. Moreover, whatever the difficulties were that prevented Russia from maximizing its overseas commercial potential, according to von Münnich, it was not so much the insufficient number of Russian built ships as the "insufficiency of skilled seamen" that had to be overcome.[39]

---

36. LeDonne, p. 271; and Daniel, "Russian Attitudes," pp. 66–69; and idem, "Grigorii Teplov," p. 425.
37. Quoted in Daniel, *Teplov: A Statesman*, p. 104.
38. Clendenning, quotation on p. 458, and see p. 447.
39. Daniel, "Russian Attitudes," pp. 71–72; idem, *Teplov: A Statesman*, p. 103; and Clendenning, pp. 458–59.

Another member of the Commission, Senator Ivan Ivanovich Nepliuev (1693–1773), dated his service back to when Peter I sent him abroad in 1716. His diplomatic missions at the Porte (from 1721) and his governorship of Orenburg (1742–1758) served him well as vice president of the College of Commerce. Nepliuev took a dim view of Russia's recent performance in commerce, reasoning that state revenues had declined because Peter I's initiatives had not been continued. He argued that more vigorous support should be given to the Russian merchantry by creating more favorable credit facilities and conditions for commerce and manufactures. Because high duties burdened both the merchantry and the population, he wanted the tariff revised and custom duties reduced. He also advocated building a domestic mercantile fleet to transport Russian goods abroad, thereby emancipating Russia from its dependency upon foreign carriers.[40]

The most gifted member of the Commission was State-Secretary Grigorii N. Teplov (1717–1779) whose writing skills and learning made him one of Catherine's most influential and adept advisers. Wallace Daniel states that Teplov served not only as "the Commission's executive director" but also as "Catherine's commercial adviser."[41] He had been educated abroad and in 1742 was named adjunct member of the Academy of Science. His early service as tutor, estate manager and personal aide to Hetman Kyrylo Rozumovs'kyi during Elizabeth's reign were later rewarded with positions of trust under Catherine. He was the architect of Catherine's policy of tightening Russia's control over the Ukranian tletmanate.[42] He wrote detailed reports on contemporary state policies, on commerce and on the merchantry.

Teplov was well qualified for his role in the Commission on Commerce not only for his erudition but also because he had practical commercial experience. For example, in 1763, with Catherine's approval, Teplov and four Tula merchants formed a Company of Merchants valued at 90,000 rubles to conduct Russian trading operations in the Levant.[43] During Elizabeth's reign, he had become involved with Senator Peter Shuvalov to sell Ukrai-

---

40. Herbert Leventer's "Introduction" to *Zapiski Ivana Ivanovicha Nepliueva (1693–1773)* (Cambridge, MA: Oriental Research Partners, 1974; originally published in St. Petersburg, 1872), pp. i-x, passim; and Troitskii, pp. 97–98.
41. Daniel, *Teplov: A Statesman*, pp. 48, 111.
42. Daniel, "Grigorii Teplov," pp. 414–15; and Zenon E. Kohut, *Russian Centralism and Ukrainian Autonomy. Imperial Absorption of the Hetmanate 1760s-1830s* (Cambridge, MA: Harvard Ukrainian Research Institute, 1988), pp. 93–96. For additional details on Teplov's early life and career, see Daniel, *Teplov: A Statesman*, Chap. 1.
43. For facsimile documents on this company, PRO SP 91/72, ff. 163–66, 208–9.

nian tobacco to the French, and this experience undoubtedly led Catherine in 1763 to give Teplov a monopoly to improve the cultivation of Ukrainian tobacco. In 1765 he published a book on the planting of tobacco. In subsequent years Teplov pursued both the development of Ukrainian tobacco and, more important, Russian trade connections with France via the Black and Mediterranean seas.[44] Neither a mercantilist or physiocrat, according to Wallace Daniel, Teplov believed that society

> had a life of its own and its true essence lay in the process of becoming rather than in objects that already existed. In building a strong national economy, government officials thus should not proceed by inventing principles and forcing the facts into them. The principles had to emerge from the facts, which were not the same for different lands and at different stages. Required in all aspects of the economy was not a dim awareness of the laws of nature, but a precise knowledge of these laws and a clear expression of them. Finding the connection between the individual and the whole in the economy, proceeding from the particular to the general, represented the task of economic science.[45]

Teplov wanted Russia's economy to equal the wealth and productivity of advanced European countries and believed that to accomplish this Russia had to study the methods practiced in those countries. Consequently, he suggested that the government should establish in St. Petersburg a "society of economists to study the publications of the most renowned writers," and observe English methods of increasing agricultural productivity. In criticizing the present state of Russian mercantile operations, Teplov argued that in Europe "craftspeople in the towns not only support themselves, but they also provide great profit [to the state]." Russia therefore needed to establish what he vaguely described as a "middle order of people between the nobility and the peasants," who would play an active role in commerce. The state had to improve craft production and exchange, and peasants had to be permitted to engage in handicrafts in the towns. He was pessimistic that the Russian merchantry in their present ignorance and traditional economic conservatism could meet the requirements of a state dedicated to prosperity for they "are without education, good faith, and credit."

---

44. Kirchner, "Ukrainian Tobacco for France," especially pp. 183, 186–87, 302, note 40; idem, "Franco-Russian Economic Relations in the Eighteenth Century," pp. 132–75, passim; Clendenning, pp. 460–61; and Daniel, *Teplov: A Statesman*, pp. 111–16.
45. Daniel, "Grigorii Teplov," p. 416.

Teplov believed that trade was a "special science" with its own laws of conduct. Therefore, Russian merchants had to be trained in bookkeeping, Russia's mercantile fleet expanded and the transportation of goods improved upon, and trustworthy credit facilities had to be established. The merchantry should be motivated by economic reward, not simply by social perquisites.[46]

These were the men whom Catherine entrusted with developing Russia's commercial policies and practices, the Anglo-Russian trade treaty and the tariff. But the British ministry neither perceived the magnitude of Catherine's commercial aspirations nor recognized the subtle changes in her decision-making process, both of which would affect the political and commercial Anglo-Russian relationship. Although the Earl of Sandwich, one of the principal foreign secretaries, was correct in cautioning Buckinghamshire that the Russians might use the commercial treaty "as a Bait to make the Treaty of Alliance go down, upon their own Terms," he was clearly wrong in exaggerating Britain's independence from Russian commodities. "We have such Resources within ourselves & Colonies," he said, "that we could, at any Time, direct that Branch of Trade into such a Channel as would totally deprive Russia of any Share in it."[47] The Russians could hardly be bluffed by such a threat.

Catherine was less successful politically than she was commercially with Great Britain. Her two-fold strategy was to prevent France and Austria from acting against her in Poland and to gain Prussia's and Britain's support for her actions there. But when she was about to conclude her defensive alliance in April 1764 with Prussia, which called for joint action in Poland, she could not obtain Britain's assistance in the event that Turkey attacked Russia over Poland.[48]

In March 1764 Golitsyn informed Buckinghamshire that trade talks would be delayed because of "the great immediate Attention, which was necessary given to the Affairs of Poland, & to some Difference of Opinion with Regard to the Manner of setting the general Method in which the Commissioners of Commerce were to proceed."[49] A month later Panin told Buckinghamshire that it would be impossible for Russia to conclude a treaty of com-

---

46. Ibid., pp. 418–31, passim; idem, "Russian Attitudes," pp. 71–86, passim; and idem, *Teplov: A Statesman.*
47. Sandwich to Buckinghamshire, Feb. 17, 1764, PRO SP 91/73. See also Spencer, p. 48n.
48. Kaplan, *The First Partition of Poland*, pp. 5–35; Spencer, pp. 46–49; and see supra note 12 in this chapter.
49. Buckinghamshire to Sandwich, Mar. 2/13, 1764, PRO SP 91/73.

merce with England "unless there was Appearance of England's being disposed to Unite Her Interests intimately with those of Russia and that no Treaty of alliance could really cement the Friendship of the Two Countries, unless [England] entered into the Views of Russia in the North, & particularly in Poland."[50] That spring, after several Polish magnates declared their support for Catherine's policies and royal candidate, thousands of Russian troops occupied vital centers in the Polish-Lithuanian Commonwealth with tens of thousands more stationed on the frontier. On April 11 Russia concluded an alliance with Prussia which contained an article on Russo-Prussian commercial relations. Thereafter, the French, Austrian, and Ottoman courts no longer threatened to interfere in Poland, and Britain continued its policy of non-involvement.[51]

The resolution of the Polish question, however, did not speed up the Anglo-Russian trade talks. In late September the Russian court informed Buckinghamshire that crucial changes had been made in the draft of the commercial treaty and that he would be expected to accept it and forward it to his court for ratification. When the Englishman refused, Panin told him that the negotiations would subsequently be handled by the Russian envoy in London and that Russia might discontinue the temporary extension of the commercial treaty, thereby abrogating Britain's most-favored-nation privilege.[52]

Thus, within two years of Russia's withdrawal from the Seven Years' War, fortified with self-confidence in her policies both at home and abroad, Catherine appeared poised to challenge the maritime power of Great Britain. The increasing annual volume of Russian commodity exports to Great Britain had resulted in favorable trade balances for Russia and revealed Russia's greater potential value to the British. The Russians had found the British market not only lucrative but eager for more Russian goods to satisfy its own economic growth. If its commercial policy with Great Britain was successful, Russia could hope someday to command as much respect at sea as it had on land.

---

50. Ibid., Apr. 2/13, 1764.
51. Kaplan, *The First Partition of Poland*, pp. 25–35.
52. Buckinghamshire to Sandwich, May 4/15, 12/23, 14/25, May 21/June 1, June 15/26, and July 2/13, 1764, PRO SP 91/73, Aug. 17/28, Sept. 14/25, Sept. 24/Oct. 5, Sept. 28/Oct. 9, and Nov. 9/20, 1764, PRO SP 91/74. See also Sandwich to Buckinghamshire, Aug. 3, 1764, PRO SP 91/73.

# Chapter III
# The Anglo-Russian Commercial Treaty of 1766 and the Russian Tariff of 1766/1767

Like his many other reforms, Peter the Great's tariff of 1724 was complex and far-reaching. It included many restrictions that not only protected Russia's commerce, industry and agriculture from foreign competition, but, in several ways unforeseen at the time, actually hindered some areas of Russia's economic development. Soon after the attempted implementation of the tariff, both foreign and domestic merchants complained about its regulations.

In 1731 a new tariff was proclaimed that, although still decidedly protectionist, was more straightforward and easier to execute than its predecessor. For more than twenty years this tariff guided Russia's domestic and foreign customs but it neither met the challenges of an expanding economy nor provided the revenues demanded by the state treasury.

In the 1750s Empress Elizabeth initiated a series of reforms intended to augment revenues through greater freedom in commercial transactions. In 1754 she abolished domestic customs duties that had long hindered internal economic development. In 1757 she issued a new tariff that governed both domestic and foreign customs. It was hoped that these reforms—as well as others in different branches of the economy—would accommodate the needs of the treasury to a greater extent and stimulate a more rapid growth of the economy. But the tariff contained unworkable provisions. Moreover, the Seven Years' War further burdened state finances and Elizabeth had to resort to additional expedients to solve the short-term problems facing Russia.[1]

---

1. There are several works dealing with Russian tariffs during the first half of the eighteenth century, among them; R. I. Kozintseva, "Ot tamozhennogo tarifa 1724 g. k tarifu 1731 g.," *Voprosy genezisa kapitalizma v Rossii* (Leningrad: Leningrad University Press, 1960), pp. 182–216; M. Ia. Volkov, "Tamozhennaia reforma 1753–1757 gg.," IZ 71 (1962): pp.

Soon after coming to the throne, Catherine II requested her newly appointed Commission on Commerce to devise a new tariff on the basis of five principles. First, smuggling should be prevented. Second, foreign goods needed by Russia should have a low duty while superfluous luxury goods should bear high duties or be prohibited. To encourage domestic manufacturing the third principle stated that raw materials should be preferred for import. Fourth, exportation of Russian materials, particularly manufactured products, should be encouraged while prohibiting or highly taxing the export of goods needed by Russia. The final principle advocated light or no duties on "every thing useful to the preservation of Health, & necessary for the convenience of Life."[2]

Russia had extreme difficulty in restricting the smuggling of luxury goods. The Commission, in the hope of reducing the flow of smuggled articles, decided to impose only a 10 percent duty on diamonds, jewelry, white lace, and certain drugs. Fruits, such as lemons, and chestnuts, would not be assessed a duty. It was assumed that the consumer who could afford such items would be more inclined to purchase them through legal, rather than clandestine, channels in order to avoid fines or imprisonment.

There were about 150 additional luxuries that were assessed a 20 percent duty. Those goods for domestic ornament or pleasure carried a 100 percent imposition. Among them were wines, gastronomic goods and exquisite edibles, expensive carpets, and large-sized mirrors. Those luxury articles that were in abundance in Russia were either assessed a duty of 200 percent or prohibited entirely. The list of prohibited imports was extensive.

This tariff appeared protectionist though it could be argued that it was much less so than those of 1724, 1731, and 1757. The majority of imported goods carried a duty of either 12, 20, or 30 percent, with the last impost applied to about 25 percent of the total imports. There were, of course, several categories of imports that carried lesser imposts. Articles that could encourage the production of agricultural goods or handicrafts—seeds, roots, and dyes—could be imported duty free. Merchandise not produced in Russia but indispensable to Russians would be mildly taxed at 4 percent or would be duty free. Included in this category were edibles and drugs. Imported goods that would eventually be re-exported to Asia would have an impost of 6 percent. A 12 percent

---

134–57; and Konstantin N. Lodyzhenskii, Chaps. IV-VII in *Istorii russkago tamozhennago tarifa* (St. Peterburg: V. S. Balashev, 1886).
2. TARIFF, f. 218.

duty was applied to indispensable products that could be, but as yet were not, produced in Russia. All unsorted merchandise, such as worsteds, broad cloths, colored and uncolored woolen yarns, blankets, and unfinished hats would be imported with an imposition of 15 percent. A 20 percent duty was levied on imports that were of either poorer quality or of higher quality but of lower price than their supposed Russian equivalents. Those imported goods that were already being produced effectively in Russia were assessed a 30 percent duty to encourage domestic production. There were 206 articles in this category alone.[3]

Russian export policy was more difficult to execute because the dynamics of the international market required a more sensitive approach. Those domestic goods considered indispensable to Russia, but of insufficient supply, were either prohibited from export, controlled by specific export regulations or assessed a duty of 200 percent.[4] Generally speaking, raw, rather than processed, agricultural products would be assessed higher export duties. However, if there were sufficient supplies in Russia of a particular grain for which there was a considerable foreign demand, a modest duty would be imposed. This principle also applied to any other agricultural product that might be competitive on the international market. Russia also exported large quantities of manufactured goods, and the rate of duty charged would vary with the quality of the products themselves and the effective foreign demand for them. Naval stores, metallurgical products, caviar, salted fish and meat, skins, and furs were in abundance in Russia and in great demand overseas.[5]

The tariff was flexible when it came to the currency with which to pay duties. Duties would be collected in rixdollars but in lieu of rixdollars, one half could be paid in silver money and the other half in any current coin of Russia, with one rixdollar equivalent to 125 kopeks. However, a special privilege was accorded to Russian subjects "if they export, or import their goods in Russian ships, either for sale in Russia, or for re-exportation to Persia, they shall pay at a rate of 90 Cop. only per Rixdollar."[6] This amounted to a 28 percent abatement and was intended to encourage both the development of the Russian merchantry and the Russian maritime

---

3. Lodyzhenskii, pp. 110–19 et seq. A full discussion of Russia's imports is provided in Chap. XIII, below.
4. TARIFF, f. 263; and Lodyzhenskii, pp. 110–19 et seq.
5. TARIFF, ff. 263–64; and Lodyzhenskii, pp. 116–17 et seq.
6. TARIFF, ff. 264–65.

fleet. This provision not only set a precedent, it also undermined the Anglo-Russian treaty of commerce.

The Commission submitted its recommendation along with a supplementary report of its proceedings to Catherine in May 1766. After making some minor alterations Catherine sent the tariff to the Senate for promulgation with the Commerce College responsible for periodically reviewing the tariff. The tariff applied to all custom houses of the Russian empire except those of Astrakhan, Orenburg, Troitskii, Siberia, Fredrikshamn, Viborg, and Baltic Sea ports which were governed by special ordinances. The new tariff was published on September 1/12, 1766 and was officially translated into Dutch, English, French, and German. The tariff would take effect beginning March 1/12, 1767 and was intended to change one important branch of the Russian economy, commerce.[7]

The Commission had discussed the impact that British trade would have on the Russian economy, and two of the authors of the tariff, Ernst von Münnich and Grigorii N. Teplov, were plenipotentiaries to the Anglo-Russian treaty. In January 1765 Baron Heinrich von Gross, Russia's envoy to Great Britain, delivered Catherine's New Year's greeting to George III in the form of a new draft of the commercial treaty. By its direct language the commercial treaty intended to circumvent Britain's long-established maritime policy and practice and loosen Britain's grip on Russian trade. Gross said that henceforth his court intended to treat Russian merchants preferentially because they were responsible for Russia's stronger trade position in the world market.[8]

Two salient features of the 1734 treaty were conspicuously absent from the Russian draft: the requirement that both Russian and British merchants pay the same duty rates on exports (Article IV), and the permission for British merchants to engage in the Persian trade across Russian territory (Article VIII).[9] Concerning the former, Ambassador Buckinghamshire had informed his court that the Russian plenipotentiaries had been surprised to learn that the 1734 treaty provided for equal payment of export duties because, in the Russian view, that was "directly contrary to a positive law of Peter the Great's." The Englishman took this as Russia's intention not only "To abate the Dutys when Russ Merchants exported Goods in Russ Vessels; but now . . . they mean to shew that

---

7. TARIFF, ff. 217–65; and Lodyzhenskii, pp. 108–19.
8. Sandwich to Board of Trade, Jan. 11, 1765, PRO CO 388/52.
9. "Projet d'un nouveau Traité de Commerce avec la Cour d'Angletterre," PRO SP 91/75, ff. 1–8.

Indulgence to their own Subjects, even when they make use of English Vessels." This policy, he said, "must ultimately ruin the English Factory at Petersburg."[10]

The omission of the British privilege to trade with Persia across Russia would affect Great Britain differently. Prior to the treaty of 1734 the British had competed in this trade with Armenian merchants who were paying Russia only a 3 percent transit duty while British merchants were paying from 25 to 75 percent. Armenians exported raw silk and other products from Persia to the United Provinces where they, in turn, purchased all their woolen goods. Although British complaints were redressed in the 1734 treaty which gave them a most-favored-nation status and allowed them to pay only a 3 percent ad valorem charge, that arrangement did not last. Captain John Elton, who initiated the Anglo-Persian trade connection, entered the Shah's services and built a fleet for him on the Caspian. Russia's security was threatened and Empress Elizabeth canceled the British trading privilege. Now the Russians were apprehensive that there might be a recurrence of British political adventurism if they should again grant this commercial favor; also, they had committed themselves to keeping as much of this trade in the hands of their own merchants as possible. It was undoubtedly the latter reason that caused the Russians to reject the French in 1761 when they had requested the Persian trading privilege for themselves.[11]

Sandwich forwarded the Russian draft to the Board of Trade and requested from it an "extremely explicit" statement "both as to what may be wav'd, as well as what must be insisted upon," so that "His Majesty may judge, whether It will be more eligible, by dropping the Treaty entirely to subject the British Trader in Russia to the uncertain regulations of that Government, or to stipulate for them such Advantages, as we may be able to procure by dint of Negociation."[12] On March 1, after receiving the recommendations

---

10. Buckinghamshire to Sandwich, Sept. 14/25, Sept. 24/Oct. 5, Sept. 28/Oct. 9, and Nov. 9/20, 1764, PRO SP 91/74.
11. Sandwich to Board of Trade, Jan. 11, 1765, PRO CO 388/52. For general background see Douglas K. Reading, *The Anglo-Russian Commercial Treaty of 1734* (New Haven: Yale University Press, 1938); Sandys et al. to Bute, May 18, 1762, PRO CO 389/31, pp. 26–61, and PRO SP 91/69, ff. 234–45; and Frank Fox, "French-Russian Commercial Relations in the Eighteenth Century and the French-Russian Commercial Treaty of 1787" (Ph.D. diss., University of Delaware, 1968), p. 99. For background to Armenian trade, see Frédérick Mauro, "Merchant Communities, 1350–1750," in *The Rise of Merchant Empires. Long-Distance Trade In The Early Modern World, 1350–1750*, ed. James D. Tracy (Cambridge: Cambridge University Press, 1990), pp. 270–74; and Roger P. Bartlett, *Human Capital. The settlement of foreigners in Russia 1762–1804* (Cambridge: Cambridge University Press, 1979), pp. 149–54, passim.
12. Sandwich to Board of Trade, Jan. 11, 1765, PRO CO 388/52.

of the Russia Company and the Advocate General, the Board returned a lengthy, detailed report to Sandwich.¹³ On March 15, 1765, Sandwich instructed George Macartney, the newly appointed envoy-extraordinary to Russia, to negotiate a commercial treaty with Russia.¹⁴

Contrary to the Board of Trade's recommendation, Sandwich would not insist on Russian merchants paying the same export duty rates that British merchants had to pay, provided the Russians permitted British merchants to sell wholesale to each other when they could not otherwise dispose of goods destined for export: "after Thirty Years Advancement in Commerce since the Conclusion of that Treaty, this beneficial Concession is more to be wished for, than expected from the Court of Russia." Concerning the Persian trade, Sandwich believed that British merchants could be prevented from entering into "military" adventurism.¹⁵

In Articles X and XI the Russian court provoked Great Britain over principles governing neutral shipping and contraband of war. The British rejected those articles because they would establish the policy, "Free Ships Make Free Goods." Russia had excluded naval stores from the contraband list and this was seen as a way for Russia to trade with Britain's enemies. The British insisted that the original 1734 treaty articles (XI and XII) covering these matters (with the addition of *masts* as "*Munitions de Guerre*") be reinserted into the new agreement.¹⁶ Great Britain also wanted the designation "Russian Empire" in the treaty to include the previously conquered Baltic territories of Narva and Riga because their continued exclusion would deprive British merchants in those ports of the same protection and trading privileges their brethren enjoyed in St. Petersburg.¹⁷

---

13. Hillsborough et al. to Sandwich, Mar. 1, 1765, PRO CO 389/31, pp. 281–338; John Pownall to Robert Nettleton, Jan. 15 and 25, 1765, and to James Marriott, Jan. 25, 1765, PRO CO 389/31, pp. 241–45, and 245–46, respectively; "The Memorial of the Governor Consuls and the Court of Assistants of the Russia Company" to Board of Trade, Feb. 1, 1765, and James Marriott to John Pownall, Feb. 4 and 12, 1765, PRO CO 388/52, and PRO SP 91/75.
14. Sandwich to Macartney, Mar. 15, 1765, PRO SP 91/75, ff. 87–193; and King George III to Macartney, Oct. 24, 1764, PRO SP 91/74, f. 88.
15. Sandwich to Macartney, Mar. 15, 1765, PRO SP 91/75, ff. 87–193; Hillsborough et al. to Sandwich, Mar. 1, 1765, PRO CO 389/31, pp. 296, 298, 304, 306; and Martin Mierop to Samuel Swallow, British Consul General, St. Petersburg, and Board of Trade, Dec. 21, 1762, Nov. 30, 1763, PRO CO 388/50, 51, respectively.
16. Hillsborough et al. to Sandwich, Mar. 1, 1765, PRO CO 389/31, pp. 285–87, 308, 310, 312, 314, 316.
17. Sandwich to Macartney, Mar. 15, 1765, PRO SP 91/75, ff. 90–91; and Buckinghamshire to Sandwich, Sept. 14/25, 1764, PRO SP 91/74. For details of the restrictions imposed on British merchants in Riga see British Factory, Riga, to Swallow, May 1/12, 1765, in Swallow

But for all these meticulous instructions, Macartney failed to live up to his ministry's confidence. The Russians ignored whatever objections the English envoy had raised and presented him with an ultimatum and a declaration explaining Catherine's steadfast position on the new trade treaty. Remarkably, Macartney signed the treaty on August 4/15, 1765, and sent it to London for ratification.[18] By doing this he contravened Sandwich's explicit instructions: "When You have brought the Treaty to the form and Substance, which shall be finally agreed to by the Court of Russia, You will transmit it to me without loss of Time, in Order to receive His Majesty's Pleasure with Respect to its absolute Conclusion."[19]

What could have caused Macartney to do this? Macartney's mission to Russia has been the subject of considerable study. Michael Roberts believes Macartney's original appointment, failures and problematic successes stem from the conventions of English politics.[20] Indeed Macartney was "a beginner" who was "unfamiliar with the forms of diplomatic business, and unschooled in the etiquette of a great court. Of commercial and economic matters it does not appear that he had any other grasp." But at the same time Roberts tells us that Macartney certainly was better suited to his task than Buckinghamshire had been: "His forthrightness, his businesslike enthusiasm, and his undeniable social talents proved more congenial to the Russians than the pompous self-sufficiency of the previous ambassador."[21] Yet it was Buckinghamshire who had the common sense and diplomatic forbearance to reject the previous Russian draft treaty, which he knew was incompatible with his court's policies.

Macartney's own, somewhat self-serving, explanation reveals why he allowed himself to become intimidated and why he felt compelled to contravene his instructions. He stated that if he had not signed the treaty there would have been "an immediate Revocation of the Privileges of our factory."[22] In signing he had been able to gain some but not all of the more important provisions the

---

to Grafton, Aug. 8/19, 1765, PRO SP 91/76, ff. 194–99, and Macartney to Grafton, Sept. 13/24, 1765, and enclosure, PRO SP 91/76, ff. 270–71. See also Bartlett, pp. 85–94, passim; and Edward C. Thaden, Chaps. 1 and 2 in *Russia's Western Borderlands, 1710–1870* (Princeton: Princeton University Press, 1984).
18. Macartney to Grafton, Aug. 8/19, 1765, PRO SP 91/76. See also Macartney to Sandwich, May 20/31, June 14/25, June 24/July 5, and July 5/16, 1765, PRO SP 91/75; and Macartney to Grafton, July 29/ Aug. 9, 1765, No. 6, PRO SP 91/76, ff. 119–21.
19. Sandwich to Macartney, Mar. 15, 1765, PRO SP 91/75, ff. 93–94.
20. Michael Roberts, "Macartney in Russia," EHR Supplement 7 (1974): p. 2.
21. Ibid., p. 17.
22. Macartney to Grafton, Aug. 8/19, 1765, No. 1, PRO SP 91/76.

British had wanted to include in the treaty. British merchants could pay their import and export duties in either rubles or rixdollars, or any other acceptable foreign currency that would be equal in value to a rixdollar at 125 kopeks. But the Persian trade would require yet a separate negotiation; British merchants in Narva and Riga could not be placed on an equal footing with their compatriots in St. Petersburg; and British merchants would not be allowed to sell wholesale to each other. Concerning the latter, although such exchanges might be carried on clandestinely, Macartney stated that the Russians would "by no means admit it as a legal Permission, fearing that we might become absolute Masters of the Market."[23]

Articles X and XI were rectified to satisfy British views on contraband of war and neutral shipping. Military stores—that is, specifically enumerated arms and munitions—were determined to be contraband of war but there was absolutely no mention of masts or naval stores as contraband in the treaty as subsequently ratified. However, during the Seven Years' War, the British had ignored the provisions of the treaties they had concluded governing neutral shipping, blockades, and contraband of war. The true test of the value of Articles X and XI would come some fifteen years later during the crisis of the Armed Neutrality when Catherine demanded that Great Britain honor every word in these two articles.[24]

Article IV also caused grief. After successfully obtaining the stipulation that Russian and British merchants would both pay the same custom duty rates on exports, Macartney allowed the Russians to insert the following reservation clause: *"Mais alors on se reserve de la Russie, en reciprocité de l'Acte de Navigation de la Grande Bretagne, la liberté de faire dans l'interieur tel arrangement particulier qu'il sera trouvé bon, pour encourager et etendre la Navigation Russienne [sic]."*[25] Macartney had also been given a declaration, authorized by Catherine but not part of the treaty document itself, stating that the British merchants residing in Russia would participate in all future internal commercial arrangements and would derive the same advantages as Russian merchants. The declaration, however, failed to elaborate just how that would be accomplished.[26]

---

23. Ibid., enclosures 4, 6, and 8.
24. Ibid., enclosures 4 and 8; CT, III, pp. 228–29. Cf., Roberts, p. 19 and note 9.
25. Macartney to Grafton, Aug. 8/19, 1765, No. 1, enclosures 4 and 8 (ff. 130–31).
26. Ibid., enclosures 4, 8, and 6 (Declaration dated in St. Petersburg, July 25/Aug. 5, 1765, ff. 119–21).

Macartney correctly perceived the ambitious course that Catherine had embarked upon in international commerce: "The Intentions of this Ministry are to render Russia, if possible, a Marine and a Commercial Power." He went on to explain how the Russian court hoped to succeed in this goal, and, therefore, he must have known of the tariff deliberations of the Commission on Commerce:

> To encourage their Navigation they propose to grant an Exemption from certain Duties, or rather a Diminution of the present Duties paid upon Merchandise exported from St. Petersburg on Board their own Ships.... This was a Scheme of Peter the First, and there is now actually an Ukaze of His dated 1718 in force granting such Encouragement; but it has been found hitherto without Effect, and I am persuaded will continue so in spite of all their Efforts of this Nature. So many interior Causes, not to mention others, such as Want of Money, Want of Industry, Want of Genius, etc. contribute to frustrate all their Plans, that all our intelligent Merchants here are convinced, it will be impossible to accomplish them.[27]

The British ministry was unanimous that the reservation clause would give Russia sole power to make unspecified domestic mercantile arrangements that would make the treaty ineffectual. The ministry denied the value of the explanatory declaration because it neither protected the interests of British merchants sufficiently nor had the legal force of the treaty, and therefore did not want to ratify the treaty as the Russians had presented it. The British were further perturbed that the Russians had, to borrow Frank Spencer's characterization, "desecrated the British mercantilist holy-of-holies, the Navigation Acts."[28] But due to the difficulties of the negotiations the ministry judged that, with the exclusion of Article IV, the Russian draft treaty was the best that Great Britain could obtain, for it met Britian's minimum requirement and it was better than no treaty at all.

Thus, the ministry was prepared to ratify the treaty immediately provided the Russian court would accept its revision of the declaration. First, the British wanted a guarantee that in any future commercial arrangement in Russia, as a consequence of the reservation clause of Article IV, British subjects would not only participate in such an arrangement but also would derive from it the same advantages accruing to Russian subjects. Second, that in

---

27. Ibid., enclosure 4.
28. Frank Spencer, ed., *The Fourth Earl of Sandwich: Diplomatic Correspondence, 1763–1765* (Manchester, 1961), pp. 57, 299.

such an arrangement the commerce of British subjects would not in any manner be limited or restrained. Finally, they wanted the declaration to have the same legal force as if it were inserted into the treaty itself—as Lord Chancellor Northington put it, the *sine qua non* for the treaty's ratification.[29]

Macartney continued to insist that his signing of the treaty was wise, for he had gained more in the face of a recalcitrant Russian court than anyone else had believed possible. If the British merchants in St. Petersburg were persuaded of that—with their independent "Public Acknowledgement of the intire & unreserved approbation of every article in this Treaty"—then why did the merchants in London not approve? Macartney trusted in the Russian declaration, which was for him as valid as the treaty itself, and further claimed that London's hesitancy would make the British vulnerable to the machinations of other nations. A refusal to ratify the treaty would leave the field wide open to Britain's competitors: "An Ukase would be immediately issued, by which the English factory would be deprived of their privileges & put upon the same footing as the other traders."[30]

The British revision of the declaration incensed Panin. Macartney described how Panin became offended when he learned what Britain wanted. Separating himself from the other plenipotentiaries, Panin said that only ministers signed declarations and he himself was "the Minister and he alone." He could not tolerate Britain's distrust of him or the Empress. If he or Catherine were "capable of breaking their word or departing from their engagements in any forms, whatever, it was not likely that any Declaration would bind them, be it even so solemnly made or unequivocally worded." If the British did not like the treaty with his accompanying explanatory letter, they could take such measures as they thought proper. However, there was one thing they could definitely count on: "The British factory here would be put exactly upon the same footing with the Merchants of other nations," that is, the British would lose their most-favored-nation status.[31]

Macartney had submitted a formal explanatory *Mémoire* along

---

29. Grafton to Lord Chancellor Northington, Sept. 27, 1765, PRO SP 91/76, ff. 211–12; and Grafton to Macartney, Sept. 29, 1765, and enclosure, PRO SP 91/76, ff. 234–48, especially f. 246. See also Northington to Grafton, Sept. 28, 1765, PRO SP 91/76, ff. 214–15.
30. Macartney to Grafton, Oct. 20/31, 1765, PRO SP 91/76, ff. 305–11; Swallow et al. to Macartney, Oct. 28/Nov. 8, 1765, in Macartney to Grafton, Nov. 5/16, 1765, No. 2, PRO SP 91/76, ff. 316–17.
31. Macartney to Grafton, Nov. 5/16, 1765, Nos. 2 and 3, PRO SP 91/76, ff. 312, 318, respectively.

with the British declaration to Panin, but the responding *Pro Memoria* (signed by Golitsyn and Panin under Catherine's authority) merely reflected the Russian court's earlier stand and Macartney did not accept it. Subsequently, on November 4/15, 1765, Panin responded with another *Pro Memoria* along with a formal declaration signed by Golitsyn and himself which, although closer to the sense of the British declaration, did not meet the requirement that it have the legal force of the treaty itself.[32] Macartney sent this declaration to London along with another perceptive evaluation of Russia's evolving self-image:

> This court rises hourly higher & higher in her Pride, & dazzled by her present Prosperity looks with less deference upon other powers & with more admiration on herself. Strengthened as they are by the alliances of Denmark & Prussia, Proud of having imposed a Monarch upon Poland & elated by their recent Success in Sweden, I am persuaded that we shall every day find them less moderate in their pretensions & more difficult in Negociations. It is therefore my Lord, my humble opinion . . . that the ratifications shou'd be exchanged as soon as possible.[33]

The British ministry was still dissatisfied. The whole matter had now become a matter of principle; Britain would only exchange ratifications if all four plenipotentiaries signed the declaration.[34]

In February 1766 rumors spread among the merchants in St. Petersburg that cancellation of the statutory commercial tie with the British was imminent. A clerk in Panin's office let it be known that an *ukaz* was being readied to revoke Empress Elizabeth's declaration that had prolonged the privilege of British merchants, and that Russian soldiers would soon be quartered in English houses. When confronted by Macartney, Panin told him that "it was impossible ever to Negotiate on equal Terms with the British Nation"; that "he must now look upon the Treaty of Commerce as absolutely annulled"; and, most revealing, "that many Persons had, very loudly, expressed Their Resentment of [Great Britain's] Attempt to enchain their Commerce. . . and that the general Opinion, of the whole Empire, was, to leave the Trade equally free to all Nations."[35]

---

32. Ibid., No. 2, ff. 289–90, and No. 3, ff. 319, 321–22, 325–27, 329.
33. Ibid., ff. 319–20.
34. Grafton to Macartney, Dec. 24, 1765, PRO SP 91/76, ff. 351–54.
35. Macartney to Grafton, Feb. 11/22 and Mar. 3/14, 1766, PRO SP 91/77, ff. 30–31, and 58, respectively.

In late March the British ministry gave up its demand that the four Russian plenipotentiaries sign the declaration, provided that the Russian court would strike the reservation clause from Article IV. Furthermore, Macartney was to make clear to the Russian court that its favorable balance of trade was dependent upon British commerce and that Great Britain could replace Russian commodities with North American goods. Finally, whatever the outcome of future discussions, Macartney was under no circumstances to allow the Russians to annul the 1734 treaty before the British ministry had one last opportunity to reconsider its present position.[36]

Catherine II's reported response was that George III did not trade with Russia *"pour ses beaux Yeux"* and that he would be compelled to buy from Russia "with or without a Treaty."[37] Catherine then ordered Panin to prepare a *Mémoire* that would revoke Empress Elizabeth's 1759 declaration as soon as the first British vessel arrived in Kronstadt for the opening of the spring trade.[38]

Finally, in the early part of May 1766, Macartney stumbled over a device for bringing the tedious negotiation to a conclusion. At the end of a protracted conversation with Panin, Macartney folded a piece of paper and on one side wrote out the contentious reservation clause as the Russian court would have it and, on the other side, the one favored by his court. Panin indicated that it might be admissible and promised to take it up with Catherine.[39] Another six weeks elapsed, however, before Panin delivered to Macartney a similarly double-columned document that contained the British version of the reservation clause on one side and a revised Russian version on the other.[40]

Despite their differences, Macartney concluded that the reworded Russian version did not undermine the spirit or substance of the British version. Panin said that his court would not wait any longer to settle the matter of the commerce treaty and he wanted Macartney to sign the treaty. Panin would forego the issuance of new plenipotentiary powers. Once again, and contrary to his explicit instructions, Macartney signed the treaty on June 20/July 1, 1766, with Panin, Golitsyn, Münnich, and Teplov signing for Russia. "Had I not seized this opportunity," Macartney wrote to his

---

36. Grafton to Macartney, Mar. 25, and Apr. 18, 1766, PRO SP 91/77, ff. 50–54, and 70–73, respectively.
37. Macartney to Grafton, Apr. 10/21, 1766, PRO SP 91/77, ff. 101–2.
38. Ibid., Apr. 15/26, 1766, No. 1, PRO SP 91/77, ff. 103–5.
39. Macartney to Grafton, May 2/13, 1766, PRO SP 91/77.
40. Macartney to Conway, June 23/July 4, 1766, No. 5, PRO SP 91/77, ff. 176–77.

ministry, "had I not signed the new treaty . . . the whole would have been lost to us forever."[41]

The British ministry acquiesced but did not forgive Macartney for disobeying his orders and yielding "to so unreasonable a Proposition." It informed him that he would soon be replaced by "a Minister of the first Rank."[42] At the same time the British ministry did not fully comprehend the implication the reservation clause would have on the future Anglo-Russian trade. It did not foresee how Catherine might someday choose to use it and the Russian tariff to undermine the British in that trade. The pertinent section of Article IV appeared in the treaty as follows:

> to preserve a just equality between the Russian and British merchants, with regard to the exportation of provisions and other commodities, it is farther stipulated, that the subjects of Russia shall pay the same duties on exportation, that are paid by the British merchants on exporting the same effects from the ports of Russia; *but then each of the high contracting parties shall reserve to itself the liberty of making, in the interior parts of its dominions, such particular arrangements as it shall find expedient for encouraging and extending its own navigation.*[43]

Thus, with the conclusion of the Anglo-Russian commercial treaty and the publication of the tariff, Catherine had progressed along the path of economic development that she had envisioned when she had ascended the throne. Great Britain's commodity trade deficit with Russia had grown considerably.[44] But Catherine would have preferred its composition to have been arranged differently, that is, she would have liked Russian merchants to have had a greater share in the value of that trade turnover. From 1764 through 1766, the years for which we have complete data, the distribution of the value of the overseas commodity trade turnover of about twenty different merchant nationals who shipped goods via St. Petersburg/Kronstadt was as follows. Russian merchants held the largest share of imports with an annual average of 30.2 percent—greater than that of the French (2.2 percent); the Dutch (14.0 percent); the Hanse merchants from Hamburg, Lübeck and

---

41. Ibid.
42. Conway to Macartney, Aug. 1, 1766, PRO SP 91/77. The British ratification reached St. Petersburg in late August (Conway to Macartney, Aug. 5, 1766 and Macartney to Conway Aug. 15/26, 1766, PRO SP 91/77), and at the beginning of September Macartney was in possession of the Russian ratification (PRO SP 108/440, and Macartney to Conway, Aug. 24/ Sept. 4, 1766, PRO SP 91/77).
43. CT, III, pp. 226–27 (italics mine); and full text of treaty, CT, III, pp. 215–34.
44. See SOT, ff. 94, 180.

Rostock (15.3 percent); and the British merchants whose share was 24.2 percent. However, Russian merchants had only 9.4 percent of exports which put them ahead of the French (2.0 percent) and the Dutch (6.3 percent) but behind the Hanse merchants (combined 12.9 percent). British merchants dominated exports with a huge 62.9 percent. The combined Russian and British merchant trade turnover equaled 63.6 percent of the total with Russian merchants accounting for 19.5 percent and British merchants accounting for 44.1 percent.[45]

Although similar comprehensive data are not available for the other Russian ports for these early years in Catherine's reign, the information we do have permits the general view that British merchants prevailed in the Riga trade, which was second in overall commercial importance only to St. Petersburg/Kronstadt. Although British merchants did not enjoy the preferential treatment accorded their compatriots in the capital port, as provided for under the Anglo-Russian treaty of commerce, by their own admission they dominated the export trade of that port: "It will appear that near one half of all the Exports of Riga to all Parts, are shipt by the British residing here." Because British merchants handled only a small share of Riga's import trade, the balance of overseas commodity trade was heavily in favor of Russia:

> The Ballance of our Trade is Yearly near a Million of Crowns or Dollars Alberts in favour of Russia; and that such Ballance is for the most Part Paid for, in Species Dollars and Ducats, brought in by the Brittish residing here and the rest paid for with Bills of Exchange.[46]

A comparison of the estimated minimum value of the British merchant overseas commodity trade turnover can be made between St. Petersburg/Kronstadt, Riga, and Narva for the years 1763–1764. British merchant trade turnover in the capital port was nearly four times greater than in Riga and nearly eight times greater than in Narva (see Table MN–I).[47]

The significance of the value of the British merchant overseas commodity trade turnover in Riga and Narva alone is most notable

---

45. Compiled and computed from the records of the St. Petersburg Custom House, PRO BT 6/231; SP 91/73, 76, 107; CO 388/95; and VTR, *Vedomosti*, Nos. 1–2.
46. British Factory, Riga, to Swallow, May 1/12, 1765, in Swallow to Grafton, Aug. 8/19, 1765, PRO SP 91/76, f. 195.
47. Table MN-I is compiled and computed from the records of the St. Petersburg Custom House: for 1763, PRO BT 6/231, and SP 91/73, 107; for 1764, PRO BT 6/231, SP 91/107, CO 388/95, and VTR, *Vedomost'*, No. 1.

### TABLE MN-I. COMPARISON BETWEEN THE VALUE OF BRITISH MERCHANT OVERSEAS COMMODITY TRADE TURNOVER IN ST. PETERSBURG, RIGA, AND NARVA: ANNUAL AVERAGE 1763–1764[47]

[Measured in Current Rubles]

|  | Exports | Imports | Total Trade Turnover |
|---|---|---|---|
| St. Petersburg | 3,526,819 | 1,203,791 | 4,730,610 |
| Riga | 1,224,803 | 71,264 | 1,296,067 |
| Narva | 607,956 | 11,038 | 618,994 |

### TABLE MN-II. COMPARISON BETWEEN THE VALUE OF THE BRITISH MERCHANT OVERSEAS COMMODITY TRADE TURNOVER IN RIGA AND NARVA AND THE RUSSIAN MERCHANT OVERSEAS COMMODITY TRADE TURNOVER IN ST. PETERSBURG IN 1764[48]

[Measured in Current Rubles]

|  | Exports | Imports | Total Trade Turnover |
|---|---|---|---|
| Russian Merchants St. Petersburg | 565,607 | 1,782,103 | 2,347,710 |
| British Merchants Riga | 1,257,069 | 62,806 | 1,319,875 |
| British Merchants Narva | 630,340 | 12,606 | 642,946 |

when compared to the value of the Russian merchant overseas commodity trade turnover in St. Petersburg/Kronstadt. The available data are for the year 1764. The minimum estimated value of the combined British merchant trade turnover in Riga and Narva was 84 percent of the value of the Russian merchant trade turnover in St. Petersburg/Kronstadt (see Table MN–II).[48]

Catherine would seek, by whatever means available, to reverse the fortunes of the Russian and British merchantry. Catherine intended, at every future opportunity, to conclude trade treaties with as many countries as would benefit Russia's commercial and economic development, and to manipulate her tariffs to obtain the

---

48. Table MN-II is compiled and computed from the records of the St. Petersburg Custom House, PRO BT 6/231, SP 91/107, and CO 388/95; and VTR, *Vedomost'*, No. 1.

most beneficial concessions for Russian merchants. Catherine wanted to emancipate Russian overseas commerce from the concentration of British mercantile influence, to extricate it from its dependence upon the British trade, and in so doing to make Russia a competitive and formidable maritime power in its own right. The published propositions and maxims in her Instruction (*Nakaz*) to the Legislative Commission in 1767 provide clues to her future commercial policies. It suffices to cite but three examples:

> 324. England has no fixed Book of Rates, or Tariff, settled with other Powers; Her Tariff varies (to use the Expression) at every Session of Parliament by particular duties, either imposed or removed; Being always exceedingly jealous of the Trade carried on with her, she seldom binds herself by Treaties with other States, and depends upon no other Laws than her own.

> 327. Much less should a state subject itself to the Inconveniency of selling all its Goods to one Nation only, under pretence that they will take the whole at a certain price.

> 328. The true Maxim is, to exclude no People from your Trade without very important Reasons.[49]

---

49. *Catherine the Great's Instruction (NAKAZ) to the Legislative Commission, 1767*, ed., Paul Dukes (Newtonville, MA, Oriental Research Partners, 1977), pp. 84–85; and see also Basil Dmytryshyn, "The Economic Content of the 1767 *Nakaz* of Catherine II," *The American Slavic and East European Review* 19:1 (1960): pp. 1–9.

# Part II
# Russian Commodity Exports and Great Britain's Economic Development: From the Seven Years' War to the Armed Neutrality

# INTRODUCTION:
## The Anglo-Russian Overseas Commodity Trade Balance

In the decades following the Seven Years' War, Great Britain became Russia's most lucrative trading partner. The factors that created the increasing demand for goods in Great Britain and led to the enormous growth of its economy during that period also had an extraordinary effect on Anglo-Russian trade. Although recent scholarship does not stress the influential role played by foreign trade in Britain's economic growth, Britain's foreign trade grew faster than the economy itself. As a result Great Britain was far richer than it might otherwise have been. British industrial output during the eighteenth century almost quadrupled while home consumption trebled.[1]

It is reasonable to speculate that Russia's overseas trade, especially its exports, must have played a similar role in that country's economic growth. The value of Russia's overseas commodity trade turnover grew about fivefold with about a tenfold increase in exports and a sixfold increase in imports during Catherine II's reign.[2] It could be expected that the resulting expansion of Russia's export industries would encourage the growth of other industries and, as a consequence of spending more and more of the income from the export sector, a positive multiplier effect would take place in other parts of the economy. It also would be reason-

---

1. For the earlier view see Phyllis Deane and H. J. Habakkuk, "The Take-Off in Britain," in *The Economics of Take-Off Into Sustained Growth*, ed. W. W. Rostow (New York, 1963), pp. 77–80. For recent reviews of the problem see W. A. Cole, "Factors in demand 1700–80," in *The Economic History of Britain since 1700. Volume 1: 1700–1860*, ed. Roderick Floud and Donald McCloskey (Cambridge: Cambridge University Press, 1981), pp. 36–65, passim; François Crouzet, "Towards an export economy: British exports during the Industrial Revolution," Chap. 6 in *Britain ascendant: Comparative studies in Franco-British economic history* (Cambridge: Cambridge University Press, 1990); and Carl-Ludwig Holtfrerich, "Introduction: The Evolution of World Trade, 1720 to the Present," in *Interactions in the World Economy: Perspectives from International Economic History*, ed. Carl-Ludwig Holtfrerich (New York: New York University Press, 1989), p. 4.
2. See Chap. XIV below for details.

able to expect that "this would induce higher levels of investment, employment and income, and might also stimulate innovation which would make possible still higher levels of output and income in the long run."[3]

The increase in Russia's overseas commodity trade with Great Britain soon after the conclusion of the Seven Years' War represented both a rebound from the decline during the war years and a new increase in Britain's demand for goods. During the first four years of Catherine's reign (1763–1766) compared to the effective period of Russia's military involvement in the war (1756–1762), Great Britain increased its imports from Russia by about 43 percent, while its exports to Russia grew by only about 35 percent ("Official Values") and, therefore, deepened its overseas commodity trade balance deficit with Russia. This was prior to the conclusion of the Anglo-Russian Commercial Treaty of 1766 and the enactment of the Russian tariff of 1766/67. During the years of war between Great Britain and the American Colonies, France, Spain, and the United Provinces (1776–1782), Britain's trade deficit with Russia worsened. Thus, the wars had the effect of making Britain even more dependent on Russia's economic and strategic material resources than before.[4]

During the period following the Seven Years' War Russia exported some forty different raw and manufactured goods valued at several million rubles per annum to more than sixty ports in England, Scotland, and Ireland. The economic and social values that these goods had for Great Britain were significant. For example, from 1764 through 1782 Great Britain imported about 43 percent of its cordage, about 14 percent of its pitch and tar, a substantial amount of its bristles, isinglass, and 58 percent of its tallow from Russia.[5] But there were several Russian commodities in particular that made enormous contributions to Great Britain's economic growth because they tended to have multiplier effects and because they provided British entrepreneurs with a means of exploiting advances in productivity. A close examination of the ex-

---

3. Cole, p. 42. For an interesting discussion of the implications of international trade on economic development, see Gerald M. Meier, "Theoretical Issues Concerning the History of International Trade and Economic Development," in *Interactions in the World Economy*, ed. Holtfrerich, pp. 33–58.
4. See SOT, ff. 94, 180.
5. For cordage, pitch and tar, bristles, isinglass, and tallow, see records of the St. Petersburg Custom House, PRO SP 91/73, 75–76, 78–79, 83, 88, 91–92, 94–96, 99–104, 107; BT 6/231; FO 65/1, 6, 9, 13, 15; FO 97/340; CO 388/54, 56, 95; SRO CO/18; and BT 6/232, ff. 7–11, 18–22; BT 6/240, ff. 97, 101, 103, 122, 124, 126, 130–32, 134–36; CU 3 and 14; and *Vedomosti*, Nos. 3–5.

change of goods between the two countries provides not only a valuable index into what the British needed and wanted most from Russia, but also what surpluses Russia was capable of generating beyond its own domestic consumption. And because exports tend to generate imports, much can be learned about Russia's economy and social values from a study of Russia's imports from Great Britain, an empire that traded with the whole world.

Chapters IV, V, and VI examine Russia's bar iron, hemp, flax, linen, and timber exports to Great Britain from the Seven Years' War to the Armed Neutrality. In Chapter XIII, Great Britain's exports to Russia are described, examined and analyzed.

# Chapter IV
# Bar Iron

Peter Mathias has stated that Russian bar iron exports to Great Britain were "by far the greatest direct contribution of Russian commerce to the industrial revolution."[1] Certainly no one would deny the significance that imported Russian bar iron had in meeting the expanding demand, both at home and abroad, for British metals and their manufacture during the second half of the eighteenth century. In 1788, the first year for which there is an agreed upon figure for British bar iron output, Great Britain produced 32,000 tons of bar iron and imported 51,000 tons, of which almost 30,000 tons came from Russia. Thus, about 35 percent of the total bar iron Great Britain imported and domestically produced that year (slightly more than 83,000 tons) was Russian.[2]

Next to textiles, British metals were the largest group of British exports. Cast iron made the pots, pans, cannon, shot, stoves, and anchors but it was the more malleable bar iron that went into the manufacture of wire, nails, horseshoes, bolts, gates, and fences. In the first half of the eighteenth century, Great Britain was without abundant and concentrated resources of ore, fuel, and water power. Therefore, its iron industry could not dramatically increase its output to meet the demand for iron wares, which had effectively outstripped domestic production. It thus became necessary to import large quantities of bar iron. Reliable data on British iron output are meager but it is clear that while domestic pig iron production met home demand, bar iron did not. During the period 1715–1720, 16,000 tons of bar iron were domestically produced. This was equal to the amount of bar iron imported during 1700–1709 and was about 4,000 fewer tons than were imported between 1720 and 1729. In 1750 the production of domestic bar

---
1. MATHIAS, p. 4.
2. Charles K. Hyde, *Technological Change and the British Iron Industry 1700–1870* (Princeton: Princeton University Press, 1977), pp. 92–94, 113; and BAR IRON, BT 6/230.

iron was about 19,000 tons, but imported bar iron was nearly 35,000 tons.³

Only in the first part of the nineteenth century, when it had substantially increased its productive capacity, would Great Britain emancipate itself from dependence upon foreign bar iron. But this would happen, according to the late Ralph Davis, only after "a long succession of economies in the production of iron" had taken place, only after the "improvements in the design of puddling furnaces and blast furnaces and a great increase in their average size, the introduction of the hot blast in 1828, [and] the close integration of processes from smelting to the production of fabricated pieces."⁴

It was the increased use of coke in both smelting and refining during the second half of the eighteenth century that helped to reduce costs and to increase British iron production. From 1750 to 1788 pig iron production grew by about 150 percent and bar iron increased by about 70 percent. But most British iron smelting and forging operations continued to depend heavily on charcoal, a fuel source that Britain could not economically supply in sufficient quantity. Moreover, the prime cost of either Swedish or Russian bar iron at the forge was such that even with the additional charges of freight, handling, and customs, the average quality of imported bar iron competed successfully with the British product. Between the 1740s and the 1770s imported bar iron almost doubled from an annual average of 22,500 tons to 44,100 tons.⁵ These bar iron imports had a salutary effect; their absence "would certainly have imposed higher costs upon the British economy for a key material."⁶

---

3. "An Account of the Quantity of Iron Imported into England from Xmas 1742 to Xmas 1762 distinguishing each Year & how much from Foreign Parts & how much from America," Inspector-General, Custom House, London, PRO CO 390/9; Hyde, "British Iron Output, 1715–1750," Appendix A, pp. 213–20; idem, "Imports of Iron Products into England and Wales 1720–1749, Annual Average [in tons]," Table 3.1, pp. 42, 45; and H. S. Kent, *War and Trade in Northern Seas: Anglo-Scandinavian Economic Relations in the Mid-Eighteenth Century* (Cambridge: Cambridge University Press, 1973), p. 186.
4. Ralph Davis, *The Industrial Revolution and British Overseas Trade* (Leicester: Leicester University Press, 1979), pp. 27–28.
5. Beginning in the 1760s there was also the employment of the "potting and stamping process." British pig iron output moved from 28,000 tons in 1750 to 70,000 tons in 1788, of which about 50,000 tons went into castings or the production of bar iron. The output of bar iron grew from 18,800 tons in 1750 to 32,000 tons in 1788. Hyde, pp. 7–8, 42–48, 51, 53–54, 56–68, 76–83, 92–94, 213–20, 234; Kent, pp. 9–11, 59–66, 70–72, 76–77, 79, 184, 186; and Davis, pp. 27–28.
6. MATHIAS, p. 8. Mathias has also argued that "the debate about the relative scarcity of charcoal for iron making in eighteenth-century England has ignored the likely effect which the absence of rising imports of Russian bar-iron doubtless would have had upon its price; given the multiplier involved between the output of bar-iron in tonnage and the necessary

During the second half of the eighteenth century Russia became the principal beneficiary of Great Britain's decision to redirect its purchases of bar iron away from Sweden. This decision was prompted by the relative unavailability of British bar iron imports from the North American colonies, the frequent breakdown of Anglo-Swedish commercial relations, and the increased production of Russian bar iron for the export market. American bar iron production and technology had not advanced sufficiently to provide the mother country with what it needed, despite preferential treatment and the alleviation of import duties. Transportation costs were also a deterrent. Sweden's frequent anti-British trade policies prevented it from maximizing its iron exports to Great Britain.[7] Undoubtedly, Sweden's economic crisis and political turmoil following the Seven Years' War accelerated Russia's successful penetration of the British market. Sir John Goodricke, the British envoy to Sweden, remarked that the Swedes believed "that their Iron is so absolutely necessary to us, that let them make what Regulations they will, we must have it. . . . The measures taken upon this Doctrine, have raised them up a Rival in Russia, that will in short demonstrate the falsity of it to their cost."[8]

Russia had already revealed its competitive potential in shipping bar iron to England during the 1750s when England's annual average imports were about 30,000 tons and Sweden's share was about 19,000 tons. In 1750 Russia supplied England with about 15,000 tons; in the following year, 5,000 tons; and, in 1755, 10,000 tons.[9] During the Seven Years' War Russia exported via St. Petersburg/Kronstadt an annual average of 10,000 tons of bar iron in British ships alone, the overwhelming number of which departed for ports in Great Britain and Ireland.[10]

As soon as the war was over, Russia's bar iron shipments to Great Britain significantly increased both relatively and absolutely over the previous decade. Russia surpassed Sweden in supplying suitable bar iron to England in 1765, 1767, 1768 and 1769. With the exception of 1775, Russia held the predominant position

---

inputs of charcoal tonnage." Ibid., p. 4. Cf., M. W. Flinn, "Technical Change as an escape from Resource Scarcity: England in the Seventeenth and Eighteenth Centuries," in *Natural resources in European history. A conference report*, ed. Antoni Mączak and William Parker (Washington, DC: Resources for the Future, 1978), pp. 139–59.
7. Kent, pp. 9–11, 59–66, 70–72, 76–77, 79, 184, 186.
8. Quoted in Michael Roberts, *British Diplomacy and Swedish Politics, 1758–1773* (Minneapolis: University of Minnesota Press, 1980), p. 255. See also pp. 251–56, passim.
9. Kent, pp. 62–63, 186.
10. Compiled and computed from the records of the St. Petersburg Custom House, PRO FO 65/15.

### TABLE BI-I. GREAT BRITAIN'S BAR IRON IMPORTS FROM 1764 THROUGH 1782[12]

[Annual Average, Measured in Tons]

| Years | Total Imports | Swedish Imports | Russian Imports | Swedish Percentage of Total | Russian Percentage of Total |
|---|---|---|---|---|---|
| 1764–66 | 45,140 | 22,912 | 18,842 | 50.8 | 41.7 |
| 1767–69 | 46,500 | 18,993 | 24,032 | 40.8 | 51.7 |
| 1770–74 | 49,388 | 17,167 | 29,204 | 34.8 | 59.1 |
| 1775–79 | 44,527 | 19,357 | 23,957 | 43.5 | 53.8 |
| 1778–82 | 43,292 | 15,261 | 27,435 | 35.3 | 63.4 |
| 1764–82 | 46,450 | 18,092 | 25,760 | 38.9 | 55.5 |
| 1767–82 | 46,695 | 17,188 | 27,058 | 36.8 | 57.9 |

within the British import market well beyond the period of technological breakthrough with the "puddling process" in 1783–1784 and its subsequent general adoption in the mid-1790s.[11]

For the entire nineteen-year period from 1764 through 1782 Russian bar iron accounted for an annual average of 55.5 percent of Great Britain's total bar iron imports. During the three years prior to the Anglo-Russian Commercial Treaty of 1766 and the Russian tariff of 1766/1767, Russian bar iron imports represented an annual average of about 42 percent of Great Britain's total bar iron imports. But in the following three years (1767–1769) they jumped to about 52 percent and thereafter represented a majority of bar iron imports. From 1778 through 1782, in particular, when Great Britain was at war with the American Colonies, France, Spain, and the United Provinces, Russia's share of Great Britain's bar iron imports increased to an annual average of about 63 percent (see Table BI–I).[12]

Given the centrality of St. Petersburg/Kronstadt in the overseas commerce of the Russian Empire with its waterway connection to the Ural metallurgical industry, it is not surprising that the capital port held an overwhelming position in Russia's international iron trade. Even in the absence of complete archival data it is clear that almost 90 percent of Russia's total iron exports departed from

---

11. BAR IRON, PRO BT 6/240 and 230.
12. Table BI-I is compiled and computed from BAR IRON, PRO BT 6/240.

St. Petersburg/Kronstadt during the period 1764–1782.[13] Therefore what can be said about the nature of bar iron exports via St. Petersburg/Kronstadt is in effect a statement about almost the whole of Russia's overseas trade in bar iron.

Archival data are available for bar iron exports from St. Petersburg/Kronstadt in British ships from 1753 and in all ships from 1760. In the seven years prior to the conclusion of the Anglo-Russian Commercial Treaty and the publication of the Russian tariff (1760–1766), British ships nearly doubled their annual average transport of bar iron over the previous seven-year period (1753–1759). This accounted for about 87 percent of the total bar iron exported from St. Petersburg/Kronstadt in the 1760–1766 period. Once the treaty and tariff were in effect the volume of Russian bar iron exports significantly increased, particularly to Great Britain and Ireland. From 1767 through 1782 British ships transported an annual average of about 30,000 tons of bar iron, which accounted for about 88 percent of the total bar iron shipped from St. Petersburg/Kronstadt. The total bar iron tonnage on British ships departing St. Petersburg/Kronstadt exceeded the total amount of bar iron that Great Britain imported from the whole of the Russian Empire, but because 96 percent of the British ships sailed for ports in Great Britain and Ireland, Great Britain must have been the destination for the overwhelming amount of this bar iron (see Table BI–II).[14]

The monetary value of Russian bar iron exports was considerable. Although the archival data are not comprehensive, and are at times even inconsistent, they are sufficient to provide reasonable estimates for most of the period under discussion. The value of bar iron exports relative to the total value of all commodity exports from St. Petersburg/Kronstadt can be determined. Further com-

---

13. This estimate is derived from the data for total iron—bar, old, nail, rod, hoop, etc.—exports in the records of the St. Petersburg Custom House, PRO SP 91/73, 75–76, 78–79, 83, 88, 91–92, 94–96, 99–104, 107; BT 6/231; FO 65/1, 6, 9, 13, 15; FO 97/340; and CO 388/54, 56, 95; and is compared to the total iron exports from Russia in "Tablitsa 50. Otpusk russkogo zheleza za granitsu v 1722–1822 gg.," in S. G. Strumilin, *Izbrannye proizvedeniia. Istoriia chernoi metallurgii v SSSR* (Moskva: Nauka, 1967), p. 197. "Tablitsa 50," however, is derived from College of Commerce archival data for only the years 1745–1765 and 1797–1822. Cf., "Tablitsa 8. Otpusk zheleza iz Rossii v raznyia gody XVIII stoletiia," in V. I. Pokrovskii, ed., *Sbornik svedenii po istorii i statistike vneshnei torgovli Rossii* (St. Peterburg: M. P. Frolov, 1902), I, p. 235.

14. Table BI-II is compiled and computed from the records of the St. Petersburg Custom House, PRO SP 91/69, 73, 75–76, 78–79, 83, 88, 91–92, 94–96, 99–104, 107; BT 6/231; FO 65/1, 6, 9, 13, 15; FO 97/340; and CO 388/54, 56, 95. For confirmation of the destination of British ships via Sund, see Hans. Chr. Johansen, *Shipping and Trade between the Baltic Area and Western Europe 1784–95* (Odense, Denmark: Odense University Press, 1983), in particular, pp. 82–95. See "Weights and Measures," p. xiii.

## TABLE BI-II. BAR IRON EXPORTS FROM ST. PETERSBURG FROM 1753 THROUGH 1782[14]

[Annual Average, Measured in Tons]

| Years | Bar Iron Exports All Ships | Bar Iron Exports British Ships | British Percentage of Total |
|---|---|---|---|
| 1753–59 | N/A | 8,860 | N/A |
| 1760–66 | 17,403 | 15,078 | 86.6% |
| 1764–69 | 32,590 | 29,923 | 91.8% |
| 1770–74 | 35,780 | 32,221 | 90.1% |
| 1775–79 | 33,036 | 28,230 | 85.5% |
| 1778–82 | 35,669 | 29,439 | 82.5% |
| 1764–82 | 32,140 | 28,156 | 87.6% |
| 1767–82 | 34,246 | 30,039 | 87.8% |

parisons can be made between the value of bar iron exports in British ships and all ships and the former can be compared to the total value of all commodity exports by British merchants in St. Petersburg/Kronstadt.

During the period 1763–1782 prices for exported bar iron fluctuated from a low of slightly more than 68 kopeks per pood in 1763 to a high of almost 87 kopeks per pood in 1773. Prices stabilized at 80 kopeks per pood during the latter 1770s.[15] In 1763–1764 the estimated annual average value of bar iron exports was 791,495 rubles, of which about 682,819 rubles, or slightly more than 86 percent, was exported in British ships. Since British merchants overwhelmingly transported their exports in British ships, bar iron exports represented about 19 percent of the total value of exports by British merchants for those two years.[16]

Compared to the total value of all commodity exports from St. Petersburg/Kronstadt from 1768 through 1779, the annual average value of bar iron exports was about 19 percent for all ships and 16.5 percent for British ships. The latter represented no less than 27 percent of the total value of the commodity exports by British merchants. During this period the annual average value

---

15. Compiled and computed from the records of the St. Petersburg Custom House, PRO SP 91/69, 73, 75–76, 78–79, 83, 88, 91–92, 94–96, 99–104, 107; BT 6/231; FO 65/1, 6, 9, 13, 15; FO 97/340; and CO 388/54, 56, 95.
16. Compiled and computed from the records of the St. Petersburg Custom House, PRO BT 6/231; SP 91/73, 75–76, 83, 94, 107; FO 65/15; CO 388/95; and VTR, *Vedomost'*, No. 1.

### Figure 1

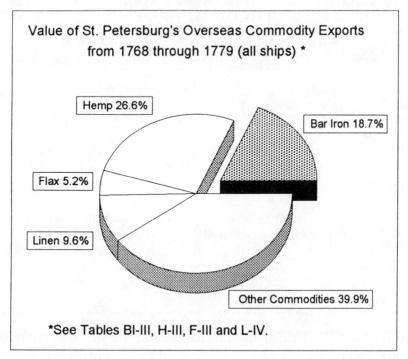

of bar iron exported from St. Petersburg/Kronstadt was about 1.7 million rubles (see Figure 1, Table BI–III).[17]

It is obvious that the volume and value of Russian bar iron exports were enormously important not only to the overall British mercantile operation but also to the overall foreign trade profile of Russia's capital port. Although archival data for the export of bar iron from other Russian ports during the period 1763–1782 are fragmentary, the evidence that is available leaves no question that St. Petersburg/Kronstadt dominated Russia's overseas commerce in bar iron.[18] As long as Great Britain would find it necessary to import Russian bar iron to supplement its own domestic

---

17. Table BI-III is compiled and computed from the records of the St. Petersburg Custom House, PRO SP 91/83, 88, 91–92, 94–96, 99–104, 107; BT 6/231; FO 65/1, 6, 9, 13, 15; and FO 97/340; and VTR, *Vedomosti*, Nos. 3–5.
18. Data for bar iron exports from Archangel, Narva, and Riga are available in the records of the St. Petersburg Custom House, PRO SP 91/75, 79, 102–4, 107; FO 65/1, 6, 9; BT 6/231; and CO 388/56, 95.

### TABLE BI-III. VALUE OF ST. PETERSBURG'S COMMODITY AND BAR IRON EXPORTS FROM 1768 THROUGH 1779[17]

[Exclusive of Custom Duties and Sundry Charges]
(Annual Average, Measured in Current Rubles, in Thousands)

| Years | Value of Commodity Exports | Value of Bar Iron Exports | | | |
|---|---|---|---|---|---|
| | | All Ships | Percentage | British Ships | Percentage |
| 1768–69 | 7,135 | 1,456 | 20.4% | 1,337 | 18.7% |
| 1770–74 | 8,269 | 1,751 | 21.2% | 1,576 | 19.1% |
| 1775–79 | 9,922 | 1,671 | 16.8% | 1,423 | 14.3% |
| 1768–79 | 8,945 | 1,668 | 18.7% | 1,473 | 16.5% |

| Years | Value of Commodity Exports by British Merchants | Percentage Value of Bar Iron Exports in British Ships |
|---|---|---|
| 1768–69 | 4,855 | 27.5% |
| 1770–74 | 5,049 | 31.2% |
| 1775–79 | 5,970 | 23.8% |
| 1768–79 | 5,401 | 27.3% |

production, the British mercantile community in St. Petersburg would insure that the capital port played the dominant role in that trade. The British and the Russians had everything to gain from such a mutually remunerative relationship.

# Chapter V
# Hemp, Flax and Linen

Great Britain's importation of Russian flax, hemp and linen goods was essential to the British way of life, industry and economy. The products into which the basic fibers of flax and hemp were transformed affected the entire fabric of British society. Because of hemp's special strength, durability and resistance to the action of water, it was highly commercial when made into cables, cordage, rope, fine threads and fishing nets or, when combined with the more flexible and elastic flax, into sailcloth and a variety of coarse linens. The cultivation and preparation of both flax and hemp were especially suited to Russia's climate and labor-intensive capabilities, which annually produced tens of thousands of tons for domestic and foreign consumption. While it might be impossible to calculate precisely the contribution that the growth and sale of hemp and flax made to Russia's overall economy, their huge overseas exports suggest that their contribution must have been significant. There is no question that Great Britain's increasing strategic and commercial demand for naval stores, as well as its expanding linen industry, provided a lucrative market for Russia's hemp and flax. The British also purchased millions of yards of Russian linens annually for either home consumption or re-export.

More than other textiles, linen met a wider and deeper need for all groups of English society. According to some authorities linen fell into a category between outright necessities and luxuries termed "decencies." In the home, linen had many uses for the table and bed; in industry, there was canvas, sacking, and sailcloth; and, of course, there was linen clothing. Because linen goods so thoroughly permeated English society it has been suggested that during the middle decades of the eighteenth century demand rose faster than the rate of population growth. Yet, until recently, historians have not paid as much attention to the linen industry as to the woolen or cotton industries because linen's growth was not as spectacular and because greater emphasis has been placed on the export market than on imports, for which the linen trade has been known.

According to N. B. Harte, the noted English historian of the textile industry, "linen was, in fact, the most important manufactured import into pre-industrial England. Until the end of the eighteenth century import of 'linens' ranked second only to imported 'groceries' in total value." Groceries—mainly sugar, coffee, tea, and several tropical and semi-tropical foodstuffs which because of climate were commodities heavily imported into Britain— moved from 17 percent of total imports ("official values") in 1700 to 35 percent in 1800. Harte asserts that "there was no climatic or other straightforward reasons why so much linen was imported. Reasons outside the realm of mere geography have to be sought to discover why an economy — especially one in the course of generating an industrial revolution — should be so dependent on foreign industry for an important manufactured commodity." Linen accounted for some 15 percent of the total imports for about half the century and then dropped to about 5 percent by 1800. The explanation for this is seen in the dramatic fall in imported continental European linen, while imports of Irish and Scottish linen increased greatly and England's own linen production rapidly accelerated. Nevertheless, despite its relative decline among other English imports, the value of linen imported during the century nearly doubled.[1]

During the first half of the eighteenth century, the linen industries of Ireland, Scotland, and England developed unevenly, England being the least advanced in output and quality. To meet the growing demand for linen goods at home and abroad a series of governmental measures was employed to increase output and improve quality. The government encouraged and subsidized hemp and flax cultivation at home and in the North American colonies, English linen manufacturers received export bounties, and higher tariffs were imposed on imported continental European linens. Since 1696 Irish linens had entered England duty-free and beginning in 1705 Ireland could export its linens directly to the colonies. With the Act of Union in 1707 Scottish linens also entered England duty-free. To further assist these industries so-called development corporations were established: in 1711 in Ireland with

---

[1]. N. B. Harte, "The Rise of Protection and the English Linen Trade, 1690–1790," *Textile History and Economic History*, ed. N. B. Harte and K. G. Ponting (Manchester: Manchester University Press, 1973), pp. 74–76, 108–9.

the Board of Trustees for the Linen and Hempen Manufacturers and in 1727 in Scotland with the Commissioners and Trustees for Improving Fisherys and Manufactures. In 1746 the British Linen Company was incorporated for the purpose of stimulating the Scottish linen industry and trade, especially in coarse linens. By mid-century the English market became the largest free-trade area for Irish and Scottish linens.[2]

At the outbreak of the Seven Years' War Great Britain and the colonies consumed 80 million yards of linen, of which approximately 32.2 million were imported from abroad with a little less than 7 percent re-exported to foreign countries. Of the remaining approximately 50 million yards, about 12 million were imported from Ireland, 12 million were manufactured in Scotland and 25.8 million were produced in England.[3]

During the 1760s Great Britain's retained European linen imports had fallen to an annual average of 17.8 million yards. By the late 1760s Irish linen imports into England were already greater than those retained imports from Europe. In 1770 England retained 18.6 million yards imported from Europe while Irish and Scottish linens on the English market amounted to about 32.7 million yards, of which Ireland's share was approximately 19.7 million. This trend was the result of a gradual removal from the English market of the traditionally heavy continental European linen suppliers.[4] First the French, then the Dutch and Flemish, and finally the German linen producers progressively withdrew from a market no longer as profitable for them. This was due to the increasing improvement in the quantity and quality of domestic linen, the continued use of export bounties for linen manufacturers (until 1832), the abolition of the duties on imported flax (in 1731) and yarn (in 1756) and a series of discriminatory protective tariff acts against foreign linen imports (in 1767). Only Russia im-

---

2. Ibid., pp. 91–96; John Horner, *The Linen Trade of Europe during the Spinning-Wheel Period* (Belfast, 1920), pp. 222–26, 230, 274, 285; Henry Hamilton, *An Economic History of Scotland in the Eighteenth Century* (Oxford: Clarenden Press, 1963), pp. 134–54; W. G. Rimmer, *Marshalls of Leeds Flax-Spinners 1788–1886* (Cambridge: Cambridge University Press, 1960), pp. 1–8; Alastair J. Durie, *The Scottish Linen Industry in the Eighteenth Century* (Edinburgh: John Donald Publishers, 1979); and *States of the Annual Progress of the Linen Manufacture, 1727–1754. From the Records of the Board of Trustees for Manufactures etc., in Scotland preserved in the Scottish Record Office*, ed. R. H. Campbell (Edinburgh, 1964).
3. These figures are cited in Horner on p. 233 from a report before Parliament in 1756.
4. Harte, pp. 91–94, especially Table 4.2 "Irish and Scottish Linen Production for the English Market," p. 93; L. M. Cullen, *Anglo-Irish Trade 1660–1800* (New York: August M. Kelley, 1968), p. 60; Hamilton, pp. 404–5; and P. P. Linen, 1773, 101–11 et seq., and Appendixes 12 and 13, pp. 119–30.

proved its absolute and relative position among continental suppliers of linen to Great Britain.[5]

It is important to note that Great Britain could not attain self-sufficiency in the production of the raw materials to sustain its linen industry and trade and that during the second half of the eighteenth century Britain had become dependent on imports of flax and hemp from Russia. Even Ireland, itself a major producer of these essential raw materials, imported flax and hemp from Russia.[6] Indicative of this situation is R. G. Rimmer's statement that "West Riding merchants from Knaresborough and Barnsley rode regularly to Hull to buy Russian flax."[7] A vivid picture and extensive discussion of how dependent the British were on the importation of Russian flax and hemp is to be found in *The Letters of Thomas Langton, Flax Merchant of Kirkham, 1771–1788*. Langton became so involved in this trade that, in 1787, he sent his youngest son, Tom, who was then seventeen years of age, to join the firm of Messrs. Thorley Morison and Co. in Riga.[8]

## HEMP

The volume and value of Russian hemp imported into Great Britain was impressive. Although the anomalous categories such as "hemp sundry" or just "hemp" were occasionally included, at least four qualities of Russian hemp for export were usually distinguished: "clean," "outshot," "half clean," and "codilla." When hemp was imported into Great Britain, however, British customs recorded it under one heading as "hemp rough,"[9] making it impossible to correlate the quality and quantity of hemp imports from Russia.

Just prior to the outbreak of the Seven Years' War Russian hemp represented an annual average of about 88 percent of the total hemp England imported. During the course of the war it rose to about 90 percent. After the war was over, although the total

---

5. Harte, pp. 77–86, 94–102; and Rimmer, p. 5. See also RC, Mar. 22, 1764, VII, pp. 175–76, and passim.
6. See the records of St. Petersburg Custom House, PRO SP 91/69, 73, 75–76, 78–79, 83, 88, 91–92, 94–96, 99–104, 107; BT 6/231; CO 388/54, 56, 95; FO 97/340; and FO 65/1, 6, 9, 13, 15.
7. Quoted in Rimmer, p. 3. See also Hamilton, pp. 134–54; Horner, pp. 234–35, 291–94; and Harte, pp. 77–86, 94–102.
8. Edited by Joan Wilkinson (Manchester: Printed for the Cheltham Society by Carnegie Publishers, 1994). I am indebted to Professor Jeremy Black who brought this work to my attention.
9. See PRO CU 3 (1753–1780), passim.

### TABLE H-I. GREAT BRITAIN'S HEMP IMPORTS FROM 1764 THROUGH 1782[10]

[Annual Average, Measured in Tons]

| Years | Total Imports | Russian Imports | Russian Percentage of Total |
|---|---|---|---|
| 1764–66 | 15,560 | 14,455 | 92.9 |
| 1767–69 | 15,473 | 14,048 | 90.8 |
| 1770–74 | 19,592 | 18,796 | 95.9 |
| 1775–79 | 18,877 | 18,216 | 96.5 |
| 1778–82 | 22,434 | 21,960 | 97.9 |
| 1764–82 | 18,602 | 17,771 | 95.5 |
| 1767–82 | 19,172 | 18,392 | 95.9 |

volume of hemp imports remained at about prewar levels, the proportion of Russian hemp increased. In the 1770s, despite the recession in the linen industry and the war with the American colonies, Great Britain significantly increased its hemp imports over the preceding decade. Total hemp imports rose by 25 percent but Russian hemp increased by 30 percent. After Britain's war widened to include France, Spain and the United Provinces (1778–1782), Russian hemp accounted for about 98 percent of total hemp imports. For the entire period from 1764 through 1782 Russian hemp accounted for an annual average of about 96 percent of the total hemp Great Britain imported (see Table H–I).[10]

Nearly all of Great Britain's hemp imports from the Russian Empire came from St. Petersburg; any disruption in its supply caused an immediate reaction in London. During the summer of 1761 a fire destroyed the old hemp warehouses in St. Petersburg, which contained the property of 98 Russian and foreign proprietors. The result was a loss of hemp valued at more than one million rubles—about 10 percent of St. Petersburg's total overseas trade turnover that year—and the destruction of 327 warehouses valued at just under 100,000 rubles.[11] Because of this the commissioners to Great Britain's Royal Navy placed an order for the immediate

---

10. Table H-I is compiled and computed from reports of the Inspector–General, Custom House, London: for the years 1753–1762, PRO CO 390/9; for the years 1764–1782, HEMP, BT 6/240.
11. The fire took place on June 29/July 10 and the College of Commerce compiled the list of losses to merchants on Aug. 20/31, 1761, in Keith to Weston, Sept. 4/15, 1761, PRO SP 91/68, f. 323.

purchase of 2,000 tons of hemp at St. Petersburg—about 15 percent of Britain's total annual average procurement of Russian hemp during the Seven Years' War. Given "the Circumstances of the Market," the British firm authorized to make this hemp purchase, A. Thomson & G. Peters Company of London and St. Petersburg, was "to secure as much as You can of It upon the Spot; and contract for the rest, to be delivered in or before May [1762], if you can, or as early as possible; for what Contracts you may make, you have Liberty to advance as far as a Tenth Part of the Value upon Account." This purchase was to be kept "Secret" in order "not to raise the Price unnecessarily." A. Thomson & G. Peters Company of St. Petersburg was authorized to draw for its "Disbourse upon Messrs. Pels, Hope, Cliffords, Muilman and Crop's Houses in Amsterdam."[12] In this case, as in so many others that would follow, British merchants drew on bills of exchange in Amsterdam for payment of their imports from the Russian Empire.[13]

From 1764 through 1782 the annual average tonnage of hemp shipped from St. Petersburg/Kronstadt in British vessels—about 96 percent of which were designated as sailing for ports in Great Britain and Ireland—was approximately equal to the total amount of "hemp rough" Great Britain imported from the whole of Russia. Contrary to what might have been expected, the Anglo-Russian Commercial Treaty (1766) and the Russian tariff (1766/67 which encouraged Russian shipping) had no appreciable effect on the Anglo-Russian hemp trade. In the previous seven years (1760–1766) British ships accounted for an annual average of about 72 percent of the total hemp tonnage exported overseas from St. Petersburg/Kronstadt, but thereafter through 1782 their share fell to 64 percent. During this period British vessels transported less hemp but more bar iron.[14] Nevertheless, by 1782 St. Petersburg's total hemp tonnage exported to all destinations had doubled and

---
12. A. Thomson & G. Peters, London, to Messrs. Thomson, Peters & Company, St. Petersburg, Nov. 24, 1761, in Bute to Keith, Nov. 25, 1761, PRO SP 91/68, f. 342.
13. For additional comment on bills of exchange, see J. A. S. L. Leighton-Boyce, *Smith's the Bankers 1658-1958* (London: National Provincial Bank Ltd., 1958), pp. 88, 184–88, 192–213, 236, and passim; Jennifer Newman, "Anglo-Dutch commercial cooperation and the Russia trade in the eighteenth century," in *The interactions of Amsterdam and Antwerp with the Baltic region, 1400–1800* (Leiden: Martinus Nijhoff, 1983), pp. 95–103; and see Chap. XIV, below.
14. Cf. Table H-II and Table BI-II. Detailed archival evidence on the commodities exported in Russian ships to Great Britain and Ireland is meager. For example, in 1778, two Russian ships bound for London did not carry any hemp; in 1780, 38 Russian ships exported 364 tons of hemp but none to British ports; and in 1782, 14 Russian ships exported 375 tons to three ports in Great Britain and Ireland. Compiled and computed from the records of the St. Petersburg Custom House: for 1778, PRO SP 91/102; for 1780, FO 65/1; for 1782, FO 97/340.

### TABLE H-II. HEMP EXPORTS FROM ST. PETERSBURG FROM 1753 THROUGH 1782[15]

[Annual Average, Measured in Tons]

| Years | Hemp Exports All Ships | Hemp Exports British Ships | British Percentage of Total |
|---|---|---|---|
| 1753–59 | N/A | 14,719 | N/A |
| 1760–66 | 18,316 | 13,252 | 72.4% |
| 1764–69 | 21,937 | 15,002 | 68.4% |
| 1770–74 | 26,912 | 17,958 | 66.7% |
| 1775–79 | 30,365 | 19,099 | 62.9% |
| 1778–82 | 34,444 | 20,514 | 59.6% |
| 1764–82 | 27,177 | 17,675 | 65.0% |
| 1767–82 | 28,441 | 18,216 | 64.0% |

the tonnage carried in British ships had increased by nearly 60 percent (see Table H–II).[15]

"Clean hemp" was the most desirable, the most exported, and the most expensive of the qualities of hemp exported. The annual average amount of "clean hemp" exported from St. Petersburg/ Kronstadt from 1764 through 1782 in all ships was more than 21,000 tons (about 80 percent of all hemp exported). Nearly 15,000 tons were transported in British ships (that is, almost 84 percent of the hemp of all kinds exported in British ships). Thus, not only was the amount of "clean hemp" exported in British ships equal to about 68 percent of the total "clean hemp" exported in all ships from St. Petersburg/Kronstadt, but it was also equal to about 83 percent of the total "hemp rough" that Great Britain imported from Russia during that period.[16]

---

15. Table H-II is compiled and computed from the records of the St. Petersburg Custom House, PRO SP 91/69, 73, 75–76, 78–79, 83, 88, 91–92, 94–96, 99–104, 107; BT 6/231; FO 65/1, 6, 9, 13, 15; FO 97/340; CO 388/54, 56, 95; SRO CO/18; and HEMP, BT 6/240.
16. The volume of "clean hemp" exports are compiled and computed from the records of the St. Petersburg Custom House, PRO SP 91/73, 75–76, 78–79, 83, 88, 91–92, 94–96, 99–104, 107; BT 6/231; FO 65/1, 6, 9, 13, 15; FO 97/340; CO 388/54, 56, 95; and SRO CO/ 18. Great Britain's hemp imports are compiled and computed from HEMP, BT 6/240.

The price of "clean hemp" fluctuated from a low of 1.20 rubles per pood (1770) to a high of 1.65 rubles per pood (1778). The price of "outshot hemp" moved between 1.00 ruble per pood (1770) and 1.50 rubles per pood (1771, 1779), with British ships transporting about 43 percent of it. "Half clean hemp" fell in price once below a ruble per pood to 90 kopeks (1770) but reached 1.40 rubles per pood twice (1777, 1779), with 35 percent of it shipped in British bottoms. Finally, the price of "hemp codilla," although hovering mostly around 40 kopeks per pood (dipping to 35 kopeks in 1770), jumped in price to 60 and 80 kopeks per pood during the last four years of the 1770s, with British ships transporting about 53 percent of it. Ibid.

Although the archival data are not complete and occasionally are inconsistent, reasonable estimates can be determined for the relative monetary value of hemp exports to the total value of all commodity exports, in all ships, in British ships and by British merchants in St. Petersburg/Kronstadt.

The monetary value of "clean hemp" exports made up a significant share of the total value of all commodity exports from St. Petersburg/Kronstadt. During the immediate postwar years of 1763–1764, the estimated annual average value of "clean hemp" exported from the capital port was a little more than 1.3 million rubles, of which almost 1.1 million rubles or about 80 percent was exported in British ships. The latter represented nearly 30 percent of the total value of commodity exports by British merchants for those two years.[17]

From 1768 through 1779 the total value of all kinds of hemp exported from St. Petersburg/Kronstadt amounted to an annual average of nearly 27 percent of the total value of all commodity exports—"clean hemp" exports accounted for 22 percent. In British ships, the value of all kinds of hemp exported amounted to slightly more than 17 percent of the total value of all commodity exports and was equal to 29 percent of the total value of all commodity exports by British merchants—"clean hemp" exports accounted for about 16 and almost 26 percent, respectively. Finally, the proportionate value of "clean hemp" to all kinds of hemp exported in British ships (89.5%) was greater than it was in all ships (about 82%) (see Figure 2, Tables H–III and H–IV).[18] [19]

The annual average value of hemp exported in all ships from St. Petersburg/Kronstadt during the period 1768–1779 was about 40 percent greater (2.4 million rubles) than the value of bar iron (1.7 million rubles). Their combined value amounted to about 45 percent of the total value of commodity exports; in British ships it was about 34 percent. The value of hemp (1.6 million rubles) and bar iron (1.5 million rubles) exports in British ships together amounted to 56 percent of the total value of exports by British merchants.[20]

The archival data from the other ports of Russia, while frag-

---

17. For the value and volume of "clean hemp" exported from St. Petersburg/Kronstadt from 1763 through 1766 see the records of the St. Petersburg Custom House, PRO SP 91/73, 75–76, 78, 83, 94, 107; BT 6/231; FO 65/15; CO 388/54, 56, 95; and VTR, *Vedomosti*, Nos. 1–2.
18. Table H-III is compiled and computed from the records of the St. Petersburg Custom House, PRO SP 91/83, 88, 91–92, 94–96, 99–104; BT 6/231; FO 65/1, 6, 9, 13, 15; FO 97/340; SRO CO/18; and VTR, *Vedomosti*, Nos. 3–5.
19. Table H-IV, ibid.
20. Cf., Table H-III with Table BI-III.

## Figure 2

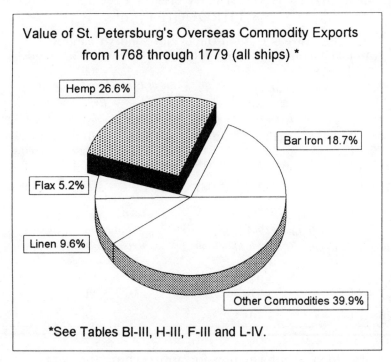

Value of St. Petersburg's Overseas Commodity Exports from 1768 through 1779 (all ships) *

- Hemp 26.6%
- Bar Iron 18.7%
- Flax 5.2%
- Linen 9.6%
- Other Commodities 39.9%

*See Tables BI-III, H-III, F-III and L-IV.

## TABLE H-III. VALUE OF ST. PETERSBURG'S COMMODITY AND HEMP EXPORTS FROM 1768 THROUGH 1779[18]

[Exclusive of Custom Duties and Sundry Charges]
(Annual Average, Measured in Current Rubles, in Thousands)

| Years | Value of Commodity Exports | Value of Hemp Exports | | | |
|---|---|---|---|---|---|
| | | All Ships | Percentage | British Ships | Percentage |
| 1768–69 | 7,135 | 1,904 | 26.7% | 1,261 | 17.7% |
| 1770–74 | 8,269 | 2,087 | 25.2% | 1,414 | 17.1% |
| 1775–79 | 9,922 | 2,867 | 28.9% | 1,817 | 18.3% |
| 1768–79 | 8,945 | 2,381 | 26.6% | 1,557 | 17.4% |

| Years | Value of Commodity Exports by British Merchants | Percentage Value of Hemp Exports in British Ships |
|---|---|---|
| 1768–69 | 4,855 | 26.0% |
| 1770–74 | 5,049 | 28.0% |
| 1775–79 | 5,970 | 30.4% |
| 1768–79 | 5,401 | 28.8% |

### TABLE H-IV. VALUE OF ST. PETERSBURG'S COMMODITY AND CLEAN HEMP EXPORTS FROM 1768 THROUGH 1779[19]

[Exclusive of Custom Duties and Sundry Charges]
(Annual Average, Measured in Current Rubles, in Thousands)

| Years | Value of Commodity Exports | Value of Clean Hemp Exports | | | |
|---|---|---|---|---|---|
| | | All Ships | Percentage | British Ships | Percentage |
| 1768–69 | 7,135 | 1,563 | 21.9% | 1,109 | 15.5% |
| 1770–74 | 8,269 | 1,812 | 21.9% | 1,324 | 16.0% |
| 1775–79 | 9,922 | 2,318 | 23.4% | 1,574 | 15.9% |
| 1768–79 | 8,945 | 1,981 | 22.1% | 1,393 | 15.6% |

| Years | Value of Commodity Exports by British Merchants | Percentage Value of Clean Hemp Exports in British Ships |
|---|---|---|
| 1768–69 | 4,855 | 22.8% |
| 1770–74 | 5,049 | 26.2% |
| 1775–79 | 5,970 | 26.4% |
| 1768–79 | 5,401 | 25.8% |

mentary, do yield information for appreciating the degree of the capital port's hegemony in Russia's overseas hemp trade. Riga was the second most important port in Russia's overseas commerce and was second in the hemp trade. In 1763–1764, the British merchants alone in Riga exported an annual average of 3,350 tons of "sundry hemp" at a value of approximately 293,000 rubles. This equaled 25 percent of the total volume and value of hemp exported in British ships from St. Petersburg/Kronstadt. During the wartime period from 1778 through 1782 Riga's export of all kinds of hemp was an annual average of about 13,000 tons, which was about 38 percent of the total hemp exported from St. Petersburg/Kronstadt. About 25 percent (that is, almost 3,300 tons, about the same quantity that British merchants had exported in 1763–1764) was designated for ports in Great Britain and Ireland. This was equivalent to about 16 percent of the hemp tonnage shipped in British vessels from St. Petersburg/Kronstadt, almost all of which were designated for ports in Great Britain and Ireland.[21]

---

21. The accounts of Riga's hemp exports in 1763–1764 are compiled and computed from the records of the St. Petersburg Custom House, PRO SP 91/75, 107; BT 6/231; and CO 388/95; in 1778–1782, ibid., PRO SP 91/103–4; and FO 65/1, 6, 9. During 1778–1782,

### TABLE F-I. GREAT BRITAIN'S FLAX IMPORTS FROM 1764 THROUGH 1782[23]

[Annual Average, Measured in Tons]

| Years | Total Imports | Russian Imports | Russian Percentage of Total |
|---|---|---|---|
| 1764–66 | 7,063 | 5,111 | 72.4 |
| 1767–69 | 6,917 | 5,526 | 79.9 |
| 1770–74 | 6,509 | 5,459 | 83.9 |
| 1775–79 | 6,988 | 5,793 | 82.9 |
| 1778–82 | 6,089 | 5,040 | 82.8 |
| 1764–82 | 6,694 | 5,419 | 81.0 |
| 1767–82 | 6,625 | 5,477 | 82.7 |

## FLAX

Like Russian hemp, flax was exported in several qualities inclusive of "flax sundry" and simply "flax." Flax exports were frequently labeled "12-head" ("clean flax"), "9-head," "6-head," and "codilla" at St. Petersburg/Kronstadt and Archangel. At Riga the sorted qualities of flax went by different names and therefore are not easily comparable to those at other ports. And, as in the case of hemp imports into Great Britain, British customs recorded all flax imports as simply "flax rough."[22]

From 1764 through 1782 Great Britain imported about 81 percent of its total flax from Russia. In the three years prior to the conclusion of the Anglo-Russian Commercial Treaty in 1766 and the promulgation of the Russian tariff in 1766/67, Great Britain's Russian flax imports represented an annual average of about 72 percent of its total flax imports. During the remainder of the period 1767–1782 this amount increased to about 83 percent (see Table F–I).[23]

Flax exports from St. Petersburg/Kronstadt nearly doubled from 1764 through 1782. Exports in British ships accounted for an annual average of 73 percent of the total which amounted to 43 percent of Great Britain's total flax imports from Russia. It was only during the late 1770s, when Great Britain was at war with

---

Archangel exported an estimated annual average of 655 tons of "sundry hemp," about 42 percent of which was shipped to ports in Great Britain and Ireland. Ibid.
22. See PRO CU 3 (1753–1780), passim.
23. Table F-I is compiled and computed from FLAX, BT 6/240.

### TABLE F-II. FLAX EXPORTS FROM ST. PETERSBURG FROM 1753 THROUGH 1782[24]

[Annual Average, Measured in Tons]

| Years | Flax Exports All Ships | Flax Exports British Ships | British Percentage of Total |
|---|---|---|---|
| 1753–59 | N/A | 1,223 | N/A |
| 1760–66 | 2,144 | 1,509 | 70.4% |
| 1764–69 | 1,955 | 1,613 | 82.5% |
| 1770–74 | 2,649 | 2,025 | 76.5% |
| 1775–79 | 4,525 | 3,150 | 69.6% |
| 1778–82 | 4,124 | 2,734 | 66.3% |
| 1764–82 | 3,163 | 2,308 | 73.0% |
| 1767–82 | 3,434 | 2,500 | 72.8% |

the American colonies, France, and Spain, that British shipments reached 54 percent of the total flax that Britain imported from Russia (see Table F–II).[24]

The unit export prices of flax were about 50 percent higher than hemp at St. Petersburg/Kronstadt. The best quality and most expensive flax was "12-head flax" or "clean flax" which sold for 2.80 rubles per pood at its high, while "clean hemp" sold for 1.65 rubles per pood at its high. Of the total amount of all varieties of flax exported from 1764 though 1782, 72 percent was "12-head flax." About 83 percent of the latter was exported in British ships and accounted for 82 percent of the total flax exported in British ships.[25] Despite its higher price the total value of flax exported from St. Petersburg/Kronstadt from 1768 through 1779 amounted to only slightly more than 5 percent of the total value of commodities exported. During that same period the total value of hemp exports was five times greater than the total value of flax exports

---

24. Table F-II is compiled and computed from the records of the St. Petersburg Custom House, PRO SP 91/73, 75–76, 78–79, 83, 88, 91–92, 94–96, 99–104, 107; BT 6/231; FO 65/1, 6, 9, 13, 15; FO 97/340; CO 388/54, 56, 95; and SRO CO/18.
25. The export price at St. Petersburg/Kronstadt for "12-head flax" or "clean flax" fluctuated from a low of 1.80 rubles (1770) to a high of 2.80 rubles per pood (1766). The price of "9-head flax" moved from 1.60 rubles (1770) to a high of 2.45 rubles per pood (1766), with British ships transporting about 57 percent of it. "6-head flax" sold for as low as 1.40 rubles (1770) and as high as 2.15 rubles per pood (1766), with British ships transporting about 27 percent of it; and finally, "flax codilla" went for 60 kopeks per pood for most of the period under discussion and British ships transported about 48 percent of it. Compiled and computed from the records of the St. Petersburg Custom House, PRO SP 91/73, 75–76, 78–79, 83, 88, 91–92, 94–96, 99–104, 107; BT 6/231; FO 65/1, 6, 9, 13, 15; FO 97/340; CO 388/54, 56, 95; and SRO CO/18.

## Figure 3

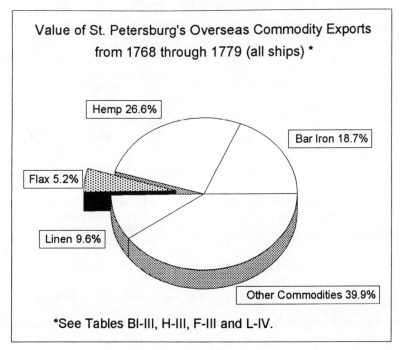

Value of St. Petersburg's Overseas Commodity Exports from 1768 through 1779 (all ships) *

Hemp 26.6%
Bar Iron 18.7%
Flax 5.2%
Linen 9.6%
Other Commodities 39.9%

*See Tables BI-III, H-III, F-III and L-IV.

because of the greater volume shipped. (see Figure 3, Tables H–III and F–III.)[26]

Narva and Riga competed successfully with St. Petersburg/ Kronstadt in the overseas flax trade, particularly in regard to British participation. In 1763–1764 British merchants in Narva and Riga exported an annual average of more flax (4,140 and 3,172 tons, respectively) than did all merchants in St. Petersburg/Kronstadt (2,779 tons). In terms of the monetary value, flax exports by British merchants in Narva and Riga amounted to an annual average of about 39 percent of the total value of their commodity exports during 1763–1764.[27]

---

26. Table F-III is compiled and computed from the records of the St. Petersburg Custom House, PRO SP 91/ 83, 88, 91–92, 94–96, 99–104; BT 6/231; FO 65/1, 6, 9, 15; FO 97/340; and VTR, *Vedomosti*, Nos. 3–5.
27. The export and import custom accounts of British merchants in Narva and Riga for the years 1763–1764 are compiled and computed from the records of the St. Petersburg Custom House, PRO SP/73, 75, 107; BT 6/231; and CO 388/95. For St. Petersburg/Kronstadt in 1763–1764, ibid.; and PRO SP 91/76, 83; and FO 65/15. For Great Britain's flax imports, see FLAX, BT 6/240.

### TABLE F-III. VALUE OF ST. PETERSBURG'S COMMODITY AND FLAX EXPORTS FROM 1768 THROUGH 1779[26]

[Exclusive of Custom Duties and Sundry Charges]
(Annual Average, Measured in Current Rubles, in Thousands)

| Years | Value of Commodity Exports | All Ships | Percentage | Value of Flax Exports British Ships | Percentage |
|---|---|---|---|---|---|
| 1768–69 | 7,135 | 286 | 4.0% | 251 | 3.5% |
| 1770–74 | 8,269 | 363 | 4.4% | 284 | 3.4% |
| 1775–79 | 9,922 | 639 | 6.4% | 473 | 4.8% |
| 1768–79 | 8,945 | 465 | 5.2% | 358 | 4.0% |

| Years | Value of Commodity Exports by British Merchants | Percentage Value of Flax Exports in British Ships |
|---|---|---|
| 1768–69 | 4,855 | 5.2% |
| 1770–74 | 5,049 | 5.7% |
| 1775–79 | 5,970 | 7.9% |
| 1768–79 | 5,401 | 6.6% |

In 1766, when flax was fetching its highest price during the period 1763–1782, Narva and Riga exported a total of more than 9,000 tons of flax. This was more than seven times the amount of flax exported from St. Petersburg/Kronstadt. Sixty-three percent of Narva's and Riga's flax exports were designated for ports in Great Britain and Ireland.[28]

Finally, during the wartime period from 1778 through 1782, Narva and Riga exported an annual average of more than 10,000 tons of flax, about two and one-half times as much flax as was exported from St. Petersburg/Kronstadt. The percentage of flax designated for ports in Great Britain and Ireland from Narva and Riga was about the same (54%) as from St. Petersburg/Kronstadt.[29]

There can be no doubt that Great Britain's linen manufacture and trade were heavily dependent upon imports of Russian flax and hemp. Even though the exact total of Great Britain's linen

---

28. Narva and Riga exports in 1766 are compiled and computed from the records of the St. Petersburg Custom House, PRO CO 388/54. For exports from St. Petersburg/Kronstadt in 1766, ibid.; PRO SP 91/78, 83; FO 65/15; and BT 6/231. For Great Britain's flax imports in 1766, see FLAX, BT 6/240.
29. Narva and Riga flax exports for 1778–1782 are compiled and computed from the records of the St. Petersburg Custom House, PRO SP 91/103–4; FO 65/1, 6, 9; and BT 6/231. For exports from St. Petersburg/Kronstadt for 1778–1782, see ibid.; PRO SP 91/102; and FO 97/340. For Great Britain's flax imports, see FLAX, BT 6/240.

### TABLE L-I. THE APPROXIMATE LINEN EQUIVALENT OF ENGLAND'S RUSSIAN FLAX IMPORTS: 1765–1785[30]

| Years | Imported Russian Flax [cwt.] | Total Approximate Linen Equivalent ['000 yards] | Harte's Total Approximate Linen Equivalent ['000 yards] | Russian % of Harte's Linen Equivalent |
|---|---|---|---|---|
| 1765 | 84,349  | 10,607 | 21,256 | 49.9% |
| 1770 | 104,550 | 13,148 | 42,761 | 30.7% |
| 1775 | 138,702 | 17,442 | 44,902 | 38.8% |
| 1780 | 112,774 | 14,182 | 44,343 | 32.0% |
| 1785 | 167,647 | 21,082 | 45,947 | 45.9% |

output is not known there are crude estimates that have been generally accepted. For example, in 1738 England's linen production amounted to about 21.5 million yards; in 1756, 25.8 million yards. N. B. Harte proposes a computation for estimating the "minimum levels" of English linen production for specific years. The formula he uses is that 1 cwt. [hundredweight = 112 lb.] of flax yielded 50 lbs. of potential yarn and that 3,550 cwt. of yarn produced 1 million yards of linen. These "minimum levels" make no allowance, however, for changes in the average quality of linen produced from the different qualities of the flax and yarn which went into its production. Nor do they account for the unknown amount of imported flax that went into the manufacture of textiles other than linen or how much imported hemp might have been used in the production of linen. By using Harte's computation the approximate Russian contribution to England's linen output can be estimated by converting imported Russian flax into equivalent linen yards. Between 1765 and 1785 the Russian contribution fluctuated between 30 and 50 percent (see Table L–I).[30]

The Russian contribution to English linen output can be further appreciated by comparing the imported Russian flax-linen equivalent to the quantity of Irish and Scottish linen imported into England. For the year 1770 Scottish and Irish linen imports into England amounted to approximately 13 and 19.7 million yards, respectively. The Russian flax-linen yard equivalent in 1770 was

---

30. Table L-I is compiled and computed from Harte, pp. 103–7, especially Table 4.3, "English imports of flax and yarn and the approximate linen equivalent, 1700–90," p. 104; and from FLAX, BT 6/240.

about 13.1 million yards, that is, approximately equal to the Scottish linen imports and two-thirds of the Irish linen imports into England. In 1780 the Scottish and Irish linen imports into England were 13.4 and 18.3 million yards, respectively. The Russian flax-linen equivalent for that year was about 14.2 million yards, more than likely equal to the Scottish imports and about three-quarters of the Irish linen imports into England.[31]

## LINEN

Russia produced great quantities of linens during the second half of the eighteenth century, although the precise amount is not known.[32] It exported a wide variety of these in significant quantities to Great Britain. But this contribution to the British market is difficult to estimate because of the problems associated with nomenclature, measurement, and monetary value in the custom sources. N. B. Harte has summarized the difficulties and complexities in the British sources that have confounded scholars for quite some time:

> The problems of extracting the figures for linen from the ledgers of the Inspector-General of Imports and Exports defeated even the energy of Mrs. Schumpeter. The absence of figures for the linen import trades is the most conspicuous omission from her *English Overseas Trade Statistics, 1697–1808*. The ledgers (P.R.O., Customs 3) distinguished over a hundred different types of linen, differentiated by length, breadth, quality, colour, value, use or place of origin. To overcome the problems of comparability arising from different types of linen being measured by pieces, bolts, yards, and ells of varying lengths (and the "hundred ells" of either 100 or 120 ells), the Inspector-General used conventionalised "official values" which were quite distinct from the official rates of the Book of Rates. But the official

---

31. Compiled and computed from FLAX, BT 6/240 and Table 4.2, "Irish and Scottish linen production for the English market ('ooo yards)," in Harte, p. 93. I am accepting Harte's statement at the bottom of p. 93: "The Scottish figures include a certain amount of linen destined for the Scottish domestic and export markets, but the greater part was for the English market." Harte is apparently also interpreting "Irish exports to Great Britain" to mean "for the English market," which, for the purposes of this approximate comparison, I have also accepted. See also Cullen, p. 60; and Hamilton, pp. 404–5.
32. The history of the Russian linen industry during the second half of the eighteenth century is wanting, although there are several articles and books that treat this subject. For example, G. S. Isaev, *Rol' tekstil'noi promyshlennosti v genezise i razvitii kapitalizma v Rossii 1760–1860* (Leningrad: Nauka, 1970), in particular, pp. 118–32; E. I. Rubinshtein, *Polotnianaia i bumazhnaia manufaktura Goncharovykh vo vtoroi polovine XVIII veka* (Moskva: Sovetskaia Rossiia, 1975); and Evg. Diubiuk, *Polotnianaia promyshlennost' Kostromskogo kraia vo vtoroi polovine XVIII i pervoi polovine XIX veka* (Kostroma: Sovetskaia, 1921).

values also mask the significant quality [of the linen] and do not get behind the misleading entries produced by the complexities of the duty structure.[33]

British merchants who imported foreign linens into Great Britain were also challenged by British Customs. A Parliamentary report published at the end of the nineteenth century reviewed the regulations governing customs during the reign of George III and identified a general problem. It quoted a competent authority who wrote in 1751:

> What a maze our merchants must be in: if we consider the many exceptions, and exceptions from exceptions, the many regulations, and regulations of regulations for collecting the Customs and for paying the drawbacks upon goods re-exported, we must conclude it impossible for any merchant in this country to be master of his business if he be what we call a general merchant; consequently he must trust to those honest gentlemen called Custom House Officers for the duties he is to pay upon importation, and the drawbacks he is entitled to upon exportation. Can we wonder at the decay of our commerce under such circumstances? Should we not rather wonder that we have any left?[34]

The report illustrated its point with a specimen of the customs fate of 2,000 ells of Russian linen imports in the year 1784. The total duty of 69 pounds and 17 shillings to be paid by the merchant "had to be painfully built up" from no less than ten enumerated "branches" of duty and this specimen was considered "a simple one."[35]

Sources from Russian custom records also frustrate research. In the data available from the records of the St. Petersburg Custom House there are more than a dozen kinds of Russian linens for export that are measured in either arshins or pieces and occasionally in both. The Russian arshin was equal to .77 yard, or 2.33 feet, or 27.96 inches [which the merchants in the Anglo-Russian trade accepted as 28 inches]. The Russian piece was equal to 45 arshins or 35 yards. While it is possible to convert the volume of Russian linens measured in arshins and or pieces into British linen yards or even into British ells (1 ell = 1.25 yards),[36] this would not resolve all the problems of nomenclature, measurement, and mone-

---

33. Harte, pp. 85n-86n.
34. P. P. Customs, 1897, pp. 13–14.
35. Ibid., p. 14.
36. See "Weights and Measures." p. xiii.

tary value of imported Russian linens into Great Britain. A commodity shipped from a Russian port by name or description is not usually recorded in the custom ledger at the port of destination by that same name, description, or measurement. It is therefore almost impossible to correlate the Russian export and the British import of the same linen commodity (and the reverse situation exists when researching British commodity imports into Russia). An example of this is found in Great Britain's importation of Russian sailcloth which, when exported from Russia, was recorded in "pieces" only.[37] But in two different sources comprising official British customs, sailcloth is recorded differently. One source denotes sailcloth imports in "hundreds" with separate listings for Russia, East Country, et cetera (but without indicating that sailcloth might have been imported from Russian territories in East Country). The other source records imported sailcloth in "yards" as simply "Sails Foreign made" without noting the country of origin.[38]

Russia's opportunity to penetrate the British sailcloth market was largely owing to the Anglo-Irish competition in that trade. In 1746 the Irish Parliament granted an export bounty on sailcloth which, by 1750, adversely affected England's sailcloth industry enough for England to retaliate. England imposed not only a duty equal to the Irish bounty on its Irish sailcloth imports but also provided its own sailcloth makers with a bounty when they exported to foreign countries. Although the Irish bounty was subsequently discontinued, the English duty and bounty remained and continued to undermine the Irish sailcloth industry and trade.[39] Nevertheless, Great Britain still could not satisfy its growing demand for sailcloth and thus imported large quantities of Russian sailcloth. From 1764 though 1782 an annual average of about 70 percent of Britain's imported sailcloth came from Russia, mostly during the years prior to Britain's war with the American colonies, France, Spain, and the United Provinces. The overwhelming amount of Britain's Russian sailcloth came from St. Petersburg/Kronstadt.[40]

Despite the obstacles noted above there are customs data that

---

37. Cf., Isaev, who states it in arshins, "Tablitsa 10, Eksport polotna za 1760–1769 gg. cherez Peterburgskii port," p. 123.
38. Cf., SAILCLOTH, BT 6/240 and 230; and P. P. Linen, 1773, Appendix 12. For problems dealing with the ambiguity of the term "East Country," see Chap. 1, note 10.
39. Horner, pp. 233–34; and Robert Stephenson's testimony to the parliamentary investigatory committee, P. P. Linen, 1773, p. 108.
40. Compiled and computed from SAILCLOTH, BT 6/240. For sailcloth exports from St. Petersburg/Kronstadt and other Russian ports, see the records of the St. Petersburg Custom House PRO SP 91/73, 75–76, 78–79, 83, 88, 91–92, 94–96, 99–104, 107; BT 6/231; CO 388/54, 56, 95; FO65/1, 6, 9, 13, 15; FO 97/340; and SRO CO/18.

permit reasonable comparisons to be made, however crude the estimates might be. There is a 1773 parliamentary report that included England's imports, re-exports and retained imports of foreign linens from 1752 through 1771. The report also included testimony about the conditions of the linen trade and industry. First, there was a lament about the recent recession, which was attributed to insufficient governmental support of the domestic linen manufacture and the effect that low import custom duties and re-export drawbacks had on the importation of foreign linens. Second, it was noted that the industry was unable to cope with foreign competition. This resulted in a significant drop in production, prices, and employment during the late 1760s. Third, the report stated that the archaic customs system and the official rates upon which the duties were based created additional problems because the import and re-export custom declarations on specific foreign linens frequently did not reflect reality. For example, the report noted that the designation of "Narrow Germany" did not necessarily mean linen imported from the Germanies with a specific measurement and quality for which a modest duty was normally paid. In fact, better quality linens entered and exited customs as "Narrow Germany." Finally, Russian linens were cited as having frequently undercut the domestic market because of their original lower ratings and duties (regardless of their improved quality) and their cheaper prices.[41]

Indeed, Russian linen did relatively well on the English market. During the twenty-year period from 1752 through 1771 an estimated annual average of 19 percent of England's total foreign linen imports was Russian linen. For the period after the Seven Years' War (1763–1771) Russian linen made up 22.6 percent. Of the total amount of England's foreign linen imports that were subsequently re-exported, Russian linen amounted to 6.6 percent for the entire period, increasing to nearly 9 percent after the war. Finally and significantly, the amount of England's total retained foreign linen imports estimated as Russian linen was 24 percent for the entire twenty-year period, increasing to 28.6 percent after the war (see Table L–II).[42]

Russia's contribution to England's linen manufacture and trade can be demonstrated further. From 1752 through 1771 England's Russian linen imports were an estimated annual average of

---

41. P. P. Linen, 1773, pp. 101–11.
42. Table L-II is compiled and computed from P. P. Linen, 1773, Appendixes 12 and 13, pp. 119–30.

TABLE L-II. RUSSIA'S ESTIMATED SHARE OF ENGLAND'S TOTAL FOREIGN LINEN IMPORTS, RE-EXPORTS, AND RETAINED IMPORTS FROM 1752 THROUGH 1771[42]

[Annual Average, Measured in '000 Yards]

| Years | Total Linen Imports | Russian Linen Imports | % | Total Linen Re-Exports | Russian Linen Re-Exports | % | Total Linen Retained Imports | Russian Linen Retained Imports | % |
|---|---|---|---|---|---|---|---|---|---|
| 1752–55 | 31,512 | 5,469 | 17.4% | 7,290 | 322 | 4.4% | 24,222 | 5,147 | 21.2% |
| 1756–62 | 27,466 | 4,393 | 16.0% | 8,315 | 425 | 5.1% | 19,151 | 3,967 | 20.7% |
| 1763–66 | 26,462 | 5,967 | 22.5% | 7,376 | 572 | 7.8% | 19,087 | 5,396 | 28.3% |
| 1767–71 | 24,988 | 5,635 | 22.6% | 8,251 | 783 | 9.5% | 16,737 | 4,852 | 29.0% |
| 1752–71 | 27,455 | 5,234 | 19.1% | 7,906 | 523 | 6.6% | 19,549 | 4,710 | 24.1% |
| 1763–71 | 25,643 | 5,783 | 22.6% | 7,862 | 689 | 8.8% | 17,781 | 5,094 | 28.6% |

approximately 5.2 million yards. Ireland's linen production destined for the English market was an annual average of approximately 14.5 million yards, and Scotland's was 11 million yards. Therefore, England's Russian linen imports were a little more than one-third of the Irish and almost one-half of the Scottish imports during this period.[43] In 1770 England imported approximately 6.2 million yards of linen from Russia, which, when added to the crude estimate of the Russian flax-linen equivalent for that year of about 13.1 million yards, made Russian linen imports about equal to Irish linen imports into England (19.7 million yards) and approximately 50 percent greater than Scottish linen imports into England (13 million yards).[44]

These estimates of England's Russian linen imports compiled from British customs reports are closely supported by the data taken from Russian customs records. From 1763 through 1771, according to the former, England's Russian linen imports amounted to an estimated annual average of about 5.8 million yards. According to the latter a minimum estimate for a selected number of Russian linens—broad and narrow diaper and linen, crash, drilling, flem, and ravenduck—exported in all ships from St. Petersburg/Kronstadt amounted to an annual average of about 7.3 million yards.[45] Approximately 6.4 million yards (about 88 percent) were exported in British ships, almost all designated for ports in Great Britain and Ireland. In 1770 the minimum estimate of these Russian linens exported in all ships from the capital port amounted to about 7.4 million yards with approximately 6.3 million yards exported in British ships, 6.2 million yards of which England imported (see above).[46]

Nevertheless, whatever stimulus the Anglo-Russian Commercial Treaty might have given to increasing Russian linen exports to Great Britain, it was not sufficient to overcome Britain's reduced demand for Russian linens. This was owing in part to the recession in the linen industry and trade during the early 1770s

---

43. Estimates are derived from calculations of data on Ireland's and Scotland's linen production and exports provided in Cullen, p. 60; Hamilton, pp. 404–5; Harte, p. 93; Horner, pp. 201–3; and P. P. Linen, 1773, Appendix 12.
44. Estimates compiled and computed from P. P. Linen, 1773, Appendix 12; and Table L-I; Cullen, p. 60; Hamilton, pp. 404–5; Harte, p. 93; and Horner, pp. 201–3.
45. Cf., Isaev, "Tablitsa 10, Eksport polotna za 1760–1769 gg. cherez Peterburgskii port," p. 123.
46. Compiled and computed from P. P. Linen, 1773, Appendix 12; Table L-II; and the records of the St. Petersburg Custom House, PRO SP 91/69, 73, 75–76, 78–79, 83, 88, 91, 107; BT 6/231; CO 388/54, 56, 95; and FO 65/15. Although archival data for the other Russian ports are not comprehensive during the period 1763–1771, those that are available suggest that the amount of linen goods exported was negligible.

## TABLE L-III. MINIMUM ESTIMATES OF SELECTED LINENS EXPORTED FROM ST. PETERSBURG FROM 1763 THROUGH 1782[47]

[Broad & Narrow Diaper & Linen, Crash, Drilling, Flem, & Ravenduck]
(Annual Average, Measured in Yards)

| Years | Linen Exports | | British Percentage of Total |
|---|---|---|---|
| | All Ships | British Ships | |
| 1763–66 | 7,289,309 | 6,573,733 | 90.2% |
| 1767–69 | 7,149,053 | 5,987,666 | 83.8% |
| 1770–74 | 6,338,645 | 5,618,838 | 88.6% |
| 1775–79 | 7,075,361 | 5,590,288 | 79.0% |
| 1778–82 | 7,311,682 | 4,872,329 | 66.6% |
| 1763–82 | 7,005,326 | 5,695,443 | 81.3% |
| 1767–82 | 6,934,330 | 5,475,871 | 79.0% |

and to Britain's increased domestic production. During the entire period from 1763 through 1782 the total of the above-mentioned Russian linens exported from St. Petersburg/Kronstadt fluctuated widely—from an annual average low of about 6.3 million yards in 1770–1774 to a high of about 7.3 million yards in 1763–1766 and 1778–1782. The amount exported in British ships, however, declined continuously from a high of about 6.6 million yards in 1763–1766 (about 90 percent) to a low of about 4.9 million yards in 1778–1782 (about 67 percent) when Britain was at war at first with France and Spain and later with the United Provinces. However, during the entire period from 1763 through 1782 British ships accounted for about 80 percent of the total linen exported from St. Petersburg/Kronstadt (see Table L–III).[47]

---

47. Table L-III is compiled and computed from the records of the St. Petersburg Custom House, PRO SP 91/73, 75–76, 78–79, 83, 91–92, 94–96, 99–104, 107; BT 6/231; CO 388/54, 56, 95; FO 65/1, 6, 9, 13, 15; and SRO CO/18.

These estimates reinforce the observation made by scholars concerning the underlying trend in Russia's long-term relationship with the British linen industry and trade. For example, Peter Mathias has stated:

Higher duties for linen in the course of the century encouraged the growth of the Irish, Scottish and English production so that imports fell relative to total output and the import of flax and yarn. These increased four-fold (in terms of linen equivalent) during the century while indigenous production also may have quadrupled by 1780. The value of foreign (i.e. not Irish) finished linen imports fell from £ .85m to £ .52m over the century. The position of Russian flax and yarn imports was evidently a function of relative price. . . . When the eighteenth century closed, Russia was principally relegated to the role of primary

Significant as these data appear, they do not reveal all we should know about Russian linen on the British market. Professor Peter Mathias graciously reminded me several years ago that the customs trade data themselves probably understated the actual Russian component in British linen imports:

> Re-routing of Baltic produce, through Holland and some of the German ports was not uncommon. British ships picked up an increasing share of the carrying trade. . . . Russian flax, linen yarn and linens "in their white" (unfinished and unbleached) featured in German and Dutch imports and, when worked up, exported as finished products to Great Britain. In that sense, the Russian component in British linen imports was greater than Professor Kaplan's direct data reveal.[48]

The monetary value of the Russian linens exported in British ships was substantial and it reinforced the importance of British participation in the commodity export trade of St. Petersburg/Kronstadt. This is obvious from an examination of just broad and narrow linen, crash, drilling, flem, and ravenduck exports from 1768 through 1779. Their annual average estimated minimum value alone amounted to 9.6 percent and 8.1 percent of the total value of commodity exports in all ships and in British ships, respectively. When the value of sailcloth is added, those percentages increase to 12.9 in all ships and to 8.7 in British ships. The value of the aforementioned linen exports in British ships as a percentage of the estimated total value of commodity exports by British merchants amounted to 13.5 percent. This percentage increases to 14.4 percent when sailcloth is added (see Figure 4, Table L–IV).[49] More than one million rubles were received annually from the sale of linens at St. Petersburg/Kronstadt. The evidence for

---

produce - or intermediary product - supplier to the industries of other countries.
  MATHIAS, p. 7.
48. MATHIAS, p. 14.
49. Table L-IV is compiled and computed from the records of the St. Petersburg Custom House, PRO SP 91/83, 88, 91–92, 94–96, 99–104; BT 6/231; FO 65/1, 6, 9, 13, 15; FO 97/340; and VTR, *Vedomosti*, Nos. 3–5. The relative prices of Russian linens calculated per arshin during the 1760s and 1770s were as follows: crash linen stayed very much within the range of 3 to 4 kopeks; narrow and broad linen fluctuated from about 5 kopeks to slightly more than 13 kopeks; although not shown in Table L-IV because of the fragmentary nature of the data, narrow and broad diaper moved within a range from about 7 kopeks to as much as 14 kopeks but stayed closer to 11 kopeks; ravenducks fluctuated from 10 to on occasion slightly more than 14 kopeks; drillings moved from 11 to 14 kopeks, but stayed very much at 12 kopeks; and flems dipped to a low of 11 and 12 kopeks and rose to a high of 17 kopeks. Sailcloth, which was much more expensive than any other linen export, commanded from 14 to 20 kopeks per arshin. Ibid.

### Figure 4

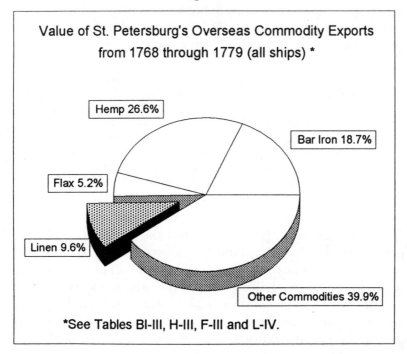

Value of St. Petersburg's Overseas Commodity Exports from 1768 through 1779 (all ships) *

- Hemp 26.6%
- Bar Iron 18.7%
- Flax 5.2%
- Linen 9.6%
- Other Commodities 39.9%

*See Tables BI-III, H-III, F-III and L-IV.

linen exports from the other ports of Russia is too fragmentary for meaningful comparison.

As with bar iron, Great Britain's huge and valuable imports of Russian hemp, flax and linens had a significant contributory effect on the expansion of British manufacture and trade from the end of the Seven Years' War to the peace in 1783 by dampening domestic prices and by sustaining the demand for labor. Together they accounted for just over 40 percent of St. Petersburg's overseas commodity exports alone.

These Russian imports had ramifications as well for Great Britain's political and military-strategic policies. Because of their steady supply Great Britain was able to pursue its imperial foreign policy at a time when its own resources of naval stores proved inadequate to meet current challenges. Whatever hopes Great Britain had for achieving military and foreign policy successes during this period, they depended upon the importation of Russian goods to keep the British imperial naval and mercantile fleet afloat. In no case was this more true than with timber.

### TABLE L-IV. VALUE OF ST. PETERSBURG'S COMMODITY AND SELECTED LINEN EXPORTS FROM 1768 THROUGH 1779[49]

Broad and Narrow Linen, Crash, Drilling, Flem and Ravenduck
[Exclusive of Custom Duties and Sundry Charges]
(Annual Average, Measured in Current Rubles, in Thousands)

| Years | Value of Commodity Exports | Value of Linen Exports | | | |
|---|---|---|---|---|---|
| | | All Ships | Percentage | British Ships | Percentage |
| 1768–69 | 7,135 | 957 | 13.4% | 780 | 10.9% |
| 1770–74 | 8,269 | 711 | 8.6% | 668 | 8.1% |
| 1775–79 | 9,922 | 976 | 9.8% | 769 | 7.8% |
| 1768–79 | 8,945 | 863 | 9.6% | 729 | 8.1% |
| | | *1,150 | *12.9% | *778 | *8.7% |

| Years | Value of Commodity Exports by British Merchants | Percentage Value of Linen Exports in British Ships |
|---|---|---|
| 1768–69 | 4,855 | 16.1% |
| 1770–74 | 5,049 | 13.2% |
| 1775–79 | 5,970 | 12.9% |
| 1768–79 | 5,401 | 13.5% |
| | | *14.4% |

*The value of sailcloth added.

# Chapter VI
# Timber

The alleged scarcity of timber resources in both Great Britain and Russia during the second half of the eighteenth century troubles historians today as much as it did policy makers of both countries two centuries ago. In 1926 Robert Greenhalgh Albion framed the historical perspective and argument for the scarcity of timber resources in Great Britain in the opening paragraph of his study, *Forests and Sea Power. The Timber Problem of the Royal Navy 1652–1862:*

> From the days when Cromwell ruled England till the battle of Hampton Roads sounded the knell of wooden ships of war, the heads of the English Navy worried over its timber shortage. Not only were the woodlands of England becoming less able to supply oak for the hulls of the King's ships, but in all Britain there grew no trees suitable for the best masts. Dependence on foreign lands for masts, therefore, had to be recognized from the outset; but it was only with the greatest reluctance that England looked to forests beyond the seas for timber to replace her own oak.[1]

Thomas S. Ashton followed Albion a generation later and added a cogent generalization to this discourse: "Wisely or unwisely, Englishmen had chosen to use their land to produce grain and livestock rather than trees, and to meet their need for timber largely by import from Norway and the coastal areas of the Baltic."[2] Both these studies reflect contemporary eighteenth-century allegations and discussions in the British marketplace and in the House of Commons about the scarcity of timber resources. The shipbuilding and construction industries had placed such demands on domestically produced timber that prices were pushed higher and the importation of timber became imperative.

---

1. Robert G. Albion, *Forests and Sea Power. The Timber Problem of the Royal Navy 1652–1862* (Cambridge: Harvard University Press, 1926), p. vii.
2. Thomas S. Ashton, *Economic Fluctuations in England, 1700–1800* (Oxford: Clarendon Press, 1959), p. 85.

The relative scarcity of shipbuilding timber threatened British national security. In 1771, a parliamentary committee was "appointed to consider how His Majesty's Navy may be better supplied with TIMBER." Among the points of enquiry were:

> First, the State of Timber fit for the Supply of His Majesty's Navy, its Sufficiency or Insufficiency for that Purpose; Secondly, in Case of Insufficiency, to what Causes it might be imputed; Thirdly, its Operation on the Prices; Fourthly, the Means of further Supply at Home, and by Importation; and under this Head, naturally come the Nature and Quality of the different Sorts of Timber.

What the committee subsequently learned and historians have savored was that Britain's dependency on foreign timber had definitely increased during the post-Seven Years' War period because "there was a great Scarcity of Timber for Ship-building in *England* . . . that there is not a sufficient Quantity of Timber in *England* to be purchased at any Price."[3] This view was reinforced by the Fourth Earl of Sandwich when he declared in 1774 in the House of Commons that on becoming First Lord of the Admiralty (1771–1782) he found that "the Navy was in a ruinous state, with not six months' timber of any in stock."[4]

But our knowledge about the semantics, the perceptions, and the genuineness of the scarcity of timber that prevailed in Great Britain during the second half of the eighteenth century has been clarified and refined by new details and analyses in the studies of M. J. Williams and R. J. B. Knight. While Williams states that "Sandwich's claims were somewhat exaggerated" he does admit that there was sufficient evidence to support the Navy Board's complaints about the shortage of critical shipbuilding timber and the neglect of the Royal Forests. According to Williams, "during

---

3. P. P. Timber, 1771, p. 15.
4. Quoted in M. J. Williams, "The Naval Administration of the Fourth Earl of Sandwich" (D. Phil. diss., Jesus College, University of Oxford, 1962), p. 283.

When Sandwich assessed the dockyards in 1774 he worried about shipbuilding productivity and wrote:

> We set out in the year 1756 with 69 ships of the line, we ended in the year 1762 with 81, 12 were taken from the enemy and employed during the war in the King's service, & 'tho we built as fast as possible in the King's Yard, and employed all the Merchant Yards that could build a ship for us, it is to be observed that we finished with exactly the same number with which we set out.

Quoted in Knight, "The Building and Maintenance of the British Fleet during the Anglo-French Wars 1688–1815," in *Les Marines de guerre Européennes XVII–XVIII<sup>e</sup> siècles*, ed. Martine Acerra, José Merino, and Jean Meyer (Paris: Presses de l'Université de Paris-Sorbonne, 1985), p. 36.

the period 1763–1769 the purchases of English and Foreign timber . . . had scarcely kept pace with consumption, so that at the end of 1770 the stocks were dangerously low . . . being considerably less than one year's consumption." Moreover, according to a later parliamentary report, the "seriously decayed condition" of the timber in the Royal Forests was still apparent by 1783.[5] Knight argues that Albion "failed to realize that there was a steady increase in the size of ships of the line through the century" and that Albion also overemphasized "the problem of supply rather than administration." Knight asserts that: "Lack of materials was not a problem, although shortages sometimes led to building with unseasoned timber." "At the root of the problem," he avers, "was a lack of skilled labour, insufficient dockyard facilities to maintain the fleet, administrative shortcomings and a failure of political will."[6]

Nevertheless, it is still fair to say that although Great Britain's timber stands in the mid-eighteenth century were still richly endowed, they no longer met the increasing demands of the shipbuilding and construction industries. Not enough good domestic woods were marketed. Inland transportation costs were at times equal to the cost of the timber itself and were exorbitant when compared to shipping by sea. Moreover, the British hand-sawn timber industry could not compete with the production of foreign water-powered mills, particularly those in the East Baltic.[7]

Great Britain's fear of domestic timber insufficiency and the implications for its foreign policy made it necessary for the nation to buy more and more of its timber supplies from Russia. During Catherine II's reign Russia became one of Great Britain's larger suppliers of timber, especially in building construction and strategic shipbuilding timber and masts. This had the dual effect of alleviating British anxieties and exacerbating Russian perceptions of timber scarcity. From 1740 onward Britain placed more demands

---

5. Quotations in Williams, pp. 283, 291, 294, passim, et seq.
6. R. J. B. Knight, "New England Forests and British Seapower: Albion Revised," *American Neptune* 46:4 (Fall 1986): p. 223; idem, "The Building and Maintenance of the British Fleet," pp. 36, 42, and 46n; and Williams, p. 286. Knight's credentials are impressive. Upon completion of his thesis, "The royal dockyards in England at the time of the American War of Independence" (University of London, unpublished, 1972), he joined the National Maritime Museum. In 1977 he became Custodian of Manuscripts and in 1983 he headed the Museum's computer group. He is now head of the Documentation and Research Division.
7. P. P. Timber, 1771, pp. 15–29, passim; and Sven-Erik Åström, who provides a summary of the advanced technology and development of the sawmilling industry in the Baltic, "Foreign Trade and Forest Use in Northeastern Europe, 1660–1860," *Natural resources in European history. A conference report*, ed. Antoni Mączak and William N. Parker (Washington, DC: Resources For the Future, 1978), pp. 43–48, 53, 56–64, passim.

on its domestic timber production by enlarging the number and sizes of ships for mercantile and military use. Moreover, the supply of colonial-built ships on British account was curtailed after the American Revolution. From 1755 to 1815 royal naval ships of the line were 40 percent bigger than those built between 1689 and 1755. During the Seven Years' War, an accelerated maritime building program necessitated using unseasoned timber. After the war additional timber went to repair or rebuild older warships (optimally every ten years), to build many new ones and to meet the considerable requirements of the East India Company. From midcentury on, the East India Company increased the size and number of its own ships, making it a strong competitor to the navy for timber. It was plainly stated to the Parliamentary committee in 1771 that: "the Consumption of Timber since 1762, in building and repairing Ships, has been as great as would have built 60 Ships of the Line; and that in Fact, since the Year 1762, 34 Ships of the line have been actually built and launched."[8]

The British demand for shipbuilding timber was part of a wider consumption pattern. Ashton has shown that during the first half of the eighteenth century increased timber imports into Great Britain coincided with the expansion of construction. After the Seven Years' War that correlation continued with a marked increase in construction, if not the "biggest building boom of the century." From 1762 to 1769, for example, the value of timber imports doubled. When construction eased in 1770 there was a fall in timber imports, but they sharply recovered in 1771, reflecting a return to boom building conditions. Subsequently, in the years running up to the American War, building activity fluctuated greatly. A significant expansion occurred from July 1776 to July 1777, but the war years that followed (1778–1782) witnessed a sharp fall in timber imports owing to increasing freight and insurance rates. This British building frenzy prompted Ashton to quote Horace Walpole: "Rows of houses shoot out every way like a polypus, and so great is the rage of building everywhere, that if I stay here a fortnight without going to town, I look about to see if no new house is built since I last went. America and France must tell us how long this embarrassment of opulence is to last."[9]

The timber needs of Great Britain were so varied that no single country could easily satisfy all its requirements in quality, quan-

---

8. P. P. Timber, 1771, p. 16 et seq., and Appendixes; and Knight, "The Building and Maintenance of the British Fleet," passim.
9. Ashton, pp. 91–100, quotation from p. 100.

tity, size, shape, and price. Although peculiar political and commercial policies occasionally influenced the pattern of the timber trade, more frequently it was freight costs—and consequently the distance traveled—that determined the direction of the British trade in timber as well as the import price. Ships carrying timber from Baltic ports to Great Britain were not only more secure than those plying the stormy Atlantic in the American trade, they also could make more voyages a year. The disadvantage of making the long haul is all the more important when viewing this bulky cargo, especially in proportion to its value, as part of the total shipping tonnage of the English import trade. During the early 1750s a little more than half the tonnage was taken up by timber and forest products. Despite the parliamentary encouragement given to that trade through bounties and preferential treatment, Atlantic transportation costs alone made timber imports prohibitive.[10] North America could not readily be considered an alternative source of timber supplies. Great Britain necessarily had to reach across the North Sea and also into the Baltic for more of its timber.[11]

The economic value of Riga timber to Great Britain's construction and building trades cannot be emphasized enough, as testimony before a parliamentary committee revealed:

---

10. "In 1721 it was found that the high rates of duty had operated to prevent the importation of Timber from British North America, where there was an inexhaustible supply, whilst on the other hand European Timber kept entire possession of the English market at high prices. The heavy cost of freight from America was the principal cause of the inability of the colonists to compete for the English trade, and therefore Parliament decided to allow the importation of Lumber from British America free of duty for 21 years. And this permission was subsequently extended until 1809." P. P. Customs, 1897, p. 261. According to H. S. K. Kent, "lifting the duties on colonial timber could not compensate for the high freight-rates from America." *War and Trade in Northern Seas: Anglo-Scandinavian economic relations in the mid-eighteenth century* (Cambridge: Cambridge University Press, 1973), p. 12. Peter Mathias remarks that

> despite a continuum of comment during the later seventeenth and eighteenth centuries that British colonial north America (whether before or after 1776) could and should have provided the iron and forest products for which Britain depended on the Baltic countries, in fact, despite certain bounties and other encouragements of a minor and/or temporary nature, no significant imperial preference was introduced in the eighteenth century by way of differential import duties to off-set the higher initial development costs and freight charges associated with new-world colonial output, let alone give colonial supplies a price advantage.

MATHIAS, p. 8.
11. Charles F. Carroll, *The Timber Economy of Puritan New England* (Providence: Brown University Press, 1973), pp. xi–xii, 18–20, 149, 151, 156–57; Kent, pp. 11–13, 39–46; Arthur R. M. Lower, *Great Britain's Woodyard. British America and the Timber Trade, 1763–1867* (Montreal and London, 1973), pp. 9–11, 13–18, 43; Bernard Pool, *Navy Board Contracts 1660–1832: Contract Administration under the Navy Board* (Connecticut: Archon Book, 1966), pp. 78–79, 88–89, 99–101; and P. P. Timber, 1835, pp. 337–40, 342–46, 351–53, 374–75.

> Riga ... the port from which almost all the timber in the log, from 12 to 13 inches square, required for building and other purposes in this country, was imported. From 1757 to 1778, nine-tenths of all the fir timber 12 inches square, and upwards, entered as purchased or imported in the mercantile books ... is Riga timber. It has the valuable property for timber in the log, for the materials out of which joists, and girders are made, of being very rigid—of bending little under great weights. Moreover, it is very regularly squared, very straight, very clear of knots, straight in the grain, and very durable.... Amongst the uses to which I have known Riga timber to be applied, on account of its stiffness, and freedom for knots, is the making of the arms for carrying the sails of windmills.[12]

Indeed, according to John Summerson, it has been well established that

> After bricks, the London builder's chief material was timber ... Baltic fir and English oak.... Georgian builders in London used fir for nearly everything ... finding it adequate for most structural purposes and ... for ornament since they habitually painted and gilded it.... The fir logs were shipped from Danzig, Riga and Memel to the port of London and sold cheaper than oak.[13]

Thus, in 1776, when the construction of Bedford Square ("the only intact eighteenth-century square remaining in London") was undertaken, the building agreement enumerated "the dimensions of each story, and the quality of materials for the different parts of the houses. For example, only the 'best Memel or Riga timber,' might be used, the floors were to be laid with 'good yellow seasoned deals free from sap.'"[14]

About the same time that the British became exercised over their own domestic timber insufficiency, loud and repetitive voices were heard in Russia concerning the shortage of economical timber for fuel, shelter and construction. Such a clamor would appear contrary to what most students of Russia's natural resources would expect because, by any eighteenth century standard of appreciation, Russia's timber stands were enormously rich. It would be difficult to imagine Russia having a scarcity of timber. Soviet research, however, has ascertained that from the end of the seventeenth to

---

12. P. P. Timber, 1835, "Minutes in Evidence," Aug. 7, 1835, Henry Warburton, p. 371.
13. John Summerson, *Georgian London* (London: Penguin Books, 1978), pp. 80–81.
14. Donald J. Olsen, *Town Planning in London: The Eighteenth and Nineteenth Centuries* (New Haven and London: Yale University Press [Yale Historical Publications: Miscellany v. 80], 1964), p. 46.

the beginning of the twentieth century—a time when the total land area expanded but woodlands decreased—there was a significant absolute and relative decline in forestation. But at the beginning of Catherine II's reign, especially in the northern and northwestern provinces, which were proximate to ports and from which much of the export trade in timber would be supplied, deforestation had been relatively minor.[15]

Yet soon after Catherine came to the throne many village peasants, entrepreneurs, and governmental policy makers complained about the shortage of readily available timber for a variety of purposes. To many of them this was a consequence of Catherine's forceful efforts to stimulate Russia's economic recovery by increasing the production, the exportation, and therefore the consumption of timber. These dissident voices railed against what they considered the practice of excessive, if not ruthless, timber exploitation. Although the government did eventually take measures to reduce deforestation on both state and private lands, their effect was at best realized over the long term. The continued complaints over timber scarcity, even though many of them focused on local or regional problems, sounded a national alarm of distress.

The incremental deforestation of Russia was rooted in the historical processes of population growth and economic expansion. For example, the canal projects initiated during the first half of the century such as the Vyshnii Volochek system resulted in the improvement of the internal waterway infrastructure and facilitated the marketing of goods during the second half of the century but they used up huge amounts of timber. Equally significant was the manufacture each year of thousands of *barki* (boats, barges, and rafts) that carried goods to market. The critical role these carriers played in the consumption of timber resources is manifested by their being worthy, with rare exception, to make the difficult voyage only one way, that is, one-time use. After the voyage the battered carriers were converted into firewood. One estimate of the number of trees consumed in the manufacture of these boats and barges ranges from 300,000 to 500,000 annually; moreover, from 1764 to 1794 the price of one carrier increased tenfold.[16] In the

---

15. M. A. Tsvetkov, *Izmenenie lesistosti evropeiskoi Rossii s kontsa XVII stoletiia po 1916 god* (Moskva: Akademii Nauk, 1957), pp. 9–74, 201, and especially pp. 123–33. See also R. A. French, "Russians and the Forest," in *Studies in Russian Historical Geography*, ed. James H. Bater and R. A. French (London: Academic Press, 1983), I, pp. 23–44, especially pp. 38–41; and M. A. Tseitlin, *Lesnaia promyshlennost' Rossii i SSSR* (Leningrad, 1940).
16. Robert E. Jones, "Getting the Goods to St. Petersburg: Water Transport from the Interior 1703–1811," SR 43:3 (Fall 1984): pp. 413–33, passim; and idem, *Provincial Development*

heavily forested Novgorod *guberniia* where the waterway transportation network connected with Baltic commerce via St. Petersburg, Narva, and Riga, the trade in wood was especially profitable. Opposition to timber exports were loud and frequent there because "the local forests provided the inhabitants . . . with most of their building material and virtually all of their fuel. They also supplied the raw materials for the tools and implements of daily life as well as for the handcrafted goods sold for cash."[17]

Russian officialdom reacted without much success. In 1763 Senator Ivan Nepliuev, a member of the Commission on Commerce, vigorously criticized the awarding of timber concessions to foreigners. Their exports, according to him, had not only depleted wood stands in the Narva region but also had unduly caused significant price increases of timber and other domestic goods in St. Petersburg.[18] Governor Jacob Sievers of Novgorod *guberniia* recommended in his "On Lands, Woods, Peat, Coal, the Rural Economy, and a Society of Economy or Agriculture" (1764) that peat and coal be substituted for wood as a fuel. However, sufficient supplies of these substitutes could not be discovered. His suggestion concerning the appointment of foresters to regulate the use of the *guberniia*'s timber resources also did not materialize.[19] Perhaps the most representative and widespread opposition to deforestation came during the deliberations of the Great Commission (Legislative Commission, 1767–1768) where one delegate ominously summed up the protestations: "our descendants will have to burn straw."[20] But Catherine II, like her predecessors, chose to subordi-

---

*in Russia. Catherine II and Jacob Sievers* (New Brunswick: Rutgers University Press, 1984), p. 149.
17. Jones, *Provincial Development*, p. 63.
18. P. H. Clendenning, "William Gomm: A Case Study of the Foreign Entrepreneur in Eighteenth Century Russia," JEEH 6:3 (Winter 1977): pp. 533–48, particularly p. 539; idem, "The Economic Awakening of Russia in the Eighteenth Century," JEEH 14:3 (Winter 1985): p. 455; N. N. Firsov, *Pravitel'stvo i obshchestvo v ikh otnosheniiakh k vneshnei torgovle Rossii v tsarstvovanie imperatritsy Ekateriny II. Ocherki iz istorii torgovoi politiki* (Kazan': Imperatskii Universitet, 1902), pp. 59–61; and French, I, pp. 36–38.
19. Jones, *Provincial Development*, pp. 148–50, and idem, "Getting the Goods to St. Petersburg," p. 427.
20. Quoted in SIRIO, VIII, p. 50. For other protestations by delegates from several parts of Russia: SIRIO, IV, pp. 104–5, 113–14, 127; VIII, 88, 424–25, 442–43; XIV, 101, 166, 177, 187–88, 256, 274, 278, 280, 325, 330, 348, 354, 370–71, 416, 421–22; XXXII, 368, 495; XLIII, 128, 227, 557, 602; LXVIII, 4, 11, 16–17, 28–29, 38–40, 89, 107, 114, 255, 266, 278, 287, 322, 370, 407, 428, 432, 484–85, 558–60, 582; XCIII, 8, 72, 108, 124, 158, 236, 280–83, 313, 319–20, 344, 347–49, 362, 374, 411, 518; CVII, 84–85, 95, 100–2, 138, 190, 238–39, 245, 279–80, 286–87, 303–4, 342, 358, 371–72, 394–95, 408, 429, 464, 520, 526, 618; CVIII, 207, 244, 261, 364, 458, 536; CXXIII, 52–53, 64–65, 85, 147–48, 230, 268, 277, 292; CXXXIV, 26, 42, 52, 190, 192, 203, 208, 231–33, 261, 284, 323, 342, 398; and CXLVII, 72–73, 78–79, 81. For an excellent discussion of the deliberations of the Great

nate the attempts to halt the expropriation of timber resources to her principal goal of accelerating Russia's economic growth.[21]

It was during her reign that Great Britain, owing to domestic economic needs and strategic military requirements, increased its procurement of Russian timber products and this, according to Sven-Erik Åström, was facilitated by "several important structural changes" that were taking place in the Northern European timber trade.[22] For centuries the Baltic, with its generous climate, soil, and beneficial river network, had sustained a major timber exporting industry with abundant hard- and softwoods of the right kind of texture and size. The Baltic *pinus sylvestris,* for example, because of its strength, resilience and high resistance to decay, had wide and important uses and was ideal for masts. Yet the region's natural wealth was unevenly distributed and by the middle of the eighteenth century the less-endowed Sweden, having ceded to Russia its best saw-milling areas, had lost its significance in the timber trade. Because of its proximity, Denmark-Norway continued to supply Britain heavily (even during wartime) with most of its sawn timber and smaller masts, but gradually its share in the British timber market declined from "almost 50 percent to between 20 and 25 per cent."[23] During the latter part of the eighteenth century Great Britain's chief suppliers of timber were Christiania in Norway, Memel (Klaipéda) in Prussia, Gdańsk (Danzig) in Poland, and Viborg-Fredrikshamn [Finnish: Viipuri-Hamina, in Old Finland], Narva, Riga, and St. Petersburg in Russia.[24]

Several variables determined which foreign port would take

---

Commission see Wilson R. Augustine, "Notes Toward a Portrait of the Eighteenth-Century Russian Nobility," CASS/CSS 4:3 (Fall 1970): p. 379.

21. For a review of Russian legislation and other governmental efforts that sought to regulate and prevent the expropriation of timber stands, see Tsvetkov, pp. 32–33, 43, 69–74, and passim; and Denise Eeckaute, "La legislation des forêts au XVIII$^c$ siècle," *Cahiers du monde russe et soviétique,* 9:2 (1968): pp. 194–208. In 1777 Governor Sievers reported to Catherine that St. Petersburg's demand for more and more timber was causing a dangerous depletion of forests in the Novgorod *guberniia,* an inflationary upswing in the cost of boats, and erosion along the waterway system. In the 1780s Alexander Romanovich Voronstov, president of the Commerce College, recommended that logs be floated through the waterway system instead of shipping them by the slower moving barges. This idea was dropped when it became evident that such activity would cause significant damage to the boats, locks, and wharves, and also interfere with the transport of commodities along the waterway system. Jones, "Getting the Goods to St. Petersburg," p. 427; and idem, *Provincial Development,* pp. 149–50. See also RC, Nov. 15, 1782, VIII, pp. 94–95.
22. Sven-Erik Åström cites the findings of B. Nüchel Thomsen, B. Thomas, and J. W. Oldam, "Dansk-engelsk Samhandel 1661–1963," in *Erhversvhistorisk Årbog 1965* (Aarhus, 1966), pp. 52–54, in "North European timber exports to Great Britain, 1760–1810," in *Shipping, Trade and Commerce. Essays in memory of Ralph Davis,* ed. P. L. Cottrell and D. H. Aldcroft (Liecester: Liecester University Press, 1981), p. 86.
23. Ibid., p. 81. See also Kent, pp. 39–40, 43–45, and Appendix I, Figure 1, p. 178.
24. Åström, "North European timber exports to Great Britain," p. 81.

the lead at a given time in the general, or in a particular branch of the timber trade with Great Britain. Because of the variation in sizes, shapes, quantities, and different kinds of woods from each exporting port, it is nearly impossible to develop an accurate comparison between Great Britain's timber imports from Russia and those from other countries. Nevertheless, certain correlations will be attempted because there are sufficient data to indicate the degree of importance that Russia had for the developing British construction and shipbuilding industries.

Several factors influenced the transport of timber to market, from the felling of the tree to its retail sale. The key determinants of the price of imported timber were the prime cost of the article, freight and insurance costs, and the import duty. Until the British Consolidation Act of 1787,[25] import duties on timber had been nominal and unchanged and did not thwart the continued flow of sawn boards (deals and battens), masts, and other wood cargoes into Great Britain. Insurance costs were relatively stable during that time. But war had an unpredictable effect on not only the cost of timber cargoes but also on the direction of the timber trade. A helpful example of the complexities that determined the cost of timber is provided in the testimony before Parliament by people whose knowledge of the timber trade had been acquired over decades of practical experience. During the Seven Years' War the average prime cost of Riga timber was 16s. 7d., with the average freight cost per load at 30s. 2d. making the average cost of import (inclusive of incidentals) 48s. 4d. But during the peacetime years of 1763–1766, while the average prime cost at Riga increased to 19s. 9d., the average freight cost per load dropped to 20s. 4d. and the average cost of import decreased to 41s. 6d.[26]

These and other factors affected Riga's leadership in the timber trade of Northern Europe. Albion tells us that Riga was famous for "some of the best timber of European commerce," having "the reputation of exporting the best masts in the world." Its balks of squared logs of fir timber and its oak, "Riga Wainscot," were renowned.[27] In 1763 about 230,000 rubles, that is, about 20 percent of the total value of all exports by British merchants in Riga, were spent on "Wooden Goods" (2,885 masts and spars, 30,115 balks, and oak wood et cetera), most of which undoubtedly ended up in

---

25. In 1787 Mr. Pitt introduced "a Bill in which he repealed all the existing compound duties and consolidated them into single rates for each article." P. P. Customs, 1897, p. 15.
26. P. P. Timber, 1835, "Minutes in Evidence," Aug. 7, 1835; and Henry Warburton, p. 346.
27. Albion, p. 141.

Great Britain and Ireland. About the same distribution took place in the following year. In 1766 about 71 percent of the wainscot logs, 40 percent of the planks, 34 percent of the fir balks, and 17.5 percent of the masts (556 of various diameters) exported from Riga went to England.[28] From 1761 through 1770 an annual average of at least 61 percent of England's imported oak plank and 13.5 percent of its oak timber came from East Country and Russia.[29] This orientation in Great Britain's procurement policy continued after Sandwich came to the Admiralty. According to Williams, "the Navy Board was certainly not convinced that the Baltic oak was in any way superior to the native product. Sandwich's Administration, however, appreciably increased their purchases of Baltic timber."[30] From 1757 to 1780 the British navy had obtained a good deal of its wainscot from Holland or Ostend but after the Anglo-Dutch War broke out Riga became its principal supplier of wainscot logs. During the second half of the eighteenth century the price of Riga wainscot logs increased by nearly 200 percent but the increase in freight costs exceeded those of price.[31] Until the first partition of Poland (1773), Riga had been the principal port in Northern Europe supplying squared logs of fir timber to the British building and construction trades. Even after Memel superseded it in the 1770s Riga still continued to ship large quantities of these valuable high quality logs to Britain. During the wartime period (1778–1782) Riga exported an annual average of 69,000 logs to both belligerents and neutrals, with 21 percent being shipped to Great Britain and Ireland.[32]

It is impossible to develop a comprehensive, equivalent and accurate comparison of manufactured deal boards and battens exported from northern European ports and their importation into Great Britain during the second half of the eighteenth century. Deal boards and battens varied not only in size and quality but also

---

28. Riga exports, 1763–1764, 1766, compiled and computed from the records of the St. Petersburg Custom House, PRO SP 91/73, 75, 107; BT 6/231; and CO 388/54, 95.
29. P. P. Timber, 1771, Appendix 17, pp. 50–51, and discussion on pp. 15–24, passim.
30. Williams, pp. 306–7 et seq., where he also discusses relative costs of English and British oak.
31. Although it is fair to say that there were mixed opinions about the quality of American oak, it was certainly considered a bad substitute for wainscot because it was susceptible to warping. Memel wainscot was larger and produced more boards of greater breadth than those from Riga, but the Memel wood was plainer and less mellow. Riga wainscot logs came down the Dwina from Poland, from Volhynia and from other parts of Ukraine. P. P. Timber, 1835, "Minutes in Evidence," Aug. 7, 1835; Henry Warburton, pp. 351–53, 374–75, Appendix, 1, p. 379; and Kent, p. 40.
32. Riga exports, 1778–1782, compiled and computed from the records of the St. Petersburg Custom House, PRO SP 91/103–4; FO 65/1, 6, 9; and BT 6/231.

in the manner in which they were counted for export and import: by the piece, dozen (which was not always 12), hundred, or Great Hundred (either 100 or 120 pieces). But as Ralph Davis tells us, even "these categories concealed significant variations in size. A hundred of deals contained 120 pieces of wood, and the dimensions of each piece might be as little as 8 ft. x 7 in. x 1/2 in. or as much as 20 ft. x 11 in. x 3 1/4 in."[33] Despite these drawbacks there are some relationships and trends in the sawn wood trade that are evident from the data. First, Russia figured prominently among northeastern European suppliers of deals and battens to western and southern Europe. Second, although it may never have reached the higher volume levels attained by Denmark-Norway as an exporter of wood boards to Great Britain, Russia surely was far ahead of its other competitors, Sweden and Prussia, by the latter part of the century.[34]

Two prominent early sawn wood exporting ports of the Russian Empire were Viborg and Fredrikshamn. Originally these were the Finnish towns of Viipuri and Hamina, formerly in the part of Sweden (southeastern Finland) that had become part of the Russian Empire by the Peace of Åbo in 1743. Finnish archival scholarship shows that these towns had benefited early in the century from capital investment and from the introduction of Dutch fine-blade sawmilling techniques, which had already appeared in Riga and Narva and would later be found in St. Petersburg.[35] By the mid-1760s it is estimated that Viborg had become the largest producer and exporter of sawn timber in the Russian Empire. By then, too, the British had eclipsed the Dutch as the principal timber buyers and are credited with having created the first "boom" for the Finnish sawn timber industry. During the period 1765–

---

33. Ralph Davis, *The Rise of the English Shipping Industry in the Seventeenth and Eighteenth Centuries* (Newton Abbot, Great Britain: David and Charles, 1972), p. 182. The St. Petersburg Standard = 12 planks of 1 and 1/2 in. by 11 in. by 12 ft., or 16.5 cubic feet. Åström, "Foreign Trade and Forest Use in Northeastern Europe," Table 1, p. 45. "The unit of measurement, the dozen, was subject to fluctuation. It was not a unit of volume but one of pieces. In the early eighteenth century a standardized dozen was adopted in which 12 planks measured 1.5 in x 11 in x 12 ft. This was known as a reduced dozen, equalling 0.47 cubic metres." Jorma Ahvenainen, "Britain as a Buyer of Finnish Sawn Timber 1760–1860," in *Britain and the Northern Seas. Some Essays*, ed. Walter Minchinton (West Yorkshire: Lofthouse Publications, 1988), p. 150. The Russian tariff of 1766/67 included a table of equivalents for the number of deals by length and thickness. TARIFF, f. 262.
34. See "Account of the Quantities of Fir Timber, Deals and Battens, Imported into the United Kingdom from 1788 to 1834, both Years inclusive," P. P. Timber, 1835, Appendix 5, pp. 385–86.
35. Åström, "Foreign Trade and Forest Use in Northeastern Europe," pp. 43–48, 53, 56–64, passim; idem, "North European timber exports to Great Britain, 1760–1810," pp. 86–87, 90; and Ahvenainen, pp. 149–55 et seq.

1779, the five-year running average of British ships arriving at Viborg exceeded 200, a significant number by any measure of shipping in the Baltic, and about one third of the deal exports went to the British market. Viborg's deal exports in 1765 were more than three times those from St. Petersburg, and by 1780 had grown by more than 30 percent. Fredrikshamn was less productive than Viborg. In 1765 Fredrikshamn's deal exports equaled the number of deals exported from St. Petersburg, but by 1780 it also experienced a 30 percent increase in deal exports.[36]

The Russian archival data for the volume of deal exports from St. Petersburg/Kronstadt are complete from 1760 to the end of the century. During the first full four years of Catherine II's reign (1763–1766) deal exports were an annual average of 56,888 pieces, with British ships transporting more than 51,000, that is, almost 90.7 percent. But after the conclusion of the Anglo-Russian Commercial Treaty and the promulgation of the Russian tariff, the total deal exports increased from an annual average of about 140,000 pieces (1767–1769) to about 182,000 (1770–1774). The number of deals exported in British vessels during these years grew at an even greater rate, from about 116,000 (1767–1769) to about 173,000 (1770–1774). Although the number of deal exports from St. Petersburg/Kronstadt declined to about 107,000 during the wartime years (1778–1782), British vessels still transported about 91 percent of them. It is interesting to note that deal exports from Riga during this wartime period were greater than those from the capital port with a total annual average of almost 132,000, of which 21 percent were shipped to Great Britain.[37] Overall, from 1763 through 1782, British ships exported almost 92 percent of the total annual average number of deals from St. Petersburg/Kronstadt (see Table T–I).[38]

Riga and other Russian ports had achieved their niche in the mast trade in the early part of the eighteenth century[39] and during and after the Seven Years' War they became principal suppliers to Great Britain and other countries. During the second half of the

---

36. Ibid.
37. Riga exports, 1778–1782, compiled and computed from the records of the St. Petersburg Custom House, PRO 91/102–4; FO 65/1, 6, 9; and BT 6/231.
38. Table T-I is compiled and computed from the records of the St. Petersburg Custom House, PRO SP 91/69, 73, 75–76, 78–79, 83, 88, 91–92, 94–96, 99–104, 107; BT 6/231; FO 65/1, 6, 9, 13, 15; FO 97/340; CO 388/54, 56, 95; and SRO CO/18.
39. Joseph J. Malone, "England and the Baltic Naval Stores Trade in the Seventeenth and Eighteenth Centuries," MM 58:1 (Feb. 1972): pp. 383–84, 389; and idem, *Pine Trees and Politics. The Naval Stores and Forest Policy in Colonial New England, 1691–1775* (Seattle: University of Washington Press, 1964), pp. 55–56.

## TABLE T-I. ST. PETERSBURG'S DEAL EXPORTS FROM 1763 THROUGH 1782[38]

[Annual Average, Measured in Pieces]

| Years | Deal Exports All Ships | Deal Exports British Ships | British Percentage of Total |
|---|---|---|---|
| 1763–66 | 56,888 | 51,583 | 90.7% |
| 1767–69 | 140,142 | 116,380 | 83.0% |
| 1770–74 | 182,097 | 172,946 | 95.0% |
| 1775–79 | 134,597 | 122,185 | 90.8% |
| 1778–82 | 107,395 | 98,272 | 91.5% |
| 1763–82 | 128,265 | 117,659 | 91.7% |
| 1767–82 | 146,109 | 134,178 | 91.8% |

century Denmark-Norway dominated Great Britain's imports in smaller masts and the Russian Empire progressively captured the extraordinarily valuable "Great Mast" market.

The outfitting of a major ship with masts required a great variety of woods. For example, an 800-ton ship required 17 New England and Riga middle and great masts and an additional 49 smaller "Norway masts" for booms and spars.[40] "A Royal Navy Masting Table" for the year 1743 called for main masts to have the following lengths and diameters: 99 feet by 33 1/2 inches (70-gun ship), 102 feet by 35 inches (80-gun ship), 108 feet by 36 1/2 inches (90-gun ship), and 114 feet by 38 inches (100-gun ship).[41] However, for the Royal Navy to actually obtain masts over 32 inches in diameter for one of its ships of the line was an exceptional occurrence. Of the total 42 New England masts with diameters of 31 inches and above used by the Royal Navy during the period 1763–1770, only 10 were "the exceptionally big masts" over 32 inches, and only one mast each of 34 and 35 inches was used.[42]

The British customs practice of categorizing and recording masts into "Great," "Middle," and "Small" for most of the second half of the eighteenth century presents obstacles to precise evaluations and comparisons among mast suppliers to Great Britain be-

---

40. Kent, pp. 43–44.
41. Malone, *Pine Trees and Politics*, p. 150. For other masting tables see P. P. Timber, 1771, Appendix 4, "Account of the Dimensions and Scantlings of the Frame, Beams, Knees, &c. used in a Ship of each Class, from 100 Guns to Frigates of 28 Guns, inclusive," pp. 32–34; and Albion, "The Timber Problem of the Royal Navy, 1652–1862," MM 38:1 (Feb. 1952): p. 5 (during the Napoleonic period).
42. Williams, pp. 317–18.

cause the sizes associated with each category are not universally agreed upon. The scholarship on this varies with the sources consulted. Albion writes about "the loose use of the term 'great masts.' In the customs accounts, all masts over twenty inches in diameter were included under that designation."[43] However, Joseph J. Malone, who studied the colonial New England timber industry, states "that the Book of Rates established that small masts were from eight to twelve inches diameter, 'middling' from twelve to eighteen inches, and anything over eighteen inches was a 'great' mast."[44] Kent, whose specialty is the Norwegian timber trade, differs from both Albion and Malone and writes that: "Few 'great masts,' that is, masts of a diameter above 12 inches, came from the Norwegian forests, but 'middle' and 'small' masts of respectively 8 to 12 inches, and 6 to 8 inches diameter, were obtained in larger quantities than from any other source, to be used in naval construction, and in industry and trade for derricks and hoists."[45] At least from 1786, Kent correctly describes what customs accounts recorded for masts but from 1795 masts were recorded in "Loads."[46]

There was an interdependent and, at times, apparently an inverse relationship in the mast trade between the availability and marketability of the New England *pinus strobus* and the Baltic *pinus sylvestris*. The *pinus sylvestris*, the fir along the Baltic, and the *pinus strobus*, the white pine of the forest of the northern American colonies, were the chief sources of great masts and were much sought after for major naval ships. Although the New England mast was inferior in strength and durability to those from Riga, its size and relatively lighter weight made it highly attractive as a main mast for the largest ships of the Royal Navy. This monarch of the forest could grow as tall as two hundred feet and as wide as six feet across at the butt.[47] Prior to the American War, Great Britain had access to and could purchase both masts conveniently. However, according to Albion, after centuries of forest exploitation Baltic *pinus sylvestris* over 27 inches in diameter were no longer abundant for the market in masts; that made the bigger New England *pinus stro-*

---

43. Albion, *Forests and Sea Power*, p. 284.
44. Malone, *Pine Trees and Politics*, p. 56.
45. Kent, p. 43. According to sources used by Williams the largest Norway mast used by the Royal Navy was 16 inches in diameter. Williams, p. 317n.
46. See "An Account of the Goods, Wares and Merchandize imported into England [Scotland] from Russia . . . 5th January 1786 to the 5th January 1800 . . .," a report from the Inspector–General, Custom House, London, PRO BT 6/230, ff. 45–54, and MASTS, BT 6/230.
47. Albion, *Forests and Sea Power*, pp. 30–32; idem, "The Timber Problem," p. 6; Malone, *Pine Trees and Politics*, p. 148; and Lower, pp. 30, 42.

*bus* more desirable and more valuable. It should be obvious, he says, that "in the largest size, an additional inch in diameter made a tremendous difference in cost."[48]

Albion further argues that with the outbreak of the American War the supply of the *pinus strobus* for the British market was severely curtailed and that this created an immense *matériel* problem for the Royal Navy, which hindered its strategic military response to the rebellious colonies. British procurers had therefore to find adequate replacements in the Riga mast. It is especially at this point that both Williams and Knight critique Albion severely. Knight had the benefit of Williams's study and his comments represent a more encompassing rebuttal: "In no other aspect was Albion more emphatic than when he stressed the critical dependence of the British navy on the very large mast sticks, which were only available from New England forests. . . . Albion, however, was simply not correct in this assertion."[49] Williams states that Albion "did not mention however . . . that any mast over 30" was rare and that masts over 27" whether from New England or not, were not plentiful."[50] Williams adds: "The period in which American masts were no longer imported and during which the Baltic masts were all important, has been much misinterpreted,"[51] because Albion dated the British switch to the Riga masts as late as 1775. In general agreement—but not in every detail—Knight adds: "What happened, therefore, when the New England sticks were no longer in the mast ponds, which they certainly were not between 1778 and 1780?" There was no "shortage of large masts; there were plenty in the yards," there were "masts from the Baltic forests, mostly shipped out from Riga."[52]

Knight details this point further: "There had been many more Russian masts than American from the time of the Seven Years War. At the time of the Falkland Islands Crisis of 1770 and 1771, American masts remained at under a thousand, while Riga, much more flexible to British demand, supplied over three thousand." Moreover, Knight says: "When it came to the American War, Riga was well able to supply the British dockyards. By 1779, Baltic masts imported were slightly under five thousand, while there were no

---

48. Albion, "The Timber Problem," p. 6.
49. Knight, "Albion Revised," p. 222.
50. Williams, p. 317.
51. Ibid., p. 320.
52. Knight, "Albion Revised," p. 224. According to the table based on Admiralty sources provided by Williams, the Royal Navy consumed New England masts 27 inches and up as late as 1781 when their supply became exhausted. Williams, p. 319.

American masts at all."⁵³ The Admiralty sources that Williams consulted support these contentions unequivocally. In comparing the consumption of masts by the Royal Navy during wartime Williams calculates that the British consumed 1,271 Riga masts (26–16 inches) during 1759–1760 and nearly twice as many masts, 2,506, during 1779–1780. During the entire Seven Years' War (1756–1762) the British consumed 4,274 Riga masts (26–16 inches) but during the five-year period 1777–1781 the number consumed increased to 5,317.⁵⁴

Knight buttresses his argument by discussing the "made mast," something that other scholars apparently either do not know or accept. The "made mast" was constructed of more than one stick by piecing together Baltic firs to achieve the maximum dimension. According to Albion the "made mast" was "in Great Britain a lost art," but Knight provides ample citation that it was an art ongoing since the 1690s. For example, William Falconer in his 1769 *An Universal Dictionary of the Marine* made it plain that: "the lower masts of the largests ships are composed of several pieces united into one body [as] a mast, formed by this assemblage, is justly esteemed much stronger than one consisting of any single trunk, whose internal solidity may be very uncertain."⁵⁵ Regarding the comparative advantage of the quality and cost of the Riga mast over the American mast, Knight cites British shipwright testimony. In 1780 the Portsmouth shipwrights responded to "how savings could be made in timber consumption. Since masts were now made of Riga fir, they replied, they could be made one inch less in diameter, 'as they will be superior in strength, much cheaper and of less weight aloft.'" Knight concludes that: "Quality was, therefore, not a problem."⁵⁶

Both Russian and British customs accounts detailing the number of masts exported from the Russian Empire and imported into Great Britain, which will be discussed below, substantiate the contentions of Knight and Williams concerning British dockyard inventories of masts and their consumption by the Royal Navy. However, all this does not mean that the British did not become "anxious about the supply of large masts" from time to time. A "scare" occurred in November 1777 when the Navy Board re-

---

53. Knight, "Albion Revised," p. 224.
54. Williams, pp. 334 and 318n, respectively, and see also Appendix 3, "Masts: Consumption," pp. 582–86.
55. Knight, "Albion Revised," p. 223; and Albion, "The Timber Problem," pp. 10–11.
56. Knight, "Albion Revised," p. 224 et seq.

ported that although "His Majesty's Magazines are at present *well supplied* with masts, there are but few larger masts due upon contract, and the present contractor apprehended difficulty in procuring further supplies."[57] This prompted the Admiralty to dispatch agents to Riga (they also went to other Baltic ports including St. Petersburg) to purchase "a sufficient number of the largest masts that can be procured." By the end of March 1778 their mission had been satisfactorily completed.[58]

The reports from that trip are extremely informative about the complexities relating to the identification of agents, merchants and factors involved in the determination of marketable timber resources, their cutting and production, their delivery to port and their sorting, and finally their overseas shipment as masts from Riga and other Baltic ports to Great Britain. The Admiralty agents were "looking for masts of from 24" to 28" and down to 20." But these were hard to come by in Danzig as the custom was not to separate large masts from small."[59] At Memel masts of 24 inches were not available. At Riga they learned that the heavy masts due that spring were already "bespoke," that the French and the Spaniards were buying up available stocks, and that "the forests were greatly exhausted." Riga merchants working on commission used Riga burghers as middlemen who would then contract with Polish and Ukrainian proprietors. The largest masts came from the region of the Polish Ukraine and the Russian Ukrainian area bordering the Dniepr river. It took about two years to deliver them to Riga. The admiralty agents found no large masts of 24 inches and up available for purchase. It was hoped that the Russian government would permit felling in their Ukrainian forests. The Admiralty agents, after spending some time in St. Petersburg, returned to Riga where they contracted for the future delivery of a sufficient number of masts.[60] However, in March 1780 the Navy Board sounded another alarm about their "very great anxiety . . . from the apparent scarcity of great masts in Europe and of the difficulty of procuring a supply equal to the consumption of the fleet."[61] There was now also a shortage of shipping and Williams avers that the "Armed Neutrality doubtless had a destructive effect on the Baltic Timber and Mast supplies."[62]

---

57. Quoted in Williams, p. 321.
58. Ibid., pp. 321–22.
59. Ibid., p. 323.
60. Ibid., pp. 324–26.
61. Ibid., pp. 326–27.
62. Ibid., p. 327.

The British and Russian customs sources only occasionally provide the size in length and/or diameter of the masts exported from the Russian Empire and imported into Great Britain. The spotty Russian data reveal that primarily Riga and secondarily St. Petersburg made significant contributions in supplying Great Britain with masts. For example, 2,885 Riga masts and spars (recorded together) were exported in 1763 by British merchants who undoubtedly shipped most of them to Great Britain.[63] In 1766, 556 Riga masts of various diameters, representing 17.5 percent of the total masts exported, were shipped to England.[64] Of the annual average of 3,587 masts exported from Riga to European countries during the period 1777–1782, 722 and 1,935 masts were shipped to Great Britain in 1777–1779 and 1781–1782, respectively.[65]

Russian and British sources provide not only the number but also occasionally the price of masts. For example, from a low average price of 47 rubles each, 21 (probably small) masts were exported in British ships from St. Petersburg in 1773; at other times the price was three to four times greater. In 1763 the average price was 187 rubles (33 masts); in 1766, 138 rubles (86 masts); and, in 1778, 200 rubles (59 masts).[66] In Great Britain, prior to the war with France in 1778, the price of Riga masts 21 inches in diameter was £3 and nearly £8 for one 26 inches in diameter. In 1781, prices in Great Britain for masts had increased by about 50 percent over what they had been in 1775.[67]

Of the total number of "Great Masts" that Great Britain imported from 1764 through 1782, an annual average of about 71 percent came from the Russian Empire. The small and middle category masts from Russia were appreciable but not significant (12.6 and 17.2 percent, respectively). Prior to the outbreak of the American War, Russia supplied about 57 percent of the total annual average number of "Great Masts" to Great Britain. In 1776 that proportion increased to 67 percent; during the period 1775–1779, it jumped to more than 81 percent; and, from 1778 through 1782,

---

63. Riga exports, 1763, compiled and computed from the records of the St. Petersburg Custom House, PRO SP 91/73, 107.
64. Riga exports, 1766, compiled and computed from the records of the St. Petersburg Custom House, PRO CO 388/54.
65. Riga exports, 1777–1782, compiled and computed from the records of the St. Petersburg Custom House, PRO SP 91/102–4; FO 65/1, 6, 9; and BT 6/231.
66. For St. Petersburg's mast exports from 1760 through 1782 see Custom House records, PRO SP 91/69, 73, 75–76, 78–79, 83, 88, 91–92, 94–95, 99–104, 107; FO 65/1, 6, 9; FO 97/340; CO 388/54, 56, 95; and BT 6/231.
67. Williams, p. 333, Appendix 5; "Prices of Naval Stores," pp. 591–94, passim. See also "An Account of the Average Prices [1764–1785], Navy Office, Mar. 21, 1787, PRO BT 6/231, ff. 34–35.

### Figure 5

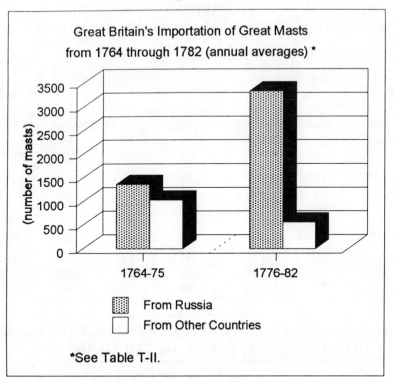

Russian masts represented more than 87 percent of the total "Great Masts" imported into Great Britain (see Figure 5, Tables T–II–IV).[68] [69] [70]

The deficiencies in both British and Russian archival sources prevent, except for isolated brief periods, a correlation of the flow of the volume of other timber exports from Russia to Great Britain and even a comparison of timber exports among Russian ports. Furthermore, given the diversity in nomenclature, size, and quality et cetera of the numerous timber products, it is probably impossible to determine in current monetary values the contribution of Russian timber imports into Great Britain, although such bold and inconclusive attempts have been made.[71] But simply on the basis of the volume of Russian timber products imported into Great

---

68. Table T-II is compiled and computed from MASTS BT 6/240.
69. Table T-III, ibid.
70. Table T-IV, ibid.
71. See Jorma Ahvenainen, pp. 155 ff; and Åström, "North European timber exports to Great Britain, 1760–1810," Appendix, pp. 96–97.

### TABLE T-II. GREAT BRITAIN'S IMPORTATION OF GREAT MASTS FROM 1764 THROUGH 1782[68]

[Annual Average, in Pieces]

| Years | Total From All Sources | America | Denmark-Norway | East Country | Russia | Russian % of Total |
|---|---|---|---|---|---|---|
| 1764–66 | 2,731 | 958 | 105 | 129 | 1,499 | 54.9% |
| 1767–69 | 1,912 | 730 | 21 | 93 | 1,067 | 55.8% |
| 1770–74 | 2,626 | 762 | 47 | 185 | 1,606 | 61.2% |
| 1775 | 1,575 | 866 | 31 | 117 | 561 | 35.6% |
| 1776 | 1,789 | 49 | 159 | 378 | 1,200 | 67.1% |
| 1775–79 | 3,164 | 211 | 107 | 247 | 2,577 | 81.4% |
| 1778–82 | 4,694 | 32 | 120 | 412 | 4,102 | 87.4% |
| 1764–75 | 2,386 | 812 | 54 | 142 | 1,357 | 56.9% |
| 1776–82 | 3,884 | 42 | 122 | 368 | 3,327 | 85.7% |
| 1764–82 | 2,938 | 528 | 79 | 225 | 2,082 | 70.9% |

### TABLE T-III. GREAT BRITAIN'S IMPORTATION OF MIDDLE MASTS FROM 1764 THROUGH 1782[69]

[Annual Average, in Pieces]

| Years | Total From All Sources | America | Denmark-Norway | East Country | Russia | Russian % of Total |
|---|---|---|---|---|---|---|
| 1764–66 | 2,223 | 385 | 1,229 | 81 | 404 | 18.2% |
| 1767–69 | 2,654 | 466 | 1,766 | 26 | 385 | 14.5% |
| 1770–74 | 3,118 | 506 | 1,924 | 137 | 526 | 16.9% |
| 1775 | 3,287 | 153 | 2,747 | 144 | 230 | 7.0% |
| 1776 | 3,089 | 265 | 2,417 | 143 | 245 | 7.9% |
| 1775–79 | 3,201 | 114 | 2,445 | 183 | 420 | 13.1% |
| 1778–82 | 4,085 | 35 | 2,793 | 261 | 962 | 23.5% |
| 1764–75 | 2,792 | 436 | 1,797 | 96 | 435 | 15.6% |
| 1776–82 | 3,852 | 80 | 2,760 | 234 | 745 | 19.3% |
| 1764–82 | 3,183 | 305 | 2,152 | 147 | 549 | 17.2% |

Britain from the end of the Seven Years' War to the peace of 1783, it seems to me that Russia's contribution to both Great Britain's overall economic growth in domestic building and strategic military security was extraordinarily important, if not crucial, during times of British national emergency.

### TABLE T-IV. GREAT BRITAIN'S IMPORTATION OF SMALL MASTS FROM 1764 THROUGH 1782[70]

[Annual Average, in Pieces]

| Years | Total From All Sources | America | Denmark-Norway | East Country | Russia | Russian % of Total |
|---|---|---|---|---|---|---|
| 1764–66 | 4,261 | 687 | 2,431 | 76 | 957 | 22.5% |
| 1767–69 | 3,888 | 501 | 2,576 | 106 | 670 | 17.2% |
| 1770–74 | 4,205 | 240 | 3,202 | 306 | 420 | 10.0% |
| 1775 | 5,109 | 283 | 4,228 | 211 | 343 | 6.7% |
| 1776 | 4,867 | 144 | 4,101 | 207 | 297 | 6.1% |
| 1775–79 | 4,702 | 94 | 3,880 | 248 | 449 | 9.5% |
| 1778–82 | 4,468 | 14 | 3,665 | 234 | 503 | 11.3% |
| 1764–75 | 4,215 | 421 | 2,938 | 191 | 610 | 14.5% |
| 1776–82 | 4,573 | 43 | 3,822 | 225 | 444 | 9.7% |
| 1764–82 | 4,347 | 277 | 3,264 | 203 | 549 | 12.6% |

# Part III
# War and Commerce on the High Seas: Conventions, Commodities, Contraband

# Chapter VII
# Neutrality Unarmed

From the beginning of her reign Catherine II was determined to broaden Russia's international commercial relations. Without denying the valuable association Russia had with Great Britain, the latter's prevailing position in Russia's overseas trade appeared to Catherine to limit Russia's commercial and political flexibility and potential revenues. During the 1770s and 1780s, therefore, Catherine embarked upon an ambitious overseas trade and navigation effort in the hope that several countries would conclude treaties of commerce with Russia.[1]

In doing this Catherine brought to fruition policies initiated by her predecessors. In March and April 1756 the Conference at the Imperial Court of Elizabeth had formulated a general and systematic state plan, which included Russia's intention "to unite the commerce of the Baltic with the Black Sea and through that have in our hands almost all the Levant Commerce."[2] Upon ascending the throne Catherine confirmed Peter III's commercial policy and declared that Russian grain would be exported through the Black Sea.[3] By its partition of the Polish-Lithuanian Commonwealth in 1773 and its military victory over the Turks in 1774, Russia in-

---

1. With Denmark-Norway (July 9, 1780 and Oct. 8/19, 1782); with Sweden (July 21/Aug. 1, 1780); with the United Provinces (Dec. 24, 1780/Jan. 4, 1781); with Prussia (May 8/19, 1781); with the Ottoman Empire (June 10/21, 1774 and June 10/21, 1783); with Austria (Oct. 9, 1781 and Nov. 1/12, 1785); with Portugal (July 13/24, 1782 and Dec. 9/20, 1787); with the Kingdom of Two Sicilies (Feb. 10/21, 1783 and Jan. 6/17, 1787); and with France (Dec. 31, 1786/Jan. 11, 1787).
2. *Tsentral'nyi Gosudarstvennyi Arkhiv Drevnikh Aktov*, Fond. 178, *Konferentsiia pri vysochaishem dvore*, delo 1/75–95, quoted in Herbert H. Kaplan, *Russia and the Outbreak of the Seven Years' War* (Berkeley, California: University of California Press, 1968), pp. 55–56. Already in 1745 the Commerce College had proposed the commercial development of Russia's southern provinces and had sought permission from the Ottoman Empire for Russian merchantmen to navigate the Black Sea. This effort failed. Mose Lofley Harvey, "The Development of Russian Commerce on the Black Sea and Its Significance" (Ph.D. diss., University of California, 1938), pp. 10–11.
3. Patricia Ann McGahey Herlihy, "Russian Grain and Mediterranean Markets, 1774–1861" (Ph.D. diss., University of Pennsylvania, 1963), p. 28; Carol S. Leonard, *Reform and Regicide. The Reign of Peter III of Russia* (Bloomington: Indiana University Press, 1993), pp. 110–12; and PSZ/XVI/11,630/July 31/Aug. 10, 1762.

creased its territory and population in the west and the south, reinforced its access to the Baltic and Black seas and enhanced its potential revenues and trading opportunities. In 1775 Russia's total foreign overland and overseas commodity trade turnover was 31 million rubles and by 1783 it would increase to almost 43 million rubles.[4] Even Great Britain, the greatest sea and economic power in the world, greatly depended upon Russia's natural and manufactured resources and repeatedly sought its diplomatic and political support. Catherine's prestige had grown to the point where she considered herself a player, if not an arbiter-mediator, in greater European affairs.

The Treaty of Küçük Kaynarca of 1774, which concluded the Russo-Turkish War (1768–1774), not only assured Russia free navigation in the Black Sea, thereby facilitating the trade link with the Baltic, but also altered the balance of commerce in the Black and Mediterranean seas.[5] In 1775 Russia decreed a special tariff for the Black Sea ports, which lowered custom duties there by approximately 25 percent below the general tariff.[6] By 1782 Kherson became an active trading port. In June 1783 a Treaty of Commerce was concluded between the Empire of Russia and the Ottoman Porte which, among other liberal provisions, permitted Russian ships to pass through the Dardanelles into the Mediterranean.[7] Henceforth, Aktiar (later Sevastopol'), the best natural harbor in the area, became the haven for the Black Sea fleet. Nevertheless, repeated political turmoil and military combat in the Crimea created unstable and untenable conditions that threatened its fragile independent integrity. Finally, toward the end of 1783, the Ottoman Empire acquiesced in Russia's annexation of the peninsula. The Treaty of Küçük Kaynarca was reaffirmed, exclusive of the provision of an independent Crimea, in a Convention signed at

---

4. See Chap. XIV, below.
5. Treaty of Küçük Kaynarca of July 10/21, 1774 (ratifications exchanged, Istanbul, January 13/24, 1775, modified by a *convention explicative* of March 10/21, 1779) is quoted in J. C. Hurewitz, *Diplomacy in the Near And Middle East. A Documentary Record: 1535–1914* (Princeton: D. Van Nostrand Company, 1958), I, pp. 54–61; and see also Alan W. Fisher, *The Russian Annexation of the Crimea 1772–1783* (Cambridge: Cambridge University Press, 1970), pp. 55–56. For a review of the documentary sources on Russia's commerce in the region of the Black and Mediterranean seas, see Herlihy, pp. 1–14.
6. PSZ/XX/14,355/193–6/Aug. 4/15, 1775; V. I. Pokrovskii, ed., *Sbornik svedenii po istorii i statistike vneshnei torgovli Rossii* (St. Peterburg: M. P. Frolov, 1902), I, p. xxvii; N. D. Chechulin, *Ocherki po istorii russkikh finansov v tsarstvovanie Ekateriny II* (St. Peterburg: Senatskaia Tipografiia, 1906), p. 213; Mikhail D. Chulkov, *Istoriia kratkaia rossiiskoi torgovli* (Moskva, 1788), pp. 293–301, passim; and Frank Fox, "French-Russian Commercial Relations in the Eighteenth Century and the French-Russian Commercial Treaty of 1787" (Ph.D. diss., University of Delaware, 1968), pp. 193–200.
7. For text of treaty, CTS 48/pp. 333–65, and PRO FO 65/12.

Aynali Kavak in January 1784. Additional Ottoman losses resulted from the Treaty of Jassy in January 1792 whereby all the territory between the Bug and Dnestr rivers was ceded to Russia.[8]

Catherine appreciated the commercial challenges facing her country's merchants, who were relatively inexperienced in the Levant trade. She also was mindful of the well-established interests that France and Great Britain had in this region and sought to accommodate them, offering Britain a partnership of sorts in the Levant trade. The British, however, were more concerned about their French competitors.[9] France already had a considerable share of the commerce of the Ottoman Empire and during the 1770s they had increased their trade with Russia. In May 1775 the French ministry informed its envoy to Russia that the British should not be allowed to acquire the same monopolistic rights on the Black Sea as they had maintained in the Baltic trade, and supplied him with a copy of the unsuccessful French commercial treaty project of 1761.[10] In July 1783 King Louis XVI granted a subsidy in ships, money, and commercial privileges to a French firm experienced in the Levant trade, to establish itself in Kherson. In early 1784 French ships arrived there and began exporting Russia's commodities—including grain—to the West.[11] In February 1784 Catherine opened the trade of the Black Sea region to all foreigners from countries friendly to Russia, offering them Russian citizenship and granting them generous trading and navigation privileges:

> If any foreigner conceives the wish to settle in ... Our towns and settlements, and to take Our citizenship [he would be given] limitless freedom to set up factories, handicrafts, and all else permitted for his own and common advantage, and he shall enjoy all these benefits and privileges granted to Our other subjects of equal status with him.[12]

---

8. Fisher, pp. 128–57; M. S. Anderson, "The Great Powers and the Russian Annexation of the Crimea, 1783–4," SEER 37:88 (December 1958): p. 19; idem, "Great Britain and the Russian Fleet, 1769–70," SEER 31:76 (1952/53): pp. 148–63; idem, "Great Britain and the Russo-Turkish War of 1768–74," EHR 69 (1954): pp. 39–58; CTS 49/pp. 11–15; and CTS 51/pp. 279–85.
9. Suffolk to Gunning, Feb. 14 and Apr. 28, 1775, PRO SP 91/98, ff. 24–26 and 106, respectively; and Gunning to Suffolk, Feb. 26/Mar. 9, Mar. 9/20, 12/23, and 16/27, 1775; PRO SP 91/98, ff. 75, 92, 97, and 100, respectively.
10. Fox, pp. 193, 200–1 et seq.
11. Fox, pp. 213–15; and Herlihy, pp. 28–34. Austrian ships from Trieste also passed through the Dardanelles in 1784 to avail themselves of the new trading opportunities with Russia in the Black Sea. Herlihy, p. 28.
12. Quoted in Roger P. Bartlett, *Human Capital. The settlement of foreigners in Russia 1762–1804* (Cambridge: Cambridge University Press, 1979), p. 129; and PSZ/XXII/15,935/50–51,

Catherine fully intended to make this southern region attractive to increased immigration, and during the remainder of the century there was indeed "the meteoric development of Russia's southern borderlands."[13]

Thus, what the imperial Russian leadership had envisioned in the middle of the eighteenth century—a Baltic-Black Sea trade connection—became a reality a quarter century later through the force of arms. Although it would be well into the nineteenth century before Russia's commerce in the Levant would mature, the Anglo-French competition for Russia's favor in the south had an influence on Russia's northern trade. It not only alerted Great Britain to safeguard its dominant trading position there by seeking the renewal of its commercial treaty and long-delayed political alliance with Russia, but also accelerated France's effort to conclude a trade treaty with Russia.

Meanwhile, in the Atlantic, the Anglo-French competition exploded into a war. In 1775 the War of Independence in America sealed off much of that trade to Great Britain and made the latter more dependent than ever on Baltic and Russian supplies. It also created a modest flow of commercial traffic between America and Russia. In 1778 the outbreak of the naval war between Great Britain and France, and then Spain in 1779, created severe difficulties for neutral countries seeking to continue their peacetime trading practices.

Catherine boldly insinuated herself into this widening and deepening international crisis by issuing the "Declaration of the Empress of Russia regarding the Principles of Armed Neutrality." Addressed to the courts of London, Versailles, and Madrid on February 28/March 10, 1780, the Declaration stated "the regard she has for the rights of neutrality and the liberty of universal commerce." But these principles had not

> prevented the subjects of Her Imperial Majesty from being often molested in their navigation, and stopped in their operations, by those of the belligerent Powers.
> These hindrances to the liberty of trade in general, and to that of Russia in particular, are of a nature to excite the attention of all neutral nations. The Empress finds herself obliged

---

Feb. 22, 1784 (o.s.) and enclosure to Fitzherbert to Carmarthen, Mar. 5/16, 1784, No. 12, PRO FO 97/340.
13. Quoted in Bartlett, pp. 116–17; and see Chap. 4, "Southern Russia 1764–1796" in idem., pp. 109–42.

therefore to free it by all the means compatible with her dignity and the well-being of her subjects.[14]

The principal historical question is why did Catherine become involved? How did she expect to benefit politically and commercially? The answer lies more in overconfidence than in policy. By the time of the 1775–1778 crisis, it seems to me, Catherine had become supremely confident in Russia's commercial capabilities and overseas trading policies and had become persuaded that Great Britain, without the continuous flow of Russian raw and manufactured goods, could neither achieve its goal of domestic economic growth nor succeed in its strategy of political-military hegemony in the Atlantic. Her recent political successes against the Poles and the Turks—Russia's ability to mobilize hundreds of thousands of troops and send a naval fleet of warships into the Mediterranean Sea—reinforced this self-esteem and tempted her to risk further advancing Russia's power and prestige by intervening in what was essentially a distant European-Atlantic Great Power conflict.

It cannot be accepted that Russia, as Isabel de Madariaga has written, "held the balance of power in Europe between the great powers and was looked up to as the possible leader of the small powers."[15] On the contrary, if Catherine did seek to achieve such high status among the greater powers of Europe by using the smaller countries to achieve her goal, the evidence suggests that they, rather than looking to her as their leader, used her and finessed her. The greater and smaller powers of Europe were both far more experienced in that kind of diplomatic gambit than Catherine. If anything, by intervening in these crises so far away from the territories with which she was much more familiar, she exposed herself as being naive and overreaching.

Further, Madariaga's contention that "The detailed domestic history of Britain and the actual conduct of the war in America impinged but little on the diplomatic relations between Britain and Russia"[16] misses the point of the commercial relationship between those two countries. The fundamental reason for the Armed Neutrality was Great Britain's attempt to deprive its enemies of strategic resources. When Great Britain went to war against France in

---

14. AN, p. 273.
15. See Betty Behrens's review of Isabel de Madariaga's *Britain, Russia and the Armed Neutrality of 1780. Sir James Harris's Mission to St. Petersburg during the American Revolution* (London: Hollis & Carter, 1962), in EHR 79 (1964): p. 864; quotation from Madariaga, p. 439.
16. Madariaga, p. xii.

1778 it sought to prevent the French from securing vital *matériel* for its navy from any overseas source under any condition. On July 29 the British Royal Navy was ordered "to capture or destroy all French goods and ships encountered on the high seas." On August 21 the Royal Navy was further charged "to intercept, seize, and bring into British ports any neutral vessels found to be carrying 'Naval or Warlike Stores' to French ports." In the course of executing these orders the Royal Navy could not avoid violating long-established international maritime principles and laws.[17] The British motive guiding this policy was as obvious as the language to express it was unambiguous. On October 27, 1778, Lord Suffolk, the British northern secretary, summed up British policy to James Harris, the British envoy to Russia: "The great & unanswerable Principle of Self-defense indispensably obliges His Majesty to prevent, as far as possible, his enemies from being supplied with Naval or Warlike Stores."[18]

From the end of 1778 and throughout 1779 the *démarches* of Great Britain gave every appearance that it would go to war with any state that would not acquiesce to its policies on the high seas. The King of Sweden indicated that he was not willing to accept these policies for such seizures would not only violate Anglo-Swedish treaties but also would ruin Sweden's economy, which depended heavily on the export of naval stores. In December 1779, when the Swedish frigate *Trolle* convoyed two merchant ships carrying naval stores to France, the Royal Navy seized them, causing Sweden to withdraw the *Trolle*. Sweden was not prepared to go to war to defend the maritime principle underlying its national integrity.[19]

The Dutch provinces had divided allegiances toward Great Britain and France, which prevented them from formulating a unified policy against the British seizure of their merchantmen. Two different historic treaties governed an ambiguous Anglo-Dutch relationship. The first was the Commercial and Maritime Treaty between Charles II of Great Britain and the States-General of the United Provinces. Concluded September 11, 1674, it defined contraband of war as being warlike materials—arms and munitions—but not naval stores; in fact, Article Four went further

---

17. David Syrett, *Neutral Rights and the War in the Narrow Seas, 1778–82* (Fort Leavenworth, KS: Combat Studies Institute, U. S. Army Command and General Staff College, 1985), pp. 2 and 14.
18. Suffolk to Harris, Oct. 27, 1778, PRO SP 91/102, No. 35, f. 511. For Russia Company concerns about Royal Navy protection of its merchantmen, see RC, VIII, passim.
19. Syrett, pp. 9–12.

than most treaties by listing what was *not* contraband of war. Under this treaty, if one power were at war, the other had the right to trade with both the warring powers.[20]

However, the second treaty governing Anglo-Dutch relations, the Anglo-Dutch Treaty of Alliance of 1678, stipulated: "When the States are attacked his Britannic Majesty shall assist them with 10,000 foot, and the States his Majesty with 6,000 foot, and with twenty men of war well fitted."[21] Although Great Britain, at the beginning of its naval war with the French, did not ask the Dutch to comply with the Alliance of 1678, Joseph Yorke, its ambassador to The Hague, on September 1, 1778, informed the Prince of Orange-Nassau, the Stadtholder of the United Provinces, that the British would seize Dutch merchantmen if they carried naval or warlike materials to the French. On December 14 the Royal Navy was directed to execute this policy.[22]

France had not remained idle while British naval squadrons executed their intercept, search and seizure operations. In February 1778 it signed two treaties with the American colonies.[23] During the summer it continued to fit out its ships of war and also declared erratic and inconsistent policies. These ranged from subjecting neutral merchantmen to seizure if they transported arms, munitions and contraband (the latter could be interpreted to include even naval stores), to proclaiming that "free ships make free goods" provided they did not carry warlike materials. In early 1779 the French government was prepared to apply prize law to all Dutch ships except those from the Francophile cities of Amsterdam and Haarlem. The Dutch, still divided over the direction their neutral policy should take, found themselves progressively squeezed between the demands of the British and French.[24]

The British, having received from early 1776 repeated intelligence that France and Spain were preparing their fleets for war, anticipated a combined Franco-Spanish naval strike. They also threatened to treat Amsterdam and Haarlem as French allies while

---

20. Daniel A. Miller, *Sir Joseph Yorke and Anglo-Dutch Relations, 1774–1780* (The Hague: Mouton, 1970), Appendix C, pp. 118–19.
21. Ibid., Appendix B, pp. 116–17.
22. Syrett, p. 14. The Stadtholder was also commander in chief of the Dutch armed forces, director of the East India Company, and president of the Bank of Amsterdam.
23. Samuel Flagg Bemis, *The Diplomacy of the American Revolution* (Bloomington: Indiana University Press, 1957), Chap. V, passim.
24. M. J. Williams, "The American War, 1775–1778," Chap. 2, Part 2, and "The Entry of the Bourbon Powers, 1778–1782," Chap. 2, Part 3 in "The Naval Administration of the Fourth Earl of Sandwich 1771–82" (D.Phil. diss., Jesus College, University of Oxford, 1962); Syrett, pp. 12–27, passim; Miller, pp. 60–105, passim; and Madariaga, pp. 58–59, 96–139, passim.

granting the other Dutch cities the protection of the Commercial and Maritime Treaty of 1674. However, when in the summer of 1779 (one that has been compared to 1588) the British learned that the Franco-Spanish fleet was sailing to invade, an alarmed ministry formally requested the United Provinces to come to Britain's aid with troops and warships under the provisions of the Anglo-Dutch Alliance of 1678. The Dutch did not comply and a series of inconclusive Anglo-French-American-Dutch naval confrontations followed. On November 19, 1779, the British cabinet directed the Royal Navy henceforth to interdict Dutch merchantmen carrying naval stores to the French, even if it meant war. By early 1780 these acts had sufficiently undermined the principles of international maritime law as to make legal fictions of previously ratified conventions and treaties. On April 17 Great Britain dissolved the maritime and political alliances it had with the United Provinces by suspending their "peculiar privileges . . . respecting their trade and navigation in time of war."[25]

Under these circumstances it seemed unlikely that any country could long remain "neutral" in the eyes of the principal belligerents, and Catherine's "Declaration" of February 28/March 10, 1780 must be evaluated within this context. The idea for united action to defend their overseas trade and shipping had circulated among the neutral maritime states of northern Europe for some time prior to this "Declaration."[26] It is paradoxical that the least vulnerable of these "neutral" states took the lead in such a *démarche*. Russia's maritime fleet was grossly insignificant in the Baltic-Atlantic trade. Virtually all of its supply of vital resources was paid for and transported by foreign "neutral" and "belligerent" states. To risk becoming embroiled in what was a distant maritime struggle was an extraordinarily dangerous commercial and political venture. Indeed, Russia had much to lose by getting into such a struggle, and could turn a greater commercial profit by staying out of it because Russia's vital resources would be in even greater demand during wartime than peacetime, by both neutral and belligerent states. Most notable among the latter was Great Britain, which was perceived as being strategically the most powerful of all the maritime powers and was Russia's largest overseas trading partner. For Russia to endanger its commercial and political rela-

---

25. Ibid., and AN, p. 281.
26. Ole Feldbaek, *Dansk neutralitetspolitik under krigen 1778–1783* (København: G. E. C. Gads Forlag, 1971), p. 152.

tionships with that power would appear on the surface not to be in its best interests.

It seems to me that the explanation for Russia's policy, then, must be found in Catherine's ambition and her desire to become a major player in greater European affairs. By her intervention, she sought not only to become a protector of the lesser, more vulnerable, and would-be united neutral maritime states, but to become the arbiter-mediator between them and the belligerents—Great Britain, France, and Spain—whom she addressed in her "Declaration" as her equals. She also sought to challenge their respective hegemonies in greater Europe. The question is did Catherine subordinate Russia's *raison d'État* to her own personal ambition? Was she an egoistic opportunist rather than a "brilliant" one?[27]

It cannot be said that Catherine acted without provocations, even if they pale in comparison with the repeated violations committed against other neutrals. In August 1778 the American privateer, *General Mifflin*, sank a British merchantman and seized off northern Norway seven others that were transporting Russian commodities out of Archangel. Catherine protested and recommended to Denmark-Norway that some kind of joint naval action be undertaken to prevent such illegal acts in the future. More directly wounding, however, was the British seizure, on October 10, 1778, of the first Russian ship, *Jonge Prins*, that carried flax and hemp for Nantes. Two weeks later, as we have noted, the Russian court was given to understand that it was a matter of Britain's "Self-defense" to prevent its "enemies from being supplied with Naval or Warlike Stores." Other such incidents followed. In January 1780 Spain captured the *Concordia*, which had been chartered by Russian merchants with a cargo for France and Italy; then the Russian ship, *St. Nicholas*, bound for Italy, was seized.[28]

Catherine's direct action was both swift and secret, much to the surprise of her key advisors and the resident diplomatic community. On February 8/19, 1780, she personally ordered the fitting out of 15 ships of the line and five frigates. In a similar fashion she decided upon the "Declaration" of February 28/March 10, which appeared as a spirited *démarche* issuing from the culmination of a sequence of intolerable insults to the integrity of Russia and the

---

27. Madariaga, p. 140.
28. Ibid., pp. 52–95, passim, pp. 146–50, 156–58; and Suffolk to Harris, Oct. 27, 1778, PRO SP 91/102, No. 35, f. 511.

dignity of the Empress.²⁹ Following her preamble Catherine enumerated five points reflecting her understanding of well-established principles of international maritime law and neutral shipping:

> (1) That neutral vessels may navigate freely from port to port and along the coasts of the nations at war.
> (2) That the effects belonging to subjects of the said Powers at war shall be free on board neutral vessels, with the exception of contraband merchandise.
> (3) That, as to the specification of the above-mentioned merchandise, the Empress holds to what is enumerated in the 10th and 11th articles of her treaty of commerce with Great Britain, extending her obligations to all the Powers at war.
> (4) That to determine what constitutes a blockaded port, this designation shall apply only to a port where the attacking Power has stationed its vessels sufficiently near and in such a way as to render access thereto clearly dangerous.
> (5) That these principles shall serve as a rule for proceedings and judgments as to the legality of prizes.³⁰

Catherine concluded her "Declaration" with a warning that, while she would continue to observe strict neutrality, she was "preparing a considerable part of her maritime forces." If the belligerents did not conduct themselves on the high seas in accordance with these aforesaid principles, she would take appropriate military action to protect Russia's flag, trade and navigation.³¹

The "Declaration" brought into sharper focus the rights and duties of neutral nations and questioned the belligerents' self-serving interpretations of "warlike materials," "contraband of war," "naval stores," and the doctrine "free ships make free goods." From 1778 onward, notwithstanding the treaties they may have had with other countries, the belligerents at times included "naval stores" in their definition of contraband of war. Catherine's "Declaration" made it clear that only "warlike materials" such as arms and munitions constituted contraband of war.³²

This principle was in essence what Empress Elizabeth had insinuated into the 1761 draft of the Anglo-Russian Commercial Treaty but which the British had declared "absolutely inadmissible" because it meant "The Doctrine of free Ships making free

---

29. Madariaga, pp. 158–71 et seq., and pp. 459–63.
30. AN, p. 274.
31. Ibid.
32. Williams, pp. 87–90, 120–23; Syrett, pp. 1–9, passim; and Feldbaek, pp. 145–55, passim.

Goods." They also claimed that its insertion would enable Russia to trade with Britain's enemies. Now Catherine continued Elizabeth's policy. She intended to hold Great Britain not only to the strict limitation of what Article XI of the Anglo-Russian Commercial Treaty of 1766 stipulated contraband to be but also to what Article X said about blockading the trade of a country at war with Great Britain.[33]

The political and commercial policies of the belligerents had merged and become national policies during 1778/79. Was Catherine willing to do the same for Russia in 1780? In principle Russia had as much cause as any country to complain about the ruthless interpretations the belligerents had given to what constituted contraband of war and a legal naval blockade. But given Russia's meager fleet of merchantmen plying the open seas, the belligerents presented only a small threat to Russia's overseas trade carried in Russian bottoms. "It appeared strange" to the Swedish foreign minister that "for the sake of protecting two or three merchant ships at the most" Russia was outfitting a squadron of 20 warships.[34]

If Catherine had really wanted to maintain the integrity of the universal principle of neutral trade and navigation, she had an alternative. Rather than threaten to defend neutral trade and navigation by a problematic military force, she could have temporarily merged her commercial policy—so lucrative to Russia and so necessary to the belligerents—with her political goal of greater involvement in European affairs, thereby establishing an integrated national policy. Such a policy would have prohibited shipping Russia's vital resources to Great Britain, France, and Spain until they conformed to the principles in her "Declaration." Had she been willing to do that her intervention into the Atlantic-European political arena might have resulted in making her the crucial arbiter-mediator that she seemed so anxious to become.

Instead, on April 3, 1780, less than a month after she had proclaimed the "Declaration," Catherine delivered to the States-General of the United Provinces, through Prince Dmitrii Golitsyn, her envoy extraordinary at the Hague, a "Memorandum" that not only reaffirmed the principles in her "Declaration" but also outlined the future League of Armed Neutrality. She invited the States-General and the courts of Copenhagen, Stockholm, and Lisbon "to make common cause with her" in order "to protect the

---
33. CT, III, pp. 228–29, and supra, Chaps. I and III.
34. Quoted in Syrett, p. 31.

trade and navigation, and at the same time observe a strict neutrality."[35]

During the remaining weeks of April, Spain, France, and Great Britain responded to Catherine's "Declaration." The Spanish and French courts welcomed it with alacrity for they took it to mean "free ships make free goods," which, if implemented, would go a long way toward satisfying their pressing requirements for greater supplies of naval stores. On April 18 Conde de Floridablanca, the Spanish secretary of state, on behalf of his sovereign, stated that Catherine's "principles" were those that "have always guided the King" concerning neutral shipping, contraband and blockade. Spain's offense to Russia's commerce was placed in the context of imitating Great Britain's behavior: "It has been entirely owing to the conduct of the English navy, both in the last and the present war (a conduct wholly subversive of the received rules among neutral Powers) that His Majesty has been obliged to follow their example; since the English paying no respect to a neutral flag."[36]

France responded as favorably as Spain but did so in a more elevated manner, suggesting specific behavior for neutral countries on the high seas. In his April 25th reply the King of France protested his attachment "to the freedom of the seas" and "could not but with the truest satisfaction see the Empress of Russia adopt the same principle and resolve to maintain it" for she has declared "in favor of a system which the King is supporting at the price of his people's blood, and that Her Majesty adopts the same rights as he would wish to make the basis of the maritime code."[37]

Great Britain reacted to the mounting coalition of opposition in two ways. First, on April 17 the subjects of the United Provinces, who carried most of the naval stores to both French and Spanish ports, were suspended in their "peculiar privileges . . . respecting their trade and navigation in time of war" and were placed on the same footing as subjects of other neutral states. On April 19 British warships and privateers were instructed to execute their operations accordingly.[38] Second, in a letter to Catherine dated April 23, the British dismissed the "Declaration." Notwithstanding their language of expression, there could be no doubt that the British

---

35. AN, pp. 275–76.
36. AN, pp. 279–80.
37. AN, pp. 284–86.
38. AN, p. 281.

would continue the policy of intercept and seizure of Russian merchantmen carrying naval stores to a hostile power. His Britannic Majesty had given "the most precise orders respecting the flag of Her Imperial Majesty, and the commerce of her subjects" and those "orders to this intent have been renewed, and the utmost care will be taken for their strictest execution. . . . but in case any infringements, contrary to these repeated orders, take place, the courts of the admiralty . . . will redress every hardship in so equitable a manner, that Her Imperial Majesty shall be perfectly satisfied."[39]

By a Resolution on April 24, 1780, the States-General of the United Provinces affirmed the principles and intentions of Catherine's "Declaration" and "Memorandum," bringing Russia and the Netherlands one step closer to forming the League of Armed Neutrality.[40]

On May 8/19 Russia published an "Ordinance on Commerce and Navigation" that not only prohibited Russian merchantmen from participating or giving aid to belligerent powers, but also specified what comprised contraband of war; that is, arms and munitions and not naval stores or other provisions. In addition to setting forth a detailed set of restrictions and obligations for both Russian and non-Russian merchants Catherine extended the protection of the Russian flag to foreigners: "These prerogatives shall be enjoyed not only by our born subjects but also by foreigners, who are domiciled under our rule and pay public taxes like our own subjects; that is to say, as long as they dwell in our country, since in any other case they would not be permitted to use the merchant flag of Russia."[41] This latter provision led to considerable confusion and irritation among those foreign merchants in Russia who did not avail themselves of the privileges that accompanied what some called a kind of double mercantile citizenship. As will be shown in a later chapter, it would be difficult to distinguish statistically between the foreign merchant who became a subject of the Russian Empire under this provision and the native-born imperial subject when both participated in Russia's overseas trade.

The British ministry remained apprehensive that Catherine would use military force to defend her neutrality policies despite the contrary view held by James Harris, the British ambassador in

---
39. AN, p. 282.
40. AN, pp. 283–84.
41. AN, pp. 291–94.

Russia. Because British nationals held commands in the Russian navy, Harris was pleased to report, on May 15/26, 1780, that his countrymen had informed Catherine that they could not fight against their mother country and would quit her service should they be commanded to do so. Harris had also learned that the Baltic fleet would be "employed for the protection of the *Russian Trade Alone.*" But should Russian merchantmen be stopped by the British, Catherine "would lay an embargo on all our trading Vessels here, & annul our Treaty of commerce." The conclusion of a commercial treaty with the French would then follow.[42]

On the evening of June 21 the Russian Baltic fleet of three squadrons, comprising 15 battleships and four frigates, sailed from Kronstadt for the purpose of protecting *only* Russian merchant ships and property.[43] Ironically, Russia formally requested British port facilities for its ships even though those same squadrons might have to defend against British naval interception and seizures of Russian merchantmen. Consistent with the provisions of the 1766 Anglo-Russian Commercial Treaty, the British agreed, and subsequently provided such assistance.[44] Similar Russian naval expeditions took place in 1781 and 1782. The Russian naval patrols did not engage belligerents and, except for sailing mishaps, experienced uneventful sojourns until the Treaty of Paris in 1783. Although British warship commanders and privateers had been ordered to respect Articles X and XI of the 1766 treaty of commerce, they continued to intercept and detain Russian merchantmen.[45]

During the summer of 1780 Great Britain and Denmark-

---

42. Harris to Stormont, May 15/26, 1780, Nos. 49 and 52, PRO SP 91/105, ff. 37, 40–41, and 78–79, respectively; and see, Nos. 50 and 51, ff. 53–61, and 66–72, respectively. See R. C. Anderson, "British and American Officers in the Russian Navy," MM 33 (1947): pp. 17–27, which is a corrective to Fred. T. Jane, "British and American officers in or connected with the Russian service," pp. 714–24, Appendix C in *The Imperial Russian Navy, its Past, Present, and Future* (London: W. Thacker & Co., 1899); R. J. Morda Evans, "Recruitment of British Personnel for the Russian Service 1734–1738," MM 47 (1961): pp. 126–37; and M. S. Anderson, "Great Britain and the Growth of the Russian Navy in the Eighteenth Century," MM 42 (1956): pp. 132–46.
43. Harris to Stormont, June 9/20, and 12/23, 1780, Nos. 70 and 71, PRO SP 91/105, ff. 155, 159, respectively.
44. Stormont to Harris, July 14 and 18, 1780, Nos. 39, 43, and enclosures; PRO SP 91/105, ff. 176–82, 195–98, 221–23; and Admiralty communications of Aug. 26 and 30, 1780, ibid., ff. 284, 286, 289 et seq.
45. R. C. Anderson, *Naval Wars in the Baltic during the Sailing-Ship Epoch, 1552–1850* (London: C. Gilbert-Wood, 1910), pp. 237–40; Madariaga, p. 382; and AN, pp. 328–29. Andreas Bode, *Die flottenpolitik Katharinas II und die konflikte mit Schweden und der Türkei (1768–1792)* (Wiesbaden: Otto Harrassowitz, 1979), pp. 243–85, provides a comprehensive ship listing for the Russian naval fleets of the Baltic, Black, and Caspian seas, and captured ships—

Norway produced a brilliant diplomatic gambit—a *coup* of sorts—which finessed Catherine and undermined the *raison d'être* of the emerging union of neutral states. From the end of May, Great Britain had secretly and successfully negotiated with Denmark-Norway an Explanatory Article to replace Article 3 of their 1670 treaty of alliance and commerce. This substantially changed the language and meaning of contraband of war from arms and munitions and other warlike materials to include, among others, naval stores: "also ship-timber, tar, pitch and rosin, sheet copper, sails, hemp and cordage, and generally whatever immediately serves for the equipment of vessels; unwrought iron and deal planks, however, excepted." The Explanatory Article was dated July 4, 1780, in Copenhagen, five days prior to the Armed Neutrality Convention Denmark-Norway signed with Russia, and it swept away everything for which Catherine had thus far worked. She had been finessed by the country she had intended to lead.[46]

It is immediately clear why Great Britain would engage in such a negotiation. The underlying motive of Denmark-Norway was to protect a complex overseas trading operation that reached as far away as the West Indies, now a major naval war zone. During wartime in the Atlantic there were opportunities from which Denmark-Norway could benefit greatly provided it were permitted the freedom of the seas; more so if there should be a decline in Dutch shipping which, at this time, could be anticipated. The influence of the American War of Independence and the outbreak of war between Great Britain and the Bourbon Powers afforded Denmark-Norway "a prosperity far in excess of what previous wars had brought about. This was due, not only to the men of shipping themselves, but also to a wise neutrality policy on the part of the state."[47]

The League of Armed Neutrality formally came into existence with the signing in Copenhagen, on July 9, 1780, of a Convention between Denmark-Norway and Russia, a prototype of conventions

---

information on category of ship, shipyard, dates of construction and commission, modification, and decommission.

46. AN, pp. 295–96; and Feldbaek, pp. 145–55, passim. Cf. Madariaga, pp. 187–89.
47. Lauritz Pettersen, "The influence of the American War of Independence upon Danish-Norwegian shipping. A survey based on late Danish and Norwegian research," in *The American Revolution and the Sea, Proceedings of the 14th International Conference of the International Commission for Maritime History* (London: National Maritime Museum, 1974), pp. 70–73, quotation, p. 73. Hermann Kellenbenz, "The Armed Neutrality of Northern Europe and the Atlantic Trade of Schleswig-Holstein and Hamburg during the Seven Years War and the War of Independence," ibid., pp. 74–81; and Feldbaek, pp. 145–55, passim.

subsequently agreed to by several other powers. Article 2 defined contraband of war: for Russia, as in Articles X and XI of the Anglo-Russian Commercial Treaty of 1766; and for Denmark-Norway, as in Article 3 of the treaty of commerce between Denmark-Norway and Great Britain of July 11, 1670, which had been modified by the secret explanatory article of July 4. Article 3 of the Armed Neutrality Convention defined a legally blockaded port as: "where the attacking Power has stationed its vessels sufficiently near and in such a way as to render access thereto clearly dangerous." Both parties agreed in Article 4: "to fit out, separately, a proportionate number of ships of the line and frigates . . . as convoys to the trading ships of their respective subjects, wherever the commerce and navigation of each nation shall require it." When ships of war of one party would not be readily available, Article 5 provided that such ships of the other party may be called into service for protection. If either contracting party's neutrality were infringed, and appropriate and speedy compensation for such violation was not forthcoming, then both signatories "will make use of reprisals towards that Power that refuses to do them justice, and will immediately unite, in the most efficacious means, to execute these just reprisals." In "Separate Articles," signed that same day, each party agreed to facilitate and provide the ships of the other with a haven port; that the Baltic was a "closed" and neutral sea; and that both parties would maintain the "tranquility" of the North Sea.[48] The following year the King of Denmark and Norway incorporated the principle of a neutral Baltic Sea into a declaration: "His Majesty could not admit thereto armed vessels of the Powers at war for the purpose of committing acts of hostility against any one whatsoever."[49] Given the strategic position of the Öresund, no one would question that he had the ability to back up that policy.

Sweden joined the League when it signed a Convention with Russia on July 21/August 1, 1780, in St. Petersburg. The doctrine of "free ships make free goods" was supposedly embodied in Article 3 which vaguely stated "that all other commerce shall be and remain free." Yet, the definition of contraband of war was anchored to an ambiguous phrase in Article 11 of the Anglo-Swedish Treaty of Commerce of October 21, 1661, that included "all other things necessary for warlike use." Like Denmark-Norway, Sweden

---

48. AN, pp. 299–307, passim.
49. AN, p. 290, whose date should read May 8, 1781.

employed legalistic devices to avoid offending either Great Britain or Russia.[50]

All this left the Dutch in an exposed position. They played a poor diplomatic hand and wasted valuable time by seeking Russia's protection of their East and West Indies overseas possessions, a rather extraordinary request given the qualifications of Russia's naval power. Stormont did not like that negotiation and in August speculated that if "the Court of Petersburg does take such a Step, it will be a Measure little short of direct Hostility, and, as pernicious in its Consequences as actual War."[51] In September Catherine informed the Dutch that neither she nor membership in the League of Armed Neutrality would provide them with such a guarantee.[52] Finally, on November 20 the States-General acceded to the League by a Resolution.[53] On December 20 and 21, 1780, Great Britain ordered that all Dutch ships and cargoes found on the high seas be captured and destroyed. In effect, Britain had declared war on the Netherlands.[54] On January 4, 1781, the States-General officially signed the respective League Conventions between Denmark-Norway, Sweden, and Russia.[55] However, when the United Provinces formally requested that the League of Armed Neutrality come to its aid against Great Britain none of the signatories complied. Nor did France or Spain come to the aid of the United Provinces against their common enemy. The Fourth Anglo-Dutch War reduced the Dutch from their former significant position in overseas trade to a level from which they would never recover.[56]

During the winter of 1780–1781, the British were still in a quandary as to what Catherine might do in these troubled waters, and Stormont took steps to avoid provoking her. On November 20, 1780, he informed the commanders of British warships and

---

50. AN, pp. 311–17 and Syrett, p. 33. Denmark-Norway acceded to the Convention of Armed Neutrality between Russia and Sweden on September 7, 1780, and Sweden acceded to the Convention between Russia and Denmark-Norway, on September 9, 1780. For texts, see AN, pp. 321–22, and 322–23, respectively. Russia informed the belligerent powers of these accessions on November 7, 1780, AN, pp. 324–25. For America's accession to Catherine's "Declaration" on October 5, 1780, see AN, pp. 323–24, and Bemis, pp. 164–71, passim.
51. Stormont to Harris, Aug. 8, 1780, No. 45, PRO SP 91/105, f. 225.
52. Harris to Stormont, Aug. 25/Sept. 5, No. 99, Aug. 28/Sept. 8, No. 100, and Sept. 1/12, 1780, No. 102; PRO SP 91/106, ff. 4, 7–8, and 11–12, respectively.
53. AN, pp. 325–28.
54. AN, pp. 330–45.
55. AN, pp. 346–49, 350, 351, 352–53, respectively.
56. For a concise narrative of these events see Syrett, pp. 35–42.

privateers that "some inconvenience has arisen from an ignorance of the nature and extent of our engagements with our good sister the Empress of all the Russias." They were ordered to seize only those Russian merchantmen whose cargoes comprised warlike materials such as arms and munitions, but *not* naval stores and other provisions, and they were to conform to the enclosed Articles 10 and 11 of the Anglo-Russian Commercial Treaty.[57] Stormont even thought he could buy Russia off by ceding it Minorca in exchange for an alliance between the two countries.[58] How Russia fit into the affairs of greater Europe, and Great Britain in particular, was deliberated at a Cabinet Council meeting on January 19, 1781, during which Lord Sandwich, the First Lord of the Admiralty, received the ministry's unanimous support for his views:

> Russia . . . is determined to gain some considerable advantage to herself in consequence of the part she takes with regard to the present war in Europe. . . .
> What will our situation be if she takes a part with our enemies? If Minorca is her object, they will readily promise it to her if she will decide against us; and if she accepts their offer, she will as readily engage that Spain shall have Gibraltar; and no power of this country can prevent the execution of this arrangement, if the fleets of Russia and of the Northern Alliance are joined to those of the House of Bourbon.
> But this will not be the only cession we shall be obliged to make if Russia declares against us, for we shall then literally speaking be in actual war with the whole world. . . .
> Powers united against us will dismember our State and make such partition among them as they shall think fit. . . .
> We shall never again figure as a leading Power in Europe, but think ourselves happy if we can drag on for some years a contemptible existence as a commercial State.[59]

Sandwich attributed to Catherine a power that some historians have denied her. They have been critical of her great leap forward into the Atlantic crisis, seeing it as a diplomatic bluff from which she sought only greater prestige for herself and Russia. Even this might be giving Catherine more credit than she is due, for it seems more likely that she was simply confused in her policies and *démarches*. Sandwich overestimated her power to make or

---

57. AN, pp. 328–29.
58. Stormont to Harris, Oct. 20, 1780, No. 61, PRO SP 91/106, ff. 73–74; SP 91/107, passim; and "Russia Refuses Minorca: The Dutch Mediation," Chap. 12 in Madariaga.
59. Quoted in G. R. Barnes and J. H. Owen, eds., *The Private Papers of John, Earl of Sandwich, First Lord of the Admiralty, 1771–1782* (London: Publications of the Navy Records Society, 1932–38), IV, pp. 23–26.

break Great Britain. As before, it was the dominant sea power of Great Britain that determined the reality of maritime principles and law. Although Russia's trade with Great Britain flourished during the period 1778–1782,[60] in making her leap into the Atlantic arena of greater European affairs Catherine badly stubbed her toe, for in the end, whatever her motivations, she did not achieve commercially or politically what she had declared to be Russia's legal prerogatives in international trade and navigation.

---

60. See Chap. XI, below.

# Chapter VIII
# The Expiration of the Anglo-Russian Commerical Treaty of 1766

Great Britain's defeat in the American War of Independence made it commercially more dependent upon Baltic and, in particular, Russian vital resources. Consequently, it was to be expected that Britain would be more accommodating to Russia when negotiating the renewal of the Anglo-Russian Commercial Treaty that was to expire in 1786. Catherine, for her part, recognized Russia's improved bargaining position and became more adamant and insistent in this negotiation and refused to give up her failed policy of Armed Neutrality.

The Russian tariff of 1782/83 was one of several reforms that Catherine had designed to improve Russia's chronically poor economic and financial condition. "Public finance was a state secret in Russia," writes John LeDonne, but that "did not necessarily imply that exact figures were known even to the leadership." According to him, while the first "revenue budget" had been compiled in 1769, it was only from 1781 onward that revenue and expenditure tables, from which the approximate basis of the Russian Empire's budget could be determined, were produced annually. LeDonne summed up the state of affairs by quoting a phrase attributed to Catherine: "No one knew how much was collected and how much was spent."[1]

Catherine was aware that her revenues could not support her ambitions, but rather than moderate those ambitions she chose to inflate the money supply. She hoped that such action would stimulate the production and turnover of commodities, enhance taxable resources, and contribute to the overall prosperity of the country.

---

1. John P. LeDonne, *Ruling Russia: Politics and Administration in the Age of Absolutism, 1762–1796* (Princeton: Princeton University Press, 1984), pp. 226–27, 245–51; A. Kulomzin, "Gosudarstvennye dokhodi i raskhodi v tsarstvovanie Ekateriny II," SIRIO 1870–1871, V, pp. 219–94, VI, pp. 219–304; James A. Duran, Jr., "The Reform of Financial Administration in Russia during the Reign of Catherine II," CASS/CSS 4:3 (Fall 1970): pp. 485–96; and James E. Hassell, "Catherine II and Procurator General Vjazemskij," JGO 24:1 (1976): pp. 23–30.

The anticipated cost of the Russo-Turkish War (1768–1774) had caused her to establish the Assignat Bank in 1768/69 to print paper rubles and, unlike Empress Elizabeth, who had been unable to do so during the Seven Years' War, to borrow on the international market. Between 1769 and 1773 Catherine had borrowed 10 million Dutch guilders or slightly less than 5 million rubles. But the Russo-Turkish War had cost Russia an estimated 47.5 million rubles and, when Russia's loans fell due beginning in 1779, Catherine borrowed again and added to Russia's debt between 6 and 9 million guilders or from 3 to 4.5 million rubles. Despite this infusion of funds "deficits were still being covered chiefly out of assignat issues. Bank note volume reached 40.1 million assignat rubles in 1784 against foreign liabilities of 9.7 million, or 19 percent of the total state budget." Catherine nonetheless continued to pay for her foreign adventures by increased borrowing and printing of assignats.[2]

The tariff of 1782/83 was intended to improve Russia's international commercial position by further stimulating the production of agricultural and industrial goods for foreign markets. It also signaled a policy of freer trade. Custom rates were generally lower than those in the tariff of 1766/67, the duties on many exported and imported commodities were now lifted and the regulations affecting timber exports were considerably liberalized. On the average, import duties were about 10 percent ad valorem. While luxury items continued to be taxed at much higher levels, raw materials were taxed at about 2 percent. Merchants of foreign states that did not have a trade treaty with Russia could now pay half their duties in rixdollars and half in rubles. The duties on the majority of goods at Black and Azov sea ports remained 25 percent lower than those at other Russian ports. The new tariff coincided with Russia's administrative-territorial reform and the lands that had been conquered from Sweden were now to a greater extent integrated into the customs system of the Empire.[3] The tariff had

---

2. James C. Riley, *International Government Finance and the Amsterdam Capital Market 1740–1815* (Cambridge: Cambridge University Press, 1980), pp. 153–57, and especially the footnotes on pp. 304–6, passim; and LeDonne, pp. 205–65, passim. See Klaus Heller, *Die Geld–und Kreditpolitik des russischen Reiches in der Zeit der Assignater: (1768–1839/43)* (Wiesbaden, 1983); and M. P. Pavlova-Sil'vanskaia, "K voprosu o vneshnikh dolgakh Rossii vo vtoroi polovine XVIII v.," *Problemy genezisa kapitalizma. K Mezhdunarodnomu kongressu ekonomicheskoi istorii v Leningrade v 1970 g. Sbornik statei* (Moskva: Nauka, 1970), pp. 301 ff.

3. The Commission on Commerce completed the Tariff on September 27, 1782 (o.s.), and it went into effect on January 1, 1783 (o.s.), PSZ/XXI/15,520; Konstantin N. Lodyzhenskii, *Istoriia russkago tamozhennago tarifa* (St. Peterburg: V. S. Balashev, 1886), pp. 137–40; N. D. Chechulin, *Ocherki po istorii russkikh finansov v tsarstvovanie Ekateriny II* (St. Peterburg: Senat-

a far-reaching effect on Anglo-Russian relations because its principles and provisions were incorporated into the Russian draft proposal for the renewal of the Anglo-Russian Commercial Treaty.

In early 1785 Alleyne Fitzherbert, Great Britain's ambassador to Russia, reported that before the Russians would agree to another treaty of commerce, they would propose several "material Innovations" and endeavor "to force upon us, in some Shape or other the ... Doctrine of the Neutral League." To forestall any interruption in the Anglo-Russian trade, he recommended that Britain initiate discussions "immediately."[4] The British ministry's reply was prompt, brief and traditional. It would seek to determine whether the present treaty should simply be renewed or be re-negotiated to include alterations deemed necessary for the improvement of British trade with Russia.[5] Thus, a tedious negotiation similar to the one in the 1760s began.

While the British deliberated, the French and Austrians acted. In early September 1785 Russia agreed to join with France in a treaty of commerce on the condition that France would first accede to the League of Armed Neutrality. The French would be willing to do this if the Russians could exact the same provision from the British.[6] In November the Habsburg Emperor Joseph II of Austria and Catherine II agreed to make the Black Sea and the Danube River the principal arteries for the flow of goods between the two countries for the next twelve years.[7]

At the end of 1785, soon after Fitzherbert received his new instructions, Catherine appointed four plenipotentiaries to negotiate a commercial treaty with him. They were Vice Chancellor Ivan A. Osterman, State Secretary Alexander A. Bezborodko, Peter V.

---

skaia Tipografiia, 1906), pp. 214–16; and Edward C. Thaden, *Russia's Western Borderlands, 1710–1870* (Princeton: Princeton University Press, 1984), Chap. 2.
4. Fitzherbert to Carmarthen, Mar. 7/18, 1785, No. 12 and Feb. 21/Mar. 4, 1785, No. 10, PRO FO 65/13. For Fitzherbert's instructions, see PRO FO 65/11 (Aug. 20, 1783).
5. Carmarthen to Fitzherbert, Apr. 12, 1785, PRO FO 65/13.
6. Fitzherbert to Carmarthen, Aug. 22/Sept. 2, 1785, No. 34, PRO FO 65/13. For the negotiations leading to the commercial treaty between France and Russia, see Frank Fox, "Negotiating with the Russians: Ambassador Ségur's Mission to Saint Petersburg, 1784–1789," *French Historical Studies* 7:1 (Spring 1971): pp. 47–71; J. L. van Regemorter, "Commerce et Politique: Préparation et négociation du traité franco–russe de 1787," *Cahiers du Monde russe et soviétique* 4:3 (Juillet Septembre 1963): pp. 230–57; and Orville T. Murphy, *Charles Gravier, Comte de Vergennes: French Diplomacy in the Age of Revolution: 1719–1787* (Albany: State University of New York Press, 1982), pp. 447–58, passim. For the reflections of the French ambassador who negotiated the treaty, see Count Louis Philippe de Segur, *Memoirs and Recollections of Count Segur, Ambassador from France to the Courts of Russia and Prussia, &c. &c. Written by Himself* (London: Henry Colburn, 1826; reprint, New York: Arno Press and *The New York Times*, 1970), vol. II: pp. 271–351, passim.
7. CTS 49/pp. 391–440. Fitzherbert to Carmarthen, Sept. 26/Oct. 7, 1785, No. 39, and Apr. 22/May 3, 1786, No. 20, PRO FO 65/13 and 14, respectively.

Bakunin, a member of the College of Foreign Affairs, and Alexander R. Vorontsov, President of the Commerce College and principal author of the tariff of 1782/83. They provided Fitzherbert with a formidable new draft treaty which substantially revised the 1766 treaty. If the British signed it, they would be accepting the principles of the Armed Neutrality, the modification of the Act of Navigation in favor of Russian shipping and the deletion in Article IV of the 1766 treaty of the provision that British and Russians subjects pay the same custom rates. Fitzherbert could not entertain this negotiation without first receiving new instructions; the Russian court extended the provisions of the 1766 treaty to the end of the year.[8]

The reorganized British Board of Trade initiated formal inquiries in the fall of 1786 and the Board's minutes, together with customs data, provide a clear and comprehensive view of the nature and extent of the British dependence on particular articles of Russian production. The minutes also reveal how misguided the British had been about the provisions of the Anglo-Russian trade treaty of 1766 and the Russian tariff of 1766/67 that deliberately undercut it.

The Board noted that Britain continued to import bar iron principally from Russia, followed by Sweden, with much smaller quantities from Denmark-Norway and Spain. While the best quality came from Sweden, it constituted only one-sixth of the total Swedish imports. Next in quality was Russian bar iron, which greatly satisfied British entrepreneurs. As to price, the best Swedish bar iron sold at £23 per ton; the best Russian and the second-quality Swedish went for £18 per ton (inferior sorts from £16.10 to £17); Norwegian iron from £16 to £17; and Spanish sold at about £17. As to quality, the "inferior sorts of the Swedish and Russian" compared favorably with English iron and the price of these imports competed favorably with "the best English Iron [which was] fixed at £19 per Ton."[9]

---

8. "Projet d'un Traité de Commerce et de Navigation entre Sa Majesté l'Impératrice de toutes les Russies, et Sa Majesté Le Roi de la Grande Bretagne," delivered to Fitzherbert in February and received in London on March 14, 1786, in Camarthen to Fitzherbert, Nov. 17, 1786, No. 11, PRO FO 97/341. See also Carmarthen to Fitzherbert, Nov. 25, 1785, No. 16, PRO FO 65/13, and Fitzherbert to Carmarthen, Dec. 9/20, 12/23, and 19/30, 1785, Nos. 50–52, respectively, PRO FO 65/13; Fitzherbert to Carmarthen, Jan. 6/17, and Feb. 9/20, 1786, Nos. 3 and 7, PRO FO 97/340, respectively; and Fitzherbert to Carmarthen, May 5/16, and June 26/July 7, 1786, Nos. 22 and 31 (with enclosure of "Note" from Russian court of June 24/July 5, 1786), respectively, PRO FO 65/14.

9. "Minute," Aug. 11 (inclusive of information produced and inserted on Aug. 17), 1786, PRO BT 6/226, and "Minute," Sept. 4, 1786, PRO BT 5/4, pp. 25–27.

Great Britain had recently concluded a trade treaty with France and wanted to increase the sale of British manufactured iron wares to France by making them more competitive with similar French imports from Germany, Sweden, and Russia. The Board considered reducing the duties on imported Russian bar iron by 2 1/2 percent to encourage more Russian imports. Specialists in the international iron trade testified that in the past when Britain had increased the duties on imported Russian bar iron, imports actually increased. Moreover, should Great Britain reduce the 5 percent duty on imported Russian bar iron the Russians would likely take advantage of that reduction and increase their price of exported bar iron to Great Britain. They stated that when "the Duty payable in Russia on the Export of assorted Bar Iron was taken off by the last tariffe [1782/83], the price continues the same to the Importer here." In their opinion this was owing to "the Rapacity of the Russian Iron-Masters."[10]

> The Russian Iron Masters are few, but very rich; the Swedish numerous, and generally not very opulent.... The Russian Government have lately exempted all assorted Iron from Duty, which is evidently to rival the Swedes; it has tended to enrich a few Individuals, who are thereby enabled to keep a larger stock, and have enormously increased their Demands and Profits.[11]

It appears that only with its investigation of the Russian tariff of 1782/83 did the Board become aware of how the tariff of 1766/67 had undermined the provisions of the Anglo-Russian Commercial Treaty; how those responsible for concluding that treaty had failed to understand what they had actually agreed to; and how Russia had intertwined its regulatory definitions of "naturalized subjects" and "Russian ships" into its commercial and tariff policies and practices. The Board was startled to learn that the tariff of 1782/83 had considerably increased the import custom duties on several British cloth manufactures by at least 100 percent, from about 15 to 30 percent ad valorem.[12] Moreover, the Board believed

---

10. "Minutes," Aug. 24 and Sept. 4, 1786, PRO BT 5/4, pp. 6–7, 26–27, respectively.
11. "Minute," Sept. 4, 1786, PRO BT 5/4, pp. 26–27. Testimony before the Board on hemp, flax, and tallow imports reaffirmed the trend of the past twenty years. Russia continued to supply Britain with almost all of its hemp, flax, and tallow. "Minutes," Aug. 24 and Sept. 27, 1786, PRO BT 5/4, pp. 7, 47–48, respectively.
12. Article XXIV of the tariff of 1766/67 provided that imported British soldiers' cloth and coarse Yorkshire cloth be assessed a duty of two kopeks, broad flannels, one kopek, and narrow flannels three-quarters of a kopek. The 1782/83 tariff, however, increased the duty on soldiers' cloth to twenty-one kopeks, coarse Yorkshire cloth to seventeen kopeks, and flannels by 30 percent of their value. A. Greenwood, secretary, Russia Company, Court of

that these increased duties affected only the British trade. For example, the tariff of 1782/83 made a minor but significant change regarding silk imports. Prior to that tariff the duty was laid on the arshin measure of silk, but after 1783 it was placed on the weight of the silk "which gave an immense Advantage to the light silks of France and Italy, so as to render the Importation of heavy solid British Manufacture almost nothing, as they were too dear to be sold in any Quantity; before they were preferred by the Middling People."[13]

The Board was further vexed that "Russian subjects" obtained a reduction in the duties they would pay—3/8 on exports and 1/8 on imports—if they used "Russian ships," which did not necessarily mean "built in Russia," and which the Board believed violated Article IV of the 1766 treaty. However, Russian subjects had actually been accorded preferential treatment in the form of tariff abatements and the payment of custom duties since the Russian tariff had gone into effect in 1767, and the British ministry had been alerted to these violations of the treaty.

In 1771 Samuel Swallow, the Russia Company representative and British General Consul in St. Petersburg, had sent to the British ministry a facsimile of a Senate edict that modified the procedures under which foreign merchants without the protection of a trade treaty were to pay their custom duties. In so ordering, the edict had significantly reaffirmed Russian subjects in their comparative trading advantage.[14] Robert Gunning, the British envoy to Russia, had reported in 1773 "that in this very Tariff there are actual Infractions of the Treaty of Commerce" and enumerated several of them.[15]

In August 1786 Walter Shairp, the Russia Company representative and British General Consul in St. Petersburg, wrote the Board that, while the Russians had enjoyed the abatement in imports during the past twenty years, they had not used the abatement when exporting Russian goods to Great Britain. He added

---

Assistants, Aug. 23, 1785, PRO FO 65/13; and "Comparison of the Duties paid on the following Articles in Russia by the Tariffe of 1767 and that of 1782," PRO FO 65/17, f. 309; PRO BT 6/231 and 232 (f. 33). See also RC, Jan. 17, 1783, VIII, pp. 97–99. The Russia Company appeared unaware of the practices of actual payment by the British Factory in St. Petersburg. Forster to Hawkesbury, Oct. 6, 1786, PRO FO 97/340.
13. "Mr. Eton's Observations on the income of Duties in Russia on British Articles," in "Mr. Eton's Remarks on the present State of the Importation Trade from Great Britain to Russia, 1794," PRO FO 97/342.
14. Swallow to Suffolk, Nov. 29/Dec. 6, 1771, PRO SP 91/88, f. 295.
15. Gunning to Suffolk, Aug. 13/24, 1773, PRO SP 91/94, No. 74, ff. 1–6.

that it was now so ordered at the St. Petersburg Custom House.[16] But, in 1790 this view was corrected by Edward Forster, Governor of the Russia Company, when he informed Lord Hawkesbury, President of the Board of Trade "that the Allowance of 3/8 of the Duty on Goods exported from St. Petersburgh to different Ports of Great Britain, in Russian ships, for Russian Account, was actually made before the Expiration of the Treaty of 1766." Forster provided the names of the ships involved and added "Tho' their ultimate Destination may have been well known to have been to Great Britain, they may have cleared out as for Elsignore."[17]

The Russian court had developed *pari passu* both the Anglo-Russian treaty of 1766 and the Russian tariff of 1766/67. Article IV of the treaty provided "that the subjects of Russia shall pay the same duties on exportation, that are paid by the British merchants on exporting the same effects from the ports of Russia." But since there had been such a fuss about the language in the various drafts of the infamous reservation clause, Article IV continued, "but then each of the high contracting parties shall reserve to itself the liberty of making, in the interior parts of its dominions, such particular arrangements as it shall find expedient for encouraging and extending its own navigation."[18] Apparently the Russian court presumed that the provisions of the tariff were consistent with the intention of the reservation clause, in particular the language in its original draft—inclusive of the critical phrase *"en réciprocité de l'Acte de Navigation de la Grande Bretagne"*—and that it further believed the reservation clause permitted the modification of the clause preceding it in Article IV.

Although the British ministry had blamed Macartney for failing to obey his instructions he should have been given credit for informing the British ministry what the Russians were really up to in August 1765: "To encourage their Navigation they propose to grant an Exemption from certain Duties, or rather a Diminution of the present Duties paid upon Merchandise exported from St. Petersburg on Board their own Ships... this Immunity is to extend to British Subjects, as well as to Russians."[19] The Russian tariff of 1766/67 provision giving Russian subjects an abatement of 28

---

16. Shairp to Carmarthen, Aug. 11/22, 1786, PRO FO 97/340; and "Minutes," Aug. 24, Sept. 27, and Nov. 10, 1786, PRO BT 5/4, pp. 6–8, 45–49, 72–73, respectively.
17. Forster to Hawkesbury, May 28, 1790, and Hawkesbury's reply of May 30, 1790, PRO FO 65/18.
18. Macartney to Grafton, Aug. 8/19, 1765, No. 1, enclosures 4 and 8 (ff. 130–1), supra, Chap. III.
19. Ibid., enclosure 4.

percent "if they export, or import their goods in Russian ships, either for sale in Russia, or for re-exporation to Persia, they shall pay at a rate of 90 Cop. only [instead of 125] per rixdollar,"[20] was not only an unvarnished and absolute contradiction of Article IV, it was an important precedent to the tariff of 1782/83.

In 1786 the official rate of exchange at the St. Petersburg Custom House was still one rixdollar to 125 kopeks. On the open market the rixdollar fluctuated between 130 and 155 kopeks; and, in August 1786, when Shairp wrote his report, one rixdollar went for 142 kopeks. The relative disadvantage under the tariff of 1782/83 to the non-Russian subject or the foreigner unprotected by a treaty in 1786, according to Shairp's calculations, was equal to 1/4 to 3/4 percent on the prime cost of the goods imported. The advantage accruing to the Russian subject over the non-Russian subject from the abatement awarded solely on the duty of exportation was, according to Shairp, between 5 and 7 percent profit on goods exported from Russia to Britain.[21]

One consequence of the tariff of 1782/83 was to encourage several British merchants of the British Factory to become "naturalized" Russian burghers of St. Petersburg. Such Russian "naturalization" efforts had been publicly advertised in the "Ordinance on Commerce and Navigation" of May 1780 during the Armed Neutrality crisis,[22] the "Manifesto" of February 1784, which opened the trade in the Black Sea region to foreigners,[23] and the "Charter of Rights and Privileges Granted to the Cities of the Russian Empire" of April 1785.[24] The British Board of Trade and the Russia Company were familiar with their own country's practice of naturalizing foreigners who wanted to avail themselves of the privileges of the Anglo-Russian Commerical Treaty. Because they were convinced that such a practice would at times be harmful to their own commercial interests, the Russia Company and members of Parliament sought to prevent foreign merchants from becoming natural-

---

20. TARIFF, f. 264, italics added; and supra, Chap. III.
21. Shairp to Carmarthen, Aug. 11/22, 1786, PRO FO 97/340.
22. See Chap. VII, supra.
23. See Chap, VII, supra.
24. "Gramota na prava i vygody gorodam Rossiiskoi imperii," PSZ/XXII/16,187 [sic] [16,188 ?]/358–84, April 21, 1785 [o.s.], article 66, p. 364; B. N. Mironov, "K voprosu o roli russkogo kupechestva vo vneshnei torgovle Peterburga i Arkhangel'ska vo vtoroi polovine VXIII–nachale XIX veka," *Istoriia SSSR* 6 (1973): p. 131; and for a discussion of the Charter, see A. A. Kizevetter, *Gorodovoe polozhenie Ekateriny II 1785 g.* (Moskva: Tipografiia Imperatorskago Moskovskago Universiteta, 1909); and J. Michael Hittle, Chap. 10 in *The Service City. State and Townsmen in Russia, 1600–1800* (Cambridge, MA: Harvard University Press, 1979).

ized.[25] The Board now worried that the Russian naturalization policy might have a detrimental effect on British shipping and trade.

The issue of the naturalization of foreigners as Russian burghers has a historical, methodological and historiographical significance that goes beyond the simple but important defection of a number of British merchants in Russia. It goes to the heart of the question who dominated the Russian overseas trade during Catherine II's reign and will be treated in detail in a subsequent chapter of this study. For the present, two opinions on this matter presented to the Board of Trade are worth noting. In the first Shairp lamented that British merchants in Russia would choose to follow the lead of other foreigners and become temporary burghers of Russia:

> The advantages considerable as they are would be of less consequence were they enjoyed merely by the Native Russians—but the Priviledge of Burghership & Naturalization are so easily procured in Russia, that Germans & other Foreigners becoming temporary Burghers of any City in the Empire or associating themselves with persons who are such are entitled to a participation of every Immunity and might Trade to all Europe or direct to Great Britain on a better footing than even the British Factory.[26]

The Russia Company, aware of the fact that in 1783 thirteen ships (the average cost of which was 14,444 rubles) were built in Archangel on British account,[27] further exacerbated the fears of the Board that the benefits of Russian naturalization

> will prove a Temptation so strong as to induce many of His Majestys Subjects, to interest themselves in the Building Ships in Russia, and as owners of them, and Russian Subjects, procure to themselves very great comparative advantages, both in the immediate Commerce between His Majesty's Dominions and Russia, and in that of other European States. That habituating themselves to Foreign Connections, they will become less

---

25. M. Sierra, Secretary, Russia Company, to Suffolk, Mar. 17, 1774, PRO SP 91/95, f. 118; Suffolk to Gunning, Mar. 29, 1774, No. 12, PRO SP 91/95, f. 140; Suffolk to Swallow, June 17, 1774, PRO SP 91/96, f. 68; Swallow to Suffolk, June 5/16, 1775, PRO SP 91/99, ff. 5–6; and Suffolk to Swallow, July 14, 1775, PRO SP 91/99, f. 7. For instances of naturalized foreigners seeking the freedom of the Russia Company to trade with Russia, see RC, VII, passim.
26. Shairp to Carmarthen, Aug. 11/22, 1786, PRO FO 97/340.
27. PRO BT 6/231, ff. 133 and 180, and Shairp to Carmarthen, May 2, 1784, No. 11, FO 97/340.

interested in the introduction of British Manufactures in Russia.[28]

The Board had also to find a way of redressing the inequality in the payment of duties by British and Russian subjects in the Anglo-Russian trade. The Russia Company advised against retaliating by imposing countervailing duties. First, such duties would be difficult to execute and, second, Russia could counter this action by either reducing the duties payable by Russian merchants or by increasing the duty on merchandise exported in British ships by non-Russian subjects. The Russia Company also rejected the idea that "Ships of the two Countries should be placed on an equal footing in all respects" because Great Britain would have to rescind its differential custom duty of eleven shillings per ton on iron imported in all foreign ships. It recommended that the new treaty with Russia provide that British subjects in Russia "when importing in Russian Ships enjoy the advantages granted to Russian Subjects, as in Great Britain, Russian Subjects, when importing in British Ships, receive every advantage of a British Subject as to national Duties. The distinction being made in Great Britain between the *Ships* carrying and not *the persons* importing."[29]

But in reality this distinction could not work because Russia's definition of a "Russian ship" was not the same as Great Britain's. Shairp identified this problem as follows: "Russian Subjects, whether Natives of Russia or foreigners naturalized for Life, or only of a limited time, exporting the Produce of this Country in *Russian Ships, navigated by half the Crew & Captain natives or naturalized Subjects of Russia* have an allowance of three Eighths of the Duty paid by foreign Merchants on the same Goods so exported in Russian, or in foreign vessels."[30] Furthermore, a foreign ship became Russian by virtue of the naturalization of its owner or master. As late as 1794 the Board of Trade received reports that the ships of

---

28. "Minute," Nov. 10, 1786, PRO BT 5/4, pp. 72–73.
29. "Minutes," Aug. 24, Sept. 27, and Nov. 10, 1786, PRO BT 5/4, pp. 8, 48–49, and 72–73, respectively.
30. Shairp to Carmarthen, Aug. 11/22, 1786, PRO FO 97/340. The relative benefits to Russian subjects (naturalized or native born) extended also to tonnage or lastage charges:

> Russian Ships navigated according to the beforementioned Regulations pay but half the duty of Tonnage or Lastage paid by British & other foreign Ships, thus a British Ship of 100 Lasts on entering the harbour pays Ro. 8 Lastage & Ro. 4.75 tonnage toward cleansing & Repairing the Narva harbour, a Russian Ship pays but Ro. 4, Lastage and the full Ro. 4.75 cop. toward the Narva Harbour; the same Lastage is again levied on both on clearing for Sea: the Expence of taking out their Passes is the same.

Ibid.

naturalized Russian burghers were Russian ships: "A Lubeck Master of a vessel becomes a Burgher, and all his Sailors, or the half—and this is a Vessel entitled to all the Advantages."[31]

On these two points—the ethnic/national origin and total number of the crew and master of the ship, and the origin of the place where the ship was built—the Russian maritime practice was contrary to the 8th Section of the Act of Navigation:

> That no Goods, or Commodities of the growth, production or Manufacture of Muscovy . . . shall be imported into this Kingdom or Ireland *in any Ships or Vessels whatsoever, but in such as do truly and without fraud belong to the people thereof and where of the master and three fourths of the Mariners at least are British, or in vessels, that are of the built of the Country or place of which the Goods are the growth, production or Manufacture . . . and whereof the Master and three fourths of the Mariners at least are of the said Country or place.*[32]

The continuation of this policy into the 1790s explains, in part, why Catherine had chosen not to make major capital investments in Russia's own commercial shipbuilding. Apparently it was in the short term far more expedient for Catherine and less expensive for Russia to "naturalize" the ships belonging to and/or whose masters were foreigners.

In November 1786 the Board of Trade submitted its recommendations to Foreign Secretary Marquis of Carmarthen (later created the Duke of Leeds), who sent Alleyne Fitzherbert a massive instruction meticulously reviewing each article of the Russian treaty proposal. Following in the tradition of his predecessors Carmarthen argued that Russia was dependent upon Great Britain's commerce, manufacture, and market, and demonstrated this by comparing Russia's small positive trade balance with France during the past fifteen years to the large one it had with Great Britain. With only modest burden to itself Great Britain could substitute imports from Russia with those from North America. And if Russia were prevented from selling its natural resources and manufactures to Great Britain, those goods "must either remain on Hand, without a Purchaser, or if they are purchased at all by the Merchants of any other Nation, it would be only with a View of introducing them into this Country, where alone they can ulti-

---

31. Quoted from "Mr. Eton's Observations on the increase of Duties in Russia on British Articles," in "Mr. Eton's Remarks on the present State of the Importation Trade from Great Britain to Russia, 1794," in PRO FO 97/342.
32. William Wood to John Pownall, Sept. 19, 1761, PRO CO 388/49, italics added; and supra, Chap. I.

mately be consumed. . . . The great Quantities of Hemp, Iron and Tallow, annually imported into this Country, can find a Market no where else." The Russians would not receive as good a price for their produce outside Britain; nor would they be paid as punctually. But Carmarthen also admitted that Britain was "indebted to the Produce of Russia, for a Regular Supply of Naval Stores, which are so essential to the Support of its Maritime Force."[33]

Carmarthen held the view that given the value Russia attached to its trade with Great Britain, it should agree to reciprocity in the new treaty of commerce: "that the shortest Way of bringing this Business to a Conclusion, is, to revert to the Treaty of 1766 and to make only such additions to it, as the Interests of the Two Countries, and the Change of Circumstances may be thought to require." What he feared was an interminable negotiation. He told Fitzherbert to "avoid all fruitless Discussions . . . confine the Negotiation to Points of absolute Necessity, or high Utility . . . throw aside & banish from it, every Consideration on which the two Courts are not likely to agree."[34]

Carmarthen recognized Catherine II's right under the reservation clause of Article IV of the 1766 treaty to make such changes as would be beneficial to Russia. But although George III could retaliate in kind[35] Carmarthen averred that "a warfare of Duties is very unpleasant & even unbecoming between two Nations, who are making Treaties of Commerce, and who wish to remain with each other, on Terms of the most strict commercial Alliance."[36]

In Articles 7 and 8 of their treaty proposal the Russians had reduced the duties on British imported cotton and woolen goods from about 30 to 20 percent ad valorem. But the British considered this rate still too high and there were several varieties of cloth and manufacture missing from the enumeration. What the British wanted was what they had recently gained in the Anglo-French commercial treaty: that the upper limit on duties should be from 10 to 12 percent ad valorem.[37]

Great Britain had lowered the import duty on Russian iron 5 percent below that on iron imported from Sweden and from other countries with whom Britain did not have or was not presently negotiating a most-favored-nation arrangement. This was done in

---

33. Carmarthen to Fitzherbert, Nov. 17, 1786, No. 11, PRO FO 97/341.
34. Ibid.
35. Ibid., Nos. 12 and 14.
36. Ibid.
37. Ibid., Nos. 11–12.

reciprocity for the stipulations in the Russian draft that British merchants would continue to pay their duties in the current coin of Russia (except in Riga where even Russian subjects paid duties in rixdollars), and in the future pay 25 percent less in duties at the Russian ports in the Black and Azov seas than they would pay at other Russian ports.[38]

However, once again the issues concerning contraband, neutral shipping, and armed neutrality became matters of contention between Russia and Great Britain, striking as they did at the heart of the two nations' histories, traditions and laws. In addition to their reservations about the language and the procedures involving search and seizure, the British objected to the proposition in Article 18 that, excepting contraband only, the property belonging to the subjects of belligerents would remain free on board neutral vessels. Only that part of Article 20 that conformed to the definition of contraband in Article XI of the 1766 treaty would be acceptable. Its final clause and those similar clauses in Articles 22 and 23 were inadmissible because they were based on the supposition that "all goods not contraband shall be free."[39]

Under no circumstance would the British admit the principle of armed neutrality into the commercial treaty or into maritime law in any manner. Article 27 of the Russian draft proposal stated that British law would define a "British ship" and Russian law a "Russian ship." Carmarthen recognized in the latter *"the Danger* of this Stipulation to the Commerce of His Majesty's Dominions," and the King found it *"wholly inadmissible"* because it was contrary to the "ancient & favorite Law."[40] Article 26 extended naturalization to foreigners even during wartime. This convinced Carmarthen "that the Court of Petersburgh proposes to convoy, under the sanction of the Russian Flag, the enjoyment of this Privilege to all other neutral Nations, and even to the very Subjects of the Power with whom His Majesty may be engaged in War."[41]

Carmarthen believed that the Russians would in the end give

---

38. Ibid.
39. Ibid., Nos. 11 and 13, Article 18 stated that, in addition to the damages normally paid for an unjust seizure, "complete satisfaction" would be made for insulting the flag of the ship so seized. According to the British this would make every capture an affair of state and would produce endless disputes between the two countries. The British found Article 24 inconvenient, since it required that the captor, who finds contraband on board a neutral ship, must take it on board his own ship and dismiss the neutral ship if its master offers to deliver up the contraband. The British argued that this could not always be done and also might interfere with the mission of the captor. Article 25 addressed courts of the admiralty and needed additional words of explanation. Article 19 required clarification. Ibid.
40. Ibid., Nos. 11–12.
41. Ibid., Nos. 11 and 13.

up their demands and conclude a treaty agreeable to the British. But in the meantime, to prevent interruptions in British trading rights, Fitzherbert was to obtain from the Russians an extension of the 1766 treaty by a declaration similar to what Empress Elizabeth had given to the British in September 1759. Under such an arrangement British merchants would enjoy their trading rights until a new treaty was concluded.[42] In mid-December 1786, soon after Fitzherbert received his instructions, he met with the Russian plenipotentiaries to inform them of his ministry's decision. He was confident that Catherine would "open her Eyes to the Danger of entering into any precipitate Measures."[43]

However, Catherine publicly appeared no longer to depend solely upon the good commercial graces of Great Britain and took a decisive measure to widen Russia's international commercial opportunities. The Orthodox New Year of 1787 celebrated Russia's signing of a commercial treaty with France, encouraging the ambassador of France to boast of how it foreshadowed a closer union with Russia. Fitzherbert, in seeking to discount it, said it was simply a way of importing French wines into Russia at greatly reduced custom duties. He was correct in that sarcasm because the most detailed article in the treaty concerned wines. However, he was wrong if he believed he could make light of the value the Russians attached to the importation of French wines and brandies. In 1783 and 1787 French wines and brandies ranked sixth in value among all commodity imports into St. Petersburg/Kronstadt with more than 400,000 and about 650,000 rubles, respectively. In 1792 the value of imported French wines and brandies into the capital port would be over a million rubles, ranked fourth behind imported linens, sugar, and cottons, and ahead of British woolens.[44] But the Franco-Russian commercial treaty was important in another way because about half of the treaty was devoted to maritime rights founded upon the principles of the Armed Neutrality.[45]

---

42. Ibid., No. 15; "Note pour Monsieur de Keith . . . ," Sept. 8/19, 1759, enclosure in No. 15; and supra, Chap. I.
43. Fitzherbert to Carmarthen, Dec. 1/12 and 4/15, 1786, Nos. 59–60, PRO FO 97/341.
44. See Table CI-III, Chap. XIII, below.
45. Fitzherbert to Carmarthen, Jan. 2/13, 1787, No. 6, PRO FO 65/15. See Regemorter; Segur; Murphy; and Fox, and idem, "French-Russian Commercial Relations in the Eighteenth Century and the French-Russian Commercial Treaty of 1787" (Ph.D. diss., University of Delaware, 1968), pp. 321 ff. "Treaty of Commerce and Navigation between France and Russia," signed at St. Petersburg, Dec. 31/Jan. 11, 1787, and ratified on Apr. 3/14, 1787, in CTS, 50/pp. 103–38; and PRO FO 65/15, ff. 299–320. This treaty was soon followed by the "Treaty of Commerce between Russia and the Two Sicilies, signed at Czarskoe-Selo, 6 (17) January 1787" and ratified on May 27/June 7, 1787, CTS 50/pp. 147–66; and PRO FO 97/341. France's trade to Russia did not depend exclusively on its own ships carrying such

Nonetheless, Fitzherbert could take pride that Catherine had modified her former position regarding the renewal of the Anglo-Russian Commercial Treaty and had authorized in a new Russian draft treaty that three-fourths of the crew on board a Russian vessel must be Russian subjects. Her ambassador to Great Britain, Semen R. Vorontsov, brother to Catherine's chief negotiator, would soon take up with Carmarthen the matters of definition of a "Russian ship" and the mercantile privileges accruing to naturalized Russian subjects. Catherine also increased the number of British goods stipulated in the treaty to receive reduced Russian import duties. But she remained firm on three points: that while arms and munitions constituted contraband of war aboard neutral vessels, all other goods would remain free; that Russian and British subjects exporting from Russia would not pay equal duties; and that the new treaty would include the controversial reservation clause in Article IV of the 1766 treaty.[46] Catherine extended the provisions of the 1766 treaty to April 1/12, 1787, when the trading season would open. On the condition that Great Britain would categorically accept the new Russian proposals, Semen Vorontsov would inform Carmarthen that British merchants would immediately enjoy their privileges under the new draft treaty as if it had already been signed. Because Fitzherbert would accompany Catherine to Kiev he could present her with his court's formal answer there.[47]

Fitzherbert did not underestimate the Russian "ultimatum." Although Alexander R. Vorontsov had told him that, despite Catherine's proclamations, Russia could not "succeed in [its] Projects for the Formation of a Mercantile Marine," Fitzherbert was mindful also that Vorontsov had for some time spoken "violently against" Nikita Panin and the other plenipotentiaries for having signed the 1766 treaty, and that Vorontsov considered the British

---

goods to Russian ports because a great deal was transhipped via northern Europe. For an insightful discussion of this, see Paul Butel, "France, the Antilles, and Europe in the Seventeenth and Eighteenth Centuries: Renewals of Foreign Trade," *The Rise of Merchant Empires. Long-Distance Trade in the Early Modern World, 1350–1750*, ed. James D. Tracy (Cambridge: Cambridge University Press, 1990), pp. 153–73, in particular, pp. 164–68.

46. Fitzherbert to Carmarthen, Dec. 29, 1786/Jan. 9, 1787, No. 2, PRO FO 97/341; and Jan. 2/13, Nos. 3, 5–6, PRO FO 65/15. "Observations des Plénipotentiaires de Sa Majesté Impériale sur cette Note ainsi que sur les remarques qui s'y trouvent insérées [sic]" and "Copie de la Note verbale et confidentielle remise par Monsieur de Fitzherbert aux Plénipotentiaires de L'Impératrice [sic]," idem, enclosed in No. 3, ff. 20–38; and "Projet d'un Traité de Commerce et de Navigation entre Sa Majesté L'Impératrice de toutes les Russies et Sa Majesté le Roi de la Grande Bretagne [sic]," idem, enclosed in No. 3, ff. 40–80. See also James W. Marcum, "Semen R. Vorontsov: Minister to the Court of St. James's for Catherine II, 1785–1796" (Ph.D. diss., University of North Carolina, 1970), pp. 79 ff.

47. Fitzherbert to Carmarthen, Jan. 2/13, 1787, Nos. 3–5, PRO FO 65/15.

Factory an obstacle to the advancement of Russia's trade by Russia's merchants.[48]

A perturbed Carmarthen received Semen Vorontsov's reaffirmation of "Free Ships, Free Goods" also as an "ultimatum." In March, 1787, he wrote to both Walter Shairp and Fitzherbert that the British Factory would have to trade without a treaty.[49] In April the treaty expired, as Catherine had promised it would, and the British Factory merchants henceforth had to pay half their duties in rixdollars and open their homes to quartering by Russian soldiers and government officials.[50] By mid-1787 several British merchants in St. Petersburg had become naturalized Russian burghers, *Inostrannye gosti,* and continued to transact their commercial affairs as Russian merchants. The British Factory subsequently barred the defectors from its meetings and membership and sought assistance from London.[51] The social base of the British merchant community had been split.[52]

The Board of Trade's initial response was pompously ineffectual and wrong: "It is to be supposed that these Persons are not of much Note, or they would not have been tempted by such small Advantages to have deserted so very respected a Body." The Board, nevertheless, wanted the names of the deserters.[53] By the end of 1787 the Russia Company had come to take these matters more seriously. It had learned from a questionnaire it had sent to the British Factory that there were indeed considerable advantages accruing to British merchants who had become naturalized burghers. A naturalized burgher would be exempt from the burdens imposed on his colleagues at the Factory. He could trade both wholesale and retail and would no longer be obliged to pay half his duties in rixdollars but entirely in rubles. He would receive a one-eighth abatement on the import duty and three-eighths on the export duty when shipping in Russian vessels. To the question:

---

48. Ibid.
49. Carmarthen to Shairp and Fitzherbert, Mar. 16 and Apr. 6 (No. 3), 1787, PRO FO 97/341, respectively.
50. Charles Henry Fraser and Walter Shairp to Carmarthen, Apr. 13/24, No. 16, and Apr. 26/May 7, 1787, PRO FO 65/15, respectively.
51. Shairp to Forster, May 21/June 1, with enclosure, and June 4/15, 1787; Fraser's dispatches of May 11/22, June 22/July 3 and June 29/July 10, 1787, Nos. 20, 27, and 28; and Forster to Fraser, June 16, 1787, PRO FO 65/15. See also RC, VIII, pp. 239–51, et seq. British merchants in Riga were repeatedly threatened with the cancellation of their trading privileges unless they became naturalized Riga Burghers. Whitworth to Carmarthen, Jan. 19/30, 1789, No. 6, PRO FO 65/17.
52. See Frédérick Mauro, "Merchant Communities, 1350–1750," in *The Rise of Merchant Empires. Long-Distance Trade in the Early Modern World,* pp. 255–56.
53. Board of Trade to Fraser, June 15, 1787, No. 2, PRO FO 65/15.

"If the practice of assuming the Burghership of St. Petersburg should become more general, what would be the probable Effects on the British Trade and Navigation [?]" the Russia Company received the answer: "The Effects would . . . be both important and highly detrimental."[54]

---

54. Forster to Carmarthen, Dec. 13, 1787, No. 15 and enclosure, PRO FO 65/15, ff. 357–60. See also RC, Dec. 7, 1787, VIII, p. 239.

# Chapter IX
# International Politics and Commerce in Crisis

In August 1787 Turkey declared war on Russia, providing Sweden with a handsome opportunity to attack Russia in the hope of regaining Swedish territories previously lost to Russia. Catherine II deployed the greater part of her land forces to the south and directed part of her northern fleet into the Mediterranean to shore up the recently established Black Sea squadron. In February 1788 Emperor Joseph II, an ally of Russia since 1781, reluctantly entered the war against the Ottoman Empire. In June Sweden's fleet set sail to invade Russian Finland and Russia became engaged in a war on two fronts.[1]

During this period, British commercial policy toward Russia did not change, although new men came to represent Great Britain in St. Petersburg. Walter Shairp died in July 1787 and was succeeded by John Cayley as consul; in late 1788 Charles Whitworth succeeded Fitzherbert as ambassador.[2] Subtle changes, however, did take place in Britain's European alliance system. In September 1787 Frederick William II of Prussia, with British support, restored to his brother-in-law the authority of the Stadtholder of the United Provinces. In April 1788 both Britain and Prussia signed treaties with the United Provinces, and in August Britain concluded a defensive alliance with Prussia.

At the same time, Catherine presumed she could obtain British naval and navigation assistance in this new war, as she had during the first Russo-Turkish War, and she took steps to purchase

---

1. Andreas Bode, *Die Flottenpolitik Katharinas II und die konflikte mit Schweden und der Türkei (1768–1792)* (Wiesbaden: Otto Harrassowitz, 1979), pp. 113–30; R. C. Anderson, *Naval Wars in the Baltic during the Sailing-ship Epoch. 1522–1850* (London: C. Gilbert-Wood, 1910), pp. 241–93; idem, *Naval Wars in the Levant, 1559–1853* (Liverpool: University Press, 1952), pp. 318–47; and see also Ia. Zutis, *Ostzeiskii vopros v XVIII veke* (Riga: Knigoizdatel'stvo, 1946), Chap. XVI, pp. 61–125.
2. Correspondence on these matters is in PRO FO 65/15 and 97/341, passim. A "Treaty of Commerce between Portugal and Russia" was signed at St. Petersburg, December 9/20, 1787, CTS 50/pp. 253–76; and PRO FO 97/341.

supplies and hire ships and British seamen to transport Russian troops to the Levant. But this time the British chose "to observe an exact Neutrality"[3] in order to contain Russia's aggressive activities. William Pitt, the First Lord of the Treasury and Chancellor of the Exchequer, made the telling point to Semen Vorontsov that "during the last war with Turkey, England had friendly and commercial ties with Russia."[4] Semen Vorontsov understood Pitt's meaning and considered the Armed Neutrality the principal obstacle to a *rapprochement* with Great Britain. In August 1788 he wrote his brother Alexander: "The hatred that England feels for us has no other cause than the armed neutrality,"[5] and he repeatedly argued that it should be rescinded.

On April 21/May 2, with the opening of the 1789 shipping season, Catherine informed the diplomatic community in St. Petersburg that Russia would offer its protection to all neutral flags trading in the Baltic.[6] However, Swedish naval vessels subsequently intercepted, searched, and seized the non-contraband cargoes of British ships in the Russian trade. Russian privateers were reported to have committed similar acts although their original commission had been directed against Swedish vessels.[7]

The Russo-Turkish-Swedish War created additional burdens for Russia's already fragile financial situation. In 1787 Russia had printed 53.8 million assignat rubles that depressed the market rate from 97.1 in 1787 to 92.2 in 1788. From 1787 Russia borrowed heavily in the United Provinces and reputedly also in Antwerp, Genoa, Venice, and perhaps even in Lucca and Leghorn. The ruble had declined against the Dutch guilder from an annual average of 47.5 stuivers in 1750, to 39 in 1787, and to 34 in 1788. The first Russo-Turkish War had cost Russia an estimated 47.5 million rubles and the Russian court must have been anxious about how much the current war would cost.[8]

---

3. Carmarthen to Fraser, Mar. 18, 1788, No. 1, PRO FO 97/341. See also Fraser to Carmarthen, Apr. 1/12 and 14/25, 1788, PRO FO 65/16.
4. Quoted in James W. Marcum, "Semen R. Vorontsov: Minister to the Court of St. James's for Catherine II, 1785–1796" (Ph.D. diss., University of North Carolina, 1970), p. 99, and see pp. 94–103, passim.
5. Quoted ibid., p. 137.
6. Whitworth to Leeds, May 4/15, 1789, No. 27, PRO FO 65/17, f. 111.
7. Ibid., Oct. 29/Nov. 9, 1789, No. 56, PRO FO 65/17; Richard Crawshay to William Fawkener, secretary to the Board of Trade, Aug. 27, 1788; Fawkener to Carmarthen, Aug. 28, 1788, PRO FO 65/16; and Marcum, pp. 139–40, 159.
8. James C. Riley, *International Government Finance and the Amsterdam Capital Market 1740–1815* (Cambridge: Cambridge University Press, 1980), pp. 155–59, passim, and notes on pp. 304–7; Marten G. Buist, *At Spes Non Fracta. Hope & Co. 1770–1815. Merchant Bankers and*

By the end of 1789 Russia's international situation was critical. The prospect of Prussia's intervention in the war dampened Catherine II's and Joseph II's hope for a conclusive military and political victory over Turkey and Sweden. Frederick William II had already stirred up anti-Russian sentiment in Poland at the Four Years' Diet (1788–1791) and he appeared close to forming an alliance with the Turks and the Poles. Catherine wanted Great Britain to keep Prussia from acting precipitously. She instructed Semen Vorontsov that she was prepared to conclude a defensive alliance and a commercial convention with Great Britain. She had earlier that year sent to him the essential provisions of the latter, stripped of the demands the British had previously found so objectionable. As soon as the convention on commerce was signed in London negotiations could proceed to conclude a formal Anglo-Russian treaty of navigation and commerce.[9]

But Great Britain could not have prevented the Prussian alliances even if it had wanted to. At the end of January 1790, within days of Semen Vorontsov's receipt of Catherine's instructions, Prussia signed a treaty with the Ottoman Empire and at the end of March it concluded a defensive alliance with Poland.[10] On February 9, 1790, eleven days before Joseph II died, the British ministry denied Catherine its support for a peace that would permit "such considerable exchanges of territory," and would allow Russia's frontier to expand further into Turkish territory. Carmarthen, now the Duke of Leeds, informed Semen Vorontsov that such a

---

*Diplomats at Work* (The Hague: Martinus Nijhoff, 1974), "Baltic Affairs," pp. 73–154, passim, "Russian Loans," Appendix D, "Loans issued by Hope & Co. on behalf of Russia between 1787 and 1793," Appendix D-1, p. 497; "Russia Loans. By de Smeth in Amsterdam . . . De Wolff in Antwerp . . . Hope & Co. in Amsterdam . . . A Loan . . . in Genoa . . . ," in Whitworth to Leeds, Jan. 10/21, 1791, No. 2, PRO FO 65/20; Konstantin N. Lodyzhenskii, *Istoriia russkago tamozhennago tarifa* (St. Peterburg: V. S. Balashev, 1886), pp. 149–54, passim; N. N. Firsov, *Pravitel'stvo i obshchestvo v ikh otnosheniiakh k vneshnei torgovle Rossii vo tsarstvovanie imperatritsy Ekateriny II* (Kazan': Imperatskii Universitet, 1902), pp. 86–123, passim; Klaus Heller, *Die Geld–und Kreditpolitik des russischen Reiches in der Zeit der Assignaten (1768–1839/43)* (Wiesbaden: 1983); and M. P. Pavlova-Sil'vanskaia, "K voprosu o vneshnikh dolgakh Rossii vo vtoroi polovine XVIII v.," *Problemy genezisa kapitalizma. K Mezhdunarodnomu kongressu ekonomicheskoi istorii v Leningrade v 1970 g. Sbornik statei* (Moskva: Nauka, 1970), pp. 301–33, passim.

9. "Traduction du Rescript de l'Impératrice au Comte de Woronzow du 8/19, Decbr. 1789" [received by Semen Vorontsov on January 29, 1790], PRO FO 65/18, ff. 26–29; "Kopiia. Na podlinnom podpisano Sobstvennoiu Eiammperatorskago velichestva rukoiu tako: Byt' po Semy v S. P. Burga. Marta 4go 1789go" [sic], "Projet de Convention avec la Cour de Londres" (Reçu par le Courier Tripolski, le 10 Avril N. S. 1789), PRO FO 65/18, ff. 30–33; "Projet de Convention avec la Cour de Londres," "Delivered by Woronzov 30 Jan. 1790", "Copy transmitted to the Board of Trade with letter from the D. of Leeds requesting Report Thereupon," PRO FO 65/20; and Leeds to Vorontsov, Feb. 9, 1790, PRO FO 65/18.

10. CTS 50/pp. 473–83 and 489–95, respectively.

settlement would "affect the interests of various European Powers, and [would] appear to the King as calculated either to Prolong, & even to Extend the War, than to put an End to its Calamities."[11]

Furthermore, at about this time the British began to calculate an alternative to their commercial treaty with Russia. There is enough evidence to warrant the speculation that Great Britain, greatly encouraged by its recent closeness with Prussia, contemplated substituting for some of the strategic commodities it imported from Russia, those from Prussia and the region of non-partitioned Poland-Lithuania. Following the partition of Poland, Prussian ports had become more active in their trade with Great Britain. James Durno, a timber merchant and British Consul in Memel [Klaipéda], was well connected with the Prussian government and the wealthy Radziwiłł family whose forest products figured prominently in the timber trade. The British ministry solicited his opinion about the advisability of negotiating an Anglo-Prussian commercial treaty and the exploitation of the resources of Poland-Lithuania proper.[12]

In March 1790 the Board of Trade responded to Carmarthen about the Russian proposal for a Convention on Commerce as a preliminary to a *bona fide* treaty. Although it was pleased "that those principles of Maritime Law, detrimental to the Navigation of this Country [the armed neutrality], are at present laid aside and will no longer embarrass this Negociation," the Board would not recommend "to renew the Treaty of 1766, until some Explanation is given of the sense in which the Ministers of Her Imperial Majesty understand certain Articles in this Treaty."[13] Specifically, the Board was more critical of the distinctions in the abatement provision of the 1782/83 tariff, the payment of export custom duties by both British and Russian merchants and the reservation clause in Article IV of the 1766 treaty. Before such a Convention on Commerce could be concluded as a preliminary to a Treaty of Commerce and Navigation with Russia, the Russian court must first place British subjects and British ships in the export trade from Russia on "the

---

11. Leeds to S. Vorontsov, Feb. 9, 1790, PRO FO 65/18, f. 49.
12. Dietrich Gerhard, *England und der Aufstieg Russlands* (München und Berlin: Verlag von R. Oldenbourg, 1933), "Polens wirtschaftliche Lage" and "Der englisch–polnische Handelsplan," pp. 275–308 et seq., passim; John Ehrman, *The British Government and Commercial Negotiations with Europe 1783–1793* (Cambridge: Cambridge University Press, 1962), pp. 125–26, 112–20, passim; and James Durno's reports, PRO FO 65/17 and PRO BT 6/244.
13. Leeds to Board of Trade, Feb. 26, 1790, PRO FO 65/18, and Stephen Cottrell, secretary to the Board of Trade, to Leeds, Mar. 20, 1790, PRO BT 6/180, pp. 99, 101. Copies of this lengthy document are also in PRO BT 3/2, pp. 213–35; and PRO FO 65/18, ff. 121–35.

same footing in all respects with the Subjects and Ships of the Russian Dominions."[14]

The Board argued against the British ministry dwelling on Russia's abatement of one-eighth the duty on imports into Russia by Russian subjects in Russian ships. The reason it gave was in itself a somewhat remarkable admission—that for decades Great Britain had knowingly vitiated the most-favored-nation principle provided for in the 1766 treaty:

> Russian Goods imported into this Country in Russian Ships, are by the Laws of Great Britain subject to a Distinction which equally operates in favour of British Ships; and it is worth remarking that by the 9th Clause of the Act of Navigation of 1660, several Articles imported from Russia in Russian Ships now pay more than the like Articles imported from other Foreign Countries in the Ships of those Countries.[15]

As the spring warmed to summer in 1790 it appeared that British political and commercial considerations were merging into one national policy. The decisive reason for Great Britain's refusal to conclude a Convention on Commerce with Russia lay in Pitt's changing attitude toward Russia. Commercial imperatives involving Russia were sacrificed to international political considerations. According to M. S. Anderson the British "idea of a commercial treaty with Poland and a great expansion of Anglo-Polish trade" had "become one of the pivots of British policy," and had "more than anything else led Pitt to take up an openly anti-Russian attitude."[16]

From the early months of 1790, while Leopold II of Austria (brother and successor to Joseph II) sought to mend his political

---

14. Cottrell to Leeds, Mar. 20, 1790, PRO BT 6/180, pp. 106–8. See also, pp. 102–5, 114. Moreover, the Board stipulated that because it is nowhere mentioned in the 1766 treaty "Ireland shall be understood to be included in this Treaty" and "Le Sujets de la Grande Bretagne" would encompass its population. Ibid., pp. 109–10, ff.
15. Ibid., p. 114.
16. M. S. Anderson, *Britain's Discovery of Russia 1553–1815* (London: Macmillan & Co., 1958), pp. 147 and 148–49. See also Gerhard, pp. 275–383, passim; Ehrman, pp. 128–34; and John Harold Clapham's insightful brief article, "The Project for an Anglo-Polish Treaty (1782–1792)," *Baltic Countries* 1 (1935): pp. 33–35.

Leeds asked Whitworth secretly to find someone specialized in the cultivation of hemp "from those parts of Russia or Poland Livonia. . . . who would be willing, on receiving proper Encouragement, to emigrate to Canada" to instruct "the Canadians in the best and cheapest method of preparing it." Leeds to Whitworth, Sept. 21, 1790, No. 8, PRO 65/19. Whitworth employed an experienced hemp planter originally from Brunswick, and subsequently others, but two of them were later arrested and imprisoned by the Russians. Whitworth to Leeds, Oct. 18/29, 1790, No. 62, and April 4/15, 1791, PRO FO 65/19 and 20, respectively; and Grenville and Leeds to Whitworth, Dec. 20 and 24, 1790, PRO FO 65/19, respectively.

fences with his rival Frederick William II of Prussia, Pitt aimed to prevent Catherine II from expanding Russia's frontier further into Ottoman territory and even attempted to make her withdraw to a line *ante bellum*. Catherine II refused British mediation but Leopold II subsequently accepted it to extricate Austria from the war with Turkey. Britain collaborated with Prussia in subsidizing Sweden to continue the war against Russia. The international tension built to the point that when Great Britain mobilized its navy for the avowed purpose of protecting its interests in North America against the perceived threat from Spain, it fueled speculation that Britain would soon turn that fleet against Russia. But the international political arrangement once again changed when Leopold II and Frederick William II concluded the Convention of Reichenbach in June 1790 and when Sweden made peace with Russia in August.

By the early months of 1791 the Anglo-Prussian policy against Russia's expansion had taken shape. In late March the British ministry drafted an ultimatum to Catherine, inclusive of a deadline, that insisted that she make peace with the Ottoman Empire to the satisfaction of Britain and Prussia; that is, Russia should give up Ochakov. Pitt's request in Parliament for a partial mobilization of the fleet provoked a public debate over Great Britain's alleged war aims and commercial policies. Semen Vorontsov played an active role defending his country and Empress by encouraging the opposition to Pitt, supplying it with information and arguments. Politicians and the public alike believed Russia "a power whom we could neither attack, nor be attacked by." No one appeared ready to go to war for Ochakov, a place they could neither pronounce nor find on the map. The press went against Pitt. The opposition did not view Russia's activities against the Ottoman Empire as a threat to the balance of power.[17]

In early April, soon after Pitt had dispatched Britain's ultimatum to Whitworth to be given to Catherine, Pitt changed his mind and put a stop to its delivery. William Fawkener, a Secretary of the Board of Trade, was sent to St. Petersburg to arrange a compromise with Catherine. On May 3, 1791, the Poles proclaimed their celebrated Constitution. At the end of July both the British and the Prussian envoys at St. Petersburg agreed to Russia's annexation of Ottoman territory between the Bug and the Dniestr rivers, which

---

17. For Semen Vorontsov's activities, see E. A. Smith, *Lord Grey 1764–1845* (Oxford: Clarendon Press, 1990), p. 33; Marcum, pp. 136, 178–80 ff.; and A. G. Cross, *"By the Banks of the Thames." Russians in Eighteenth Century Britain* (Newtonville, Mass.: Oriental Research Partners, 1980), pp. 23–27.

included Ochakov, and a preliminary peace was signed at Jassy in August 1791.

From the expiration of the Anglo-Russian Commercial Treaty to the end of the Russo-Turkish-Swedish War it had become increasingly expensive for the merchants of the British Factory to conduct their trade. In a periodic report to London, John Cayley wrote in March 1791: "the difference of paying one half the Amount of the duties in Rix dollars, or in Russ Money (which latter mode, the Nations, foreign Guests or Burghers & other Nations under Treaties of Commerce Enjoy the privilege of) is become from the present low & variable course of Exchange, an object of 30 to 32 Per cent upon the Amount of the Duties to the prejudice of the English Nation."[18]

In London, on April 1, 1791, Edward Forster wrote the British ministry about the Russia Company's concern for the safety of the valuable cargoes on the more than 200 ships preparing to sail from Britain to Russia, and also about the several ships that had wintered in Kronstadt, which were about to sail for Britain. Moreover, British merchants with their families in several parts of Russia had become exercised, Forster wrote, over the consequences of not having the protection of a treaty. Since the British merchants sold their products to the Russians on long credit and frequently paid considerable advances to the Russians for their goods prior to delivery, it was extremely important to the Russia Company that their ships reach their destinations on schedule to prevent a serious balance of payments crisis.[19] In May the British ministry assured For-

---

18. Cayley to Leeds, Mar. 7/18, 1791, PRO FO 65/20.
19. Forster to Leeds, Apr. 1, 1791, PRO FO 65/20; and see also RC, Apr. 1, 1791, VIII, pp. 327–29. An example of British merchants extending long-term credit to their Russian counterparts is noted by John Morewood of St. Petersburg, a partner in and factor for Longsdon & Morewood, a small English cotton manufacturing firm. John Morewood, with his younger brother Thomas, arrived in St. Petersburg in 1783 and expected "either to sell for ready money or barter for saleable Russian produce and so make returns immediately." But these expectations were not fulfilled because, as he reported in May 1784: "I have this week sold near R1,500 amount of velverets and white goods, but . . . I am compelled to give twelve months credit." In July Andrew Morewood Jr. of Manchester wrote:

> My brother writes that the buyers from the country have been down two days, that they are all unanimous in giving our goods the preference and that he has bargained for nearly the whole of our stock at fair prices, but that the credit they take is twelve months.

In 1785 John Morewood reported "that he has sold half the white goods and a few velverets, and says his prospects still wear a favourable appearance, and that if it was not for our circumscribed capital he should be filled with very gay hopes." Quoted in Stanley Chapman, "James Longsdon (1745–1821), Farmer and Fustian Manufacturer: The Small Firm in the Early English Cotton Industry," *Textile History* 1:3 (Dec. 1970): pp. 272–75. In 1787 John Morewood declared 26,513 rubles and 25 kopeks of imports at St. Petersburg. VTR, *Vedomost'*, No. 6.

ster that no precipitous action against Russia would take place to endanger the trade between the two countries. It remained silent, however, on the possibility of concluding a treaty of commerce with Russia.[20]

In January 1792 Edward Forster wrote privately to Foreign Secretary William Grenville, successor to the Duke of Leeds, that "several very respectable" members of the Russia Company, whose mercantile interests in Great Britain and in Russia were considerable, had provided him with a confidential representation entitled "For Information" to be given to him, with copies only for Lord Hawkesbury, President of the Board of Trade, and William Pitt. This document at first addressed well-known matters; namely, that from the expiration of the treaty of commerce the importance of the British Factory had gradually been declining; that its members had lost their personal privileges and perquisites; and that they paid about one-third more duties on their merchandize than either Russian merchants or other merchant nationals who were protected by treaties of commerce with Russia.[21]

"For Information" then went on to state that several British Factory merchants had chosen not to complain publicly for fear of undermining "the progress of Political measures." But others had chosen to become naturalized Russian burghers, acquiring the privilege of paying their duties with Russian currency, which "gives them a most essential advantage over those of the Factory." In the present circumstances it remained doubtful that merchants of the Factory could continue "their trade under such great disadvantages & such unequal terms." If the situation did not change for them "they will have no alternative than to . . . follow the example that has been shewn them by accepting the benefits of Russian Naturalization & render the British Factory totally useless." It was the opinion of "For Information" that Russia had long sought "To Form a Commercial Marine, & to introduce a spirit of Foreign Commerce," and "to this end Naturalization with all its benefits is

---

20. See exchange of notes between Grenville and Forster dated May 3, 1791, with enclosures, PRO FO 65/20. Similar concerns and reassurances were expressed in Forster's correspondence with Grenville, July 25, Aug. 14, and 15, 1791, PRO FO 65/22 and PRO FO 97/341, respectively.
21. "For Information," dated Jan. 17, 1792, in Forster to Grenville, Jan. 19, 1792, PRO FO 97/341. Charles Jenkinson, first earl of Liverpool (1727–1808), was given the title Lord Hawkesbury when he was appointed president of the Board of Trade in 1786. William Wyndham Grenville (1759–1834), youngest son of George Grenville and cousin of William Pitt, had been vice president of the Board (1786–1789) and was created Baron Grenville of Wotton in 1790.

freely granted: The Russian Flag is easily obtained for Foreign built ships, & an abatement is allowed of 1/8 of the duties upon goods imported & 3/8 upon goods exported in Russian ships & for account of Russian Subjects." But Russia had advanced much beyond these modest steps:

> The Russians possess in great abundance every material for building & equipping ships; & in fact do build & completely equip them at less expence than most other European nations can do; many of them in the last war have been sold to England, Holland, France, & Spain; several of such burthen as to be employed in the East & West India Trades & even in his Majesty's service.
> To carry into full effect this view of the Russian Government nothing seems wanting but the abilities, the activity, the Experience, the Fortunes & the enterprising spirit of British Traders and Navigators. Already Russian built, & foreign naturalized, ships navigated by Crews at least half Russians are become Carriers to & from Ports where they were formerly unknown.[22]

If the British government would not immediately assist the British Factory, the authors of "For Information" apprehended that merchants of the Factory would be "driven to the necessity of becoming Russian Burghers" and they would promote and improve "the Russian Navigation."

> Artificers of every kind, in every branch connected with that art, may be allured from Great Britain by the liberal encouragement of their countrymen. British Seamen may be tempted into their service; Instruction & example may join to accelerate the accomplishment of this grand object, &, in no great length of time, the chief part of the Russia trade, not only with other foreign nations, but even with Great Britain itself, may be carried on in Russian Ships & cease to be what it has long been one of the most useful nurseries of hardy seamen.[23]

Whitworth was less optimistic about Russia's maritime future. "Whatever Progress towards Perfection has been made in the Imperial Navy, that, in the Merchant Service," he states, "has not, by any Means, kept Pace with it; and there are, at this Moment, but very few Russian vessels employed in the Carrying Trade. The Idea of employing Russian Subjects in this Service, which would

---

22. Ibid.
23. Ibid.

necessitate their more immediate Intercourse with other Countries, is too contrary to the Spirit of this Government ever to be adopted by it."[24]

Notwithstanding the conflicting views stated above, Catherine's subjects had not fulfilled her hope of developing an independent Russian maritime fleet and mercantile community capable of competing effectively with other powers in overseas commerce. But Catherine had created the impression that Russia could do this with the help of foreigners, not only from among British Factory merchants but also, and in particular, from among Germanic merchant nationals. They had become naturalized Russian burghers and they knew much more about mercantile affairs than her native-born subjects. This impression was enough to threaten the maritime powers. But by doing this Catherine had also wasted valuable time and resources that could have been better put to use for longterm mercantile and maritime purposes. This temporary expedient was a weakness in her policies and a cause of her procrastination in coming to terms with the British over the commercial treaty.

Whitworth's views about the *mentalité* of the Russian court in early March 1792 were insightful: "The Imagination of Her Imperial Majesty and Her Ministers are too exalted, at this Moment, on Commercial, as well as on every other Subject, to leave Room to hope that we should find them more reasonable than when we last abandoned the [commercial] Negotiation." His views were also predictive:

> They are sensible of this Necessity we are under of coming to their Market for Objects we cannot so conveniently find else where as their Importation from Great Britain has considerably increased within these few years, so much so, indeed, as to bring the Balance nearer Par than was ever supposed likely to be the Case—they are unwilling to give up the Advantages they now enjoy of receiving the Duties on our Importation in Silver Money, and which, were it not evaded in some Instances, by English Merchants entering their Goods under Russian Names, a Fraud which it is intended to counteract, would be very considerable indeed.

Whitworth was pessimistic: "Unless, therefore, her Imperial Majesty should have some immediate Inducement to cultivate His Majesty's good Will, I see no Reason to suppose, that any Proposal

---

24. Whitworth to Grenville, Feb. 19/Mar. 2, 1792, No. 10, PRO FO 65/23.

will be made on Her Part, towards an Arrangement on this Point; nor, indeed, is it possible to establish any System 'till the Business of Poland which Certainly is at present the Object of all the Attention of this Court shall have been determined."[25]

But in April 1792, responding to an appeal from the King of England, Catherine, through her vice chancellor Osterman and Platon Alexandrovich Zubov, her current influential young favorite, told Whitworth that she wanted to reopen formal commercial negotiations with Great Britain. For that purpose Semen Vorontsov had been authorized to activate the instructions and plenipotentiary powers he had been given two years earlier.[26] In early May, Alexander Vorontsov told Whitworth that his brother's instructions "were so drawn up, as to remove all, or at least the greatest Obstacles, which stood in the Way of that Negotiation," and made clear what that meant. "The Empress felt herself obliged . . . in Honour not to abandon a System which she had announced with so much Publicity, and without the Adoption of which She had pledged Herself with other Powers, never hereafter to subscribe to a Treaty of Commerce." However, because of her avowed friendship for the King of England the "Principle of Armed Neutrality might be omitted" in the Convention, "reserving its Discussion to the Negotiation for a Treaty." Moreover, as a means of removing other obstacles to such an agreement, preferential customs duties would, henceforth, be awarded only to Russian subjects born in the Empire.[27]

Nevertheless, events in Europe overshadowed the commercial negotiation. In France the Revolution had abolished the monarchy and declared the birth of the First Republic and in Poland-Lithuania tumult reigned. The hostilities that spilled over France's frontiers toward the end of 1792 persuaded Catherine that Great Britain's friendship was necessary to execute her ambition in Poland-Lithuania and to thwart revolutionary France. The correspondence of both Semen Vorontsov and Charles Whitworth testifies to Catherine's willingness (notwithstanding her previously declared principle of the Armed Neutrality) to use her influence in Denmark and Sweden to assure the safety of British shipping in the event that France went to war against Great Britain.[28] January

---
25. Ibid.
26. Ibid., Apr. 13/24, 16/27, Apr. 20/May 1, and Apr. 23/May 4, 1792 (Nos. 18–20 and 22), respectively, and Grenville to Whitworth, Mar. 27, 1792 (No. 6), PRO FO 65/23.
27. Whitworth to Grenville, Apr. 27/May 8, 1792, No. 23, PRO FO 65/23.
28. Marcum, pp. 258–62 ff.; and Whitworth to Grenville, Jan. 27/Feb. 7, 1793, No. 8, PRO FO 65/24.

1793 marked the execution of King Louis XVI of France, the signing of the Russo-Prussian treaty for the Second Partition of Poland and the declaration of war by the French Convention against Great Britain. Russian-French relations had already been strained for several months and in February Catherine formally ended them.[29]

This international political crisis precipitated an "extraordinarily convenient occasion" for Catherine to introduce new financial and commercial measures.[30] It had become patently obvious to everyone at the Russian court that the cost of the Russo-Turkish-Swedish War had exacerbated Russia's already deteriorating financial situation. From 1787 Russia had repeated its practice of international borrowing and assignat rubles were printed to pay its bills. In 1787, 53.8 million assignat rubles were printed and from 1788 through 1793 an additional 24 million were issued. From 1787 to 1793 Dutch loans alone raised f. 53.5 million. The market rate of the assignat ruble continued to fall: 97.1 in 1787; 92.2 in 1788; 78.7 in 1792; and to 73.5 in 1793. The exchange rate of stuivers to the ruble continued to erode: 39 in 1787; 34 in 1788; 27 in 1792; and 24.5 in 1793.[31] Moreover, Russia's favorable balance of overseas commodity trade was in jeopardy in 1790 and 1792 and was actually unfavorable in 1791. St. Petersburg/Kronstadt suffered unfavorable balances in all three years.[32] In mid-March 1793 John Cayley, the British General Consul, reported from St. Petersburg that the Russians were in "Agitation" to "Reduce the increasing Amount of the Imports, [and] many Prohibitions of foreign Articles of Luxury ... will speedily be issued."[33] The tariff of 1782/83 was reexamined. Alexander Vorontsov, its principal author, whose reputation had recently been tarnished by the perception of his complicity in the "Radishchev Affair," had already fallen and taken leave from the Russian court.[34] The challenge confronting Catherine was whether she could stem Russia's sinking rate of exchange and improve its balance of international trade.

---

29. Whitworth to Grenville, Feb. 11/22, 1793, No. 13, PRO FO 65/24. "Traduction de l'Edit Émané de SA MAJESTÉ IMPÉRIALE de toutes les Russies, et addresse a SON Sénat le 8 [19] Fevrier 1793," and "Formule de Serment," in Whitworth to Grenville, Feb. 15/26, 1793, No. 15, PRO FO 65/24.
30. Lodyzhenskii, pp. 151–52.
31. Supra, footnote 8.
32. See Chap. XIV, below.
33. Cayley to Grenville, Mar. 7/18, 1793, PRO FO 65/24.
34. Lodyzhenskii, pp. 149–54; Riley, pp. 158–59; Whitworth to Grenville, Mar. 3/14, 1793, No. 18, PRO FO 65/24; Marcum, pp. 244–49; and Robert E. Jones, "Opposition to War and Expansion in Late Eighteenth Century Russia," JGO 32:1 (1984): pp. 46–49.

On March 14/25, 1793, Great Britain and Russia signed in London a Commercial Convention[35] and a Convention for Concerted Action Against France.[36] With modifications on the jurisdiction of the College of Commerce and the tariff reduction at the Black and Azov Sea ports, the Commercial Convention renewed the expired 1766 treaty of commerce for six years "till a Definitive Arrangement for a Treaty of Commerce can be agreed upon between the two courts."[37] The defensive alliance contained interesting provisions, which are best summed up by Isabel de Madariaga: "to close their ports to all French ships and to prohibit French trade with neutral countries. This last provision ran totally counter to the principles of the armed neutrality proclaimed by Catherine with such pomp in 1780."[38]

But within a month after signing the Convention on Commerce the Russian court resorted to its tariff policy to effect ad hoc changes in the Anglo-Russian commercial relationship. However, the goal this time was significantly different from past policy. These tariff edicts beginning in April 1793 were not designed, as the earlier ones had been, to give a simple preference to Russian merchants who would transport their commodities in Russian ships and thereby encourage the development of a Russian mercantile fleet. Nor were they enacted merely to increase the rate of custom duties on the importation of British merchandise into Russia. On April 14/25, 1793, Catherine II issued a detailed edict that canceled all commercial intercourse between Russia and France. Included in that edict, however, was also a list, dated April 8/19 but only published on April 26/May 7, of presumed French merchandise that Russia now prohibited from importation and which would be treated as contraband after June.[39] But the edict had a

---

35. *Convention between His Britannick Majesty and the Empress of Russia. Signed at London, the 25th of March, 1793* (London: Edward Johnston, in Warwick-Lane, 1793), 8 pp., PRO BT 6/233, ff. 97–101; CTS 51/pp. 491–97; and "Minute," Whitehall, Mar. 15, 1793, PRO FO 97/342.
36. "Convention for Concerted Action against France between Great Britain and Russia, signed at London, 25 March 1793," CTS/pp. 52/1–6; Osterman to Semen Vorontsov, Jan. 27/Feb. 7, 1793, PRO FO 65/24; and Grenville to Semen Vorontsov and Whitworth, Mar. 20 and 26, No. 4, 1793, PRO FO 65/24, respectively.
37. *Convention . . . the 25th of March, 1793.*
38. Isabel de Madariaga, *Russia in the Age of Catherine the Great* (New Haven: Yale University Press, 1981), p. 443.
39. PSZ/XXIII/17,111/414–17, Apr. 8/19, 1793, published Apr. 26/May 7, 1793, "Rospis tovaram . . . Smotri knigu Tarifov"; "Traduction de l'Edit de SA MAJESTÉ IMPÉRIALE, Émané du Sénat dirigeant le 14 [25] Avril 1793," and "Traduction. L'Original est confirmé par Sa Majesté Impériale dans ces termes: Soit fait ainsi. St. Petersbourg le 8 [19] Avril 1793. LISTE des marchandises et effets des pays étrangers, dont l'importation tant par mer que par terre est défendue dans l'Empire de Russie," in Whitworth to Grenville, May 17/

broader purpose. Many of the goods listed came from countries other than France, for example, metalwares, which Russia normally obtained from Great Britain and not France, were included. The situation was further complicated by the fact that France transhipped significant quantities of goods to northern European states which in turn re-exported them to Russia. Whitworth later concluded that because "no distinction had been made between friend and foe" the creation of the list "might possibly have been done with a view to the Finances of Russia."[40]

Several results were intended from the enactment of these measures. By putting a halt to a significant number of foreign imports Russia's trade balance was expected to improve and regicide France would be punished at the same time. Under this policy, Russia's ally Great Britain was also punished. The British Factory and other foreign mercantile houses had long before contracted for their imports. Many of the enumerated prohibited goods were already en route to Russia.[41]

The Russian court published additional edicts with the intention of bringing order out of the chaos that was engulfing the Custom House. These new edicts revised the earlier prohibition list and extended the deadline for the mandatory sale or re-export of such merchandise. However, these new enactments only made matters worse, especially for the British trade, as the edict of December 13/24, 1793 actually expanded the list of prohibited goods to include British merchandise that had previously not been prohibited.[42]

Whitworth complained to the Russian court that he had, during the summer, been given "a Verbal assurance" by the Custom

---

28, 1793, No. 32, PRO FO 65/24 and PRO BT 1/14, ff. 391–96; copy of "Translation of the List of Prohibited Goods by the Empress's Ukase of the 14th April o. s. 1793. The List is dated the 8th April o. s. 1793 . . . The original signed Count Alex. Bezborodko," PRO FO 97/342; Whitworth to Grenville, Apr. 12/23 (No. 26), Apr. 18/29, May 17/28 (No. 32), and May 30/June 10, 1793, in June 3/14, 1793 (No. 36), PRO FO 65/24; and Lodyzhenskii, pp. 152–54.
40. Whitworth to Grenville, Oct. 8/19, 1793, No. 62, PRO FO 65/25; and see Paul Butel, "France, the Antilles, and Europe in the Seventeenth and Eighteenth Centuries: Renewals of Foreign Trade," *The Rise of Merchant Empires. Long-Distance Trade in the Early Modern World, 1350–1750,* ed. James D. Tracy (Cambridge: Cambridge University Press, 1990), pp. 153–73, in particular, pp. 164–68.
41. Whitworth to Grenville, dated May 30/June 10, 1793, in June 3/14, 1793, No. 36, PRO FO 65/24.
42. PSZ/XXIII/17,169/476–77, Dec. 13/24, 1793; PSZ XXIII/17,170/477–78, Dec. 17/28, 1793; "Translation of Her Imperial Majesty's Edict of the 18th [O. S.] December 1793 under her Sign Manual [sic]," and "Translation of an Ukase issued by the Senate printed and published under date of the 30th Dec. 1793 [O. S.] [sic]," in Whitworth to Grenville, Dec. 23/Jan. 3, No. 1 (and enclosure), and Jan. 13/24, 1794, No. 4, PRO FO 65/26, respectively.

House that particular categories of British cloths were exempt from prohibition. He also stated that those exempted cloths had been imported "during the whole Season without the smallest difficulty." According to him, one consequence of the edict of December 17/28, which mandated the sale or re-export of prohibited merchandise by July 1794, was that "the retail merchants on the pretext of not having a sufficient time to dispose of [them], return them on the hands of their Commissioners." In January 1794 Whitworth estimated that "the amount of British Manufactured Goods now prohibited, is . . . near a million Sterling."[43]

The Board of Trade assessed the potential damage and concluded that the edicts extended the prohibition "indiscriminately to all Woollen Goods whatever . . . by which means some of the most considerable branches of the Importation Trade from Great Britain are at once totally cut off, and several of our principal Manufacturing towns exposed to a most heavy loss."[44] John Cayley reported from St. Petersburg that the recently prohibited British goods were not luxuries but had "always been considered as articles of necessity in Russia, and by no means interfering with any useful manufacture of its own. Indeed, they cannot be made in Russia at all." Cayley's petition to the Russian court was ignored.[45] In March 1794 Whitworth estimated that the value of British goods in Russia that had to be disposed of under the law amounted to 1.2 million pounds sterling.[46]

It seems clear that the intent of the 1793 edicts was to reduce the British trade to Russia by prohibiting the importation of specific categories of British merchandise that had traditionally been sold to Russia. This prohibition would not only have a direct negative impact on the volume of imported British goods transported in British ships to Russia, it would also burden those British manufacturers who produced for the Russian market with "The goods being totally unsaleable in any other Market."[47]

Certainly that became the result. Complaints from British

---

43. Complaint submitted to Russian court on Jan. 1/12, 1794, enclosure in Whitworth to Grenville, Jan. 6/17, 1794, No. 3, PRO FO 65/25.
44. Quoted in "Mr. Eton's Remarks on the present State of the Importation Trade from Great Britain to Russia, 1794," PRO FO 97/342 and PRO BT 1/14.
45. "Consul Cayley's Representation, 1794," in "Mr. Eton's Remarks on the present State of the Importation Trade from Great Britain to Russia, 1794," PRO FO 97/342 and PRO BT 1/14, ff. 387–88.
46. Whitworth to Grenville, Mar. 18/29, 1794, No. 20, PRO FO 65/26.
47. Quotation from "The Memorial of the Merchant Manufacturers of the City of Norwich . . ," Aug. 16, 1794, in Cottrell to Burges, Aug. 20, 1794, PRO FO 65/28. See also Cayley's report, "What is the just and reasonable preference the Russian Government can give to their own Subjects?" dated April 24, 1794, in "Mr. Eton's Remarks on the present State of

manufacturers became so persistent that no cabinet official could ignore them. A petition from the Sheffield and Hallamshire cutlery manufacturers, whose "families are nearly in want of the common necessaries of Life," exposed the vulnerability of the Anglo-Russian commercial connection. Their cutlery had been manufactured from imported Russian iron and now was forbidden to enter Russia.[48] The longterm market orientation for British manufacturers was in jeopardy.

Grenville ordered Whitworth not to sign either the political-military alliance or the commercial treaty with Russia until Russia revoked the prohibition edicts: "the Security and protection of the Commerce of His Majesty's Subjects in The Russian Empire can afford the only solid basis of friendship or union between the two Courts."[49] But after having repeatedly failed to persuade the Russian court to revoke the prohibitory edicts and having gained the support of the British Factory, Whitworth recommended to Grenville "retaliation on the part of England" by refusing to import Russian goods, which he believed would have an immediate effect on the manufactories in Russia. "The Empress and her Ministry are convinced that in a commercial point of view we cannot do without them, and that they may deal with our Imports as they please, lopping it off branch by branch without risk or fear of reprisals." According to Whitworth, "unless we can prove to them that we are not so much in their power as they imagine, they will not stop until they have gradually reduced the importation from Great Britain to nothing." The only recourse, therefore, was "A war of prohibition."[50]

---

the Importation Trade from Great Britain to Russia, 1794," PRO FO 97/342 and PRO BT 1/14, ff. 397–99.
48. "The Memorial of the Master Wardens, Searchers and Assistants of the Company of Cutlers, within Hallamshire in the County of York, under their Common Seal; and of the Merchants, Manufacturers, and principal Traders of Sheffield," in Cottrell to Burges, Mar. 29, 1794, PRO FO 97/342. See also, "At a General Meeting of the Manufacturers of Norwich, at the Guildhall in that City on Wednesday the 12th February 1794," in Grenville to Whitworth, Feb. 22, 1794, No. 9, PRO FO 65/26; "The Memorial of the Merchant Manufacturers of the City of Norwich . . ," Aug. 16, 1794, in Cottrell to Burges, Aug. 20, 1794, PRO FO 65/28, and PRO BT 5/9, pp. 261–64; and "The Memorial of the Merchants and Manufacturers of the Town of Birmingham, concerned in the Russian Trade," March 18, 1794, in Grenville to Whitworth, Mar. 28, 1794, PRO FO 65/26.
49. Grenville to Whitworth and Semen Vorontsov, Feb. 22, 1794, No. 9, PRO FO 65/26. See also Grenville to Whitworth, Jan. 17, 28, and 31, 1794, Nos. 2, 7–8, PRO FO 65/26. For Great Britain's political-military goals, see John M. Sherwig, *Guineas And Gunpowder: British Foreign Aid in the Wars with France 1793–1815* (Cambridge, MA: Harvard University Press, 1969), pp. 69–70; and Whitworth-Grenville correspondence on the negotiation for a political-military-commercial alliance in PRO FO 65/26–27, passim.
50. See several dispatches from Whitworth to Grenville, Mar. 10/21, (No. 16, with enclosures of Mar. 4/15 and 6/17), Mar. 14/25, Mar. 18/29, Mar. 24/Apr. 1, Mar. 28/Apr. 8, Apr.

In June 1794 Catherine published yet another edict that allowed for a wide interpretation in setting new deadlines of the sale, re-export, and/or the destruction of British merchandise. The edict permitted the continued sale of British steel and iron wares in Russia until April 1, 1795, but within six weeks thereafter all unsold inventories had to be re-exported, during which time they would not be subject to duties and charges. However, British manufactures that had been imported prior to publication of their official prohibition were to be sold no later than April 1, 1796, and within six weeks thereafter all unsold inventories had to be re-exported, during which time they would not be subject to duties or charges.[51] As with the edicts of 1793 the ultimate intention of this edict was to foreclose the Russian market to British manufactures.

When the Anglo-Russian Defensive Alliance was finally concluded on February 7/18, 1795, Article XXI stated that in the future both countries would conclude a treaty of commerce[52] but Catherine II chose not to accommodate the British in commercial affairs until the day she died. In fact, had Catherine not died in November 1796, the British might have suffered another devastating blow to its commerce with Russia. In September 1796 Catherine published a revision of the tariff of 1782/83 that was to take effect in January 1797.[53] It reflected Catherine's final judgment about the future of Russia's overseas commerce and, in particular, that with Great Britain. Catherine correctly recognized that the prices of both imported and exported merchandise had risen considerably since 1783 and that the tariff of 1782/83 had lost its meaning and had to be revised to meet current conditions and challenges. Therefore, the 1796 tariff added to the import prohibitions of the edict of April 8/19, 1793.[54]

---

18/29, (Nos. 19–22, and 26), and June 23/July 4, 1794 (No. 38), PRO FO 65/26 and 27, respectively.

51. PSZ/XXIII/17,215/517–18, June 14/25, 1794; "Translation. Her Imperial Majesty's Edict of the 14th June 1794. Published by the Senate the 18th. [O. S.]," in Whitworth to Grenville, June 23/July 4, No. 38, 1794, PRO FO 65/27; and "Extract from the Empress's Ukase (or Edict) dated 14th June 1794 [O. S.] published the 18th. [O. S.] (Translation)," in "Mr. Eton's Remarks on the present State of the Importation Trade from Great Britain to Russia, 1794," PRO FO 97/342 and PRO BT 1/14. See also Whitworth to Grenville, Apr. 29/May 9, 1794, No. 29, PRO FO 65/27 and PRO 97/342, with enclosure, Whitworth to Osterman, Apr. 26/May 6, 1794.

52. CTS 52/pp. 315–26; and for the conclusion and ratification of the Defensive Alliance, see Whitworth's correspondence to Grenville, PRO FO 65/29–30, passim.

53. PSZ/XXIII/17,511/935–40, Sept. 16/27, 1796, [effective Jan. 1/12, 1797]; and for discussions on the implication for Anglo-Russian overseas commerce, see PRO FO 65/34–35; and Lodyzhenskii, pp. 154–57.

54. Ibid.

Moreover, experience had shown Catherine that the provision that accorded abatements to Russian subjects who exported or imported goods in Russian ships (three-eighths on exports and one-eighth on imports) had not produced the desired result. Rather, in addition to the abuses it created, it also reduced significantly Russia's custom revenues. The abatements were deleted from the tariff of 1796, wherein it was stipulated that everyone shall henceforth pay full duty. Finally, and most important for the intended future of Anglo-Russian commercial relations, the tariff of 1796 doubled the duties on many articles of export and imposed even greater duties on some articles of import. In general, the ad valorem duties on British manufacturers were to increase to about 50 percent; on British cotton goods to about 70 percent; and from two to three times the value on British ale and porter. Soon after Paul ascended the throne he suspended the tariff of 1796.[55]

Catherine's death in November 1796 brought to an end the British hope that Russia's diplomatic and military power would be used against France. Immediately on ascending the throne Paul recalled Catherine's commitment of an auxillary force of 60,000 men.[56]

A disappointed William Pitt offered a bitter observation: "It is difficult to say whether one ought to regret the most that she had not died sooner or lived longer."[57]

---

55. Ibid.; and Shairp to Grenville, Nov. 25/Dec. 6, 1796, PRO FO 65/35.
56. Roderick E. McGrew, *Paul I of Russia 1754–1801* (Oxford: Clarendon Press, 1992), pp. 150, 282, 308.
57. Quoted in Sherwig, p. 83.

# Part IV
# The Balance of Trade and Culture

# Chapter X
# Merchant Nationals

At the time of Catherine II's death Russia was still fundamentally dependent upon foreign merchants, foreign carriers, foreign seamen and foreign mercantile services to conduct its overseas trading operations. This represented a failure of the first order for Catherine's commercial, maritime and political policies, for which she must bear the primary responsibility.

Some of the factors that prevented Russia from taking control of its own overseas trade were historical in nature and had persisted despite Peter the Great's efforts to overcome them during the first quarter of the eighteenth century. When Catherine ascended the throne Russia's internal communications network—the waterway system and infrastructure—was still in need of improvement to provide speedier and safer delivery of goods to and from market areas. The inadequacy of this system increased the cost of transporting merchandise, particularly the heavy and bulky commodities such as iron and timber. The primitive level of Russian financial institutions, which could not offer easy credit facilities to traders, meant that the foreign merchants frequently had to lend their Russian counterparts funds, sometimes for over a year, to insure delivery of goods on schedule. Under such an arrangement Russian merchants lost some of their flexibility and independence when contracting orders. In addition Russians still needed to be better educated in order to apply to Russia the trading methods employed by advanced maritime states.

There is no doubt that during the thirty-four years of her reign Catherine had the power to mobilize the human and *matériel* resources required to develop a mercantile program that could have made Russia a competitive and formidable maritime state. Her failure to do this was inconsistent with her intention to maximize Russia's potential profit from overseas commerce.

Russia had the timber and naval stores necessary to build a substantial overseas maritime fleet of its own, as it was already a principal supplier of such commodities to maritime states and also built ships for them in Russia. If the necessary construction and nautical skills were not readily and sufficiently available, these could have been imported and taught to native Russians who subsequently would have found steady employment in an expanding shipbuilding industry. Native Russians could have been intensively trained in mercantile services as well.

Catherine's failure cannot be explained away by the wars that punctuated her reign; wars were endemic to all maritime countries during the second half of the eighteenth century. In fact overseas commercial rapaciousness frequently served as a cause for countries to go to war.

Had she chosen to do so, Catherine could have emulated England's Act of Navigation by decreeing one for her own Empire. Such an act would have been well understood and accepted within the historical context of her ambition for Russia's overseas commerce. Early in her reign Catherine had published substantial decrees on trade and navigation from which she could have compiled such a Russian Act of Navigation. The corpus would have been drawn from the Anglo-Russian Commercial Treaty of 1766, the Tariff of 1766/67, the *Nakaz* to the Legislative Commission of 1767, the Tariff of 1782/83, several commercial treaties concluded in the 1780s, the prohibitive tariff regulations of 1793, and the abortive Tariff of 1796. Catherine had committed Russia to a maritime code during the crisis of the Armed Neutrality. However, she could not sustain that code because she could not develop Russia's naval power sufficiently to make other countries respect it.

Soon after the conclusion of the Seven Years' War Catherine recognized that the old way of doing things would not significantly increase Russia's overseas trading potential. But instead of inaugurating a mercantile program that would have produced a large maritime fleet, built and equipped in Russia and staffed and supervised by native-born subjects of the Empire, or encouraging the development of entrepreneurial skills that would have made Russia's overseas mercantile operations more productive and profitable, Catherine chose two policies that could result in short-term benefits at best. First, she decreed that Russian merchants be subsidized. They would receive an abatement when transporting their merchandise on Russian ships. For this reason the Tariff of 1766/67 included an abatement that amounted to 28 percent on both exports and imports and the Tariff of 1782/83 included an abatement of 3/8 on exports and 1/8 on imports. More than twenty-five years later the Russian court declared this abatement policy a failure and abolished it in the abortive Tariff of 1796. Second, Catherine merged the abatement policy with the policy of naturalizing foreign merchants and owners and masters of foreign ships. In this manner she created a Potemkin-like mercantile marine. The chief beneficiary of this failed policy was Great Britain, because by supplying much of the maritime facilities and mercantile services it was able to assume the prevalent position in Russia's overseas trade. This deprived Russia of an enormous sum of money in transportation and insurance costs alone.

In a pioneering study the Soviet scholar N. L. Rubinshtein argued that Russia's relatively small maritime fleet was one reason why native Russian merchants had a smaller share of Russia's overseas commodity trade turnover.[1] But he also stated that the evidence showing Britain's dominance in Russia's overseas export trade warranted serious reexamination and verification, asserting that the "true nature" of native Russian involvement in the empire's overseas trade was obscured. He argued that because fewer native Russian merchants were recorded in the port registry books than should have been, their historic role in Russia's overseas commerce has been undervalued. According to Rubinshtein the overseas trade data should have reflected not only a higher number of native Russian merchants, but also a greater value for their merchandise in the total distribution of Russia's overseas commodity foreign trade. In support of this view he identified several dozen native-born Russian merchants, proprietors of manufactories and landed estates who purchased and sold their commodities through foreign merchants and agents but whose names were omitted from the port registry books.[2]

Rubinshtein stated that some foreigners became *Inostrannye gosti* (naturalized Russian burghers) to benefit from commercial concessions that would accrue only to Russian merchants, but that he could not specify their number because the extant sources are incomplete. During the second half of the eighteenth century the central Russian officialdom repeatedly failed to obtain regularly from the St. Petersburg Custom House administration information

---

1. N. L. Rubinshtein, "Vneshniaia torgovlia Rossii i russkoe kupechestvo vo vtoroi polovine XVIII v.," IZ 54 (1955): pp. 343–61.
2. Had it been available to him, Rubinshtein would have found comfort in the April 1794 testimony of John Cayley, the British Consul General and British Factory representative in St. Petersburg, to the British Board of Trade: "At least 2/3 of the Goods imported from Great Britain are ordered for Account of the Russian Merchants who alone profit from the Gains made thereon." Cayley's report: "What is the just and reasonable preference the Russian Government can give to their own Subjects?" dated April 24, 1794, in "Mr. Eton's Remarks on the present State of the Importation Trade from Great Britain to Russia, 1794," PRO FO 97/342 and PRO BT 1/14, ff. 397–99.

on both Russian and foreign merchant nationals who participated in Russia's overseas commodity trade. This is probably one reason why the extant archival repository in Russia appears to be incomplete.[3]

Such opportunism by foreigners, including English merchants from the British Factory in St. Petersburg, is generally well known and is not peculiar to the Russian Empire during the eighteenth century. The implication of it for determining the actual number of native Russian merchants and their contribution to Russia's overseas trade turnover is clear. To ascertain the number of native Russian merchants and the value of their commerce in Russia's trade turnover, the recorded number of naturalized Russian burghers and the value of their trade turnover would have to be subtracted from the total number of recorded Russian merchants and their trade turnover. If this computation could be accomplished historians would have a more accurate bookkeeping system for Russia's trade accounts. But would these resultant numbers and percentages reduce or increase the incidence of native Russian merchant involvement in Russia's overseas trade turnover and would they reflect as well the value of their participation? The answer to this question lies in being able first to determine the original ethnic or national identity of the naturalized Russian burghers, and second to recalculate their trade turnover within the merchant national group to which they had originally belonged. To a significant degree this can be accomplished by using the extant customs and commercial sources which are in British and Russian archives.

In his 1974 study on Russian grain exports during the second half of the eighteenth century the Soviet historian B. N. Mironov stated that among the archival sources available to him one category related to the role played by merchant nationals and that it was in two parts. One part contained data on merchant nationals who participated in Russia's overseas trade turnover but did not list their names; the other contained similar lists with names. From the latter, two lists had already been published for St. Petersburg in 1795 and 1804 and several for Riga. In manuscript, according to Mironov, lists existed of merchant nationals who traded via St. Petersburg/Kronstadt in the years 1772–1775, and 1787; and for Archangel in the years 1773–1774, 1783, 1797–1801.[4]

---

3. VTR, p. 5; and B. N. Mironov, "Eksport russkogo khleba vo vtoroi polovine XVIII-nachale XIX v.," IZ 93 (1974): pp. 149–54.
4. Mironov, p. 151.

In 1981 the Soviet historian E. I. Indova and her colleagues published for "the first time" the extant archival manuscripts—*Vedomosti* [Lists, Registers]—containing the names of both Russian and foreign merchants who traded via the port of St. Petersburg/Kronstadt inclusive of the current ruble valuations of their exports and imports for the years 1764, 1765, 1772, 1773, 1775, 1787, and 1792. Indova also republished the lists for St. Petersburg in 1795 and 1804.[5]

The Russian archival sources concerning merchant nationals reposing in Russia and in Great Britain complement and supplement each other. Similar to the *Vedomosti* published by Indova for nine years, that part of the Russian manuscript collection in Great Britain that specifically relates to merchant nationals trading via St. Petersburg/Kronstadt is grouped according to the merchants' ethnic/national origins but does not contain the names of individual merchants, although it is far more discretely and consistently categorized throughout and contains information that the *Vedomosti* do not for thirty-four years during the period 1764–1801. The British collection records information under at least nineteen different groups of merchant nationals—such as Russian, British, Dutch, Mecklenburger et cetera—plus "Passengers and Merchants," "Ship Captains," and "Sundry Nations." The manuscripts published by Indova occasionally include the additional categories of "Merchants of Various Nations" (which lists merchants from different parts of the Russian Empire—from Reval [Tallinn], Narva, Novorossia, et cetera, as well as from Central Asia), "Foreign Merchants," and "Russian and Foreign Merchants of Various Towns."

The *Vedomosti* must be used with caution because the manuscripts are fraught with internal inconsistencies. For example, as with the materials in the British archives, there are mistakes in the arithmetic that require that the *Vedomosti* be recalculated. Moreover, because the *Vedomosti* are for several random years only it would be risky to attempt to ascertain a trend from their data.[6] The three *Vedomosti* for 1787 are the most problematic and are numbered No. 6, No. 7, and No. 8. No. 6 for Imports is the only List that includes a separate category for *Inostrannye gosti*. No. 7

---

5. VTR, in particular, pp. 3–8, passim. Indova was unable to locate the list for the year 1774 to which Mironov had referred.
6. Cf., B. N. Mironov, "K voprosu o roli russkogo kupechestva vo vneshnei torgovle Peterburga i Arkhangelska vo vtoroi polovine XVIII-nachale XIX veka," *Istoriia SSSR* 6 (1973): pp. 129–40; idem, "Sotsial'naia mobil'nost' rossiiskogo kupechestva x XVIII-nachale XIX veka (opyt izucheniia)," *Problemy istoricheskoi demografii SSSR. Sbornik Statei*, ed. R. N. Pullat (Tallinn: Akademiia Nauk Estonskoi SSR, 1977), pp. 207–17.

lists Exports and, while both No. 6 and No. 7 are printed in Russian, No. 8 is printed in French but is entitled incorrectly in Russian by Indova as "Imports" when in fact it comprises "Exports." Although No. 7 and No. 8 are identical in their grand totals, their sub-totals are not and the arithmetic is incorrect. Under "St. Petersburg Merchants" No. 8 records the names of only about one third of the number of the individual merchants who are listed in No. 7 and excludes the separate listings of Russian merchants from three different towns that are included in No. 7. Although No. 7 and No. 8 do not include a separate category for *Inostrannye gosti*, both record former members of the British Factory who, we have determined from other sources, had in fact become naturalized Russian burghers of St. Petersburg. Moreover, a few of the British Factory merchants who had become naturalized Russian burghers are recorded in No. 7 and No. 8 under both St. Petersburg Russian merchants and English merchants, perhaps reflecting their transactions before and after becoming *Inostrannye gosti*. Finally, the names of the merchants are listed alphabetically in the republished *Vedomosti* for 1795 and 1804, the former taken from a publication in German and the latter from one in English, and therefore cannot be compared to other *Vedomosti* by categories of merchant national groups. Moreover, the list for 1795 is incorrect because the total value given for exports excludes the value of Kronstadt's exports.

The methodological approach I have employed to determine the contribution made to the total value of the overseas commodity trade turnover of St. Petersburg/Kronstadt by merchant nationals during Catherine II's reign is in two parts. Part one provides an analysis of the *prima facie* evidence, which includes *Inostrannye gosti* and those foreign merchants who temporarily traded under the rubric of Russian merchants even though they were not specifically identified as naturalized burghers of St. Petersburg. The second part provides a similar analysis and identifies the respective contribution made by merchant nationals as representing their original ethnic/national merchant community although they had become Russian naturalized burghers or had traded temporarily under the rubric of Russian merchants. This reconstruction of merchant national overseas trading was accomplished by comparing the names of the foreigners listed in the *Vedomosti*, before and after the naturalization or the change of trading status that took place, with information on merchant nationals in collateral archival documents and in scholarship based on archival data that contain the names of merchant nationals. For example, there are the names of those

British Factory and Hanse merchants who traded via St. Petersburg/Kronstadt but who became naturalized Russian burghers or who temporarily traded under the rubric of Russian merchants.

The British merchants' share of the *prima facie* monetary value of the total overseas commodity trade turnover of St. Petersburg/Kronstadt amounted to an annual average of about 39 percent during the thirty years between 1764 and 1796 covering all but four years of Catherine's reign (1762–1763, 1780, and 1794). Russian merchants accounted for an annual average of about 44 percent. British merchants had an annual average of about 57 percent of the value of exports and 20 percent of the value of imports; Russian merchants had an annual average of 30 percent of the value of exports and almost 61 percent of the value of imports. Russian and British merchants together accounted for an annual average of nearly 87 percent of the total value of exports and 81 percent of the total value of imports (see Figures 6 and 7, Tables MN–III and IV).[7][8]

From 1764 through 1796, except for Russian merchants, no other merchant national group was a serious competitor to British merchants in either exports or imports. This was most striking during the period 1764–1782 when British merchants sustained a fairly steady share of about 60 percent of the total value of exports and about 24 percent of the total value of imports. Russian merchants during this period gradually augmented their share of the increasing total value of exports from about 9 percent to about 20 percent, and from about 30 percent to about 48 percent in imports.

However, after the peace was concluded in 1783 and the Russian tariff of 1782/83 went into effect Russian merchants significantly increased their share of the total value of the overseas commodity trade turnover. This continued to the end of Catherine's reign. For example, from 1783 through 1787, while British merchants had an annual average of about 61 percent of the value of exports and about 21 percent of imports, Russian merchants increased the value of their exports to an annual average of about 25 percent while their imports jumped to about 62 percent. During the period 1788–1792 the Russian merchants' share of the

---

7. Table MN-III is compiled and computed from the records of the St. Petersburg Custom House, PRO BT 6/231–33, 141, 266; SP 91/73, 75–76, 94, 102–4, 107; CO 388/95; FO 65/6, 9, 13–15, 17–18, 20, 24, 26, 33, 35, 52; FO 97/340–42; and VTR, *Vedomosti*, Nos. 1–5, 7–10.

8. Table MN-IV is compiled and computed from the records of the St. Petersburg Custom House; ibid.; and VTR, *Vedomosti*, Nos. 1–6, 9–10.

**Figure 6**

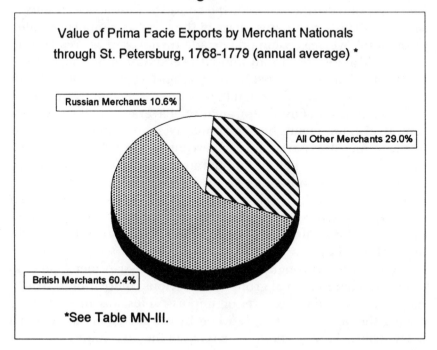

Value of Prima Facie Exports by Merchant Nationals through St. Petersburg, 1768-1779 (annual average) *

Russian Merchants 10.6%
All Other Merchants 29.0%
British Merchants 60.4%

*See Table MN-III.

value of exports peaked when it doubled to about 50 percent, exceeding the British for the first time, and their imports increased by a third to 83.5 percent. During the same time the value of the British merchants' share of exports fell to about 41 percent and imports plummeted to 7.5 percent. Though British merchants had regained their earlier position of strength in both exports and imports toward the end of Catherine's reign, by that time Russian merchants had achieved a level in exports at more than half that of the British and they dominated imports.

The reason the Russian merchants' share of the value of the total trade turnover increased significantly from the 1780s onward is because an appreciable number of foreign merchants had become naturalized Russian burghers, or their trade transactions via St. Petersburg/Kronstadt were recorded in the custom ledgers as Russian merchants. The value of their trade turnover was not entered into the custom ledgers under their former merchant national identity. However, the *prima facie* data presented in Tables MN-III and MN-IV do not reflect these important merchant national shifts and therefore give a false impression of the real distribution of the value of the total trade turnover. For instance, until

## TABLE MN-III. VALUE OF PRIMA FACIE EXPORTS BY MERCHANT NATIONALS AS PERCENTAGE OF THE TOTAL PRIMA FACIE VALUE OF ST. PETERSBURG'S OVERSEAS COMMODITY EXPORTS: 1764–1796[7]

[Exclusive of Custom Duties and Sundry Charges]
[Annual Average, Measured in Current Rubles, in Thousands]

| Years | Total Exports | Russian Merchants | % | British Merchants | % | Combined Percentage |
|---|---|---|---|---|---|---|
| 1764–66 | 5,978 | 564 | 9.4 | 3,757 | 62.9 | 72.3 |
| 1768–72 | 7,352 | 575 | 7.8 | 4,988 | 67.8 | 75.7 |
| 1773–77 | 9,416 | 1,142 | 12.1 | 5,536 | 58.8 | 70.9 |
| 1768–79 | 8,945 | 952 | 10.6 | 5,401 | 60.4 | 71.0 |
| 1783–87 | 13,193 | 3,304 | 25.0 | 8,057 | 61.1 | 86.1 |
| 1788–92 | 20,965 | 10,503 | 50.1 | 8,639 | 41.2 | 91.3 |
| 1793 | 23,758 | 9,938 | 41.8 | 13,122 | 55.2 | 97.1 |
| 1795–96 | 34,521 | 12,479 | 36.1 | 21,101 | 61.1 | 97.3 |
| 1764/1796 | 13,703 | 4,067 | 29.7 | 7,656 | 55.9 | 85.6 |

## TABLE MN-III. Continued

| Years | Total Exports | Dutch Merchants | % | French Merchants | % | Combined Percentage |
|---|---|---|---|---|---|---|
| 1764–66 | 5,978 | 379 | 6.3 | 120 | 2.0 | 8.3 |
| 1768–72 | 7,352 | 359 | 4.9 | 115 | 1.6 | 6.4 |
| 1773–77 | 9,416 | 434 | 4.6 | 326 | 3.5 | 8.1 |
| 1768–79 | 8,945 | 393 | 4.4 | 292 | 3.3 | 7.7 |
| 1783–87 | 13,193 | 143 | 1.1 | 314 | 2.4 | 3.5 |
| 1788–92 | 20,965 | 132 | .6 | 725 | 3.5 | 4.1 |
| 1793 | 23,758 | 100 | .4 | 151 | .6 | 1.1 |
| 1795–96 | 34,521 | 1 | .0 | 3 | .0 | .0 |
| 1764/1796 | 13,703 | 256 | 1.9 | 321 | 2.3 | 4.2 |

| Years | Total Exports | Hamburg Merchants | % | Lübeck Merchants | % | Rostock Merchants | % | Combined Percentage |
|---|---|---|---|---|---|---|---|---|
| 1764–66 | 5,978 | 318 | 5.3 | 319 | 5.3 | 137 | 2.3 | 12.9 |
| 1768–72 | 7,352 | 158 | 2.1 | 329 | 4.5 | 121 | 1.6 | 8.3 |
| 1773–77 | 9,416 | 293 | 3.1 | 370 | 3.9 | 322 | 3.4 | 10.5 |
| 1768–79 | 8,945 | 281 | 3.1 | 325 | 3.6 | 218 | 2.4 | 9.2 |
| 1783–87 | 13,193 | 77 | .6 | 56 | .4 | 30 | .2 | 1.2 |
| 1788–92 | 20,965 | 0 | .0 | 15 | .1 | N/A | .0 | .1 |
| 1793 | 23,758 | 4 | .0 | 7 | .0 | N/A | .0 | .0 |
| 1795–96 | 34,521 | N/A | .0 | 5 | .0 | N/A | .0 | .0 |
| 1764/1796 | 13,703 | 172 | 1.3 | 183 | 1.3 | 118 | .9 | 3.4 |

**Figure 7**

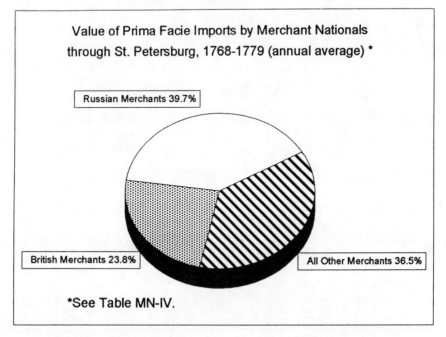

1779 the value of the combined trade turnover for the Hanse merchants of Hamburg, Lübeck, and Rostock was about 10 percent. But from the early 1780s onward many Hanse merchants either became *Inostrannye gosti* and/or traded as Russian merchants. Their trade transactions no longer appeared under their former ethnic/national identity but were recorded in the custom ledgers under Russian merchants.

Table MN-V illustrates the comparative value of the trade transactions by Russian, British, and Hanse merchant nationals in 1787. In the *Vedomosti* for imports in 1787 are the names of those foreign merchants whose trade transactions were recorded either as naturalized Russian burghers or as Russian merchants without the designation of *Inostrannye gosti*. Several of these foreign merchants were also recorded as Russian merchants in the *Vedomosti* comprising exports in 1787.

British Factory reports provide the names of those British merchants who became *Inostrannye gosti* in 1787. In some cases their different trade transactions were recorded in the *Vedomosti* not only as *Inostrannye gosti* or Russian merchants but also as British merchants, possibly reflecting transactions made prior to their for-

## TABLE MN-IV. VALUE OF PRIMA FACIE IMPORTS BY MERCHANT NATIONALS AS PERCENTAGE OF THE TOTAL PRIMA FACIE VALUE OF ST. PETERSBURG'S OVERSEAS COMMODITY IMPORTS: 1764–1796[8]

[Exclusive of Custom Duties, Sundry Charges, and Specie Imported]
[Annual Average, Measured in Current Rubles, in Thousands]

| Years | Total Imports | Russian Merchants | % | British Merchants | % | Combined Percentage |
|---|---|---|---|---|---|---|
| 1764–66 | 5,639 | 1,703 | 30.2 | 1,366 | 24.2 | 54.4 |
| 1768–72 | 6,782 | 2,504 | 36.9 | 1,713 | 25.3 | 62.2 |
| 1773–77 | 7,680 | 3,119 | 40.6 | 1,933 | 25.2 | 65.8 |
| 1768–79 | 7,310 | 2,903 | 39.7 | 1,740 | 23.8 | 63.5 |
| 1783–87 | 12,194 | 7,599 | 62.3 | 2,569 | 21.1 | 83.7 |
| 1788–92 | 20,449 | 17,078 | 83.5 | 1,541 | 7.5 | 91.0 |
| 1793 | 14,581 | 10,340 | 70.9 | 2,879 | 19.7 | 90.7 |
| 1795–96 | 24,688 | 15,763 | 63.9 | 7,607 | 30.8 | 94.7 |
| 1764/1796 | 11,777 | 7,173 | 60.9 | 2,326 | 19.7 | 80.6 |

TABLE MN-IV. Continued

| Years | Total Imports | Dutch Merchants | % | French Merchants | % | Combined Percentage |
|---|---|---|---|---|---|---|
| 1764–66 | 5,639 | 790 | 14.0 | 126 | 2.2 | 16.2 |
| 1768–72 | 6,782 | 718 | 10.6 | 137 | 2.0 | 12.6 |
| 1773–77 | 7,680 | 525 | 6.8 | 197 | 2.6 | 9.4 |
| 1768–79 | 7,310 | 606 | 8.3 | 146 | 2.0 | 10.3 |
| 1783–87 | 12,194 | 291 | 2.4 | 76 | .6 | 3.0 |
| 1788–92 | 20,449 | 182 | .9 | 295 | 1.4 | 2.3 |
| 1793 | 14,581 | 111 | .8 | 34 | .2 | 1.0 |
| 1795–96 | 24,688 | 3 | .0 | 163 | .7 | .7 |
| 1764/1796 | 11,777 | 431 | 3.7 | 155 | 1.3 | 5.0 |

| Years | Total Imports | Hamburg Merchants | % | Lübeck Merchants | % | Rostock Merchants | % | Combined Percentage |
|---|---|---|---|---|---|---|---|---|
| 1764–66 | 5,639 | 393 | 7.0 | 426 | 7.6 | 43 | .8 | 15.3 |
| 1768–72 | 6,782 | 362 | 5.3 | 435 | 6.4 | 25 | .4 | 12.1 |
| 1773–77 | 7,680 | 252 | 3.3 | 297 | 3.9 | 41 | .5 | 7.7 |
| 1768–79 | 7,310 | 290 | 4.0 | 336 | 4.6 | 38 | .5 | 9.1 |
| 1783–87 | 12,194 | 216 | 1.8 | 108 | .9 | 25 | .2 | 2.9 |
| 1788–92 | 20,449 | 6 | .0 | 14 | .1 | N/A | .0 | .1 |
| 1793 | 14,581 | 5 | .0 | 15 | .1 | N/A | .0 | .1 |
| 1795–96 | 24,688 | 7 | .0 | 12 | .0 | N/A | .0 | .1 |
| 1764/1796 | 11,777 | 234 | 2.0 | 212 | 1.8 | 32 | .3 | 4.1 |

mal naturalization. Moreover, the different transactions of Richard Sutherland, banker to the Russian court, are listed separately under both British and Russian merchants, either as an individual or jointly with another merchant national, reflecting a partnership arrangement. John Henry [Iuogan Gindrikh] Schneider was a naturalized British merchant and a member of the Russia Company who arrived in St. Petersburg in mid-1775 and opened a merchant house with a non-British foreigner. Although the British Factory report of 1787 did not name him as becoming a naturalized Russian burgher, the *Vedomosti* recorded his imports as an *Inostrannyi gosti* and his exports as a Russian merchant. In the *Vedomosti* of 1792 he and his company are listed as British merchants.[9]

The overseas commercial transactions of Hanse merchants were recorded in one or more categories at the St. Petersburg Custom House: either as *Inostrannye gosti,* as Russian merchants, as merchants of their respective ethic/national heritage, or as merchants of another merchant national group with whom they had become associated. It is possible to recalculate some of their custom house entries and determine at minimum the estimated value of their trade turnover via St. Petersburg/Kronstadt for the year 1787.

The *prima facie* figures in Table MN-V. A represent foreign merchant national *Inostrannye gosti* and/or foreign merchants who had their trade turnover included in the Russian merchant trade turnover in 1787. The Russian merchants' share of exports was about 39 percent, more than twice the *prima facie* percentage in each of the two preceding years. The British merchants' share of the *prima facie* value of exports was about 51 percent, significantly less than the immediate preceding years (1784 = 64.8%; 1785 = 66.9%; 1786 = 67.3%). During the period 1764–1786 the British merchants' share had never been less than 51 percent.[10] The spread in the distribution of the 1787 *prima facie* value of imports was greater than in exports. The Russian merchants' share in 1787

---

9. For the list of British Factory merchants who became *Inostrannye gosti,* see the British Factory letter and list dated May 31/June 11, 1787, in Shairp to Forster, June 1/12, 1787, PRO FO 65/15; for the particulars of John Henry [Iuogan Gindrikh] Schneider, see Samuel Swallow to Suffolk, June 1/16, 1775, PRO SP 91/99, ff. 5–6; and VTR, *Vedomosti,* Nos. 6–9. Indova has published a table from 1775 on manufactories that includes the names of British and other foreign merchants who had also become proprietors of manufactories. E. I. Indova, ed., "O Rossiiskikh Manufakturakh vtoroi polovini XVIII v.," *Istoricheskaia geografiia Rossii XII–nachalo XX v. Sbornik statei k 70–letiiu professora Liubomira Grigor'evicha Beskrovnogo* (Moscow: Nauka, 1975), pp. 284–345.

10. For the period covering 1764–1787, see the records of the St. Petersburg Custom House, PRO BT 6/231–232, 141, 266; SP 91/73, 76, 94, 102–4, 107; FO 65/6, 9, 13–15; FO 97/340; CO 388/95; and VTR, *Vedomosti* Nos. 1–5, 7–8.

## TABLE MN-V. COMPARISON BETWEEN THE VALUE OF THE TOTAL OVERSEAS COMMODITY TRADE TURNOVER BY RUSSIAN, BRITISH, AND HANSE MERCHANTS AT ST. PETERSBURG: 1787[12]

[Exclusive of Custom Duties, Sundry Charges, and Specie Imported]
[Measured in Current Rubles, in Thousands]

A. *INCLUSIVE OF FOREIGN MERCHANT NATIONAL INOSTRANNYE GOSTI AND/OR THOSE WHO HAD THEIR TRADE TURNOVER INCLUDED IN RUSSIAN MERCHANT TRADE TURNOVER*

| Total Exports | Russian Exports | % | British Exports | % | Hanse Exports | % |
|---|---|---|---|---|---|---|
| 15,950 | 6,266 | 39.3 | 8,187 | 51.3 | 66 | .4 |

| Total Imports | Russian Imports | % | British Imports | % | Hanse Imports | % |
|---|---|---|---|---|---|---|
| 15,314 | 12,084 | 78.9 | 1,569 | 10.2 | 239 | 1.6 |

B. *EXCLUSIVE OF BRITISH AND HANSE MERCHANT NATIONAL INOSTRANNYE GOSTI AND/OR THOSE WHO HAD THEIR TRADE TURNOVER INCLUDED IN THE RUSSIAN MERCHANT TRADE TURNOVER*

| Total Exports | Russian Exports | % | British Exports | % | Hanse Exports | % |
|---|---|---|---|---|---|---|
| 15,950 | 1,632 | 10.2 | 11,414 | 71.6 | 1,474 | 9.2 |

| Total Imports | Russian Imports | % | British Imports | % | Hanse Imports | % |
|---|---|---|---|---|---|---|
| 15,314 | 8,554 | 55.9 | 3,242 | 21.2 | 2,096 | 13.7 |

jumped to nearly 79 percent, nearly double the annual average share for the period 1768–1779 and about 18 percent greater than the peak level reached in 1785. Conversely, the British merchants' share of imports fell precipitously to about 10 percent in 1787 from about 26 percent in 1786.[11]

The decline of the British merchants' share in the *prima facie* value of the total trade turnover in 1787 was owing to several factors: the expiration of the Anglo-Russian trade treaty (just as spring shipping opened that season) that removed the protection previously accorded to British merchants, especially their most-favored-nation privilege; the Russian government's encourage-

---

11. Ibid.; and VTR, *Vedomosti*, Nos. 1–6.

ment to foreign merchants to become *Inostrannye gosti* and be accorded the beneficial privileges under the 1782/83 Russian tariff; and the defection of some British merchants who became naturalized Russian burghers, *Inostrannye gosti,* or who traded as Russian merchants. Conversely, these factors also largely explain why the Russian merchants' share increased so dramatically in 1787.

Counter-factually, Table MN–V. B[12] illustrates the resultant recalculation of the distribution of the 1787 trade turnover, reflecting the respective merchant national shares as they might have been exclusive of the conversion of non-Russian merchant nationals to the status of *Inostrannye gosti* or who traded as Russian merchants. Thus, the exports and imports of Russian merchants drop to about 10 percent and 56 percent, respectively; the exports and imports of British merchants increase significantly to about 72 and about 21 percent; and the exports and imports of Hanse merchants jump to about 9 and almost 14 percent. These recalculated estimates are about equal to those of the two previous decades as represented in Tables MN–III and –IV.

Five years later, in 1792, most of the British merchants who had become *Inostrannye gosti* in 1787 had returned to their former legal status as British merchants. Their trade turnover, accounts via St. Petersburg/Kronstadt were once again listed under English merchants. Nevertheless the estimated trade turnover of the few British merchants who remained *Inostrannye gosti* was significant and amounted to about 66 percent of the trade turnover of those British merchants who in 1787 had traded as Russian merchants.[13] In 1792 the *prima facie* value of exports and imports by Russian merchants amounted to almost 57 and nearly 85 percent of the total value of exports and imports respectively. The *prima facie* value of exports and imports by British merchants amounted to about 36 and about 7 percent of the total exports and imports.

---

12. Table MN-V is compiled and computed from the records of the St. Petersburg Custom House for 1787, PRO BT 6/231–32, 141; VTR, *Vedomosti,* Nos. 6–8; Shairp to Forster, June 1/12, 1787, and subsequent correspondence of the Board of Trade and Russia Company during June–August, 1787, PRO FO 65/15; and, for the verification of the names of Hanse merchant nationals, see Christoph Friedrich Menke, whose data are from the Archiv Reval, "Die wirtschaftlichen und politischen Beziehungen der Hansestädte zu Russland im 18. und frühen 19. Jahrhundert" (Ph.D. diss., Göttingen Universität, 1959), pp. 136–38. There is a discrepancy between the data in the British archival manuscripts and the manuscript collection published by Indova. There is a difference of 1.7 percent between the higher valued British data and the lower valued Russian data for the estimated value of the overseas commodity trade turnover for Russian and British merchant nationals in 1787 at St. Petersburg/Kronstadt. The estimated value of the turnover trade represented in Table MN-V for 1787 is an average of the two.

13. VTR, *Vedomost',* No. 9, and the list of British Factory defectors in Shairp to Forster, June 1/12, 1787, PRO FO 65/15. See also RC, Feb. 24, 1789, VIII, p. 260.

### TABLE MN-VI. COMPARISON BETWEEN THE VALUE OF THE TOTAL OVERSEAS COMMODITY TRADE TURNOVER BY RUSSIAN, BRITISH, AND HANSE MERCHANTS AT ST. PETERSBURG: 1792[14]

[Exclusive of Custom Duties, Sundry Charges, and Specie Imported]
[Measured in Current Rubles, in Thousands]

A. *INCLUSIVE OF FOREIGN MERCHANT NATIONAL INOSTRANNYE GOSTI AND/OR THOSE WHO HAD THEIR TRADE TURNOVER INCLUDED IN RUSSIAN MERCHANT TRADE TURNOVER*

| Total Exports | Russian Exports | % | British Exports | % | Hanse Exports | % |
|---|---|---|---|---|---|---|
| 21,053 | 11,964 | 56.8 | 7,620 | 36.2 | 7 | .0 |

| Total Imports | Russian Imports | % | British Imports | % | Hanse Imports | % |
|---|---|---|---|---|---|---|
| 23,296 | 19,778 | 84.9 | 1,698 | 7.3 | 7 | .0 |

B. *EXCLUSIVE OF BRITISH, HANSE, AND DANISH MERCHANT NATIONAL INOSTRANNYE GOSTI AND/OR THOSE WHO HAD THEIR TRADE TURNOVER INCLUDED IN RUSSIAN MERCHANT TRADE TURNOVER*

| Total Exports | Russian Exports | % | British Exports | % | Hanse Exports | % |
|---|---|---|---|---|---|---|
| 21,053 | 7,176 | 34.1 | 9,254 | 44.0 | 3,162 | 15.0 |
|  | 6,703 | 31.8 |  |  |  |  |

| Total Imports | Russian Imports | % | British Imports | % | Hanse Imports | % |
|---|---|---|---|---|---|---|
| 23,296 | 15,862 | 68.1 | 2,516 | 10.8 | 3,105 | 13.3 |
|  | 14,630 | 62.8 |  |  |  |  |

The *prima facie* value of exports and imports by Hanse merchants was neglible (see Table MN–VI. A).[14]

However, the distribution radically changes when the estimated trade turnover value for British and Hanse merchants who traded as Russian merchants is recalculated and reestablished according to their own ethnic/national identity (see Table MN-VI. B). The Russian merchant exports and imports drop to about 34 and about 68 percent; the British merchant exports and imports increase to 44 and almost 11 percent; and the Hanse merchant

---

14. Table MN-VI is compiled and computed from the records of the St. Petersburg Custom

exports and imports increase dramatically to 15 and about 13 percent. Moreover, in 1792 one Danish merchant can be absolutely identified as trading as a Russian merchant. His exports and imports amounted to 473,186 and 1,231,436 rubles respectively. When those amounts are subtracted from the Russian merchants' *prima facie* trade turnover, the value of their exports and imports is further reduced to 31.8 and 62.8 percent.[15]

While researching Hanse merchant participation in Russia's overseas trade Christoph Friedrich Menke discovered Reval [Tallinn] archival lists of merchant nationals trading at St. Petersburg/Kronstadt in the years 1793 and 1795. Menke recalculated their trade turnover according to their ethnic/national origins and his findings for those years reaffirm the direction and magnitude of my estimates that were calculated for the years 1787 and 1792 (see Tables MN–V–VI). The Hanse merchants, whom Menke refers to in this case as "German," had a large share of the value of the trade turnover in 1793 and 1795. In 1793 their share of exports and imports were 29.8 and 50.2 percent, and in 1795 their share of exports and imports was 35.9 and 50.3 percent.[16]

Given Menke's calculations, the Russian and British merchants' share of the value of total overseas commodity trade turnover via St. Petersburg/Kronstadt changed dramatically in 1793 and 1795. After the value of the trade turnover of those foreign merchants trading as *Inostrannye gosti* or simply as Russian merchants is recalculated and restored to their former ethnic/national origins, the Russian merchants' share dropped in 1793 from nearly 42 percent of the *prima facie* exports to 7 percent and in 1795 it collapsed altogether falling from about 37 percent to 0.2 percent. The Russian merchants' share of *prima facie* imports fell precipitously in 1793 from nearly 71 percent to nearly 18 percent and in 1795 it declined considerably from nearly 66 percent to about 13 percent. The British merchants in 1793 and 1795, according to Menke's calculations, rebounded to their pre-1787 lev-

---

House for 1792, PRO BT 6/232; FO 65/24; VTR, *Vedomost'*, No. 9; and Menke for the verification of the Hanse merchants, pp. 136–38. There is a discrepancy in the data between the British archival manuscripts and the manuscript collection published by Indova. There is a difference of .3 percent between the higher valued British data and the lower valued Russian data for the estimated value of the overseas commodity trade turnover for Russian and British merchant nationals in 1792 at St. Petersburg/Kronstadt. The estimated value of the turnover trade represented in Table MN-VI for 1792 is an average of the two.
15. PRO BT 6/232; FO 65/24; VTR, *Vedomost'*, No. 9; and Menke for the verification of the Hanse merchants, pp. 136–38.
16. Menke, p. 130; and Tables MN-V-VI, supra.

TABLE MN-VII. COMPARISON BETWEEN THE VALUE OF THE TOTAL OVERSEAS COMMODITY TRADE TURNOVER BY RUSSIAN AND BRITISH MERCHANTS AT ST. PETERSBURG: 1793[17]

[Exclusive of Custom Duties, Sundry Charges, and Specie Imported]
[Measured in Current Rubles, in Thousands]

A. *INCLUSIVE OF FOREIGN MERCHANT NATIONAL INOSTRANNYE GOSTI AND/OR THOSE WHO HAD THEIR TRADE TURNOVER INCLUDED IN RUSSIAN MERCHANT TRADE TURNOVER*

| Total Exports | Russian Exports | % | British Exports | % |
|---|---|---|---|---|
| 23,758 | 9,938 | 41.8 | 13,122 | 55.2 |

| Total Imports | Russian Imports | % | British Imports | % |
|---|---|---|---|---|
| 14,581 | 10,340 | 70.9 | 2,879 | 19.7 |

B. *EXCLUSIVE OF MERCHANT NATIONAL INOSTRANNYE GOSTI AND/OR THOSE WHO HAD THEIR TRADE TURNOVER INCLUDED IN RUSSIAN MERCHANT TRADE TURNOVER*

| Total Exports | Russian Exports | % | British Exports | % |
|---|---|---|---|---|
| 23,758 | 1,663 | 7.0 | 14,397 | 60.6 |

| Total Imports | Russian Imports | % | British Imports | % |
|---|---|---|---|---|
| 14,581 | 2,610 | 17.9 | 3,689 | 25.3 |

els. In 1793 their share rose from about 55 percent in *prima facie* exports to about 61 percent, and in 1795 it leveled off at about 60 percent. In 1793 the British merchants' share increased from nearly 20 percent in the *prima facie* imports to about 25 percent, and in 1795 it remained steady at about 28 to 29 percent (see Tables MN–VII–VIII).[17] [18]

The *prima facie* data for the distribution of the value of the total overseas commodity trade turnover via St. Petersburg/Krons-

---

17. Table MN-VII. A is compiled and computed from the records of the St. Petersburg Custom House, PRO BT 6/233 and FO 65/26; and Table MN-VII. B is compiled from Menke, whose data are from the Archiv Reval, p. 130.
18. Table MN-VIII. A is compiled and computed from the records of the St. Petersburg Custom House, PRO BT 6/233; FO 65/33, 35; and VTR, *Vedomost'*, No. 10; and Table MN-VIII. B is compiled from Menke, whose data are from Archiv Reval, p. 130.

## TABLE MN-VIII. COMPARISON BETWEEN THE VALUE OF THE TOTAL OVERSEAS COMMODITY TRADE TURNOVER BY RUSSIAN AND BRITISH MERCHANTS AT ST. PETERSBURG: 1795

[Exclusive of Custom Duties, Sundry Charges, and Specie Imported]
[Measured in Current Rubles, in Thousands]

**A. INCLUSIVE OF FOREIGN MERCHANT NATIONAL INOSTRANNYE GOSTI AND/OR THOSE WHO HAD THEIR TRADE TURNOVER INCLUDED IN RUSSIAN MERCHANT TRADE TURNOVER**

| Total Exports | Russian Exports | % | British Exports | % |
|---|---|---|---|---|
| 31,932 | 11,881 | 37.2 | 19,124 | 59.9 |

| Total Imports | Russian Imports | % | British Imports | % |
|---|---|---|---|---|
| 23,019 | 15,148 | 65.8 | 6,670 | 29.0 |

**B. EXCLUSIVE OF MERCHANT NATIONAL INOSTRANNYE GOSTI AND/OR THOSE WHO HAD THEIR TRADE TURNOVER INCLUDED IN THE RUSSIAN MERCHANT TRADE TURNOVER**

| Total Exports | Russian Exports | % | British Exports | % |
|---|---|---|---|---|
| 31,932 | 64 | 0.2 | 19,124 | 59.9 |

| Total Imports | Russian Imports | % | British Imports | % |
|---|---|---|---|---|
| 23,019 | 3,039 | 13.2 | 6,491 | 28.2 |

tadt by merchant nationals can no longer be considered valid. Although the influence of the recalculated data for the years 1787, 1792, 1793, and 1795 might appear moderate when substituted for *prima facie* data in the limited number of aggregated averages that are illustrated in Tables MN–IX–XI,[19][20][21] it demonstrates how exaggerated the Russian merchants' share of the value of the total overseas commodity trade turnover has been. During Catherine's reign, in comparison to the *prima facie* data, the Russian merchants'

---

19. Table MN-IX is compiled and computed from the records of the St. Petersburg Custom House, PRO BT 6/231–33, 141, 266; SP 91/73, 75–76, 94, 102–4, 107; CO 388/95; FO 65/6, 9, 13–15, 17–18, 20, 24, 26, 33, 35, 52; FO 97/340–42; VTR, *Vedomosti*, Nos. 1–10; and Menke, p. 130.
20. Table MN-X, ibid.
21. Table MN-XI, ibid.

## Figure 8

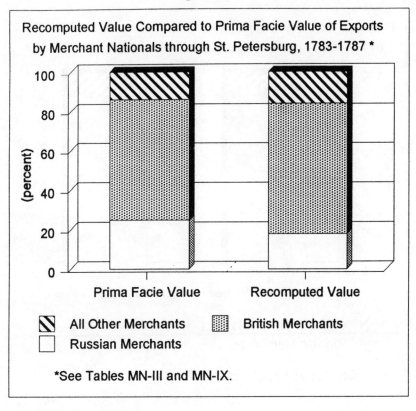

**Figure 9**

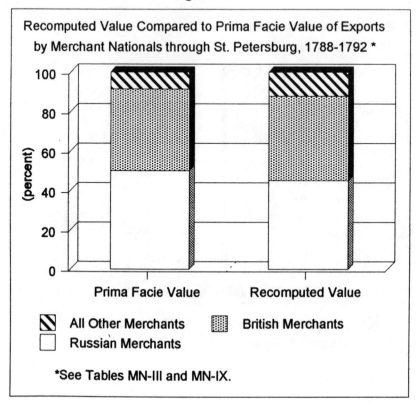

### Figure 10

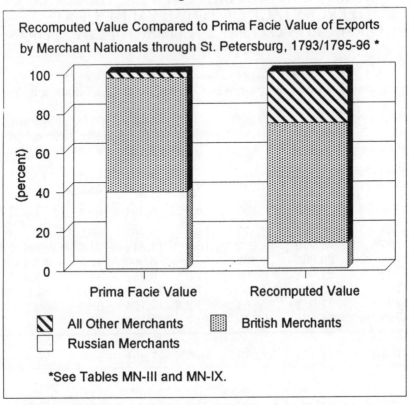

Recomputed Value Compared to Prima Facie Value of Exports by Merchant Nationals through St. Petersburg, 1793/1795-96 *

*See Tables MN-III and MN-IX.

## TABLE MN-IX. RECOMPUTED VALUE OF EXPORTS BY RUSSIAN AND BRITISH MERCHANTS AS PERCENTAGE OF TOTAL VALUE OF ST. PETERSBURG'S OVERSEAS COMMODITY EXPORTS: 1764–1796[21]

[Exclusive of Custom Duties and Sundry Charges]
[Annual Average, Measured in Current Rubles, in Thousands]

| Years | Total Exports | Russian Merchants | % | British Merchants | % | Combined Percentage |
|---|---|---|---|---|---|---|
| 1764–66 | 5,978 | 564 | 9.4 | 3,757 | 62.9 | 72.3 |
| 1768–72 | 7,352 | 575 | 7.8 | 4,988 | 67.8 | 75.7 |
| 1773–77 | 9,416 | 1,142 | 12.1 | 5,536 | 58.8 | 70.9 |
| 1768–79 | 8,945 | 952 | 10.6 | 5,401 | 60.4 | 71.0 |
| 1783–87 | 13,193 | 2,377 | 18.0 | 8,702 | 66.0 | 84.1 |
| 1787 | 15,950 | 1,632 | 10.2 | 11,414 | 71.6 | 81.8 |
| 1788–92 | 20,965 | 9,451 | 45.1 | 8,966 | 42.8 | 87.8 |
| 1792 | 21,053 | 6,703 | 31.8 | 9,254 | 44.0 | 75.8 |
| 1793 | 23,758 | 1,663 | 7.0 | 14,397 | 60.6 | 67.6 |
| 1795 | 31,932 | 64 | .2 | 19,124 | 59.9 | 60.1 |
| 1795–96 | 34,521 | 6,570 | 19.0 | 21,101 | 61.1 | 80.2 |
| 1764/1796 | 13,703 | 2,675 | 19.5 | 7,861 | 57.4 | 76.9 |

share of the value of exports in the recalculated data is reduced from nearly 30 to about 20 percent; in imports the Russian merchants' share is reduced from almost 61 to almost 53 percent, and the Russian merchants' share of the total overseas commodity trade turnover is reduced from about 44 percent to almost 35 percent. In the recalculated data the British merchants' share is increased only slightly: within the range of 57 to 58 percent in exports, between 20 and 22 in imports, and between 40 and 42 percent in the total turnover trade (cf., Tables MN-I-II and VIII-IX, Figures 6-13).

It is a matter of speculation what the effect might be if recalculated data for the years other than 1787, 1792, 1793 and 1795 were to become available. It must also remain a matter of speculation, in the absence of comprehensive data concerning merchant national participation, what might be the British merchants' share of the value of the total overseas commodity trade turnover in the other ports of the Russian Empire. Yet there is enough collateral evidence to surmise that British merchants were active in the over-

## Figure 11

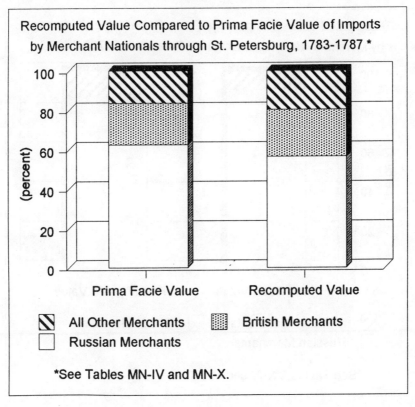

**Figure 12**

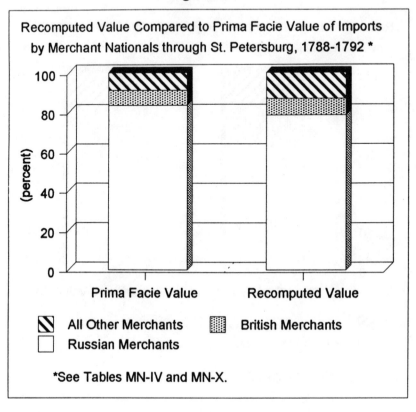

**Figure 13**

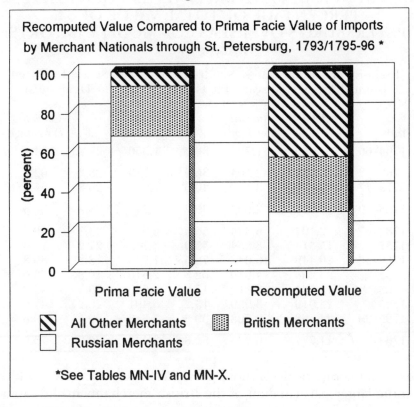

TABLE MN-X. RECOMPUTED VALUE OF IMPORTS BY RUSSIAN AND BRITISH MERCHANTS AS PERCENTAGE OF TOTAL VALUE OF ST. PETERSBURG'S OVERSEAS COMMODITY IMPORTS: 1764–1796[22]

[Exclusive of Custom Duties, Sundry Charges, and Specie Imported]
[Annual Average, Measured in Current Rubles, in Thousands]

| Years | Total Imports | Russian Merchants | % | British Merchants | % | Combined Percentage |
|---|---|---|---|---|---|---|
| 1764–66 | 5,639 | 1,703 | 30.2 | 1,366 | 24.2 | 54.4 |
| 1768–72 | 6,782 | 2,504 | 36.9 | 1,713 | 25.3 | 62.2 |
| 1773–77 | 7,680 | 3,119 | 40.6 | 1,933 | 25.2 | 65.8 |
| 1768–79 | 7,310 | 2,903 | 39.7 | 1,740 | 23.8 | 63.5 |
| 1783–87 | 12,194 | 6,893 | 56.5 | 2,904 | 23.8 | 80.3 |
| 1787 | 15,314 | 8,554 | 55.9 | 3,242 | 21.2 | 77.0 |
| 1788–92 | 20,449 | 16,048 | 78.5 | 1,705 | 8.3 | 86.8 |
| 1792 | 23,296 | 14,630 | 62.8 | 2,516 | 10.8 | 73.6 |
| 1793 | 14,581 | 2,610 | 17.9 | 3,689 | 25.3 | 43.2 |
| 1795 | 23,019 | 3,039 | 13.2 | 6,491 | 28.2 | 41.4 |
| 1795–96 | 24,688 | 9,708 | 39.3 | 7,518 | 30.5 | 69.8 |
| 1764/1796 | 11,777 | 6,223 | 52.8 | 2,430 | 20.6 | 73.5 |

seas commodity trade of those ports. When such activity is added to the larger participation of the British merchants in the capital port of the Empire it is reasonable to assume that British merchants prevailed in the total overseas commodity trade turnover of the Russian Empire during the reign of Catherine II.

This interpretation gains support when the question of Great Britain's role in the overseas trade of the Russian Empire is examined from the standpoint of whose vessels transported the commodities to and from Russia.

## TABLE MN-XI. RECOMPUTED VALUE OF TRADE TURNOVER BY RUSSIAN AND BRITISH MERCHANTS AS PERCENTAGE OF THE TOTAL VALUE OF ST. PETERSBURG'S OVERSEAS COMMODITY TRADE TURNOVER: 1764–1796[23]

[Exclusive of Custom Duties, Sundry Charges, and Specie Imported]
[Annual Average, Measured in Current Rubles, in Thousands]

| Years | Total Turnover | Russian Turnover | % | British Turnover | % | Combined Percentage |
|---|---|---|---|---|---|---|
| 1764–66 | 11,616 | 2,267 | 19.5 | 5,123 | 44.1 | 63.6 |
| 1768–72 | 14,134 | 3,079 | 21.8 | 6,701 | 47.4 | 69.2 |
| 1773–77 | 17,096 | 4,261 | 25.0 | 7,469 | 43.8 | 68.8 |
| 1768–79 | 16,255 | 3,856 | 23.8 | 7,140 | 44.0 | 67.7 |
| 1783–87 | 25,387 | 9,270 | 36.5 | 11,606 | 45.7 | 82.3 |
| 1787 | 31,263 | 10,186 | 32.6 | 14,553 | 46.6 | 79.1 |
| 1788–92 | 41,414 | 25,499 | 61.6 | 10,670 | 25.8 | 87.3 |
| 1792 | 44,349 | 21,333 | 48.1 | 11,770 | 26.5 | 74.6 |
| 1793 | 38,339 | 4,273 | 11.1 | 18,086 | 47.2 | 58.3 |
| 1795 | 54,951 | 3,102 | 5.6 | 25,615 | 46.6 | 52.3 |
| 1795–96 | 59,209 | 16,279 | 27.5 | 28,619 | 48.3 | 75.8 |
| 1764/1796 | 25,481 | 8,899 | 34.9 | 10,290 | 40.4 | 75.3 |

# Chapter XI
# Ships and Tonnage

Catherine II failed to produce a Russian maritime fleet—that is, Russian built and manned by Russian-born citizens—that was capable of emancipating Russia from its dependence on foreign carriers in its overseas commerce. Certainly after the publication in 1780 of the "Ordinance on Commerce and Navigation,"[1] Catherine's policy of naturalizing foreign ships in the process of naturalizing foreign merchants, ship captains, and owners meant that most Russian ships were Russian only in a formal sense. Indicative of this situation is the oft-quoted report in 1777 of the secretary to the French embassy in St. Petersburg that there were only 12 to 15 Russian mercantile ships capable of making long ocean voyages.[2] Nevertheless, there is convincing evidence that only a small number of trading ships (as opposed to small vessels or smacks) flying the Russian flag sailed into the North Sea and Atlantic. Hamburg served ships from the ports of the Russian Empire, in particular, those from Archangel, Riga, and St. Petersburg. During the period 1763–1796, approximately 1,000 trading ships of all flags sailing from Russian ports visited Hamburg (including 159 in 1772 which carried Russian grain to western Europe) making an annual average of about 30. In the Baltic trade with Hamburg during that period, only 14 have been identified as trading ships flying the Russian flag. The total number flying the Russian flag during the period 1763–1796 was 43, 8 of them in 1793.[3]

Although the available archival data on ship arrivals and departures are incomplete, one can estimate an annual average of

---

1. See Chaps. VII–IX supra.
2. C. Ahlström, "Aspects of the commercial shipping between St. Petersburg and Western Europe, 1750–1790," *The interactions of Amsterdam and Antwerp with the Baltic region, 1400–1800* (Leiden: Martinus Nijhoff, 1983), pp. 152–60, in particular pp. 155–56; and Giliane Besset, "Les relations commerciales entre Bordeaux et la Russie au XVIIIe siècle," *Cahiers du Monde russe et soviétique*, XXIII: 2 (Avril–Juin 1982): p. 202.
3. Compiled from the Hamburg Custom House reports in PRO CO 388/95; SP 82/84–98; FO 33/1–3, 5–9, 12; and FO 97/240; and in Staatsarchiv Hamburg, 371-2, Admiralitäts-Kollegium, F 8, Bd. 1–3; F 9; F 10, Bd. 1–15; F 11; F 12, Bd. 1–19; and F 13, Bd. 1–7.

more than 2,500 vessels passed through Russia's overseas ports between 1783 and 1801. Almost 70 percent of this traffic passed through St. Petersburg/Kronstadt and Riga, with the capital port handling slightly more traffic than Riga. British ships accounted for an estimated annual average of more than 25 percent of that number, with about 20 different mercantile groups sharing in the rest: Denmark-Norway and Sweden each accounted for an annual average of about 14 percent; Prussia for about 10 percent; Bremen, Hamburg, Lübeck and Rostock together for about 8 percent; and, according to the *prima facie* evidence, Russian ships accounted for about 8 percent.[4] British shipping actually had a greater influence in Russia's overseas commerce than these figures would suggest. Great Britain concentrated almost all of its trading activity in the northern seaports of the Russian Empire and those ports predominated in the total overseas commerce of the Empire.

Of the available shipping data for St. Petersburg/Kronstadt from 1753 through 1801, only ship departures are complete for the period 1764–1801. From 1764 to the end of Catherine II's reign in 1796 the annual average total of ship departures was 707; when Paul's reign (1797–1801) is included it was 736. British ships represented an annual average of 52 percent of the total departures during both reigns and about 96 percent of those British ships sailed from St. Petersburg/Kronstadt for ports in Great Britain and Ireland (see Table S–I).[5]

From 1753 through 1763, that is from just before the outbreak of the Seven Years' War until peace was restored, the annual average number of British ships departing St. Petersburg/Kronstadt was 163. In the subsequent eleven-year period (1764–1774) the number of British ships departing increased by nearly 62 percent to 264, accounting for slightly more than 52 percent of the total. The latter period was a time of relative peace for Great Britain, but not for Russia, which was at war in the Polish-Lithuanian Commonwealth and with the Ottoman Empire. Yet there is no evidence to suggest that those wars had an appreciable negative influence on Russia's northern shipping trade. It also appears that neither the Anglo-Russian Commercial Treaty of 1766 nor the

---

4. Data on ships entering and departing the overseas ports of the Russian Empire for the period from 1753 through 1801 are compiled and computed from the records of the St. Petersburg Custom House, PRO SP 91/69, 74, 76, 78–79, 83, 88, 91–92, 94–96, 99–104, 107; FO 65/1–2, 4–6, 9, 12–15, 17–18, 20, 22–24, 26, 33, 35, 50, 52; FO 97/340–41; CO 388/54, 56, 95; and BT 6/141, 231–33, 266.
5. Table S-I, ibid.

### TABLE S-I. SHIPS DEPARTING ST. PETERSBURG/ KRONSTADT FROM 1753 THROUGH 1801[5]

(Annual Average)

| Years | Total Ships Sailed | British Ships Sailed | Percentage British Ships |
|---|---|---|---|
| 1753–63 | N/A | 163 | N/A |
| 1764–74 | 506 | 264 | 52.2 |
| 1775–82 | 652 | 317 | 48.6 |
| 1778–82 | 668 | 306 | 45.8 |
| 1783 | 644 | 270 | 41.9 |
| 1783–87 | 750 | 351 | 46.9 |
| 1787 | 752 | 394 | 52.4 |
| 1788–92 | 976 | 532 | 54.5 |
| 1793 | 848 | 542 | 63.9 |
| 1794–96 | 1,019 | 582 | 57.1 |
| 1797–1801 | 933 | 477 | 51.1 |
| 1764–96 | 707 | 368 | 52.1 |
| 1764–1801 | 736 | 383 | 52.0 |

Russian Tariff of 1767 had any special influence on British ship sailings.

However, during the wartime period from 1775 through 1782 the share of British shipping from St. Petersburg/Kronstadt declined. While the annual average number of British ship departures increased to 317, it represented only 48.6 percent of the total number. More specifically, from 1778 through 1782, when Great Britain was at war with France, Spain and the United Provinces, the annual average number of British ship departures declined to 306 and represented only 45.8 percent of the total. Russian shipping information is incomplete for this period but Russian ships in the export trade do not appear to be significant. From 1780 through 1782 an annual average of only 32 ships that were declared to be Russian sailed from St. Petersburg/Kronstadt.[6]

As an indicator of trading activity the number of ships by itself is an incomplete measure of Great Britain's involvement in the Russian Empire's shipping trade. The tonnage those ships transported would be of equal, if not greater, importance. The weight and volume of the bulky cargoes, such as bar iron, flax, hemp, tim-

---

6. Compiled and computed from the records of the St. Petersburg Custom House, PRO BT 6/141, 231, and FO 65/1–2, 4–5.

ber et cetera, exported from Russia were greater than the British domestic manufactures, re-exports and luxury goods imported into Russia. For example, from 1771 through 1782, while the annual average tonnage per British ship in the British import trade from Russia was 198, it was 163 tons per British ship in the British export trade to Russia. That is, the British ships from Russia were about 21 percent larger than those to Russia.[7] Of course, British ships destined for the Baltic departed British ports often in ballast and sometimes they loaded cargoes along the way but their exact number is difficult to determine.[8]

The extant British archival data for Great Britain's world-wide overseas shipping are organized by region and nation-state and provide information on tonnage as well as vessels. Although this source treats East Country separately from Russia and therefore does not reveal the exact nature of the British overseas shipping trade at Riga, it nonetheless complements and amplifies the archival data from Russia on the shipping pattern at St. Petersburg/Kronstadt and at other ports of the Russian empire. This shipping source does not indicate the kind of tonnage recorded as either "registered," "measured," or "tons burthen." "Registered tonnage" appeared in the vessel's certificate of registration and was used by "ships' captains and ship owners who wanted to minimize lighthouse and port duties that were based on tonnages." It was about two-thirds the "measured tonnage" which "was the basis for building, buying, and selling vessels and was computed by means of a given formula" which was first introduced in 1695 and then revised by the Act of 1773.[9] On the other hand, "the invariable measure of a ship's size, from her owner's point of view," according to Ralph Davis, "was 'tons burthen,' deadweight tonnage, the number of tons weight which would lade a ship previously empty (except for stores) down to her minimum safe freeboard or load-

---

7. VT, ff. 225–70.
8. See, in particular, W. E. Minchinton and D. Starkey, "British shipping, The Netherlands and The Baltic, 1784–1795," and Hans Chr. Johansen, "Ships and cargoes in the traffic between the Baltic and Amsterdam in the late eighteenth century," in *The interactions of Amsterdam and Antwerp with the Baltic region*, pp. 181–91, and 161–70, respectively; and Johansen's indispensable study, *Shipping and Trade between the Baltic Area and Western Europe 1784-95* (Odense, Denmark: Odense University Press, 1983), in particular, pp. 82–95.
9. See Christopher J. French's "Note" where he provides a concise review of the problem and the literature on this subject: "Eighteenth-Century Shipping Tonnage Measurements," *Journal of Economic History* 33:2 (June 1973): pp. 434–43; and see also its application to Scotland by Gordon Jackson, "Scottish Shipping, 1775–1805," *Shipping, Trade and Commerce. Essays in Memory of Ralph Davis*, ed. P. L. Cottrell and D. H. Aldcroft (Leicester: Leicester University Press, 1981), pp. 117–36, passim.

line."[10] By the end of the eighteenth century "tons burthen" was actually greater than "measured tonnage" and therefore also "registered tonnage." Compounding the problem of determining the nature of recorded ship tonnage is the fact that the Act of 1773 was not effectively employed until the General Act of Registration of 1786, which required that all vessels henceforth be calculated in terms of "measured tons." This process of assigning ship tonnage has made scholars wary of attempting to compare ship tonnage before and after 1786.[11]

Notwithstanding these reservations, the twofold purpose of the analysis of British shipping tonnage that follows adds to our understanding of the Anglo-Russian trading relationship. Although it might be impossible to determine precisely the overall magnitude and fluctuation of shipping tonnages prior to 1786 without first knowing whether that tonnage was "registered," "measured," or "tons burthen," it is known from other sources that from 1783 onward there was an appreciable increase in the exportation of heavy Russian commodities to Great Britain. Also, in certain years after 1783, there was an increase in the amount of manufactured goods and other commodities transported in British ships to Russia. Moreover, while the exact nature of Great Britain's shipping tonnages prior to 1786 might not become known, it might also be assumed that there was a kind of homogeneous character to them, that is, there was presumably a mix of "registered," "measured," and "tons burthen" shipping tonnages similarly employed in Great Britain's trading with Russia and other countries and that a meaningful comparison between the two can be attempted. Finally, the comparison of the shipping tonnages in Great Britain's trade with Russia and with other countries both before and after 1786 does provide a meaningful assessment of the relative importance of the Anglo-Russian trade.

Great Britain's imports from the Russian Empire were significant when compared to its imports from other European countries. From 1771 through 1796, covering most of Catherine II's reign, an annual average of slightly more than 15 percent of the British ships and nearly 25 percent of the ship–tonnage Great Britain imported from "Foreign Europe" (Denmark-Norway, France, Holland, Flanders, Poland, Sweden, East Country, Russia, Ger-

---

10. Ralph Davis, *The Rise of the English Shipping Industry in the Seventeenth and Eighteenth Centuries* (Newton Abbot, Great Britain: David Charles, 1972), p. 178.
11. See French and Jackson, passim. For the character of the ship tonnage in the Russian-British trade, see Johansen, *Shipping and Trade*, pp. 82–95, passim.

many, Spain and Canaries, Portugal and Maderia, Minorca, Streights and Gibraltar, Malta, Italy and Venice, and Turkey) came from Russia. When Paul's reign is included Russia's share increases to slightly more than 17 percent of the British ships and almost 28 percent of the ship-tonnage during the period 1771–1801. In Great Britain's export trade to "Foreign Europe" an annual average of almost 7 percent of the British ships, transporting nearly 12 percent of the ship-tonnage, went to Russia during Catherine's reign. When Paul's reign is included Russia's share increased to almost 9 and 15 percent in ships and ship-tonnage respectively (see Table VT–I).[12]

The Anglo-Russian trade fluctuated considerably during the period 1771–1801, most noticeably between times of peace and war. During the seven years (1771–1777) prior to the war between Great Britain and France, Spain and the United Provinces in 1778–1782, an annual average of 385 British ships transported an annual average of almost 80,000 ship-tons of Russian goods to Great Britain. This accounted for an annual average of nearly 12 percent of the British ships and almost 22 percent of the ship-tonnage in Great Britain's import trade from "Foreign Europe." But during the five years of war (1778–1782), Great Britain became much more dependent on Russian resources, especially strategic naval stores and ship timber. Early in the naval campaign of 1778 Great Britain proclaimed "a program of preemptive buying of all naval stores,"[13] and from 1778 through 1782 an annual average of 497 British ships transported an annual average of 93,000 ship-tons of Russian goods to Great Britain. This accounted for an annual average of nearly 25 percent of the British ships and nearly 35 percent of the ship-tonnage in Great Britain's import trade from "Foreign Europe" (see Table VT-I).

In terms of Great Britain's exports to Russia during these two periods, a more exaggerated situation developed. During the period 1771–1777 an annual average of 116 British ships transported an annual average of 19,000 ship-tons of British manufactures, re-exports, and luxury goods to Russia. This accounted for an annual average of 4 percent of the British ships and 6.6 percent of the ship-tonnage in Great Britain's export trade to "Foreign Europe." But during the five years of war (1778–1782) an annual average of 170 ships transported an annual average of

---

12. Table VT-I is compiled and computed from VT, ff. 255–70.
13. David Syrett, *Neutral Rights and the War in the Narrow Seas, 1778–82* (Fort Leavenworth, KS: Combat Studies Institute, U. S. Army Command and General Staff College, 1985), p. 7.

### TABLE VT-I. GREAT BRITAIN'S OVERSEAS COMMERCE WITH "FOREIGN EUROPE" AND RUSSIA FROM 1771 THROUGH 1801: VESSELS AND TONNAGE[12]

(Annual Average)

| Years | British Ships Inwards | | | British Tonnage Inwards | | |
|---|---|---|---|---|---|---|
| | From Foreign Europe | From Russia | % From Russia | From Foreign Europe | From Russia | % From Russia |
| 1771–77 | 3,223 | 385 | 11.9 | 362,977 | 79,068 | 21.8 |
| 1778–82 | 1,998 | 497 | 24.9 | 267,018 | 93,272 | 34.9 |
| 1783 | 3,769 | 534 | 14.2 | 474,017 | 116,358 | 24.5 |
| 1783–87 | 4,546 | 575 | 12.7 | 581,369 | 123,065 | 21.2 |
| 1787 | 5,354 | 653 | 12.2 | 695,566 | 135,571 | 19.5 |
| 1788–92 | 6,043 | 766 | 12.7 | 815,751 | 178,159 | 21.8 |
| 1793 | 4,803 | 812 | 16.9 | 778,907 | 194,307 | 24.9 |
| 1794–96 | 3,407 | 912 | 26.8 | 572,804 | 197,493 | 34.5 |
| 1797–1801 | 2,524 | 817 | 32.4 | 449,420 | 173,289 | 38.6 |
| 1771–96 | 3,866 | 594 | 15.4 | 513,802 | 127,413 | 24.8 |
| 1771–1801 | 3,650 | 630 | 17.3 | 503,418 | 134,813 | 27.7 |

| Years | British Ships Outwards | | | British Tonnage Outwards | | |
|---|---|---|---|---|---|---|
| | To Foreign Europe | To Russia | % To Russia | To Foreign Europe | To Russia | % To Russia |
| 1771–77 | 2,927 | 116 | 4.0 | 285,771 | 18,986 | 6.6 |
| 1778–82 | 1,367 | 170 | 12.5 | 157,176 | 27,532 | 17.5 |
| 1783 | 2,798 | 128 | 4.6 | 276,897 | 25,058 | 9.0 |
| 1783–87 | 3,884 | 169 | 4.3 | 424,799 | 36,134 | 8.5 |
| 1787 | 4,674 | 215 | 4.6 | 551,156 | 48,981 | 8.9 |
| 1788–92 | 4,962 | 266 | 5.4 | 622,261 | 63,580 | 10.2 |
| 1793 | 2,922 | 181 | 6.2 | 418,171 | 43,666 | 10.4 |
| 1794–96 | 2,580 | 518 | 20.1 | 420,992 | 113,689 | 27.0 |
| 1797–1801 | 2,128 | 527 | 24.8 | 370,208 | 116,734 | 31.5 |
| 1771–96 | 3,162 | 214 | 6.8 | 373,182 | 44,379 | 11.9 |
| 1771–1801 | 2,995 | 265 | 8.8 | 372,702 | 56,050 | 15.0 |

28,000 ship-tons of Great Britain's exports to Russia. This accounted for an annual average of 12.5 percent of the British ships and 17.5 percent of the ship-tonnage in Great Britain's export trade to "Foreign Europe" (see Table VT-I).

The archival records indicate that Russian ships participated in the overseas commerce of St. Petersburg/Kronstadt during the war years (1778–1782); but whether they were truly Russian ships—that is, Russian built, Russian owned and operated by Russian sailors—is problematic.[14] By Article 4 of the "Ordinance on Commerce and Navigation" of May 8/19, 1780, foreigners were extended the protection and enjoyment of the Russian flag:

> these prerogatives shall be enjoyed not only by our born subjects but also by foreigners, who are domiciled under our rule and pay public taxes like our own subjects; that is to say, as long as they dwell in our country, since in any other case they would not be permitted to use the merchant flag of Russia.[15]

From 1780 through 1782 an annual average of at least 32 ships that were declared to be Russian participated in exports, and about half as many were involved in imports at St. Petersburg/Kronstadt. In 1780, 38 Russian ships, for which the names of both ships and masters are available, sailed with a full range of Russian commodities for Baltic and European ports. Nevertheless, several of those ships did not arrive at their destinations with full cargoes because they were either seized, disabled or lost at sea. Near the close of the shipping season in 1781, 32 Russian ships had already departed Kronstadt with cargoes for Baltic and European ports, including London, and the names of the ships, their masters, and owners are available. A dozen ships under Russian colors are alleged to have carried goods to Great Britain's enemies from St. Petersburg. Four Russian government frigates also sailed with cargoes. Moreover, 32 out of 36 newly-built ships at Kronstadt were designated as Russian. In 1782, 14 Russian ships exported cargoes destined for ports in Great Britain.[16]

The year 1783 inaugurated a period of peace in European affairs. The Russian annexation of the Crimea, together with the new tariff that year, made Russia's mercantile expansion all the more promising by giving it greater access to the Mediterranean Sea and the Levant trade via the Black Sea. Two hundred and seventeen ships declared as Russian ships arrived at Russian ports that year and, even though many of those ships must have sailed

---

14. During the period 1722–1780 "Russian-born shipmasters never succeeded in controlling more than a mere five per cent of the vessels" in the St. Petersburg-Amsterdam trade. J. Th. Lindblad and P. de Buck, "Shipmasters in the shipping between Amsterdam and the Baltic, 1722–1780," *The interactions of Amsterdam and Antwerp with the Baltic region*, pp. 134–52, quotation from p. 140; and see Lindblad, *Sweden's Trade with the Dutch Republic 1738–1795* (The Netherlands: Van Gorcum, ASSN, 1982).
15. AN, p. 292.
16. Compiled and computed from the records of the St. Petersburg Custom House: for 1780, PRO BT 6/231, 141, and FO 65/1–2; for 1781, FO 65/4–5; for 1782, BT 6/231.

only between Russian ports, their number is indicative of Russia's increased maritime strength. Of the total 2,637 ship arrivals, the 217 Russian ships accounted for 8.2 percent, placing them behind the ships of Great Britain (587 = 22.3%), Sweden (569 = 21.6%), and Denmark-Norway (310 = 11.8%) but ahead of the ships of the United Provinces (204 = 7.7%), of Bremen, Hamburg, Lübeck, or Rostock (which together totaled 251 = 9.5%), of Prussia (189 = 7.2%) and the Ottoman Empire (70 = 2.7%).[17]

For the whole Empire in 1783 there were 534 British ships employed in the export trade, which made them the third largest since 1771 (1781 = 580, 1782 = 542). They nevertheless transported the largest ship-tonnage ever of Russian goods to Great Britain. However, the 270 British ships that sailed from St. Petersburg/Kronstadt accounted for only 41.9 percent of the total ship departures, making it the fourth lowest percentage since 1764 (see Tables VT-I and S-I).

In 1786 the total number of ships arriving in Russian ports was 2,155, down 18.3 percent from 1783. The British again led all other carrier groups with 705 ships representing 32.7 percent of the total. Compared to 1783 this was an increase of 20 percent in number and 10 percent in share. The number of Russian ships declined to 107 and equaled 5 percent of the total. This placed them behind the ships of the United Provinces (366 = 17%), Sweden (314 = 14.6%), and Denmark-Norway (276 = 12.8%) but ahead of the ships of Prussia (92 = 4.3%), and Bremen, Hamburg, Lübeck or Rostock (which together totaled 184 = 8.5%). The ships of America, Austria, France, Portugal, and several other nations each represented less than 1 percent of the total.[18]

In 1787, despite the fact that the Russian court canceled the extension of the Anglo-Russian Commercial Treaty of 1766 at the beginning of the shipping season, Great Britain imported its greatest amount of cargo up to that time from the Russian Empire: about 136,000 ship-tons in 653 British ships. This represented about 12 percent of the total number of British ships and about 20 percent of the total ship-tonnage in Great Britain's import trade from "Foreign Europe" (see Table VT-I). From St. Petersburg/Kronstadt slightly more than 52 percent of the total ship departures (394/752) were British, representing 60 percent of the total

---

17. Compiled and computed from the records of the St. Petersburg Custom House, PRO FO 97/340.
18. Compiled and computed from a report made to the College of Commerce in the records of the St. Petersburg Custom House, PRO BT 6/231.

number of British ships in the export trade from the Russian Empire to Great Britain (see Table S-I). From Riga, of the total 702 ships that departed, 205 (29%) sailed for British ports and were most likely British vessels. Of the remaining 497 ships, 180 were Dutch (25.6%), 97 were Russian (13.8%), 83 were Swedish (11.8%), 66 were Danish (9.4%), and 30 were Prussian (4.3%). From Archangel, of the total 128 ships that departed, 51 (39.8%) sailed for British ports and were most likely British vessels.[19]

Nearly 50 percent of the total 804 ships that arrived at St. Petersburg/Kronstadt in 1787 were British. Of the remaining 404 non-British vessels nearly 16 percent were Russian (64/404). This placed Russian ships behind only those of the United Provinces (71 = 17.6%) but ahead of those of Denmark-Norway (57 = 14.1%), Sweden (46 = 11.4%), Prussia (37 = 9.2%) and Bremen, Hamburg, Lübeck, or Rostock (taken together totaled 82 = 20.3%)[20]

The Anglo-Russian trade flourished during the period of the Russo-Turkish-Swedish War and the French Revolution (1788–1792). Compared to the previous five-year period (1783–1787) the annual average number of British ships increased by 33 percent and the ship-tonnage by about 45 percent in Great Britain's import trade from the Russian Empire. In its export trade to Russia, the annual average number of British ships increased by slightly more than 57 percent and the ship-tonnage by 76 percent.[21]

When Catherine II precipitously cut off trade with revolutionary France in 1793 her customs prohibitions also had a negative effect on Britain's export trade to Russia. While Russian exports to Great Britain set a record, Russian imports from Great Britain plummeted. However, while Great Britain's overall trade with "Foreign Europe" declined, the Anglo-Russian trade component of that trade improved. Nearly 25 percent of Great Britain's total import ship-tonnage from "Foreign Europe," transported by nearly 17 percent of the total number of British ships, came from the Russian Empire. But only a little more than 10 percent of Great Britain's export ship-tonnage to "Foreign Europe," trans-

---

19. These figures are nearly identical to Johansen's calculation for the number of British ships westward bound via the Sund: 397 from St. Petersburg; 205 from Riga; 32 from Narva; 52 from other Russian ports. Johansen, *Shipping and Trade*, p. 94. Archangel and Riga ship departures are compiled and computed from records of the St. Petersburg Custom House PRO BT 6/231; and see Tables S-I and VT-I.
20. Compiled and computed from the records of the St. Petersburg Custom House, PRO BT 6/141, 231–32, and FO 65/17; and Table S-I.
21. Tables S-I and VT-I; PRO BT 6/231–33; FO 65/17–18, 20, 23–24; and FO 97/341.

ported by more than 6 percent of the total number of British ships, went to the Russian Empire (see Table VT-I).

In the crisis year of 1793 British ships dominated the trade of St. Petersburg/Kronstadt, representing about 64 percent of the total number trading there. In addition, 99 percent of the departing British ships were destined for ports in Great Britain and Ireland. The percentage of British ships destined for British ports was not insignificant at other Russian ports: 61 percent of 108 ships from Archangel, about 36 percent of 78 ships from Fredrikshamn and Viborg, about 33 percent of 70 ships from Narva, and 20 percent of the 869 ships from Riga. In sum, no less than 42 percent of the foregoing total of ships was destined for ports in Great Britain and Ireland.[22]

At war, Great Britain placed greater demands on Russian resources and set new records for Russia's share of the trade of "Foreign Europe." From 1794 through 1796 an annual average of 912 British ships (27%) transported an annual average of nearly 200,000 ship-tons (34.5%) of Russian goods to Great Britain. In Great Britain's export trade to "Foreign Europe" an annual average of 518 British ships (20%) transported an annual average of almost 114,000 ship-tons (27%) of British manufactures, re-exports, and luxury goods to Russia despite Russia's continuing restrictions on British imports into Russia (see Table VT-I).

During Emperor Paul's reign (1797–1801) the pattern of shipping at Russian ports changed significantly. The turnover of ships at Russian ports declined, the number and share of British ships in the Russian trade declined, and the distances sailed by some of those ships trading with Russia contracted. In 1796, the final year of Catherine II's reign, a high of 3,444 ships entered and departed Russian ports. However, from 1797 through 1800 that number progressively declined to an annual average of 2,886. In 1796 the number of British ships that arrived at Russian ports was 1,317, representing 38.3 percent of the total. In 1797 that number dropped to 814, representing 25.4 percent of the total. Comparing 1797 to 1796, the ships of Denmark-Norway entering Russian ports declined 19 percent (from 572 to 462) and their share slipped from 16.4 to 14.4 percent. However, entering Swedish ships jumped 39 percent (280/390) and increased their share from 8.1 to 12.2 percent; Prussian ships advanced by 35 percent (391/529) and raised their share from 11.4 to 16.5 percent; and the

---

22. Compiled and computed from the records of the St. Petersburg Custom House, PRO FO 65/26.

ships of Bremen, Hamburg, Lübeck and Rostock together gained 11 percent (217/241), improving their share from 6.3 to 7.5 percent. The number of Russian ships increased slightly from 301 to 308, resulting in a modest increase in their share of the total from 8.7 to 9.6 percent. In the Black Sea the ships from the Ottoman Empire jumped 67 percent (238/398) and nearly doubled their share from 6.9 to 12.4 percent.[23]

The annual average number of ship departures from St. Petersburg/Kronstadt during Paul's reign fell to the level of 1793, below the high of more than 1,000 achieved during the last three years of Catherine's reign. The British share declined from 57 to 53.7 percent (see Table S-I). Comparing 1797 to 1796, the total turnover of ships at the capital port dropped 24 percent, from 2,316 to 1,763. The number of British ships fell 35 percent, from 1,362 to 880, and its share of the turnover of all ships declined from nearly 59 percent to 50 percent. Even the total number of Russian ships declined from 123 to 112 although its share of the total turnover increased from 5.3 to 6.4 percent. The number of ships from America plummeted 44 percent (118/52) along with their share of the total (from 5.1 to 2.9 percent); and the number of ships from Denmark-Norway dropped about 24 percent (260/198) but sustained their former share at 11 percent. However, the number of departing Swedish ships rose about 26 percent (112/141) and increased their share from 4.8 to 8.0 percent. The ships from Bremen, Hamburg, Lübeck, and Rostock together gained 25 percent (172/215) and lifted their share from 7.4 to 12.2 percent. The number of Prussian ships increased by 17 percent (70/82) and its share rose from 3.0 to 4.7 percent.[24]

But the Anglo-Russian trade, as part of Great Britain's trade with "Foreign Europe," increased dramatically while Great Britain was at war in Europe. During the period 1797–1801, although the annual average number of British ships in Great Britain's trade with "Foreign Europe" fell to its lowest level since the wartime period of 1778–1782, the proportion of British ships in the Anglo-Russian trade of "Foreign Europe" was the highest since 1771, the first year for which the archival data are available. Slightly more than 32 percent of the total number of British ships in the import trade from "Foreign Europe" came from Russia, and almost 25

---

23. Compiled and computed from the records of the St. Petersburg Custom House, PRO BT 6/233; and for ships at Black Sea Ports, see FO 97/340, BT 6/231, and FO 65/33, 50, 52.
24. Compiled and computed from the records of the St. Petersburg Custom House, PRO BT 6/231, 233, and FO 65/33.

percent of the total number of British ships in Great Britain's export trade with "Foreign Europe" went to Russia. While Great Britain's total shipping tonnage fell to its lowest level with "Foreign Europe" since 1783, its import tonnage from Russia reached a high with more than 38 percent of the total and its export tonnage to Russia reached more than 31 percent of the total (see Table VT-I).

These figures and percentages in ships and tonnage once again demonstrate Great Britain's extraordinary dependence upon Russia's resources for its economic growth and strategic capability during Catherine's reign. This dependence was especially pronounced during wartime and persisted even when Russia denied Great Britain favorable tariffs and the protection of a commercial treaty, inclusive of the privilege of the most-favored-nation.

These figures and percentages substantiate the leading role that the British mercantile community played in Russia's overseas commerce. Russia was absolutely dependent upon foreign and, in particular, British shipping to sell its natural resources and manufactured products on the international market. Without such foreign shipping assistance Russia could not have sustained its overseas commerce because Russian shipping, however much subsidized or however defined, could not by itself have accomplished this national task. Because of this it should have been clear to everyone at the Russian court that Catherine II's mercantile and maritime policies, which had been designed to put an end to this dependence, had been failures.

# Chapter XII
# Russian Commodity Exports and Great Britain's Economic Development:
From the Peace of Paris to the end of Catherine II's Reign

Great Britain's campaign of buying Russian natural and manufactured resources accelerated soon after the restoration of peace in 1783 and coincided with the greater expansion of Britain's economy. This was also a period when the value of the ruble declined dramatically against English and Dutch currency and when British shipping increased sharply in the Russian trade. From 1783 through the end of Catherine's reign in 1796, Great Britain remained Russia's foremost trading client for a wide assortment of Russian goods and, although their volumes and values increased over the preceding decades, their composition did not change. Great Britain continued to import from Russia bar iron and other irons, flax, linseed and linseed oil, hemp, hempseed and hemp oil, cordage, linens and sailcloth, saltpeter and potash, rozin, pitch and tar, tallow and tallow candles, soap, wax and wax candles, timber boards, wainscot and squared logs, masts, hides, sole leather, bristles, animal hair and manes, feathers, skins and furs, mats, tobacco, caviar and isinglass, and a variety of cereal grains, including wheat and rye. For example, from 1783 through 1796 Great Britain imported 58 percent of its cordage, nearly 32 percent of its pitch and tar, a substantial amount of its bristles, isinglass, and potash, and 90 percent of its tallow from Russia.[1]

The aforementioned Russian commodities imported into

---

1. For cordage, pitch and tar, bristles, isinglass, potash, and tallow, see records of the St. Petersburg Custom House, PRO FO 65/12–15, 17–18, 20, 23–24, 26, 33, 35, 50, 52; FO 97/340–41; BT 6/141, 230–33, 240, 266; VTR, *Vedomosti*, Nos. 7–8; BT 6/230, ff. 11, 13, 15, 17, 29, 31, 45–49, 52–53; BT 6/232, ff. 7–11, 18–22; BT 6/240, ff. 97, 101, 103, 122, 124, 126; and CU 3, 14, and 17.

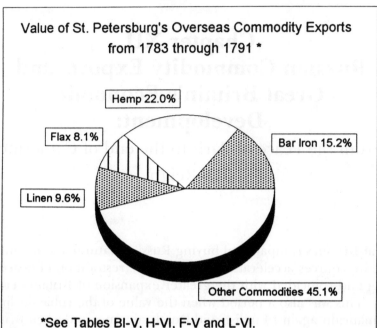

**Figure 14**

Value of St. Petersburg's Overseas Commodity Exports from 1783 through 1791 *

Hemp 22.0%
Flax 8.1%
Bar Iron 15.2%
Linen 9.6%
Other Commodities 45.1%

*See Tables BI-V, H-VI, F-V and L-VI.

Great Britain during the latter part of Catherine II's reign are impressive for their volume, value, and multiplier effect in the British economy and because they suited special industrial or consumer purposes.

While it is doubtful that an English citizen would take pride in advertising the use or consumption of a good imported from Russia as much as a Russian would a good imported from Great Britain, the fact remains that Russian imports were well known to those who used them, depended on them, or made a living wage from them. This was indeed true regarding the economic and strategic value the British placed on the importation of Russian bar iron, hemp, flax, linen and timber.

## BAR IRON

Although British bar iron output increased during the last quarter of the eighteenth century, Great Britain remained dependent on foreign sources to meet the increasing demand for metal products made of bar iron. Compared to the period 1764–1782 there was an annual average increase of nearly 10 percent in imported bar

## TABLE BI-IV. GREAT BRITAIN'S BAR IRON IMPORTS FROM 1783 THROUGH 1796[2]

(Annual Average, Measured in Tons)

| Years | Total Imports | Russian Imports | % | St. Petersburg's Bar Iron Exports All Ships | St. Petersburg's Bar Iron Exports British Ships | % |
|---|---|---|---|---|---|---|
| 1783 | 47,914 | 31,270 | 65.3 | 29,754 | 25,647 | 86.2 |
| 1783–87 | 48,657 | 29,732 | 61.1 | 34,579 | 30,340 | 87.7 |
| 1787 | 46,748 | 23,097 | 49.4 | 26,998 | 24,843 | 92.0 |
| 1788 | 51,496 | 29,541 | 57.4 | 39,365 | 35,020 | 89.0 |
| 1788–92 | 53,329 | 28,282 | 53.0 | 37,880 | 33,443 | 88.3 |
| 1793 | 58,962 | 34,984 | 59.3 | 41,750 | 37,001 | 88.6 |
| 1794 | 42,479 | 23,045 | 54.3 | 34,492 | 27,974 | 81.1 |
| 1794–96 | 48,427 | 27,400 | 56.6 | 36,599 | 29,801 | 81.4 |
| 1783–96 | 51,013 | 29,089 | 57.0 | 36,703 | 31,809 | 86.7 |

iron from all sources and nearly 13 percent from Russia during the period 1783–1796. Russian bar iron amounted to an annual average of 57 percent of Great Britain's total bar iron imports during 1783–1796, slightly higher than during the previous period. St. Petersburg/Kronstadt continued to be the largest Russian supplier of bar iron. British ships transported an annual average of about 87 percent of that port's exported bar iron tonnage (see Table BI–IV).[2] Bar iron exports from other Russian ports to British ports were negligible.

In terms of sheer volume, and despite the arguments and fears of British politicians, the absence of a treaty does not appear to have had an effect on Great Britain's imports of Russian bar iron, which, notwithstanding rising domestic production, remained absolutely vital for the British iron and steel industry. These imports peaked in 1793, and had been contracted for prior to the conclusion of the Anglo-Russian commerce convention in March 1793. The sharp drop in Great Britain's bar iron imports

---

2. Table BI-IV is compiled and computed from BAR IRON, PRO BT 6/240 and 230; and the records of the St. Petersburg Custom House, PRO FO 65/12–15, 17–18, 20, 23–24, 26, 33, 35, 50, 52; FO 97/340–41; and BT 6/141, 231–33, 266. Cf., S. G. Strumilin, *Izbrannye proizvedeniia. Istoriia chernoi metallurgii v SSSR* (Moskva, 1967), pp. 196–97.

Archangel was the only other Russian port that exported an appreciable amount of bar iron. During the period 1783–1796 it exported slightly more than 5 percent of the total exported from St. Petersburg/Kronstadt of which about 85 percent was shipped to Great Britain. See the records of the St. Petersburg Custom House, PRO BT 6/231–33, 266; FO 65/13–15, 33; and FO 97/340.

in 1794 was probably due to the weakness of the British economy in 1793, to the recent outbreak of the European war, and to the contribution that the puddling process was having on British domestic bar iron output which, in turn, had the effect of somewhat dampening Great Britain's dependence upon foreign imports of bar iron.

To appreciate the significance that Russian bar iron had in the British economy, one need only recall that in 1788 (the first year during the second half of the eighteenth century for which there is an agreed-upon figure) British bar iron output was 32,000 tons. In that year Russian bar iron imports into England and Scotland were almost 30,000 tons—an amount almost equal to the entire British domestic output. The total amount of bar iron imported and domestically produced that year was more than 83,000 tons, of which Russian bar iron imports represented slightly more than 35 percent. In 1794 (the only other year for which there is an agreed-upon figure for British domestic output), 50,000 tons of bar iron were produced in Britain and 42,000 tons were imported. Of the amount imported, Russian bar iron amounted to just over 23,000 tons, the lowest during the period 1783–1796, representing 46 percent of the British output. Of the total bar iron tonnage that was domestically produced and imported in 1794, Russian bar iron accounted for nearly 25 percent. Thus, about 30 percent of all the British products made from bar iron during the period 1783–1796 was Russian in origin.[3]

There are archival data on the monetary value of commodity exports from St. Petersburg/Kronstadt for ten of the fourteen years during the period 1783–1796: from 1783 through 1791 and for the year 1793. In comparing the two periods for which there are continuous export prices in current rubles—1768–1779 and 1783–1791—the annual average value of Russian commodity exports increased by 86 percent and bar iron by 52 percent (see Figure 14, Table BI-V[4] and cf. Figures 1, Table BI-III). In 1793 the value of commodity exports increased again by 13 percent, and bar iron increased by 16 percent over the period 1788–1791. The current price of a pood of Russian bar iron increased annually

---

3. For Great Britain's bar-iron output see Charles K. Hyde, *Technological Change and the British Iron Industry 1700–1870* (Princeton, 1977), pp. 92–94, 113.
4. Table BI-V is compiled and computed from the records of the St. Petersburg Custom House, PRO BT 6/141, 231–33, 266; FO 97/340–41; and FO 65/13–15, 17–18, 20, 26; and VTR, *Vedomosti*, Nos. 7–8.

## TABLE BI-V. VALUE OF ST. PETERSBURG'S COMMODITY AND BAR IRON EXPORTS: 1783–1791, 1793[4]

[Exclusive of Custom Duties and Sundry Charges]
(Annual Average, Measured in Current Rubles, in Thousands)

| Years | Value of Commodity Exports | Value of Bar Iron Exports | | | |
|---|---|---|---|---|---|
| | | All Ships | Percentage | British Ships | Percentage |
| 1783–87 | 13,193 | 1,983 | 15.0 | 1,744 | 13.2 |
| 1788–91 | 20,942 | 3,289 | 15.7 | 2,860 | 13.7 |
| 1793 | 23,758 | 3,814 | 16.1 | 3,380 | 14.2 |
| 1783–91 | 16,637 | 2,537 | 15.2 | 2,240 | 13.5 |

| Years | Value of Commodity Exports by British Merchants | Percentage Value of Bar Iron Exports in British Ships |
|---|---|---|
| 1783–87 | 8,702 | 20.0 |
| 1788–91 | 8,894 | 32.2 |
| 1793 | 14,397 | 23.5 |
| 1783–91 | 8,787 | 25.5 |

from 80 kopeks in 1783 (and its high during the period 1768–1779) to 1 ruble 45 kopeks in 1793.

In proportion to the total value of all commodity exports from St. Petersburg/Kronstadt from 1783 through 1791, the annual average value of bar iron was about 15 percent in all ships and 13.5 percent in British ships. The latter represented 25.5 percent of the total value of the commodity exports by British merchants. These percentages were each slightly less than during the period 1768–1779 but they continue to demonstrate the significant economic importance of bar iron exports to the British merchant community.

## HEMP

To a much greater extent Great Britain was dependent on hemp from Russia. From 1783 through 1796 Great Britain increased its annual average importation of hemp from all sources by about 27 percent and by almost 29 percent from Russia over the previous period 1764–1782. The nearly 23,000 annual average tons of imported Russian hemp accounted for 97 percent of Great Britain's total hemp imports during 1783–1796, and slightly exceeded the

### TABLE H-V. GREAT BRITAIN'S HEMP IMPORTS FROM 1783 THROUGH 1796[5]

(Annual Average, Measured in Tons)

| Years | Total Imports | Russian Imports | % | St. Petersburg's Hemp Exports All Ships | St. Petersburg's Hemp Exports British Ships | % |
|---|---|---|---|---|---|---|
| 1783 | 13,355 | 13,319 | 99.7 | 21,080 | 10,263 | 48.7 |
| 1783–87 | 16,377 | 16,119 | 98.4 | 26,740 | 15,741 | 58.9 |
| 1787 | 18,990 | 18,499 | 97.4 | 23,376 | 16,671 | 71.3 |
| 1788–92 | 26,218 | 25,217 | 96.2 | 36,754 | 22,882 | 62.3 |
| 1793 | 27,692 | 26,911 | 97.2 | 30,138 | 21,810 | 72.4 |
| 1794–96 | 29,598 | 28,873 | 97.6 | 32,834 | 22,866 | 69.6 |
| 1783–96 | 23,533 | 22,872 | 97.2 | 31,865 | 20,251 | 63.6 |

percentage imported during the period 1764–1782. St. Petersburg/Kronstadt was again the largest supplier of Russian hemp. British ships transported an annual average of about 64 percent of that port's hemp exports, about the same proportion as during the period 1764–1782. This was about 89 percent of Great Britain's total hemp imports from Russia during 1783–1796 and represented a decline of about 10 percent from the period 1764–1782 (see Table H–V).[5]

Riga was the only other Russian port that exported substantial quantities of hemp between 1783 and 1796, with an annual average of about 13,000 tons (an amount equal to the annual average during the wartime period 1778–1782), that is, about 40 percent of the volume exported from St. Petersburg/Kronstadt. Of that amount, slightly more than 4,000 tons, about 33 percent, was shipped to British ports, amounting to an increase of about 700 tons over the previous period.[6]

Between 1783 and 1791 the value of hemp exports from St. Petersburg/Kronstadt was an annual average of about 54 percent greater than it was during the period 1768–1779. However, in 1793, unlike the increase that took place in the value of bar iron exports, the value of hemp exports declined considerably from an annual average of about five million current rubles during 1788–

---

5. Table H-V is compiled and computed from HEMP, PRO BT 6/240 and 230; and the records of the St. Petersburg Custom House, PRO FO 65/12–15, 17–18, 20, 23–24, 26, 33, 35, 50, 52; FO 97/340–41; and BT 6/141, 231–33, 266.
6. For Riga exports between 1783 and 1796 see the records of the St. Petersburg Custom House, PRO BT 6/231–33, 266; FO 97/340; and FO 65/13–15, 33.

## TABLE H-VI. VALUE OF ST. PETERSBURG'S COMMODITY AND HEMP EXPORTS: 1783–1791, 1793[7]

[Exclusive of Custom Duties and Sundry Charges]
(Annual Average, Measured in Current Rubles, in Thousands)

| Years | Value of Commodity Exports | Value of Hemp Exports | | | |
|---|---|---|---|---|---|
| | | All Ships | Percentage | British Ships | Percentage |
| 1783–87 | 13,193 | 2,587 | 19.6 | 1,558 | 11.8 |
| 1788–91 | 20,942 | 5,001 | 23.9 | 3,110 | 14.9 |
| 1793 | 23,758 | 3,745 | 15.8 | 2,723 | 11.5 |
| 1783–91 | 16,637 | 3,660 | 22.0 | 2,248 | 13.5 |

| Years | Value of Commodity Exports by British Merchants | Percentage Value of Hemp Exports in British Ships |
|---|---|---|
| 1783–87 | 8,702 | 17.9 |
| 1788–91 | 8,894 | 35.0 |
| 1793 | 14,397 | 18.9 |
| 1783–91 | 8,787 | 25.6 |

1791 to 3.7 million rubles (see Table H-VI). Moreover, again unlike bar iron, the price of a pood of hemp in its four varieties— "clean," "outshot," "half clean," and "codilla" at St. Petersburg/Kronstadt fluctuated considerably during the period 1783–1791. For example, in 1785 the current price of a pood of "clean hemp," the best quality and the most desired, fell to a low of 1 ruble 30 kopeks, which was its price during the late 1760s and early 1770s. In 1789 it was at its high at 2 rubles 50 kopeks; in 1790 it declined to 2 rubles 20 kopeks and, in 1791, to 1 ruble 90 kopeks. In 1793 a pood of "clean hemp" increased to 2 rubles.

Between 1783 and 1791 the total value of exported hemp of all kinds from St. Petersburg/Kronstadt was an annual average of 22 percent of the value of all commodity exports, and that which was transported in British ships was 13.5 percent (representing a decline of about 4 percent for each from the period 1768–1779). The latter was about 26 percent of the total value of commodity exports by British merchants, a slight decline from the earlier period 1768–1779 (see Figure 14, Table H–VI).[7]

---

7. Table H-VI is compiled and computed from the records of the St. Petersburg Custom House, PRO BT 6/141, 231–33, 266; FO 97/340–41; and FO 65/13–15, 17–18, 20, 26; and VTR, *Vedomosti*, Nos. 7–8.

The value of hemp exports from St. Petersburg/Kronstadt was again greater than that of bar iron; together their annual average value accounted for 37 percent of the total annual average value of commodity exports from the capital port. This was a sharp decline from 45 percent in the earlier period. In British ships the annual average value of bar iron and hemp was about the same, and together accounted for 27 percent of the total value of commodity exports for 1783–1791, a significant decline from 34 percent in the earlier period. The combined total annual average value of bar iron and hemp transported in British ships amounted to 51 percent of the total value of commodity exports by British merchants but this too was sharply down from 56 percent in the earlier period.

In the face of the enormous and growing demand for hemp from the Russian Empire—at a time when the shipping in the Russian trade had sharply increased and when the value of the ruble had steadily declined against the Dutch guilder and English pound Sterling—the export price of hemp at the capital port did not increase as much as bar iron. This occurrence was undoubtedly owing to the prolonged deterioration in the quality of the hemp for export at St. Petersburg, a situation complained about for several years by the British, who purchased about two-thirds of the capital port's hemp exports.[8] Moreover, the overall fall in total hemp receipts had an undermining effect on the precarious balance of trade at the capital port, which resulted in crisis in 1793, a subject discussed in greater detail in Chapters XIII and XIV below.

## FLAX

Compared to the period 1764–1782, Great Britain's annual average amount of imported flax increased by about 78 percent from all sources and about 68 percent from Russia during the period 1783–1796. Of the total flax imported during 1783–1796, Russian flax represented an annual average of about 77 percent, down from about 82 percent during 1764–1782. But during 1783–1796 St. Petersburg/Kronstadt played a greater role in the Anglo-Russian flax trade than during the earlier period. British ships transported 79 percent of the exported flax from the capital port during 1783–1796, an increase over the 73 percent during 1764–

---

8. See Admiralty and Navy Board letters from 1791–1794, FO 97/342.

## TABLE F-IV. GREAT BRITAIN'S FLAX IMPORTS FROM 1783 THROUGH 1796[9]

(Annual Average, Measured in Tons)

| Years | Total Imports | Russian Imports | % | St. Petersburg's Flax Exports All Ships | St. Petersburg's Flax Exports British Ships | % |
|---|---|---|---|---|---|---|
| 1783 | 6,135 | 5,246 | 85.5 | 5,811 | 4,186 | 72.0 |
| 1783–87 | 9,527 | 7,742 | 81.3 | 7,000 | 4,848 | 69.3 |
| 1787 | 13,484 | 10,173 | 75.4 | 7,778 | 5,825 | 74.9 |
| 1788–92 | 12,101 | 9,201 | 76.0 | 8,031 | 6,503 | 81.0 |
| 1793 | 13,562 | 10,437 | 77.0 | 7,338 | 6,662 | 90.8 |
| 1794–96 | 14,924 | 11,285 | 75.6 | 8,515 | 7,310 | 85.8 |
| 1783–96 | 11,891 | 9,108 | 76.6 | 7,717 | 6,096 | 79.0 |

1782. Moreover, while British ships from the capital port transported about 43 percent of Great Britain's total Russian flax imports during 1764–1782, they transported nearly 67 percent of it during 1783–1796 (see Figure F–IV, Table F–IV).[9]

Between 1783 and 1796 the annual average flax exports from Narva and Riga together exceeded the flax exports from St. Petersburg/Kronstadt by about 30 percent. Forty percent of Narva's and Riga's exports were shipped to British ports. The flax exports from Narva, Riga, and St. Petersburg/Kronstadt made up the total amount of Great Britain's flax imports from the Russian Empire.[10]

The annual average value of flax exported from St. Petersburg/Kronstadt between 1783 and 1791 was nearly three times the value exported during the period 1768–1779. But, in 1793 the value of flax declined, not as severely as did that of hemp but undoubtedly for the same reason, from the annual average 1.7 million current rubles during 1788–1791 to nearly 1.4 million (see Figure 14, Table F–V).[11] The price of a pood of the several varieties of flax fluctuated widely. For example, "clean or 12-head flax," the best and the most desired, in 1783 sold as low as 2 rubles and 30 kopeks per pood, a price available during the 1760s. In 1789 the

---

9. Table F-IV is compiled and computed from FLAX, PRO BT 6/240 and 230; and the records of the St. Petersburg Custom House, PRO FO 65/12–15, 17–18, 20, 23–24, 26, 33, 35, 50, 52; FO 97/ 340–41; and BT6/ 141, 231–33, 266.
10. For Narva's and Riga's exports between 1783 and 1796 see the records of the St. Petersburg Custom House, PRO BT 6/231–33, 266; FO 97/340; and FO 65/13–15, 33.
11. Table F-V is compiled and computed from the records of the St. Petersburg Custom House, PRO BT 6/141, 231–33, 266; FO 97/340–41; and FO 65/13–15, 17–18, 20, 26; and VTR, *Vedomosti*, Nos. 7–8.

## TABLE F-V. VALUE OF ST. PETERSBURG'S COMMODITY AND FLAX EXPORTS: 1783–1791, 1793[11]

[Exclusive of Custom Duties and Sundry Charges]
(Annual Average, Measured in Current Rubles, in Thousands)

| Years | Value of Commodity Exports | Value of Flax Exports All Ships | Percentage | British Ships | Percentage |
|---|---|---|---|---|---|
| 1783–87 | 13,193 | 1,044 | 7.9 | 743 | 5.6 |
| 1788–91 | 20,942 | 1,728 | 8.3 | 1,431 | 6.8 |
| 1793 | 23,758 | 1,389 | 5.8 | 1,289 | 5.4 |
| 1783–91 | 16,637 | 1,348 | 8.1 | 1,049 | 6.3 |

| Years | Value of Commodity Exports by British Merchants | Percentage Value of Flax Exports in British Ships |
|---|---|---|
| 1783–87 | 8,702 | 8.5 |
| 1788–91 | 8,894 | 16.1 |
| 1793 | 14,397 | 9.0 |
| 1783–91 | 8,787 | 11.9 |

price reached a high of 4 rubles but subsequently declined. In 1790 the price dropped to 3 rubles 80 kopeks; in 1791, to 3 rubles 40 kopeks; and in 1793, to 3 rubles 20 kopeks.

Between 1783 and 1791 the total value of exported flax from St. Petersburg/Kronstadt was an annual average of 8.1 percent of the value of all commodity exports, and that which was transported in British ships amounted to 6.3 percent. Both figures represent an increase of more than 50 percent over the period 1768–1779. The annual average value of flax transported in British ships between 1783 and 1791 was nearly 12 percent of the total value of commodity exports by British merchants, nearly twice as much as during the period 1768–1779.

Russian flax made a significant contribution to England's linen output. Following N. B. Harte's schema (and the output years he provides in cwt. already discussed in Chapter V), in 1785 the 167,647 cwt. of Russian flax imported into England was equivalent to approximately 21 million yards of linen. That represented about 45.9 percent of England's estimated total linen output of approximately 45.9 million yards. In 1790 the 110,153 cwt. of imported Russian flax was equivalent to approximately 13.9 million yards of linen and represented about 34 percent of England's estimated total linen output of approximately 40.7 million yards.[12]

## LINEN

It is nearly impossible to estimate the quantity of Russian linen that Great Britain imported, re-exported and retained for the period 1783–1796, as was the case for 1753–1771 which has been demonstrated in Chapter V. The data for such a comparison are not readily available. However, archival data are available to estimate the minimum quantity of several Russian linens—broad and narrow diaper and linen, crash, drilling, flem, and ravenduck—that were exported from St. Petersburg/Kronstadt destined for the British market. From 1783 through 1796 an annual average of nearly six million yards of these linens was transported by British ships destined for British ports. This represented 79 percent of the total quantity exported, about the same as during the period 1763–1782 (see Table L–V).[13] It also indicates the upper limits of marketable Russian linens in competition with England's increasing domestic production.

The significance of Russian linen on the British market may be appreciated by comparing it with England's linen output. In terms of Harte's estimate, the 5.4 million yards of Russian linens in 1785 and the 6 million yards in 1790 exported from St. Petersburg/Kronstadt and destined for British ports represented 11.7 and 14.5 percent of England's approximate minimum linen output.[14] Moreover, in 1790 these 6 million yards of Russian linens were equal to almost 18 percent of the Irish and about 33 percent of the Scottish linen production destined for the English market.[15]

Between 1783 and 1796 Great Britain had become less dependent on Russian sailcloth and only 44 percent of its foreign sailcloth came from Russia, a decline from 70 percent during the

---

12. N. B. Harte, Table 4.3, "English imports of flax and yarn and the approximate linen equivalent, 1700–90," in "The Rise of Protection and the English Linen Trade, 1690–1790," *Textile History and Economic History*, ed. N. B. Harte and K. G. Ponting (Manchester, 1973), p. 104, and FLAX, PRO BT 6/230.
13. Table L-V is compiled and computed from the records of the St. Petersburg Custom House, PRO FO 65/12–15, 17–18, 20, 23–24, 26, 33, 50, 52; FO 97/340–41; and BT 6/141, 231–33, 266. Cf., G. S. Isaev, *Rol' tekstil'noi promishlennosti v genezise i razvitii kapitalizma v Rossii 1760–1860* (Leningrad: Nauka, 1970), pp. 123–24.
14. Harte, Table 4.2, "Irish and Scottish linen production for the English market," p. 93.
15. Harte, Table 4.3, "English imports of flax and yarn and the approximate linen equivalent, 1700–90," p. 104.

## TABLE L-V. MINIMUM ESTIMATE OF SELECTED LINENS EXPORTED FROM ST. PETERSBURG FROM 1783 THROUGH 1796[13]

[Broad and Narrow Diaper and Linen, Crash, Drilling, Flem, and Ravenduck]
(Annual Average, Measured in Yards, in Thousands)

| | Linen Exports | | British Percentage |
|---|---|---|---|
| Years | All Ships | British Ships | of Total |
| 1783 | 6,253 | 4,898 | 78.3 |
| 1783–87 | 6,228 | 5,271 | 84.6 |
| 1787 | 7,127 | 6,141 | 86.2 |
| 1788–92 | 7,669 | 6,122 | 79.8 |
| 1793 | 7,706 | 5,952 | 77.2 |
| 1794–96 | 9,025 | 6,590 | 73.0 |
| 1783–96 | 7,449 | 5,906 | 79.3 |
| | *9,109 | *5,991 | *65.8 |

*when sailcloth is added

period 1764–1782.[16] Although Archangel exported an appreciable amount of sailcloth as well as other linen products, nearly all of Great Britain's Russian sailcloth came from St. Petersburg/Kronstadt.[17]

The prices of these individual linens increased but fluctuated widely during the period 1783–1791. Compared with 1768–1779, their annual average exported value from St. Petersburg/Kronstadt increased significantly—by 84 percent in all ships and by 70 percent in British ships. As a percentage of the total value of commodity exports, however, the value of these linens did not increase above the level of 1768–1779: in all ships it was the same at 9.6 percent, and in British ships it actually declined from 8.1 to 7.4 percent. When the value of sailcloth is included there was a further decline. As a percentage of the total value of commodity exports by British merchants, exported Russian linens transported in British

---

16. For Great Britain's sailcloth imports see SAILCLOTH, PRO BT 6/240 and 230. For Russian sailcloth exports to Great Britain see the records of the St. Petersburg Custom House, PRO FO 65/12–15, 17–18, 20, 23–24, 26, 33, 35, 50, 52; FO 97/340–41; and BT 6/141, 231–33, 266.
17. For sailcloth and linen exports from Russian ports other than St. Petersburg/Kronstadt during the period 1783–1796, see the records of the St. Petersburg Custom House, PRO BT 6/231–33, 266; FO 65/13–15, 33; and FO 97/340.

## TABLE L-VI. MINIMUM ESTIMATE OF THE VALUE OF ST. PETERSBURG'S COMMODITY AND SELECTED LINEN EXPORTS: 1783–1791, 1793[18]

[Broad and Narrow Diaper and Linen, Crash, Drilling, Flem, and Ravenduck]
[Exclusive of Custom Duties and Sundry Charges]
(Annual Average, Measured in Current Rubles, in Thousands)

| Years | Value of Commodity Exports | Value of Linen Exports All Ships | Percentage | British Ships | Percentage |
|---|---|---|---|---|---|
| 1783–87 | 13,193 | 1,092 | 8.3 | 895 | 6.8 |
| 1788–91 | 20,942 | 2,214 | 10.6 | 1,662 | 7.9 |
| 1793 | 23,758 | 2,993 | 12.6 | 2,241 | 9.4 |
| 1783–91 | 16,637 | 1,591 | 9.6 | 1,236 | 7.4 |
|  |  | *2,042 | *12.3 | *1,261 | *7.6 |

| Years | Value of Commodity Exports by British Merchants | Percentage Value of Linen Exports in British Ships |
|---|---|---|
| 1783–87 | 8,702 | 10.3 |
| 1788–91 | 8,894 | 18.7 |
| 1793 | 14,397 | 15.6 |
| 1783–91 | 8,788 | 14.1 |
|  |  | *14.3 |

*when sailcloth is added

ships were 14.1 during the period 1783–1791, slightly higher than in 1768–1779 (see Figure 14, Table L–VI).[18]

## TIMBER

The challenge to researching and analyzing the Anglo-Russian trade in timber during the period 1783–1796 is similar to the one

---

18. Table L-VI is compiled and computed from the records of the St. Petersburg Custom House, PRO BT 6/141, 231–33, 266; FO 97/340–41; and FO 65/13–15, 17–18, 20, 26; and VTR, *Vedomosti*, Nos. 7–8. The fluctuating current prices of Russian linen per arshin were: crash, from a low 3.5 kopeks (1783–1786) to a high of 7 kopeks (1793); narrow linen, from a low of 7 kopeks (1783–1784) to a high of 11 kopeks (1791); broad linen, from a low of 11 kopeks (1783, 1785–1786) to a high of 15 kopeks (1790, 1793); broad diaper, from a low of 11.5 kopeks (1784–1787) to a high of 18 kopeks (1791); narrow diaper, from a low of 8 kopeks (1783–1785, 1787) to a high of 13.5 kopeks (1790); and per piece, ravenducks, from a low of 6 rubles 50 kopeks (1783, 1786) to a high of 14 rubles (1793); drillings, from a low of 4 rubles 50 kopeks (1783) to a high of 25 rubles (1789–1790); flems, from a low of 9

for linens. In addition to the problem created by the lack of comprehensive archival data, the exporting and importing ports did not employ the same nomenclature to rate the size, shape, and quality of timber products. It is therefore difficult to correlate the volume of all the timber products in the Anglo-Russian trade and even more difficult to establish monetary values that are meaningful.

Moreover, although the available data for several timber products discussed below are from authentic sources, they are derived from disparate archival sources and occasionally do not match perfectly. In at least one instance regarding Great Britain's importation of sawn boards, there is a wide margin of disagreement in the sources. Nonetheless, one source in particular facilitates the comparison of the foreign suppliers of several timber products imported into Great Britain. In 1835 a Parliamentary Committee reviewed and reported on the probable impact that the contemporary international timber trade might have on the British market, and collected trade statistics from the 1780s and 1790s for that purpose. These data are from the Custom House, London, and for the most part are supported by customs data from the St. Petersburg Custom House. The Parliamentary Committee report also includes valuable testimony from reliable people in the British timber trade.

During the 1780s and 1790s Riga continued to enjoy the reputation for the quality of the woods it exported to Great Britain, among them, the "Riga Wainscot" and the "Riga Great Masts." Subsequent to the Anglo-Dutch War Riga supplanted Holland and Ostend to become the leading supplier of wainscot logs to Great Britain. During the period of 1785–1796 almost all of Riga's annual average 20,000 wainscot logs were shipped to British ports.[19]

Riga was also the largest Russian exporter of squared logs of fir timber to Great Britain. Prior to the first partition of the Polish-Lithuanian Commonwealth and even after Memel [Klaipéda] had superseded it during the mid-1770s, Riga shipped an annual average of 21 percent of its exported 69,000 logs to Great Britain during the wartime period 1778–1782.[20] And between 1783 and 1796 Riga shipped about 30 percent of its annual average 60,000 ex-

---

rubles 50 kopeks (1785) to a high of 22 rubles (1793); and the most expensive, sailcloth, from a low of 80 rubles (1787) to a high of 160 rubles (1793).

19. For Riga's exports of wainscot logs see P. P. Timber, 1835, Appendix 1, p. 379; and the records of the St. Petersburg Custom House, PRO BT 6/231–33, 266; FO 97/340; and FO 65/13–15, 33.

20. See Chap. VI, Timber, supra.

## TABLE T-V. GREAT BRITAIN'S IMPORTS OF "FIR TIMBER EIGHT INCHES SQUARE AND UPWARDS" FROM 1788 THROUGH 1796[22]

(Annual Average, Measured in Loads)

| Years | Total Imports | From Prussia | % | From Norway | % | From Russia | % |
|---|---|---|---|---|---|---|---|
| 1788 | 205,087 | 161,282 | 78.6 | 20,329 | 9.9 | 20,355 | 9.9 |
| 1788–92 | 223,959 | 169,518 | 75.7 | 25,470 | 11.4 | 22,646 | 10.1 |
| 1793 | 199,963 | 137,092 | 68.6 | 30,403 | 15.2 | 19,932 | 10.0 |
| 1794–96 | 175,772 | 126,592 | 72.0 | 25,652 | 14.6 | 18,917 | 10.8 |
| 1788–96 | 205,230 | 151,606 | 73.9 | 26,079 | 12.7 | 21,101 | 10.3 |

ported logs to British ports.[21] But during the 1780s and 1790s it was Prussia that dominated the trade to Great Britain of "fir timber eight inches square and upwards," and from 1788 through 1796 it supplied an annual average of 151,606 loads of these logs, that is, nearly 74 percent of the total 205,230 loads Great Britain imported from all countries. Norway was a distant second with almost 13 percent and Russia was third with slightly more than 10 percent (see Table T–V).[22]

Although Norway continued to supply Great Britain with most of its ship masts, the Russian or Riga mast was renowned for its best-quality larger masts, the Great Masts, exported to Great Britain for its largest ships. In British customs import data during the 1780s the designation of "Masts Great . . . Middle . . . Small" (without diameter dimensions) gives way to diameter inches of: "Masts 6 to 8 . . . 8 to 12 . . . 12 Upwards." The latter "12 Upwards" is then the Great Mast. Moreover, until 1795 British customs recorded imported masts by the piece but from 1795 on masts were recorded by the load (the latter however can be converted into pieces). Russian customs records provide even less information about the size of masts exported.[23]

---

21. For Riga exports of squared logs of fir timber see P. P. Timber, 1835, Appendix 1, p. 379; and records of the St. Petersburg Custom House, PRO BT 6/231–33, 266; FO 97/340; and FO 65/13–15, 33.
22. Table T-V is compiled and computed from P. P. Timber, 1835, Appendix, No. 5, p. 384. For a discussion of the Baltic square fir timber log trade for selected years during this period, see Sven-Erik Åström, "North European timber exports to Great Britain, 1760–1810," in *Shipping, Trade and Commerce. Essays in Memory of Ralph Davis*, ed. P. L. Cottrell and D. H. Aldcroft (Leicester: Leicester University Press, 1981).
23. Cf., annual mast entries in CU 3; "An Account of the Quantity . . . of the principal Articles . . . imported into England [Scotland] from Russia, every fifth Year, between the 5th January, 1765 and the 5th January. 1786. . . ." PRO BT 6/240; "An Account of the Quantity

**Figure 15**

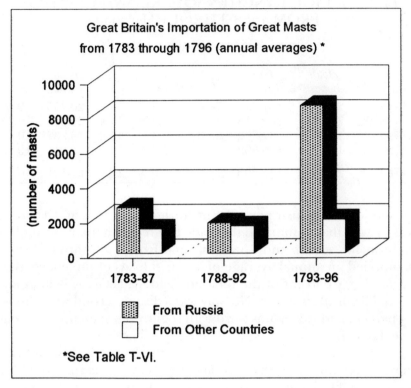

In 1783, when peace was restored, Great Britain's store of Great Masts was immediately re-supplied. Britain procured more than 11,000 Great Masts from all sources, more than twice the annual average number it had obtained during the wartime years 1778–1782. Nearly 70 percent of them came from Russia. During the period 1783–1787 the annual average number declined to four thousand, with 65 percent of them coming from Russia. During 1788–1792 the number contracted further to slightly more three thousand, with about 52 percent coming from Russia. But, in 1794, the year after war broke out with France, Great Britain imported slightly more than 5,700 Great Masts from all sources, more than double the number procured in 1793, with Russia sup-

---

of Timber imported into Great Britain since the Year 1783...." PRO BT 6/241; "An Account of the Goods, Wares, and Merchandize, imported into England [Scotland] from Russia from the 5th January 1786 to the 5th January 1800...." PRO BT 6/230; Elizabeth Boody Schumpeter, Table XVII, "Quantities and Value of Selected British Imports from 1772 to 1808," in *English Overseas Trade Statistics 1697–1808* (Oxford: Clarendon Press, 1960); and Chap. VI, Timber, supra.

## TABLE T-VI. APPROXIMATE NUMBER OF GREAT MASTS IMPORTED INTO GREAT BRITAIN FROM 1783 THROUGH 1796[24]

(Annual Average, Measured in Pieces)

| Years | Total Imports From All Sources | Total Imports From Russia | % |
|---|---|---|---|
| 1783 | 11,310 | 7,904 | 69.9 |
| 1783–87 | 4,062 | 2,645 | 65.1 |
| 1787 | 2,425 | 1,692 | 69.8 |
| 1788 | 2,863 | 1,497 | 52.3 |
| 1788–92 | # 3,345 | 1,764 | 52.7 |
| 1793 | 2,496 | * 1,437 | 52.6 |
| 1794 | 5,706 | 4,087 | 71.6 |
| 1795 | 12,247 | 10,781 | 88.0 |
| 1796 | 21,419 | 17,739 | 82.8 |
| 1783–96 | 5,630 | 4,006 | 71.2 |

\# 1789–1791 = not including Scotland
\* 1793 = not including Scotland

plying about 72 percent of them. As the war progressed, Great Britain's strategic requirement in Great Masts increased spectacularly and so did Russia's supply of them. In 1795 Britain imported more than 12,000 Great Masts from all sources with more than 10,000 of them coming from Russia for 88 percent of the total. In 1796 Britain imported more than 21,000 Great Masts and more than 17,000 came from Russia for almost 83 percent of the total. Overall, between 1783 and 1796 Great Britain procured an annual average of more than 5,600 Great Masts from all sources, of which 4,000 came from Russia for 71 percent of the total (see Figure 15, Table T–VI).[21]

Although Great Britain continued to import most of its sawn boards from Norway there is a problem with the data supporting its share of that trade. According to the information in the Parliamentary Committee report for the period 1788–1796, an annual average of 67 percent of Great Britain's imported "deals and deal ends" and "battens and batten ends" came from Norway, amounting to an annual average of 49,407 Hundreds or 5,928,840 pieces (one Hundred = 120 pieces). Another 22 percent came

---

24. Table T-VI is compiled and computed from ibid., and MASTS, PRO BT 6/240 and 230; and the records of the St. Petersburg Custom House, PRO FO 65/12–15, 17–18, 20, 23–24, 26, 33, 35, 50, 52; FO 97/340–41; and BT 6/141, 231–33, 266.

### TABLE T-VII. GREAT BRITAIN'S IMPORTS OF "DEALS AND DEAL ENDS" AND "BATTENS AND BATTEN ENDS" FROM 1788 THROUGH 1796[25]

(Annual Average, Measured in Hundreds)

| Years | Total Imports | From Norway | % | From Russia | % |
|---|---|---|---|---|---|
| 1788 | 73,129 | 51,683 | 70.7 | 14,573 | 19.9 |
| 1788–92 | 75,900 | 48,701 | 64.2 | 16,659 | 21.9 |
| 1793 | 72,178 | 45,740 | 63.4 | 17,734 | 24.6 |
| 1794–96 | 70,336 | 48,472 | 68.9 | 14,936 | 21.2 |
| 1788–96 | 73,713 | 49,407 | 67.0 | 16,204 | 22.0 |

from Russia and amounted to 16,204 Hundreds or 1,944,480 pieces (see Table T–VII).[25]

But according to the data from records of the St. Petersburg Custom House the Russian share is understated in the British Parliamentary Committee report. Between 1783 and 1796 the annual average number of exported sawn boards, referred to simply as deals, from St. Petersburg/Kronstadt was 11 times greater than it was during the previous period 1763–1782. This is not in doubt. An annual average of 95 percent of the total 2.4 million (20,146 Hundreds) exported deal boards were transported in British ships, nearly all of which were destined for British ports. And this is not in doubt. But compared to the Parliamentary Committee report during the period 1788–1796 the St. Petersburg records show an excess of an annual average of about 900,000 sawn boards that were not recorded as having been imported into Great Britain from Russia.

While the magnitude of such a discrepancy is the first of its kind encountered in Russia's trade with Great Britain, there are reasons why the Russian sawn boards might not have found their way into British customs data. The most obvious one is expressed by Ralph Davis: "Timber came in a large number of standard categories.... A hundred deals contained 120 pieces of wood, and the dimensions of each piece might be as little as 8 ft. x 7 in. x 1/2 in.

---

25. Table T-VII is compiled and computed from P. P. Timber, 1835, pp. 385–86. For a discussion of the Baltic trade in sawn timber boards for selected years during this period see Åström.

or as much as 20 ft. x 11 in. x 3 1/4 in."[26] That is, the number of pieces, however measured or computed in St. Petersburg's export data, might not have been recorded in the same way in British customs import data. We know from the 1767 Russian tariff that Russian deals were exported in a variety of sizes and the tariff actually provides a conversion table showing how many deals by such and such dimensions were the equivalent of 100 deals by such and such dimensions.[27] Moreover, in his study on Norwegian timber exports, Steiner Kjaerheim provides a cautionary alert to this problem when he states: "All data are here given in 'pieces' rather than deals, since cargoes were made up of deals as deals, but the term was also used to indicate the number of planks, one plank counting as two deals."[28] But there are also known cases of fraudulent practices and smuggling in the Norwegian timber trade with Great Britain and this could also have happened in the Russian trade. For example, H. S. K. Kent describes incidents where so-called Norwegian timber destined for Great Britain had originated elsewhere in the Baltic "under coloured papers"; where "Direct smuggling of Norwegian timber is a further factor modifying calculations of imports"; and where "Norwegian timber entering England unrecorded in the customs ledgers was due to the policy of bringing up neutral timber-ships bound for enemy ports, and purchasing their cargo for the use of the royal dockyards. Evidence suggests that the quantities were very considerable."[29] Therefore, if the approximate 900,000 Russian sawn boards in fact entered Great Britain, but for some reason had not been recorded as imported and were not included in the Parliamentary Committee report in 1835 as coming from Russia, the Russian share of Great Britain's total imported "deals and deal ends" and "battens and batten ends" during the period 1788–1796 would increase from 22 percent to 32 percent (cf. Tables T–VIII and VII).[30]

Once again, from 1783 through 1796, and for all the reasons

---

26. Ralph Davis, *The Rise of the English Shipping Industry* (Newton Abbot, Great Britain: David & Charles, 1972), p. 182.
27. TARIFF, ff. 259–62, especially f. 262.
28. Steiner Kjaerheim, "Norwegian Timber Exports in the 18th Century: A Comparison of Port Books and Private Accounts," *The Scandinavian Economic History Review* 5:2 (1957): pp. 188–201, quotation on p. 191.
29. H. S. K. Kent, "The Anglo-Norwegian Timber Trade in the Eighteenth Century," *Economic History Review*, 2nd series, 8:1 (Aug. 1955): pp. 62–74, quotations on pp. 65–66.
30. Table T-VIII is compiled and computed from the records of the St. Petersburg Custom House, PRO FO 65/12–15, 17–18, 20, 23–24, 26, 33, 35, 50, 52; FO 97/340–41; and BT 6/141, 231–33, 266.

## TABLE T-VIII. ST. PETERSBURG'S DEAL EXPORTS FROM 1783 THROUGH 1796[30]

(Annual Average, Measured in Pieces and Hundreds)

| Year | Total Exports in All Ships | Total Exports in All Ships | Total Exports in British Ships | Total Exports in British Ships | % |
|---|---|---|---|---|---|
| | Pieces | Hundreds | Pieces | Hundreds | |
| 1783 | 1,075,406 | 8,962 | 1,012,148 | 8,435 | 94.1 |
| 1783–87 | 1,428,685 | 11,906 | 1,348,521 | 11,238 | 94.4 |
| 1787 | 1,492,491 | 12,437 | 1,352,925 | 11,274 | 90.6 |
| 1788 | 2,339,065 | 19,492 | 2,248,086 | 18,734 | 96.1 |
| 1788–92 | 2,864,069 | 23,867 | 2,720,228 | 22,669 | 95.0 |
| 1793 | 3,725,258 | 31,044 | 3,615,292 | 30,127 | 97.0 |
| 1794–96 | 2,885,423 | 24,045 | 2,793,058 | 23,275 | 96.8 |
| 1783–96 | 2,417,521 | 20,146 | 2,309,872 | 19,249 | 95.5 |
| 1788–96 | 2,966,875 | 24,724 | 2,843,956 | 23,700 | 95.9 |

stated earlier in Chapter VI, which are even more applicable during the post Independence period of America, Great Britain demonstrated how indispensable the timber resources of Russia were in sustaining its economic growth in the building trades, in shipbuilding and, especially during wartime, its strategic military security.

# Chapter XIII
# The Material of Culture

Each year during Catherine II's reign Russians regularly imported from overseas hundreds of different commodities that can be described as necessities, decencies, and luxuries. The greater number of them were far more expensive than most of the commodities Russia exported. The list of these imports reads like a consumers' guide to what the wealthiest Russians deemed indispensable to their lifestyle, which, for the lack of a more attractive if not a more accurate expression, has been termed "Westernized." Although the notion of being "Westernized" involved much more than the simple transfer and acquisition of material from abroad (even if much of it did not originate in Western Europe), to be "Westernized" however surely meant much more than the simple transfer and acquirement of the manners, the ideas and the *mentalité* that the elites and the wealthy in the West practiced and propagated.

The extravagant expenditure on these imports reflected the socio-cultural values of the wealthiest Russians as much as it did the economic and commercial imperatives of the Empire. In addition to those items that were either consumed or worn soon after their arrival, there were several other imports that acted as economic multipliers. These imports either sustained or created business transactions in the wholesale and retail trades and were the inputs in the crafts, the arts, and the manufactories of Russia. In this manner the cultural influence of this foreign merchandise trickled down to the people.

This analysis of Russia's overseas import trade focuses primarily on the value of its merchandise and to a lesser degree on its volume or weight. This approach, it seems, provides greater insights into the socio-economic-cultural makeup of Russia's wealthiest elites. It also informs us about Russia's maritime trade. By its nature the composition of Russia's import trade in contrast to its export trade, its comparatively higher ratio of value to volume or weight per item, required far less tonnage and ship space, and thereby fewer ships, than the export trade. Moreover, because only several Russian exports were valued as high per unit as were so many of Russia's imports, an enormous amount of export tonnage was required to pay for those imports. When this could not be achieved precarious trade balances resulted in 1790, 1791, and 1792, which contributed to the commercial and economic crisis in 1793.

This analysis also focuses on the overseas imports into St. Petersburg/Kronstadt and closely examines its commodity composition and value, although it should be noted that other Russian ports imported essentially the same range of commodities as the capital port.[1] The reasons for doing this are that the extant data for this port are far more plentiful than for any other port of the Russian Empire and also that St. Petersburg/Kronstadt accounted for an annual average of 79 percent of the total value of overseas imports of the principal northern ports during the latter part of Catherine II's reign.[2]

The wide range of merchandise the Russians imported into St. Petersburg/Kronstadt is detailed among two dozen extant annual data lists during the period 1763–1796. However, not all of the lists are complete either in terms of the ships that transported those imports or comprehensive in listing both volumes and values per commodity imported. From the data available the total expenditure for individual commodity imports for several years which were representative of Russia's overseas commerce (1764, 1766, 1783, 1787, 1792, 1796) can be calculated, but, because of the lack of comprehensive data, only the minimum expenditure can be stated for the period 1768–1779. Several imports are clustered either because they appeared that way in the archival documents themselves or because of their similarity of purpose. Although most merchandise was imported under the same nomenclature on both non-British and British ships, occasionally their quality and consequently their value per unit varied from each other. The value of imported merchandise was recorded at the St. Petersburg Custom House by weight, volume, size, or piece per unit and by lump sum ruble value.[3]

---

1. For data on the imports into Archangel, Fredrickshamn, Narva, Pernau, Reval, Riga, and Viborg, see the records of the St. Petersburg Custom House, PRO SP 91/73, 75, 107; BT 6/231–33; CO 388/95; and FO 65/33, 48;
2. A detailed discussion of the overall monetary value of Russia's overseas commodity imports is provided in Chap. XIV.
3. Compiled and computed from the records of the St. Petersburg Custom House PRO SP 91/73, 75, 78–79, 94, 96, 107; CO 388/54, 95; BT 6/231–33; FO 65/3, 16; FO 97/340; Whitworth's report to Grenville on sugar imports from 1788 through 1791 taken from St. Petersburg Custom House, Sept. 30, 1794, No. 53, FO 65/28; and Mr. Eton's report on

Imports are arbitrarily divided into eight major merchandise categories to focus attention on what the wealthiest Russian elites chose to buy with their money: textiles and apparel; groceries; wines and spirits; colors and dyes; metals; drugs and medicines; minerals, salts and substances; and toys and trinkets (which can be characterized as knick knacks). However, several items served more than one purpose, for example as a color dye, as a spice or a condiment, or as a medicine, and therefore could have been included in another category. Moreover, a few of the categories have an affinity for each other and could have been amalgamated. For example, colors and dyes and the mordants among the minerals, salts and substances are frequently used together in the processing of textiles; groceries marry well with wines and spirits; and certain minerals, salts, and substances are used as drugs and medicines. Between 1764 and 1796 textile and apparel merchandise commanded about 38 percent of the total expenditure on imports; groceries about 25 percent; wines and spirits and colors and dyes each slightly more than 8 percent; metals slightly more than 2 percent; and drugs and medicines, minerals, salts, and substances, and toys and trinkets each slightly more than 1 percent (see Table CI–I).[4]

There were only 13 items that appeared among the top 10 most valued commodities imported into St. Petersburg/Kronstadt in the years 1783, 1787, 1792, and 1796. Among them, cottons, linens, silks and woolens figured prominently but sugar consistently ranked second in each of those years. These were followed by indigo, French brandy and wine, coffee, fresh and dried fruit, Spanish and Portuguese wines, olive oil, cochineal, and tea (see Table CI–II).[5]

Textile imports into St. Petersburg/Kronstadt alone led the market with approximately 34 percent of the overall value of commodity imports between 1764 and 1796 (in particular years, 1783 = 34%, 1787 = 33%, 1792 = 33%, 1796 = 36%). Although

---

the "Importation of British Manufactured Goods in 1792," FO 97/342. For a discussion of intercontinental trade dating from the preceding period but bearing on Russia's commodity imports such as coffee, logwood, pepper, sugar, fine spices and textiles, see Niels Steensgaard, "The Growth and Composition of the Long-Distance Trade of England and the Dutch Republic before 1750," *The Rise of Merchant Empires. Long-Distance Trade in the Early Modern World 1350–1750*, ed. James D. Tracy (Cambridge: Cambridge University Press, 1990), pp. 102–52.
4. Sources for Table CI-I are compiled and computed from the records of the St. Petersburg Custom House cited in footnote 3 above; and Tables MN-X, Chap. X, supra, and VRT II, Chap. XIV, below.
5. Table CI-II, ibid.

## TABLE CI-I. VALUE OF SELECTED CATEGORIES OF IMPORTS AS PERCENTAGE OF TOTAL VALUE OF COMMODITY IMPORTS INTO ST. PETERSBURG: 1764-1796[4]

[Exclusive of Custom Duties, Sundry Charges, and Specie Imports]
(Annual Average, Measured in Current Rubles, in Thousands)

| Years | Total Imports | Textiles and Apparel | % | Groceries | % |
|---|---|---|---|---|---|
| 1764/66 | 5,472 | 2,227 | 40.7 | 1,069 | 19.5 |
| 1768-79 | 7,310 | 2,370 | 32.4 | 934 | 12.8 |
| 1783 | 11,674 | 4,768 | 40.8 | 2,651 | 22.7 |
| 1787 | 15,314 | 5,935 | 38.8 | 4,366 | 28.5 |
| 1792 | 23,296 | 8,797 | 37.8 | 5,423 | 23.3 |
| 1796 | 26,356 | 10,063 | 38.2 | 7,858 | 29.8 |

| Years | Wines and Spirits | % | Colors and Dyes | % |
|---|---|---|---|---|
| 1764/66 | 508 | 9.3 | 589 | 10.8 |
| 1768-79 | 462 | 6.3 | 536 | 7.3 |
| 1783 | 950 | 8.1 | 999 | 8.6 |
| 1787 | 1,068 | 7.0 | 1,298 | 8.5 |
| 1792 | 2,524 | 10.8 | 1,625 | 7.0 |
| 1796 | 2,078 | 7.9 | 2,587 | 9.8 |

Russia was itself a significant linen producer and although it exported large amounts of coarse linens, in particular to Great Britain, in the latter part of Catherine II's reign it also became a significant importer of quality linens. Russia imported more than two million rubles of linens from overseas in 1787, more than four million in 1792 and more than six million in 1796. The British had to compete with the French, Dutch, and other European suppliers. The value of British linens imported into St. Petersburg/Kronstadt, in 1792, for example, accounted for about half the total value of imported linens. The British dominated overseas cotton imports into St. Petersburg/Kronstadt but while their value increased—one million rubles in 1787, 1.5 million in 1792, and 1.6 million in 1796—compared with the value of linen imports cottons lost ground.[6] During the period 1764-1796, the value of imported

---

6. See ibid. For Great Britain's exports of linens, cottons, and silks to Russia see "An Account of the [British] Goods . . . to Russia . . ," BT 6/232, ff. 12-16, 23-25; and BT 6/230, ff. 55-60; "An Account of the Goods, Wares & Merchandize being Foreign Produce or

## TABLE CI-I. Continued

| Years | Metals | % | Drugs and Medicines | % |
|---|---|---|---|---|
| 1764/66 | 135 | 2.5 | 113 | 2.1 |
| 1768–79 | 206 | 2.8 | 18 | .2 |
| 1783 | 255 | 2.2 | 101 | .9 |
| 1787 | 457 | 3.0 | 280 | 1.8 |
| 1792 | 401 | 1.7 | 349 | 1.5 |
| 1796 | 603 | 2.3 | 424 | 1.6 |

| Years | Minerals, Salts, and Substances | % | Toys and Trinkets | % |
|---|---|---|---|---|
| 1764/66 | 103 | 1.9 | 96 | 1.7 |
| 1768–79 | 43 | .6 | 75 | 1.0 |
| 1783 | 143 | 1.2 | 134 | 1.2 |
| 1787 | 314 | 2.1 | 281 | 1.8 |
| 1792 | 221 | .9 | 225 | 1.0 |
| 1796 | 434 | 1.6 | 194 | .7 |

silks more than tripled: in 1792, it peaked at a little over 900,000 rubles, but fell back to 550,000 rubles in 1796. Great Britain did not acquire a significant niche in Russia's overseas imports of silks.

The value of woolen imports, which had led all other textiles during the first two decades of Catherine's reign and amounted to 2.4 million rubles in 1783, lagged behind both imported linens and cottons in 1787 and in 1792. The 1793 tariff prohibitions had significantly curtailed British imports of woolens; in 1796 its value declined to less than 800,000 rubles.[7] In terms of quantity, woolens imported into the capital port between 1763 and 1779 amounted to an annual average of approximately 1.5 million arshins or 1.2 million yards. Most of it came from Great Britain in British ships. In 1780 during the height of the war British ships transported nearly 2 million arshins of woolens (1.5 million yards) to St. Petersburg/Kronstadt. But soon after the peace was restored in 1783 the

---

Manufacture Exported from England [Scotland] to Russia between 5th January 1764 and 5th January 1786 [and between 1786 and 1800] . . ," BT 6/232, ff. 0–4, 22–23; BT 6/230, ff. 61–64, 67–68; and see CU 3, 14 and, 17 for these years.

7. Ibid. For Russia's domestic production of textiles during Catherine II's reign, see the survey by G. S. Isaev, *Rol' tekstil'noi promyshlennosti v genezise i razvitii kapitalizma v Rossii 1760–1860* (Leningrad: Nauka, 1970), pp. 118–78, passim.

## TABLE CI-II. MOST VALUED COMMODITIES IMPORTED INTO ST. PETERSBURG: 1783–1796[5]

[Exclusive of Custom Duties and Sundry Charges]
(Measured in Current Rubles, in Thousands)

| Rank | 1783 | 1787 | 1792 | 1796 |
|---|---|---|---|---|
| First | Woolens 2,367 | Linens 2,118 | Linens 4,327 | Linens 6,583 |
| Second | Sugar 1,639 | Sugar 1,969 | Sugar 2,849 | Sugar 5,381 |
| Third | Silks 732 | Cottons 1,165 | Cottons 1,460 | Indigo 1,600 |
| Fourth | Linens 641 | Woolens 1,115 | French Wine and Brandy 1,181 | Cottons 1,599 |
| Fifth | Indigo 600 | Silks 760 | Woolens 1,119 | Coffee 834 |
| Sixth | French Wine and Brandy 432 | French Wine and Brandy 651 | Silks 908 | Spanish Wine 816 |
| Seventh | Cottons 382 | Indigo 595 | Indigo 776 | Woolens 778 |
| Eighth | Coffee 249 | Coffee 558 | Fruit 746 | Silks 550 |
| Ninth | Fruit 241 | Fruit 425 | Spanish Wine 580 | Portuguese Wine 523 |
| Tenth | Cochineal 220 | Tea 377 | Olive Oil 567 | Cochineal 484 |

quantity of woolen imports began falling: 1783 = 3.1 million arshins (2.4 million yards); 1787 = 2.9 million arshins (2.2 million yards); 1792 = 2.4 million arshins (1.9 million yards); 1796 = 1.1 million arshins (.9 million yards).[8]

The following table represents the approximate total value

---

8. See sources compiled and computed from the records of the St. Petersburg Custom House cited in footnote 3 above. For British woolen exports to Russia, see "An Account of the Quantity and Value of all Woollen Goods exported from England to Foreign Kingdoms and States of Europe in the Years 1764, 1765, 1766 and 1785, 1786, 1787, distinguishing each Species of Goods and the Countries to which exported . . ," BT 6/241, f. 73; "An Account shewing the Increase or Decrease of the Quantity and Value of Woollen Goods exported to all Parts upon a Comparative View of the Years 1764, 1765 & 1766 with 1785, 1786 & 1787 . . ," BT 6/240, ff. 146–47; "An Account of the Goods, Wares and Merchandize being British Produce or Manufacture exported from England [Scotland] to Russia between 5th January 1764 and 5th Jan. 1786 . . ," BT 6/232, ff. 12–16, 23–25; "An Account of the Goods, Wares and Merchandize being British Produce or Manufacture exported from England [Scotland] to Russia from the 5th January 1786 to the 5th January 1800 . . ," BT 6/230, ff. 55–60; and see CU 3, 14 and 17 for these years.

### TABLE CI-III. ST. PETERSBURG'S OVERSEAS TEXTILE IMPORTS: 1783–1796[9]

[Exclusive of Custom Duties and Sundry Charges]
(Measured in Current Rubles; Arshins and Equivalent Yards, in Thousands)

|         | 1783  | 1787  | 1792  | 1796  |         |
|---------|-------|-------|-------|-------|---------|
| Cottons | 382   | 1,165 | 1,460 | 1,599 | Rubles  |
|         | 764   | 3,020 | 1,310 | 3,868 | Arshins |
|         | 595   | 2,349 | 1,019 | 3,009 | Yards   |
| Linens  | 641   | 2,118 | 4,327 | 6,583 | Rubles  |
|         | 2,226 | 1,188 | 2,131 | 3,433 | Arshins |
|         | 1,731 | 924   | 1,658 | 2,670 | Yards   |
| Silks   | 732   | 760   | 908   | 550   | Rubles  |
|         | 652   | 456   | 223   | 40    | Arshins |
|         | 507   | 355   | 174   | 31    | Yards   |
| Woolens | 2,367 | 1,115 | 1,119 | 778   | Rubles  |
|         | 3,120 | 2,891 | 2,435 | 1,103 | Arshins |
|         | 2,427 | 2,248 | 1,894 | 858   | Yards   |

and quantity of these four imported textiles into St. Petersburg/Kronstadt in the years 1783, 1787, 1792, and 1796. Imported textiles at the Custom House there were almost always recorded in rubles and arshins, but they were also denoted in pieces, poods, and in lump sums of rubles. This was frequently the case with imported silks, which prevents us from knowing the quantity of silk that was imported (see Table CI–III).[9]

British manufactures comprised about 71 percent, and foreign merchandise comprised about 29 percent, of the total value, in English pounds sterling ("official values"), of Great Britain's exports to Russia during the periods 1769–1774 and 1784–1789.[10] In addition to textiles, which undoubtedly accounted for the largest share of the value of manufactures, the British exported dozens of other important commodities that either came directly from Great Britain or were reexported to Russia in British ships.

---

9. Sources for Table CI-III are compiled and computed from the records of the St. Petersburg Custom House cited in footnote 3 above.

10. Compiled and computed from "A Comparative View of the Value of the British Manufactures Exported from Great Britain to all parts in the following Years distinguishing each Year and the Countries," and ". . . also of the Foreign Merchandize . . ," PRO BT 6/241, ff. 93, 150–52, and SOT ff. 94, 180. For itemized lists of exports from England and Scotland of both British and non-British merchandise to Russia from 1764 through 1799, see the several archival documents in above footnotes entitled "An Account of the Goods, Wares and Merchandise . . . to Russia," and see CU 3, 14 and 17 for these years.

For example, among the metals more than 70 percent of the lead and more than 90 percent of the tin bars imported into St. Petersburg/Kronstadt arrived in British ships. The Russians also spent hundreds of thousands of rubles annually on imported furs and skins, about 80 percent of which were beaver and otter skins that the British brought to them.[11]

Textile imports had a direct influence on the importation of dyes and the mordants (from among the minerals, salts, and substances) that were indispensable to the processing of those fabrics. During the second half of the eighteenth century all colored fabrics were dyed with natural dyes, which, although time-consuming and labor-intensive, suited Russia's social and economic structure. Most fabric dyeing frequently involved a two-step process in which the fiber had to be prepared by a mordant, a metallic salt that combined chemically with both the fiber and the dye—the process of mordanting—so that the fabric could absorb the color of the dye. Without the mordant, the dyed fabric would wash out or fade. Because different mordants produce different shades of color from the same natural dye substance—a tin mordant brightens while an iron mordant darkens the color—the process of mordanting is crucial to the dyeing of fabrics. Moreover, an unevenly mordanted fabric could not be an evenly dyed fabric. Wool fibers are easy to mordant because they are porous, but those of cotton and linen are not; they inhibit the penetration of the mordant. Raw silk requires the removal of its sticky gum prior to mordanting.[12]

Russia's importation of foreign dyes and mordants was expensive when compared with Russia's commodity exports, but the dyes and mordants were indispensable to the conversion of imported textiles into attractive fabrics for consumption by the

---

11. See sources to Tables CI-I-III above.
12. For a survey of the history of dyes and mordants used in textile processing, see Agnes Geijer, *A History of Textile Art. A Selective Account* (1979; reprint, London: Pasold Research Fund in association with Sothby Parke Bernet, 1982); Stuart Robinson, *A History of Dyed Textiles* (London: Studio Vista Limited, 1969), especially pp. 20–32; idem, *A History of Printed Textiles* (Cambridge: M.I.T. Press, 1969); R. D. Harley, *Artists' Pigments c. 1600–1845. A Study in English Documentary Sources* (New York: Elsevier, 1970; London: Butterworth Scientific, 2nd ed., 1982); William F. Leggett, *Ancient and Medieval Dyes* (Brooklyn, NY: Chemical Publishing, 1944); J. J. Hummel, *Textile Fabrics and Their Preparation for Dyeing*, new and revised edition, ed. Paul N. Hasluck (London: Cassell, 1906); William Partridge, *A Practical Treatise on Dying of Woollen, Cotton, and Skein Silk with the Manufacture of Broadcloth and Cassimere including the Most Improved Methods in the West of England*, introduction by J. de Mann and technical notes by K. G. Ponting (Edington, Wiltshire: Pasold Research Fund, 1973); K. G. Ponting, "Logwood: An Interesting Dye," JEEH 2:1 (Spring 1973): pp. 109–19; and Arthur M. Wilson, "The Logwood Trade in the Seventeenth and Eighteenth Centuries," in *Essays in the History of Modern Europe*, ed. Donald C. McKay (New York: Harper & Brothers, 1936), pp. 1–15.

wealthiest elites who could afford them. The combined value of the dyes, the mordants, and the textiles imported into St. Petersburg/Kronstadt during the latter part of Catherine II's reign was equal to more than 40 percent of the total value of commodity imports.[13] Only a few of these dyes and mordants will be discussed here.

Alum (potassium aluminum sulphate) is the most common mordant used in dyeing. It is frequently combined with cream of tartar (purified argol), which is not strictly speaking a mordant itself but rather an assistant, a chemical that assists in creating the bonding between dye and fiber to brighten and even the color applied. About 21,000 poods of alum were imported annually into St. Petersburg/Kronstadt, about 22 percent of which came from Great Britain. Its price ranged mostly between two and three rubles per pood although in 1796 a pood of alum sold for nearly five rubles. Tartar/argol was twice as expensive as alum but only about one tenth as much was imported.[14]

Cochineal was one of the more expensive dye substances. Its import price in St. Petersburg/Kronstadt was frequently more than 200 rubles per pood, and in 1796 its price increased to 243 rubles. Although only about 1,000 poods were imported annually, of which about 15 percent arrived on British ships, cochineal ranked among the top 10 most valuable imports.[15] Cochineal is prepared from the dried bodies of the insects *Coccus cacti* which are found in South America. It is estimated that 50,000 insects weigh only two pounds and that 70,000 make one pound of cochineal.[16] Depending on the mordant used, cochineal produces a variety of reds, violets, and purples: the mordant alum produces magenta but alum with cream of tartar makes crimson; tin crystals (stannous chloride) and cream of tartar produce bright scarlet; and iron (copperas, green vitriol) makes purple-grey colors.

Curcuma (turmeric, korkum) is a yellow powder extracted from the root of the plant *Curcuma tinctoria,* which is found in eastern lands. Without a mordant it produces orange but will not remain fast. Although only about 1,600 poods per annum were imported during the latter part of Catherine II's reign, its price increased significantly from about 7.5 rubles in 1787 to about 21 rubles in 1796.[17]

---

13. See sources to Tables CI-I-III above.
14. Ibid.
15. Ibid.
16. Cited in R. D. Harley (1970), p. 125; Harley (1982), p. 136; and Robinson, *Dyed Textiles*, p. 25; but originally from Leggett, p. 84.
17. See sources to Tables CI-I-III above.

Indigo was the second most expensive dye imported. Its price increased steadily during Catherine's reign to reach 155 rubles per pood in 1796. About 5,000 poods of indigo were imported annually into St. Petersburg/Kronstadt, of which about 30 percent came from Great Britain, and it ranked high among the 10 most valued imports. Indigo comes from the plant *Indigofera tinctoria,* grown in the West Indies, India, Asia, and Egypt, and reaches the market as a blue powder which is insoluble in water. Indigo therefore requires a reducing agent to convert it into its leuco compound (Indigo White), making it soluble in water and allowing the fiber to absorb the Indigo Blue dye.[18]

Logwood is from the heartwood of *Hematoxylon campeachianum,* a tree grown in South America. Because it entered the market in large blocks, it was named logwood or blockwood, but it also became known as Campechy or St. Martin's Wood. These names must have denoted differences in quality, because the custom records of Archangel and St. Petersburg list logwood, Campechy, and St. Martin's Wood imports, which differ somewhat in price. At St. Petersburg/Kronstadt both logwood and St. Martin's Wood were priced at about two rubles per pood during the first two decades of Catherine's reign, but after 1783 the price of St. Martin's Wood increased to three and four rubles per pood. About 25,000 poods of both woods were imported annually into St. Petersburg/Kronstadt with modest amounts arriving on British ships. Logwood with alum produces a dark slate-blue and purplish blue; with tin and cream of tartar purple; and with iron, rich black which is especially attractive in silks and woolens.[19]

Madder is the dried plant root of *Rubia tinctorium* cultivated in France and especially in the Netherlands. With alum and cream of tartar it produces brownish red; with tin and cream of tartar, a bright red; and with iron, a purplish brown. A substantial amount of madder, about 13,000 poods, was imported annually during the latter part of Catherine's reign and its price fluctuated between 6 and 11 rubles per pood.[20]

Next to textiles sugar was the second most valuable commodity imported into St. Petersburg/Kronstadt in 1783, 1787, 1792, and 1796. In the latter year it accounted for almost 20 percent of the total value of commodity imports. Several kinds of sugar were imported: refined, raw or powdered, melis, candy, lump, and ad-

---

18. Ibid.
19. Ibid., and also Ponting and Wilson.
20. Ibid.

ditional sorted grades. When compared to the price of Russian commodity exports, the average price of each kind of sugar was expensive. For example, from 1764/66 to 1796 the price per pood of refined sugar increased from almost 7 to about 19 rubles; raw or powdered sugar from 3 to almost 12 rubles; candy from 10 to slightly more than 24 rubles; and melis from 9 to slightly more than 17 rubles. Although most of Russia's sugar imports originally came from the French West Indies, Brazil, the Danish islands, and the British West Indies were also sources of supply. As intermediaries, the Netherlands and Hamburg provided Russia with most of the French sugar. Although Russia had a small number of sugar refineries,[21] the refineries of the Netherlands and Hamburg supplied Russia with most of its imported refined sugar. The amount of sugar imported into St. Petersburg/Kronstadt was an annual average of about 89,000 poods in 1764/66, about 157,000 poods during 1768–1776, 260,000 poods in 1783, 335,000 poods during 1787–1792, and 300,000 poods in 1796. Modest amounts of these sugar imports were transported in British ships.[22] However, when France lost its source of sugar following the Haitian revolt and declared war against Great Britain in 1793, French sugar reexports plummeted. In 1792, French reexported sugar to Hamburg was less than half of what it had been in 1791; in 1793, it fell even further until it was hardly visible in 1794–96. British reexported sugar to Hamburg nearly trebled during the latter period and a significant amount of it was shipped to Russia.[23]

Much less coffee than sugar was imported into St. Petersburg/Kronstadt, but its price in the early decades of Catherine's reign was already substantial at 10 rubles per pood. During that time about 10,000 poods were imported annually, but in the latter part of her reign coffee imports more than tripled and its price almost

---

21. See Indova, ed., "O Rossiiskikh Manufakturakh vtoroi polovini XVIII v.," *Istoricheskaia geografiia Rossii XII–nachalo XX v. Sbornik statei k 70-letiiu professora Liubormira Grigor'evicha Beskrovnogo* (Moscow: Nauka, 1975), pp. 284–345, passim; and K. K. Zlobin, ed., "Vedomost' sostoiashchem v St. Peterburge fabrikam, manufakturam i zavodam na Sentiabr' 1794 g.," SIRIO (1867), Vol. I, pp. 352–61.
22. For sugar imports see sources compiled and computed from the records of the St. Petersburg Custom House cited in footnote 3 above; and "An Account of the Quantity of Refined Sugar Exported [from Great Britain in British Bottoms for the years 1783–1785]," BT 6/240, f. 143.
23. See Hamburg Custom House reports, Staatsarchiv Hamburg, 371–2, Admiralitäts-Kollegium, F 10, Bd. 1–10, especially Bd. 5–10. See also Robert Louis Stein, *The French Sugar Business in the Eighteenth Century* (Baton Rouge: Louisiana State University Press, 1988), p. 100, and passim. See also idem, "The French Sugar Business in the Eighteenth Century: A Quantitative Study," *Business History* 22:1 (Jan. 1980): p. 317; and idem, "The State of French Colonial Commerce on the Eve of the Revolution," JEEH 12:1 (Spring 1983): pp. 105–17.

doubled, making it one of the more expensive commodities. Only modest amounts were transported to St. Petersburg/Kronstadt in British ships.[24]

Both dried and fresh imported fruits were, along with orange and lemon juice, of considerable importance to Russians who could afford these luxuries, and all fruits together ranked among the higher priced imports. There was a wide variety of dried fruits which included apples, cherries, currants, dates, figs, pears, prunes and prunellos, and raisins. Taken together, an annual average of 72,000 poods of dried fruit alone was imported into St. Petersburg/Kronstadt during the years 1783, 1787, 1792, and 1796.[25]

During these years other foodstuffs such as olive oil were imported in considerable quantity (an annual average of 46,000 poods) and cost between 6 and 10 rubles per pood, putting them among the higher priced commodity imports entering St. Petersburg/Kronstadt. To the olive oil must be added 1,000 poods per annum of olives. Although certain spices and condiments were generally imported in smaller quantities they nevertheless commanded prices that were extraordinarily high per pood: ginger fluctuated between approximately 6 and nearly 18 rubles; mustard between about 8 and 18 rubles; pepper between about 12 and 19 rubles; nutmeg between 25 and 80 rubles; cinnamon between approximately 55 and 90 rubles; cloves between 62 and 127 rubles; and mace between 101 and 512 rubles.[26]

To wash all this down, the Russians imported considerable quantities of wines and spirits although Russia produced plentiful amounts of vodka. France led the way with the most valued brandy and wines—Bordeaux, Burgundy and Champagne. When Russia prohibited French imports in 1793 Spanish and Portuguese wines replaced the French products although they had earlier established their niche in the wine trade with Russia. Great Britain's contribution to the drinking habits of the Russians was almost as significant because hundreds of thousands of rubles worth of ale and porter entered St. Petersburg/Kronstadt annually.[27]

Imports from Great Britain had a cultural impact on Russians. "The English have replaced the French: nowadays women and men are falling over themselves to imitate anything English; every-

---

24. For coffee imports see sources compiled and computed from the records of the St. Petersburg Custom House cited in footnote 3 above.
25. For imports of dried and fresh fruit and juices see ibid.
26. For imports of olive oil, olives, spices and condiments see ibid.
27. For imports of wines and spirits see ibid.

thing English now seems to us good and admirable and fills us full of enthusiasm." With these words by Nikolay Novikov in 1772, Anthony G. Cross introduced his informative study "The British in Catherine's Russia: A Preliminary Survey."[28] The cultural image created by Novikov and the others whom Cross quotes and describes bears out the fascination that wealthy Russians felt about English imported goods during Catherine II's reign. This is not surprising since the imitative and emulous Russians had sufficient occasion to observe the "distinctive life style" of the nearly 1500 members of the British colony in cosmopolitan St. Petersburg.

> there was an English church, a club, a coffee house, an inn, and, for a short period, an English masonic lodge and an English theatre. British visitors, for business or pleasure, arriving by land or on one of the numerous British ships that plied the Baltic during the ice-free months, found "English grates, English coats, English coal, and English hospitality," . . . and opportunities for moving in the best Russian society.[29]

Those imports are, in fact, the material basis for the acculturation of the English style of life to which wealthy Russians had pretensions. While Cross cautions his readers that the evaluation made in the year 1800 by the Cambridge don Edward Daniel Clarke might be exaggerated, it seems to me that Clarke was perfectly reasonable in making the following statement: "Whatever they [the Russians] possess useful or estimable comes to them from England. Books, maps, prints, furniture, clothing, hardware of all kinds, horses, carriages, hats, leather, medicine, almost every article of convenience, comfort, or luxury, must be derived from England or it is of no estimation."[30]

By virtue of the commodity imports from Great Britain the English political, commercial, and cultural presence in Russia, and especially in St. Petersburg, was fortified. As Cross correctly points out in his reading of the *Sanktpeterburgskie vedomosti* (St. Petersburg News), that newspaper did carry numerous and repeated items of information on English activity in the capital city.[31] Such notices must have been annoying in times of crisis. But it is a mistake to

---

28. Anthony G. Cross, "The British in Catherine's Russia: A Preliminary Survey," in *The Eighteenth Century in Russia*, ed. J. G. Garrard (Oxford: Clarendon Press, 1973), pp. 233–63, quotation on p. 233.
29. A. G. Cross, ed., *An English Lady at the Court of Catherine the Great. The Journal of Baroness Elizabeth Dimsdale, 1781* (Cambridge: Crest Publications, 1989), Introduction, p. 3.
30. Cross, "The British in Catherine's Russia," pp. 242–43.
31. Ibid., p. 243. A further reading of the *Sanktpeterburgskie vedomosti* for the years 1763, 1771, 1784, and 1790 bears out Cross's findings.

say that "it was a campaign to 'buy British' that received critical support from Catherine's ukaz of May 1793, which, following the rupture in Franco-Russian relations, prohibited the import of French goods."[32] As already discussed in Chapter IX, the Russian court not only turned against the French and their imports in 1793, it also restricted and prohibited British imports. The cost of the British importation had become too much for the Russian court in the years running up to the crisis of 1793. Compared to the value of exports, the growing amount of imports had endangered the traditional favorable balance of overseas commodity trade of St. Petersburg and even the Empire. However, if indeed there was a campaign to "buy British" that could somehow be attributed to Catherine's prohibitory commercial policies beginning in 1793, it was that British merchants could no longer import certain British goods as they had in the past and that local merchants were ordered to dispose of the British goods they had in their shops and storehouses. If during the remaining years of Catherine's reign the wealthy Russians rushed to buy more and more British goods, it was undoubtedly encouraged by the fear of their growing scarcity on the market.

---

32. Cross, "The British in Catherine's Russia," p. 243; and see Chap. IX, supra.

# Chapter XIV
# Russia's Foreign Commodity Trade Turnover:
# The Principal Ports and Great Britain

In 1775, following Catherine II's partition of the Polish-Lithuanian Commonwealth and her defeat of the Ottoman Turks in the Crimea, the value (measured in current rubles) of the Russian Empire's overseas and overland foreign commodity trade turnover was approximately 31 million. In 1783, after peace was restored in North America and Europe and the Crimea was annexed by Catherine, Russia's foreign trade turnover had increased by nearly 38 percent to almost 43 million rubles. At the time of Catherine's death in 1796 this figure had grown by two and one-half times the previous amount and peaked at nearly 109 million rubles. During Paul's reign (1797–1801) Russia's foreign trade turnover leveled off at an annual average value of about 108 million rubles (see Figure 16, Table VRT–I).[1]

Between 1783 and 1801 the value of overseas trade comprised the overwhelming share of the combined value of Russia's overseas and overland foreign commodity trade turnover: 81 percent in 1783, almost 89 percent in 1796, and 80 percent during Paul's reign. Although about two dozen ports were active in the overseas trade during this period, the principal northern ports of Archangel, Fredrikshamn (Hamina), Narva, Pernau (Pärnu), Reval (Tallinn), Riga, St. Petersburg/Kronstadt and Viborg (Viipuri) together comprised the largest share of the combined value of Russia's overseas and overland foreign commodity trade turnover: 78 percent in 1783, almost 84 percent in 1796, and nearly 74 percent during Paul's reign.

St. Petersburg/Kronstadt led all other ports in trade turnover.

---

1. Table VRT-I is compiled and computed from the records of the St. Petersburg Custom House, PRO FO 97/340–42; FO 65/13–15, 17–18, 20, 23–24, 26, 28, 30, 33, 35, 37, 40, 42, 48, 50, 52; BT 6/141, 231–33, 266; and VTR, *Vedomosti*, Nos. 5–10.

**Figure 16**

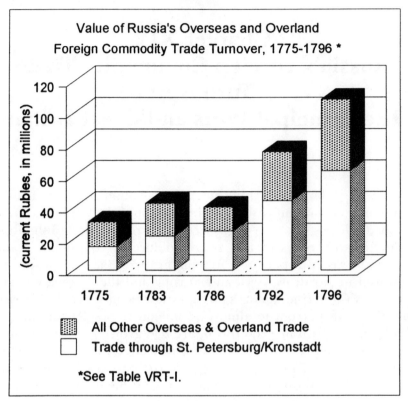

The value of its overseas commodity trade turnover in 1775 amounted to 49 percent of Russia's overseas and overland foreign commodity trade turnover: 51 percent in 1783, 58 percent in 1796, and almost 53 percent during Paul's reign. Between 1783 and 1796/1801 it amounted to an annual average of approximately 72 percent of the value of the overseas trade turnover at the principal ports (see Tables VRT–I–II).[2]

Although Riga was the second port of importance in the Empire, it was a distant second in trade turnover. Between 1775 and 1801 the value of its overseas commodity trade turnover was about 15 percent of Russia's foreign overseas and overland commodity trade turnover. Between 1783 and 1801 Riga accounted for an annual average of approximately 18 percent and Archangel for approximately 5 percent of the total value of the trade turnover at

---

2. Table VRT-II, ibid.

## TABLE VRT-I. APPROXIMATE VALUE OF RUSSIA'S OVERSEAS AND OVERLAND FOREIGN COMMODITY TRADE TURNOVER: 1775–1801[1]

[Exclusive of Custom Duties, Sundry Charges, and Specie Imports]
(Annual Average, Measured in Current Rubles, in Thousands)

| Years | Russian Empire | Overseas | % | Overland | % |
|---|---|---|---|---|---|
| 1775 | 31,027 | N/A | N/A | N/A | N/A |
| 1783 | 42,755 | 34,624 | 81.0 | 8,131 | 19.0 |
| 1786 | 40,308 | 37,207 | 92.3 | 3,101 | 7.7 |
| 1792 | 75,592 | 68,141 | 90.1 | 7,451 | 9.9 |
| 1794–96 | 91,872 | 80,929 | 88.1 | 10,943 | 11.9 |
| 1796 | 108,803 | 96,524 | 88.7 | 12,278 | 11.3 |
| 1797–1801 | 108,119 | 86,885 | 80.4 | 21,214 | 19.6 |

| Years | St. Petersburg/ Kronstadt | % | Riga | % | Principal Ports | % |
|---|---|---|---|---|---|---|
| 1775 | 15,192 | 49.0 | 6,571 | 21.2 | N/A | N/A |
| 1783 | 21,837 | 51.1 | 7,553 | 17.7 | 33,467 | 78.3 |
| 1786 | 25,187 | 62.5 | 5,976 | 14.8 | 35,891 | 89.0 |
| 1792 | 44,349 | 58.7 | 11,075 | 14.7 | 65,620 | 86.8 |
| 1794–96 | 55,295 | 60.2 | 13,441 | 14.6 | 77,600 | 84.5 |
| 1796 | 63,466 | 58.3 | 16,220 | 14.9 | 91,086 | 83.7 |
| 1797–1801 | 56,957 | 52.7 | 15,737 | 14.6 | 79,781 | 73.8 |

the principal ports. Therefore, the value of the trade turnover at St. Petersburg/Kronstadt was about four times greater than Riga's and about 14 times greater than Archangel's (see Tables VRT-I-II).

The data for the value of the overseas trade turnover are not complete for every port in every year during the period 1783–1796/1801. Nevertheless, it is fair to say that the annual average value of exports from St. Petersburg/Kronstadt represented approximately two-thirds of the total from the principal ports. This was almost three times that of Riga (23%) and about 9 to 10 times that of Archangel (7.0/6.5%). The capital port participated to an even greater extent in imports, accounting for an annual average of approximately 79 percent of the value of imports into the principal ports. Riga's share of imports was an annual average of about 7.7 to 9.5 percent and Archangel's share was about 2.9 to 2.5 percent (see Figures 17 and 18, Table VRT-II).

These numbers and percentages testify to the strength of Russia's overall foreign commodity trade, especially its overseas commerce, during the latter part of Catherine's reign. However, on

# TABLE VRT-II. APPROXIMATE VALUE OF PRINCIPAL PORTS' OVERSEAS COMMODITY TRADE TURNOVER FROM 1783 THROUGH 1801[2]
## ARCHANGEL, FREDRIKSHAMN, NARVA, PERNAU, REVAL, RIGA, ST. PETERSBURG/KRONSTADT, AND VIBORG

[Exclusive of Custom Duties, Sundry Charges, and Specie Imports]
(Annual Average, Measured in Current Rubles, in Thousands)

| Years | Principal Ports Trade Turnover | St. Petersburg/ Kronstadt Trade Turnover | % | Riga Trade Turnover | % | Archangel Trade Turnover | % |
|---|---|---|---|---|---|---|---|
| 1783–87 | 35,425 | 25,387 | 71.7 | 6,777 | 19.1 | 2,011 | 5.7 |
| 1788–92 | 57,240 | 41,414 | 72.4 | 8,337 | 14.6 | 2,918 | 5.1 |
| 1789 | 48,674 | 37,107 | 76.2 | 5,745 | 11.8 | 3,267 | 6.7 |
| 1790 | 61,689 | 44,606 | 72.3 | 8,519 | 13.8 | 3,133 | 5.1 |
| 1791 | 63,943 | 45,181 | 70.7 | 10,496 | 16.4 | 2,644 | 4.1 |
| 1792 | 65,620 | 44,349 | 67.6 | 11,075 | 16.9 | 3,315 | 5.1 |
| 1793 | 55,369 | 38,339 | 69.2 | 10,783 | 19.5 | 3,010 | 5.4 |
| 1794–96 | 77,600 | 55,295 | 71.3 | 13,441 | 17.3 | 3,953 | 5.1 |
| 1797–1801 | 79,781 | 56,957 | 71.4 | 15,737 | 19.7 | 3,375 | 4.2 |
| 1783–96 | 53,678 | 38,445 | 71.7 | 9,048 | 16.9 | 2,823 | 5.3 |
| 1783–1801 | 60,547 | 43,316 | 71.6 | 10,808 | 17.9 | 2,968 | 4.9 |

| Years | Principal Ports Exports | St. Petersburg/ Kronstadt Exports | % | Riga Exports | % | Archangel Exports | % |
|---|---|---|---|---|---|---|---|
| 1783–87 | 20,914 | 13,193 | 63.1 | 5,277 | 25.2 | 1,563 | 7.5 |
| 1788–92 | 30,716 | 20,965 | 68.3 | 6,461 | 21.0 | 2,092 | 6.8 |
| 1789 | 29,669 | 21,736 | 73.3 | 4,437 | 15.0 | 2,372 | 8.0 |
| 1790 | 31,595 | 21,642 | 68.5 | 6,691 | 21.2 | 2,209 | 7.0 |
| 1791 | 31,288 | 20,041 | 64.1 | 7,998 | 25.6 | 1,844 | 5.9 |
| 1792 | 33,154 | 21,053 | 63.5 | 8,642 | 26.1 | 2,336 | 7.0 |
| 1793 | 35,862 | 23,758 | 66.2 | 8,986 | 25.1 | 2,549 | 7.1 |
| 1794–96 | 47,994 | 31,590 | 65.8 | 11,646 | 24.3 | 3,230 | 6.7 |
| 1797–1801 | 50,676 | 34,180 | 67.4 | 11,796 | 23.3 | 2,889 | 5.7 |
| 1783–96 | 31,285 | 20,665 | 66.1 | 7,329 | 23.4 | 2,180 | 7.0 |
| 1783–1801 | 36,388 | 24,222 | 66.6 | 8,505 | 23.4 | 2,366 | 6.5 |

## TABLE VRT-II. Continued

| Years | Principal Ports Imports | St. Petersburg/ Kronstadt Imports | % | Riga Imports | % | Archangel Imports | % |
|---|---|---|---|---|---|---|---|
| 1783–87 | 14,510 | 12,194 | 84.0 | 1,500 | 10.3 | 448 | 3.1 |
| 1788–92 | 26,524 | 20,449 | 77.1 | 1,876 | 7.1 | 826 | 3.1 |
| 1789 | 19,005 | 15,371 | 80.9 | 1,309 | 6.9 | 895 | 4.7 |
| 1790 | 30,093 | 22,965 | 76.3 | 1,828 | 6.1 | 924 | 3.1 |
| 1791 | 32,656 | 25,141 | 77.0 | 2,498 | 7.7 | 800 | 2.4 |
| 1792 | 32,466 | 23,296 | 71.8 | 2,433 | 7.5 | 979 | 3.0 |
| 1793 | 19,507 | 14,581 | 74.7 | 1,797 | 9.2 | 461 | 2.4 |
| 1794–96 | 29,607 | 23,705 | 80.1 | 1,795 | 6.1 | 723 | 2.4 |
| 1797–1801 | 29,105 | 22,777 | 78.3 | 3,940 | 13.5 | 486 | 1.7 |
| 1783–96 | 22,393 | 17,780 | 79.4 | 1,719 | 7.7 | 643 | 2.9 |
| 1783–1801 | 24,159 | 19,095 | 79.0 | 2,304 | 9.5 | 602 | 2.5 |

### Figure 17

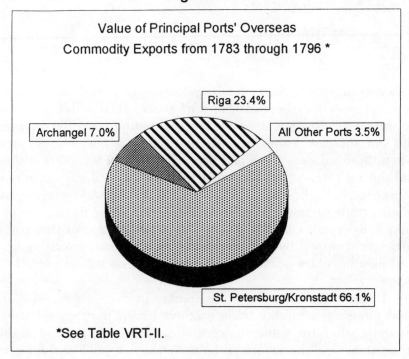

Value of Principal Ports' Overseas Commodity Exports from 1783 through 1796 *

Riga 23.4%
Archangel 7.0%
All Other Ports 3.5%
St. Petersburg/Kronstadt 66.1%

*See Table VRT-II.

**Figure 18**

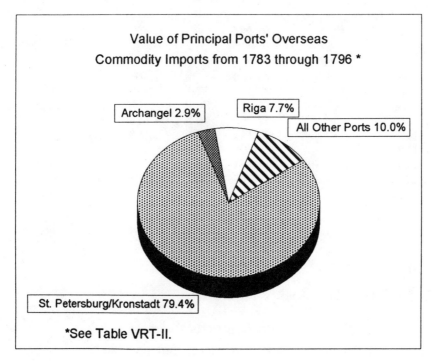

closer examination they reveal weaknesses that had far-reaching consequences for the financial and commercial stability of the country. First, Russia's favorable foreign commodity trade balances did not increase annually and they fluctuated widely. Second, when those balances remained precarious over a few years as they did during 1790–1792, a commercial and financial crisis resulted. Moreover, just when Russia's overseas and overland foreign commodity trade turnover more than doubled in value during the period 1783–1796, customs revenue as a percentage of that trade turnover dropped by half (depriving the Russian treasury of several million rubles per annum) for reasons that will be explained below.

Russia experienced wide fluctuations in its overseas and overland foreign commodity trade balances between 1775 and 1801. In 1775, when the value of exports was nearly 60 percent of the trade turnover, the favorable balance was six million rubles; in 1783 it was four million rubles and exports represented 55 percent of the trade turnover; in 1792, the year prior to the crisis of 1793, it was less than two million rubles and the value of exports declined

## TABLE VRT-III. APPROXIMATE VALUE OF RUSSIA'S OVERSEAS AND OVERLAND COMMODITY TRADE BALANCE: 1775–1801[3]

[Exclusive of Custom Duties, Sundry Charges, and Specie Imports]
(Annual Average, Measured in Current Rubles, in Thousands)

| Years | Total Trade Turnover | Total Exports | Total Imports | Balance | % Exports |
|---|---|---|---|---|---|
| 1775 | 31,027 | 18,557 | 12,469 | 6,088 | 59.8% |
| 1783 | 42,755 | 23,504 | 19,251 | 4,252 | 55.0% |
| 1786 | 40,308 | 23,244 | 17,064 | 6,181 | 57.7% |
| 1792 | 75,592 | 38,746 | 36,846 | 1,900 | 51.3% |
| 1794–96 | 91,872 | 54,917 | 36,955 | 17,961 | 59.8% |
| 1796 | 108,803 | 67,332 | 41,471 | 25,861 | 61.9% |
| 1797–1801 | 108,119 | 63,602 | 44,517 | 19,086 | 58.8% |

to 51 percent of the trade turnover. In 1796, the year of Catherine's death, the favorable balance jumped to almost 26 million rubles and the value of exports peaked at nearly 62 percent of the trade turnover. During Paul's reign, the favorable balance declined to an annual average 19 million rubles and exports declined to nearly 59 percent of the foreign commodity trade turnover (see Table VRT–III).[3]

As already indicated, between 1783–1801 the value of the trade turnover of the principal ports amounted to more than 80 percent of Russia's foreign commodity trade turnover. This figure experienced wide fluctuations. During the periods 1783–1787 and 1788–1792, the favorable trade balances amounted to an annual average of six million and four million rubles, respectively, and the value of exports amounted to 59 and 53 percent of the total trade turnover, respectively. In 1789 both the trade balance and the percentage value of exports increased: the former to 10 million rubles and the latter to 61 percent. However, in the following three years the trade balances fell precipitously: to a positive 1.5 million rubles in 1790; to a negative 1.3 million rubles in 1791; and to a positive balance of only about seven hundred thousand rubles in 1792. The value of exports as a percentage of the total trade turnover was 51.2 percent in 1790, 50.5 percent in 1792, and 48.9 percent in 1791. From 1793 onward, owing in part to the restrictions and prohibitions placed on imports, the principal ports experienced

---

3. Table VRT-III, ibid.

only favorable balances of trade: 16 million rubles in 1793; an annual average of 18 million rubles during 1794–1796; and an annual average of 21 million rubles during Paul's reign. The value of exports as a percentage of the total trade turnover exceeded 60 percent during these years, and it reached nearly 65 percent in 1793. Despite this, for the latter part of Catherine's reign (1783–1796) the favorable trade balance was only an annual average of about nine million rubles (see Table VRT-II).

The eight northern principal ports had decidedly different fluctuations and magnitudes in their overseas commodity trade balances during the latter part of Catherine's reign. Archangel had consistently favorable balances, which annually averaged 1.5 million rubles, and the value of its exports was 77 percent of this trade turnover. Narva had a trade turnover of an annual average of more than 400,000 rubles and the value of its exports was 83 percent of its total trade turnover. Riga had consistently favorable trade balances with an annual average of 5.6 million rubles, while the value of its exports was 81 percent of its total trade turnover. The smaller ports of Fredrikshamn and Viborg in southern Finland suffered deficit and near deficit trade turnover years: 1789–1793 for Fredrickshamn, and for Viborg, 1789–1791, 1794–1796, and, while the data are unclear, probably also 1792–1793. Pernau had unfavorable balances almost yearly with an annual average deficit of more than a half million rubles between 1783 and 1795. The value of its exports was about 30 percent of its total trade turnover. Reval had consistently unfavorable balances of trade with an annual average deficit of nearly 1.8 million rubles. The value of Reval's exports was less than 10 percent of its total trade turnover.

The huge trade turnover of St. Petersburg/Kronstadt dwarfed those of the other ports and its wide fluctuation considerably influenced the trade turnover curve of the principal ports as a whole. Had it not been for the consistently favorable trade balances of Riga, Archangel and Narva, which amounted to an annual average of 7.5 million rubles during 1783–1796, Russia would have experienced more than the one shock of 1793.

During 1783–1787 the balance of trade was in the capital port's favor by an annual average of only approximately one million rubles, falling to as little as a half million rubles during 1788–1792. The value of St. Petersburg/Kronstadt's exports as a percent of its trade turnover during those periods was 52 and 50.6 percent, respectively. The favorable trade balances achieved by Archangel (an annual average of more than one million rubles in each period), Riga (an annual average of 3.8 and 4.6 million rubles, respectively) and by even the more modest port of Narva (an annual average of a half million and four hundred thousand rubles, respectively) calls into serious question the trade profitability of St. Petersburg/Kronstadt. At all three ports (Archangel, Narva and Riga) during these two periods the value of exports as a percent of trade was well above 70 percent.

In 1789 the capital port's six-million-ruble favorable trade balance was only one million more than the combined balances of Archangel, Narva and Riga. However, for the next three years (1790–1792) St. Petersburg/Kronstadt suffered unfavorable balances of trade with a deficit of 1.3 million rubles in 1790, 6.4 million in 1791, and 2.2 million in 1792. Although the value of its exports as a percent of trade turnover was approximately 58.6 percent in 1789, it slid to 48.5 percent in 1790, dropped to 44.4 in 1791, and rebounded to 47.5 percent in 1792. In comparison, Archangel, Narva, and Riga each had consistent favorable trade balances which together amounted to 6.5 million rubles in 1790, 7.1 million rubles in 1791, and 8.0 million rubles in 1792. The value of their exports to their trade turnover was 70 percent or more.

In 1793, owing no doubt to the restrictions and prohibitions imposed on imports that year, the favorable trade balance of St. Petersburg/Kronstadt increased to approximately 9.2 million rubles, and the value of its exports increased to 62 percent of the trade turnover. During the final three years of Catherine's reign the capital port's favorable trade balance slipped to an annual average of 7.9 million rubles with the value of its exports declining to 57 percent of total trade turnover.

From 1783 through 1796 the favorable balance of St. Petersburg/Kronstadt's trade amounted to an annual average of 2.9 million rubles and the value of its exports was approximately 54 percent of its trade balance. While this trade balance was slightly less than twice that of Archangel's and six times greater than Narva's, it was only about 48 percent of Riga's 5.6 million ruble annual average trade balance.

An expected consequence of Russia's growing foreign commodity trade turnover should have been the generation of significant custom revenue—duties and charges. In fact the opposite was the case and took place just when the value of the trade turnover more than doubled. Although the archival data on custom revenue for the Russian Empire are available for only several random years, they are illustrative of what transpired. In 1775 custom revenue was approximately 3.3 million rubles and represented 10.7 per-

### TABLE VRT-IV. CUSTOM REVENUE AS A PERCENT OF THE VALUE OF RUSSIA'S FOREIGN COMMODITY TRADE TURNOVER: 1775–1801[4]

(Annual Average, Measured in Current Rubles, in Thousands)

| Years | Total Trade Turnover | Custom Revenue | % | Overland Trade Turnover | Custom Revenue | % |
|---|---|---|---|---|---|---|
| 1775 | 31,027 | 3,326 | 10.7 | N/A | N/A | N/A |
| 1783 | 42,755 | 5,182 | 12.1 | 8,131 | 756 | 9.3 |
| 1786 | 40,308 | 5,049 | 12.5 | 3,101 | 271 | 8.7 |
| 1795–96 | 99,589 | 5,947 | 6.0 | 12,269 | 833 | 6.8 |
| 1796 | 108,803 | 6,471 | 5.9 | 12,278 | 913 | 7.4 |
| 1797–1801 | 108,119 | 8,208 | 7.6 | 21,214 | 1,035 | 4.9 |

| Years | Overseas Trade Turnover | Custom Revenue | % |
|---|---|---|---|
| 1775 | N/A | N/A | N/A |
| 1783 | 34,467 | 4,426 | 12.8 |
| 1786 | 37,207 | 4,778 | 12.8 |
| 1795–96 | 87,320 | 5,114 | 5.9 |
| 1796 | 96,524 | 5,558 | 5.8 |
| 1797–1801 | 86,885 | 7,173 | 8.3 |

cent of the value of the trade turnover. In 1783 and 1786 custom revenue was slightly more than five million rubles and represented slightly more than 12 percent of the value of the trade turnover. However, in 1796, when the value of the overseas and overland foreign commodity trade turnover reached its highest level at nearly 109 million rubles, more than two and one-half times the 1786 trade turnover, custom revenue increased to only about 6.5 million rubles and represented only 5.9 percent of the total trade turnover (see Table VRT–IV).[4] Russia therefore lost several million rubles per annum of expected custom revenue during the period 1787–1796. Moreover, the circumstances that caused that loss also contributed to the commercial and financial crisis of 1793.

As a percentage of the trade turnover, the fluctuation in the amount of custom revenue collected on Russia's overseas commerce at the principal ports, at St. Petersburg/Kronstadt and at Riga are all similar to that of the Empire's for those years in which

---

4. Table VRT-IV, ibid.

## TABLE VRT-IV. Continued

| Years | Principal Ports' Total Trade Turnover | Custom Revenue | % | St. Petersburg/ Kronstadt's Trade Turnover | Custom Revenue | % |
|---|---|---|---|---|---|---|
| 1775 | N/A | N/A | N/A | 15,192 | 1,697 | 11.2 |
| 1783 | 33,467 | 4,316 | 12.9 | 21,773 | 2,810 | 12.9 |
| 1786 | 35,891 | 4,669 | 13.0 | 25,187 | 3,346 | 13.3 |
| 1795–96 | 83,464 | 4,816 | 5.8 | 59,209 | 3,367 | 5.7 |
| 1796 | 91,086 | 5,217 | 5.7 | 63,466 | 3,505 | 5.5 |
| 1797–1801 | 79,781 | 6,340 | 7.9 | 56,957 | 4,430 | 7.8 |

| Years | Riga's Trade Turnover | Custom Revenue | % |
|---|---|---|---|
| 1775 | 6,571 | 588 | 9.0 |
| 1783 | 7,553 | 1,018 | 13.5 |
| 1786 | 5,976 | 706 | 11.8 |
| 1795–96 | 14,508 | 670 | 4.6 |
| 1796 | 16,220 | 779 | 4.8 |
| 1797–1801 | 12,857 | 1,205 | 9.4 |

the data are available (see Table VRT-IV). During those years the custom revenue collected at St. Petersburg/Kronstadt amounted to more than half that collected in the Empire. The archival data for the capital port are complete from 1768 through 1801. There are aggregate data for the two twelve-year periods from 1768 through 1779 and from 1780 through 1791 and data for individual years between 1768 and 1801. By charting the annual average fluctuation in the custom revenue collected at the capital port the course of the fluctuation for the Empire can be approximated (see Table VRT–V).[5]

From 1768 through 1779 the custom revenue collected at St. Petersburg/Kronstadt was nearly 22 million rubles, an annual average of slightly more than 1.8 million rubles representing 11.3 percent of the value of that port's commodity trade turnover. Data are available for 1775 for both St. Petersburg/Kronstadt and the Empire. The custom revenue collected at the capital port was 1.7 million rubles (51 percent of the Empire's), which represented 11.2 percent of the value of that port's trade turnover. The custom

---

5. Table VRT-V, ibid.

### TABLE VRT-V. COMPARISON OF CUSTOM REVENUE AS A PERCENT OF RUSSIA'S AND ST. PETERSBURG'S FOREIGN COMMODITY TRADE TURNOVER: 1768–1801[5]

(Annual Average, Measured in Current Rubles, in Thousands)

| Years | Russia's Trade Turnover | Custom Revenue | % | St. Petersburg/ Kronstadt's Trade Turnover | Custom Revenue | % |
|---|---|---|---|---|---|---|
| 1768–79 | | | | 16,231 | 1,829 | 11.3 |
| 1775 | 31,027 | 3,326 | 10.7 | 15,192 | 1,697 | 11.2 |
| 1780–91 | | | | | 3,419 | 11.5 |
| 1783 | 42,755 | 5,182 | 12.1 | 21,773 | 2,810 | 12.9 |
| 1786 | 40,308 | 5,049 | 12.5 | 25,187 | 3,346 | 13.3 |
| 1787 | | | | 31,263 | 4,023 | 12.9 |
| 1788 | | | | 35,825 | 4,109 | 11.5 |
| 1789 | | | | 37,107 | 3,898 | 10.5 |
| 1790 | | | | 44,606 | 4,762 | 10.7 |
| 1791 | | | | 45,181 | 4,515 | 10.0 |
| 1792 | | | | 44,349 | 4,109 | 9.3 |
| 1793 | | | | 38,339 | 2,796 | 7.3 |
| 1794 | | | | 47,469 | 2,972 | 6.3 |
| 1795–96 | 99,589 | 5,947 | 6.0 | 59,209 | 3,367 | 5.7 |
| 1796 | 108,803 | 6,471 | 5.9 | 63,466 | 3,505 | 5.5 |
| 1797–1801 | 108,119 | 8,208 | 7.6 | 56,957 | 4,430 | 7.8 |

revenue collected for the Empire was 3.3 million rubles, which represented 10.7 percent of the value of the Empire's overseas and overland foreign commodity trade turnover.

From 1780 through 1791 the custom revenue collected at St. Petersburg/Kronstadt was approximately 41 million rubles for an annual average of 3.4 million rubles. Although the trade turnover for the capital port is not yet available for 1780 it is reasonable to estimate that custom revenue as a percentage of trade turnover for the period 1780–1791 was an annual average of about 11.5 percent. In 1783 it was 12.1 percent for the Empire and 12.9 percent for the capital port; in 1786 it was 12.5 percent for the Empire and 13.3 percent for the capital port; in 1795–1796 it was 6.0 percent for the Empire and 5.7 percent for the capital port; in 1796 it was 5.9 percent for the Empire and 5.5 percent for the capital port; and during Paul's reign (1797–1801) the annual average was 7.6 percent for the Empire and 7.8 percent for the capital port.

More significantly, not only did custom revenue as a percentage of trade turnover decline by about half between 1783–1796 but, except for one year in which it increased by two-tenths of one percent at St. Petersburg/Kronstadt, it also declined annually between 1787–1796. This decline in custom revenue as a percentage of trade turnover was one reason why Catherine substantially increased custom duties in her aborted tariff of 1796.

The principal cause for the decline in custom revenue as a percentage of trade turnover is to be found neither in the inefficient administration of custom collection nor in smuggling. Rather, it is to be found in two separate but converging aspects of Russia's foreign trading operation. One question is how did the value of imports fluctuate, and the other is who actually paid the custom duties and how much were they?

Between 1783–1787 and 1788–1792 the total value of the overseas commodity trade turnover at the principal ports increased by an annual average of about 62 percent, increasing again by almost 26 percent between 1788–1792 and 1793–1796. But from 1789 through 1793 the value of the trade turnover fluctuated significantly: in 1789 it was 48.7 million rubles; in 1790 it jumped to 61.7 million; in 1791 it rose to 63.9 million; in 1792 it peaked at 65.6 million; but in 1793 it tumbled to 55.4 million rubles. In the three remaining years of Catherine's reign, 1794–1796, it catapulted to an annual average of 77.6 million rubles. During Paul's reign (1797–1801) it increased only slightly to an annual average of 79.8 million rubles.

The value of exports from the principal ports grew by an annual average of about 47 percent between 1783–1787 and 1788–1792, and by another 46 percent between 1788–1792 and 1793–1796. The value of exports experienced modest fluctuation from 1789 through 1793: in 1789 it was 29.7 million rubles; in 1790 it increased to 31.6 million; in 1791 it declined slightly to 31.3 million; in 1792 it increased to 33.2 million; and in 1793 it peaked at 35.9 million rubles. But in the three remaining years of Catherine's reign, 1794–1796, it jumped to an annual average of 48 million rubles. During Paul's reign it increased only slightly to an annual average of 50.7 million rubles.

Between 1783–1787 and 1788–1792 the value of imports into the principal ports skyrocketed by an annual average of nearly 83 percent. Subsequently, however, between 1788–1792 and 1793–1796 it increased by as little as 2 percent. In 1789 the value of imports was 19 million rubles; in 1790 it jumped to 30.1 million; in 1791 and 1792 it increased slightly to more than 32 million; but in 1793 it plummeted to 19.5 million rubles. The favorable balance of commodity trade that the principal ports had annually enjoyed was in jeopardy in 1790 and 1792 and was actually unfavorable in 1791. St. Petersburg/Kronstadt had an unfavorable balance of commodity trade in 1790, 1791, and 1792. This adverse trend contributed to the commercial-financial crisis of 1793 that in turn provoked the draconian reduction of imports but with the corollary effect of further reducing custom revenue. In the remaining three years of Catherine's reign, 1794–1796, the value of imports rose to an annual average of 29.6 million rubles, and during Paul's reign it declined slightly to an annual average of 29.1 million rubles.

However, because the slide in custom revenue as a percentage of trade turnover actually began in 1787, prior to the wider fluctuations in imports, and came to a rest only at the beginning of Paul's reign, not in 1794 after the rebound in the value of the trade turnover, there is a compelling explanation for the shortfall in expected custom revenue and its corresponding drop in percent.

In 1787 the expiration of the Anglo-Russian Commercial Treaty caused several British merchants to become naturalized Russian merchants, *Inostrannye gosti,* and continue their overseas trading operations in St. Petersburg/Kronstadt. Their new status gave them considerable commercial benefits including generous abatements on the payment of export and import custom duties. When these benefits are added to those provided to other *Inostrannye gosti,* a large loss in custom revenue resulted. For example, in 1787, there was a shift in the value of the trade turnover of slightly more than eight million rubles from former British and Hanse merchants to *Inostrannye gosti;* in 1792, the amount of the shift increased to almost nine million rubles; and those sums were greater still in 1793 and 1795. Although the exact amount of similar shifts in the value of the trade turnover is not yet known for the other years during the period 1787–1796, foreign merchants undoubtedly continued to trade as *Inostrannye gosti* in millions of rubles.[6] The custom duties were therefore collected on those amounts but at a lower rate because of the abatements. Ironically, Catherine's decision to emancipate Russia from the British hegemony in her country's overseas commerce by refusing in 1787 to continue Britain's privileged position actually led to the diminution of Russia's custom revenue. This was just another demonstration of how the

---

6. See Chap. X, supra.

British commercial presence in Russia continued to influence that country's overseas trade.

St. Petersburg/Kronstadt, Riga, Narva, Archangel, and Reval participated to a greater extent than other Russian ports in the Anglo-Russian trade. Although there are gaps in the record, the Russian archival data yield enough information to provide a meaningful indication of the terms of that trade during the latter part of Catherine's reign (1783–1796). For example, during 1788–1791 an annual average of about 22 percent of the value of Archangel's total trade turnover was with Great Britain: almost 27 percent of its exports and almost 11 percent of its imports. In 1793 the total trade turnover increased to about 34 percent: almost 39 percent of its exports and almost 6 percent of its imports. Reval's trade with Great Britain concentrated in imports with about 23 percent of its total imports coming from that country during the period between 1788 and 1796. Narva's trade with Great Britain was more active than Archangel's or Reval's. During the period 1788–1791 trade with Great Britain averaged annually at least 37 percent of Narva's total trade turnover: almost 41 percent of its exports and at least 21 percent of its imports. In 1795 about 52 percent of Narva's exports went to Great Britain and about 27 percent of its imports came from that country.[7]

Riga's overseas commerce was as diverse as that of St. Petersburg/Kronstadt and it had an ongoing trade with Bremen, Danzig, Denmark, Flanders, France, Great Britain, Hamburg, Holland, Italy, Lübeck, Portugal, Prussia, Rostock, Spain, Sweden, and other Baltic ports.[8] Riga's trade was primarily in exports. The British were Riga's largest clients and were followed by the Dutch. Between 1783 and 1796, while the value of Riga's imports from Great Britain amounted to about 9.5 percent of its total imports, the value of its exports to that country ranged from 18 percent in 1783 to 46 percent in 1796. During the period 1787–1791 the annual average value of Riga's exports to Great Britain was 1.8 million rubles—almost 32 percent of its total exports. This resulted in a British trade deficit of 1.6 million rubles. In 1793 Riga's more than three million rubles of exports to Great Britain amounted to 33

---

7. For Archangel's, Narva's, and Reval's balance of commodity trade with Great Britain, see the records of the St. Petersburg Custom House, PRO BT 6/141, 231–33, 266; FO 97/340; and FO 65/13–15, 18, 20, 23–24, 26, 33, 35.
8. In addition to trade data mentioned herein, see Bibliography entries for V. V. Doroshenko and Elisabeth Harder-Gersdorff.

percent of its total exports, resulting in a British trade deficit of almost 2.7 million rubles.[9]

Although there is much more extant archival information on the overseas trading operation of St. Petersburg/Kronstadt than there is on any other Russian port, only an approximate calculation of the value of its trade turnover with Great Britain can be attempted and is based on the following assumptions. Because all of the British merchants in St. Petersburg/Kronstadt traded primarily in merchandise shipped to and from British ports, their annual average trade turnover would be close to the trade turnover of the capital port with Great Britain. This assumption is further supported by the occasional archival data that show nearly coincident values of the British merchant national trade turnover, and the commodity trade turnover of St. Petersburg/Kronstadt with Great Britain. It is assumed, moreover, that a relatively small amount of the trade with Great Britain was undertaken by non-British merchant nationals, perhaps equal to the amount of trade British merchants transacted with countries other than Great Britain. And finally, for the several years in which information is available and where it is in fact known that British merchant nationals had temporarily become naturalized Russian burghers, they are herein still counted as British merchants and their trade transactions are calculated as part of the British merchant national trade turnover with Great Britain. The analysis that follows is for the eleven-year period 1783–1793 and for the years 1795–1796. Because the data for merchant national trade are not extant for the year 1794 only a rough estimate will be provided for the overall period between 1783 and 1796.

As a percentage of the total value of the overseas commodity trade turnover of St. Petersburg/Kronstadt during the period 1783–1787, the value of the trade turnover of British merchant nationals annually averaged approximately 45 percent. During 1788–1792 it fell to about 25 percent. This is probably because there were an undetermined number of British merchants who had become naturalized Russian burgers, about whom there is yet no extant information during the years 1788–1791. Although the Russian government began imposing restrictions and prohibitions on British imports beginning in 1793, the trade turnover of British merchants that year still amounted to about 47 percent of the total

---

9. For Riga's balance of commodity trade with Great Britain, see records of the St. Petersburg Custom House, PRO BT 6/141, 231–33, 266; FO 97/340; and FO 65/13–15, 18, 20, 23–24, 26, 33, 35.

## TABLE MN-XII. APPROXIMATE VALUE OF BRITISH MERCHANTS' OVERSEAS COMMODITY TRADE TURNOVER AT ST. PETERSBURG: 1783–1796[10]

[Exclusive of Custom Duties, Sundry Charges, and Specie Imports]
(Annual Average, Measured in Current Rubles, in Thousands)

| Years | St. Petersburg/ Kronstadt Total Exports | British Merchants' Total Exports | % | St. Petersburg/ Kronstadt Total Imports | British Merchants' Total Imports | % |
|---|---|---|---|---|---|---|
| 1783–87 | 13,193 | 8,702 | 66.0 | 12,194 | 2,904 | 23.8 |
| 1788–92 | 20,965 | 8,966 | 42.8 | 20,450 | 1,705 | 8.3 |
| 1793 | 23,758 | 14,397 | 60.6 | 14,581 | 3,689 | 25.3 |
| 1795–96 | 34,521 | 21,101 | 61.1 | 24,688 | 7,518 | 30.5 |
| 1783–93 | 17,686 | 8,782 | 49.7 | 16,163 | 2,130 | 13.2 |

| Years | St. Petersburg/ Kronstadt Total Trade Turnover | British Merchants' Total Trade Turnover | % |
|---|---|---|---|
| 1783–87 | 25,387 | 11,606 | 45.7 |
| 1788–92 | 41,414 | 10,670 | 25.8 |
| 1793 | 38,339 | 18,086 | 47.2 |
| 1795–96 | 59,209 | 28,619 | 48.3 |
| 1783–93 | 33,850 | 10,912 | 32.2 |

trade turnover. In 1795–1796 the annual average was about 48 percent. Therefore, the estimated value of the trade turnover of St. Petersburg/Kronstadt with Great Britain annually averaged approximately 32 percent during the period 1783–1793, and it is reasonable to assume that it would increase to roughly 42 percent between 1783 and 1796 (see Table MN–XII).[10]

The value of exports as a percentage of total trade turnover by British merchants at St. Petersburg/Kronstadt comprised an annual average of about 66 percent during the period 1783–1787, about 42 percent during 1788–1792, about 60 percent in 1793, and about 61 percent in 1795–1796. Thus during the period 1783–1793 the value of exports from St. Petersburg/Kronstadt to Great Britain annually averaged approximately 49 percent; but

---

10. Table MN-XII is compiled and computed from the records of the St. Petersburg Custom House, PRO BT 6/141, 231–33, 266; FO 97/340–42; FO 65/13–15, 17–18, 20, 23–24, 26, 28, 30, 33, 35; and VTR *Vedomosti*, Nos. 6–10.

## TABLE BT-I. APPROXIMATE VALUE OF GREAT BRITAIN'S OVERSEAS COMMODITY TRADE BALANCE DEFICIT WITH ST. PETERSBURG: 1783–1796[11]

[Exclusive of Custom Duties, Sundry Charges, and Specie Imports]
(Annual Average, Measured in Current Rubles, in Thousands)

| Years | Total Trade Turnover | Britain's Exports to St. Petersburg | Britain's Imports from St. Petersburg | Britain's Trade Balance |
|---|---|---|---|---|
| 1783–87 | 11,606 | 2,904 | 8,702 | (−5,798) |
| 1788–92 | 10,670 | 1,705 | 8,966 | (−7,261) |
| 1793 | 18,086 | 3,689 | 14,397 | (−10,708) |
| 1795–96 | 28,619 | 7,518 | 21,101 | (−13,583) |
| 1783–93 | 10,912 | 2,130 | 8,782 | (−6,652) |

between 1783 and 1796 it is reasonable to assume it increased to about 57 percent. The value of imports by British merchants annually averaged about 23 percent during 1783–1787, only about 8 percent during 1788–1792, about 25 percent in 1793, and about 30 percent in 1795–1796. The value of imports from Great Britain into St. Petersburg/Kronstadt therefore annually averaged about 13 percent during the period 1783–1793, probably increasing to approximately 22 percent between 1783 and 1796.

By using the aforementioned data the approximate value of Great Britain's overseas commodity trade balance deficit with St. Petersburg/Kronstadt can be estimated. During the period 1783–1787, it annually averaged approximately 5.8 million rubles; during 1788–1792 it increased to about 7.3 million rubles; in 1793, the year when the Russian court restricted and prohibited imports into Russia from Great Britain, the trade balance deficit shot up to about 10.7 million rubles; and during the years 1795–1796 the trade balance deficit increased again to about 13.6 million rubles. For the eleven-year period from 1783 through 1793 the trade balance deficit amounted to an annual average of approximately 6.7 million rubles and was probably slightly greater for the period 1783–1796 (see Table BT–I).[11]

Great Britain met its considerable trade deficit with Russia with bills of exchange and precious metals and to a lesser extent

---

11. Table BT-I is compiled and computed from the records of the St. Petersburg Custom House, ibid.

with the mercantile services it provided Russia. The Amsterdam, London and Hamburg markets played a significant role in all these transactions. Regarding mercantile services, Hans Chr. Johansen, who has prodigiously quantified the Baltic Sund sea traffic, has stated that the British not only "must have profited from their shipping" but also profited from "the *insurance* business, which included both ship and cargo" and he has estimated those sums.[12]

The bill of exchange was extensively used in the Russian trade. Clearly stated by Dr. John Orbell

> ... The Bill of exchange was a well established widely used instrument for making payments between international centres, but it was an agreement to receive payment after delivery, and possibly sale, of goods. The seller's confidence in the standing of the buyer was essential but the bill's usefulness broke down when such confidence did not exist or where the standing of the two parties was unknown to one another. In such cases a third party, in whose creditworthiness the seller had confidence, and who had confidence in the ability of the buyer to meet his obligations, would accept the bill, i.e. he would guarantee to meet payment of the bill when it fell due, the buyer placing him in funds immediately prior to this ... .[13]

The late Artur Attman has provided a summary of how bills of exchange and precious metals worked in the Anglo-Russian trade. Owing to Great Britain's "overwhelming" positive trade balances with the United Provinces, British merchants were able not only

> to transfer Dutch coins to England or to the Baltic ports but also, more significantly, to draw bills of exchange on the English credits in Holland. These bills were used as cash in Baltic ports, they were payable in Amsterdam and they were ultimately redeemed there with precious metals which were sent to Baltic ports. The English importers availed themselves of this opportunity, and it became of ever growing significance as English imports from the Russian and Polish Lithuanian markets

---

12. Hans Chr. Johansen, "How to Pay for Baltic Products," in *The Emergence of a World Economy 1500–1914*, Part I: 1500–1850, ed. Wolfram Fischer, R. Marvin McInnis and Jürgen Schneider (Stuttgart: Steiner–Verlag-Wiesbaden-GmbH, 1986), pp. 123–42, quotation from pp. 138–39. See also, Jürgen Schneider and Oskar Schwarzer, "International Rates of Exchange: Structures and Trends of Payments Mechanism in Europe, 17th to 19th Century," ibid., pp. 143–70; and Frank C. Spooner, *Risks at Sea. Amsterdam insurance and maritime Europe, 1766–1780* (Cambridge: Cambridge University Press, 1983).
13. John Orbell, *Baring Brothers & Co., Limited. A History to 1939* (London: Baring Brothers & Co., Limited, 1985), p. 5. For a lengthy, detailed and authoritative contemporary description of the bill of exchange, see Joshua Montefiore, *A Commercial Dictionary: containing The Present State of Mercantile Law, Practice, and Custom. Intended for the Use of the Cabinet, the Counting House, and the Library* (London: Printed by the Author, 1804).

increased in the course of the 18th century. In this respect Riga became an important financial entrepot for the whole of the Russian market during the 18th century. Consignments of coins were channelled to Riga, either directly or through bill transactions; these consignments came mainly from Amsterdam, but Hamburg also sent considerable quantities of precious metals to pay for the Russian export surpluses, above all in the trade with Great Britain.[14]

Nevertheless, there is sufficient evidence to indicate that the British merchants in the Russian trade could not always depend on Amsterdam to facilitate their bills of exchange, as was the case in 1763–1764 when Dutch houses were failing and would not accept these bills.[15] This caused the Russia Company to petition that a "Course of Exchange should be established between Petersbourg and London by Bills payable in London."[16] It was thought at the time that such an exchange would result in a "saving of ten or twelve thousand Pounds sterling yearly, which is now paid to the Dutch by the English Merchants, for the negotiating these Bills." The Russia Company further believed that "once an Exchange between Russia and Great Britain directly is effected, the English Merchants may, and in all probability will, become the remitters between the Russian and the Merchants of many other parts of Europe."[17]

It is therefore noteworthy to read in Stanley Chapman's *Merchant Enterprise in Britain* that "the economy of Britain revolved round London, and the bill of exchange of a London merchant, factor or warehouseman provided both the credit and the cur-

---

14. Artur Attman, *Dutch Enterprise in the World Bullion Trade 1550–1800* (Göteborg, Sweden: Acta Regiae Societatis Scientiarum et Litterarum Gothoburgensis, Humaniora 23; Kungl. Vetenskaps-och Vitterhets-Samhället, 1983), quoted from p. 52 but see also Chaps. IV and V, pp. 45–91; idem, "The Russian Market in World Trade 1500–1860," *The Scandinavian Economic History Review* 29 (1981): pp. 177–202; and idem, "Precious Metals and the Balance of Payments in International Trade 1500–1800," *The Emergence of a World Economy*, pp. 113–21.

Considerable research on this subject and Riga's place in it has been done by the late V. V. Doroshenko and Elizabeth Harder-Gersdorff (see Bibliography). Most recently, for example, Professor Doroshenko was to present a paper entitled "Merchant-bankers in Riga Sea Trade in the 18th century," wherein he intended to show that Riga bankers were few in number and that the banking "houses of Amsterdam, Hamburg, Lübeck etc.: their credits layed [sic] the financial foundation of the Riga export and the inclusion of its Hinterland in the system of European economic links." Quoted from Doroshenko letter to Kaplan, Nov. 19, 1990.
15. See RC, Sept. 8, 1763, VII, p. 158; and RC, Jan. 27 and May 29, 1764, VII, pp. 163, 181, respectively.
16. RC, Sept. 8, 1763, VII, p. 158.
17. RC, May 29, 1764, VII, p. 181.

rency for wholesale buying and selling operations," and in particular that

> Henry Thornton, a London merchant in the Russian trade, wrote in 1802 that London had become, 'especially of late, the trading metropolis of Europe and, indeed, of the whole world; the foreign drafts (i.e. bills of exchange) [sic] on account of merchants living in our outports and other trading towns, and carrying on business there, being made, with scarcely any exceptions, payable in London'. Thornton wrote from personal experience, for twenty years earlier his firm had discounted bills in Amsterdam but were now themselves acting as guarantors for the payment of Russian bills of exchange.[18]

The archival data from the records of the St. Petersburg Custom House on specie—foreign gold and silver coin and bullion—imported into Russia are available for St. Petersburg/Kronstadt, Riga, and other Russian ports. Although St. Petersburg/Kronstadt was the capital port of Russia, during the period 1783–1796, its favorable balance amounted to an annual average of only 2,885,801 rubles compared to Riga's 5,610,550 rubles. It was Riga that imported the overwhelming amount of specie from all sources. Therefore, Attman suggests, "the export surpluses [of St. Petersburg] were probably settled through Riga to a certain extent."[19] For example, during 1783–1787, Riga imported specie the equivalent of which was an annual average of 2,633,640 rubles while St. Petersburg/Kronstadt imported only 155,296 rubles; in 1790–1791, Riga imported an annual average of 2,439,795 rubles and St. Petersburg/Kronstadt only 100,896 rubles; and in 1795–1796, Riga imported an annual average of 4,827,124 rubles while St. Petersburg/Kronstadt only 247,548 rubles. Taken together, the specie imports into St. Petersburg/Kronstadt and Riga comprised almost the total amount of specie imported into the Russian Empire by land and by sea.[20]

---

18. Stanley Chapman, *Merchant Enterprise in Britain: From the Industrial Revolution to World War I* (Cambridge: Cambridge University Press, 1992), p. 42. Chapman cites the 1802 edition of Henry Thornton's work at page 59; in the American edition the quotation is found on page 48, *An Inquiry into the Nature and Effects of the Paper Credit of Great Britain* (Philadelphia: James Humphreys, 1807). For additional commentary on bills of exchange and the British participation in their use in the Baltic and Russian trade, see J. A. S. L. Leighton-Boyce, *Smith's the Bankers 1658–1958* (London: National Provincial Bank Ltd., 1958), pp. 88, 184–88, 192–213, 236, and passim; and Stuart Ross Thompstone, "The Organisation and Financing of Russian Foreign Trade before 1914" (Ph.D. thesis, University of London, 1991), chapter 5, passim.
19. Quoted from Attman, *Dutch Enterprise*, p. 83.
20. For Russia's specie imports between 1783 and 1796, see records of the St. Petersburg Custom House, PRO BT 6/141, 231–33, 266; FO 97/340–342; and FO 65/13–15, 17–18,

There is only one extant source that provides data for Great Britain's overseas commodity trade balance deficit with the Russian Empire for the entire period from 1783 through 1796. However, it has one essential drawback because it is calculated in English pounds sterling, "official values" and not in "current real values."[21] Notwithstanding the static nature of the "official values," they do provide a guideline to the magnitude of the value of Britain's trade with Russia. Earlier in this study it was shown, according to calculations in "official values," that Russia's share of England's and Scotland's total overseas commodity imports made it the leader among eleven continental European countries who exported merchandise to Great Britain. These figures revealed that during Peter III's and Catherine II's reigns (1762–1796) Russia's percentage of England's and Scotland's total world imports was an annual average of 8.5 and 17.3 percent, respectively.[22]

These figures must now be revised. Ralph Davis's computation of the "current real values" for the years 1784–1786 and 1794–1796 not only supports in magnitude the direction of the "official values" but also suggests that Russia played an even greater role in Great Britain's total importations. The annual average percentage of Britain's total imports from Denmark-Norway, Sweden, and Russia is just slightly more than 10 percent in "official values" of Britain's total world imports during the years 1784–1786 and 1794–1796. Russia alone accounts for an annual average of 8.5 and 7.9 percent, respectively, for these years. In other words about 80 percent of the total "official values" of British imports from Denmark-Norway, Sweden, and Russia combined belonged to Russia. However, Davis's annual average percentage of "current real values" of Britain's total imports from a region which is very similar to the above-mentioned countries and which he calls

---

20, 23–24, 26, 28, 30, 33, 35, 37, 40, 42, 48, 50, 52; VRT, *Vedomosti*, Nos. 5–10; and Table VRT-II, supra. See also, Ian Blanchard, *Russia's 'Age of Silver.' Precious-metal production and economic growth in the eighteenth century* (London: Routledge, 1989), pp. 167–68, 201–5.

21. "The rates of duty were charged on a list of valuations of commodities made in 1660, and rarely altered." However, in 1696 the Inspector-General of Imports and Exports, Custom House, London, made his own estimate of the average price at which each commodity was imported and exported by consulting published price lists and merchants. During the eighteenth century very little adjustment was made in commodity valuation. It was not until 1798 that the Inspector-General readjusted the "official values" in line with the real export value based on merchants' declarations and on current prices. See Ralph Davis, *The Industrial Revolution and British Overseas Trade* (Leicester, England: Leicester University Press, 1979), pp. 77–86; and Thomas S. Ashton's introduction to Elizabeth B. Schumpeter, *English Overseas Trade Statistics, 1697–1808* (Oxford: Clarendon Press, 1960).

22. Compiled and computed from SOT, ff. 81–181.

"Northern Europe (Baltic States except Germany; Scandinavia, Iceland, Greenland, Russia)"[23] are 16.7 and 18.0 percent, respectively.

Therefore, it is reasonable to assume that approximately the same proportion in Davis's computed "current real values" also belonged to Russia for the following reasons. First, Davis has provided a list of commodities imported from "Northern Europe" during the years 1784–1786 and 1794–1796 in their computed "current real values." The overwhelming majority of those commodities also consists of those that Russia traditionally supplied to Britain in great quantities. Second, there is a significant increase in their worth according to his computed "current real values"— nearly double the annual average from 1784–1786 to 1794–1796. Third, the valuation of both Iceland's and Greenland's exports to Great Britain were relatively negligible. Fourth, the exact definition of the "Baltic States" is absent from Davis's work much the same as there is no clear indication of what "East Country" meant in either the British official custom records or in contemporary compilations of those records. Consequently, Davis's region of "Northern Europe" is almost identical to Denmark-Norway, Sweden, and Russia. Thus, since the volume of the commodities Britain imported from Russia largely dominated the composition of the commodities imported from Davis's "Northern Europe" for the respective periods 1784–1786 and 1794–1796, it seems more than reasonable to assume that Russia's *proportion* in Davis's computed "current real values" must in reality be approximately the same as that in the "official values." That is, Russian exports to Great Britain amounted to approximately 80 percent of Davis's total computed "current real values."

A further and very important implication follows. Since Davis's computed "current real values" of Great Britain's imports from "Northern Europe" exceed the "official values" by more than half, the whole region had a greater significance than hitherto believed. In proportion, this must be true of Russia's contribution to Britain's total importation from the whole world. In other words, instead of Russia accounting for an annual average of 8.5 and 7.9 percent for the years 1784–1786 and 1794–1796, respectively, in "official values" of Britain's total world imports, in Davis's computed "current real values," it probably accounted for 14.2 percent

---

23. Davis, p. 82.

for each period. Thus, Britain's commodity trade deficit with Russia was much greater than the official figures indicate.[24]

Should similar computed "current real values" be established for other years and for earlier periods during the second half of the eighteenth century, they would perhaps demonstrate conclusively how profoundly important Russia was to Great Britain's economy during the entire reign of Catherine II.

But there is a paradox here. Despite Russia's enormous commercial expansion, and although Great Britain was in significant trade deficit to Russia throughout her entire reign, the Anglo-Russian commercial relationship was such a mutually dependent one that it presented Catherine II with enormous difficulties in her quest to emancipate Russia from its continuing dependence on the British market. In fact, this was a goal she never achieved.

---

24. See Herbert H. Kaplan, "Russia's Impact on the Industrial Revolution in Great Britain During the Second Half of the Eighteenth Century: The Significance of International Commerce," *FOG* 29 (Berlin: Otto Harrassowitz Wiesbaden, 1981), pp. 38–40, and SOT, ff. 76–77, 81, 94, 97, 171, 180–81, 208; and Davis, pp. 110–13.

# Conclusion

There was a symbiotic economic relationship between Great Britain and Russia during the second half of the eighteenth century; and the contributions each country made to the development of the other can no longer be ignored. Russia serviced the Royal Navy and mercantile marine and enhanced Britain's growing industrial export sector. In the absence of viable alternative markets of supply, Russia's contribution was more than just important to Great Britain—during its wars with the American colonies, France, Spain and the United Provinces, it was crucial. Britain's demand for Russian raw materials and manufactures was great enough to offset the intermittent lack of a commercial treaty with Russia that would have protected British merchants and guaranteed them a most-favored-nation privilege.

By the time of the Napoleonic Wars Great Britain commanded about one-third of the world's total foreign trade. Russia's contribution to Britain's ascendancy was not insignificant. During the two decades preceding the Napoleonic Wars Britain's imports from Russia amounted to about 14 percent (in computed current real value) of Britain's total worldwide imports. Included among the more than forty different Russian imports were several especially important ones that made enormous contributions to Great Britain's economic growth because they tended to have multiplier effects and because they provided British entrepreneurs with a means of exploiting advances in productivity. A few examples should suffice. More than 55 percent of Great Britain's total annual average bar iron imports was Russian and, therefore, about 30 percent of all British products made from bar iron was Russian in origin. Russian hemp accounted for about 96 percent of Britain's total hemp imports. Not only did imported Russian flax amount to about 80 percent of Britain's total flax imports, but when converted into linen it represented a significant portion of England's estimated total linen output, for example, 46 percent in 1785. During the latter part of Catherine's reign, Russian tallow imports accounted for 90 percent of Britain's total tallow imports. Finally, Britain imported an annual average of 71 percent of the critically strategic Great Masts from Russia and 85 percent during the war years of 1795 and 1796.

The Russian trade and navigation figured prominently in Great Britain's overall European trade and navigation. More than 15 percent of the British ships employed in that trade transported from Russia more than 25 percent of Britain's total import ship-tonnage; and more than 7 percent of the British ships carried to Russia more than 12 percent of Britain's total export ship-tonnage. Britain's Russian trade and navigation peaked during the last three years of Catherine II's reign, 1794–1796. About 27 percent of the British ships in the European trade transported about 35 percent of Britain's total import ship-tonnage from Russia, and 20 percent of the British ships carried 27 percent of Britain's total export ship-tonnage to Russia.

Britain's Russian trade and navigation contributed to Britain's economy in still another way. Although Catherine had very much to do with the enormous expansion of Russia's overseas commerce, she nevertheless yielded a considerable part of it to British ships, British merchants, and British mercantile services. This had the effect of reducing Britain's annual trade deficit with Russia.

It also must be emphasized that, although Russia traded with all of Europe and even modestly with America, the British market was Russia's principal outlet for its merchandise, a situation that not only sustained Russia's foreign trade expansion but also tied Russia and Great Britain closer together economically. That tie was further strengthened by the Russian market for Great Britain's domestic production and re-exports. Among the wide range of merchandise that Russia imported annually, British textiles, in particular, consistently figured among the ten most valued imports. The importance of the Russian market to certain sectors of the British economy was clearly in evidence when Russia, during its commercial and financial crisis, began in 1793 to restrict British imports. This created a furor in Britain's manufacturing and mercantile establishment. The Russian crisis of 1793 also provides convincing evidence of how Russia's wealthiest elites jeopardized their country's favorable overseas commodity trade balance by insistently demanding more foreign merchandise to support their extravagant lifestyle.

Neither Great Britain nor Russia succeeded completely in its commercial and political goals regarding the other. The British ministry failed to appreciate fully Catherine II's commercial and political ambitions, although its envoy in Russia reported on them quite early in Catherine's reign. The renewal of the commercial treaty with Russia ultimately did not provide Great Britain with greater advantages because it was repeatedly undermined by Russia's tariff policies. *Pari passu* with the tedious commercial treaty negotiations, Great Britain pursued a political alliance with Russia in the hopes of making its political interests in greater Europe more secure. However, it was only at the end of her reign, after revolutionary France had gone to war in Europe, that Catherine allied herself to Great Britain.

Early in her reign Catherine continued the commercial policies of her predecessors, Elizabeth and Peter III, and acknowledged the mutually dependent commercial relationship Russia had with Great Britain. However, she wanted to emancipate Russia from Great Britain's hegemony in Russia's overseas trade. To accomplish this she concluded several commercial treaties with other European countries, and delayed the renewal of the Anglo-Russian Commercial Treaty. She also wanted to become an arbiter-mediator in greater European political and commercial affairs, and therefore embarked upon an Armed Neutrality policy, portraying herself as the leader of smaller maritime states. In doing so Catherine challenged Great Britain's maritime and mercantile authority on the high seas. She failed in her pursuit of this policy, but not because she overestimated Russia's political and commercial value to Great Britain. Rather, she overestimated her ability to influence a set of advanced European political entanglements in the distant Atlantic arena. These were not only unfamiliar to her and to Russia, but were also far more complex than those she and Russia had traditionally encountered on the nearby and less sophisticated political playing fields of the Polish-Lithuanian Commonwealth and the Ottoman Empire. However, had she threatened to deprive Great Britain of the essential supplies that it could obtain only from Russia during its war with France, Spain, the United Provinces, and North America, Catherine might have made Great Britain accommodate her commercial and political aspirations.

Catherine also lacked the domestic statecraft to pursue the long-term goal of making Russia an independent and formidable mercantile and maritime power. Had she from early in her reign invested Russia's financial and material wealth into developing a major shipbuilding and mercantile program employing and training Russia's native-born subjects, Russia could have become competitive with other maritime powers in the Baltic-North Sea-Atlantic trade and would have been better prepared to meet the challenges of the Levant trade following Russia's annexation of the

Crimea. Instead, she promoted an expedient policy of naturalizing foreigners—merchants, ship masters, and their ships. In making them Russian she treated them preferentially in their custom duty payments and in their trading operations. This policy created a false impression of Russia's true maritime and mercantile strength and could produce only transient and ultimately negative results. For example, Russia actually experienced a severe decline in custom revenue during the last decade of Catherine's reign, when its total overseas commodity trade turnover more than doubled. Catherine's short-sighted policy ultimately reinforced Russia's dependence upon foreign merchants and foreign carriers, with Great Britain becoming its principal beneficiary.

The Anglo-Russian interdependent commercial relationship could last only so long as Great Britain continued to import Russian resources to supply the engines of its economic growth, and the material required to build and repair its warships and mercantile fleet. During the latter years of Catherine's reign, when Great Britain was again at war, there were no obvious signs that Britain could dispense with Russian imports despite the technological and industrial advancements within its economy. Nothing dampened the British enthusiasm "to buy Russian." Catherine clearly recognized this situation as an opportunity; just before she died she promulgated a new, although subsequently aborted, tariff that would have placed greater burdens on Great Britain's trade with Russia.

However much the Russian trade contributed to Great Britain's economic well being and strategic security during the decades following the Seven Years' War, Great Britain for its part greatly stimulated the intensity of Russia's overseas commerce. St. Petersburg's overseas commodity trade turnover more than quintupled during the course of Catherine II's reign. From 1783 through 1796, in particular, Russia's total overseas commodity trade turnover grew 2.8 times (exports by 3.1 and imports by 2.3). It is worth noting that the tempo of Russia's foreign trade expansion was comparable to what Great Britain and France experienced.

But this growth could not be sustained following Catherine's death. In the course of the following two decades Europe rapidly changed politically and economically. Russia's overseas commerce stagnated and declined as a result of the breakdown in Russian-British commercial intercourse, the Continental Blockade, and the Napoleonic Wars. After 1815 Russia and Great Britain followed different economic paths and experienced different trading relationships. The British economy continued to change and grow rapidly; it no longer depended upon Russian resources for its economic growth and strategic military security as it had during the previous half century. Russia, however, did not change. It remained for the most part rooted in its traditional economy and did not adapt to new economic developments until well into the nineteenth century.

# Bibliography

PRIMARY SOURCES

*Archives*

Great Britain, Public Record Office, London
   State Papers, Foreign
                Russia 91/61–107
                Hamburg 82/84–98
   *Foreign Office*
                Russia 65/1–52
                Russia 97/340–342
                Hamburg 33/1–3, 5–9, 12
                Hamburg 97/240

   *Board of Trade*

| Class No. | Dates | Description | Volumes |
|---|---|---|---|
| | | GENERAL | |
| 1 | 1791–1863 | In-Letters and Files, General | 3, 10, 12–14, 17 |
| 3 | 1786–1863 | Out-Letters, General | 2, 5 |
| 5 | 1784–1940 | Minutes | 2–5, 9–11 |
| 6 | 1697–1876 | Miscellanea | |
| | | Hemp, Flax and Cordage | 97 |
| | | Custom House Accounts | 104 |
| | | Custom House Accounts | 105 |
| | | Trade with Foreign Nations | 110 |
| | | Custom House Accounts | 141 |
| | | Custom House Accounts | 142 |
| | | Reports of the Committee of Trade | 180 |
| | | Reports of the Committee of Trade | 181 |
| | | Tables of Trade and Navigation | 185 |
| | | Tobacco | 225 |

|  |  | Spanish, Russian and Portuguese Trade, Navigation, etc. | 226 |
|  |  | Russian Trade, Whale Fishery, Custom Account | 230 |
|  |  | Russia Accounts | 231 |
|  |  | Russia Accounts | 232 |
|  |  | Russia Accounts | 233 |
|  |  | Custom House Accounts | 241 |
|  |  | Custom House Accounts | 242 |
|  |  | Vermillion, Gum, Hemp, Glass, Cambricks, Paper, etc. | 244 |
|  |  | Miscellaneous Papers | 262 |
|  |  | Miscellaneous Papers | 266 |

*Colonial Office*

| Class No. | Dates | Description | Volumes |
|---|---|---|---|
|  |  | Board of Trade |  |
| 388 | 1654–1792 | Original Correspondence | 49–52, 54, 56, 58, 95 |
| 389 | 1660–1803 | Entry Books | 31–32 |
| 390 | 1654–1799 | Miscellanea | 9 |

*Customs*

| Number | Descriptions | Volumes |
|---|---|---|
| 3 | Ledgers of Imports and Exports | 50–80 |
| 14 | Ledgers of Imports and Exports (Scotland) | 1a–14 |
| 17 | States of Navigation, Commerce and Revenue | 1–22 |

*Scottish Record Office, Edinburgh*
    Colonial Office 18

*Guildhall, London*
    Russia Company Papers. The Court Minute Books. MSS. 11,741, volumes 7–9

*Staatsarchiv Hamburg*
    371–2

*Admiralitäts-Kollegium*
    F 8, Bd. 1–3
    F 9
    F 10, Bd. 1–15
    F 11
    F 12, Bd. 1–19
    F 13, Bd. 1–7

*Published*

*A Collection Of All The Treaties Of Peace, Alliance, And Commerce, Between Great Britain And Other Powers, From the Treaty signed at Munster in 1648, to the Treaties signed at Paris in 1783. To which is prefixed, A Discourse on the Conduct of the Government of Great-Britain in Respect to Neutral Nations, by the Right Hon. Charles Jenkinson.* 3 vols. London: 1785. Reprint, New York: Augustus M. Kelley, 1969.

*A Tariff, or Book of Rates, of the Customs to be levied, on Goods imported & exported, at the Port and Frontier Custom-Houses.* Translated from the Russ. PRO SP 91/78. St. Petersburg: Imperial Academy of Sciences, 1767.

Barnes, G. R., and J. H. Owen, eds. *The Private Papers of John, Earl of Sandwich, First Lord of the Admiralty, 1771–1782.* 4 vols. London, Publications of the Navy Records Society, 1932–38.

Brough, Anthony. *A View of the Importance of the Trade Between Great Britain and Russia.* London, 1789.

Campbell, R. H., ed. *States of the Annual Progress of the Linen Manufacture, 1727–1754. From the Records of the Board of Trustees for Manufactures etc., in Scotland preserved in the Scottish Record Office.* Edinburgh, 1964.

*Convention between His Britannick Majesty and the Empress of Russia. Signed at London, the 25th of March, 1793,.* London: Edward Johnston, in Warwick-Lane, 1793. PRO BT 6/233.

Great Britain
  Parliament, House of Commons
  "A Report From The Committee Appointed To Consider How His Majesty's Navy may be better supplied with TIMBER. The 6th May 1771." *Reports From Commitees Of The House of Commons.* Volume III, pp. 15–53. London, 1803.

Great Britain
  Parliament. House of Commons
  "Report From The Committee Appointed to Enquire Into The Present State Of The Linen Trade In Great Britain And Ireland. Reported on the Twenty-fifth Day of May 1773." *Reports From Committees Of The House of Commons.* Volume III, pp. 99–133. London, 1803.

Great Britain
  Parliament. House of Lords
  "First Report from The Select Committee of the House of Lords, appointed to inquire into the means of extending and securing the Foreign Trade of the Country, and to report to

the House; together with the Minutes of Evidence taken before the said Committee—3 July 1820." *Reports From Select Committees Of The House of Lords* (269). Volume III. London, 1820.

Great Britain
Parliament. House of Commons
"Report From Select Committee on Timber Duties Together With Minutes of Evidence, An Appendix, And Index," *Reports From Committees*. XIX (519), London, 1835.

Great Britain
Parliament. Presentation to both Houses
"Customs Tariffs of the United Kingdom, from 1800 to 1897. With some notes upon the history of the more important branches of receipt from the year 1660." *Accounts and Papers*. XXXIV [c. 8706.], London, 1898.

Indova, E. I., ed. "O Rossiiskikh Manufakturakh vtoroi polovini XVIII v." In *Istoricheskaia geografiia Rossii XII–nachalo XX v. Sbornik statei k 70-letiiu professora Liubormira Grigor'evicha Beskrovnogo*. Moscow: Nauka, 1975, pp. 284–345.

———., et al., eds. *Vneshniaia torgovlia Rossii cherez Peterburgskii port vo votoroi polovine XVIII - nachale XIX v. Vedomosti o sostave kuptsov i ikh torgovykh oborotakh*. Moskva: Akademii Nauk, 1981.

*Journal of the Commissioners for Trade and Plantations*. London, 1933, 1938. Vols. from 1754–1758, 1759–1763, 1764–1767, 1768–1775, and Jan. 1776–May 1778.

Kaidonov, N., ed. *Sistematicheskii katalog delam kommissii o kommertsii i o poshlinakh, khraniashchimsia v Arkhive Departmenta Tamozhennykh Sborov*. St. Peterburg: V. Kirshbaum, 1887.

Kulomzin, A., ed. "Gosudarstvennye dokhodi i raskhodi v tsarstvovanie Ekateriny II." In *Sbornik Imperatorskago Russkago Istoricheskago Obshchestva*. Vols. V, pp. 219–94, VI, 219–304. St. Peterburg, 1870–1871.

Ovchinnikov, R. V., and M. A. Sidorov, eds. *Russko-Indiiskie otnosheniia v XVIII v. Sbornik dokumentov*. Moskva: Nauka, 1965.

Parry, Clive, ed. *The Consolidated Treaty Series*. Dobbs Ferry, NY: Oceana Publications, 1969.

Partridge, William. *A Practical Treatise on Dying of Woollen, Cotton, and Skein Silk with the Manufacture of Broadcloth and Cassimere including the Most Improved Methods in the West of England*. Introduction by J. de Mann and technical notes by K. G. Ponting. Edington, Wiltshire: Pasold Research Fund, Ltd., 1973.

*Polnoe Sobranie Zakonov Rossiiskoi Imperii*. Sankt Peterburg, 1830.

Vols.: XVI/11,630; XX/14,355; XXII/15,935, 16,187 [sic] 16,188 [?]; XXIII/17,215, 17,111, 17,511, 17,169, 17,170,
*Sanktpeterburgskie vedomosti*, 1763, 1771, 1784, and 1790.
*Sbornik Imperatorskago Russkago Istoricheskago Obshchestva*. St. Petersburg 1867–1916. Vols. I, IV-VI, VIII, XIV, XXXII, XLIII, LXVIII, XCIII, CVII-CVIII, CXXIII, CXXXIV, and CXLVII.
Scott, James Brown, ed. *The Armed Neutralities Of 1780 and 1800. A Collection of Official Documents Preceded by the Views of Representative Publicists*. Carnegie Endowment for International Peace. New York: Oxford University Press, 1918.
Segur, Count Louis Philippe de. *Memoirs and Recollections of Count Segur, Ambassador from France to the Courts of Russia and Prussia, &c. &c. Written by Himself.* London: Henry Colburn, 1826–27. Reprinted, New York: Arno Press and *New York Times*, 1970, vols. II-III.
Thornton, Henry. *An Inquiry into the Nature and Effects of the Paper Credit of Great Britain*. Philadelphia: James Humphreys, 1807.
Whitworth, Sir Charles. *A Register of the Trade of the Port of London; specifying the Articles Imported and Exported, arranged under the respective Countries; with A List of the Ships Entered Inwards and Cleared Outwards. Number I. For January, February, and March, 1776. Number II. April, May, and June, 1776.* London, 1777.
―――. *State of the Trade of Great Britain in its Imports and Exports, Progressively from the Year 1697: also the Trade of each particular Country, during the above Period, distinguishing each Year. In two parts with a Preface and Introduction, setting forth the Articles whereof each Trade consists.* London, 1776 [the hand-revised and amended edition to 1801, Public Record Office, London, Board of Trade 6/185].
Zlobin, K. K., ed. "Vedomost' sostoiashchem v St. Peterburge fabrikam, manufakturam i zavodam na Sentiabr' 1794 g." In *Sbornik Imperatorskago Russkago Istoricheskago Obshchestva*. Vol. I, pp. 352–61. St. Peterburg, 1867.

### SECONDARY WORKS

Ahlström, C. "Aspects of the commercial shipping between St. Petersburg and Western Europe, 1750–1790." In *The interactions of Amsterdam and Antwerp with the Baltic region, 1400–1800*. Leiden: Martinus Nijhoff, 1983, pp. 152–60.
Anderson, Adam. *An Historical and Chronological Deduction of the Ori-*

gins of Commerce, From the Earliest Accounts. Containing An History of the Great Commercial Interests of the British Empire*. Vols. 3–4. London: Printed for J. White et al., 1801.

Ahvenainen, Jorma. "Britain as a Buyer of Finnish Sawn Timber 1760–1860," in *Britain and the Northern Seas. Some Essays*, edited by Walter Minchinton. West Yorkshire: Lofthouse Publications, 1988, pp. 149–61.

Albion, Robert G. *Forests and Sea Power. The Timber Problem of the Royal Navy 1652–1862*. Cambridge: Harvard University Press, 1926.

———. "The Timber Problem of the Royal Navy, 1652–1862." *The Mariner's Mirror* 38:1 (Feb. 1952): pp. 4–22.

Anderson, M. S. *Britain's Discovery of Russia 1553–1815*. London: Macmillan & Co., 1958.

———. "Great Britain and the Russian Fleet, 1769–70." *The Slavonic and East European Review* 31:76 (1952/53): pp. 148–63.

———. "Great Britain and the Russo-Turkish War of 1768–74." *The English Historical Review* 69 (1954): pp. 39–58.

———. "Great Britain and the Growth of the Russian Navy in the Eighteenth Century." *The Mariner's Mirror* 42 (1956): pp. 132–46.

———. "The Great Powers and the Russian Annexation of the Crimea, 1783–4." *The Slavonic and East European Review* 37:88 (December 1958): pp. 17–41.

Anderson, R. C. *Naval Wars in the Baltic during the Sailing-Ship Epoch, 1552–1850*. London, C. Gilbert-Wood, 1910.

———. "British and American Officers in the Russian Navy." *The Mariner's Mirror* 33 (1947): pp. 17–27.

———. *Naval Wars in the Levant. 1559–1853*. Liverpool, University Press, 1952.

Ashton, Thomas S. *Economic Fluctuations in England, 1700–1800*. Oxford: Clarendon Press, 1959.

Åström, Sven-Erik. *From Stockholm to St. Petersburg. Commercial Factors in the Political Relations between England and Sweden 1675–1700*. Helsinki: Studia Historica 2, Finnish Historical Society, 1962.

———. "English Timber Imports from Northern Europe in the Eighteenth Century." *The Scandinavian Economic History Review* 18:1 (1970): pp. 12–32.

———. "Foreign Trade and Forest Use in Northeastern Europe, 1660–1860." In *Natural resources in European history. A conference report*, edited by Antoni Maczak and William N. Parker. Washington, DC: Resources For the Future, 1978, pp. 43–64.

———. "North European timber exports to Great Britain, 1760–1810," in *Shipping, Trade and Commerce. Essays in memory of Ralph Davis*, edited by P. L. Cottrell and D. H. Aldcroft. Leicester: Leicester University Press, 1981, pp. 81–97.

Attman, Artur. "The Russian Market in World Trade 1500–1860." In *Scandinavian Economic History Review* 29 (1981): pp. 177–202,

———. *Dutch Enterprise in the World Bullion Trade 1550–1800*. Göteborg, Sweden: Acta Regiae Societatis Scientiarum et Litterarum Gothoburgensis, Humaniora 23; Kungl. Vetenskaps- och Vitterhets-Samhället, 1983.

———. *American Bullion in the European World Trade 1600–1800*. Göteborg: Acta Regiae Societatis Scientiarum et Litterarum Gothoburgensis, Humaniora 26, Kungl. Vetenskaps-och Vitterhets-Samhället, 1986.

———. "Precious Metals and the Balance of Payments in International Trade 1500–1800." In *The Emergence of a World Economy 1500–1914*, Part I: 1500–1850, edited by Wolfram Fischer, R. Marvin McInnis, and Jürgen Schneider. Stuttgart: Steiner-Verlag-Wiesbaden-GmbH, 1986, pp. 113–21.

Augustine, Wilson R. "The Economic Attitudes and Opinions Expressed by the Russian Nobility in the Great Commission of 1767." Ph.D. diss., Columbia University, 1969.

———. "Notes Toward a Portrait of the Eighteenth-Century Russian Nobility." *Canadian-American Slavic Studies/Canadian Slavic Studies* 4:3 (Fall 1970): pp. 373–425.

Bamford, Paul W. *Forests and French Sea Power, 1660–1789*. Toronto, 1956.

Bartlett, Roger P. *Human Capital. The settlement of foreigners in Russia 1762–1804*. Cambridge: Cambridge University Press, 1979.

Beales, Derek. *Joseph II*, vol. 1 of *In the Shadow of Maria Theresa 1741–1780*. Cambridge: Cambridge University Press, 1987.

Beck, Ludwig. *Die Geschichte des Eisens. Dritte abteilung. Das XVIII Jahrhundert*. Braunschweig: Friedrich Vieweg und Sohn, 1897.

Behren, Betty. Review of *Britain, Russia and the Armed Neutrality of 1780. Sir James Harris's Mission to St. Petersburg during the American Revolution* by Isabel de Madariaga. *The English Historical Review* 79 (October 1964): p. 864.

Bemis, Samuel Flagg. *The Diplomacy of the American Revolution*. Bloomington: Indiana University Press, 1957.

Besset, Giliane. "Les relations commerciales entre bordeaux et la Russie au XVIIIe siècle." *Cahiers du Monde russe et soviétique*, XXIII: 2 (Avril-Juin 1982): p. 202.

Blanchard, Ian. "Resource Depletion in the European Mining and

Metallurgical Industries, 1400–1800." In *Natural resources in European history. A conference report,* edited by Antoni Mączak and William Parker. Washington, DC, 1978, pp. 85–113.

———. *Russia's 'Age of Silver.' Precious-Metal Production and Economic Growth in the Eighteenth Century.* London: Routledge, 1989.

Bode, Andreas. *Die flottenpolitik Katharinas II und die konflikte mit Schweden und der Türkei (1768–1792).* Wiesbaden: Otto Harrassowitz, 1979.

Bogucka, Maria. "Northern European Commerce as a Solution to Resource Shortage in the Sixteenth-Eighteenth Centuries." In *Natural resources in European history. A conference report,* edited by Antoni Maczak and William Parker. Washington, DC, 1978, pp. 9–42.

Brenner, Robert. *Merchants and Revolution. Commercial Change, Political Conflict, and London's Overseas Traders, 1550–1653.* Princeton: Princeton University Press, 1993.

Buist, Marten G. *At Spes Non Fracta. Hope & Co. 1770–1815. Merchant Bankers and Diplomats at Work.* The Hague: Martinus Nijhoff, 1974.

Bushkovitch, Paul. *The Merchants of Moscow 1580–1650.* Cambridge, Cambridge University Press 1980.

Butel, Paul. "France, the Antilles, and Europe in the seventeenth and eighteenth centuries: renewals of foreign trade." In *The Rise of Merchant Empires. Long-Distance Trade in the Early Modern World, 1350–1750,* edited by James D. Tracy. Cambridge: Cambridge University Press, 1990, pp. 153–73.

Butterfield, Herbert. "Review Article: British Foreign Policy, 1762–5." *The Historical Journal* 6:1 (1963): pp. 131–40.

Campbell, R. H. *Carron Company.* Edinburgh, 1961.

Carroll, Charles F. *The Timber Economy of Puritan New England.* Providence: Brown University Press, 1973.

Chapman, Stanley. *Merchant Enterprise in Britain. From the Industrial Revolution to World War I.* Cambridge: Cambridge University Press, 1992.

———. "James Longsdon (1745–1821), Farmer and Fustian Manufacturer: The Small Firm in the Early English Cotton Industry." *Textile History* 1:3 (December 1970): pp. 265–92.

———., and Serge Chassagne. *European Textile Printers in the Eighteenth Century. A Study of Peel and Oberkampf.* London: Heinemann Educational Books, Ltd., The Pasold Fund, 1981.

Chechulin, N. D. *Ocherki po istorii russkikh finansov v tsarstvovanie Ekateriny II.* St. Peterburg: Senatskaia Tipografiia, 1906.

———. *Vneshniaia politika Rossii v nachale tsarstvovanie Ekateriny II, 1762–1774*. St. Peterburg, 1895.

Chulkov, Mikhail D. *Istoricheskoe opisanie rossiiskoi kommertsii*. 7 vols. St. Petersburg, 1781–1788.

———. *Istoriia kratkaia rossiiskoi torgovli*. Moskva, 1788.

Ćieslak, Edmund. "Aspects of Baltic Sea-borne Trade in the Eighteenth Century: The Trade Relations between Sweden, Poland, Russia and Prussia." *The Journal of European Economic History* 12:2 (Fall 1983): pp. 239–70.

Clapham, John Harold. "The Project for an Anglo-Polish Treaty (1782–1792)," *Baltic Countries* 1 (1935): pp. 33–35.

Clark, George N. *Guide to English Commercial Statistics, 1696–1782*. London: Royal Historical Society, 1938.

Clarke, Mary Patterson. "The Board of Trade at Work." *The American Historical Review* 17 (1912): pp. 17–43.

Clendenning, P. H. "Eighteenth Century Russian Translations of Western Economic Works," *The Journal of European Economic History* 1:3 (1972): pp. 745–53.

———. "The Anglo-Russian Trade Treaty of 1766." Ph.D. dissertation, Cambridge University, 1975.

———. "William Gomm: A Case Study of the Foreign Entrepreneur in Eighteenth Century Russia." *The Journal of European Economic History* 6:3 (Winter 1977): pp. 533–48.

———. "The Economic Awakening of Russia in the Eighteenth Century." *The Journal of European Economic History* 14:3 (Winter 1985): pp. 443–71.

———. "The Anglo-Russian Trade Treaty of 1766—An Example of Eighteenth-Century Power Group Interests." *The Journal of European Economic History* 19:3 (Winter 1990): pp. 475–520.

Cole, W. A. "Factors in demand 1700–80." In *The Economic History of Britain since 1700. Volume 1: 1700–1860*, edited by Roderick Floud and Donald McCloskey. Cambridge: Cambridge University Press, 1981, pp. 36–65.

Coleman, D. C. "Textile Growth." In *Textile History and Economic History*, edited by N. B. Harte and K. G. Ponting. Manchester: Manchester University Press, 1973, pp. 1–21.

———. *History and the Economic Past. An Account of the Rise and Decline of Economic History in Britain*. Oxford: Clarendon Press, 1987.

———., and Peter Mathias, eds. *Enterprise and history. Essays in honor of Charles Wilson*. Cambridge: Cambridge University Press, 1984.

Cottrell, P. L., and D. H. Aldcroft, eds. *Shipping, Trade and Commerce. Essays in Memory of Ralph Davis.* Leicester: Leicester University Press, 1981.

Coxe, William. *Travels in Poland, Russia, Sweden, and Denmark.* Fifth ed., 5 vols. London: Printed for T. Cadell, Jun. and W. Davies, in the Strand, 1802.

Cross, A. G. *"By the Banks of the Thames." Russians in Eighteenth Century Britain.* Newtonville, Mass.: Oriental Research Partners, 1980, pp. 23–27.

———., ed. *An English Lady at the Court of Catherine the Great. The Journal of Baroness Elizabeth Dimsdale, 1781.* Cambridge: Crest Publications, 1989, Introduction, p. 3.

Cross, Anthony C. "The British in Catherine's Russia: A Preliminary Survey." In *The Eighteenth Century in Russia,* edited by J. G. Garrard. Oxford: Clarendon Press, 1973, pp. 233–63.

Crouzet, François. *L'Economie Britanique et blocus continental.* Second edition. Paris: Economica, 1987.

———. "Opportunity and Risk in Atlantic Trade During the French Revolution." *Interactions in the World Economy. Perspectives from International Economic History.* Edited by Carl-Ludwig Holtfrerich. New York: New York University Press, 1989, pp. 90–150.

———. *Britain ascendant: Comparative studies in Franco-British economic history.* Cambridge: Cambridge University Press, 1990.

Cullen, L. M. *Anglo-Irish Trade 1660–1800.* New York: August M. Kelley, 1968.

Daniel Jr., Wallace L. "Russian Attitudes Toward Modernization: The Merchant-Nobility Conflict in the Legislative Commission, 1767–1774." Ph.D diss., University of North Carolina, 1973.

———. "The Merchantry and the Problem of Social Order in the Russian State: Catherine II's Commission on Commerce." *The Slavonic and East European Review* 60:2 (April 1977): pp. 185–203.

———. "Grigorii Teplov and the Conception of Order: The Commission on Commerce and the Role of the Merchantry in Russia." *Canadian-American Slavic Studies/Canadian Slavic Studies* 16:3-4 (Fall-Winter 1982): pp. 410–31.

———. *Grigorii Teplov: A Statesman at the Court of Catherine the Great.* Newtonville: Mass.: Oriental Research Partners, 1991.

Davis, Ralph. "English Foreign Trade, 1700–1774." *The Economic History Review,* second series, 15:2 (December 1962): pp. 285–303.

———. *The Rise of the English Shipping Industry in the Seventeenth and*

*Eighteenth Centures*. Newton Abbot, Great Britain: David and Charles, 1972.

———. *The Industrial Revolution and British Overseas Trade*. Leicester: Leicester University Press, 1979.

Deane, Phyllis, and W. A. Cole. *British Economic Growth 1688–1959*. Cambridge: Cambridge University Press, 1962.

———., and H. J. Habakkuk. "The Take-Off in Britain." In *The Economics of Take-Off Into Sustained Growth*, edited by W. W. Rostow. New York, 1963, pp. 63–82.

Deerr, Noel. *The History of Sugar*. London: Chapman and Hall, Ltd., 1950. Two volumes.

Diubiuk, Evg. *Polotnianaia promyshlennost' Kostromskogo kraia vo vtoroi polovine XVIII i pervoi polovine XIX veka*. Kostroma: Sovetskaia, 1921.

Dixon, G. Graham. "Notes on the Records of the Custom House, London." *Economic History Review* 34 (January 1919): pp. 71–82.

Dmytryshyn, Basil. "The Economic Content of the 1767 *Nakaz* of Catherine II." *American Slavic and East European Review* 19:1 (1960): pp. 1–9.

Doroshenko, V. V. *Ekonomicheskie sviazi pribaltiki s Rossei. Sbornik statei*. Riga: Zinatne, 1968.

———. "Rigas tirdznieciba 18s." In *Feodala Riga*, edited by T. Zeids. Riga: Zinatne, 1978, pp. 257–80.

———. "Torgovlia Rigi v period kontinental'noi blokady." *Istoriia* (Izvestiia Akademii Nauk Latviiskoi CCR) 7 (1979): pp. 23–32.

———. "Quellen zur Geschichte des Rigaer Handels im 17.-18. Jahrhundert und Probleme ihrer Erforschung." In *Seehandel und Wirtschaftswege Nordeuropas im 17. und 18. Jahrhundert*, edited by Klaus Friedland and Franz Irsigler. Ostfildern: Scripta Mercaturae Verlag, 1981, pp. 3–25.

———., and Elisabeth Harder-Gersdorff. "Ost-Westhandel und Wechselgeschäfte zwischen Riga und westlichen Handelsplätzen: Lübeck, Hamburg, Bremen und Amsterdam (1758/59)." *Zeitschrift des Vereins für Lübeckische Geschichte und Altertumskunde* 62 (1982): pp. 103–53. Verlag Max Schmidt-Römhild, Lübeck.

———. "Merchant-bankers in Riga Sea Trade in the 18th century." Doroshenko letter to Kaplan, Nov. 19, 1990.

Dukes, Paul, ed. *Catherine the Great's Instruction (NAKAZ) to the Legislative Commission, 1767*. Newtonville, MA, Oriental Research Partners, 1977.

Duran Jr., James A. "The Reform of Financial Administration in Russia during the Reign of Catherine II." *Canadian-American*

*Slavic Studies/Canadian Slavic Studies* 4:3 (Fall 1970): pp. 485–96.

Durie, Alastair J. *The Scottish Linen Industry in the Eighteenth Century.* Edinburgh: John Donald Publishers, 1979.

Eeckaute, Denise. "La legislation des forêts au XVIII$^e$ siècle." *Cahiers du monde russe et soviétique*, 9:2 (1968) pp. 194–208.

Ehrman, John. *The British Government and Commercial Negotiations with Europe 1783–1793.* Cambridge, Cambridge University Press, 1962.

Eicher, Carl, and Lawrence Witt, eds. *Agriculture in Economic Development.* New York, 1964.

Evans, R. J. Morda. "Recruitment of British Personnel for the Russian Service 1734–1738." *The Mariner's Mirror* 47 (1961): pp. 126–37.

Feldbaek, Ole. *Dansk neutralitetspolitik under krigen 1778–1783.* Kobenhavn: G. E. C. Gads Forlag, 1971.

Firsov, N. N. *Pravitel'stvo i obshchestvo v ikh otnosheniiakh k vneshnei torgovle Rossii v tsarstvovanie imperatritsy Ekateriny II. Ocherki iz istorii torgovoi politiki.* Kazan': Imperatskii Universitet, 1902.

Fisher, Alan W. *The Russian Annexation of the Crimea 1772–1783.* Cambridge: Cambridge University Press, 1970.

Flinn, M. W. "Revisions in Economic History. XVII. The Growth of the English Iron Industry, 1660–1760." *The Economic History Review*, second series, 11:1 (August 1958): pp. 144–153.

———. *The Origins of the Industrial Revolution.* New York, 1966.

———. "Technical Change as an escape from Resource Scarcity: England in the Seventeenth and Eighteenth Centuries." In *Natural resources in European history. A conference report*, edited by Antoni Mączak and William Parker. Washington, DC: Resources for the Future, 1978, pp. 139–59.

Floud, Roderick, and Donald McCloskey, eds. *The Economic History of Britain since 1700. Volume I: 1700–1860.* Cambridge: Cambridge University Press, 1981.

Foust, Clifford M. *Rhubarb. The Wondrous Drug.* Princeton: Princeton University Press, 1992.

———. *Muscovite and Mandarin: Russia's Trade with China and Its Setting, 1727–1805.* Durham: The University of North Carolina Press, 1969.

———. "Russia's Peking Caravan, 1689–1762." *The South Atlantic Quarterly* 67:1 (Winter 1968): pp. 108–24.

Fox, Frank. "French-Russian Commercial Relations in the Eighteenth Century and the French-Russian Commercial Treaty of 1787." Ph.D. diss., University of Delaware, 1968.

———. "Negotiating with the Russians: Ambassador Ségur's Mission to Saint Petersburg, 1784-1789." *French Historical Studies* 7:1 (Spring 1971): pp. 47-71.

Francis, A. D. *The Wine Trade*. London: Adam & Charles Black, 1972.

French, Christopher J. "Eighteenth-Century Shipping Tonnage Measurements." *Journal of Economic History* 33:2 (June 1973): pp. 434-443.

French, R. A. "Russians and the Forest." In *Studies in Russian Historical Geography*, edited by James H. Bater and R. A. French. London: Academic Press, 1983, I, pp. 23-44.

Gayer, Arthur D., Walt W. Rostow, and Anna Jacobson Schwartz. *The Growth and Fluctuation of the British Economy 1790-1850: An Historical, Statistical, and Theoretical Study of Britain's Economic Development*. 2 vols. New York: Barnes and Noble, 1975.

Geijer, Agnes. *A History of Textile Art. A Selective Account*. 1979. Reprint. London: Pasold Research Fund in association with Sotheby Parke Bernet, 1982.

Gerhard, Dietrich. *England und der Aufstieg Russlands*. München und Berlin: Verlag von R. Oldenbourg, 1933.

Gleason, Walter J. *Moral Idealists, Bureaucracy, and Catherine the Great*. New Brunswick, NJ: Rutgers University Press, 1981.

Griffiths, David M. "Russian Court Politics and the Question of an Expansionist Foreign Policy under Catherine II, 1762-1783." Ph.D. diss., Cornell University, 1967.

———. "The Rise and Fall of the Northern System: Court Politics and Foreign Policy in the First Half of Catherine II's Reign." *Canadian-American Slavic Studies/Canadian Slavic Studies* 4:3 (Fall 1970): pp. 547-69.

Grigorova-Zakharova, S. P. "Torgovlia zhelezom Golitsynykh vo II polovine XVIII veka i ee ekonomicheskie usloviia." Candidate diss., Moscow University, 1953.

Hamilton, Henry. *An Economic History of Scotland in the Eighteenth Century*. Oxford: Clarendon Press, 1963.

Hamilton, Henry. *The Industrial Revolution in Scotland*. London: Frank Cass, 1966.

Harder, Elisabeth. "Seehandel zwischen Lübeck und Russland im 17./18. Jahrhundert nach Zollbüchern der Novgorodfahrer. Erster Teil, 41 (1961), pp. 43-114; Zeiter Teil, 42 (1962), pp. 5-53; *Zeitschrift des Vereins für Lübeckische Geschichte und Altertumskunde*. Lübeck: Verlag Max Schmidt-Römhild.

Harder-Gersdorff, Elisabeth. "Mitteleuropäische Gerwerbezonen und ostbaltischer Handel im 18. Jahrhundert." In *Seehandel und*

*Wirtschaftswege Nordeuropas im 17. und 18. Jahrhundert,* edited by Klaus Friedland an Franz Irsigler. Ostfildern: Scripta Mercaturae Verlag, 1981, pp. 26–37.

———. "Zwischen Rigar und Amsterdam: die Geschäfte des Herman Fromhold mit Frederik Beltgens & Comp., 1783–1785." In *The interactions of Amsterdam and Antwerp with the Baltic region, 1400–1800.* Leiden: Martinus Nijhoff, 1983, pp. 171–80.

Harley, R. D. *Artists' Pigments c. 1600–1845. A Study in English Documtary Sources.* New York: Elsevier, 1970; London: Butterworth Scientific, 2nd ed., 1982.

Harper, Lawrence A. *The English Navigation Laws. A Seventeenth-Century Experiment in Social Engineering.* New York, 1964.

Harte, N. B. "The Rise of Protection and the English Linen Trade, 1690–1790." *Textile History and Economic History,* edited by N. B. Harte and K. G. Ponting. Manchester: Manchester University Press, 1973, pp. 74–112.

———., and K. G. Ponting, eds. *Textile History and Economic History. Essays in Honour of Miss Julia de Lacy Mann.* Manchester: Manchester University Press, 1973.

Hartley, Janet M., ed. *The study of Russian history from British archival sources.* London: Mansell Publishing, 1986.

Harvey, Mose Lofley. "The Development of Russian Commerce on the Black Sea and Its Significance." Ph.D. diss., University of California, 1938.

Hassell, James E. "Catherine II and Procurator General Vjazemskij." *Jahrbücher für Geschichte Osteuropas,* Neue Folge, 24:1 (1976): pp. 23–30.

Hatton, Ragnhild, and M. S. Anderson, eds. *Studies in Diplomatic History. Essays in Memory of David Bayne Horn.* London: Longman, 1970.

Hautala, Kustaa. *European and American Tar in the English Market During the Eighteenth and Early Nineteenth Centuries.* Helsinki: Suomalainen Tiedeakatemia, 1963.

Heaton, Herbert. *The Yorkshire Woollen and Worsted Industries.* 2nd ed. Oxford: Clarendon Press, 1965.

Heckscher, Eli F. "Multilateralism, Baltic Trade, and the Mercantilists." *The Economic History Review,* second series, 3:2 (1950): pp. 219–28.

Heckscher, Eli F. *An Economic History of Sweden.* Cambridge, MA: Harvard University Press, 1954.

Heeres, W. G. et al. *From Dunkirk to Danzig. Shipping and Trade in the North Sea and the Baltic, 1350–1850.* Hilversum, Netherlands:

Verloren Publishers, Amsterdamse Historische Reeks Nederlands Economisch-Historisch Archief, 1988.

Heller, Klaus. *Die Geld-und Kreditpolitik des russischen Reiches in der Zeit der Assignaten (1768–1839/43)*. Wiesbaden: Franz Steiner Verlag GMBH, 1983.

Herlihy, Patricia Ann McGahey. "Russian Grain and Mediterranean Markets, 1774–1861." Ph.D. diss., University of Pennsylvania, 1963.

———. *Odessa. A History, 1794–1914*. Cambridge, MA: Harvard Ukrainian Research Center, 1986.

Hildebrand, K. G. "Foreign Markets for Swedish Iron in the 18th Century." *The Scandinavian Economic History Review* 6:1 (1958): pp. 3–52.

Hittle, J. Michael. *The Service City. State and Townsmen in Russia, 1600–1800*. Cambridge, MA: Harvard University Press, 1979.

Hollingsworth, Barry. "Some Aspects of Anglo-Russian Trade in the Eighteenth Century" [Synopses of Paper Read at 15–16 December 1973 Meeting of Study Group on Eighteenth-Century Russia, University of Leeds]. *Study Group on Eighteenth-Century Russia Newsletter* 2 (Sept. 1974): pp. 12–15.

Holtfrerich, Carl-Ludwig. "Introduction: The Evolution of World Trade, 1720 to the Present." *Interactions in the World Economy. Perspectives from International Economic Hisory*. Edited by Carl-Ludwig Holtfrerich. New York: New York University Press, 1989, pp. 1–30.

———. editor. *Interactions in the World Economy. Perspectives from International Economic Hisory*. New York: New York University Press, 1989.

Horn, David Bayne. "The Board of Trade and Consular Reports, 1696–1782." *The English Historical Review* 54 (July 1939): pp. 476–80.

Horner, John. *The Linen Trade of Europe during the Spinning-Wheel Period*. Belfast, 1920.

Hummel, J. J. *Textile Fabrics and Their Preparation for Dyeing*. New and revised edition. Edited by Paul N. Hasluck. London: Cassell and Company, Ltd., 1906.

Hurewitz, J. C. *Diplomacy in the Near And Middle East. A Documentary Record: 1535–1914*. Vol. 1. Princeton, D. Van Nostrand Company, 1958, pp. 54–61.

Hyde, Charles K. "Technological Change in the British Wrought Iron Industry, 1750–1815: A Reinterpretation." *The Economic History Review*, second series, 27:2 (May 1974): pp. 190–206.

———. *Technological Change and the British Iron Industry 1700–1870.* Princeton: Princeton University Press, 1977.
Iakovtsevskii, V. N. *Kupecheskii kapital v feodal'no-krepostnicheskoi Rossii.* Moskva: Nauk, 1953.
Isaev, G. S. *Rol' tekstil'noi promyshlennosti v genezise i razvitii kapitalizma v Rossii 1760–1860.* Leningrad: Nauka, 1970.
Jackson, Gordon. *Hull in the Eighteenth Century. A Study in Economic and Social History.* London: Oxford University Press, 1972.
———. *The British Whaling Trade.* Hamden, CT: Archon Books, 1978.
———. "Scottish Shipping, 1775–1805." In *Shipping, Trade and Commerce. Essays in Memory of Ralph Davis*, edited by P. L. Cottrell and D. H. Aldcroft. Leicester: Leicester University Press, 1981, pp. 117–136.
———., with Kate Kinnear. *The Trade and Shipping of Dundee, 1780–1850.* Dundee: Abertay Historical Society Publication No. 31, 1991.
Jane, Fred T. *The Imperial Russian Navy, its Past, Present, and Future.* London: W. Thacker & Co., 1899.
Johansen, Hans Chr. *Shipping and Trade between the Baltic Area and Western Europe 1784–95.* Odense, Denmark: Odense University Press, 1983.
———. "Ships and cargoes in the traffic between the Baltic and Amsterdam in the late eighteenth century." In *The interactions of Amsterdam and Antwerp with the Baltic region, 1400–1800.* Leiden: Martinus Nijhoff, 1983, pp. 161–70.
———. "How to Pay for Baltic Products." In *The Emergence of a World Economy 1500–1914,* Part I: 1500–1850, edited by Wolfram Fischer, R. Marvin McInnis, and Jürgen Schneider. Stuttgart: Steiner-Verlag-Wiesbaden-GmbH, 1986, pp. 123–42.
Jones, Robert E. "Introduction" to *Zapiski Kniazia Iakova Petrovicha Shakhovskago, 1705–1777.* Cambridge, MA: Oriental Research Partners, 1974. Originally published in St. Peterburg, 1872.
———. *Provincial Development in Russia. Catherine II and Jacob Sievers.* New Brunswick: Rutgers University Press, 1984.
———. "Getting the Goods to St. Petersburg: Water Transport from the Interior 1703–1811." *Slavic Review* 43:3 (Fall 1984): pp. 413–33.
———. "Opposition to War and Expansion in Late Eighteenth Century Russia." *Jahrbücher für Geschichte Osteuropas* 32:1 (1984): pp. 34–51.
Kahan, Arcadius. *The Plow The Hammer and The Knout. An Economic*

*History of Eighteenth-Century Russia.* Chicago: University of Chicago Press, 1985.

Kaplan, Herbert H. *The First Partition of Poland.* New York: Columbia University Press, 1962.

———. *Russia and the Outbreak of the Seven Years' War.* Berkeley: University of California Press, 1968.

———. "Russian Commerce with Great Britain in the Second Half of the Eighteenth Century: Early Diplomatic Considerations." In *State and Society in Europe from the Fifteenth to the Eighteenth Century.* Warsaw: University of Warsaw Press, 1981, pp. 215–232.

———. "Russia's Impact on the Industrial Revolution in Great Britain During the Second Half of the Eighteenth Century: The Significance of International Commerce." *Forschungen zur osteuropäischen Geschichte* 29 (Berlin: Otto Harrassowitz Wiesbaden, 1981): pp. 7–59.

———. "Observations on the Value of Russia's Overseas Commerce with Great Britain During the Second Half of the Eighteenth Century." *Slavic Review* 45:1 (1986): pp. 85–94.

Kellenbenz, Hermann. "The Economic Significance of the Archangel Route (from the late 16th to the late 18th century)." *The Journal of European Economic History* 2:3 (Winter 1973): pp. 541–81.

———. "The Armed Neutrality of Northern Europe and the Atlantic Trade of Schleswig-Holstein and Hamburg during the Seven Years War and the War of Independence." In *The American Revolution and the Sea, Proceedings of the 14th International Conference of the International Commission for Maritime History.* London, National Maritime Museum, 1974, pp. 74–81.

Kent, H. S. K. "The Anglo-Norwegian Timber Trade in the Eighteenth Century." *Economic History Review,* second series, 8:1 (August 1955): pp. 62–74.

———. *War and Trade in Northern Seas: Anglo-Scandinavian Economic Relations in the Mid-Eighteenth Century.* Cambridge: Cambridge University Press, 1973.

Kirchner, Walther. *Commercial Relations between Russia and Europe 1400 to 1800. Collected Essays.* Bloomington: Indiana University Press, 1966.

Kizevetter, A. A. *Gorodovoe polozhenie Ekateriny II 1785 g.* Moskva: Tipografiia Imperatorskago Moskovskago Universiteta, 1909.

Kjaerheim, Steiner. "Norwegian Timber Exports in the 18th Century: A Comparison of Port Books and Private Accounts." *The Scandinavian Economic History Review* 5:2 (1957): pp. 188–201.

Knight, R. J. B. "The Performance of the Royal Dockyards in England during the American War of Independence." *The American Revolution and the Sea. Proceedings of the 14th International Conference of the International Commission for Maritime History.* London: National Maritime Museum, 1974, pp. 139–44.

———. "The Building and Maintenance of the British Fleet during the Anglo-French Wars 1688–1815." In *Les Marines de guerre Européennes XVII-XVIII<sup>e</sup> siècles*, edited by Martine Acerra, José Merino, and Jean Meyer. Paris: Presses de l'Université de Paris-Sorbonne, 1985, pp. 35–50.

———. "New England Forests and British Seapower: Albion Revised." *American Neptune* 46:4 (Fall 1986): pp. 221–29.

Knoppers, Jake V. Th. *Dutch Trade with Russia from the Time of Peter I to Alexander I: A Quantitative Study in Eighteenth Century Shipping.* 3 vols. Montréal: Interuniversity Centre for European Studies, 1976.

Kohut, Zenon E. *Russian Centralism and Ukrainian Autonomy. Imperial Absorption of the Hetmanate 1760s-1830s.* Cambridge, MA: Harvard Ukrainian Research Institute, 1988.

Koutaissoff, E. "Essays in Bibliography and Criticism. XVIII. The Ural Metal Industry in the Eighteenth Century." *The Economic History Review*, second series, 4:2 (1951): pp. 252–55.

Kozintseva, R. I. "Ot tamozhennogo tarifa 1724 g. k tarifu 1731 g." In *Voprosy genezisa kapitalizma v Rossii*. Leningrad: Leningrad University Press, 1960, pp. 182–216.

Kresse, Walter. *Materialien zur Entwicklungsgeschichte der Hamburger Handelsfloote 1765–1823.* Hamburg: Museum für Hamburgische Geschichte, 1966.

Kroll, Adam. *A Commercial Dictionary in the English and Russian Languages; with A Full Explanation of the Russian Trade, etc.* London: S. Chappel, Royal Exchange, 1800.

Kulisher, I. M. *Ocherk istorii russkoi torgovli.* St. Petersburg: Atenei, 1923.

Kutz, Martin. *Deutschlands Aussenhandel von der französischen Revolution bis zur Gründung des Zollvereins.* Beiheft 61. *Vierteljahrschrift für Sozial-und Wirtschaftsgeschichte.* Wiesbaden: Franz Steiner Verlag, 1974.

Lappo-Danilevskii, A. S. "Die russische Handelskommission von 1763–1796." In *Beiträge zur russischen Geschichte*, edited by Otto Hötzsch. Berlin, 1907.

Lauber, Jack M. "The Merchant-Gentry Conflict in Eighteenth-Century Russia." Ph.D. diss., University of Iowa, 1967.

LeDonne, John P. *Ruling Russia: Politics and Administration in the Age of Absolutism 1762–1796*. Princeton: Princeton University Press, 1984.

LeDonne, John P. *Absolutism and Ruling Class. The Formation of the Russian Political Order 1700–1825*. New York: Oxford University Press, 1991.

Lee, Sir Frank., *The Board of Trade (The Stamp Memorial Lecture delivered before the University of London on 11 November 1958)*. London, 1958.

Leggett, William F. *Ancient and Medieval Dyes*. Brooklyn, NY: Chemical Publishing, 1944.

Leighton-Boyce, J. A. S. L. *Smith's the Bankers 1658–1958*. London: National Provincial Bank Ltd., 1958.

Leonard, Carol S. *Reform and Regicide. The Reign of Peter III of Russia*. Bloomington: Indiana University Press, 1993.

———. "The Reputation of Peter III." *The Russian Review* 47 (1988): pp. 263–92.

Leventer, Herbert. "Introduction" to *Zapiski Ivana Ivanovicha Nepliueva (1693–1773)*. Cambridge, MA: Oriental Research Partners, 1974. Originally published in St. Peterburg, 1872.

Lindblad, J. Th. *Sweden's Trade with the Dutch Republic 1738–1795*. The Netherlands: Van Gorcum, ASSN, 1982.

———., and P. de Buck. "Shipmasters in the shipping between Amsterdam and the Baltic, 1722–1780." In *The interactions of Amsterdam and Antwerp with the Baltic region, 1400–1800*. Leiden: Martinus Nijhoff, 1983, pp. 134–52.

Lingelbach, Anna Lane. "The Inception of the British Board of Trade." *The American Historical Review* 30 (1925): pp. 701–27.

Lodyzhenskii, Konstantin N. *Istoriia russkago tamozhennago tarifa*. St. Peterburg: V. S. Balashev, 1886.

Lower, Arthur R. M. *Great Britain's Woodyard. British America and the Timber Trade, 1763–1867*. Montréal and London, 1973.

Lukowski, Jerzy. *Liberty's Folly. The Polish-Lithuanian Commonwealth in the Eighteenth Century, 1697–1795*. London: Routledge, 1991.

McCulloch, J. R. *A Dictionary, Practical, Theoretical, and Historical of Commerce and Commercial Navigation*, edited by Henry Vethake. 2 vols. Philadelphia, 1843.

McCusker, John J. *European Bills of Entry and Marine Lists: Early Commercial Publications and the Origins of the Business Press*. Cambridge, Mass.: Harvard University Library, 1985.

———., and Cora Gravesteijn. *The Beginnings of Commercial and Financial Journalism. The Commodity Price Currents, Exchange Rate*

*Currents, and Money Currents of Early Modern Europe.* Amsterdam: NEHA, 1991.

McGrew, Roderick E. "Dilemmas of Development: Baron Heinrich Friedrich Storch (1766–1835) on the Growth of Imperial Russia." *Jahrbücher für Geschichte Osteuropas,* Neue Folge, 24:1 (1976): pp. 31–71.

———. *Paul I of Russia 1754–1801.* Oxford: Clarendon Press, 1992, pp. 150, 282, 308.

McKay, Derek, and H. M. Scott. *The Rise of the Great Powers 1648–1815.* London: Longman, 1983.

Macmillan, David S. "The Scottish-Russian Trade: Its Development, Fluctuations, and Difficulties 1750–1796." *Canadian-American Slavic Studies/Canadian Slavic Studies* 4:3 (Fall 1970): pp. 426–42.

———. "Paul's 'Retributive Measures' of 1800 Against Britain: The Final Turning-Point in British Commercial Attitudes towards Russia." *Canadian-American Slavic Studies/Canadian Slavic Studies* 7:1 (Spring 1973): pp. 68–77.

Macpherson, David. *Annals of Commerce, Manufactures, Fisheries, and Navigation, with Brief Notices of the Arts and Sciences with them containing the Commercial Transactions of the British Empire and Other Countries.* 4 vols. London: Printed for Nichols and son et al., 1805.

Madariaga, Isabel de. *Britain, Russia and the Armed Neutrality of 1780. Sir James Harris's Mission to St. Petersburg during the American Revolution.* London: Hollis & Carter, 1962.

———. *Russia in the Age of Catherine the Great.* New Haven: Yale University Press, 1981.

———. Review of *The Well-Ordered Police State: Social and Institutional Change Through Law in the Germanies and Russia, 1600–1800* by Marc Raeff. The *London Times* publication, *Times Literary Supplement,* November 25, 1983, p. 1326.

Malone, Joseph J. *Pine Trees and Politics. The Naval Stores and Forest Policy in Colonial New England, 1691–1775.* Seattle: University of Washington Press, 1964.

———. "England and the Baltic Naval Stores Trade in the Seventeenth and Eighteenth Centuries." *The Mariner's Mirror* 58:1 (Feb. 1972): pp. 375–95.

Marcum, James W. "Semen R. Vorontsov: Minister to the Court of St. James's for Catherine II, 1785–1796." Ph.D. diss., University of North Carolina, 1970.

Mathias, Peter. "Russia and the British Industrial Revolution in the Eighteenth Century." A commentary on a paper by Professor

H. H. Kaplan delivered at the 93rd Annual Meeting of the American Historical Association, 28 December 1978, San Francisco.

Matthews, R. C. O. "The Trade Cycle in Britain, 1790–1850." In *British Economic Fluctuations 1790–1939*, edited by Derek H. Aldcroft and Peter Fearon. London, 1972, pp. 97–130.

Mauro, Frédérick. "Merchant Communities, 1350–1750." In *The Rise of Merchant Empires. Long-Distance Trade in the Early Modern World, 1350–1750*, edited by James D. Tracy. Cambridge: Cambridge University Press, 1990, pp. 255–86.

Meier, Gerald M. "Theoretical Issues Concerning the History of International Trade and Economic Development." *Interactions in the World Economy. Perspectives from International Economic History.* Edited by Carl-Ludwig Holtfrerich. New York: New York University Press, 1989, pp. 33–58.

Menke, Christoph Friedrich. "Die wirtschaftlichen und politischen Bezhiehungen der Hansestädte zu Russland im 18. und frühen 19. Jahrhundert." Ph.D. diss., Göttingen Universität, 1959.

Metcalf, Michael F. *Russia, England and Swedish Party Politics 1762–1766.* Stockholm: Almqvist & Wiksell, 1977.

Meyer, Jean, and John Bromley. "The Second Hundred Years' War (1689–1815). In *Britain and France. Ten Centuries.* Edited by Douglas Johnson, François Crouzet and François Bedarida. England: Wm. Dawson & Son Ltd., 1980. pp. 137–72.

Miller, Daniel A. *Sir Joseph Yorke and Anglo-Dutch Relations, 1774–1780.* The Hague, Mouton, 1970.

Minchinton, W. E., ed. *The Trade of Bristol in the Eighteenth Century.* Bristol: England, 1957.

———., ed. *The Growth of English Overseas Trade in the Seventeenth and Eighteenth Centuries.* London: Metheun, 1969.

———., and D. Starkey. "British shipping, The Netherlands and The Baltic, 1784–1795." In *The interactions of Amsterdam and Antwerp with the Baltic region, 1400–1800.* Leiden: Martinus Nijhoff, 1983, pp. 181–91.

Mironov, B. N. "Eksport russkogo khleba vo vtoroi polovine XVIII-nachale XIX v." *Istoricheskie zapiski* 93 (1974): pp. 149–54.

———. "K voprosu o roli russkogo kupechestva vo vneshnei torgovle Peterburga i Arkhangel'ska vo vtoroi polovine VXIII-nachale XIX veka." *Istoriia SSSR* 6 (1973): pp. 129–40.

———. "Sotsial'naia mobil'nost' rossiiskogo kupechestva v XVIII-nachale XIX veka (opyt izucheniia)." In *Problemy istoricheskoi demografii SSSR. Sbornik Statei,* edited by R. N. Pullat (Tallinn, 1977), pp. 207–17.

Mitchell, B. R., with the collaboration of Phyllis Deane. *Abstract of British Historical Statistics.* Cambridge: Cambridge University Press, 1962.

Murphy, Orville T. *Charles Gravier, Comte de Vergennes: French Diplomacy in the Age of Revolution: 1719–1787.* Albany: State University of New York Press, 1982.

Neal, Larry. *The Rise of Financial Capitalism. International Capital Markets in the Age of Reason.* Cambridge: Cambridge University Press, 1990.

Newman, Jennifer. "Anglo-Dutch commercial cooperation and the Russia trade in the eighteenth century." In *The interactions of Amsterdam and Antwerp with the Baltic region, 1400–1800.* Leiden: Martinus Nijhoff, 1983, pp. 95–103.

North, Douglass. "Ocean Freight Rates and Economic Development 1750–1913." *Journal of Economic History* 18:4 (December 1958): pp. 537–55.

O'Brien, Patrick, and Francois Crouzet. "Economic Growth in Britain and France." In *Britain and France. Ten Centuries.* Edited by Douglas Johnson, Francois Crouzet and Francois Bedarida. England: Wm. Dawson & Son Ltd., 1980. Pp. 173–95.

O'Brien, Patrick K. and Robert Quinault, eds. *The Industrial Revolution and British Society.* Cambridge University Press, 1993.

Oddy, J. Jepson. *European Commerce, Shewing New and Secure Channels of Trade with the Continent of Europe: Detailing the Produce, Manufactures and Commerce, of Russia, Prussia, Sweden, Denmark, and Germany; as well as the Trade of the Rivers Elbe, Weser, and Ems; with a General View of the Trade, Navigation, Produce and Manufactures, of the United Kingdom of Great Britain and Ireland; and its Unexplored and Improvable Resources and Interior Wealth.* London, 1805.

Öhberg, Arne. "Russia and the World Market in the Seventeenth Century. A Discussion of the Connection between Prices and Trade Routes." *The Scandinavian Economic History Review* 3:2 (1955): pp. 123–62.

Oliva, Jay L. *Misalliance: A Study of French Policy in Russia during the Seven Years' War.* New York: New York University Press, 1964.

Olsen, Donald J. *Town Planning in London: The Eighteenth and Nineteenth Centuries.* New Haven and London: Yale University Press (Yale Historical Publications: Miscellany v. 80), 1964.

Pavlova-Sil'vanskaia, M. P. "K voprosu o vneshnikh dolgakh Rossii vo vtoroi polovine XVIII v." In *Problemy genezisa kapitalizma. K Mezhdunarodnomu kongressu ekonomicheskoi istorii v Leningrade v 1970 g. Sbornik statei.* Moskva: Nauka, 1970.

Pettersen, Lauritz. "The influence of the American War of Independence upon Danish-Norwegian shipping. A survey based on late Danish and Norwegian research." In *The American Revolution and the Sea, Proceedings of the 14th International Conference of the International Commission for Maritime History.* London, National Maritime Museum, 1974, pp. 70–73.

Phipps, Geraldine M. *Sir John Merrick English Merchant-Diplomat in Seventeenth-Century Russia.* Newtonville, MA, 1983.

Pohl, Hans. *Die Beziehungen Hamburgs zu Spanien und dem spanischen Amerika in der Zeit von 1740 bis 1806.* Beiheft 45. *Vierteljahrschrift für Sozial-und Wirtschaftsgeschichte.* Wiesbaden: Franz Steiner Verlag, 1963.

Pokrovskii, V. I., ed. *Sbornik svedenii po istorii i statistike vneshnei torgovli Rossii.* St. Peterburg: M. N. Frolov, 1902, vol. I.

Ponting, K. G. "Logwood: An Interesting Dye." *The Journal of European Economic History* 2:1 (Spring 1973): pp. 109–19.

Pool, Bernard. *Navy Board Contracts 1660–1832: Contract Administration under the Navy Board.* Connecticut: Archon Book, 1966.

Price, Jacob M. "Multilateralism and/or Bilateralism: The Settlement of British Trade Balances with 'The North,' c. 1700." *The Economic History Review,* second series, 14:2 (December 1961): pp. 254–74.

———. *The Tobacco Adventure to Russia. Enterprise, Politics, and Diplomacy in the Quest for a Northern Market for English Colonial Tobacco, 1676–1722.* Transactions of the American Philosophical Society. New Series, Volume 51, part 1. Philadelphia: The American Philosophical Society, 1961.

Raeff, Marc. "Uniformity, Diversity, and the Imperial Administration." In *Osteuropa in Geschichte und Gegenwart: Festschrift für Günther Stökl zum 60. Geburtstag,* edited by Hans Lemberg, et al. Köln Wien: Bohlau Verlag, 1977, pp. 97–113.

———. *The Well-Ordered Police State: Social and Institutional Change Through Law in the Germanies and Russia, 1600–1800.* New Haven and London: Yale University Press, 1983.

Ramsay, G. D. *English Overseas Trade During the Centuries of Emergence.* London: Macmillan, 1957.

———. Review of *The Growth of English Overseas Trade in the Seventeenth and Eighteenth Centuries,* edited by W. E. Minchinton. *The Economic History Review,* 2nd series, 23:3 (December 1970): pp. 571–72.

Ransel, David L. "Nikita Panin's Role in Russian Court Politics of the Seventeen Sixties: A Critique of the Gentry Opposition Thesis." Ph.D. diss., Yale University, 1968.

———. "The 'Memoirs' of Count Münnich." *Slavic Review* 30:4 (Dec. 1971): pp. 843–52.

———. *The Politics of Catherinian Russia: The Panin Party.* New Haven: Yale University Press, 1975.

Rasch, A. A. "American Trade in the Baltic 1783–1807." *The Scandinavian Economic History Review* 13:1 (1965): pp. 31–64.

Reading, Douglas K. *The Anglo-Russian Commercial Treaty of 1734.* New Haven: Yale University Press, 1938.

Regemorter, J. L. van. "Commerce et Politique: Préparation et négociation du traité franco-russe de 1787." *Cahiers du Monde russe et soviétique* 4:3 (Juillet-Septembre 1963): pp. 230–57.

Riley, James C. *International Government Finance and the Amsterdam Capital Market 1740–1815.* Cambridge: Cambridge University Press, 1980.

———. *The Seven Years War and the Old Regime in France. The Economic and Financial Toll.* Princeton: Princeton University Press, 1986.

Rimmer, W. G. *Marshalls of Leeds Flax-Spinners 1788–1886.* Cambridge: Cambridge University Press, 1960.

Roberts, Michael. *Splendid Isolation, 1763–1780.* Reading: University of Reading Press, 1970).

———. "Macartney in Russia." *The English Historical Review* Supplement 7 (1974).

———. *British Diplomacy and Swedish Politics, 1758–1773.* Minneapolis: University of Minnesota Press, 1980.

Robinson, Stuart. *A History of Dyed Textiles.* London: Studio Vista Limited, 1969.

———. *A History of Printed Textiles.* Cambridge: M.I.T. Press, 1969.

Röhlk, Frauke. *Schiffahrt und Handel zwischen Hamburg und den Niederlanden in der Zweiten Hälfte des 18. under zu Beginn des 19. Jahrhunderts.* Beiheft 60, Teil I und II *Vierteljahrschrift für Sozial-und Wirtschaftsgeschichte.* Wiesbaden: Franz Steiner Verlag, 1973.

Rubinshtein, N. L. "Vneshniaia torgovlia Rossii i russkoe kupechestvo vo vtoroi polovine XVIII v." *Istoricheskie zapiski* 54 (Moskva, 1955): pp. 343–61.

Rubinshtein, E. I. *Polotnianaia i bumazhnaia manufaktura Goncharovykh vo vtoroi polovine XVIII veka.* Moskva: Sovetskaia Rossiia, 1975.

Samuelsson, Kurt. "International Payments and Credit Movements by the Swedish Merchant-Houses, 1730–1815." *The Scandinavian Economic History Review* 3:2 (1955): pp. 163–202.

Schmidt, K. Rahbek. "Wie ist Panins Plan zu einem Nordischen

System entstanden?" *Zeitschrift für Slavistik* 2:3 (Berlin, 1957): pp. 406–22.

Schneider, Jürgen, and Oskar Schwarzer. "International Rates of Exchange: Structures and Trends of Payments Mechanism in Europe, 17th to 19th Century," In *The Emergence of a World Economy 1500–1914*, Part I: 1500–1850, edited by Wolfram Fischer, R. Marvin McInnis and Jürgen Schneider. Stuttgart: Steiner-Verlag-Wiesbaden-GmbH, 1986, pp. 143–70.

Schumpeter, Elizabeth Boody. *English Overseas Trade Statistics 1697–1808*. Oxford: Clarendon Press, 1960.

Scott, H. M. "Great Britain, Poland and the Russian Alliance, 1763–1767." *The Historical Journal* 19:1 (1976): pp. 53–74.

———. "Frederick II, the Ottoman Empire and the Origins of the Russo-Prussian Alliance of April 1764." *European Studies Review* 7 (1977): pp. 153–75.

Scrivenor, Harry. *A Comprehensive History of the Iron Trade, Throughout the World, From the Earliest Records to the Present Period*. London: Smith, Elder and Co., 1841.

Segur, Count Louis Philippe de. *Memoirs and Recollections of Count Segur, Ambassador from France to the Courts of Russia and Prussia, &c. &c. Written by Himself*. London: Henry Colburn, 1827. Reprint, New York: Arno Press & The New York Times, 1970, II.

Semenov, A. *Izuchenie istoricheskikh svedenii o rossiiskoy vneshney torgovle i promyshlennosti s poloviny XVII-go stoletiya po 1858 god*. St. Petersburg, 1859. Reprint, 3 parts in 2 vols. Newtonville, MA: Oriental Research Partners, 1977.

Sherwig, John M. *Guineas And Gunpowder: British Foreign Aid in the Wars with France 1793–1815*. Cambridge, MA: Harvard University Press, 1969.

Smith, E. A. *Lord Grey 1764–1845*. Oxford: Clarendon Press, 1990.

Spencer, Frank. "Lord Sandwich, Russian Masts, and American Independence." *The Marriner's Mirror* 44:2 (1958): pp. 116–26.

Spencer, Frank, ed. *The Fourth Earl of Sandwich: Diplomatic Correspondence, 1763–1765*. Manchester: Manchester University Press, 1961.

Sperling, J. "The International Payments Mechanism in the Seventeenth and Eighteenth Centuries." *The Economic History Review*, second series, 14:3 (April 1962): pp. 446–68.

Spooner, Frank C. *Risks at Sea. Amsterdam insurance and maritime Europe, 1766–1780*. Cambridge: Cambridge University Press, 1983.

Steensgaard, Niels. "The Growth and Composition of the Long-Distance Trade of England and the Dutch Republic before

1750." In *The Rise of Merchant Empires. Long-Distance Trade in the Early Modern World 1350–1750*, edited by James D. Tracy. Cambridge University Press, 1990, pp. 102–52.

Stein, Robert Louis. "The French Sugar Business in the Eighteenth Century: A Quantitative Study." *Business History* 22:1 (January 1980): pp. 3–17.

———. "The State of French Colonial Commerce on the Eve of the Revolution." *The Journal of European Economic History* 12:1 (Spring 1983): pp. 105–17.

———. *The French Sugar Business in the Eighteenth Century*. Baton Rouge: Louisiana State University Press, 1988.

Storch, Heinrich Friedrich von. *Historicsh-statistiches Germälde des russischen Reichs am Ende des achtzehnten Jahrhunderts*. Vols. 1 and 2, Riga, 1797. Vols. 3–8, Leipzig: Johann Friedrich Hartknoch, 1799–1803; and *Supplementband*. Leipzig: Johann Friedrich Hartknoch, 1803.

———. *The Picture of Petersburg*. London: Longman & Rees, 1801.

Strumilin, S. G. *Izbrannye proizvedeniia. Istoriia chernoi metallurgii v SSSR*. Moskva: Nauka, 1967.

Summerson, John. *Georgian London*. London: Penguin Books, 1978.

Sutherland, Lucy S. Review of *The British Government and Commercial Negotiations with Europe 1783–1793*, by John Ehrman. *English Historical Review* 79 (October 1964): pp. 864–66.

Syrett, David. *Neutral Rights and the War in the Narrow Seas, 1778–82*. Fort Leavenworth, KS: Combat Studies Institute, U. S. Army Command and General Staff College, 1985.

Szostak, Rick. *The Role of Transportation in the Industrial Revolution. A Comparison of England and France*. Montreal and Kingston: McGill-Queen's University Press, 1991.

Thaden, Edward C. *Russia's Western Borderlands, 1710–1870*. Princeton: Princeton University Press, 1984.

Thomas, R. P., and D. N. McCloskey. "Overseas trade and empire 1700–1860." Chap. 5, in *The Economic History of Britain since 1700. Volume 1: 1700–1860*, edited by Roderick Floud and Donald McCloskey. Cambridge: Cambridge University Press, 1981.

Thompstone, Stuart Ross Thompstone. "The Organisation and Financing of Russian Foreign Trade before 1914." Ph.D. thesis, University of London, 1991.

Tooke, William. *The Life of Catherine II. Empress of Russia*. 4th ed. 3 vols. London, 1800.

Topolski, Jerzy. "Reflections on the first partition of Poland (1772). *Acta Poloniae Historica* 27 (1973): pp. 89–104.

Troitskii, S. M. *Finansovaia politika russkogo absoliutizma v XVIII veke*. Moskva: Nauka, 1966.

Tseitlin, M. A. *Lesnaia promyshlennost' Rossii i SSSR*. Leningrad, 1940.

Tsvetkov, M. A. *Izmenenie lesistosti evropeiskoi Rossii s kontsa XVII stoletiia po 1916 god*. Moskva: Akademii Nauk, 1957.

Unger, W. S. "Trade through the Sound in the Seventeenth and Eighteenth Centuries." *The Economic History Review*, second series, 12:2 (December 1959): pp. 206–21.

Vanstone, J. Henry. *The Raw Materials of Commerce*. 2 vols. London, 1929.

Viatkin, M. P., ed. *Ocherki istorii Leningrada*. Vol. 1. Moskva: Izdatel'stvo Akademii Nauk SSSR, 1955.

Volkov, M. Ia. "Tamozhennaia reforma 1753–1757 gg." *Istoricheskie zapiski* 71 (Moskva, 1962): pp. 134–57.

Warden, Alex. J. *The Linen Trade, Ancient and Modern*. London: Longman, Green, Longman, Roberts & Green, 1864.

Wilkinson, Joan. *The Letters of Thomas Langton, Flax Merchant of Kirkham*. Manchester: Printed for the Chetham Society by Carnegie Publishers, 1994.

Willan, T. S. *The Muscovy Merchants of 1555*. Manchester: Manchester University Press, 1953.

Willan, T. S. *The Early History of the Russia Company 1553–1603*. Manchester: Manchester University Press, 1956.

Williams, J. E. "Whitehaven in the Eighteenth Century." *The Economic History Review* 8:3 (April 1956): pp. 394–404.

Williams, M. J. "The Naval Administration of the Fourth Earl of Sandwich." D.Phil. diss., Jesus College, University of Oxford, 1962.

Wilson, Arthur M. "The Logwood Trade in the Seventeenth and Eighteenth Centuries." In *Essays in the History of Modern Europe*. Edited by Donald C. McKay. New York: Harper & Brothers, 1936. Pp. 1–15.

Wilson, Charles. *Anglo-Dutch Commerce and Finance in the Eighteenth Century*. Cambridge: Cambridge University Press, 1941.

———. "Treasure and Trade Balances: The Mercantilist Problem." *The Economic History Review*, second series, 2:2 (1949): pp. 152–61.

———. *England's Apprenticeship 1603–1763*. New York: St. Martin's Press, 1966.

———. *Economic History and the Historian. Collected Essays*. London: Weidenfeld and Nicolson, 1969.

Wilson, R. G. "The Supremacy of the Yorkshire Cloth Industry in the Eighteenth Century." In *Textile History and Economic History*, edited by N. B. Harte and K. G. Ponting. Manchester: Manchester University Press, 1973, pp. 225–46.

Wright, Charles, and C. Ernest Fayle. *A History of Lloyd's*. London: Macmillan, 1928.

Wyrobisz, Andrzej. "Resources and Construction Materials in Preindustrial Europe." In *Natural resources in European history. A conference report*, edited by Antoni Mączak and William Parker. Washington, DC, 1978, pp. 65–84.

Zimmerman, Jules S. "Alexander Romanovich Vorontsov, Eighteenth Century Enlightened Russian Statesman, 1741–1805." Ph.D. diss., City University of New York, 1975.

Zupko, Ronald Edward. *A Dictionary of English Weights and Measures. From Anglo-Saxon Times to the Nineteenth Century*. Madison, WI: University of Wisconsin Press, 1968.

———. *French Weights and Measures before the Revolution. A Dictionary of Provincial and Local Units*. Bloomington: Indiana University Press, 1978.

Zutis, Ia. *Ostzeiskii vopros v XVIII veke*. Riga: Knigoizdatel'stvo, 1946.

# Index

Act of Navigation of 1660 153
agricultural products 24
  importance to other countries 24
Aktiar 114
Albion, Robert Greenhalgh 88, 102
  on importance of timber 103
  on masts 104
Amsterdam
  and bills of exchange 68, 263–264 and n. 14
Anglo-Persian trade 36
Anglo-Russian Commerce Convention (1793) 161, 213
Anglo-Russian Defensive Alliance 165
Anglo-Russian relationship 18, 30
  commerce 18
  Commercial Treaty 36, 37, 38, 39, 40, 41, 122, 123, 126, 128, 132, 138, 145, 170, 199
    and rixdollar 39
    Article VIII 35
    Article IV 35, 40, 43, 44, 143
    Articles X and XI 39, 123, 126, 128
    British revision 40–42
    contraband 39
    duty rates 35
    neutral shipping 39
    expiration of 258
    extension canceled 206
    proposal 35
    reservation clause 40
    signed 43
    wording 44
  political alliance 18
Anglo-Russian trade 23, 31, 44, 51, 59
  beginnings 7
  British dependence 210, 211
  Commission on Commerce and 23
  exports vs imports 202, 203, 206
  imports 211 See bar iron; flax; hemp; linen; timber
  origin 7
Archangel 73 and n. 21, , 121, 140, 172, 207, 213, 259
  bar iron exports 213 n. 2
  linen exports 222
  overseas trade 245, 246, 247, 253
  fluctuations 252
Armed Neutrality 12, 132, 139, 144, 145, 150, 152, 159, 161, 170 See Catherine II
  blockaded port defined 128
  economic effects 105

Armenians 36
Arshin See linen
Astrakhan 6, 35
  duty free port 6
Ashton, Thomas
  on consumption of timber 91 See Timber
Åstrom, Sven-Erik
  on timber 96
Attman, Artur 263–264
Austria 19, 30
Aynali Kavak convention 115

Bakunin, Peter V. 135
Baltic-Black Sea trade 115, 116
Baltic Sea ports 35, 116
Bank of the Nobility 5
bar iron
  American 57
  Board of Trade 135
  British dependence on 55, 56, 58, 135, 212, 214
  coke use in producing 56
  duties on 136, 143
  monetary value 59
  Norwegian 135
  percent imports 213
  pig iron 56 n. 5
  prices 60, 135, 214–215
  puddling furnaces and 56, 58
  Russian exports during war 57
  salutary effect of imports 56 and n. 6
  St. Petersburg/Kronstadt 59
  Swedish 57, 135
  tonnage on British ships 59, 214
  volume and value 61
Bestuzhev-Riumin, Aleksei P. 16
Bezborodko, Alexander A. 134
bills of exchange 68, 262, 263–264
blockade defined 128
British Board of Trade 10 and n. 20, 13, 20, 135, 136, 137, 141
  on Convention on Commerce 152
  on naturalizing foreigners 139–140, 147
  on payment of duties 141, 152, 153
  response to prohibited goods 163
  silk imports 137
  Tariff of 1766/67 136 and n. 12
British Factory 147, 155, 156, 157, 158, 162, 172
  naturalized burgher records 175, 179
British-French conflict 117–124

Buckinghamshire, Earl of 20, 22, 30, 35, 38
Burke, Edmund 10
Bute, Earl of 10, 11
  on privileges 11

Carmarthen, Marquis of 144, 145, 146, 147, 153 n. 16
  on Russian trade dependence 142–143
Catherine II 15, 17, 18, 20, 22, 28, 30, 33, 35, 39, 42, 43, 44, 45
  appointments 23
  arbiter-mediator 114, 117, 121
  Armed Neutrality 132, 159, 170
  commercial policy 24, 46, 47, 117, 146, 159, 160
    edict against France 161–162
    edict on British goods 165
    effects on British economy 163–64
    failure of 169
  death 166
  Declaration . . . Armed Neutrality 116, 120, 121, 122, 123, 124, 125
  deforestation 94, 96
  domestic policy 15, 33
    money supply 132–133
  economic policies 17
  foreign trade policy 15, 22–23, 44, 46, 47, 96, 113, 210, 244
    Black Sea region 115–116
    France 207
    grain 113
    See Anglo-Russian Commercial Treaty; 1734 Treaty; Tariff of 1766/67; Tariff of 1782/83
  Instruction 25, 47
  maritime fleet 198
  "Memorandum" 123
  merchantmen protection 123, 125
  merchant subsidies 170
  money supply 132
  naval power 170
  neutral shipping declaration 122
  Northern System 18, 19
  personality 16
  Panin and 16
  policy towards England 20
  political aspirations 22, 30
  pragmatist 17
  ruling policy 16
  southern alliance 18, 19
  supreme legislator 15
  timber exports 90
Cayley, John 149, 155, 160, 163
  on payment of duties 155
  on Russian merchants 171 n. 2
Chapman, Stanley 264
Charles II 118
Clarke, Edward Daniel 243
Clendenning, P. H. 27
Commerce Collegium 24, 35, 113 n. 2
  Black Sea navigation 113 n. 2
  Teplov on 24
Commission on Commerce 5, 23, 36
  as a policy-formulating body 23
  goals of 23
commodity
  trade turnover 114
  valuation 266 n. 21
contraband 37
  Articles X and XI 39
  Catherine's declaration and 122
  war 12, 125, 127, 128
Convention of Armed Neutrality 129 n. 50
Convention of Reichenbach 154
Cross, Anthony G. 243
custom houses 7 n. 11, 35
  London 7 n. 11
  St. Petersburg 79, 138, 139, 163, 171, 224, 228, 232
    categories of records kept 182
    chaos at 162
    records See notes and tables throughout, esp. chaps. IV, V, VI, X, XI, XII, XIII
  specie 265

Daniel, Wallace L. 25, 28, 29
  view on Teplov 28
Davis, Ralph 56
  and "current real values" 266, 267
  and ship's measure 201
  and timber measurements 99 and n. 33
deforestation in Russia 94 See Timber
Denmark-Norway 42
  Explanatory Article 127
    motive for 127
  mast exports 101
  timber exports 96, 99
Durno, James
  and Prussian government 152
Dutch 21
  allegiances 118, 119
  Anglo-Dutch Treaty 119
  Anglo-Dutch War 129
  Armed Neutrality and 129
  policy on seizure 118
  proposals 21
  trade with Russia 21–22
duties 33, 133
  bar iron 136
  British Board of Trade and 137, 138
  collection 34
  currency 34, 156
  exemptions 138
  free goods 33
  goods for the wealthy 33
  high 33
  imported goods 33
  life necessities not taxed 33
  low 33, 34
  luxuries 33
  manufactured goods 34
  merchant rates 34, 37

duties (*continued*)
  payment 34, 37, 39, 144, 146, 155
    equality of 152
  Tariff of 1796 166
  timber 97 and n. 25
  unsorted merchandise 34
  used as retaliation 141
  "warfare of Duties" 143

Elizabeth 3, 9, 10, 13, 15, 16, 17, 32, 36, 42, 43
  1759 declaration 43
  new tariff 32
  death 13
Elton, Captain John 36
exports 27, 55–62
  and British economy 52
  multiplier effects of 52
  Russian expansion of 51–52
  value of 257, 261
  *See* raw materials; Riga; St. Petersburg/Kronstadt; bar iron; flax; hemp; linen; timber

Falconer, William
  on masts 104
Fawkener, William 154
Fitzherbert, Alleyne 134, 135, 142, 146, 149
  extension of treaty 145
flax
  effect on Russian economy 63
  export prices 74
  importance to British life 63, 218
  monetary value 75
  prices 219–220
  qualities:
    6-head 73
    9-head 73
    12-head 73, 74 and n. 25
    Codilla 73
  *See* Chapter V
Floridablanca, Conde de 124
forests, deforestation of 6, 94, 96
Forster, Edward 138, 155, 156
  "For Information" 156
France 21, 30
  diplomatic relations with Russia 21
  relations with Britain 119
  Teplov on 29
  trade competition 115
  trading privilege 36
  treaty of commerce 134, 145
Frederick II 13, 19
Frederick William II 149, 151, 154
Fredrikshamn (Hamina) 35, 99, 100
  overseas trade 245
    fluctuations 252
  timber exports 96, 99, 100
"Free Ships Make Free Goods" 37, 122, 124, 128

George III 19, 35, 43
  customs regulations 79
Golitsyn, Alexander M. 20, 24, 42, 43
Goltisyn, Dmitrii 123
Goodricke, Sir John 57
"Great Mast" market 101 *See* Timber
Grenville, William 156 and n. 21, 164
Gross, Baron Heinrich von 35

Hamburg
  bills of exchange 263, 264 n. 14
  and Russian ships 198
Hanse merchants 44, 175, 179, 184, 185, 186
Harris James 118, 225
Harte, N. B. 64
  on flax's importance 220
  on linen's importance 64, 221
  minimum levels of linen production 77
Hawkesbury, Lord *See* Jenkinson, Charles
hemp
  Carmarthen on 153 n. 16
  crisis of 1793 218, 254
  effect on British economy 67
  effect of fire 67–68
  effect on Russian economy 63
  importance to British life 63, 215
  import percentages 67
  monetary value of exports 70
  prices 216–217
  qualities:
    clean 66, 69 and n. 15, 70
    Codilla 66
    half clean 66
    outshot 66
  shortages 68
  tonnage 215
  *See* Chapter V

imports
  British
    parliamentary report on 81
  Russian
    analysis 231–232 *See* chapter XIII
  categories:
    textiles 233–237
      dyes and mordants:
        alum 239
        cochineal 239
        curcuma 239
        indigo 240
        logwood 240
        madder 240
        value of 236–237, 262–263
      sugar 240–241
      coffee 242–242
      fruits 242
      olive oil 242
      wines 242
      cultural impact 242–243

imports (*continued*)
  data lists 232
  effect of restrictions 252–252
  extravagant nature of 231
  value of 232, 257–258
  See *prima facie* value
Indova, E.I.
  *Vedomosti* 172
Industrial Revolution 55
  bar iron and 55
infrastructure, Russian
  effect on commerce 169
  inadequacy of 169
*Inostrannye gosti* 147, 171, 179, 182, 184, 186, 258
  In *Vedomosti* 173–174, 179
  See merchants
Ireland
  linen industry 64, 65, 77, 78 and n. 31, 83, 84 n. 47
Ivan the Terrible 16

Jenkinson, Charles 138, 156 and n. 21
Johansen, Hans Chr. 263
Joseph II of Austria 134, 151
  See Russo-Turkish-Swedish War

Keith, Robert 9, 10
Kherson 114,
  French trade to the Levant 115
Kjaerheim, Steiner
  on timber 229
Klingstedt, Timofei von 23, 26
  on Russia's relationship to Great Britain 27
Knight, R.J.B. 90 n. 6 *See* Timber
  on mast quality 104
  Royal Navy and masts 103

Langton, Thomas 66
League of Armed Neutrality 123, 127, 128, 129, 134
LeDonne, John
  on Russian budgets 132
Leeds, Duke of *See* Carmarthen
Leopold II 153, 154
linen
  and British society 63
  British consumption of 65
  British customs 79
  British Linen Company 65
  comparison of sources 77
  conditions of British linen trade 81
  effect on manufacture 86
  effect on politics and military 86
  export-import correlation 80, 221
  Irish imports 64, 65, 77, 78 and n. 31, 83, 84 n. 47
  levels of production 77
  parliamentary report on 81
  prices of Russian linens 85 n. 49, 223 n. 18
  Russian contribution 77, 81, 83
  Russian measurements 79
  Scottish imports 64, 65, 77, 78 and n. 31, 83, 84 n. 47
  types 78
  See Chapter V; Harte, N.B.
Lomonosov, M. V. 3
Lübeck 264 n. 14
luxuries 33, 43

Macartney, George 37, 39, 40, 41, 42, 43, 44,
  on Russian duty exemptions 138
de Madariaga, Isabel 117
Magistracy of Riga 12
Magistracy 24
  Teplov on 24
Malone, Joseph J. 102
manufactured goods 24
  expanded exports of 52
maritime law
  Catherine's five principles 122
maritime practice, Russian 142
masts 101, 102, 105
  Baltic 102, 103
  "Great Masts" 101, 106, 107, 224, 225, 226, 227
    described 226
  measurements 101, 102
  Memel 105
  New England 102, 103 n. 52
  Norway 225, 227
  prices 105
  Russian 103, 106
Mathias, Peter 5
  and customs trade data 85
Memel (Kkaipéda)
  log exportation 98
  masts 105
merchants
  bar iron shipments 60
    Catherine and 17, 170
  credit 155
  duty payments 133
  English 7
    Earl of Bute on 10
    privileges 11
    percentage of trade 183, 184 n. 12, 186, 187, 192
    regulations on 79
    Riga 72
    St. Petersburg/Kronstadt 72
  incentives 26
  naturalization of foreign 139–140, 147 and n. 51, 156, 171 See *Inostrannye gosti*
    ethnic origins 172
    Germanic nationals 158
  and price fixing 5
  ordinance on 125

merchants (*continued*)
  Russian
    records of 171, 182, 186, 192
    subsidies 170
    in Riga 12
    in St. Petersburg 12
    Teplov on 29–30
    trade turnover 44–45
Mironov, B.N.
  on role of merchant nationals 172
Moscovy Company 7, 10, 12
Münnich, Burchard Christoph von 23
Münnich, Ernst von 23, 24, 27, 35, 43
  director of Customs 23
  on industry, agriculture, and commerce 27
  on Russia's future economic growth 27

Narva 37, 39, 45, 46, 259
  flax exports 75 and n. 27, 76, 219
  overseas trade 245, 253
    fluctuations 252
  timber exports 96
Nepliuev, Ivan Ivanovich 23, 28, 95
  diplomatic missions 28
  on timber concessions to foreigners 95
  Russian merchantry 28
  tariff and custom reduction 28
"Northern System" 18, 19
Northington, Lord Chancellor 41
Norway
  mast exports 101
  timber export 88, 96, 99
Novgorod wood trade 95
Novikov, Nikolay 243

Ochakov 154
Orbell, Dr. John 263
Ordinance on Commerce and Navigation 125, 198, 205
Orenburg 35
Orlov brothers 16, 24
Osterman, Ivan A. 134, 159

Panin, Nikita Ivanovich 16, 17, 24, 30, 41, 43
  favored tariffs 25
  intolerance of British mistrust 46
  on British privilege 36
Panin-Teplov memoirs 25
Panin-Teplov program 25
Parliamentary Committee
  and timber trade 224, 228
Paul I, Emperor
  and shipping 208, 209
  trade balances 245, 251
Pernau (Pärnu)
  overseas trade 245
    fluctuations 252
Persia 35, 36, 37, 39
Persian Trading Company 6, 13, 14 n. 32, 17

Peter I 28, 32, 40
Peter III 4, 13, 14 n. 32, 15, 17, 19
  armistice 13
  reform 13
Pitt, William
  and armed neutrality 150
  attitude towards Russia 153, 154
  and timber duties 97 n. 25
Poland 17, 19, 30, 42, 152
  Catherine's policies 31
  Radziwiłł family 152
    exports timber 96, 98
Polish Question 19, 31
Polish throne 19
poods, defined 21
*prima facie* value 182–188
Prussia 4, 13, 30, 42
  exports timber 96

Raeff, Marc 16
Ransel, David L. 24, 25
raw materials 24, 33, 34
  bar iron *See* Chapter IV
  expanded exports 52
  flax 76 *See* Chapter V
  hemp *See* Chapter V
  importance to other countries 24
  linen *See* Chapter V
    British insufficiency 66
  timber *See* Chapter VI
Reval (Tallinn)
  overseas trade 245, 259
    fluctuations 252
Riga 12, 37, 39, 46, 259
  British merchants in 45
  flax exports 73, 75 and n. 27, 76, 219
  hemp exports 72, 216
  log exports 224
  magistracy of 12, 13
  mast trade 100, 103, 106, 224, 225
  merchant nationals 172
  overseas trade 245, 247, 253
    customs revenue 254–255
    fluctuations 252
  shipping traffic 199, 207
  timber export 96, 100, 105, 224
  trade turnover 246–247
  "Wainscot" 97, 98 and n. 31, 224
Rimmer, R. G. 66
rixdollars 34, 39
  equivalence 34
  *See* duties
Roberts, Michael 38
Royal Navy 67, 101, 104
  and Dutch 119
  and France 118
  and New England masts 102
  masting table 101
  *matériel* problem 103
Rozumovs'kyi, Hetman Kyrylo 28
Rubinshtein, N.L.

Rubinshtein, N.L. (*continued*)
  on *inostrannye gosti* 171
  on Russia's maritime fleet 171
Russia Company 7, 10, 20, 37, 137, 139, 140, 148, 155, 264
Russo-Turkish War 114
Russo-Turkish-Swedish War 149, 150, 151, 155, 160
  Anglo-Russian trade during 207
  and Russian finances 150

St. Petersburg 37, 39, 40, 41, 42
  Custom House 228
  flax exports 73
  hemp exports 67, 68, 70, 72
  timber exports 100, 106
St. Petersburg/Kronstadt 43, 44, 45, 46, 259, 265
  bar iron 214, 215
  British ships 199, 207, 208, 209
  commodity trade turnover 174
  comparison to British cities 260–261
  export venue 57
  favorable balance 252–253
    fluctuations 252
  flax exports 73, 74, 218, 219
  hemp 69, 215, 217, 218
  imports 205–206, 231–241
  linen 222
    major port 58
  merchant nationals 172, 176, 186
  overseas trade 245
  and customs revenue 254–257
  sailcloth export 80, 222
  shipping traffic 199, 200
  timber export 96, 100, 228
  trade balances 160, 187–188, 245–246
    ratio to other cities 247
Sailcloth 80 *See* linen
  Anglo-Irish competition 80
Sandwich, Fourth Earl of 30, 36, 37, 38, 89, 98, 130
Schneider, Henry 182
Schwan, C.F. 14
Scotland
  linen industry 64, 65, 77, 78 and n. 31, 83, 84 n. 47
  seamen 27
  insufficiency of 27
Seven Years' War 3, 12, 17, 18, 21, 32, 65
  armistice 4
  economic effect 3, 52 57, 65, 66
  Russian exports during 57, 65, 66
  scarcity of timber in England 89
  timber use 97, 100
Shairp, Walter 137, 139, 140
  death 149
  defines "Russian ship" 141
Shakhovskoi, Iakov P. 23, 26
  director of the College of Audit 26
  general of the Army (1753–1760), and procurator 26

goals of 23
Shemiakin, N. P. 17
ships 89, 90, 91
  Baltic ports 92
  British 199, 206, 208, 209
  departures 199
  North America 92 n. 10
  for Russian canals 94
  Russian 205–206
  seizure of merchant ships 121
  timber used 102
shipping *See* Chapter XI; tonnage
  British 199, 200, 206, 208, 209, 210
  pattern 208
Shuvalov, Peter I. 4, 5, 21, 28
Siberia 35
Sievers, Jacob 95, 96
smuggling 33
Spencer, Frank 40
Stormont 126, 129, 130
Suffolk, Lord 118
Summerson, John
  on Riga timber 93
Sutherland, Richard 182
Sweden 21, 42, 123
  and League of Armed Neutrality 128–129
  Anglo-Swedish commercial relations 57, 118
  Catherine and 123, 124
  Swedish-Russian trade agreement 21
    proposals 21
  Swedish-Russian relations 21
  *Trolle* incident 118
Swallow, Samuel 137

tariffs, Russian 32
  1724 32
  1731 32
  1766/67 33–34, 133, 135, 136, 138, 170, 200, 229
    and duties 34
    Article IV 39, 40
    Articles X, XI 39
    effective date 35
    flax imports 73
    principles of 33
    reservation clause 40
  1782/83 132, 133, 136, 137, 139, 170
    effect on Anglo-Russian relations 134
    revision 165
  1796 165, 166
    custom houses 35
    Ernst von Münnich 35
    Grigorii N. Teplov 35
  protectionist 33
Teplov, Grigorii N. 23, 24, 28, 29, 35, 43
  book on the planting 29
  Company of Merchants 28
  favored tariffs 25
  goals for Russian economy 29
  "middle order of people" 29

Teplov, Grigorii N. (*continued*)
  Russian merchants 30
  tobacco monopoly 29
  trade 29
textiles *See* exports; flax; linen
timber
  British dependence on foreign 88, 224, 230
    construction 91, 92, 93
    shipbuilding 88, 89
  categories 228–229
  costs 90, 92 and n. 10, 97
  *pinus strobus* 102, 103
  *pinus sylvestris* 96, 102, 103
  Riga 92, 93, 224
  Russian canal projects 94
  scarcity in England and Russia 88, 89, 90, 93–94
  studies on 88, 89
  transportation of 96 n. 21
  unit of measure 99 n. 33
  value 91
  *See* masts
tobacco 28–29
tonnage
  average per British ship 201, 202
  bar iron 214
  hemp 215
  importance of 200
  kinds of 201
  records of 202
trade
  balances 31
    and customs revenue 250, 254–257
      cause for decline 257
  Russian
    annual average export value 247
    balances 26–27, 31
    bills of exchange 263
    British dominance of 171
    Catherine's effect on 170–171
      fluctuations 250–251
    shift in value 258
    turnover 245
Treaty of Commerce (Russia-Turkey) 114

Treaty of Jassy 115, 155
Treaty of Küçük Kaynarca 114
Treaty of 1734 7, 10, 11, 12, 13, 14, 18, 20, 35, 36, 37, 39, 43
  Article III 11
  Article IX 11
Troitskii 35
Turkey 30

*Vedomosti* 173
  inconsistencies 173
  *Inostrannye gosti* 173–174, 179, 182
  St. Petersburg/Kronstadt records 173
Viazemskii, Prince Alexander Alexeevich 17
  budget 17
Viborg (Viipuri) 35, 99
  overseas trade 245
    fluctuations 252
  timber exports 99, 100
Volkov, Dmitrii V. 6, 13
  major tariff reform 6
Vorontsov, Alexander Romanovich 15, 135, 146, 159
  and forest depletion 96 n. 21
Vorontsov, Mikhail I. 4, 9, 15, 16, 20, 21, 22
Vorontsov family 5
Vorontsov, Roman I. 15
Vorontsov, Semen 146, 150, 151, 154, 159

Walpole, Horace 91
  on house building 91
westernization of Russia 231 *See* Imports
Whitworth, Charles 149, 153 n. 16, 157, 158, 159
  on prohibited imports 162–63, 164
Williams, M.J. *See* Timber
  Royal Navy and masts 103

Yorke, Joseph 119

Zubov, Platon Alexandrovich 159

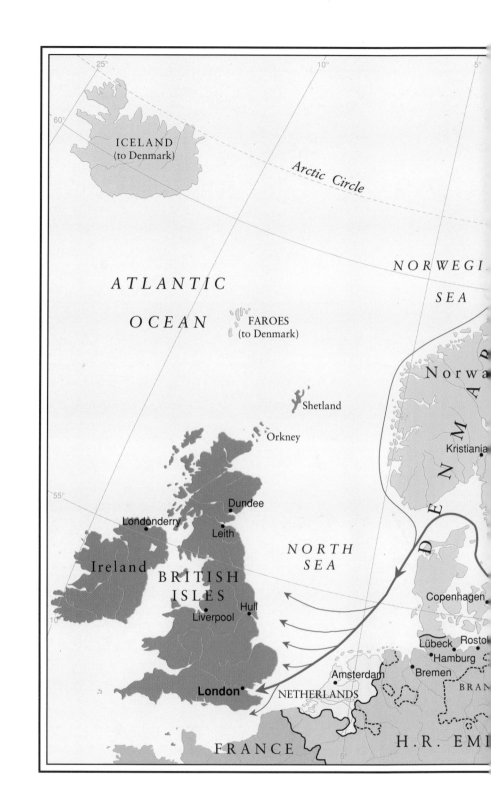